Gallium Arsenide and Related Compounds 1991

Gallium Arsenide and Related Compounds 1991

Proceedings of the Eighteenth International Symposium on Gallium Arsenide and Related Compounds, Seattle, Washington, USA, 9–12 September 1991

Edited by G B Stringfellow

CRC Press
Taylor & Francis Group
Boca Raton London New York

CRC Press is an imprint of the
Taylor & Francis Group, an **informa** business

CRC Press
Taylor & Francis Group
6000 Broken Sound Parkway NW, Suite 300
Boca Raton, FL 33487-2742

First issued in paperback 2019

ISBN-13: 978-0-85498-410-7 (hbk)
ISBN-13: 978-0-367-40289-1 (pbk)

CODEN IPHSAC 120 1–650 (1992)

British Library Cataloguing in Publication Data are available

ISBN 0-85498-410-0

Library of Congress Cataloging-in-Publication Data are available

**Visit the Taylor & Francis Web site at
http://www.taylorandfrancis.com**

**and the CRC Press Web site at
http://www.crcpress.com**

GaAs Symposium Award and Heinrich Welker Gold Medal

The Gallium Arsenide Symposium Award was initiated in 1976. Candidates for the Award are selected by the GaAs Symposium Award Committee from those who have distinguished themselves in the area of III−V compound semiconductors. The Award consists of $2500 and a plaque citing the recipient's contribution to the field. In addition, the Heinrich Welker Gold Medal, sponsored by Siemens, is also presented to the Award recipient.

The 1976 Award was presented to Nick Holonyak of the University of Illinois for his work leading to the development of the first practical light-emitting diodes and his continuing research on III−V compound semiconductors. The second Award was received by Cyril Hilsum of the Royal Signals and Radar Establishments (now at GEC Research Laboratories) for his contributions in the field of transferred-electron logic devices and the advancement of GaAs MESFETS. In 1981 the Award and Medal were presented to Gerald L Pearson, Stanford University, for his research and teaching in the field of compound semiconductor physics and new device technology. In 1982, the Award and Medal were presented to Herbert Kroemer, University of California at Santa Barbara, for his contributions to hot-electron effects, the Gunn Oscillator, and III−V heterojunction devices including the heterojunction laser. The 1984 Award and Medal were presented to Izuo Hayashi (now at Optoelectronics Joint Research Laboratories) for his contributions to the development and understanding of room temperature operation double-heterojunction lasers. In 1985, the recipient was Heinz Beneking, Technical University of Aachen, in recognition of his distinguished contributions to the development of III−V compound semiconductor technology and new structure devices. In 1986, the Award and Medal were presented to Alfred Y Cho, AT&T Bell Laboratories, for his pioneering work in the development of molecular beam epitaxy and contributions to III−V compound semiconductor research. Zhores I Alferov of the A F Ioffe Technical Institute was the recipient of the Award and Medal in 1987 for his outstanding contributions in liquid phase epitaxy, laser diodes, vapor phase epitaxy, and for innovative contributions to the technology of compound semiconductors. The 1988 GaAs Symposium Award and Welker Medal were presented to Jerry Woodall for his pioneering work in introducing the III−V alloy, AlGaAs, and his further fundamental contributions to III−V compound semiconductor physics. In 1989 the Award and Medal were presented to Don Shaw of Texas Instruments in recognition of his pioneering work in elucidating the complex mechanisms of epitaxial crystal growth using chemical vapor deposition. In 1990 the recipient was Gregory S Stillman of the University of Illinois for his outstanding contributions to the characterization of high-purity GaAs and related compounds and to the understanding and development of near infrared avalanche photodetectors.

At this meeting the 1991 GaAs Symposium Award and Welker Medal were presented to Lester F Eastman in recognition of his many contributions to III−V compound semiconductors, including the development of the concept of ballistic electron transport, the planar doping technique, the buffer layer technique, and AlInAs/GaInAs/InP heterostructures.

Dr Eastman was born in Utica, NY, and obtained B.S. (1953), M.S. (1955), and Ph.D. (1957) degrees at Cornell University. He joined the faculty of Electrical Engineering at Cornell in 1957. Since 1965 he has been doing research on compound semiconductor materials, high-speed devices, and circuits, and has been active in organizing workshops and conferences on these subjects at Cornell from 1977. In 1977 he joined other Cornell faculty members in obtaining funding and founding the National Research and Resource Facility at Cornell (now the National Nanofabrication Facility). Also in 1977 he founded the Joint Services Electronics Program and directed it until 1987. He has recently joined with others at Cornell to develop a large effort in high-frequency/high-speed optoelectronics. During the 1978–1979 year he was on leave at MIT's Lincoln Laboratory, and during the 1985–1986 year he was at the IBM Watson Research Laboratory. During 1983 he was the IEEE Electron Device Society National Lecturer. He was a member of the U.S. Government Advisory Group on Electron Devices from 1978–1988, and serves as a consultant for several industrial laboratories. He was a founder in 1985 and is now chairman of the board of directors of Northeast Semiconductors, Inc. He has been a Fellow of IEEE since 1969, a member of the National Academy of Engineering since 1986, and was aprointed the John L Given Foundation Professor of Engineering at Cornell in January, 1985.

Young Scientist Award

In 1986 the International Advisory Committee of the GaAs and Related Compounds Symposium established a Young Scientist Award to recognize technical achievements in the field of III–V compound semiconductors by a scientist under the age of 40. The Award consists of $1000 and a plaque citing the recipient's contributions. The first Young Scientist Award was presented at the 1986 Symposium to Russell D Dupuis for his work in the development of organometallic vapor phase epitaxy for III–V compound semiconductors.

In 1987 the Award was presented to Dr Naoki Yokoyama for his contributions to developing self-aligned gate technology for GaAs MESFETs and ICs and the resonant tunneling hot-electron transistor. At the 1988 conference, the Young Scientist Award was presented to Won Tsang for his distinguished contributions to the development of new MBE growth techniques and heterostructure devices. The 1989 Award was presented to Russel Fischer in recognition of his work in showing state of the art performance at DC and microwave frequencies of MESFETs, MODFETs, and HBTs using GaAs on Si. In 1990 the Award was made to Yasuhiko Arakawa for his pioneering studies of low-dimensional semiconductor lasers, particularly in regard to his prediction and demonstration of the superiority of quantum wire and quantum box configurations.

At this conference the 1991 Young Scientist Award was presented to Sandip Tiwari of the IBM Watson Research Center in recognition of contributions to the understanding and development of compound semiconductor devices, most notably the metal–semiconductor field effect transistor, the heterostructure field effect transistor, and the heterostructure bipolar transistor.

Dr Tiwari was born in 1955. He attended the Indian Institute of Technology at Kanpur, India, Rensselaer Polytechnic Institute, and Cornell University, where he received his Ph.D. degree in electrical engineering in 1980. He is a Research Staff Member at IBM Watson Research Center working in the area of Semiconductor Physics and Devices.

Dr Tiwari's contributions to the understanding and development of electronic and optical compound semiconductor devices is recorded in over 50 publications and 10 patents. These contributions have included, individually and in collaboration, new device ideas such as p-channel HFETs, SiGe HBTs, and unpinned oxide-based MOSFETs; new technologies such as p-type and n-type refractory ohmic contacts that are readily amenable to self-alignment; and an understanding of new phenomena in devices such as the alloy-potential-related high-current effect in graded hetero-structure collector HBTs.

During 1988–1989 he was a Visiting Associate Professor at the University of Michigan. He has been Associate and Guest Editor of the IEEE Transactions on Electron Devices and is the author of the textbook *Compound Semiconductor Device Physics* published by Academic Press.

Preface

This volume presents the papers from the 18th International Symposium on Gallium Arsenide and Related Compounds, held in Seattle, Washington from September 9 to 12, 1991. It includes current results on the materials, characterization, and device aspects of a broad range of semiconductor materials, particularly the III−V compounds and alloys.

The conference consisted of 3 plenary talks, 117 oral presentations, 24 poster papers, and 6 late news papers. The competition was stiff, with less than 50% of the submitted abstracts accepted for presentation at the conference. Consequently, the results included in this volume are timely and of the very highest quality.

The volume is organized with the plenary papers presented in the first chapter, the late news papers in the last chapter (in order to allow the most rapid publication schedule), and the other papers presented by topic in the remaining chapters. Naturally, some papers could be classified in several categories. Thus, the reader is encouraged to peruse the entire volume.

Finally, the editor wishes to especially thank the program committee, listed in the front of the volume, for their hard work in selecting the plenary speakers, deciding which abstracts should be accepted, and having the papers reviewed. Thanks also go to Kathryn Cantley for her invaluable assistance at the conference.

G B Stringfellow

Acknowledgments

The organizers of the Symposium gratefully acknowledge financial support provided by:

AFOSR
AIXTRON
BANDGAP TECHNOLOGY CORP
BOEING SCIENCE AND TECHNOLOGY CENTER
EPI
FISONS INSTRUMENTS
INTEVAC
IEEE
SIEMENS

Contents

Chapter 4: Processing

Chapter 5: Characterization

Chapter 6: HBTS and InGaAlAs FETS

Chapter 7: Characterization

Chapter 8: Ordering

Chapter 9: Quantum Wells

Chapter 10: Opto-electronics

Chapter 13: Late News Papers

Inst. Phys. Conf. Ser. No 120: Chapter 1
Paper presented at Int. Symp. GaAs and Related Compounds, Seattle, 1991

1

New directions for III–V structures: metal/semiconductor heteroepitaxy

J P Harbison, T Sands, C J Palmstrom, T L Cheeks, L T Florez, and V G Keramidas

Bellcore, 331 Newman Springs Road, Red Bank, N.J. 07701-7040

ABSTRACT: Epitaxial metallic films grown on semiconductor substrates are most attractive when they consist of materials which are thermodynamically stable with respect to the underlying semiconductor. This allows, under the proper growth conditions, the overgrowth of epitaxial semiconductor layers, thus forming monocrystalline semiconductor/metal/semiconductor heterostructures with exciting physics and device possibilities. This paper discusses our recent work in the growth of stable and epitaxial metallic layers which meet these stringent criteria on III–V semiconductors. Work on layers sharing either a common group III element with the III-V, such as AlAs/NiAl/AlAs or a common group V, such as GaAs/ErAs/GaAs, will be presented. The potential applications are further enhanced when the metallic films display an added functionality such as ferromagnetic behavior, demonstrated in epitaxial MnAl/AlAs/GaAs heterostructures. Examples of new directions for III-V structures made possible by such heteroepitaxy, are discussed. They include tailoring of Schottky barrier heights, integration of metallic quantum wells into high-speed three-terminal devices, and integration of magnetic storage functionality with III-V electronics.

1. INTRODUCTION

The increasingly versatile ability to successfully grow epitaxially material heterostructures which are dissimilar in terms of crystal structure, chemical bonding and even overall physical properties, has led to a wide range of possibilities in the use of this ever widening range of materials combinations. In the two decades since the First International Symposium on GaAs and Related Compounds, the concept of "related compounds" has expanded considerably. That original group, which included only a small list of closely-related lattice-matched III-V compounds, now can truthfully be said to include a broad spectrum of materials ranging from other III-V, II-VI and group IV semiconductors, to epitaxially-grown insulators, high T_c superconductors and metals. It is this last metals category that this paper addresses. We explore some of our current capabilities in forming integral metal/semiconductor monocrystalline structures, and show a few examples of how these new combinations point to new directions for III-V heteroepitaxy.

2. MOTIVATION

The motivation for exploring the concept of buried epitaxial metal layers in III-V heterostructures is multifaceted. The first reason is as an extension of the principle role metallizations already play in III-V devices, that is providing the link between the semiconductor devices and the "outside world" . Until recently, in most cases thermally unstable and polycrystalline metallizations have been adequate for most Schottky barrier and ohmic contact applications. However, we are fast reaching the limits of such schemes. As both lateral and vertical device dimensions "shrink" to the size of the individual metallic grains, the necessity for more controllable single crystal metal substitutes becomes more and more apparent. In addition, the ability to bury metal interconnect layers and ground planes provides a vehicle for increasing the flexibility of device design, leading a step closer to the long sought goal of true three-dimensional device integration. Such combinations would make possible the use of metallic films to provide not only conventional but also more demanding metal/semiconductor functions.

The second motivation is to explore the incorporation of metals into novel, non-traditional device concepts, enhancing the versatility of III-V devices. This class of applications would include metal elements which could be used as magnetic storage devices, buried heat sinks, reflectors or waveguides. As the size of a buried metal layer becomes comparable to the size of the electron wavefunction in the metal, quantum effects begin to dominate, opening up the possibility of metallic quantum wells. Incorporating these in a three terminal

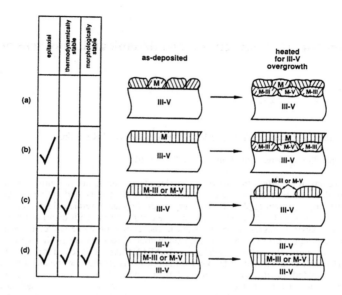

Figure 1: A schematic representation of the three criteria which must be met to allow the formation of a stable, and epitaxial buried metal layer within a III-V heterostructure.

resonant tunneling transistor with a metallic quantum well base, or using metal layers as buried gates in devices, such as an epitaxial permeable base transistor, increases the range of utilization of III-V-based devices. The challenge to the crystal grower is to conquer the complex materials issues so as to achieve the level of perfection in both the buried metal layer itself and the III-V material grown above it for the structures to meet the demanding performance requirements of these novel classes of devices.

3. STABILITY REQUIREMENTS

The rather stringent stability requirements which must be met in choosing a metal which can withstand the elevated temperatures needed to regrow III-V material above it have been thoroughly discussed by Sands et al (1990a). As shown schematically in Figure 1, the metal first of all must have the appropriate structure so as to grow epitaxially on the III-V. This is essential for subsequent monocrystalline III-V overgrowth, necessary for most device applications. Such a requirement alone, however, is still insufficient because, unless the metal is thermodynamically stable with respect to the surrounding III-V, it will eventually react with the underlying III-V when heated for overgrowth, as shown schematically on the right of case b) in the figure. Eventually it will be consumed by the reaction products and thus lose its role as an epitaxial template for the III-V overgrowth above it. If a metal meets these first two criteria, as shown in case c), it may still be unsuitable if it tends to agglomerate upon heating (as shown in the right-hand panel) because it will lose its electrical continuity, essential for most buried metal applications. Only when it is also morphologically stable to agglomeration, will it be possible to successfully achieve III-V overgrowth, retaining the metal layer intact as a continuous thin film.

The most obvious first candidates are the various elemental metals, especially those used in conventional compound semiconductor metallization applications. Although a number of elements meet at least one of the requirements listed above, only a few, such as W which is both morphologically and thermodynamically stable but not epitaxial, or Ag which is epitaxial and relatively thermodynamically stable but not morphologically stable, meet two of the three requirements. None of the elements, with the sole exception of Si, meet all three. And though GaAs on Si and Si delta-doping of GaAs both are legitimate active fields of study in the area of heteroepitaxy, Si can't be considered a viable candidate as a buried <u>metal</u> layer in GaAs. So we need to turn to metallic <u>compounds</u> in order to simultaneously meet all three buried metal layer requirements.

Again, the choices of the proper metallic compounds has been treated in detail in our previous review (Sands et al 1990a). The essence of the choice is indicated schematically in Figure 1. If one chooses M-III or M-V compounds which are the final by-products of an annealing reaction of the semiconductor with a deposited metal, one is then assured of thermodynamic stability. Selecting only those reaction products which form in a single epitaxial relationship with the underlying GaAs insures that the epitaxial requirement is met. And lastly, choosing a metallic compound with as high a melting point as possible allows the substrate to be heated to correspondingly higher temperatures, needed for III-V overgrowth, before entering a regime with high enough metal surface mobility to allow agglomeration.

The materials we have studied most closely which fall into the above categories, are the rare-earth monoarsenides such as ErAs, which form NaCl-type cubic M-V compounds (Palmstrom et al 1988, 1989) and NiAl and CoAl which form CsCl cubic M-III compounds (Harbison et al 1988, Sands et al 1990b). All three compounds have successfully been grown epitaxially (Sands et al 1990a), and though in all three cases the lattice constant of the metals is 1-2% larger than that of GaAs, exact lattice matching to the III-V can be achieved through ternary alloying of either the metal, in the case of $Sc_{.32}Er_{.68}As$ lattice matched to GaAs (Palmstrom et al 1990a), or the III-V in the case of NiAl or CoAl lattice matched to $In_{.30}Ga_{.70}As$ or $In_{.17}Ga_{.83}As$, respectively (Harbison et al 1990).

4. GROWTH BY MOLECULAR BEAM EPITAXY

The growth of these buried metal structures using molecular beam epitaxy (MBE) has been reviewed previously in detail (Sands et al 1990a). The growth sequence for the buried rare-earth monoarsenide/III-V heterostructures is similar to conventional III-V MBE in that the group V source remains on constantly. The group III and rare-earth sources are opened sequentially to build the appropriate buried metal structure. Buried ErAs layers as thin as two monolayers thick which are electrically continuous can be formed, as shown in Figure 2 (Sands et al 1990a). A problem arises, however, when one attempts GaAs overgrowth. Due to unfavorable interface surface energies, the overgrown GaAs grows in a 3-D growth mode, resulting in islands elongated along the [110] and [1$\bar{1}$0] directions, which grow to be many several nanometers thick before they laterally coalesce and

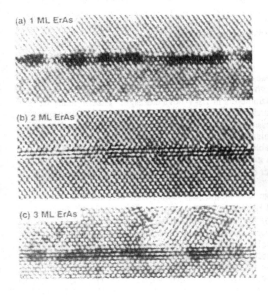

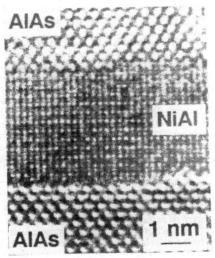

Figure 2: High-resolution cross-sectional transmission electron micrographs comparing buried layers of ErAs in GaAs with average ErAs thicknesses of 1,2 and 3 monolayers. The films are already electrically continuous over macroscopic dimensions by the middle 2 monolayer stage.

Figure 3: High-resolution cross-sectional transmission electron micrograph showing an epitaxial III-V/NiAl/III-V heterostructure fabricated by MBE (Sands et al 1990a).

completely cover the exposed ErAs. As a result, undesirable twinned GaAs variants form above the ErAs, leading to unwanted defects at the upper metal/semiconductor interface. Though a good fraction of these defects self-annihilate farther on in the overgrowth, their presence at the critical metal/III-V interface, is in general undesirable.

Realizing that this phenomenon was a wetting problem driven primarily by surface and interface energy considerations, and knowing that they were attempting to grow a polar (100) GaAs on a non-polar (100)ErAs surface, Palmstrom et al (1990b) examined the effect of growing on other crystal orientations. While ($\bar{1}\bar{1}\bar{1}$) growth provides polar-on-polar heteroepitaxy, it results in a different problem: the creation of multiple microtwin defects along the [$\bar{1}\bar{1}\bar{1}$] growth direction in both the rare-earth arsenide and overgrown GaAs layers. By performing growth in the intermediate crystal orientations, $7\bar{1}\bar{1}$, $5\bar{1}\bar{1}$, $3\bar{1}\bar{1}$, and $2\bar{1}\bar{1}$, the properties of films with a varying degree of ($\bar{1}\bar{1}\bar{1}$)- vs. (100)-like character was explored. While this work is still in progress, preliminary data indicate good growth on these intermediate directions, particularly ($3\bar{1}\bar{1}$), free from the gross defect problems discussed above with respect to straight (100) and ($\bar{1}\bar{1}\bar{1}$) growth.

MBE growth of transition metal aluminides presents difficulties in controlling background arsenic pressure which must be minimized during the intermetallic deposition step (Sands et al 1990a), although there is not as much of a problem in terms of the growth mode of the overgrown III-V. More importantly, special care must be taken to protect the exposed NiAl or CoAl surface from transition metal-arsenide formation during the time needed to raise the As flux again for III-V overgrowth. This may take several minutes. It can, however, be controlled by strategic additions of monolayers of Al during this period which just saturates the impinging As flux and slowly forms a protective AlAs layer. More recently, the advent of valved arsenic cracker MBE sources[*], vastly improves the control and response time of the As beam, alleviating many of these problems.

A second issue involves the exclusive nucleation of the (100) NiAl and CoAl variant on the (100) GaAs. Some growth conditions and shutter sequences have been found to nucleate (100) and (110) NiAl variants. By using a template process, however, of initially depositing one monolayer of Ni before codepositing Ni and Al, we can grow now the single desired (100) variant. The same holds true for Co for properly nucleating (100) CoAl. The resulting heterostructure formed when the proper steps are taken to insure the desired orientation of the metal on the III-V and then the III-V on the metal, is shown in Figure 3. Buried metallic films such as these can be made as thin as 1 nm (Sands et al 1990a) and still remain electrically continuous, making them excellent candidates for the metal quantum well devices discussed below.

5. SCHOTTKY BARRIER HEIGHT CONTROL

Having discussed the method we use to fabricate these metal/III-V heterostructures, we will now turn to some of the new directions for III-Vs made possible by such structures. The first direction lies in a more complete control of Schottky barrier height, one of the key parameters in III-V device design. Though many devices could benefit from a controllably high Schottky barrier height, Φ_B, the range of Φ_B values achievable in metallizations deposited by conventional III-V processing remains quite narrow, typically between 0.7 and 0.9 eV. Such metallizations often are not thermodynamically stable in the sense discussed earlier in this article, and they are almost never epitaxial. We have been able to exploit some of the unique properties of these new stable epitaxial metallization systems and provide a greater range of Φ_B. These modified Schottky barriers have thermally stable properties even up into the upper end of III-V processing temperatures (400 - 600 °C).

One way we have been able to tailor the effective Φ_B in the transition metal aluminide system is by controlling the thickness of an AlAs layer inserted between the GaAs and the NiAl or CoAl for chemical stability. Since AlAs has a substantially higher Φ_B for most metallizations ($\Phi_B \sim 1.1$ eV), the Φ_B of the composite metal/AlAs/GaAs heterostructure can be significantly elevated over that of metallizations directly on GaAs. This scheme is similar to using metals deposited ex situ on MBE-grown III-V structures terminated with AlGaAs layers. The advantage in the present case is that pure AlAs can be utilized despite its reactivity in air because the epitaxial metallization is deposited in situ, protecting it from air exposure. Cheeks et al (1991) have shown that the I-V barrier heights in such NiAl/AlAs/GaAs and CoAl/AlAs/GaAs heterostructures can be as high as the 1.1 eV. Furthermore, if the AlAs interlayer is made thin enough to allow electrical tunneling, this effective I-V barrier height can be varied continuously from 0.75 to 1.1 eV, yielding a controllably tailorable Φ_B, by simply varying the thickness of the AlAs tunnel barrier.

Yet another example of increased Φ_B control using the stable epitaxial structures discussed above, which relies fundamentally on the monocrystalline nature of the interface, is the work of Palmstrom et al (1990b, 1992) studying Φ_B as a function of crystal orientation. Tung et al (1989) have shown previously, in the case of monocrystalline silicide/Si heterostructures, that Φ_B can be uniquely determined by the exact atomic

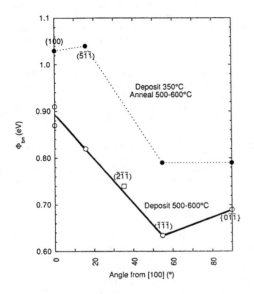

Figure 4: Φ_B for lattice matched ScErAs/GaAs diodes with different GaAs substrate orientation. Thermally stable valves over a range in excess of 0.65 - 1.0 eV are achievable.

arrangement at the single crystal interface. Palmstrom et al (1990b, 1992) rely on such a concept and explore the dependence of Φ_B on the crystallographic orientation of the underlying substrate. The study parallels the overgrowth mode work as a function of orientation discussed above in the previous section. Their results are shown in Figure 4. Because the bonding across the metal/III-V interface is so different for the (100) and ($\bar{1}\bar{1}\bar{1}$) case (nonpolar-on-polar vs. polar-on-polar) it is not surprising that Φ_B is significantly different for these two cases. Figure 4 also shows that one can obtain intermediate values between these two extremes by choosing intermediate crystallographic orientations. Furthermore, additional reproducible differences can be seen between samples deposited at low temperature and annealed at high temperature and ones deposited directly at high temperature (upper and lower curves in the figure). The different As/Ga stoichiometry at the interface determined by the substrate temperature is the probable cause for these differences. The results demonstrate quite a wide swing in Φ_B in excess of 0.65 - 1.0 eV. Furthermore, these Φ_B's are thermally stable up to the 600 °C range, making them applicable to device structures requiring relatively high-temperature processing.

6. METALLIC QUANTUM WELL DEVICES

Another new direction for III-Vs made possible by the fabrication of integral III-V/metal/III-V heterostructures arises from the discrete quantum-confined energy levels which appear in such a metallic quantum well when the metal layer becomes thin enough that it is comparable to the size of the wavefunction of the electron (Sands et al 1990a). The physics which gives rise to the quantum confinement effects is qualitatively quite similar to that in III-V heterostructure quantum wells. The differences are quantitative. Whereas a semiconductor quantum well has a relatively shallow depth of a few tenths of an eV, with intersubband spacings on the order of a tenth of an eV, metallic quantum wells are much deeper, typically 10 eV, with subband spacings on the order of an eV. In addition, the metallic wells have sheet concentrations ~100 times higher than the semiconductor wells (typically $10^{15}/cm^2$ as opposed to $10^{13}/cm^2$) with as many as 10 - 20 of the lowest subbands filled below the Fermi level. This increased carrier density results in a highly desirable decreased sheet resistance which could play a crucial role in devices in which one makes electrical contact to the well itself. Tabatabaie et al (1988) first explored the two terminal vertical transport properties of such metallic quantum wells in the 2-5 nm thickness range, cladding them with two AlAs tunnel barriers surrounded by GaAs. They showed that such a structure behaves like a canonical all-semiconductor double-barrier resonant tunneling diode. One, and on occasion two negative differential resistance regions were detected with peak-to-valley ratios at room temperature as high as 2:1, confirming metallic quantum well behavior.

Tabatabaie et al (1989) took this concept one step further by fabricating a three-terminal switching device based on a contacted metallic quantum well layer, (shown schematically in Figure 5). Processing takes advantage of the fact that the metallic well can be contacted in a straightforward manner using etches selective between the semiconductor and the metal. A low-resistance ohmic Ti-Au contact can easily be made to NiAl through these etch-exposed areas without the need for an annealing step.

The concept of the device is to modulate the current tunneling from the leftmost source through the left hand AlAs barrier into the metallic NiAl quantum well by means of a rightmost gate. Modulating the gate voltage modulates the exact shape of the right hand AlAs barrier, which in turn subtly affects the position of the confined quantum states within the metallic well. Figure 6 shows the operation conditions chosen in such a way that increasing the gate voltage, plotted on the x-axis, pulls a quantum confined state which is in resonance out of resonance, resulting in the measured monotonic decrease in the source current. The flat, near-zero gate current shown in the same figure at an expanded scale on the left, provides proof that the effect does not arise from current injection from the gate side of the device. Though the proportion of the source current modulated in this way due to these quantum effects is quite low in this first proof-of-principle device, the possibilities it suggests for yet other three-terminal metallic quantum well devices are quite rich and varied, providing exciting new directions for the future.

7. INTEGRATED MAGNETIC STORAGE AND III-V ELECTRONICS

The final example of a new direction opened up for III-V's through the use of stable epitaxial metal/semiconductor heterostructures comes in the field of magnetic materials. Many ferromagnetic thin film applications, such as dense perpendicular magnetic recording and magneto-optic recording or switching, require films with a magnetization perpendicular to the film plane. In the current generation of magneto-optic commercial films this is achieved by growth-induced stress anisotropies, not yet completely understood. In ferromagnetic epitaxial films such as Fe and Co, because of the large shape anisotropy of a thin film

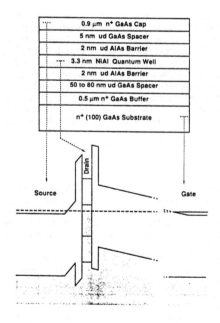

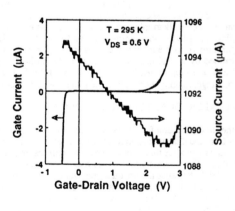

Figure 6: I-V characteristics of the device shown in Figure 5.

Figure 5: Schematic representation of a successful three-terminal metallic quantum well device dubbed a buried metal well quantum field effect device (Tabatabaie et al 1989).

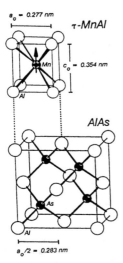

$a_o \sim 0.277$ nm

τ-MnAl

$c_o \sim 0.354$ nm

Mn

Al

AlAs

As

Al

$a_o/2 \sim 0.283$ nm

Figure 7: Epitaxial orientation of the ferromagnetic tetragonal τ-MnAl phase on the underlying (100) AlAs layer. Such an orientation leads to perpendicular magnetization desired for many applications (Sands et al 1990c).

geometry, the magnetization inevitably lies in the plane. In the case of τ-MnAl/AlAs/GaAs heterostructures which we have fabricated along the lines of the systems discussed in the above sections of this article, epitaxy has been used to orient the film magnetization perpendicular to the film surface. As shown in Figure 7, the τ-MnAl phase is a tetragonally distorted version of the cubic CsCl structure of NiAl and CoAl. This phase possesses a strong magnetocrystalline anisotropy favoring the alignment of the magnetic moment of the Mn atom along the long axis of the tetragonal cell. By growing the film epitaxially on AlAs with the tetragonal unit cell's square dimensions closely matching the (100) AlAs surface, we can "engineer" this magnetization along the proper plane-perpendicular direction. This τ-phase is the only ferromagnetic phase in the MnAl system. It is, however, a metastable phase accessed previously only by non-equilibrium techniques such as quenching or sputtering. In spite of its inherent metastability, we can use the epitaxial template's alignment energetic advantage to further stabilize this desired magnetic phase. The growth process (Harbison et al 1991), closely parallels that of the other transition metal aluminides and relies heavily, again, on the all important approach of proper initial template formation at the AlAs/MnAl interface.

The magneto-optic properties of these MnAl films are discussed more thoroughly elsewhere in this volume (Cheeks et al 1992). Preliminary measurements of very thin layer films are already within a factor of 2 or 3 of the materials currently in use, with further improvements expected for thicker films. Measurements of the perpendicular magnetic component of the films have also been made using the extraordinary Hall effect, an in-plane transport probe of the film magnetization (Leadbeater et al 1991). Not only do such measurements confirm the existence of a strong perpendicular magnetic component, but in the best films they reveal an almost ideal rectangular magnetic hysteresis loop which is optimal for many magnetic storage applications The films possess a coercive field of a few kilogauss, close to 100% remnance and a Curie temperature in the range of 250 - 400°C. One such trace is shown in the upper right-hand panel of Figure 8. The ability to have magnetic storage on chip in a form integrated with the III-V electronics is very appealing, and represents quite a significant new direction. What makes this particular measurement so appealing, in the context of new directions for III-V's, is that the MnAl being probed was part of an integrated structure, shown in the left-hand side of the figure, which included a two-dimensional electron gas (2DEG) in the underlying III-V. The simultaneous presence of a good 2DEG is evidenced by the Shubnikov de Haas oscillations displayed in the lower right, measured in the underlying semiconductor nearby. The juxtaposition of these two capabilities opens the possibility of future applications in which the non-volatile memory elements in the form of "magnetic gates" may be able to directly affect high-speed switching in underlying circuitry, based perhaps, as in this example, on a 2DEG, forming a switchable channel.

The applications of the new concepts presented here are admittedly some time off in the future. Yet they illustrate some of the powerful new directions being opened up through the newly established capability to grow crystalline metallic thin films and their subsequent use in integrated metal/III-V semiconductor heterostructures.

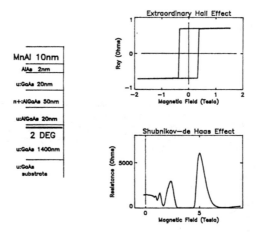

Figure 8: Results of measurements probing both the magnetic hysteresis loop of the MnAl film, using in-plane transverse resistivity referred to as the extraordinary hall effect, and the Shubnikov-de Haas oscillations revealing the presence of a two-dimensional electron gas in the underlying semiconductor portion of the heterostructure. The measurements were made on slightly different sections of the wafer whereby contacts could be made in one area to the magnetic metal film and in the other to the two-dimensional electron gas.

References

* EPI, 261 East Fifth Street, St. Paul, Minnesota, USA 55101

Cheeks T L, Sands T, Nahory R E, Harbison J P, Gilchrist H L and Keramidas V G 1991 accepted for publication in J. Electronic Mat.

Cheeks T L, Nahory R E, Sands T, Harbison J P, Brasil M J S P, Gilchrist H L, Schwartz S, Pudensi M, Allen S J Jr, Florez L T and Keramidas V G 1992 Proceedings of the 18th International Symposium on GaAs and Related Compounds held in Seattle, Washington September 9-12, 1991

Harbison J P, Sands T, Tabatabaie N, Chan W K, Florez L T and Keramidas V G 1988 Appl. Phys. Lett. 53 1717

Harbison J P, Sands T, Ramesh R, Tabatabaie N, Gilchrist H L, Florez L T and Keramidas V G 1990 J. Vac. Sc. Technol. **B8** 242

Harbison J P, Sands T, Ramesh R, Florez L T, Wilkens B J and Keramidas V G 1991 J. Cryst. Growth **III** 978

Leadbeater M L, Allen S J Jr, DeRosa F, Harbison J P, Sands T, Ramesh R, Florez L T and Keramidas V G 1991 J. Appl. Phys. **69** 4689

Palmstrom C J, Tabatabaie N and Allen S J Jr 1988 Appl. Phys. Lett. **53** 2608

Palmstrom C J, Garrison K C, Mounier S, Sands T, Schwartz C L, Tabatabaie N, Allen S J Jr, Gilchrist H L and Miceli P F 1989 J. Vac. Sci. Techol. **B7** 747

Palmstrom C J, Mounier S, Finstad T G and Miceli P F 1990a Appl. Phys. Lett. **56** 382

Palmstrom C J, Cheeks T L, Gilchrist H L, Zhu J G, Carter C B and Nahory R E 1990b Mat. Res. Soc. Extended Abstract **EA-21** 63

Palmstrom C J, Cheeks T L, Gilchrist H L, Zhu J G and Carter C B 1992 to be submitted to J. Vac. Sci. Technol.

Sands T, Palmstrom C J, Harbison J P, Keramidas V G, Tabatabaie N, Cheeks T L, Ramesh R and Silberberg 1990a Materials Science Reports **5** pp 99-170

Sands T, Harbison J P, Ramesh R, Palmstrom C J, Florez L T and Keramidas V G 1990b Mater. Sci. Eng. **B6** 147

Sands T, Harbison J P, Leadbeater M L, Allen S J Jr, Hull G W, Ramesh R, and Keramidas V G 1990c Appl. Phys. Lett. **57** 2609

Tabatabaie N, Sands T, Harbison J P, Gilchrist H L and Keramidas V G 1988 Appl. Phys. Lett. **53** 2528

Tabatabaie N, Sands T, Harbison J P, Gilchrist H L, Cheeks T L, Florez L T and Keramidas V G 1989 Technical Digest International Electron Device Meeting Dec. 3-6, 1989, Washington DC **IEDM 89** 555

Tung R T 1989 J. Vac. Sci. Technol. **A7** 598

Inst. Phys. Conf. Ser. No 120: Chapter 1
Paper presented at Int. Symp. GaAs and Related Compounds, Seattle, 1991

9

Short wavelength II–VI laser diodes

M.A. Haase, J. Qiu, J.M. DePuydt and H. Cheng
3M Company, 201-1N-35, 3M Center, St. Paul, MN 55144

Abstract   The first wide band gap II-VI semiconductor laser diodes were recently reported.   These devices emit at 490 nm (blue-green) under pulsed current injection at 77 K, and are comprised of   CdZnSe single quantum wells in ZnSe-ZnSSe waveguides.   Further advances in this technology have resulted in room temperature pulsed operation of green laser diodes.   These developments are discussed in light of the difficulties which have historically been presented by the II-VI compounds.

Introduction

Research on wide band gap II-VI materials (*i.e.*, ZnSe, which has a band gap of 2.7 eV, and related ternaries) and devices has burgeoned in the last few years.   Although there have been interesting reports of attempts to incorporate ZnSe into GaAs electronic devices, the most important applications are as visible light emitters.   Blue LEDs may find applications as indicators and in high-brightness full-color displays.   Short-wavelength (blue or green) laser diodes are expected to play a major role in next-generation optical recording systems, where the smaller spot size afforded by the short wavelength will allow significantly higher information densities.   Such lasers are also likely to be used in laser printers, digital photocopiers, and a wide variety of sensors.   Undersea optical communications are often mentioned as another possible application, since the attenuation of blue light in sea water is less than that of any other wavelength in the electromagnetic spectrum.

This paper provides a summary of the salient developments leading to the demonstration of the first wide band gap II-VI laser diodes, operating at wavelengths as short as 490 nm (Haase *et al* 1991b).   Three important obstacles stood in the way of this accomplishment.   They were—and to a large extent still *are*—1) the difficulty of attaining high *p*-type conductivity in ZnSe and related alloys, 2) the lack of a suitable ohmic contact to *p*-ZnSe, and 3) the need for an appropriate heterostructure system to provide carrier and optical confinement.

1. The quest for p-type ZnSe

Claims of p-type conductivity (albeit, extremely low conductivity) in bulk ZnSe date back at least to 1967 (Haanstra 1967). Perhaps the first truly encouraging report of p-type conductivity in ZnSe was the bulk growth of Nishizawa et al (1986). They reported hole concentrations up to 1.5×10^{15} cm^{-3} using Li acceptors. The first reports of epitaxial p-type ZnSe grown on GaAs came in early 1988. Yasuda et al (1988) used MOCVD and Li$_3$N (presumably the acceptor was Li), and Cheng et al (1988) used MBE and elemental Li. One of the more intriguing results was that of Akimoto et al (1989), who reported p-type doping and pn junctions using O acceptors from a ZnO source in MBE. Nitrogen acceptors and p-type conductivity were also reported using NH$_3$ in MOCVD (Ohki et al 1988), and MOMBE (Migita et al 1990).

Through 1990, the success of ZnSe:Li by MBE was arguably the most reproducible, as it spread to several laboratories and led to several reports of blue LEDs. Despite the widely discussed "problems" of rapid diffusion, and electromigration (Haase et al 1991a), Li doping played an important role by allowing the clear identification of the important issue of (the lack of) ohmic contacts to p-ZnSe. For lightly doped p-ZnSe, this difficulty makes Hall measurements unreliable, if not impossible. The issue of p-type contacts will be discussed in the next section. One of the important outcomes was the conclusion that C-V measurements of N_A-N_D are as informative and more reliable than Hall measurements—not to mention easier (DePuydt et al 1989). A standard mercury probe used at relatively low frequencies (<10 kHz) suffices. The real problem with ZnSe:Li (at least by MBE) was the fact that at high Li concentrations (above 10^{17} cm^{-3}) compensation occurs. Therefore, N_A-N_D in excess of 1×10^{17} cm^{-3} was never achieved (DePuydt et al 1989).

An important breakthrough came with the development of an N$_2$ plasma source for MBE by Park et al (1990), and independently by Ohkawa et al (1991). This technique employs a small helical-coil rf plasma chamber (manufactured by Oxford Applied Research) which replaces a Knudsen cell in the MBE chamber. The "active" nitrogen species is thought to be either neutral, mono-atomic N free radicals," or neutral, excited N$_2$ molecules. Park et al used the technique to achieve N_A-N_D=3.4$\times 10^{17}$ cm^{-3} and blue LEDs.

Subsequent work by Qiu et al (1991) demonstrated N_A-N_D up to 1.0×10^{18} cm^{-3} by doping with a nitrogen-plasma source. In that work we showed that the N incorporation could be increased by: 1.) increasing the rf power, 2.) increasing the II/VI flux ratio, or 3.) by decreasing the substrate temperature. We also showed that compensation occurs at N concentrations in excess of about 10^{18} cm^{-3}, thus limiting N_A-N_D. The most important discovery of that work was that higher N_A-N_D can be achieved at lower substrate temperatures. The room temperature resistivities of our ZnSe:N layers are as low as 0.7 Ω-cm. We have also made convincing variable-temperature van der Pauw-Hall measurements (DePuydt *et al* 1991).

The reproducibility of our ZnSe:N layers is excellent, and similar results have been achieved in ZnSSe alloys; particularly in ZnS$_{0.07}$Se$_{0.93}$ which is lattice-matched to GaAs at the growth temperature of 300°C.

2. The ohmic contact problem

The difficulty of making device-quality ohmic contacts to *p*-ZnSe was grossly underestimated by many of the early workers in this field. In early 1990, we (the 3M group) reported in some detail on blue LEDs which operated at over 15V at modest current densities. The problem was identified as difficulty in injecting holes into *p*-ZnSe (Haase *et al* 1990).

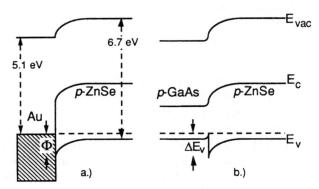

Figure 1. Band diagrams of a.) p-type ZnSe Schottky barrier, and b.) the p-GaAs/p-ZnSe interface.

The fundamental impediment to the injection of holes into *p*-ZnSe is simply that the energy of the valence band is extremely low—approximately -6.7 eV with respect to the vacuum level. Most metals form more-or-less "classical" Schottky barriers on ZnSe and other wide band gap materials. Pt and Au are among the metals with the largest work functions at 5.1 and 5.65 eV, respectively. Therefore, one would expect them to result in Schottky barrier heights Φ of 1.6 and 1.1 eV (Figure 1). In fact, the valence-band barrier height for Au-ZnSe has been measured at 1.25 eV by XPS (Xu *et al* 1988). Our experience with Pt (sputtered, evaporated, and electroplated) is that it is slightly inferior to Au for hole injection.

We have experimented with many metals (Li, Na, Mg, Ti, Cr, Mn, Ni, Pd, Pt, Cu, Ag, Au, Zn, Hg, Al, In, Sn, Pb, Sb, Bi, and many alloys thereof), surface cleaning techniques (*in situ* evaporation, wet chemical etching, sputter etching), deposition techniques (electroplating, evaporation, sputtering) and annealing techniques (including rapid thermal annealing, and laser annealing). The conclusion is that evaporated, unannealed Au provides the "least bad" contact—that is, the leakiest Schottky barrier.

The III-V materials, *e.g.*, GaAs, provide at least two advantages over ZnSe for making low resistance electrical contacts. First, several metals will alloy with GaAs at reasonable temperatures. Our experience with *p*-ZnSe has shown that the material begins to compensate, becoming more resistive, if heated to temperatures above its growth temperature (typically 300°C). Unfortunately, we have found no metals that alloy at these low temperatures. The other III-V advantage is that degenerate doping is trivial. To make a *p*-type contact, one need only to alloy with a metal that includes Zn to make very heavily *p*-doped material, and achieve large tunneling currents. With the II-VIs, of course, it is much more difficult to make p^+ material.

One might hope that hole injection could be achieved via the III-V to II-VI interface. In this case, the depth of the ZnSe valence bands presents itself as a large band discontinuity, $\Delta E_v \approx 1.0$ eV (Kowalczyk 1982, Kassel 1990), again blocking hole transport (Figure 1). Therefore, if one makes a ZnSe *pn* junction LED (or laser) on a *p*-GaAs substrate, the *p*-GaAs/*p*-ZnSe interface blocks hole flow from the substrate. If sufficient voltage is applied, current flows by avalanche injection at this interface. Filamentary (spotty) injection is observed. In the most heavily doped *p*-ZnSe, the spots are sufficiently dense to give the appearance of uniform emission.

Alternatively, we have made blue LEDs (and blue-green laser diodes) on *n*-GaAs substrates, with the *p*-ZnSe on top. In this case, the reverse-biased Schottky barrier blocks hole injection from the metal electrode. Again, at sufficiently high voltage, current flows by means of avalanche injection. Recently, with the advent of N doping *via* rf plasma, we have used heavily doped (N_A-$N_D \approx 10^{18}$ cm^{-3}) contact layers, and semitransparent Au (100 Å thick) electrodes. The current injection from such structures behaves more like tunneling current, and may be modeled with Fowler-Nordheim theory. The operating voltages of the best LEDs are typically 6 V.

3. Laser design

Recently, the first blue-green laser diodes were announced (Haase *et al* 1991b). In order to achieve sufficient carrier confinement to reach inversion, a single strained quantum well of CdZnSe was used within a ZnSe *pn* junction. CdZnSe-ZnSe quantum wells have been shown to lase by photopumping, although extremely high input powers were required (Ding *et al* 1990).

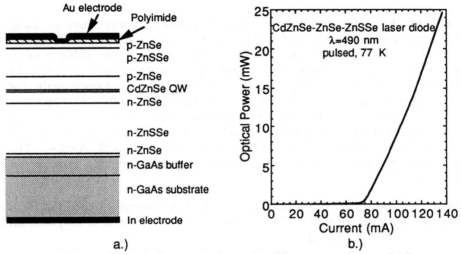

a.) b.)

Figure 2. a.) Cross-section of a blue-green laser diode.
b.) L-I characteristic of such a device.

The index of refraction of ZnSe is approximately 2.7, which can be compared to 3.9 for GaAs at the operating wavelength of 490 nm (at 77K). Therefore, light generated in the II-VI material tends to be antiguided into the substrate where it is strongly absorbed. Although a lossy waveguide is formed by the discontinuity of the imaginary part of the refractive index, better optical confinement is needed. We have used $ZnS_{0.07}Se_{0.93}$ for a cladding material. A cross section of the device is diagrammed in Figure 2. The top and bottom ZnSSe cladding layers are 1.5 and 2.5 μm thick, respectively, and the total thickness of the ZnSe light-guiding region is 1.0 μm. The $Cd_{0.2}Zn_{0.8}Se$ quantum well is nominally 100 Å thick. The *p*-type layers are doped (from an N_2 plasma source) to N_A-N_D=2×10^{17} cm^{-3}, except for the top 0.1 μm ZnSe layer which is doped to 10^{18} cm^{-3}. The lower *n*-ZnSe guiding layer is doped with Cl from a $ZnCl_2$ source to 1×10^{17} cm^{-3}, and the lower cladding and *n*-ZnSe contact layer to about 10^{18} cm^{-3}. Devices are typically patterned into 20 μm wide "gain-guided" stripes, and cleaved to cavity lengths typically on the order of 1 mm. The lower index of refraction of ZnSe gives rise to uncoated facet reflectivities of only 0.21 (compared to 0.3 for GaAs lasers).

Computer modeling of the heterostructure shown in Figure 2 shows that the optical confinement is sufficient to reduce losses in the substrate to less than 1 cm^{-1} (much less than the anticipated free carrier and scattering losses), and to provide an optical confinement factor of $\Gamma=0.013$ for a 100 Å quantum well.

Because of the persistent p-type contact problem, these devices have so far been operated only under pulsed current injection. Laser action was first seen at 77 K, at a wavelength of 490 nm (blue-green). Lasers from that first wafer have worked at temperatures as high as 200 K without facet coatings. Figure 2b shows an *L-I* characteristic for one of these devices. Output powers in excess of 100 mW per facet have been observed. Differential quantum efficiencies in excess of 20% per facet have also been measured. The threshold current for the device of Figure 2b is 78 mA which corresponds to 320 A/cm^2. The output from these devices is TE polarized, and a "speckle pattern" is clearly visible.

Figure 3 shows the characteristic electroluminescence spectra from one of these laser diodes. The devices operate in many longitudinal modes, and several lateral modes.

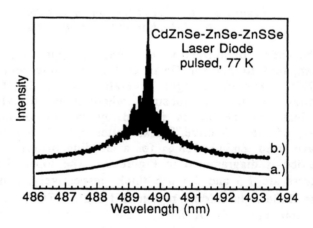

Figure 3.
Electroluminescence
spectra from a blue-
green laser diode:
a.) below threshold,
b.) above threshold.

CdZnSe-ZnSe-ZnSSe
Laser Diode
pulsed, 77 K

In order to achieve operation at room temperature, we have grown a similar structure with a deeper quantum well to enhance the carrier confinement. In this case, the well is nominally 75 Å of Cd$_{0.34}$Zn$_{0.66}$Se. By coating both facets of the devices made with this material with high reflectivity ($R \approx 0.7$) dielectric coatings (in order to reduce the gain required for stimulated emission), we have observed 77 K threshold currents as low as 13 mA for a 695 μm long device, which corresponds to 95 A/cm². These facet-coated devices also operated at room temperature at a wavelength of 535 nm (green). Figure 4a shows a room temperature *L-I* curve for one of these devices. The threshold current density is about 5000 A/cm². Other devices have had threshold current densities as low as 2800 A/cm² at room temperature.

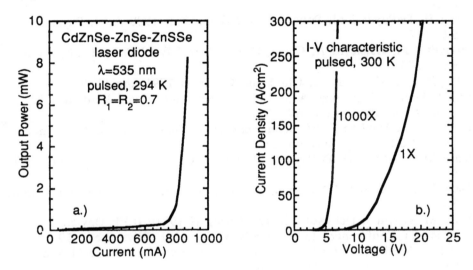

Figure 4. a.) L-I characteristic for a green laser diode at room temperature, b.) I-V characteristic for a II-VI laser diode.

The I-V characteristic for a typical laser diode is shown in Figure 4b. At low current densities, these devices act like well-behaved diodes (with turn-on voltages of 4-5 V). However, to reach the laser threshold current density, 20 V is typically required. This implies that over 10 W of input power is needed to reach threshold at room temperature. It is not surprising that the lifetimes of these lasers are only a few minutes at room temperature.

Conclusions

Recent achievements have clearly demonstrated the feasibility of blue-green laser diodes fabricated from II-VI semiconductors. However, if such devices are to become commercially viable, several outstanding challenges must be met. These include reduction of the room-temperature threshold current densities, and development of low-resistance ohmic contacts to *p*-type ZnSe.

Acknowledgements

We gratefully acknowledge the waveguide modeling of D.K. Misemer, and the technical assistance of K.A. Gleason and P.F. Baude.

References

Akimoto K, Miyajima T and Mori Y 1989b Japan. J. Appl. Phys. **28** L531

Cheng H, DePuydt J M, Potts J E, and Smith T L 1988 Appl. Phys. Lett. **52** 147

DePuydt J M, Haase M A, Cheng H and Potts J E 1989 Appl. Phys, Lett. **55** 1103

DePuydt J M, Haase M A, Qiu J and Cheng H 1991 unpublished.

Ding J, Jeon H, Nurmikko A V, Luo H, Samaranth N and Furdyna J K 1990 Appl. Phys. Lett. **57** 2756

Haanstra J H 1967 in *II-VI Semiconducting Compounds* ed D G Thomas (New York: Benjamin) pp 207-14

Haase M A, Cheng H, DePuydt J M and Potts J E 1990 J. Appl. Phys. **67** 448

Haase M A, DePuydt J M, Cheng H and Potts J E 1991a Appl. Phys. Lett. **58** 1173

Haase M A, Qiu J, DePuydt J M and Cheng H 1991b IEEE Device Research Conference; to be published in Appl. Phys. Lett. **59** (11)

Kassel L, Abad H, Garland J W, Raccah P M, Potts J E, Haase M A and Cheng H 1990 Appl. Phys. Lett. **56** 42

Kowalczyk S P, Kraut E A, Waldrop J R and Grant R W 1982 J. Vac. Sci. Technol. **21** 482

Migita M, Taike A and Yamamoto H 1990 J. Appl. Phys. **68** 880

Ohkawa K, Karasawa T and Mitsuyu T 1991 Japan. J. Appl. Phys. **30** L152

Ohki A, Shibata N and Zembutsu S 1988 Japan. J. Appl. Phys. **27** L909

Park R M, Troffer M B, Rouleau C M, DePuydt J M and Haase M A 1990 Appl. Phys. Lett. **57** 2127

Qiu J, DePuydt J M, Cheng H, and Haase M A 1991 Electronic Materials Conference

Xu F, Vos M, Weaver J H and Cheng H 1988 Phys. Rev. **B38** 13418

Yasuda T, Mitsuishi I and Kukimoto H 1988 Appl Phys. Lett. **52**, 57

Inst. Phys. Conf. Ser. No 120: Chapter 1
Paper presented at Int. Symp. GaAs and Related Compounds, Seattle, 1991

17

Two-dimensional electron optics in GaAs

H. L. Stormer, J. Spector*, J. S. Weiner, K. W. Baldwin, L. N. Pfeiffer and K. W. West
AT&T Bell Laboratories, Murray Hill, N.J. 07974

Electron mobilities in modulation-doped GaAs-AlGaAs heterostructures presently exceed values of $10^7 cm^2/Vsec$ at low temperatures, equivalent to elastic mean free paths of ~100μm (Pfeiffer *et al* 1989). These are macroscopic distances on the length scale of today's semiconductor fabrication techniques. Electronic transport in such materials can no longer be regarded as diffusive. Instead, low-energy electron propagation is ballistic and a large fraction of the carriers traverse typical device distances without a serious scattering event. On one hand, such extraordinary electronic transport can seriously alter the low-temperature performance of traditional devices that are based on high mobility 2D electron systems. On the other hand, ballistic propagation offers novel possibilities to control and manipulate electrons in such systems.

Research of the past few years shows that low-energy ballistic carriers in two-dimensional electron systems behave in many ways analogous to photons in classical geometrical optics. Ohmic contacts act as electron sources and electron absorbers (Spector *et al* 1990a). Regions totally depleted of 2D electrons such as mesa edges and under highly biased gates act as almost perfect electron reflectors (Spector *et al* 1990a). Partially depleted regions refract electrons in accordance with an equivalent of Snell's law in optics (Spector *et al* 1990b). Beams of 2D ballistic electrons can be created and they are able to intersect each other with negligible interaction (Spector *et al* 1991). Using these novel control elements we demonstrate a 2D electron lens that focuses electrons from a point source onto a point detector by means of a photolithographically defined electrostatic lens of variable "index of refraction" (Spector *et al* 1990b). In a similar arrangement, an electrostatic prism can steer a beam of ballistic electrons onto three spatially separated electron detectors (Spector *et al* 1990c). High-mobility, two-dimensional electron systems, combined with modern lithographic techniques, lend themselves at low temperatures to the implementation of a wide variety of control elements that closely resemble the devices portrayed in textbooks on geometrical optics.

References

Pfeiffer L N, West K W, Stormer H L, Baldwin K W 1989 *Appl. Phys. Lett.* **55** 1880
Spector J, Stormer H L, Baldwin K W, Pfeiffer L N, West K W 1990a *Appl. Phys. Lett.* **56** 967
Spector J, *et al* 1990b *Appl. Phys. Lett.* **56** 1290
Spector J, *et al* 1991 *Appl. Phys. Lett.* **56** 263
Spector J, *et al* 1990c *Appl. Phys. Lett.* **56** 2433

---

* Present address: Arizona State University, Tempe, Arizona 85287

Inst. Phys. Conf. Ser. No 120: Chapter 2
Paper presented at Int. Symp. GaAs and Related Compounds, Seattle, 1991

19

Facet formation observed in MOMBE of GaAs on a patterned substrate

Y.MORISHITA, Y.NOMURA, S.GOTO, Y.KATAYAMA, and T.ISU*

Optoelectronics Technology Research Laboratory,
5-5 Tohkodai, Tsukuba, Ibaraki 300-26, Japan

*Central Research Laboratory, Mitsubishi Electric Corporation,
8-1-1 Tsukaguchi-Honmachi, Amagasaki, Hyogo 661, Japan

Abstract A detailed investigation was carried out on facet formation in the metalorganic molecular beam epitaxy (MOMBE) of GaAs on a mesa-etched (100) GaAs substrate by in situ scanning microprobe reflection high-energy electron diffraction (μ-RHEED) in real time. The distribution of the growth rates on the (100) surface near the edge of the mesa-groove was also measured by μ-RHEED. Besides the initial (100) surface and a (111)A sidewall, a (411)A facet as well as (311)A and ($5\overline{1}\overline{1}$)A facets were formed at the lower and upper sidewalls, respectively, at growth temperatures between 540 and 620 °C. It was found that the formation of the ($5\overline{1}\overline{1}$)A and (411)A facets is substantially associated with the exponential variation of the growth rate on the (100) surface, due to the flow of either Ga adatoms or Ga compounds from the sidewall to the (100) surface of the mesa pattern.

1. Introduction

Epitaxial growth on patterned substrates is useful for achieving advanced structures of optoelectronic devices such as laser diodes (Dzurko et al. 1989), waveguides (Colas et al. 1989), and transistors (Asai et al. 1984). For these devices, it is important to control the shapes of the epitaxial layers, especially in the active region of the device. It is well known that crystal planes with different indices affect the growth behavior on neighboring surfaces in molecular beam epitaxy (MBE) and metalorganic chemical vapor deposition (MOCVD) on patterned substrates. It has been reported that facets are formed between different crystal planes during growth (Tsang and Cho 1977, Hersee et al. 1986).

There have been few theoretical studies so far on the fundamental growth morphology formed on patterned substrates; Wulff (Wulff 1901) has reported that an epitaxial layer tends to form a shape which minimizes the total surface energy of the crystals, and that the shape is determined by the orientation dependence of the surface energy. Jones et al. (Jones et al. 1991) have applied the Wulff constriction to the prediction of edge shapes for nonplanar epitaxial growth, and have shown that the shapes of epitaxial layers are basically predetermined by the Wulff construction replacing the energy polar diagram with a growth rate polar diagram. Ohtsuka and Suzuki (Ohtsuka and Suzuki 1989) have simulated GaAs growth on patterned substrates using a model which takes into account the effects of the incorporation, desorption and surface migration of Ga adatoms, and have demonstrated that, by assuming a certain migration length, bump-like shapes are formed on epitaxial layers grown on convex surfaces.

Although metalorganic molecular beam epitaxy (MOMBE) is a promising growth technique, the growth mechanism is still under intense investigation. We (Morishita et al. 1991) have observed facet formation during the MOMBE of GaAs on a mesa-etched (100) GaAs surface by scanning microprobe reflection high-energy electron diffraction (μ-RHEED) in real time, and have shown that the shape of the growing surface can be exactly monitored by μ-RHEED.

In this paper, we report on a detailed investigation of the effects of the growth temperature on facet formation in the MOMBE of GaAs on a mesa-etched (100) GaAs surface by μ-RHEED in real time. We also discuss the influences of the variation of the growth rate measured on the (100) surface near the edge, due to surface migration of Ga adatoms or Ga compounds, upon the shapes of epitaxial layers.

2. Experimental

GaAs MOMBE growth was carried out on (100) substrates with mesa-grooves along the [0$\bar{1}$1] direction having outward sloping sidewalls. Growth apparatus used in this study and substrate preparation were described elsewhere (Morishita et al. 1991). In the MOMBE growth, trimethylgallium (TMGa) was delivered through a low-temperature (100 °C) effusion cell. The flow rate of TMGa was controlled at 1.0 SCCM by an electronic mass flow controller. Arsenic was supplied as a molecular beam and the flux was 8×10^{-4} Pa on a beam flux monitor. The substrate temperatures (T_s) from 540 to 620 °C were monitored by an infrared pyrometer. The growth rate on the (100) surface far from the edge of a mesa-groove was about 0.4 μm/h.

For RHEED measurements, a 25 keV electron beam with a diameter of about several hundred Å was aligned along the [0$\bar{1}$1] azimuth. Fig.1 shows a schematic illustration of the epitaxial layer. The upper drawings represent the RHEED patterns observed at each point between a and f in the lower figure. The incident electron beam scanned the surface around the sidewall along a line normal to the edge of the mesa-groove (namely, parallel to the [0$\bar{1}$1] direction) at a speed of 16 sec/line. The observed RHEED patterns during growth were recorded in real time with a video system. In the present experiment, the electron beam scanned a width of about 135 Å during one frame of the video picture. The indices of the surfaces were determined from the angle of the RHEED patterns at each point on the scanned line. The widths of

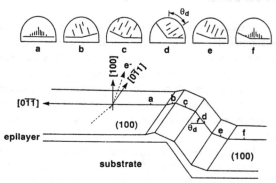

Fig. 1   Schematic diagram of the epitaxial layer. The upper drawings represent the RHEED patterns observed at each point between a and f in the lower figure.

the facets were estimated from the number of video frames on which the RHEED patterns from the respective facets were taken during one scan.

The distribution of the growth rates on the (100) surface was measured from the periods of the RHEED intensity oscillation at each point on the scanned line. For growth rate measurements, the incident electron beam scanned the (100) surface near the edge of the mesa-groove along the $[0\,\overline{1}1]$ direction for 20 msec/line. Details of the measurement system have been described elsewhere (Isu et al. 1991). This technique is very useful in simultaneously obtaining the distribution of growth rates in an observed area within a few monolayer growth sequence.

3. Results

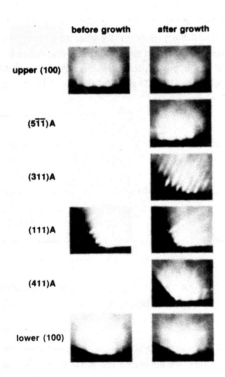

Fig. 2 Change of RHEED patterns observed at each point on the scanned line around a sidewall of a patterned substrate before and after growth.

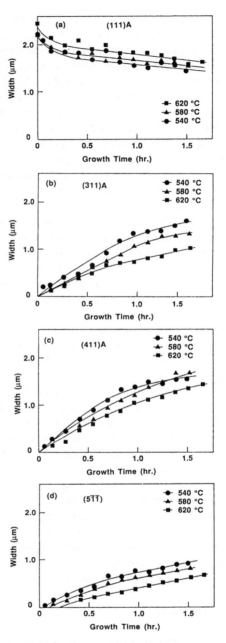

Fig. 3 The growth time dependence of the width of the (a) (111)A, (b) (311)A, (c) (411)A, and (d) (5$\overline{1}$1)A facet observed in MOMBE of GaAs for various growth temperatures.

Fig. 2 shows a set of RHEED patterns at each point on the scanned line around a sidewall of the patterned substrate before and after 1.5 hr growth at T_s of 580 °C. The RHEED patterns from a (100) surface and a sidewall with an intersecting angle of about 53° with the (100) plane were observed before growth, indicating that the initial face revealed on the sidewall was a (111)A plane. The onset of growth led to the appearance of RHEED patterns from new facets with different characteristic angles. The facet appearing at the lower sidewall had an intersecting angle of about 20° with the (100) plane; two facets at the upper sidewall had intersecting angles of about 25° and 15° with the (100) plane, respectively. These new facets appearing at the upper sidewall intersected each other at an angle of about 140°. Almost exactly the same tendency was observed with the RHEED patterns from the epilayers grown at T_s between 540 and 620 °C, indicating that the facets with the same indices are formed at the lower and upper sidewalls, respectively, under the experimental conditions studied. The facets with intersecting angles of about 20, 25, and 15° with the (100) surface correspond to the (411)A, (311)A, and (5$\overline{1}$1)A planes, respectively.

The widths of each facet observed in MOMBE of GaAs are plotted in fig. 3 for various growth temperatures as a function of the growth time. For all of the growth temperatures, the RHEED patterns from the (411)A and (311)A facets were observed immediately after the start of growth. Their width increased rapidly with growth up to about 1 h, and slightly thereafter. On the other hand, the width of the (111)A sidewall decreased rapidly with growth; the tendency then slowed down slightly. The gradient of the variation of the width for all facets increased as the growth temperature was lowered. These results indicate that (411)A and (311)A facets are quickly formed on the intersection regions between the initial (100) surface and the (111)A sidewall after the start of growth, and that the rapid development for the first 1 h of growth is enhanced at lower growth temperatures. At all growth temperatures, unlike in other cases involving different facets, the RHEED pattern from the (5$\overline{1}$1)A facet appeared after about 0.2 h of growth; the onset of the appearance increased from about 0.05 to 0.25 h as the growth temperature was raised from 540 to 620 °C. The width increased asymptotically as growth proceeded. These results suggest that a different factor affects the formation of the (5$\overline{1}$1)A facet. Details are discussed in the next section.

A set of (01$\overline{1}$) cross-sectional drawings of the epitaxial layers after 1.5 h growth which are determined by the width of each facet is shown in fig. 4. The contours of the epitaxial layers determined using μ-RHEED agree well with those obtained from secondary electron microscope observations at all growth temperatures.

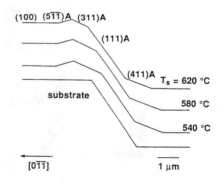

Fig. 4 A (01$\overline{1}$) cross-sectional drawing of the epitaxial layers grown at different growth temperatures.

4. Discussion

In order to explain the shapes of the epitaxial layers, we measured the growth rates on the planar (100), (111)A, (311)A, (411)A, and (511)A substrates at T_s between 540 and 620 °C. The growth rates were measured from the thicknesses of the epitaxial layers grown simultaneously on the same substrate holder. The results are summarized in table 1 for various growth temperatures. The growth rates for the (100) substrate were exactly equal to those obtained from the periods of the RHEED intensity oscillation on the (100) surface far from the edge of a mesa-groove of the patterned

Table 1. Growth rates on each substrate for various growth temperatures.

T_s (°C)	growth rate (μm/hr)				
	(100)	(111)A	(311)A	(411)A	(511)A
540	0.40	0.35	0.27	0.37	0.42
580	0.40	0.25	0.32	0.41	0.50
620	0.43	0.10	0.35	0.44	0.53

substrate at all growth temperatures. The growth rates on the (100), (311)A, (411)A, and (511)A substrates increased with raising growth temperature. On the other hand, the value for the (111)A substrate decreased as the growth temperature was raised. As has been proposed by Jones et al. (Jones et al. 1991), using the values of the growth rates summarized in table 1, the contours of the epitaxial layers are predicted for various growth temperatures by a Wulff construction, replacing the energy polar diagram with a growth rate polar diagram. The results are shown in fig. 5 for different growth temperatures.

On the other hand, the growth rate on the (100) surface near the edge of the mesa-groove was reported to vary exponentially as a function of the distance from the edge during MBE of GaAs, due to the flow of Ga adatoms from the (111)A sidewall to the (100) surface of the mesa pattern (Hata et al. 1990). The growth rates on the (100) surface near the edge were measured using the same technique during the MOMBE of GaAs. The exponential dependence of the increase of the growth rate on the distance from the edge was also obtained for the MOMBE of GaAs. In order to take into account the effect of the flow of Ga adatoms or Ga compounds from the (111)A sidewall to the (100) surface, the shapes of the epitaxial layers were modified by adding the increase of growth rates on the (100) surface near the edge of a mesa-groove to those on the planar (100) substrate. The results are also drawn in fig. 5. The lowermost and uppermost lines at each growth temperature represent the drawing of the substrate before growth and that of the contour of the epitaxial

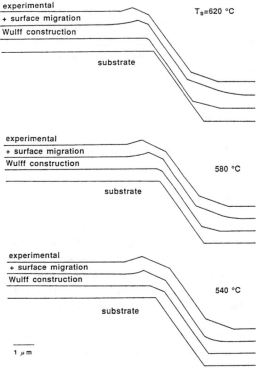

Fig. 5 A $(01\bar{1})$cross-sectional view of the epitaxial layers obtained using μ-RHEED observations and those determined by the Wulff constructions with and without taking into account the flow of Ga adatoms or Ga compounds.

layer obtained using μ-RHEED, lrespectively.

The result clearly indicates that the bump-like shape comprising of the $(5\overline{1}1)$A and (311)A facets is only obtained by taking into account

the flow of Ga adatoms or Ga compounds. On the intersection region between the (100) surface and the lower (111)A sidewall, the Wulff construction predicts no facet at T_s=540 °C; a (511)A facet is predicted at T_s>580 °C. However, the (411)A facet is actually observed by μ-RHEED. The formation of the (411)A facet is also considered to be due to the flow of Ga adatoms or Ga compounds from the (111)A sidewall to the lower (100) surface. In other words, it is considered that the surface migration of adatoms affects the formation of the (411)A facet. Although it is very difficult to measure the distribution of the growth rates on each facet, investigations about the surface migration on the facet are desired in order to exactly understand the shapes of the epitaxial layers.

5. Conclusion

A detailed observation was carried out on facet formation during the MOMBE of GaAs by μ-RHEED in real time. The distribution of the growth rates on the (100) surface near the edge of a mesa-groove was also measured by μ-RHEED. Besides the initial (100) surface and (111)A sidewall, a (411)A facet and (311)A and $(5\overline{1}1)$A facets were formed at the lower and upper sidewalls, respectively, at growth temperatures between 540 and 620 °C. It was found that the formation of $(5\overline{1}1)$A and (411)A facets is substantiality associated with the exponential variation of the growth rate on the (100) surface, due to the flow of Ga adatoms or Ga compounds from the sidewall to the (100) surface of the mesa pattern.

Acknowledgment

The authors would like to thank Dr. I.Hayashi for enlightening discussions and continuous encouragement.

References

Asai H, Adachi S, Ando S, and Oe K 1984 J. Appl Phys. 55 3868.
Colas E, Yi-Yan A, Bhat R, Seto M, and Deri R 1989 Appl. Phys. Lett. 54 1501.
Dzurko K M, Menu E P, Beyler C A, Osinski J S, and Dapkus P D 1989 Appl. Phys. Lett. 54 105.
Hata M, Isu T, Watanabe A, and Katayama Y 1990 Appl. Phys. Lett. 56 2542.
Hersee S D, Barbier E, and Blondeau R 1986 J. Crystal Growth 77 310.
Isu T, Hata M, Morishita Y, Nomura Y, and Katayama Y 1991 to be published in J. Crystal Growth.
Jones S H, Seidel L K, Lau K M, and Harold M 1991 J. Crystal Growth 108 73.
Morishita Y, Nomura Y, Goto S, Katayama Y, and Isu T 1991 to be published in Surface Sci.
Ohtsuka M and Suzuki A 1989 J. Crystal Growth 95 55.
Tsang W T and Cho A T 1977 Appl. Phys. Lett. 30 293.
Wulff G 1901 Z. Krist. 34 449.

Inst. Phys. Conf. Ser. No 120: Chapter 2
Paper presented at Int. Symp. GaAs and Related Compounds, Seattle, 1991

25

Atomically flat AlGaAs/GaAs (110) heterointerface grown by molecular beam epitaxy

G.Tanaka, K.Hirakawa, H.Ichinose and T.Ikoma

Institute of Industrial Science, University of Tokyo,
7-22-1 Roppongi, Minato-ku, Tokyo 106, Japan

Abstract : We found that one of the difficulties in growing (110) AlGaAs/GaAs hetero-structures with a high-quality interface is the lack of steps to stabilize Ga atom migration during growth. Insertion of an InAs buffer layer introduces a high density of misfit dislocations which provide equivalent steps at the growing surface. We found, however, that the hetero-interface of the structures grown by this method is very poor and neither quantum wells nor modulation doped hetero-interfaces can be grown.

We have shown that by lowering the growth temperature down to 400°C and using a high As_4 to Ga ratio (30), atomically flat (110) interface can be obtained. The FWHMs of photo luminescence spectra from quantum wells with different well widths are as narrow as those of the best (100) hetero-interface and the fluctuation of the interface roughness is less than one monolayer.

Introduction

High-quality AlGaAs/GaAs quantum wells and superlattices have been successfully grown on (100) substrates and a relatively flat hetero-interface with one mono-layer fluctuations has been obtained. At a (100) surface, anion and cation atoms sit on slightly different planes and therefore dipoles are possibly formed at the hetero-interface. The interface dipoles can contribute to the band discontinuity, which has been a hot topic discussed by many authors (Capasso et al 1985, Waldrop et al 1990, Lambrecht et al 1990, Peressi et al 1991, Sorba et al 1991).

On the other hand, at a (110) surface, anion and cation atoms sit on the same plane and the charge neutrality holds in the plane (Harrison et al 1978). Therefore, we expect no interface dipole layer to be formed (homopolar interface), and a very flat interface should be obtained on this crystal plane. However, so far only a few (Munekata et al 1987, Allen et al 1987, Zhou et al 1987) high quality interfaces have been reported on the (110) growth surface. Thus, it is important and interesting to determine the optimal growth condition to obtain a high quality (110) interface from the view point of practical applications as well as investigating the effects of interface dipoles on the hetero-interface band offset.

Munekata et al. reported that when a thin InAs layer was inserted between the substrate and an epitaxial layer, a mirror like surface was obtained on a (110) substrate. They did not, however, report the quality of the interface. Allen et al. reported that a hetero-structure with a mirror like surface was obtained on a (110) substrate tilted by 6° toward (1̄11) at temperatures higher than 550°C. On the other hand, Zhou et al. succeeded in growing a hetero-structure with relatively high electron mobility on an untilted (110) substrate when the growth temperature is low (470-500°C) and the As_4 to Ga ratio is high (20-30).

In this paper, we first test the method of Munekata et al. stressing the characterization of the hetero-interface. Secondly, we determine an optimal condition to grow AlGaAs/GaAs hetero-structures on an untilted (110) substrate without an InAs layer, in order to obtain a high quality interface. Finally, we characterized those interfaces by photoluminescence and the mobility of two dimensional electron gas.

Role of InAs buffer layer

We simultaneously grew AlGaAs/GaAs single hetero-structures on untilted (100) GaAs substrates with and without a thin InAs buffer layer by molecular beam epitaxy (MBE). The growth temperature was varied from 400°C to 600°C, and As_4 to Ga ratio from 5 to 35. The composition of Al in AlGaAs is 0.3. The surface morphologies were observed with an optical microscope and an atomic force microscope (AFM) shown in Fig. 1(a). We confirmed that the insertion of a thin InAs layer could improve surface morphologies as shown in Fig.1 (b) and (c) as reported by Munekata et al.

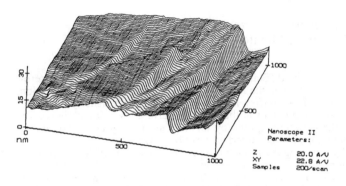

Figure 1(a)   AFM image of fasets on surface.

Figure 1(b) SEM image of surface morphology grown without an InAs buffer layer. Many facets are observed on surface.

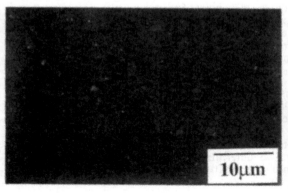

Figure 1(c) SEM image of surface morphology grown with an InAs buffer layer. By comparison with (a), a facet growth is suppressed by inserting an InAs buffer layer and a mirror-like surface is obtained.

In order to investigate the interface flatness, we fabricated quantum wells with various well widths as shown in Fig.2 and measured the photoluminescence at 77K. No clear exciton lines were observed as shown in Fig.3. There is a broad peak around 720 nm, suggesting that at the interfaces AlGaAs and GaAs were mixed and formed an alloy. The electron mobility at 77K was also very low (3000 cm^2/Vs). This indicates that the insertion of an InAs buffer layer is not a suitable method to obtain a high quality (110) interface.

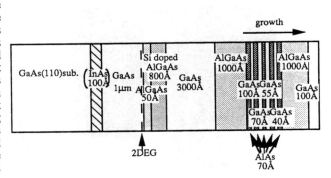

Figure 2 Quantum wells grown for characterization of interfaces

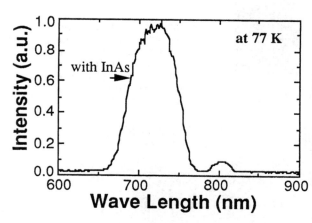

Figure 3 Photoluminescence spectrum from quantum wells at 77 K. The growth temperature is 450°C. No peaks from quantum wells are obserbed. There is a broad peak which suggests that no well-defined quantum wells are formed.

The scanning electron microscope images of cleaved surfaces of the two wafers grown under the same condition on (110) and (100) substrates are shown in Fig.4 (a) and (b). Apparently, the growth rate on (110) substrates is 0.8 times smaller than that on (100) substrates. This shows that the sticking probability of Ga atoms is lower on (110) surface than on (100) surface. This fact leads us to speculate that Ga atoms migrate for longer distance on a (110) surface than on a (100) surface to find stable positions before they stick on steps or kinks on the surface and thus have a larger probability to re-evaporate from the surface. This is reasonable since (110) surface has less steps. Moreover, a transmission electron microscope lattice image was taken on the cleaved surface (a cross sectional image) of the (110) AlGaAs/GaAs hetero-structure with an InAs buffer layer and is shown in Fig.5. As shown in the figure many dislocations are generated due to large mismatch of the lattice constants. These dislocations play an important role in providing equivalent steps where Ga atoms can stick stably before they re-evaporate. Thus, an apparent two-dimensional growth becomes possible by insertion of an InAs layer. This growth condition is similar to the case of highly misoriented substrates (Allen et al 1987), which have many steps at the surface. However, it is easily seen that such heterostructures formed with the InAs layer will give poor electronic and optical properties which was confirmed through the photoluminescence spectra and electron mobility.

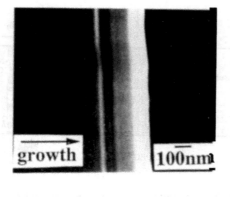

Figure 4(a)   SEM image of a cleaved surface grown on a (100) substrate. The white areas are AlGaAs layer. The black areas are GaAs layer.

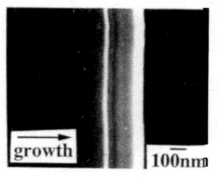

Figure 4(b)   SEM image of a cleaved surface grown on a (110) substrate. Comparing with (a), GaAs layers are thinner, showing that Ga atoms are likely to re-evaporate on a (110) surface.

Figure 5   TEM image of the interface between GaAs(110) substrate and an InAs buffer layer. At the interface many dislocations are introduced by large mismatch of lattice constance. These dislocations tilt the growth plane.

Optimal Growth Condition

From the above observations we find that one of the difficulties of growing thin (110) epitaxial layers is due to the fact that Ga atoms migrate very long distance and re-evaporate before sticking at steps or kinks on the surface. Therefore it is desirable to set up the growth condition in such a way as to suppress Ga atom migration and re-evaporation. One of the means to do so is to provide steps or kinks by introducing a strained buffer layer or by using misoriented surfaces as mentioned in the previous section. However these methods do not give a high-quality interface which is a goal of the present investigation. Another means is to lower the growth temperature to reduce the surface migration, and to increase the As_4 to Ga ratio to suppress re-evaporation of Ga atoms.

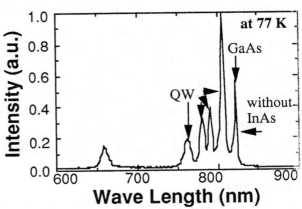

Figure 6 Photoluminescence spectrum from quantum wells at 77 K. Without an InAs buffer layer, the peaks from quantum wells are apparently obserbed.

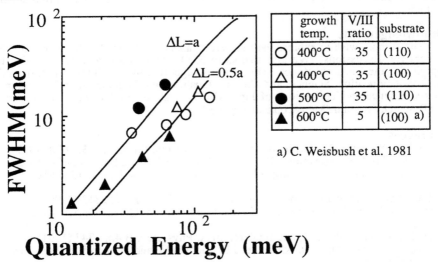

	growth temp.	V/III ratio	substrate
O	400°C	35	(110)
△	400°C	35	(100)
●	500°C	35	(110)
▲	600°C	5	(100) a)

a) C. Weisbush et al. 1981

Figure 7 FWHM of photoluminescence spectra as a function of quantized energy. Theoretical curves are obtained by assuming that fluctuations of the interface roughness are one to a half monolayer.

We grew (110) AlGaAs/GaAs quantum wells, which are schematically shown in Fig. 2, and modulation-doped single hetero-structures, in which the electron density is 5×10^{-11} cm^{-2}, at growth temperature of 400°C and an As_4 to Ga ratio of 30. The Al composition in the AlGaAs layers is 0.3.

The surfaces of these structures are all mirror like. The photoluminescence spectrum is shown in Fig.6, which shows clear excitonic peaks corresponding to each quantum well. As usual we plotted the full width at half maximum (FWHM) as a function of quantized energy in Fig.7 with growth temperature as a parameter. The FWHM is as narrow as that from one of the best (100) quantum wells (Weisbush et al 1981) and falls between the two theoretical curves which assume interface fluctuations of one-half to one monolayer as shown in the figure. The mobility of two dimensional gas was 60000 cm^2/Vs at 77K, which is comparable to that at (100) interface.

Thus, we obtained high quality hetero-interface on untilted (110) GaAs substrates without the use of an InAs layer at a low substrate temperature of 400°C and a high As_4 to Ga ratio of 30.

Conclusions

We found that one of the difficulties in growing (110) AlGaAs/GaAs hetero-structures with a high-quality interface is the lack of steps to stabilize Ga atom migration during growth. Insertion of an InAs buffer layer (Munekata et al 1987) introduces a high density of misfit dislocations which provide equivalent steps at the growing surface. We found, however, that the hetero-interface of structures grown by this method is very poor and neither quantum wells nor modulation doped hetero-interfaces can be grown.

We have shown that by the lowering growth temperature down to 400°C and using a high As_4 to Ga ratio (30), atomically flat (110) interface can be obtained. The FWHMs of quantum wells with different well widths are as narrow as those of the best (100) hetero-interface and the fluctuation of the interface roughness is less than one monolayer. These optimal conditions are similar to those of Zhou et al. but the growth temperature is smaller than theirs.

Acknowledgements

The authors wish to express their sincere thanks to T. Saito, Y. Hashimoto and M. Noguchi for discussions. We also thanks to S. Ozawa of Furukawa Electric Co. for supplying us the GaAs substrates. This work is supported by the Grant-in-Aid from the Ministry of Education, Science, and Culture, Japan, by the Foundation of Promotion of Material Science and Technology of Japan (MST), and also by the Industry-University Joint Research Program "Mesoscopic Electronics."

References

Allen L T P, Weber E R, Washburn J, and Pao Y C, Appl. Phys. Lett. **51**, 670 (1987).
Capasso F, Cho A Y, Mohammed K, and Foy P W, Appl. Phys. Lett. **46**, 664 (1985).
Harrison W A, Kraut E A, Waldrop J R, and Grant R W, Phys. Rev. B **18**, 4402 (1978).
Lambrecht W R L, Segall B, and Andersen O K, Phys. Rev. B **41**, 2813 (1990).
Munekata H, Chang L L, Woronick S C, and Kao Y H, J. of Crystal Growth **81**, 237 (1987).
Peressi M, Baroni S, Resta R, and Baldereschi A, Phys. Rev. B **43**, 7347 (1991).
Sorba L, Ceccone G, Antonini A, Walker J F, Micovic M, and Franciosi A, Phys. Rev. B **43**, 2450 (1991).
Waldrop J R, Kraut E A, Farley C W, and Grant R W, J. Vac. Sci. Technol. B **8**, 768 (1990).
Weisbuch C, Dingle R, Grossard A C, Wiegmann W, Solid State Commun. **38**, 709 (1981).
Zhou Junming, Huang Yi, Li Yongkang, and Jia Wei Yi, J. of Crystal Growth **81**, 221 (1987).

Inst. Phys. Conf. Ser. No 120: Chapter 2
Paper presented at Int. Symp. GaAs and Related Compounds, Seattle, 1991

Abrupt heterojunctions of AlGaAs/GaAs quantum wells grown on (111)A GaAs substrates by molecular beam epitaxy

T. Yamamoto, M. Fujii, T. Takebe, D. Lovell, and K. Kobayashi

ATR Optical and Radio Communications Research Laboratories,
Seika-cho, Soraku-gun, Kyoto 619-02, Japan

ABSTRACT

Morphologies and interface abruptness of AlGaAs/GaAs on (111)A GaAs substrates are studied using single quantum well (SQW) structures consisting of five GaAs layers of variable thickness. They are grown on exactly (111)A GaAs and (111)A GaAs surfaces that are misoriented towards the [100] direction by 1°, 3°, and 5°. The abruptness of AlGaAs/GaAs heterojunctions grown on GaAs (111)A surfaces has been evaluated by photoluminescence(PL) intensity measurements. It has been demonstrated that the PL peak wavelengths for SQW structures on the exactly and 5°- misoriented (111)A surfaces agree with the values calculated with the effective mass of the heavy hole of $m^*_{hh}=0.9m_0$. The abruptness of SQW structures on these surfaces is estimated as less than one monolayer from full width at half maximum(FWHM) values.

1. Introduction

Recently, epitaxial growth on patterned substrates has stimulated interest in the fabrication of lateral microstructures (Miller, Meier et al, and Kapon et al 1985, 1987, and 1987). The differences in doping type, layer thickness, and composition of ternary alloy systems between the slope and the flat are main topics of interest. With regard to epitaxial growth on (100) substrates patterned with one-dimensional features, advanced devices with lateral carrier confinement structures, such as quantum wire lasers, have attracted much attention (Kapon et al 1990). Three-dimensional confinement of carriers, such as quantum dots, may be obtained similarly by molecular beam epitaxial (MBE) growth on substrates patterned with two-dimensional features. We have made good use of the threefold rotational symmetry of the (111)A surface, and obtained an equilateral triangle pattern in which three slopes have equivalent crystallographic faces. The conductivity type of MBE grown Si-doped GaAs layers on (111)A-oriented GaAs is controlled by the growth conditions and the degree of (111)A substrate misorientation towards the (100) surface (Okano et al 1990). We have proposed and realized a new type of carrier confinement structure by the combination of these interesting properties (Fujii et al 1991).

However, it is difficult to grow GaAs and AlGaAs layers that have good surface morphology on the (111)A GaAs. Our purpose is to solve the above problem and to achieve a surface morphology and heterointerface that is satisfactory for fabricating optical devices. It has been reported that the morphology of GaAs and AlGaAs layers on (111)A substrates are dependent on growth conditions (Fuke et al 1990) and misorientation angle (Okano et al 1990). Recently, we have found that the morphology of GaAs and AlGaAs layers on (111)A substrates are strongly dependent on surface treatment conditions for the GaAs substrates before the growths (Yamamoto et al 1991).

In this paper, we investigate the dependence of the surface morphology and heterointerface quality of MBE AlGaAs/GaAs single quantum wells on (111)A GaAs substrates, upon the surface misorientation under optimized surface treatment conditions, in order to obtain mirror-like surface and high quality in epitaxial layers.

2. Experiment

Samples under study were grown by MBE on semi–insulating exactly (111)A oriented GaAs wafers and those 1°–, 3°– and 5°– misoriented towards [100]. The substrates were cleaned by organic solvents and the surface native oxide layers were removed by HCl:H_2O. Then, they were etched in NH_4OH:H_2O_2:H_2O solution and rinsed in de–ionized (DI) water. The above etchant was intentionally selected because the etch rate is very low and it leaves a smooth surface. The thermal etching temperature was 780 °C under a As$_4$ flux intensity of 3.2x10^{-5} Torr. (We have purposed the optimized surface treatments for (111)A substrates (Yamamoto et al 1991). The good surface morphology have been obtained on condition that higher thermal etching temperature under higher As$_4$ flux intensity than those of (100).) The temperature of $Al_{0.5}Ga_{0.5}$As/GaAs SQWs growth was kept at 600 °C . The As$_4$/Ga flux ratios (γ) were about 7.5 for GaAs and 5.0 for $Al_{0.5}Ga_{0.5}$As, respectively. The growth rates, corrected by reflection high energy electron diffraction (RHEED) oscillations, were about 1.0μm/h for $Al_{0.5}Ga_{0.5}$As and 0.5μm/h for GaAs. First, a GaAs buffer layer (0.2nm) and an $Al_{0.5}Ga_{0.5}$As barrier layer (50nm) were grown on (111)A GaAs substrates. Then, a SQW structure consisting of five GaAs layers of varying thickness, 2, 3, 5, 9, and 13nm, each separated by an 50nm $Al_{0.5}Ga_{0.5}$As layer, was grown. In order to investigate the dependence of optical properties on the tilt angle, all the (111)A substrates and an (100) substrate were placed side by side on a substrate holder. No growth interruption was done between the wells(GaAs) and the barriers(AlGaAs). Surface morphologies were observed by scanning electron microscopy(SEM). Photoluminescence(PL) spectra were measured at 15K using the 488nm line of an Ar ion laser.

3. Results and Discussion

Figure 1(a), (b), (c), (d), and (e) show photoluminescence spectra of SQWs on the exactly (111)A GaAs wafer, those 1°–, 3°–, 5°– misoriented, and the (100) wafer with the five quantum wells, respectively. The PL peak wavelengths of all the (111)A SQWs are longer than those of the (100) SQWs. As the misorientation from (111)A increases, the PL peaks show red shifts and the linewidths become broad. Unknown PL peaks appear in the high energy side for the (111)A 1°– and 3°– misoriented samples. The PL linewidth of the 5°–misoriented samples recover those of the exactly (111)A sample.

The peak wavelengths of PL spectra as a function of single quantum well width are shown in Fig. 2. The experimental transition wavelengths of the PL peaks are compared with theoretical values. The solid and broken lines are the calculations for (100) and (111)A orientations, respectively. It has been assumed that the recombination transition takes place between an electron and a heavy hole in the lowest eigenstates in a finite square potential well. The equation and band parameters used in the calculation

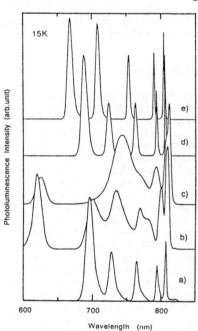

FIG. 1 PL spectra at 15K of the exactly (111)A oriented (a), 1°–, 3°–, and 5°– misoriented (111)A (b), (c), and (d), and the (100) SQWs (e). The dependence of PL peak wavelengths of AlGaAs/GaAs single quantum wells on the (111)A GaAs substrates upon the surface misorientation are shown.

Table I. The equation and band parameters used in the calculation.

Equation :
$$tan^2\left(\sqrt{\frac{m_w E L_z^2}{2h^2}}\right) = \frac{m_b(V-E)}{m_w E}$$

E: eigenvalue in the 1D finite square potential well, m_b, m_ω: barrier and well mass of the particle
L_z: well width, V: barrier height

Band parameters $m_0=9.11\times10^{-28}$ g, $E_g(T)=1.519-5.405\times10^{-4}T^2/(204+T)$

$V_c=0.63[1.247x+1.147(x-0.45)^2]$, $V_v=0.37[1.247x+1.147(x-0.45)^2]$

$m^*_{hh}GaAs[100]=(0.34)m_0$, $m^*_{hh}Al_xGa_{1-x}As[100]=(0.34+0.41x)m_0$

$m^*_{hh}GaAs[111]=(0.90)m_0$, $m^*_{hh}Al_xGa_{1-x}As[111]=(0.90+0.41x)m_0$

$m^*_e GaAs(E)=(0.0665+0.0436E+0.236E^2-0.143E^3)m_0$ (E in eV), $m^*_e Al_xGa_{1-x}As=(0.0665+0.083x)m_0$.

are listed in Table I.

The effective mass of the heavy hole for the (111) orientation is taken to be $m^*_{hh}Al_xGa_{1-x}As[111] = (0.90+0.41x)m_0$ (Hayakawa et al 1988). The dependence of the binding energy of two-dimensional electron–heavy–hole free excitons upon the quantum well width is taken into account in the calculation; 11 and 8 meV for well thicknesses of 4.2 and 14.5nm, respectively (Miller et al 1981). The valence–band discontinuity between GaAs and $Al_xGa_{1-x}As$ is taken to be 37% of the total band–gap differences ($\Delta Eg=1.247x+1.147(x-.45)^2$ (Casey et al 1978), assumed to be independent of temperature.

The PL peak wavelengths for every well width for the (100) SQWs are in good agreement with the solid line theoretically derived for the (100) orientation. This clearly shows that the desired SQW structures have been realized. The PL peak wavelengths for exactly and 5°– misoriented (111)A SQWs are consistent with the broken line theoretically derived for the (111) orientation. To our knowledge, this is first experimental demonstration that the effective mass of the heavy hole for the (111)A GaAs is $m^*_{hh}=(0.9)m_0$. On the other hands, the PL peak wavelengths for the 1°– and 3°– misoriented samples show red shift.

The dependence of the full width at half maximum (FWHM) upon the quantum

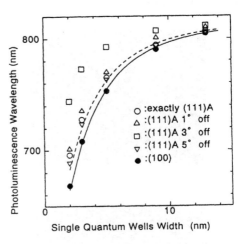

FIG. 2 Peak wavelengths of PL spectra as a function of single quantum wells width. The experimental transition wavelengths of the PL peaks are compared with theoretical values. The solid and broken lines are the calculations for (100) and (111)A orientations, respectively.

well width is shown in Fig. 3. The solid
and broken lines are the theoretical
FWHM values that are evaluated from
$[\partial E^{*}/\partial L_{z}]dL_{z}$, E^{*} is the transition energy,
assuming that the fluctuation of the well
width (dL_{z}) is one monolayer for both
(100) and (111)A orientation, respectively. For well width below 3nm, the
FWHMs of (100), exactly (111)A and
5°– misoriented (111)A samples are
narrower than the theoretical values. As
a result, the abruptness of SQW structures on these surfaces can be estimated
as less than one monolayer.

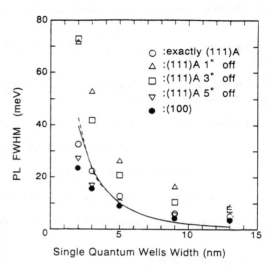

FIG. 3 Full width at half maximum (FWHM) as a
function of the quantum well width. The solid and
broken lines are the theoretical FWHM values which
were evaluated from $[\partial E^{*}/\partial L_{z}]dL_{z}$, E^{*} is the transition energy, assuming that the fluctuation of the well
width (dLz) is one monolayer for both (100) and
(111)A orientations at 0K.

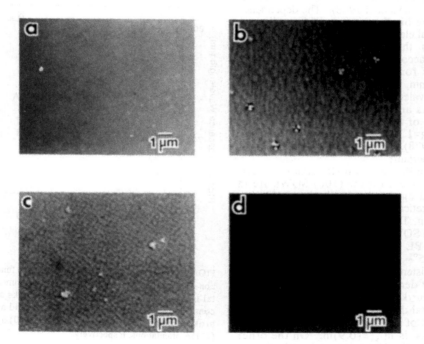

FIG.4 Surface morphology of the AlGaAs/GaAs SQW structures grown on the exactly (111)A GaAs (a), and
1°(b), 3°(c), and 5°(d)–misoriented toward [100]. The exactly (111)A surface exhibits a few defects and the 5°
misoriented surface exhibits fewer defects. On the other hands, numerous giant steps and diamond shape terraces of submicron sizes are observed on the 1°– and 3°– misoriented surfaces.

Surface morphology of the AlGaAs/GaAs SQW structures grown on the exactly (111)A and 1°-, 3°-, and 5°-misoriented (111)A surfaces are shown in Fig. 4(a), (b), (c), and (d), respectively. The exactly (111)A surface exhibits a few defects and the 5°- misoriented surface exhibits fewer defects. On the other hand, numerous giant steps of submicron sizes are observed on the 1°- and 3°- misoriented surfaces. The red shift and broadening seen in the PL spectra are probably caused by these giant steps. The mechanism of giant step formation has been explained as follows: the growth processes consist of surface migration of the growth elements and subsequent quasi-equilibrium reactions at the kinks in the steps (Suzuki et al 1990).

4. Conclusion

We have also proposed a surface emitting laser using the new structure (Fujii et al 1991).
The abruptness of AlGaAs/GaAs heterojunctions formed on (111)A GaAs surfaces has been evaluated by photoluminescence for four types of MBE-grown samples which have five single quantum wells having well thicknesses, of 2, 3, 5, 9, and 13nm, respectively. The exactly (111)A surface exhibits a few defects and the 5°- misoriented surface exhibits fewer defects. The observed peak energies from these substrates agree well with the theoretical values, which are derived assuming that the effective mass of the heavy hole for the (111) orientation are $m^*_{hh}=(0.9)m_0$. The abruptness for SQW structures on these surfaces is estimated as less than one monolayer from FWHMs values. On the other hand, a red shift and broadening of the PL spectra, an appearance of a peak in the high energy side, and the giant steps are observed on the 1°- and 3°- misoriented surfaces. Thus, it seems that giant steps and diamond shape terraces observed only on these surfaces are responsible for the change in the PL spectra.

Acknowledgements

The authors would like to thank Dr. Y. Furuhama for his encouragement throughout this work. Thanks are also due to Dr. M. Shigeta for stimulating discussions.

References

Casey H.C. and Panish M. B. 1978, Heterostructure Lasers (Academic, New York, 1978),part A.

Fujii M., Yamamoto T., Shigeta M., Takebe T., Kobayashi K., Hiyamizu S., and Fujimoto I. 1991, The 5th International Conference on Modulated Semiconductor Structures (Nara, Japan, July 8-12,) pp.149-151

Fuke S., Umemura M., Yamada N., and Kuwahara K. 1990, J. Appl. Phys 68, 997

Hayakawa T., Takahashi K., Kondo M., Suyama T., Yamamoto S., and Hijikata T. 1988, Phys. Rev. Lett. 60 349

Kapon E., Tamargo M. C., and Hwang D. M., 1987, Appl. Phys. Lett. 50(6), 347

Kapon E., Simhony S., Harbison J. P., Florez L. T. and Worland P. 1990, Appl.Phys. Lett. 56, 1825

Meier H. P., van Gieson E., Broom R. F., Walter W., Webb D. J., Harder C. and Jäckel H., 1987, Int. Symp. GaAs and Related Compounds 1987, edited by A. Christou and H. S. Rupprecht (Institute of Physics, Bristol, 1988),pp.609-612

Miller D. L. 1985, Appl. Phys. Lett. 47 ,1309

Miller R. C., Kleinman D. A., Tsang W. T., and Gossard A. C. 1981, Rhys. Rev. B24, 1134

Okano Y., Shigeta M., Seto H., Katahama H., Nishine S., and Fujimoto I. 1990, Jpn. J. Appl. Phys. 29, L1357

Suzuki T. and Nishinaga T. 1990, The 9th Symp. Record of Alloy Semiconductor Physics and Electronics, Izu-Nagaoka 1990 July p187

Yamamoto T., Takebe T., Fujii M., and Kobayashi K. 1991, The 10th Symp. Record of Alloy Semiconductor Physics and Electronics, Nagoya 1991 July 18-19 p73

Inst. Phys. Conf. Ser. No 120: Chapter 2
Paper presented at Int. Symp. GaAs and Related Compounds, Seattle, 1991

37

Confinement of excess arsenic incorporated in thin layers of MBE-grown low-temperature GaAs

J.P.Ibbetson*, C.R.Bolognesi[†], H.Weman[‡], A.C.Gossard*[†], U.K.Mishra[†]
Materials Dept.*, Dept. Electrical and Computer Engineering[†],
QUEST Center for Science and Technology[‡].
University of California, Santa Barbara, CA93106

Introduction

GaAs epilayers grown at low substrate temperatures (200-250°C) under otherwise normal MBE growth conditions are currently attracting much interest due to their unique electrical properties. Initially developed as buffer layers for power FET applications[1], Low-Temperature GaAs (LTGaAs) can reduce backgating in MESFETs [1]. Subsequently, surface layers of LTGaAs have also been employed to significantly improve MESFET gate-to-drain breakdown characteristics [2]. Most recently, indications of low-temperature (10K) superconductivity has also been reported in bulk LTGaAs [3], although this remains a controversial topic.

Material characterization studies have shown that the low growth temperature results in highly non-stoichiometric crystalline growth, with roughly 1 at.% excess As [4]. During annealing and/or postgrowth high temperature treatment the excess As redistributes itself in the crystal. In some cases metallic As precipitates have been observed [5].

Preventing out-diffusion of As-related defects into nearby active regions is very important for present and future applications of LTGaAs. Optical studies indicate that the presence of a LTGaAs buffer layer can significantly affect the quality of quantum wells ~1000Å away [6]. However, AlAs layers grown at normal growth temperatures (600°C) have been shown to prevent compensation of an n-type channel beneath LTGaAs surface layers [2]. In the present paper we present results on the optical properties of single and multiple quantum well structures incorporating thin layers (<150Å) of LTGaAs within barriers and well regions respectively. Photoluminescence (PL) is used to probe the quality of layers and interfaces in the immediate vicinity (~200Å) of LTGaAs layers,the optical properties of LTGaAs QW's, and the effectiveness of AlAs defect diffusion barriers.

MBE Growth

All samples in this study were grown in a Varian GENII solid source MBE system. Complete growth details will be published elsewhere [7]. All low-temperature (LT) growth was at a substrate temperature (T_{sub}) of 230°C as measured by a thermocouple. Otherwise growth occurred at T_{sub}=600°C.The growth rate was 0.5µm/hr, with an As_4:Ga beam equivalent pressure of 15:1.

Results

Two series of samples were grown. In the first series, a GaAs/AlGaAs single quantum well structure containing very thin layers of LTGaAs was studied. Bulk layers of LTGaAs have been shown to be optically dead, due to the large number of recombination centres they contain. In this study, the effect of these traps on the optical properties of thin layers of LTGaAs is determined. In sample A, 15Å of LTGaAs was inserted in the centre of a 150Å GaAs well. The growth was interrupted during the well growth to ramp down the temperature as described above. In sample B, just 3 monolayers of LTGaAs sandwiched by 4 monolayer thick confining barriers of AlAs grown at 600°C was inserted in the well. The structures are shown schematically in figure 1.

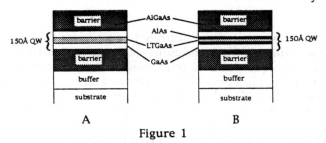

Figure 1

Schematic structure of samples A and B: The QW contains thin layers of LTGaAs.
Sample B also has AlAs diffusion barriers. No PL was observed for either sample

Neither sample was specifically annealed after the LTGaAs layer was deposited, but growth of the upper barrier and completion of the structure means that both samples saw 600°C temperatures for 5 mins. No PL due to the well was observed in either sample, indicating that the density of non-radiative traps associated with these thin LTGaAs layers is sufficient to quench any detectable amount of radiative recombination. Since 1 at.% excess As translates into an enormously high defect density ($\sim 1 \times 10^{20}$ cm^{-3}) this may not be surprising. Yet, one might expect the defects to anneal out from such a thin LT layer as T_{sub} is raised. Obviously, this is not the case. Sample C contains no LTGaAs but was grown with a long growth interruption. It shows an intrinsic QW PL, confirming that unintentional impurity incorporation is not responsible for the PL quench.

The null result of samples A and B suggests that PL can be used to monitor the degree of As confinement by optically probing wells of unequal width surrounding a LTGaAs layer. This is the basis for the second series of growths and the results of samples D-H proved to be more interesting. The primary structure consists of three decoupled GaAs wells separated by 200Å thick barrier regions, as shown in figure 2. The first well to be grown is 150Å wide, followed by a 100Å well, and a 50Å well nearest to the surface. The upper and lower AlGaAs barriers are 500Å thick.

Sample H is a control sample. In this growth all three wells are grown at 600°C. There was no temperature ramping and no extended growth interruptions. For samples D,E,F,G the 100Å well is grown at 230°C. In samples D and F, 30Å of AlAs is inserted in the 200Å thick barriers on either side of the LTGaAs layer in order to confine excess As. Samples D and E are annealed for 20 mins. at 600°C immediately following the central well in an As$_4$ flux of 5×10^{-6} torr, before proceeding with the remainder of the growth. Therefore significant redistribution of the LT-related defects is expected in these samples prior to the growth of the 50Å well. There is no anneal after the LTGaAs well in sample F or sample G. However, the remaining growth time for both samples is approximately 6 mins. at 600°C.

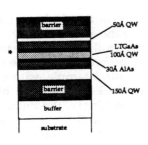

Figure 2

Growth structure for samples D-H: Three decoupled QW's, 50Å,100Å and 150Å wide. The middle (100Å) well is LTGaAs except in the control sample, H. Samples D and E are annealed for 20 mins. at 600°C following the 100Å well (*). The central barrier material in samples E and G does not include AlAs defect diffusion barriers.

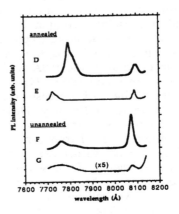

Figure 3

PL spectrum for samples D-G. D and E were annealed following the LTGaAs layer. D and F include AlAs diffusion barriers. Curves have been shifted vertically for clarity. The vertical scale has exaggerated by x5 for sample G.

The PL spectra corresponding to the four samples containing LTGaAs are shown in figure 3. Some PL due to the 150Å and 50Å wells are observed in all four samples. As expected, in no case do we detect PL due to the 100Å LTGaAs well itself. It is readily apparent that the presence of the LTGaAs layer has a significant effect on the quality of epilayers nearby.

The 50Å well double peak (due to monolayer fluctuations in the well-width) has broadened significantly in the unannealed samples which strongly suggests a degradation in the AlGaAs/GaAs interface quality. The optical efficiency of this well has also decreased compared to the annealed samples and the control sample, which suggests the introduction of non-radiative recombination centers in the well. These effects become increasingly large in sample G. In fact, the vertical scale has been exaggerated (x5) for sample G, and for this sample the peak is extremely broad (20 meV) and almost at the limit of detectability. In addition, there is a small (5-8 meV) shift in the peak position to higher energy for these two samples, with the magnitude of the shift being greater for sample G. (The seemingly large peak shift in sample E is an unfortunate artifact of barrier Al content. This sample was grown at a different time in the MBE system cycle, whereas the other samples were grown consecutively). Previous studies on quantum wells grown on LTGaAs buffer layers have shown similar trends in peak shift to higher energy [6]. In that case, it was suggested that the presence of increasing strain due to higher concentration of excess As in the buffer layers was responsible for the observed energy shift . For the present growth structures, strain is an unlikely explanation. The LTGaAs layers are very thin and it has been shown that annealing at $T_{sub} > 580°C$ for 10 mins. is sufficient to remove the lattice mismatch to GaAs even for bulk layers of LTGaAs [4]. In this case, annealing at 600°C removes the source of the peak broadening mechanism. Both annealed samples, D and E, exhibit linewidths comparable to the control sample (10 meV). However, the effect of non-radiative traps remains significant in E, which contains no AlAs barriers. Nonetheless, this effect is reduced by the anneal.

In the case of the 150Å well, whose position in the growth means that it always sees unannealed LTGaAs, the effects are different. Again the optical efficiency of the well decreases compared to the control sample, with sample G having lowest efficiency. However, now sample F exhibits the highest intensity emission, comparable to the control. Interestingly, the linewidth is fairly uniform for all four samples (5 meV), indicating there is no induced interface disorder associated with backdiffusion from the LTGaAs layer. The decreased efficiency of the two annealed samples is probably due largely to the long anneal in an As_4 overpressure, rather than the LTGaAs.

We believe that there is a simple explanation for these results based on point defect considerations. When GaAs is grown at low temperatures, excess As is incorporated in the cystal implying a high concentration of As-rich related native defects, i.e. V_{Ga}, As_{Int} and As_{Ga} type defects. Only the latter has been positively identified in LTGaAs [4], and then only for unannealed samples. As-related complexes are also likely to exist, to account for the remainder of the 1 at.% excess As. The thin LTGaAs layers in our growth structures are consequently highly localized sources of these defects in non-equilibrium concentrations..

As the lattice temperature is raised these native defects become more energetic and they will attempt to re-establish equilibrium concentrations. To do so, they must move through the crystal. Defect diffusion in III-V heterostructures has been extensively studied (readers are referred to review article by Deppe and Holonyak [8]). The observed broadening of the 50Å peak strongly suggests that the group III sublattice is involved in the forward diffusion defect migration mechanism. Therefore, we propose the following simple defect reaction as a possible first step in the redistribution mechanism for excess As in LTGaAs:

$$As_{Ga} -> V_{Ga} + As_{Int}$$

As the temperature increases many defect pairs of this type are created, with the reaction being driven by the highly non-equilibrium conditions. Typically the group III vacancy (V_{III}) defects have the highest diffusion rates of the six possible native defects in III-V crystals. The V_{Ga} defects therefore can out-diffuse from the LT layer due to the concentration gradient much faster than the As_{int} defects. These vacancies will preferentially diffuse in the direction of the growth surface which is under an As_4 overpressure, since they can decrease the free energy of the crystal by self-annihilation at the surface. As they move forward through the crystal predominantly by hopping to adjacent group III sublattice sites they induce interface intermixing of group III atoms. The average Al content of the well increases (from zero) and that of the barrier material will decrease as the V_{III} moves across the GaAs/AlGaAs interface. This will inhomogeneously shift the 2D energy levels higher. This explains the broadening of the 50Å QW PL peak and its observed shift to higher energy in the unannealed samples. The ability to prevent such an effect may be of great importance in structures that are sensitive to interface-scattering or alloy-scattering effects. There is a relatively small effect due to vacancy outdiffusion observed in samples D and E because the anneal takes place immediately following the LTGaAs layer. The growth surface is therefore less than 40 lattice sites from the V_{Ga} defects which consequently find it easily. By the time the 50Å well is deposited, the native defect concentration of this type of defect is already at its equilibrium value.

AlAs layers were initially used as diffusion barriers because of the higher Al-As bond strength compared to Ga-As bonding. Since an Al atom must break its As bonds for the vacancy to propagate through AlAs, the V_{III} diffusion rate will be much lower in AlAs than in GaAs. Although sample F demonstrates that the 30Å AlAs barriers in our structures are not completely effective at preventing vacancy outdiffusion, they do offer a significant improvement over no barriers (sample G). Nevertheless, 30Å is a very thin layer, and thicker diffusion barriers can be expected to fully protect the integrity of epilayers grown on LTGaAs.

The 150Å well is not affected by V_{Ga}-induced intermixing since it is away from the preferential diffusion direction. The PL for this well does not display significant broadening effects. However, both wells display the effects of non-radiative recombination sites, which can be attributed to As_{int} defects and related complexes. As the temperature increases, their tendency is also to reduce the local non-equilibrium concentration. In this case the diffusion path involves hopping between neighboring interstitial sites or through some intermediate configuration involving mostly group V sublattice sites. Consequently we do not expect this type of defect motion to have a large effect on group III atomic positions. However, these defects do affect the optical efficiency of quantum wells when they are present by providing carrier trap centres. From our studies it is not clear which, if any, is the preferential diffusion direction for the As_{Int} type defects, although simple thermodynamics arguments suggest that backdiffusion would dominate for a sample in an As_4 overpressure. Indeed, the 150Å well data shows the PL quenching effect most clearly. In samples F and G we see that the insertion of AlAs barriers is definitely effective at inhibiting defect backdiffusion when we compare their optical efficiency.

Summary

We have studied the optical properties of thin layers of LTGaAs using PL measurements. Quantum wells containing even a few monolayers of material grown at 230°C were found to be optically dead which suggests the defects associated with LT growth are very effective at quenching radiative recombination. Outdiffusion and backdiffusion of these defects was investigated by their effect on nearby QW's. Group III atom intermixing effects is observed in the direction of the growth surface, which suggests that V_{Ga} defects due to the LTGaAs are involved in the defect migration mechanism. The insertion of AlAs barriers was found to significantly reduce the defect diffusion problem, being most effective against backdiffusion. Annealing the sample immediately following the LTGaAs improved outdiffusion effects by allowing defects to anneal out.

This work was supported by AFOSR. UKM was supported by a NSF PYI award.

References

[1]. F.W. Smith, A.R. Calawa, Chang-Lee Chen, M.J. Manfra, L.J. Mahoney
    IEEE Elec. Dev. Lett., vol.9, 77 (1988).

[2]. L.-W. Yin, Y. Hwang, J.H. Lee, R.M. Kolbas, R.J. Trew, U.K. Mishra,
    IEEE Elec. Dev. Lett., vol.11, 561 (1990).

[3]. J.M. Baranowski, Z. Lilienthal-Weber, W.-F. Yau, E.R. Weber,
    Phys. Rev. Lett., vol.66, 3079 (1991).

[4]. M. Kaminska, Z. Lilienthal-Weber, E.R. Weber, T. George, J.B. Kortwright,
    F.W. Smith, B-Y. Tsaur, A.R. Calawa, Appl. Phys. Lett., vol. 54, 1881 (1989).

[5]. A.C. Warren, J.M. Woodall, J.L. Freeouf, D. Grischkowsky, D.T. McInturff,
    M.R. Melloch, N. Otsuka, Appl. Phys. Lett., vol 57, 1331 (1990).

[6]. K.T. Shiralgi, R.A. Puechner, K.Y. Choi, R. Droopad, G.N. Maracas,
    J. Appl. Phys., vol.69, 7942 (1991).

[7]. J.P. Ibbetson, L.-W. Yin, U.K. Mishra, *to be published*

[8]. D.G. Deppe, N. Holonyak,Jr., J. Appl. Phys., vol.64, R93 (1988).

Inst. Phys. Conf. Ser. No 120: Chapter 2
Paper presented at Int. Symp. GaAs and Related Compounds, Seattle, 1991

43

Etching of GaAs and AlGaAs by H* radical produced with a tungsten filament

R. Kobayashi, K. Fujii, and F. Hasegawa

Institute of Materials Science, University of Tsukuba,
Tsukuba Science City, 305 Japan

Abstract: It was found that GaAs was etched with a smooth surface by radical or atomic hydrogen. The etching rate was as high as 15 um/h at the substrate temperature of 850°C and tungsten(W)-filament temperature of 2000°C. Several experimental results indicate that this high etching rate is due to evaporation of Ga as gallium hydride. Reaction between atomic hydrogen and AlGaAs was different from that for GaAs. Aluminum could not be removed so easily as Ga, and an aluminum oxide layer is left on the AlGaAs surface.

1.Introduction

The effects of atomic hydrogen on epitaxial growth or semiconductor surface is one of the recent topics. Noda et al. reported that a smoother surface was obtained when a platinum catalyst was placed beside the substrate in MOVPE of GaAs using triethylarsine (Noda et al. 1989). It is reported that selective growth can be achieved in MBE growth of GaAs if radical hydrogen is supplied on the growth surface (Yamamoto et al. 1989). Low-temperature cleaning of the substrate surface by atomic or radical hydrogen has also been reported (Sugata et al. 1988, Sugaya et al. 1991).

We have studied the reaction between $GaCl_3$ and arsine (AsH_3) or arsenic vapor from arsenic metal, and found that GaAs can be grown at much lower temperatures with AsH_3 than with arsenic vapor. This is because the hydrogen in the AsH_3 molecule is much more reactive than the hydrogen of H_2 (Kobayashi et al. 1990).

The first aim of this work was, therefore, to enhance the chloride VPE of AlGaAs at lower temperatures by generating atomic hydrogen without using the very toxic AsH_3. However, we found that the atomic hydrogen etches GaAs very quickly rather than enhancing the growth.

In this paper, we report the reaction of atomic hydrogen with GaAs and with AlGaAs.

2. Experimental

Figure 1 shows the schematic diagram of the experimental setup. The (100)-oriented GaAs substrate was put at three positions A, B and C on a carbon holder (distance between A, B and C is 2 cm each). The AlGaAs substrate grown by LPE was put at only the position B. The carbon holder was heated by an infrared lamp, and its temperature was monitored by a thermocouple inserted into the carbon holder.

The atomic hydrogen was produced by a heated tungsten (W) filament. This is the simplest method and has been successfully applied to the growth of diamond films (Matsumoto et al. 1982). The diameter of the W-filament was 0.2 mm, and the length was 40 - 60 mm. The W-filament was positioned about 15-20 mm above the substrate. The temperature of the filament was estimated from the change in resistance, and was maintained at about 2000°C for most of the experiments.

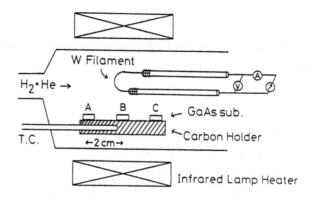

Fig.1 Schematic diagram of the experimental setup.

The experiment was performed in pure H_2 gas or (H_2 + He) gas mixture at atmospheric pressure. The gas flow rate was 180 sccm in a quartz reactor 45 mm in diameter.

Part of the substrate was covered by a small quartz chip. After etching, the chip was removed and the etched depth or the oxide thickness was measured by a surface profiler.

3. Experimental results and discussion
3.1 Etching of GaAs

A significant gas etching of GaAs substrates was observed when the W-filament was heated to 2000°C and the carbon holder temperature was kept at 850°C. The etching rate was about 15 um/h for the sample right under the W-filament, but the surface was rather smooth as shown in Fig.2(a). In order to confirm that this fast and smooth etching of the substrate is due

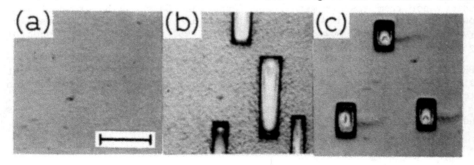

Fig.2   Surface morphology of GaAs substrates etched or annealed under various conditions ; (a) substrate temperature of 850°C and W-filament at 2000°C in pure H_2 for 60 min ;   (b) annealed at 850°C for 60 min in pure He with W-filament of 2000°C; (c) annealed at 850°C for 20 min in pure H_2 without the heated W-filament. Marker represents 20 um.

to the radical or atomic hydrogen produced by the heated W-filament, the hydrogen partial pressure was changed by adding helium gas. The resultant dependence of the etching rate on the hydrogen partial pressure is shown in Fig.3. The etching rate decreases with a decrease of the hydrogen partial pressure, and it was about 3 um/h for pure helium (zero hydrogen partial pressure). The reason why etching was observed even in pure helium, is not properly understood at the moment; it might be due to the residual hydrogen or to mere evaporation of arsenic from the surface.

The etched surface becomes rough with a decrease of the hydrogen partial pressure, and many thermal pits with a gallium droplet inside appeared as shown in Fig.2(b). This rough surface is the same as is often seen for samples annealed in pure hydrogen without a heated W-filament. An example of such a surface is shown in Fig.2(c).

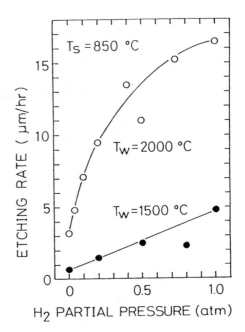

Fig.3 H_2 partial pressure dependence of the etching rate of GaAs by radical or atomic hydrogen.

The etching rate decreased very much with a decrease of the filament temperature from 2000°C to 1500°C. Still the same tendency of a decrease of the etching rate with a decrease of the hydrogen partial pressure was observed.

These facts indicate that the fast and smooth etching of GaAs is due to radical or atomic hydrogen produced by the heated W-filament: One might suspect an increase of the surface temperature due to radiation from the W-filament as the cause of the fast etching, but the temperature increase due to the radiation should be the same for different hydrogen partial pressure. Furthermore, when GaAs is heated in hydrogen or in vacuum, arsenic evaporates much faster than gallium, and gallium droplets are left on the surface as shown in Fig.2(c). There were no gallium droplets left on the surface when the sample was annealed in a hydrogen atmosphere with a heated W-filament. It was also confirmed that the gallium droplets left on the GaAs surface could be removed by annealing with the heated W-filament (Kobayashi et al. 1991). This fact suggests that the gallium is evaporated as a hydride by the radical or atomic hydrogen produced by the W-filament. Existence of gallium hydride and aluminum hydride is reported in rather old papers (Wiberg et al. 1942, Finholt et al. 1947). Therefore, it is not unreasonable to assume that gallium is removed as gallium hydride by radical or atomic hydrogen.

3.2 Transport of GaAs by atomic hydrogen

The sample right under the W-filament was always etched, but when the substrate temperature was not very high, both etching and deposition

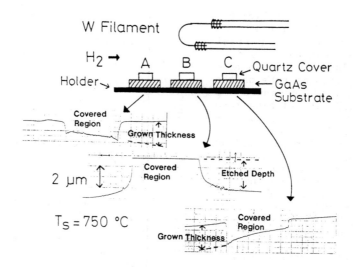

Fig.4 Surface profile of each GaAs substrate
measured by a surface profiler. The carbon
holder was about 750°C and etching or
deposition time was 60 min.

of GaAs was observed. Figure 4 shows an example of such case. The top half
of the figure shows the configuration of the experiment and the bottom
three curves show the surface profiles of the GaAs samples placed at
positions A, B and C of the carbon holder separating by about 2 cm. Center
of each profile is the portion covered by a quartz chip. The carbon
holder was 750°C, and the W-filament was 2000°C.

The sample placed at position B (right under the W-filament) was
etched, but deposition rather than etching was observed on the samples
placed at positions A and C. Since there is no GaAs source other than the
sample placed at position B, this fact indicates the transport of GaAs
from the sample at the center(B) to the samples at the both sides(A and
C). This phenomenon would be explained as follows: Density of the atomic
hydrogen must be higher at the center (position B) than at both sides
(positions A and C), therefore, arsenic and the gallium evaporated from
the center sample as gallium hydride diffuse to both sides. Since gallium
hydride is unstable, the gallium deposited in arsenic vapor and GaAs was
grown on the substrates placed at positions A and C.

Since the vapor pressure of gallium is very low even at 900°C, this
fact also indicates that the gallium is evaporated as a hydride by the
atomic hydrogen, otherwise a high rate transport of GaAs shown in Fig.4
under a pure hydrogen atmosphere can not be understood.

3.3 Etching of AlGaAs

In order to see reaction between atomic hydrogen and AlGaAs, the
same experiments were performed using LPE grown AlGaAs with Al content of
30%. Figure 5 shows the experimental results.

The results for AlGaAs was quite different from those for GaAs.

First of all, AlGaAs is never thermally etched. Even when it is annealed at 900°C in pure hydrogen gas without a heated W-filament, the surface is smooth without any thermal pit. Of course no gallium droplet is left. Instead of that, the surface is slightly oxidized.

When the atomic hydrogen is supplied, the AlGaAs layer is etched as shown at the top of the figure, but a thick aluminum oxide layer is left on the surface.

The solid circle of the figure shows thickness of the oxide layer left after etching by atomic hydrogen and the open circle shows the total etched depth after removing the oxide layer against the etching time. Both the total etched depth and the oxide layer thickness increase almost linearly with an increase of the etching time. However, thickness of the oxide layer is almost the same as the etched depth, suggesting that only gallium and arsenic are etched by the atomic hydrogen, and aluminum is left behind.

Since the reactor used here is not an ultra high vacuum type, there must be slight leakage. Therefore, when gallium is removed by the atomic hydrogen, aluminum left on the surface must have been easily oxidized and a thick oxide layer was formed. Thickness of the oxide layer increases linearly with an increase of the etching time, but it does not go through the zero point. This fact might indicate that aluminum is also slightly removed as aluminum hydride by atomic hydrogen, at least at the beginning of the etching. Further experiments in an ultra high vacuum type system is necessary to conclude whether atomic hydrogen really reacts with aluminum and evaporates as a hydride or not.

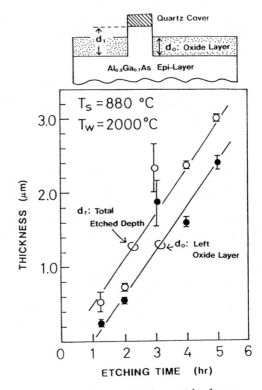

Fig. 5 Dependence of the oxide layer thickness after etching and the total depth after removal of the oxide layer on the etching time for LPE grown AlGaAs.

4. Conclusion

It was found that GaAs was etched with a smooth surface by atomic hydrogen. The etching rate was as high as 15 um/h at substrate temperature of 850°C and W-filament temperature of 2000°C. When the temperature of the carbon holder was relatively low, transport of GaAs from the sample near the W-filament to substrates distant from the filament was observed. These results suggest that gallium on the GaAs surface is removed as a hydride by radical or atomic hydrogen generated by the W-filament.

As to reactions between atomic hydrogen and AlGaAs, aluminum could not be so easily removed as gallium, and an aluminum oxide layer is left on the AlGaAs surface. Further experiments in an ultra high vacuum system are necessary to clarify the phenomenon.

5. Acknowledgments

The authors express their sincere thanks to Mr. K.Ishikura for his help on the experiments. This work is partly supported by a Grant in Aid for Scientific Research from the Ministry of Education, Science and Culture of Japan.

References

Finholt.A.E, Bond.A.C and Schlesinger, 1947 J.Amer.Chem.Soc. **69** 1199
Kobayashi.R, Fujii.K and Hasegawa.F, 1991 Jpn.J.Appl.Phys. **30** L1447
Kobayashi.R, Yamaguchi, Jin.Y and Hasegawa.F, 1990 J.Crystal Growth **102** 471
Matsumoto.S, Sato.Y, Tsutsumi.M and Setaka.N, 1982 J.Mater.Sci. **17** 3106
Noda.S, Metavikul.S, Wang.X.L, Okuda.T, Takeda.Y and Sasaki.A, 1989 Proc. 16th Int.Symp.GaAs and Related Compounds (Institute of Physics, Bristol, 1989)p.57
Sugata.S, Takamori.A, Takado.N, Asakawa.K, Miyauchi.E and Hashimoto.H, 1988 J.Vac.Sci. Technol. **B6** 1087
Sugaya.T and Kawabe.M, 1991 Jpn.J.Appl.Phys. **30** L402
Wiberg.E and Johanson.T, 1942 Angew.Chem. **55** 38
Yamamoto.N, Kondo.N and Nanishi.Y, 1989 J.Crystal Growth **96** 705

Reduction in the outdiffusion into epitaxial Ge grown on GaAs using a thin AlAs interlayer

A. L. Demirel, S. Strite, A. Agarwal, M. S. Ünlü, D. S. L. Mui, A. Rockett, and H. Morkoç

1. Introduction

The lattice-matched GaAs-Ge interface is not only important for the study of polar-nonpolar heteroepitaxy [1], but it may also prove useful for improving optical and electronic devices such as heterojunction bipolar transistors [2], metal-insulator-semiconductor field effect transistors (MISFET) and photodetectors [3]. A major obstacle to the realization of practical GaAs-Ge devices is the large amount of Ga diffusion at the Ge on GaAs (Ge/GaAs) interface which leads to uncontrolled p-type doping of the Ge [4,5]. In this paper, we compare Ge/GaAs(100) structures with similar structures incorporating a thin AlAs interlayer. Secondary ion mass spectroscopy (SIMS) measurements indicate that no measurable Al outdiffusion occurs and that Ga outdiffuses much more slowly in the presence of an AlAs interlayer. X-ray photoemission spectroscopy (XPS) shows Ga segregation in Ge grown directly on GaAs while no Ga is observed in samples incorporating an AlAs interlayer. Hall-effect measurements show lower background carrier concentrations in the layers incorporating AlAs which can not be accounted for by depletion around the Ge/AlAs heterojunction. $Au/Si_3N_4/Ge$ MIS capacitors and Ge/GaAs p-i-n diodes in which AlAs interlayers were used exhibited superior electrical characteristics compared to similar layers grown directly on GaAs. Our results indicate that a thin AlAs layer at the Ge/GaAs heterojunction suppresses Ga outdiffusion leading to a reduced background carrier concentration in the epitaxial Ge and improved performance in GaAs-Ge devices grown on GaAs substrates.

2. Experimental

Crystal growth was conducted in the University of Illinois Epicenter Facility which is described in Ref. 6. The GaAs and AlAs films were grown in a Perkin Elmer Model 430 III-V chamber with conventional solid evaporation sources including an As cracker. Ge was grown in a Perkin Elmer Si-Ge MBE system using electron beam evaporation. The Ge growth temperatures were measured by a thermocouple in direct contact with the back of the sample holder. For each sample a 0.2 μm nominally undoped GaAs buffer layer was deposited on a semi-insulating GaAs (100) substrate. In order to study its effect as a diffusion barrier, 100 Å of AlAs was grown in some samples before they were transferred under ultra-high vacuum (UHV) to the Si-Ge MBE system by the procedure described in Ref. 1. Ge growth was two dimensional exhibiting a mixed domain (2×1)+(1×2) reconstruction under all growth conditions. SIMS analyses were carried out using Cs^+ ions. XPS spectra were obtained in a Perkin Elmer 5300 spectrometer with Al K_α x-rays at an emission angle of 20°. The electron escape depth for photoelectrons with kinetic energies of 160 to 370 eV, corresponding to binding energies of 1326 to 1118 eV, is 7 to 10 Å. Thus, all impurities detected by XPS are on or very close to the sample surface. No evidence for residual gas contamination of the growth surface during transfers was found in any of the experiments. Hall-effect measurements at both room temperature and 77 K were made on samples fabricated in van der Pauw patterns.

3. Results

Outdiffusion of Ga, Al and As into epitaxial Ge grown on the AlAs interlayer at initial growth temperatures of 200 and 300°C was investigated by SIMS. The initiation temperature was maintained during the growth of 200 Å of Ge at a rate of 0.2 Å/s. The growth rate and temperature were then increased to 0.4 Å/s and 300°C (for the 200°C layer) and an additional ≈2300 Å of Ge was deposited. A SIMS scan of the 300°C layer is shown in Fig. 1 (a). SIMS scans of the 200°C layer were quite similar. The Al, Ga and As signals all fall quickly from the matrix level to the SIMS background level. There is no evidence of outdiffusion of any of the constituent elements. In contrast, a significant amount of Ga outdiffusion is observed by SIMS in a Ge/GaAs layer initiated at 300 °C (Fig. 1 (b)). In this case, the Ga extends ≈700 Å into the Ge while decreasing approximately four orders of magnitude in concentration to the background level.

Surface segregation of Ga, Al and As through epitaxial Ge was investigated by XPS. Fig. 2 (a) shows a photoelectron spectrum taken of the surface of a 500 Å Ge layer grown on an AlAs interlayer. Only the Ge $2p_{1/2}$ and $2p_{3/2}$ peaks, with binding energies of 1248.4 eV and 1217 eV respectively, and their associated satellite and energy loss peaks are evident. Fig. 2 (b) shows the XPS spectrum of an 800 Å Ge layer grown directly on GaAs. In addition to the peaks seen in Fig. 2 (a), the Ga $2p_{1/2}$ and $2p_{3/2}$ peaks are observed at 1144 eV and 1118 eV respectively. These results are consistent with the Ga outdiffusion measured by SIMS. The large surface Ga concentration observed by XPS suggests that the outdiffusion mechanism is Ga surface segregation. For a 500 Å Ge layer grown at 500°C on AlAs, As $2p_{3/2}$ peak was detected at 1326 eV. Therefore, the AlAs interlayer effectively suppresses Ga outdiffusion even at higher temperatures although the possibility of As outdiffusion arises.

Results of Hall-effect measurements taken at room temperature and 77K on 0.5 μm Ge layers grown on both AlAs

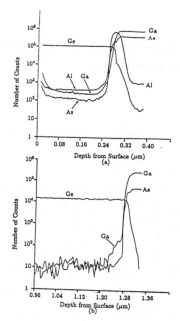

Fig. 1. SIMS profiles of Ge layers deposited on GaAs. (a) Ge growth initiated at 300°C with a 100 Å AlAs interlayer. (b) Ge growth initiated at 300 °C without an AlAs interlayer.

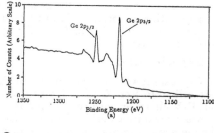

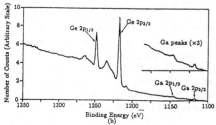

Fig. 2. XPS spectra of epitaxial Ge grown at 300°C: (a) 500 Å of Ge on AlAs. (b) 800 Å of Ge on GaAs.

TABLE 1. Summary of electrical properties of Ge epitaxial layers on GaAs(100)

AlAs interlayer	Ge growth initiation temperature (°C)	Carrier conc.(RT) ($\times 10^{17}$ cm^{-3})	μ_{RT} (cm^2 V^{-1} s^{-1})	Carrier conc.(77K) ($\times 10^{17}$ cm^{-3})	μ_{77} (cm^2 V^{-1} s^{-1})
No	150	p = 4.0-5.5	120-155	p = 0.4-0.5	425-500
No	300	p = 8.0-12.5	135-190	p = 6.0-8.7	170-235
Yes	200	n = 0.75-1.22	694-697	n = 0.29	760
Yes	300	n = 3.47-5.00	455-504	n = 3.33	315

and GaAs are summarized in Table 1. Ge grown on AlAs is observed to be n-type. When grown directly on GaAs, p-type Ge having higher absolute carrier concentrations than the layers grown on AlAs were obtained. We also note that the layers grown at 300°C exhibit a higher carrier concentration compared to the same structures initiated at the lower temperatures. From the SIMS data and the work of others [4,5], we conclude that Ga outdiffusion is the cause of the p-type conductivity of the Ge/GaAs. The observation of n-type conductivity in layers grown on AlAs suggests that the diffusion of Ga and Al has been reduced to a point that the polarity of the carriers is determined by n- type impurities, possibly As from the underlying AlAs.

We investigated the possibility that Hall-effect measurements were affected by conduction through the GaAs due to unintentional doping by Ge or another impurity. The diffusion of Ge into the underlying GaAs during growth was investigated by capacitance-voltage profiling of a p$^+$-Ge/N$^-$-GaAs diode in which the Ge growth was initiated at 300°C. The total GaAs carrier concentration attributable to Ge was found to be below 5×10^{15}/cm^3 within 80 Å of the Ge-GaAs interface. This observation suggests that the majority of the measured electron concentration in Ge/AlAs is attributable to donors (probably As) in the Ge.

The effect of the AlAs interlayer was further investigated in Au/Si$_3$N$_4$/Ge MIS capacitors grown on semi-insulating GaAs substrates. The layer structure of the p-type MIS devices grown directly on GaAs and the n-type devices incorporating AlAs are given in Fig. 3. The polarity of the MIS structures was chosen to match the carrier type of the Ge which was observed by Hall measurements. The initial 200 Å of Ge growth in both devices occurred at 300°C and a rate of 0.3 Å/s. The growth temperature and rate were then increased to 500°C and 0.5 Å/s.

Figures 4 (a) and (b) show the quasi-static and high frequency capacitance-voltage (C-V) curves for the p-type and n-type capacitors, respectively. Following Ref. 7, the ionized dopant concentration was calculated at the edge of the depletion layer from the slope of $1/C^2$ versus voltage. We obtained an n-type dopant concentration of 8.8×10^{16} cm^{-3} for the device grown on AlAs. A p-type dopant concentration of 2.7×10^{17} cm^{-3} was measured for the device grown directly on GaAs, confirming the results of the Hall-effect measurements. The high frequency C-V curve of the n-type layer resembles an ideal S-shaped profile more nearly than the p-type device. We interpret this

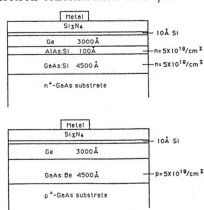

Fig. 3. Layer structures of n- type and p-type Au-Si$_3$N$_4$-Ge MIS capacitors.

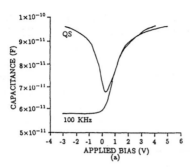
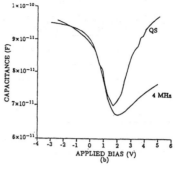

Fig. 4. Capacitance-voltage curves of (a) n-type and (b) p-type Au-Si$_3$N$_4$-Ge MIS capacitors.

to be indicative of a lower bulk trap density in the Ge grown on AlAs. This conclusion is supported by conductance-voltage measurements also performed on these devices.

One possible application of the Ge/GaAs system is the complementary MISFET. Due to the small band gap of Ge, Ge substrates conduct at room temperature and unsuitable for FET applications due to leakage current and parasitic capacitance which degrade the high frequency performance. It is desirable to grow Ge MISFETs on a semi-insulating substrate such as GaAs. Furthermore, MISFETs require a low background carrier concentration in the channel to enable the current flow to be modulated through the insulator. Our results indicate that the development of a high quality n-channel Ge MISFET may benefit from the incorporation of an AlAs interlayer.

The effect of an AlAs interlayer on the electrical characteristics of Ge p-i-n homojunction diodes grown on GaAs substrates was also investigated. For these structures, Ge growth was initiated in the same manner as for the MIS structures and after 200 Å of growth the temperature and rate were increased to 400°C and 0.4 Å/s. A 2000 Å thick B-doped p$^+$ layer was grown followed by another 2000 Å of nominally undoped Ge. Finally, 1000 Å of Sb-doped n$^+$ Ge was grown. The devices were fabricated using standard techniques [1]. Figure 5 (a) shows the current-voltage characteristics of the two Ge p-i-n diodes. The diode grown on AlAs shows a factor of thirty reduction in leakage current and a much larger breakdown voltage (Fig. 5 (b)). These results strongly suggest the advantage of a thin AlAs diffusion barrier in the fabrication of proposed high sensitivity Ge detectors operating in the 1.0-1.5 μm wavelength range.

4. Summary

We investigated the effect of a thin AlAs interlayer at the Ge/GaAs interface on the background carrier concentration of the epitaxial Ge. SIMS, XPS and Hall-effect measurements indicated that the problem of Ga outdiffusion could be circumvented by the AlAs interlayer. Ge MIS capacitors and p-i-n diodes grown on AlAs interlayer exhibited improved electrical characteristics compared to identical devices grown directly on GaAs.

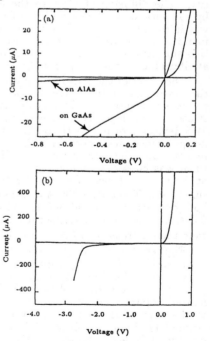

Fig. 5. Current-voltage characteristics of Ge p-i-n diodes grown on GaAs substrates with and without AlAs diffusion barrier.

5. Acknowledgements

This work was supported by the SDIO/IST division of the Office of Naval Research under grant N00014-84-K-0274 and by the United States Air Force Office of Scientific Research under contract AFOSR-89-0239. Surface and bulk analyses of the deposited layers were carried out in the Epicenter and the Center for Microanalysis of Materials and was supported by the Department of Energy under contract DEFG02-91ER45439. The authors would like to thank B. Bowdish for his tireless support in the Epicenter. One of us (S. S.) wishes to acknowledge the support of NSF and AFOSR Graduate Fellowships during different stages of this research.

References

1. S. Strite, M. S. Ünlü, K. Adomi, G.-B. Gao, A. Agarwal, A. Rockett, H. Morkoç, D. Li, Y. Nakamura, and N. Otsuka, J. Vac. Sci. Tech. B 8 (1990) 1131

2. S. Strite, M. S. Ünlü, K. Adomi, G.-B. Gao, and H. Morkoç, Elect. Dev. Lett. 11 (1990) 233

3. N. Chand, J. Klem, and H. Morkoç, Appl. Phys. Lett. 48 (1986) 484

4. M. Kawanaka, and J. Sone, J. Crystal Growth 95 (1989) 421

5. A. L. Demirel, S. Strite, A. Agarwal, M. S. Ünlü, H. Morkoç, and A. Rockett (unpublished results)

6. Tim Studt, "Grow and Analyze Complex Materials Like Never Before", in R&D Magazine, May 1990, 88

7. E. H. Nicollian, and J. R. Brews, in MOS Physics and Technology (Wiley, New York, 1982)

Inst. Phys. Conf. Ser. No 120: Chapter 2
Paper presented at Int. Symp. GaAs and Related Compounds, Seattle, 1991

Improvement of the electrical properties of MBE grown Ge layers and its application to collector-top n-GaAs/p-Ge/n-Ge HBTs

Tohru Kimura, Masafumi Kawanaka, Toshio Baba, and Jun'ichi Sone
Fundamental Research Labs., NEC Corporation, 34 Miyukigaoka, Tsukuba, Ibaraki 305, Japan

1. Abstract

This paper reports improved electrical properties in MBE-grown Ge layers for collector-top n-GaAs/p-Ge/n-Ge heterojunction bipolar transistors (c-top GaAs/Ge HBTs). Both the electron-beam-gun (E-gun) and the Knudsen-cell (K-cell) were tested for use in Ge evaporation. E-gun grown Ge pn-diodes had the problem of large leakage current in reverse bias conditions. Leakage current was sufficiently low with K-cell grown diodes, and Ga diffusion into the Ge layer from the GaAs layer was reduced by a factor of 2 as compared to E-gun growth. This is believed to be caused by the elimination of ionized Ge atoms, which tend to induce crystal imperfections.

HBTs fabricated on K-cell grown wafers had a base sheet resistance of 90 ohms/square, and a current gain was as high as 80. To the best of our knowledge, these indicate the largest ratio of current gain to base resistance among bipolar transistors yet reported.

2. Introduction

N-GaAs/p-Ge/n-Ge HBTs have the following advantages for use in high speed and low power dissipation devices.*(Jadus & Feucht 1969, Kroemer 1982, Strite et al 1990)*

1) Because Ge is a narrow band gap (Eg= 0.66eV) material, turn-on voltage can be as low as 0.5V.

2) High hole mobility, high dopant solubility, and a low Schottky-barrier height help attain low base sheet resistance and low contact resistance.

Further, when these HBTs are of a collector-top structure, fabrication is relatively easy and base-collector capacitance is reduced.

In the previous study, we presented c-top GaAs/Ge HBTs which had been grown for the first time by molecular beam epitaxy (MBE).*(Kimura et al 1989)* Because of the low vapor pressure of Ge, it was possible to get a practical growth rate (about 2.0um/hour) with an E-gun. We observed large reverse leakage current in the Ge pn-diodes, however, and unusually large Ga diffusion in the Ge layer prevented shortening of the base width. In other words, improvement in Ge crystal quality would be necessary if we were to improve HBT operational characteristics.

In the present study, we have attempted to improve Ge pn-diode characteristics by using K-cell evaporation, thus eliminating Ge ionized atoms in the molecular beam.

3. Experiment

In our experiments, by using a two growth chamber MBE system, one chamber for GaAs growth and the other for Ge growth, we were able to eliminate As flux completely during Ge layer growth. If As flux can be eliminated during Ge growth, an unintentionally-doped Ge layer, grown on Ga-stabilized GaAs surface, exhibits p-type conduction because of selective Ga diffusion from the GaAs surface.*(Kawanaka & Sone 1989)*

A GaAs buffer layer and a GaAs emitter layer were grown on a n^+-GaAs (100) substrate in the GaAs growth chamber at a growth temperature of 500 C and a growth rate of 1.0um/ hour. The wafer was then transferred to the Ge growth chamber and the Ge layer was grown sequentially.

We experimented with two types of evaporation equipment for Ge layer growth, E-gun and K-cell. For E-gun evaporation, we prepared two types of wafers to test the effects of ionized Ge atoms. The first type (type E1) of wafer was grown with a PBN crucible inserted between the Ge source and the Cu hearth. This construction of the E-gun enhanced the Ge source charge-up, and the number of ionized Ge atoms in molecular beam increased. The other wafer (Type E2) was grown by using a normal E-gun (without a PBN crucible). The growth rate of these 2 wafers is 2.0 um/hour. We also grew wafers with a K-cell, which is, essentially, a thermal effusion cell that has a PBN crucible and that generates a neutral molecular beam. Wafers grown by K-cell (Type K), therefore, are unaffected by ionized Ge atoms. In order to avoid the decomposition of PBN crucible, however, the growth rate is limited to 0.2 um/hour.

	Sub-emitter	Emitter	Base	Collector
E1	$4 \times 10^{18} cm^{-3}$ 7000A	$1 \times 10^{17} cm^{-3}$ 3000A	$1 \times 10^{19} cm^{-3} \rightarrow 1 \times 10^{16} cm^{-3}$ 4000A	$1 \times 10^{16} cm^{-3}$ 2000A
E2	$4 \times 10^{18} cm^{-3}$ 7000A	$1 \times 10^{17} cm^{-3}$ 3000A	$1 \times 10^{19} cm^{-3} \rightarrow 1 \times 10^{16} cm^{-3}$ 4000A	$1 \times 10^{16} cm^{-3}$ 2000A
K	$4 \times 10^{18} cm^{-3}$ 7000A	$1 \times 10^{17} cm^{-3}$ 3000A	$1 \times 10^{19} cm^{-3} \rightarrow 2 \times 10^{17} cm^{-3}$ 2000A	$2 \times 10^{17} cm^{-3}$ 2000A

Table.1   Layer structure of HBT wafers

The layer structures for E1, E2, and K wafers are listed in Table 1. Substrate temperature for Ge growth was 500 C in all cases.

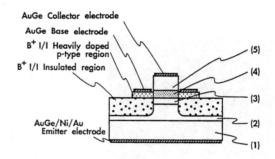

Fig.1 shows the device structure of the c-top GaAs/Ge HBTs we produced with the 3 types of wafers. The fabrication process was as follows. :

1) Collector area was covered by photo-resist (PR) and B⁺ ions were implanted. This result in a collector-top

(1) Heavily doped n-type GaAs substrate
(2) Heavily doped n-type GaAs sub-emitter layer
(3) n-type GaAs Emitter layer
(4) Unintentiory doped p-type Ge Base layer
(5) n-type Ge Collector layer

Fig. 1 Collector-top n-GaAs/p-Ge/n-Ge HBT Structure

structure. (B⁺ ion implantation into GaAs/Ge bilayers makes Ge layers the heavily-doped p-type and makes GaAs layers semi-insulating. Consequently, B⁺ ion implantation forms the extrinsic base region at the same time that it limits the emitter-base junction area.*(Ishizuka et al 1986))*

2) The Ge layer, exclusive of the collector area, was etched off down to the p-type Ge region by reactive ion etching (RIE) using CF_4 gas.

3) The external base and collector area were covered by PR and the Ge layer, which

remained outside of the device, was then etched off completely by selective etching using CF_4 RIE. Etching rates of Ge and GaAs were 290A/min. and 5A/min. respectively, at a flow rate of 30 SCCM, pressure of 5 Pa, and power of 83.3W/cm².

4) AuGe/Ni/Au alloy was deposited and alloyed at 430 C for n-GaAs emitter contact electrode.

5) Non-alloyed AuGe electrodes were deposited on both the p-Ge base and n-Ge collector layers.

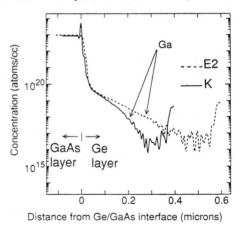

Fig. 2 Ga Doping Profile in Ge layers

4. Results

Differential interference optical microscope measurements determined that surface defect (larger than 5um) densities of E1, E2, K were $20000/cm^2$, $4000/cm^2$, $400/cm^2$, respectively. The higher the number of ionized Ge atoms in the molecular beam, the greater the surface defect density. We can assume the bulk defect density of E1 is higher than E2, and E2 is higher than K.

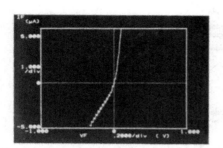

Fig. 3 I-V Characteristics of Ge pn Diode fabricated on E2

The Ga profiles in Ge layer of E2 and K are shown in Fig.2. Even though the growth time for E2 wafer is 6.7 times shorter than that for K, its degree of Ga diffusion in the Ge layer is larger. Assuming Ga diffusion to be enhanced by defects in the Ge layer, more crystal imperfections exist in E-gun grown Ge layers than in K-cell grown layer. This is consistent with the data on surface defect densities.

N-Ge/p-Ge diodes fabricated on E1 showed unstable current-voltage (I-V) characteristics. Fig.3 and Fig.4 shows the I-V characteristics of the Ge pn-diodes fabricated on E2 and K. On K wafer, however, n-Ge/p-Ge diode I-V characteristics are fine, ideality factor of n=1.2 and a turn-on voltage of 0.1V. (Fig.4) In reverse biased conditions, the leakage current density (at 0.5V) was less than one tenth that for E2. A saturation current density of $2.0 \times 10^{-4} A/cm^2$ was estimated and a reverse current density of $2.0 \times 10^{-3} A/cm^2$ (at 0.5V) can be measured from Fig.4. This suggests that the recombination current in the Ge pn-diode fabricated on the K wafer is smaller than that on the E1 and E2 wafers. The number of recombination centers is less in the K-cell grown Ge layers.

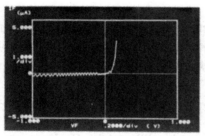

Fig. 4 I-V Characteristics of Ge pn Diode fabricated on K

From these results, we may say that the poor I-V characteristics of n-Ge/p-Ge diodes have been brought about by crystal imperfections that were introduced by Ge ionized atoms in the molecular beam.

On the other hand, n-GaAs/p-Ge diodes produced by E-gun evaporation exhibited as nearly ideal I-V characteristics as did those produced with K-cell evaporation. It is possible

to take the view that the I-V characteristics of
n-GaAs/p-Ge diodes are independent of the
number of ionized Ge atoms in the molecular
beam. This is because when almost all of the
depletion region lies within the n-GaAs emitter layer, defects in Ge layer would not, then,
seriously affect I-V characteristics.

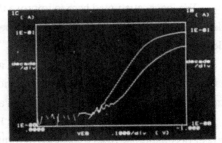

We also used a K wafer to fabricate a
collector-top n-GaAs/p-Ge/n-Ge HBT. Collector size was 125um x 50um. A Gummel plot is
given in Fig.5. The ideality factor n of collector current (Ic) is 1.02 and that of base current
(Ib) is 1.17. These are fairly good values as
compared to AlGaAs/GaAs HBTs.

Fig. 5 Gummel Plot of c-top GaAs/Ge
HBT fabricated on K (V_{BC}=0V)
Ideality factor n of Ic is 1.02
and n of Ib is 1.17

Common emitter DC characteristics are
shown in Fig.6. The HBT exhibits a current
gain (beta) of 80 under emitter-grounded conditions. The base sheet resistance (Rb) of 90
ohms/square was measured by transmission
line method. The ratio of current gain to base
sheet resistance (beta/Rb)was as large as 0.89
(square/ohms). Typical values of (beta/Rb) of
Si bipolar transistor and of GaAs HBT are 0.1
(beta=100, Rb=1000 ohms/square) and 0.17
(beta=50, Rb=300 ohms/square), respectively. To the best of our knowledge, (beta/
Rb)=0.89 is the largest value among bipolar
transistors ever reported.

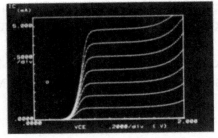

Fig. 6 Common emitter DC characteristics
of c-top GaAs/Ge HBT
(Base current step of 10uA)

Emitter grounded current gain of 80
Base sheet resistance of
90 ohms/square

Although the base doping density is 100 times higher than that of the emitter, and the
emitter-base junction area is about three times larger than the base-collector junction area,
(beta/Rb) of 0.89 was achieved. This excellent characteristics was possible because of : (1)
the large band gap difference in the GaAs/Ge heterojunction (delta Eg =764meV); (2) the
small carrier injection from the emitter to the extrinsic base region. (the current density
through the n-GaAs/i-GaAs/p-Ge diode is 2800 times smaller than that through the n-GaAs/p-
Ge diode (at Ic= 1.0mA)); and (3) the high hole mobility of the Ge which was made possible
by the refined Ge crystal quality.

5. Summary

We have attempted here to improve the properties of MBE-grown Ge layers on GaAs. Using K-cell evaporation of Ge, we have been able to get improved characteristics in p-Ge/n-Ge diodes.

We have also successfully produced a collector-top GaAs/Ge HBT with a high current gain of 80, and a low base sheet resistance of 90 ohms/square. To the best of our knowledge, (beta/Rb)=0.89 is the largest value among bipolar transistors ever reported.

These characteristics make GaAs/Ge HBTs a promising candidate for high-speed, low-power dissipation devices to be used in future LSIs.

References

Ishizuka F, Sugioka K, Itoh T, (1986) J.A.P. vol.59 pp.495

Jadus D K, Feucht D L, (1969) IEEE Trans. Electron Devices, vol.ED-16 pp.102

Kawanaka M, Sone J, (1989) J. of Crystal Growth vol.95 pp.421

Kimura T, Kawanaka M, Sone J, (1989) Proc. 47th IEEE Device Res. Conf. (Cambridge, MA), 2A-8

Kroemer H, (1982) Proc. IEEE vol.70 no.1 pp.13

Strite S, Unlu M S, Adomi K, Gao G, Morkoc H, (1990) E.D.L. vol.11 no.5 pp.233

Inst. Phys. Conf. Ser. No 120: Chapter 2
Paper presented at Int. Symp. GaAs and Related Compounds, Seattle, 1991

61

Defects in vertical zone melt (VZM) GaAs

P.E.R.Nordquist, R.L.Henry, J.S.Blakemore[1], R.J.Gorman
and S.B.Saban[1]
Naval Research Laboratory
Washington, D.C. 20375-5000
[1]Western Washington University
Bellingham, WA 98225

ABSTRACT: VZM growth of GaAs yields semi-insulating (SI) material with a low etch pit density (EPD ca. $2\text{-}4 \times 10^3$). We have used the NaOH-KOH eutectic mixture and the Abrahams-Buiocchi (AB) etch to explore the relationship between defects in GaAs (including precipitates) and crystal growth variables. The correlation of cooling regimes and arsenic overpressure during growth with the nature and distribution of etch-revealed defects and with electrical properties is discussed.

1. INTRODUCTION

GaAs is grown commercially by the liquid encapsulated Czochralski (LEC) process, by the horizontal Bridgman (HB) or by the vertical Bridgman (VB) procedure. The LEC process grows GaAs under a steep temperature gradient and yields a material with a high and geometrically characteristic pattern of dislocations (Jordan et al 1980). The Bridgman procedures, while achieving more shallow and controllable temperature gradients, yield, in the case of HB, D-shaped wafers. Crystals grown by all of these procedures follow a "normal freeze model" (Pfann 1978) resulting in an uneven axial distribution of dopants or impurities in the crystal. Vertical zone melting (VZM) combines growth in a shallow temperature gradient with solidification according to a zone melting model resulting in a much more uniform distribution of dopants along the axis of the crystal. We report here investigations to determine the effect of growth process variables (primarily arsenic overpressure and post-growth cooling regimes) on the crystalline quality and electrical properties of semi-insulating VZM GaAs measured by physical characterization and chemical etching.

2. EXPERIMENTAL

VZM growth of GaAs has been described elsewhere (Swiggard 1989, Henry et al 1991, Nordquist et al 1990). Briefly, however, a GaAs charge in a pyrolytic boron nitride (PBN) crucible is sealed in an evacuated ampoule and passed through a furnace having a quadrantly heated spike zone. Boric oxide is used as an encapsulant to prevent arsenic loss from the crystal during growth. The quadrant heaters are adjusted to give a molten zone in the GaAs of a uniform 22 mm width. Guard furnaces on either side of the spike zone also are used to control the size of the molten zone and the temperature gradient in which the crystal grows.

The pressure within the growth ampoule was maintained using titanium-gettered argon or by an amount of elemental arsenic sufficient to give the desired pressure (calculated as As_2) included in the ampoule.

(100)-oriented crystals of 29 mm full-body diameter were grown at 0.5 cm/hr under uniform thermal conditions. Following growth, the crystals were subjected to one of two soak/cooling regimes. The standard, long soak/cooling procedure involved a soak at 1200°C for 12 hr, during which time the furnace was inverted to allow the boric oxide to drain off, followed by a 32 hr cooling to room temperature(35°C/hr). An alternative short soak/cooling cycle comprised a three hour drain period followed by a seven hour cool down to room temperature (160°C/hr).

Etching studies were conducted on wafers chem-mechanically polished with bromine-methanol. Both the NaOH-KOH eutectic etch (Lessoff and Gorman 1984) and the AB etch (Abrahams and Buiocchi 1965) were used. Etch times were 45 minutes at 350°C for the eutectic etch and 3 minutes at room temperature for the AB etch. Etch pits were counted on photomicrographs taken at several locations along a (110) direction running through the center of the wafer. An intercept method (Stirland 1977) was used to measure etch pit density (EPD) on AB etched samples.

Neutral EL2 was measured by near-infrared transmittance (Dobrilla and Blakemore 1985); ionized EL2 was determined from room temperature Hall effect data on 8 mm squares using In contacts and an ambipolar correction (Blakemore et al 1989). EPD, Hall effect and EL2 measurements were made on slices of GaAs taken from just above the cone region.

3. RESULTS

3.1 EL2

Table 1 summarizes EL2 concentrations measured on VZM-grown GaAs grown under several pressure and cooling regimes. There does not appear to be an unequivocal correlation between EL2 concentration and arsenic overpressure: neutral EL2 values are 6-7 x 10^{15}/cm^3 and total EL2 is ca. 25%

higher.   In an earlier study (Nordquist et al 1990) an
apparent increase in EL2 content with increasing arsenic
overpressure was noted but these earlier results are not
confirmed in this work.   It does appear, however, that lower
EL2 values are achieved by a rapid post-growth cooling of
the crystal, a trend that suggests that arsenic may be
frozen in the crystal in an inactive site and that a
diffusion of arsenic to an active site must occur for
maximum EL2 concentrations to be realized.

Table 1   EL2 and Growth Conditions

	Pressure (atm)	Gas	soak/ cool	EL2 $N_o (10^{15})$	Total EL2 $N_t (10^{15})$
A	1.25	As_2	slow	6.3	8.5
B	1.40	As_2	slow	7.1	9.3
C	1.98	As_2	slow	7.2	8.1
D	2.02	As_2	fast	4.6	7.6
E	1.0	Ar	slow	7.5	8.6

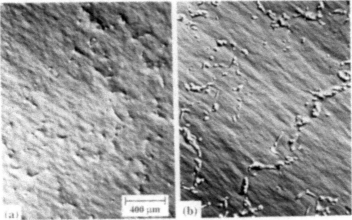

Fig.1.Typical AB etch pit distribution of
VZM GaAs: (a) slow cool; (b) rapid cool.

3.2 Etching, Dislocations and Precipitates

Etch pit densities (EPD) measured on VZM GaAs are reported
in Table 2.   EPDs derived from the NaOH-KOH eutectic etch
show no apparent correlation with arsenic overpressures in
the range 1.2 to 2.0 atmospheres.   There may be an increase

in EPD on fast cooling of the crystal but the effect is not great. As reported earlier (Nordquist et al 1990,Henry et al 1991), there appears to be no unique geometric distribution of the EPD in VZM GaAs crystals: etch pits are uniformly distributed across the crystal with both the eutectic etch and with the AB etch.The AB etch also allows other conclusions to be drawn about defects in GaAs. Because the AB etch exhibits a memory effect, it allows visualization of a three dimensional array of defects in addition to producing etch pits at the site of arsenic precipitates or inclusions in the crystal (Stirland 1977, Cullis et al 1980). Brown and Warwick (1986) have established that, in semi-insulating LEC GaAs, the AB etch produces both grooved and ridged etch figures. Ridges correspond to isolated dislocations while grooves reveal tangles of dislocations frequently in polygonized cellular arrays. Figure 1 is a photomicrograph of AB-etched wafers of VZM-grown GaAs. The material in Figure 1a (crystal C in Table 1) was cooled slowly under 2 atm arsenic pressure, while that in Figure 1b (crystal D, Table 1) was cooled rapidly under the same arsenic pressure. Essentially all of the etch figures revealed by the AB etch on VZM GaAs are of the same Nomarski contrast; physical appearance and profile measurements indicate that they are virtually all ridges. Those very few grooves that do appear to be present are all found quite near the edge of the wafer. Figure 1b also suggests that, upon fast cooling of the crystal, there is a pronounced tendency for the dislocations to form cellular arrays. These arrays, however, appear to consist entirely of ridged etch features, rather than the grooved structures found on etching of LEC material. Slow cooling (i.e. Figure 1a) appears to produce isolated dislocations which show little tendency to aggregate into cells. While the cellular arrays produced in VZM material by rapid cooling are not as dramatic as those found by Brown in LEC GaAs, they are, nonetheless, visible to the unaided eye in oblique light.

Table 2 Growth Conditions and Etch Pit Density

	Pressure (atm)	Gas	Soak/ cool	EPD ($\times 10^3$/cm^2) NaOH-KOH	AB
A	1.25	As$_2$	slow	2.5	12.0
B	1.40	As$_2$	slow	3.5	6.3
C	1.98	As$_2$	slow	2.8	8.0
D	2.02	As$_2$	fast	4.4	11.0
E	1.0	Ar	slow	4.8	12.0

A unique feature of the AB etch is that it makes possible the visualization of arsenic precipitates in GaAs. Figure 2 shows a photomicrograph of AB-etched VZM wafers. The linear features in each photograph are dislocations; the elliptical structures (arrowed), all of which are oriented in the same (110) direction, have been identified (Cullis et al 1980) as arsenic precipitates. These As-related etch pits are almost invariably associated with dislocations; we rarely observe them in dislocation-free areas. Although virtually all of the dislocations that we have observed in VZM GaAs (regardless of pressure conditions or cooling regimes) have As-pits associated with them the density of the As-pits and their size do appear to be functions of cooling regime. Thus in the slowly cooled GaAs in Figure 2a, the dislocations are decorated over essentially their entire length with As-pits ca. 6um x 3um. In contrast, in the quickly cooled material in Figure 2b, the precipitates occur along a significantly smaller fraction of the dislocation length and the precipitates themselves are smaller: ca. 2um x 1um. The difference in precipitate density and size between slowly-cooled and quickly-cooled GaAs is consistent with the "freezing in" hypothesis of arsenic being trapped in inactive sites invoked above in the discussion of the EL2 content of VZM GaAs.

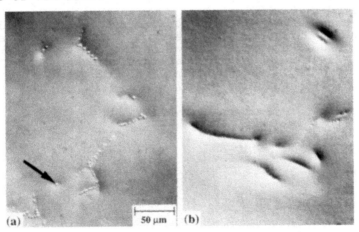

Fig. 2. Arsenic precipitates (arrow) revealed by the AB etch: (a) slow cool; (b) rapid cool.

4.SUMMARY

Post-growth thermal treatment appears to affect the electrical and crystalline properties of VZM GaAs. Measured values of EL2 are lower for rapidly cooled material compared to that cooled at a slower rate. While there is no unequivocal effect of post-growth cooling rate on etch pit density, AB etching of rapidly cooled GaAs reveals a distinctly cellular pattern of dislocations. These cells are less distinct than in LEC GaAs and consist only of ridges in contrast to LEC cells in which both grooves and ridges are present. All dislocations in VZM material incorporate arsenic precipitates but again there is a correlation with cooling regime: dislocations in rapidly cooled material are sparsely decorated with relatively small arsenic precipitates whereas those in slowly cooled GaAs are heavily decorated with larger precipitates of arsenic. The relevance of these observations to device fabrication (e.g. ion implantation) is currently under study.

REFERENCES

Abrahams M S and Buiocchi C J 1965 *J.Appl.Phys.* 36 2855
Blakemore J S, Sargent L, Tang R-S and Swiggard E 1989
    Appl.Phys.Lett. 54 2106
Brown G T and Warwick C A 1986 *J.Electrochem.Soc.*133 2576
Cullis A G, Augustus P D and Stirland D J 1980 *J.Appl.Phys.*
    51 2556
Dobrilla P and Blakemore J S 1985 *J.Appl.Phys.*58 208
Henry R L, Nordquist P E R ,Gorman R J and Quadri S B 1991
    J.Crystal Growth 109 228
Jordan A S, Caruso R and VonNeida A R 1980 *Bell Syst.*
    Technical J. 59 593
Lessoff H and Gorman R J 1984 *J.Electron. Mater.*13 733
Nordquist P E R, Henry R L, Blakemore J S, Moore W J and
    Gorman R J 1990 *Inst.Phys.Conf.Ser.*No 112 49
Pfann W G 1978 *Zone Melting* (New York:Krieger)
Stirland D J 1977 *Inst.Phys.Conf.Ser.* No 33a 150
Swiggard E 1989 *J.Cryst. Growth* 94 556

Inst. Phys. Conf. Ser. No 120: Chapter 2
Paper presented at Int. Symp. GaAs and Related Compounds, Seattle, 1991

67

Constant temperature growth of uniform-composition $In_xGa_{1-x}As$ bulk crystals by supplying GaAs

Kazuo Nakajima and Toshihiro Kusunoki

Fujitsu Laboratories Ltd. Atsugi, 10-1 Morinosato-Wakamiya, Atsugi 243-01, Japan

ABSTRACT: Ternary $In_{0.05}Ga_{0.95}As$ and $In_{0.14}Ga_{0.86}As$ bulk crystals with a uniform composition were found to be grown at a constant temperature by supplying the depleted solute elements of Ga and As to the growth melt. The pseudo-binary InAs-GaAs melts were used as the growth melts. The Liquid Encapsulated Czochralski (LEC) technique with a method of supplying GaAs source material was used to grow the ternary bulk crystals. Polycrystalline GaAs was used as a source material. 6 mm thick $In_{0.05}Ga_{0.95}As$ and 4 mm thick $In_{0.14}Ga_{0.86}As$ single bulk crystals were obtained on the GaAs and $In_xGa_{1-x}As$ ($0.05 < x < 0.08$) seeds by the constant temperature growth through only GaAs supply.

1. Introduction

Homogeneous ternary bulk crystals offer the possibility of preparing a variety of epitaxial structures without lattice-matching to a binary compound semiconductor. One of the toughest problems is how to grow ternary bulk crystals with a uniform composition. In the Liquid Encapsulated Czochralski (LEC) growth, ternary bulk crystals have been grown cooling the melts without melt replenishment by Bonner et al. (1988). In principle, the ternary crystals grown by the cooling method must have some gradient on the composition.

In the pseudo-binary phase diagram, uniform ternary crystals are expected to be able to grow at a constant temperature using supply of solute elements. However, in the LEC growth of ternary crystals, there are no reports on the growth at a constant temperature with supply of solute elements, and this theory has not been demonstrated.

One of the important materials as ternary substrates is $In_xGa_{1-x}As$. Especially $In_{0.05}Ga_{0.95}As$ can be used as a substrate for ZnSe and $In_{0.12}Ga_{0.88}As$ can be used as that for an $In_{0.57}Ga_{0.43}P/In_{0.12}Ga_{0.88}As/In_{0.57}Ga_{0.43}P$ heterostructure of 0.98 µm lasers.

In this work we have tried such a constant temperature LEC growth of $In_xGa_{1-x}As$ bulk crystals from the InAs-GaAs pseudo-binary melts using supply of solute elements, Ga and As, and we have demonstrated the above mentioned theory by obtaining the $In_xGa_{1-x}As$ bulk crystals with a uniform composition.

2. Experiments

In order to efficiently grow ternary bulk crystals, it is desirable that the crystals should be grown from pseudo-binary melts. Fig. 1 shows the InAs-GaAs pseudo-binary phase diagram. By the LEC technique, compositionally graded $In_xGa_{1-x}As$ (0.05 < x < 0.14) crystals can be grown cooling the melts along the liquidus line between Ta and Tb. These graded crystals can be used as seed crystals for the growth of uniform ternary bulk crystals. The ternary bulk crystals should be grown at a constant temperature to keep the crystal composition constant. Therefore, the technique of continuously supplying solute elements is important to grow the ternary bulk crystals at a constant temperature. For the growth of $In_xGa_{1-x}As$ from the pseudo-binary melt, the supply of GaAs source increases supersaturation in the melt and gives the driving force of the growth to the melt.

We have grown compositionally graded <111>B $In_xGa_{1-x}As$ (0.05 < x < 0.08) crystals on GaAs seeds by the LEC technique. The growth temperature was cooled during growth. The graded crystals were cut to make bars to be used as seed crystals. The LEC technique with GaAs supplying was used to grow the uniform $In_{0.05}Ga_{0.95}As$ and $In_{0.14}Ga_{0.86}As$ crystals on the GaAs and $In_xGa_{1-x}As$ (0.05 < x < 0.08) graded seeds. Fig. 2 shows the apparatus used which is a high pressure system for growing crystals at 9 atm. The GaAs source bar was dipped into the melt by moving a rod, and it was supplied during the growth. The crucible is 1 mm thick pyrolytic boron nitride (PBN) and the diameter is 6 cm. The melt compositions are $X_{Ga}^L = 0.306$, $X_{In}^L = 0.194$ and $X_{As}^L = 0.500$ for $In_{0.05}Ga_{0.95}As$, and $X_{Ga}^L = 0.18$, $X_{In}^L = 0.32$ and $X_{As}^L = 0.50$ for $In_{0.14}Ga_{0.86}As$, where X_i^L represents the atomic fraction of one element, i, in the melt. The weight of the melt was about 120 g.

In order to maintain single crystal growth, the pulling speed was fixed at 1.5 mm/h and the rotation rate of 40 rpm was used. The melt temperature was monitored at two points A and B as shown in Fig. 2, and constant during the growth. Immediately after the seed was touched to the melt, the working rod with the GaAs source was pulled down into the melt to start the supply of the solute elements and the growth of the uniform ternary bulk crystal. The melt temperature near the source material was higher than that near the seed. The melt composition of Ga near the

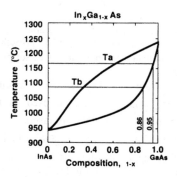

Fig. 1 Pseudo-binary InAs-GaAs phase diagram.

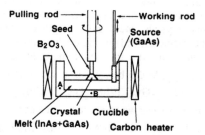

Fig. 2 LEC apparatus with supplying source material.

source was larger than that near the seed. Therefore, the solute elements were continuously supplied toward the seed by diffusion and convection.

An electron-prove microanalyzer (EPMA), employing energy dispersive X-ray detection (KEVEX DELTA I), was used to measure the crystal composition.

3. Results

Single crystals of $In_xGa_{1-x}As$ with $x = 0.05$ can be easily grown on GaAs, but it is difficult to grow single crystal of $In_xGa_{1-x}As$ with $x = 0.08$ on GaAs. In this work, it was found that and $In_xGa_{1-x}As$ single crystal with $x = 0.14$ can be grown on an $In_xGa_{1-x}As$ seed with $x = 0.08$. Therefore, the compositionally graded $In_xGa_{1-x}As$ seed crystals $(0.05 < x < 0.08)$ are required to obtain $In_xGa_{1-x}As$ single bulk crystals with $x = 0.14$. $In_xGa_{1-x}As$ bulk crystals with $x = 0.05$ and 0.14 were grown on the GaAs and graded $In_xGa_{1-x}As$ seeds by the constant temperature growth, respectively.

Fig. 3 shows the In composition, x, in $In_xGa_{1-x}As$ crystals on GaAs seeds as a function of the thickness. These crystals were grown by the ramp-cooling LEC method without supplying solute elements. The data curves a and b correspond to the results for the melt amount of 45 g and 35 g, respectively, at a constant cooling rate of 4 °C/h and a constant pulling rate of 2 mm/h. The composition variation in the crystals becomes smaller as the melt amount increases, but the crystals have some gradient on the composition.

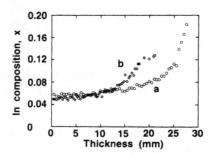

Fig. 3 In composition, x, in $In_xGa_{1-x}As$ crystals grown on GaAs by the ramp cooling LEC technique without supplying source materials.

Fig. 4 shows the outside view of $In_{0.14}Ga_{0.86}As$ bulk crystals together with the ternary seed crystals. When GaAs source was not supplied, about 1 mm thick crystal was grown on the seed as shown in Fig. 4a. This result implies that some amounts of supersaturation existed in the growth melt near the seed at the initial stage of the growth. The growth time was 1 h and 50 min. For the crystal growth of Fig. 4b, the GaAs source was dipped into the melt by 1 mm at the interval of 2 h. The single part of the bulk crystal was about 4 mm thick from the crystal/seed interface. The growth time was 5 h and 50 min.

Fig. 5 shows the In composition, x, in $In_xGa_{1-x}As$ crystals obtained by the constant temperature LEC growth with supplying GaAs solute elements as a function of the thickness. The data lines a and b show the compositional variation $In_{0.14}Ga_{0.86}As$ grown on the compositionally graded $In_xGa_{1-x}As$ seeds. The data line a is 1 mm thick single crystal without supplying GaAs and the data line b is 4 mm thick single crystal with supplying GaAs. The data lines a and b correspond to the crystals shown in Fig. 4a and 4b, respectively. The data line c is a part of 6 mm thick single $In_{0.05}Ga_{0.95}As$ grown on the GaAs seed with supplying GaAs. The In composition of the data line b and c is almost constant. This implies that in the InAs-GaAs

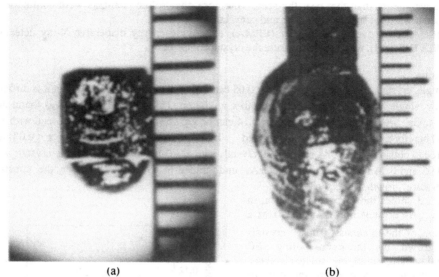

(a)   (b)

Fig. 4 Outside views of $In_{0.14}Ga_{0.86}As$ bulk crystals grown on $In_xGa_{1-x}As$ graded seeds. (a) crystal grown by only supersaturation and (b) crystal grown using the supplying technique of GaAs.

pseudo-binary system, $In_xGa_{1-x}As$ ternary crystals with a uniform composition can be grown at a constant temperature regardless of supplying of only GaAs source.

Fig. 6 shows the weight of grown $In_{0.05}Ga_{0.95}As$ crystals as a function of the weight of dissolved GaAs source in the melt. The constant temperature LEC growth was used. The weight of grown crystals increases almost proportionally as the weight of dissolved GaAs source

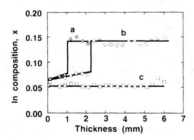

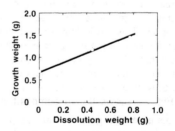

Fig. 5 In composition, x, in (a) $In_{0.14}Ga_{0.86}As/$
$In_xGa_{1-x}As$ graded seed without supplying GaAs,
(b) $In_{0.14}Ga_{0.86}As/In_xGa_{1-x}As$ graded seed
with supplying GaAs, and (c) $In_{0.05}Ga_{0.95}As/$
GaAs with supplying GaAs as a function of thickness.

Fig. 6 Weight of grown
$In_{0.05}Ga_{0.95}As$ crystals as a
function of the weight of
dissolved GaAs source in the
growth melt.

increases. From this result, it is confirmed that the constant temperature growth from the pseudo-binary melt can increase the weight of grown crystals with a uniform composition by supplying only GaAs source.

4. Discussion

The supplying technique used in this work is not sufficient for controllability of continuous supply of solute elements, so only small ternary crystals have been obtained. The source-current-controlled (SCC) method reported by Nakajima (1987) may be one of the efficient supplying methods in the point of controllability. In the SCC method, dissolved amount of solute elements in the growth melts can be controlled by electric current which is flowed through the source material. The SCC method can be included in the growth chamber of the LEC furnace to control the continuous supply of solute elements. The constant temperature growth from the pseudo-binary melts using such a LEC technique may have a possibility to obtain larger ternary bulk crystals with a uniform composition.

5. Conclusions

$In_{0.05}Ga_{0.95}As$ and $In_{0.14}Ga_{0.86}As$ single bulk crystals with a uniform composition were found to be obtained by the constant temperature growth with supplying of only GaAs to the InAs-GaAs pseudo-binary growth melt. In this growth, the LEC technique was used, and the GaAs and compositionally graded <111>B $In_xGa_{1-x}As$ ($0.05 < x < 0.08$) crystals were used as seed crystals. 6 mm thick $In_{0.05}Ga_{0.95}As$ and 4 mm thick $In_{0.14}Ga_{0.86}As$ single bulk crystals were obtained on the GaAs and $In_xGa_{1-x}As$ seeds.

Acknowledgements

The authors wish to acknowledge electron-probe microanalyzer measurements with C. Takenaka and K. Kato.

References

Bonner W A, Skromme B J, Berry E, Gilchrist H L and Nahory R E 1988 Proc. 15th Inter. Symp. on GaAs and Related Compounds, Atlanta, Georgia, Inst. Phys. Conf. Ser. 96, ed. Harris J S (Inst. Phys. Bristol-Philadelphia) pp 337-342

Nakajima K 1987 J. Appl. Phys. 61 4626

Inst. Phys. Conf. Ser. No 120: Chapter 2
Paper presented at Int. Symp. GaAs and Related Compounds, Seattle, 1991

73

In$_x$Al$_{1-x}$As/InP: organometallic molecular beam epitaxial growth and optical properties

M.J.S.P. Brasil[a], R.E. Nahory, W.E. Quinn,
M.C. Tamargo, R. Bhat and M.A. Koza
Bellcore, Red Bank, New Jersey 07701-7040

Abstract: In$_x$Al$_{1-x}$As samples are investigated using photoluminescence at 5 K. A shift of the bandedge emission energy between OM-MBE and OMCVD samples of the same composition suggests that an impurity level is dominating the photoluminescence in the OM-MBE samples. This interpretation is supported by photoreflectance measurements. The photoluminescence is shown to be a very powerful technique to investigate the quality of In$_x$Al$_{1-x}$As/InP interfaces. The interface seems to improve when we decrease its growth-halt time. A valence band offset in the range of 350 to 400 meV is estimated for this type II interface using InP quantum wells for electrons.

I. Introduction

The In$_x$Al$_{1-x}$As/InP heterostructure possesses properties that make it very promising for device applications [Kroemer 1983]. It combines the high saturation velocity of InP with a large type II conduction band offset, very suitable for two-dimensional electron gas structures. Improved performance devices such as InP-channel MESFET's and HIGFET's have already been reported [Hanson et al 1987, Fathimulla et al 1988]. In$_x$Al$_{1-x}$As has been grown previously by molecular beam epitaxy (MBE) and organometallic chemical vapor deposition (OMCVD). Relatively few studies however have been done on thin layer structures of In$_x$Al$_{1-x}$As/InP. In this paper we report the growth and properties of the In$_x$Al$_{1-x}$As/InP structure prepared by organometallic molecular beam epitaxy (OM-MBE), and compare with OMCVD structures.

II. Growth

The samples were grown using a Vacuum Generators V80-H gas source MBE system. Trimethylindium and tri-isobutylaluminum were used for the group III sources and cracked arsine and phosphine were used as the group V sources. Substrate temperature ranged from 500° C to 530° C, where the incorporation of group III elements was monotonic. Sample stoichiometry was changed by altering the flux ratio of the two group III gases. The arsine flow was fixed at 2 sccm. The InP growth rate was ~ 0.7μm/h and the InAlAs ~ 0.3 μm/h. Undoped InAlAs layers, with thicknesses varying between 500 and 10000 Å, were grown on 200 to 1000 Å

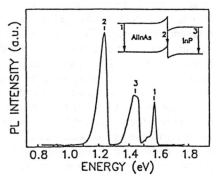

Figure 1. PL spectrum at 5K of an In$_{.48}$Al$_{.52}$As sample grown by OM-MBE. The inset shows a band diagram identifying the transitions observed.

InP buffer layers, using either n^+ or semi-insulating InP substrates. Thin InP layers were grown between thick $In_{0.52}Al_{0.48}As$ layers for electron quantum wells.

III. Results

The photoluminescence (PL) spectrum at 5 K for a thick InAlAs sample grown on n^+InP is shown in fig. 1. We observe three distinct peaks related to the three possible transitions indicated in the diagram in fig. 1. Peak **2** is assigned to the spatially indirect transition through the $In_xAl_{1-x}As$/InP interface. Peak **3** has an unusual shape due to the highly doped, degenerate, n^+ InP substrate from which the emission originates.

Peak **1** is assigned to a transition in the InAlAs layer. Some authors have used the energy of the luminescence peak as a direct measure of the bandgap energy [Wakefield et al 1984, Davies et al 1984, Chu et al 1988, Brennan et al 1989, Oh et al 1990]. The values presented in the literature for the most common reference, the bandgap of the $In_{0.52}Al_{0.48}As$ lattice-matched ternary, show however a wide variation, which gives an uncertainty about the detailed origin of this emission. Figure 2 presents the energy of our

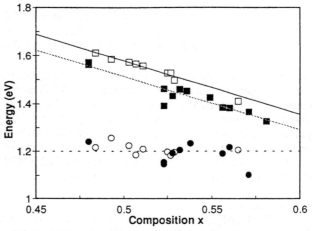

Figure 2. Energy of the observed photoluminescence peaks for OM-MBE samples (InAlAs emission, ■ and interface emission, ●) and OMCVD samples (InAlAs emission, □ and interface emission, ○).

PL peaks as a function of the composition. The $In_xAl_{1-x}As$ compositions have been measured by double-crystal X-ray diffraction. The more mismatched samples present a slightly larger error in composition due to partial relaxation. The figure shows results for several samples grown by OM-MBE and a comparison with samples grown by OMCVD [Bhat et al 1991]. The two continuous lines correspond to the empirical equation given by Wakefield et al 1984, which applies specifically to this small range of compositions, but with an added constant. The constants used were, respectively, +25 meV and -40 meV. Thus, our data exhibits a shift of 65 meV between the PL peak energies observed in OMCVD and OM-MBE samples having the same composition. This shift could be explained by the effects of an impurity level, phase ordering or phase separation. In the following discussion we conclude that the former is correct for the present samples.

Phase ordering has been observed in various alloys grown by OMCVD and MBE, such as InGaP. It changes the band structure of the material, usually inducing a decrease of the bandgap energy [Mascarenhas et al 1989]. Phase separation, sometimes also called clustering, consists of regions with compositions higher and smaller than the average. Depending on the size of these regions, the PL will present a peak only at the smallest bandgap. Oh et al. 1990 attributed an observed variation of the InAlAs band gap as a function of the growth temperature to clustering. The fact that our PL peak energies for InAlAs samples grown by OMCVD, grown between 650 and 700 C, are higher than those for InAlAs samples grown by OM-MBE, usually grown at 500 to 530 C, is consistent with the fact that high temperatures tend to inhibit phase ordering.

On the other hand, the 65 meV shift, constant for all compositions, can be easily explained by the presence of an impurity with this binding energy. Previous workers have suggested that impurities can dominate the observed PL spectra of InAlAs samples, with one study [Praseuth et al 1987] reporting a difference of 70 meV between photoluminescence and excitation spectroscopy peaks for samples grown by MBE, and another study [Gaskill et al 1990] reporting a difference of 45 meV between PL and photoreflectance measurements for samples grown by OMCVD. To clarify this question in

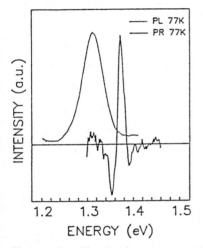

Figure 3. Photoluminescence and photoreflectance spectra at 77 K for an InAlAs sample grown by OM-MBE showing the energy difference between their features.

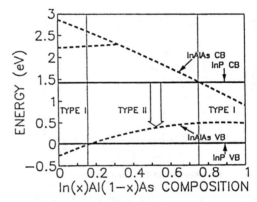

Figure 4. InAlAs (dashed) and InP (solid) band edges versus In composition. Dotted lines separate type I and type II interface regions. The arrow marks the interface transition for lattice match.

our case, we have measured the photoreflectance (PR) at 77 K. Figure 3 shows a comparison between the PL and the PR of a thick InAlAs layer grown by OM-MBE. We observe that the PL and the PR features do not line up. PR is not expected to be sensitive to extrinsic effects, such as impurities, and thus should reveal the intrinsic bandgap, or multiple bandgaps if phase separation exists. The PR indicates a bandgap energy ~ 60 meV higher than the PL peak. This difference between emission and reflectance energies is not consistent with either phase ordering or phase separation and can only be explained by an impurity level, although its chemical identification is not possible at this point. The fact that this impurity is not observed in our OMCVD samples is reasonably understood due to the differences in growth conditions. Considering our comparison of PL and PR spectra, along with our PL data for the two kinds of growth techniques, we deduce that the intrinsic bandgap for $In_{0.52}Al_{0.48}As$ lattice-matched to InP is 1.539 eV at 5K, where we have included the calculated exciton Rydberg of 6 meV.

Figure 2 also shows the energy of the PL peak related to the InAlAs/InP interface as a function of the InAlAs composition. We observe that this energy is almost constant for $0.48 < x < 0.58$. We can notice some scatter in the values, with most of the measurements ranging between 1.15 and 1.25 eV, but no systematic dependence on the composition is observed. The fact that the interface transition does not follow the same variation as the bandgap of InAlAs in that range, which is greater than 400 meV, indicates that most of this bandgap variation is in the conduction band offset and thus is not affecting the interface transition. This is also consistent with the diagram presented in

Figure 4 which helps in understanding this system, having type I and type II domains. The diagram presents the variation of the $In_xAl_{1-x}As$ bandedges as a function of the composition x using the InP bands as references. To draw the diagram we have used the band offset values of the studied heterojunctions [Aina et al 1988], InAs/InP, AlAs/InP and $In_{0.52}Al_{0.48}As$/ InP, and calculated a second degree interpolation, with no strain included. We observe that, in the range of composition of our samples (0.48 < x < 0.58), the system is always type II and the bandgap variation is primarily related to changes in the conduction band.

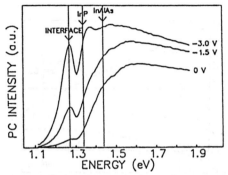

Figure 5. Photocurrent spectra for various reverse bias voltages. The dashed lines indicate the InP and InAlAs bandgap energies and the peak energy for the InP/InAlAs interface transition.

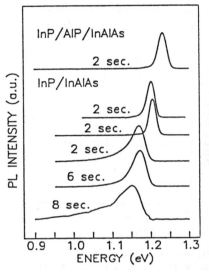

Figure 6. Photoluminescence spectra at 5 K showing the interface emission for different growth-halt times for the InP/InAlAs interface. The top-most spectrum is for an AlP-rich interface sample.

Figure 5 shows the photocurrent at 300 K for the sample from fig. 1 as a function of the reverse voltage applied between a front Schottky contact and a back ohmic contact on the InP:n^+ substrate. We observe two resolved peaks that we attribute respectively to the InP bandgap and the InP/InAlAs interface. Superimposed on these peaks, there is a broad shape, attributed to the InAlAs bandgap. For zero applied voltage, the field should be primarily in the InAlAs layer, so that the photocurrent is dominated by the InAlAs absorption whose broad shape probably originates from Franz-Keldysh effects. As we increase the applied voltage, the field extends into the InAlAs/InP interface, so that the peak related to this region increases. These results show that a substantial absorption occurs at the InAlAs/InP interface, which gives rise to a strong photocurrent as soon as there is a field to separate the photogenerated carriers. Finally, at the largest reverse voltage, the field evidently extends into the InP buffer layer since a broad peak appears close to the bandgap of InP.

The type II interface between InAlAs and InP creates a separation of the two types of carriers, so that electrons are confined in the InP layer and holes are confined in the InAlAs layer. The intense PL peak observed indicates a strong overlap of the electron and hole wave functions precisely at the interface, which makes the PL a highly sensitive technique to investigate the interface quality of this system. Figure 6 shows the PL spectra in the interface peak region for various lattice-matched samples. The layer structure is always the same: a thin (500-700 Å) undoped InAlAs layer grown on an undoped InP buffer layer (1000 Å). The difference from sample to sample comes from the time of the growth-halt in switching from InP growth to InAlAs growth or in the specific gas switching sequence. During the growth-halt, an arsenic overpressure was maintained, after which both In and Al were simultaneously introduced into the chamber. The interface emission energy varies between 1.05 and 1.25 eV, becoming narrower and stronger when the peak maximum moves to higher energies, suggesting better interface quality. As a general conclusion, the interface seems to improve when we decrease its growth-halt time, but some randomness is also seen, suggesting other factors may be affecting the interface quality. A very striking result was that the strongest and narrowest interface emission belongs to a sample grown with the flux equivalent of 2 monolayers of AlP at the interface in between the InP and the InAlAs. The emission observed for this sample is still characteristic of the InP/InAlAs interface and indicates that the electron (InP) - hole (InAlAs) overlap is strong. A detailed study of this effect will be reported elsewhere.

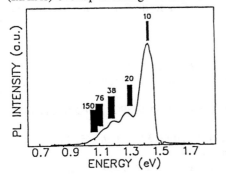

Figure 7. PL spectrum (5K) for InAlAs with five InP quantum wells. Numbers give nominal thicknesses in Angstroms; bars show energies calculated for VB offsets = 350 to 400 meV.

In order to obtain the band offset values and further understand the interface transition, we have also studied the PL for a sample with thin InP layers separated by thick InAlAs layers. The InP layers will act as quantum wells for electrons. Figure 7 shows the PL spectrum at 5 K of a multiple quantum well structure, with nominal InP quantum well thicknesses of 10, 20, 38, 76 and 150 Å, as estimated from the growth time and the growth rate deduced from thick InP layers. The InAlAs barriers were 500 Å thick. The spectrum shows 5 peaks corresponding to the 5 quantum wells. The bars represent the calculated range of transition energies for each quantum well using the nominal thicknesses and taking the valence band offset ranging between 350 and 400 meV to arrive at a best fit. The parameters used in this calculation are, for $In_{0.52}Al_{0.48}As$ (InP): bandgap energy, 1.539 eV (1.423 eV); electron effective mass, 0.08 m_0 (0.08 m_0); heavy hole effective mass, 0.68 m_0 (0.45 m_0). Our result for the valence band offset is close to the larger values reported in the literature, which are 324 meV [Bhat et al 1991], 290 meV [Aina et al 1988], 230 meV, [Hybertson et al 1991] and 160 meV [Waldrop et al 1990].

Using our valence band offset, we can analyze more carefully the interface emission presented in fig. 6, now considering an effective bandgap through the interface of 1.05 ± 0.03 eV. We observe peaks over a range of energies, up to 180 meV larger than this value. This is at least partially explained by the electron and hole confinement energies in the triangular shaped quantum wells built at the interface. In this model, "good interfaces" are expected to have a sharper, narrower band bending region with concomitantly larger confinement energies. We cannot neglect however the interface chemistry whereby certain ternary alloys may appear near the interface. Finally, the possibility of a defect or impurity related interface transition could exist for some samples with broad and low energy peaks, as pointed out previously [Caine et al 1984] where an interface transition energy of 0.96 eV was deduced.

IV - Conclusions

We observed a different bandedge PL emission energy between OM-MBE samples and samples of the same composition grown by OMCVD. Comparison of photoreflectance (PR) and PL spectra suggests that an impurity level is dominating the PL in the OM-MBE samples as opposed to phase separation or ordering effects. We deduce a value of 1.539 eV for the low temperature intrinsic bandgap of $In_{0.52}Al_{0.48}As$ lattice-matched to InP. Efficient interface emission is observed, which is a highly sensitive probe of interface quality. Changes in the interface transition energy and peak shape are observed and correlated to the growth-halt time for the interface. More efficient and higher-energy interface transitions are observed for shorter growth-halt times. InP quantum wells were also investigated, from which the valence band offset for this type II interface was estimated to be in the range from 350 to 400 meV.

Acknowledgments: The authors thank R.J. Martin for technical assistance. One of us (MJSPB) thanks Fundação de Amparo à Pesquisa no Estado de São Paulo (Brazil) for financial support during this work.

References
Aina L, Mattingly M and Stecker L 1988 *Appl. Phys. Lett.* **53**, 1620
Bhat R, Koza M A, Kash K, Allen S J, Hong W P, Schwarz S A, Chang G K and Lin P 1991 *J. Crystal Growth* **108**, 441
Brennan T M, Tsao J Y, Hammons B E, Klem J F, Jones E D 1989 *J. Vac.Sci. Technol.* **B7**, 277
Caine E J, Subbanna S, Kroemer H and Merz J L 1984 *Appl. Phys. Lett.* **45**, 1123
Chu P, Wieder H H 1988 *J. Vac. Sci. Technol.* **B6**, 1369
Davies G J, Kerr T, Tuppen C G, Wakefield B and Andrews D A 1984 *J. Vac. Sci. Technol.* **B2**, 219
Fathimulla M A, Loughran T, Stecker L, Hempfling E, Mattingly M and Aina O 1988 *IEEE Elect. Dev. Lett.* **EDL-9**, 223
Gaskill D K, Bottka N, Aina L and Mattingly M 1990 *Appl. Phys. Lett.* **56**, 1269
Hanson C M, Chu P, Wieder H H and Clawson A R 1987 *IEEE Electron Device Lett.* **EDL-8**, 53
Hybertsen M S 1991 *Appl. Phys. Lett.* **58**, 1759
Kroemer, H 1983 *IEEE Electr. Device Lett.* **EDL-4**, 20
Mascarenhas A, Kurtz S and Olson J M 1989 *Phys. Rev. Lett.* **63**, 2108
Oh J E, Bhattacharya P K, Chen Y C, Aina O and Mattingly M, 1990 *J. Electr. Materials* **19**, 435
Praseuth P R, Goldstein L, Hénoc P, Primot J 1987 *J. Appl. Phys.* **61**, 215
Wakefield B, Halliwell M A G, Kerr T, Andrews D A, Davies G J and Wood D R 1984 *Appl. Phys. Lett.* **44**, 341
Waldrop J R, Kraut E A, Farley C W and Grant R W 1990 *J. Vac. Sci. Technol.* **B8**, 768
[a]On leave from UNICAMP, Campinas, Brazil.

Inst. Phys. Conf. Ser. No 120: Chapter 2
Paper presented at Int. Symp. GaAs and Related Compounds, Seattle, 1991

79

Strained quantum well InGaSb/AlGaSb heterostructures grown by molecular beam epitaxy

J. F. Klem, J. A. Lott, J. E. Schirber, and S. R. Kurtz

Sandia National Laboratories, Albuquerque, NM 87185

ABSTRACT: Strained, modulation-doped InGaSb/AlGaSb quantum wells were grown on InP substrates by molecular beam epitaxy and characterized by magneto-transport measurements. Hole transport properties were strongly correlated with growth conditions . Shubnikov-de Haas measurements yielded a hole mass of $(0.15 \pm 0.02)\, m_0$, and hole mobilities as high as 3300 cm^2/Vs were obtained at 77K, with a density of 1.6×10^{12} cm^{-2}, thus showing promise for p-type field-effect transistor applications.

1. INTRODUCTION

The development of a fast, low-power compound semiconductor complementary logic has been hampered by the lack of a high-performance p-channel companion to the n-channel field-effect transistors (FETs) now available. Recent demonstrations of InGaAs p-channel devices have shown that strain-induced light/heavy hole splitting leads to enhanced hole transport (Osbourn 1986, Drummond *et al.* 1986, Ruden *et al.* 1989); yet the p-channel device still limits the ultimate performance of these circuits. Small band offsets, resulting in high gate leakage currents, and the conflicting needs for high hole densities and high hole mobilities (Jones *et al.* 1990) pose significant difficulties in these materials.

An alternate materials combination has recently been proposed for a complementary logic system in the compound semiconductors (Longenbach *et al.* 1990). This system relies on InAs-channel devices, which are expected to provide high electron saturation velocities, for the n-type FETs, while the p-type devices would have channels of GaSb. GaSb has a higher low-field mobility than GaAs, and has estimated valence band offsets to AlSb and AlAs$_{0.2}$Sb$_{0.8}$ of 0.4 and 0.52 eV, respectively, which has allowed fabrication of high-quality p-channel FETs (Luo *et al.* 1990a, 1990b). In this work, we examine the growth and characteristics of strained-channel InGaSb/AlGaSb p-type modulation-doped structures, which should offer improved transport due to the removal of the light/heavy hole degeneracy in a manner analogous to strained InGaAs-channel devices.

2. EXPERIMENTAL

The structures were grown on (100)-oriented semi-insulating InP substrates by molecular beam epitaxy at substrate temperatures from 400 to 550 °C. The group V species were As$_4$ and Sb$_4$, provided by conventional and "low-temperature" effusion cells, respectively. Growth rates were 0.4 to 0.6 monolayers/second. A typical structure, as shown in Figure 1, consisted of a thin AlSb buffer layer, a thicker Al$_{0.75}$Ga$_{0.25}$Sb buffer (sometimes with an AlSb/GaSb superlattice inserted in the center), a GaSb "smoothing layer", the InGaSb channel, an undoped AlSb spacer layer, a Be-doped AlSb layer, and a GaSb cap.

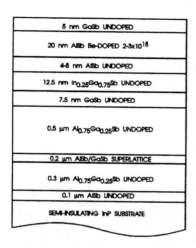

Fig. 1. Typical structure of the InGaSb/AlGaSb modulation-doped samples.

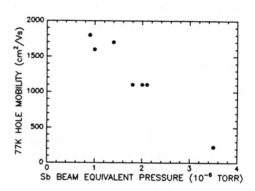

Fig. 2. Hall hole mobilities at 77K for similar GaSb-channel structures grown with various Sb beam fluxes.

Four-terminal measurements using the Van der Pauw geometry were used to determine carrier concentrations and mobilities at 300, 77, and 4K. In addition, Shubnikov-de Haas oscillations were analyzed below 4K to measure carrier masses.

3. RESULTS

The surface morphologies of these structures were generally specular, despite the 4% mismatch between the InP substrate and the AlGaSb buffer layer. The best morphologies were generally obtained with AlSb buffers, while buffers with compositions of $Al_{0.75}Ga_{0.25}Sb$ or $Al_{0.5}Ga_{0.5}Sb$ appeared slightly rough under phase-contrast microscopy. The mobilities obtained in these structures obey an opposite correlation, however, with the AlGaSb-buffer structures displaying higher mobilities. For this reason, as well as to avoid problems with the hygroscopic nature of AlSb and to achieve a reasonable valence band offset with GaSb, $Al_{0.75}Ga_{0.25}Sb$ was used for the buffer material. Surface morphologies were found to be improved considerably with the inclusion of a 0.1 μm AlSb layer at the beginning of growth, as well as the inclusion of a thin 100Å/100Å GaSb/AlSb superlattice in the buffer layer. These layers were found to have little effect on the measured Hall mobilities in the samples.

The Sb/III ratio during growth of these structures was found to significantly influence their electrical properties, in agreement with earlier studies of the optical properties of AlSb/GaSb superlattices (Suzuki *et al.* 1986). Figure 2 shows measured hole mobilities at 77K for similar AlGaSb/GaSb structures grown with various Sb fluxes. At an Sb beam equivalent pressure of approximately 7×10^{-7} torr, the growth was barely Sb-stabilized and exhibited a c(2x6) reconstruction (Chang *et al.* 1977). The highest hole mobilities were found at Sb pressures slightly above this point. At higher Sb pressures, hole mobilities and concentrations both declined rapidly. This correlates with previous observations of the degradation of GaSb/AlSb crystalline quality at higher V/III ratios, where twins and stacking faults were found to form as a result of the decreased surface mobilities due to the high Sb flux (Suzuki *et al.* 1986).

Reflection high-energy electron diffraction (RHEED) patterns were monitored during the growth, and no intensity oscillations were observed at growth temperatures corresponding to optimal electrical properties in these structures. At lower temperatures, however, clear and persistent oscillations were observed for both AlGaSb and InGaSb layers after brief growth interruptions. These interruptions were performed at temperatures near 500 °C under an Sb flux. Figure 3 shows oscillations in the specular spot intensity for an azimuth a few degrees off the two-fold reconstructed <110> direction as a function of growth temperature for an $Al_{0.75}Ga_{0.75}Sb$ layer. Oscillations are weak for temperatures of 470 °C and above, but are quite pronounced at 425 °C. This is indicative of the transition from island growth and coalescence to a step-flow growth mode in this temperature range. Extremely intense oscillations were also observed for GaSb at 400 °C, and somewhat less intense oscillations were observed for $In_{0.25}Ga_{0.75}Sb$ in the 400-425 °C range.

The electrical characteristics of these structures also are fairly sensitive to the growth temperature. Figure 4 shows Hall mobilities measured at 77K versus growth temperature. We found that the highest mobilities were obtained for growth near 500 °C for most of the structure. The InGaSb strained layers were grown at 425-460 °C. The mechanism for degradation of electrical characteristics at low temperatures is most likely the same as that encountered

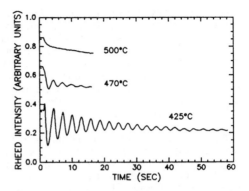

Fig. 3. Specular RHEED intensity oscillations for $Al_{0.75}Ga_{0.25}Sb$ at various growth temperatures.

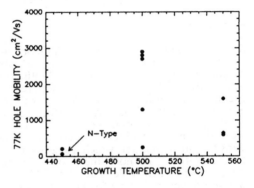

Fig. 4. Hall hole mobilities at 77K for similar InGaSb- and GaSb-channel structures grown at various substrate temperatures.

for excessive Sb overpressures, i. e. an increase in crystalline defects resulting from reduced surface mobility, which is demonstrated by the RHEED oscillation results discussed above. The reason for the degradation at higher temperatures is less clear, but may be related to higher point-defect densities. This is supported by the observation that the lower mobilities are usually accompanied by smaller measured hole concentrations, indicating the presence of a compensating defect. Although it was not attempted in this work, it is conceivable that either raising or lowering the Sb flux would lead to a shift in the growth temperature which optimizes the electrical characteristics of these structures.

The highest 77K mobilities obtained in these structures were 2200 and 3300 cm^2/Vs for GaSb- and InGaSb-channel samples, with hole densities of 2.4 and 1.6×10^{12} cm^{-2}, respectively. The corresponding 300K mobilities were 280 and 520 cm^2/Vs, at 3.9 and 1.6×10^{12} cm^{-2}. Mobilities increased monotonically with decreasing temperature between 300 to 77K, but measurements at 4K showed substantially lower mobilities than at 77K. Those samples with the highest hole concentrations (due to higher doping levels) generally showed the least degradation in mobility at 4K. In addition, as the thickness of the undoped spacer layer between the doped layer and channel was increased beyond 40Å, the mobilities remained constant or declined, as did the hole concentrations. This indicates that impurity scattering due to the doped layer is probably not the limiting factor for the mobilities in these samples. Rather, some other temperature-dependant mechanism which is screened by high hole concentrations is implicated. One plausible mechanism is dislocation scattering, which is reasonable in light of the densities of dislocations expected in these highly-mismatched structures.

Magnetoresistance measurements were performed on one sample to determine the hole mass in a strained-channel $In_{0.25}Ga_{0.75}Sb$ structure. At 4K the mobility for the p-type sample used was 2700 cm^2/Vs and the two-dimensional hole density was 2.6×10^{12} cm^{-2}. The effective mass was determined from the temperature dependence of the Shubnikov-de Haas oscillations in the magnetoresistance between 1 and 4 K. Because of the relatively low value of the mobility, data were taken at fields above 4.5 T. The mass of the holes in this material was found to be $m^* = (0.15 \pm 0.02)$ m_0 in the field range 4.5-6.5 T, thus demonstrating the light hole character expected for this strained quantum well system. To our knowledge, this is the first measurement of light hole masses in the InGaSb/AlGaSb system.

4. SUMMARY

Strained-quantum well InGaSb/AlGaSb p-type modulation-doped structures were grown by molecular beam epitaxy on InP substrates and characterized by RHEED and magnetotransport measurements. Optimal transport characteristics were obtained at temperatures high enough to suppress or nearly suppress RHEED intensity oscillations, indicating a step-flow growth mode, and at Sb overpressures barely high enough to stabilize the c(2x6) surface reconstruction. Mobilities as high as 520 and 3300 cm^2/Vs were obtained at 300 and 77K, respectively, with a hole density of 1.6×10^{12} cm^{-2}. Mobilities decreased below 77 K and were correlated with hole densities in a manner which suggests that dislocation scattering dominates at low temperatures. A hole mass of 0.15 m_0 was measured at low temperatures, demonstrating light-hole character. This material system should provide the basis for a high-performance p-type field-effect transistor technology.

5. ACKNOWLEDGEMENTS

The authors would like to thank J. Avery and D. Overmyer for technical support. This work was supported by the U. S. Department of Energy under contract No. DE-AC04-76DP00789.

REFERENCES
Chang C-A, Ludede R, Chang L L and Esaki L 1977 *Appl. Phys. Lett.* **31** 759
Drummond T J, Zipperian T E, Fritz I J, Schirber J E and Plut T A 1986 *Appl. Phys. Lett.* **49** 461
Jones E D, Biefeld R M, Klem J F and Lyo S K 1990 *Inst. Phys. Conf. Ser.* **106** 435
Longenbach K F, Beresford R and Wang W I 1990 *IEEE Trans. Electron. Dev.* **37** 2265
Luo L F, Longenbach and Wang W I 1990a *IEEE Electron Dev. Lett.* **11** 567
Luo L F, Longenbach and Wang W I 1990b *Intl. Elec. Dev. Meeting Tech. Digest* 515
Osbourn G C 1986 *J. Vac. Sci. Technol.* **B4** 1423
Ruden P P, Akinwande A I, Narum D, Grider D E and Nohava J 1989 *Intl. Elec. Dev. Meeting Tech. Digest* 515
Suzuki Y, Ohmori Y and Okamoto H 1986 *J. Appl. Phys.* **59** 3760

Inst. Phys. Conf. Ser. No 120: Chapter 2
Paper presented at Int. Symp. GaAs and Related Compounds, Seattle, 1991

83

Growth and characterization of InAs$_{1-x}$Sb$_x$ layers on GaSb substrates

A. Y. Polyakov, M. Stam, A. Z. Li* and A. G. Milnes
Carnegie Mellon University, Pittsburgh, PA 15213-3890
* Laboratory of Functional Materials for Informatics, Shanghai Institute of
Metallurgy (Academia Sinica), Shanghai, China.

ABSTRACT: InAs$_{1-x}$Sb$_x$ layers with good surface morphology have been grown by
MBE on GaSb substrates in the whole range of compositions. It is shown that
heterostructures on n-GaSb substrates behave as quasi Schottky diodes whereas
structures on p-GaSb display good ohmic behavior. The band offsets for different
compositions are calculated from C-V measurements.

1. Introduction

The InAsSb ternary system is of considerable interest for long wavelength IR sensors
because for different compositions it covers the range of 4 - 10 μm and the InAs$_{0.35}$Sb$_{0.65}$
solid solution comes closest among all the III-V materials to the band gap of HgCdTe,
(Chen et al., 1987). The cut-off wavelength can be even further increased by growing strained superlattices, (Osbourn, 1984; Kurtz et al., 1988). Moreover very recent
theoretical calculations of Wei and Zunger (1991) show that when grown on the (111)
surface, the equiatomic InAs$_{0.5}$Sb$_{0.5}$ composition might undergo an ordering transition
that would decrease the band gap even further. The InAsSb system has no suitable lattice matched substrate for growth in the composition range. There have been reported,
however, examples of successful growths of low Sb content compositions lattice matched
to GaSb by LPE (Gertner et al., 1979; Bubulac et al., 1980), MOCVD (Srivastava et
al., 1986) and MBE (Chin et al., 1988). Growth on InSb substrates was mainly done
for strained superlattice studies (Kurtz et al., 1988). Several works have been devoted
to growing InAsSb on InAs by LPE (Cheung et al., 1985; Mohammed et al., 1986) or
MBE (Chen et al., 1987; Yen et al., 1988). Despite a very large lattice mismatch to
InAs, a material with good electronic and recombination properties could be grown.
Therefore it was of interest to examine growth on GaSb which provides such apparent
advantage as the ability to use GaSb as a wide band gap window.

2. Experimental

Our InAs$_{1-x}$Sb$_x$ samples were grown by MBE in the Perkin-Elmer 400 system. The substrates were (100) oriented n- or p-GaSb with carrier concentrations in the range 10^{17}-
10^{18}cm^{-3}. Before growth, mechanically polished substrates were etched for 5 seconds

in Br_2:methanol (0.3% Br_2) and mounted by In on a molybdenum block. The substrate temperature during oxide desorption was 560°C under Sb_4 flux and the growth was at 480°C. For composition with x = 0.50, the As_4 pressure was maintained at 5 x 10^{-6} Torr; Sb_4 to In beam equivalent pressure ratio was $\sim$ 30. Lattice-matched compositions were grown at higher As_4 pressure (2 · 10^{-5} Torr) and the Sb_4/In ratio around 10. The layer composition was determined from x-ray measurements assuming that the lattice parameter follows Vegard's law. Layer thicknesses were in the range 0.3 - 1 μm and the growth rates were about 0.3 μm/h. Layers were either undoped or Be-doped. To assess concentration and mobility values of electrons or holes in the layers, Van der Pauw measurements were performed on structures grown in similar runs on semi-insulating GaAs substrates. For thicknesses larger than 0.3 μm there was no appreciable dependence of measured concentration and mobility on the layer thickness. Electrical characterization of heterojunction properties was done by C-V and I-V measurements in the 77 - 400 K temperature range. For these measurements, Au dots of approximately 1 mm in diameter were evaporated on the InAsSb layers using a shadow mask. These dots served as a mask when defining mesas by etching in H_2SO_4:H_2O (20:1) solution. The back contact to the GaSb substrate was formed by In. The same structures were also used for DLTS measurements in the 100 - 400 K range using a SULA DLTS spectrometer.

3. Results and Discussion

As indicated by x-ray measurements, the x-values of our $InAs_{1-x}Sb_x$ layers were 0.1, 0.12, 0.5, 0.6 and 1. All of the grown layers had good surface morphology. Most detailed studies were done for the lattice matched composition x = 0.1. These layers were grown either n-type (concentration 6.8 · 10^{16}cm^{-3} and mobility 12700 cm^2/V·s at room temperature as measured by Van der Pauw method on a SI GaAs substrate) or p-type with the concentration 1 · 10^{18}cm^{-3} and mobility 300 cm^2/V·s. When grown on n-GaSb (n $\sim$ 2 · 10^{17}cm^{-3}) substrates, both p- or n-type structures had practically identical I-V characteristics (in Fig. 1, the I-V curve for n-type structure is shown) with the ideality factors in the forward direction between 1.2 and 1.3. The reverse current is governed by the surface leakage in the GaSb substrate. If the saturation current is calculated from the I-V curve in the forward direction, it has an activation energy close to 0.55 eV. The capacitance-voltage characteristic of the n-$InAs_{0.9}Sb_{0.1}$/n-GaSb structure is shown in Fig. 2.

The $1/C^2$ plot is linear and gives a cut-off voltage of 0.7 V for the lattice matched (x = 0.1) composition. For the p$^+$-$InAs_{0.9}Sb_{0.1}$/n-GaSb structures, the cut-off voltage is 0.82 V. For both n- and p-type layers, the band bending in the n-GaSb substrate is larger than the band gap and a p-type inversion layer is formed at the heterointerface on the GaSb side. When n-type layers were grown on n$^+$-GaSb substrates (2 · 10^{18}cm^{-3}), the forward and reverse currents were greatly increased, the ideality factor became 2.6 and I-V curves were practically temperature independent indicating that, in this case, tunneling was prevalent. The structures n-$InAs_{1-x}Sb_x$/p-GaSb had perfectly linear I-V characteristics for all x-values at all temperatures.

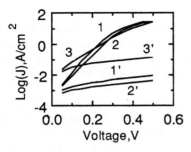

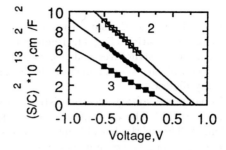

Fig. 1. Current-voltage characteristics for n-InAs$_{1-x}$Sb$_x$/n-GaSb; 1, 1'–forward and reverse currents for x = 0.1; 2, 2'–for x = 0.5; 3, 3'–for x = 1.

Fig. 2. Capacitance-voltage characteristics for n-InAs$_{1-x}$Sb$_x$/n-GaSb; (1) x = 0.1, (2) x = 0.5, (3) x = 1; S is the diode area.

Thus electrical properties of our InAs$_{0.9}$Sb$_{0.1}$/GaSb structures are quite similar to the behavior of heterojunctions involving InAs (Harrison, 1977) or InAs$_{0.95}$Sb$_{0.05}$ (Srivastava et al., 1986). The band diagram for the n-n isotype heterostructure is shown in Fig. 3a. The band bending as measured by C-V is related to the GaSb substrate. From its value, we can as usual calculate the charge in the depletion region of the n-GaSb (including the charge of the holes in the inversion layer) that should be balanced by the charge in the accumulation region of the n-InAsSb. Then the band bending in the narrow band gap material can be calculated using Poisson's equation. When doing that, we have neglected the effects of quantization which is permissable as shown for n-InGaAsSb/n-GaSb by Afrailov et al., (1990). The values of ΔE_c and ΔE_v for x = 0.1 are given in Fig. 4 and are in agreement with the results of Srivastava et al. (1986).

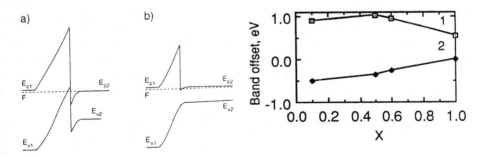

Fig. 3. Band diagrams for n-InAs$_{1-x}$Sb$_x$/n-GaSb; a) x = 0.1, b) x = 1.

Fig. 4. Band offsets ΔE_c (1) and ΔE_v (2) in InAs$_{1-x}$Sb$_x$/GaSb as a function of x.

The I-V characteristics of the lattice mismatched n-InAs$_{1-x}$Sb$_x$/n-GaSb structures with x = 0.5 and 0.6 bear no evidence of tunneling generally expected because of the very large density of misfit dislocations at the heteroboundary (selective etching reveals a very dense dislocation network in this region). The ideality factor is again close to unity (1.25) and the saturation current is even slightly lower than for the lattice-matched composition (Fig. 1). Photosensitivity of these structures is the same as for lattice-matched structures and it is twice as large as the photosensitivity of GaSb p-n junctions and an order of magnitude larger than the photosensitivity of Au/n-GaSb Schottky diodes.

Capacitance-voltage characteristics are shown in Fig. 2. The cut-off value is 0.8 eV for x = 0.5. The concentration of electrons from Van der Pauw measurements is 10^{18}cm^{-3}. Then if the Fermi level position in this degenerate sample is calculated using the effective mass values and zone parameters for InSb one arrives at the ΔE_c and ΔE_v values for x = 0.5 and 0.6 shown in Fig. 4.

The I-V curve for the n-InSb/n-GaSb heterojunction is shown in curve 3 Fig. 1. The ideality factor is 2.5, the magnitudes of forward and reverse current are increased and the temperature dependence of the current is very weak, all indicating that tunneling via interface states dominates. The cut-off voltage from the C-V curve is 0.46 eV and the corresponding band diagram is shown in Fig. 3b. It can be seen from this figure and from Fig. 4 that, in contrast to previous cases, there is a portion of band gap that is shared by GaSb and InSb and in which the interface states can be expected to operate in a "normal" way. Perhaps it is the lack of overlap between the band gaps of GaSb and InAsSb of lower x-values that makes the interface states relatively inefficient in tunneling and recombination. Another possibility to consider is the influence of screening of the dislocations by the degenerate electrons in the accumulation region (Partin et al., 1991).

DLTS spectra of lattice-matched and lattice mismatched n-InAsSb/n-GaSb structures are shown in Fig. 5. Peak A observed in the lattice matched structure is apparently related to an electron trap in bulk n-GaSb, as confirmed by comparison with the DLTS spectrum of an Au/n-GaSb Schottky diode in Fig. 6. This peak is markedly absent in the lattice mismatched structure which can be explained by the gettering action of the misfit dislocations at the heteroboundary. This depletion of deep centers penetrates to some depth into the substrate as confirmed by DLTS measurements on a Schottky diode prepared on the bulk n-GaSb substrate from which the InAs$_{0.5}$Sb$_{0.5}$ layer has been etched away by Br$_2$:methanol (Fig. 6). The band edge photoluminescence intensity at room temperature is also increased by about 3 times in the "gettered" substrate as compared to the initial n-GaSb.

The peak B is present in DLTS spectra of all the InAsSb structures. This is not the bulk electron trap in the substrate since it is not observed in the Schottky diode. It can also be hardly related to the interface traps because it is observed both in lattice matched and lattice mismatched structures. Its common occurrence in all the InAsSb samples and its apparent activation energy 0.6 eV make us think that this peak is related to the refilling of the notch on the GaSb side (Fig. 3) by the holes after emptying by the injection pulse (a process in a way similar to generation of carriers in the inversion layer of an MIS structure pulsed from accumulation to inversion).

The origin of the hole trap-like feature in the DLTS spectra of InAsSb samples (Fig. 5) is not at present understood.

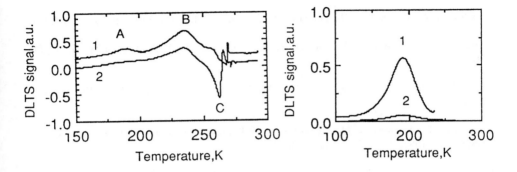

Fig. 5. DLTS spectra of (1) n-InAs$_{0.9}$Sb$_{0.1}$/ n-GaSb and (2) n-InAs$_{0.5}$Sb$_{0.5}$/n-GaSb structures.

Fig. 6. DLTS spectra of (1) Au/n-GaSb Schottky diode and (2) Schottky diode on GaSb substrate from which InAs$_{0.5}$Sb$_{0.5}$ layer has been etched away.

4. Conclusion

We have shown that epitaxial layers of InAs$_{1-x}$Sb$_x$ with good surface morphology can be grown in the whole range of compositions on GaSb substrates. When grown on n-GaSb, these heterostructures behave like quasi Schottky diodes whereas structures grown on p-substrates display an ohmic behavior. No obvious manifestations of the detrimental effects of misfit dislocations at the interface of lattice mismatched structures is observed unless there appears a portion of shared band gap between GaSb and InAsSb as for the InSb/GaSb structure. Gettering of certain deep centers in the substrate by presumably the strain field produced by a large misfit in heavily lattice mismatched compositions has been observed.

Possible applications of InAsSb/GaSb structures might include photodetectors in the GaSb wavelength range. Heterostructures on n-GaSb substrates are in a way analogous to n-i-p-i structures (Döhler, 1986) in the sense that they provide efficient separation of holes and electrons between GaSb and InAsSb (Fig. 3a) and thus increase the effective lifetime of carriers by many times. Therefore, one might expect to obtain good photosensitivity even for lattice mismatched compositions. An obvious advantage of such an approach over the n-i-p-i concept is that the absorption coefficient is the bulk absorption coefficient of low band gap InAsSb. To capitalize on the effect of carrier separation, the InAsSb layers should be grown p-type.

5. Acknowledgements

Research at Carnegie Mellon University was supported by NSF grant ECS 8915034. Professor T. E. Schlesinger is thanked for the use of his photoluminescence facilities. A. Y. Polyakov is visiting from the Institute of Rare Metals, Moscow, Russia.

6. References

Afrailov M. A., Baranov A. N., Dmitriev A. P. et al., 1990, Sov. Phys. Semicond., **24**, 876.

Bubulac L. O., Andrews A. M., Gertner E. R., Cheung D. T., 1980, Appl. Phys. Lett., **36**, 734.

Chen M. Y., Levine B. F., Bethea C. G., Choi K. K., Cho A. Y., 1987, Appl. Phys. Lett., **50**, 927.

Cheung D. T., Andrews A. M., Gutner E. R., Williams G. U., Clark J. E., Pasko J. G., Longo J. T., 1985, Appl. Phys. Lett., **30**, 587.

Chin T. H. and Tsang W. T., 1985, J. Appl. Phys., **57**, 4572.

Döhler G. H., 1986, IEEE J. Quantum Electr., **QE-22**, 1682.

Gertner E. R., Andrews A. M., Bubulac L. O., Cheung D. T., Ludowise M. J., Reidel R. A., 1979, J. Elect. Mater., **8**, 545.

Harrison W. A., 1977, J. Vac. Sci. Technol., **14**, 1016.

Kurtz S. R., Biefeld R. M., Dawson L. R., Fritz I. J., Zipperian T. E., 1988, Appl. Phys. Lett., **53**, 1961.

Mohammed K., Capasso F., Logan R. A., Van Der Ziel J. P., Hutchinson A. L., 1986, Electronics Lett., **22**, 215.

Osbourn G. C., 1984, J. Vac. Sci. Technol., **B2**, 176.

Partin D. L., Green L., Morelli D. T., Heremans J., Fuller B. K., Thrush G. M., 1991, Abstracts of Electron Materials Conference, Boulder, Colorado, late news paper in session O.

Srivastava A. K., Zyskind J. L., Lum R. M., Dutt B. V., Klingert J. K., 1986, J. Vac. Sci. Technol., **B4**, 1064.

Wei S. H. and Zunger A., 1991, Appl. Phys. Lett., **58**, 2684.

Yen M. Y., People R., Wecht K. W., Cho A. Y., 1988, Appl. Phys. Lett., **52**, 489.

Inst. Phys. Conf. Ser. No 120: Chapter 2
Paper presented at Int. Symp. GaAs and Related Compounds, Seattle, 1991

89

An investigation of the structural and insulating properties of cubic GaN for GaAs–GaN semiconductor–insulator devices

S. Strite, D. S. L. Mui, G. Martin, Z. Li, David J. Smith, and H. Morkoç

I. Introduction

Cubic GaN is a wide bandgap semiconductor ($E_g \approx 3.45$ eV [1]) with potentially useful applications as an insulator or passivation layer for GaAs-based devices. As a prospective insulator material for GaAs, GaN suffers from a 20 % lattice mismatch as well as significant thermal mismatch [1]. In its favor, GaN is isoelectronic with GaAs which simplifies the interface considerably. The SiO_2/Si system provides the model insulator-semiconductor interface despite its fairly large lattice mismatch (6.7 %) [2], so it appears possible that the lattice mismatched GaN on GaAs system may perform well. However, only two groups [3,4] have reported on GaN-GaAs insulator-semiconductor structures. In the present work, we present the first comprehensive investigation of the insulating and structural properties of cubic GaN for applications in GaN-GaAs semiconductor-insulator-semiconductor (SIS) and metal-insulator-semiconductor (MIS) diode structures.

II. Experimental

The samples reported here were grown in two identical vacuum-connected Perkin Elmer 430 molecular beam epitaxy (MBE) systems. One system was modified to permit the addition of a Wavemat MPDR microwave electron cyclotron resonance (ECR) plasma source which was used to provide a flux of ionized and atomic nitrogen. Nitrogen was taken from liquid nitrogen boil-off, filtered first for particulates and finally for impurities using a Semigas Nanochem filter. The SIS and MIS structures shown schematically in Fig. 1 were grown on Si-doped (100) GaAs substrates. After GaAs growth, the Ga shutter was closed while the sample remained under an As flux. Once a nitrogen plasma was ignited, GaN growth commenced with the simultaneous closing of the As shutter and opening of the Ga and ECR shutters. GaN was grown at 10-20 nm per hour at 580-620°C. The nitrogen overpressure was 1×10^{-4} torr with approximately 100 W of microwave energy being fed into the plasma. The final GaAs emitter growth for the SIS structures occurred in the second MBE to avoid the lengthy interruption necessary to pump out the nitrogen and reheat the As source material.

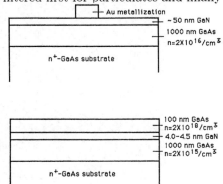

Fig. 1. Structures of GaN-GaAs MIS (above) and SIS (below) diodes.

The SIS diodes were fabricated as circular dots of radius 125 μm. The GaAs was etched with 24:1 H_2O_2:NH_4OH which exposes the nitride and stops, except for small isolated windows where the etch goes deep into the GaAs substrate. Collector contacts were made through those vias with AuGe/Ni/Au as was the top GaAs emitter contact. For the MIS diode structures, 150 μm radius Au dots were evaporated on the nitride and Au evaporated onto the back side of the substrate formed the ohmic contact.

III. Results

The quality of the GaN and its interfaces with GaAs was investigated by high resolution transmission electron microscopy (HRTEM). An HRTEM micrograph of an SIS diode structure in Fig. 2 (a) shows the epitaxial relationship at the GaN on GaAs (GaN/GaAs) interface as well as verifying the continuity of the GaN interlayer. Interface roughening between the GaN grown on the GaAs (GaN/GaAs) is evident, although the interface appears to be chemically abrupt. However, the degree of roughening, approximately 4-5 monolayers in the worst areas, is significantly less than that observed in thicker layers of cubic GaN grown on both GaAs [1] and β-SiC [6]. Also visible are some defects in the GaAs emitter which presumably formed to accomodate the large lattice mismatch between the two materials. A higher magnification micrograph is shown in Fig. 2 (b). The GaN layer is confirmed to be cubic and has a uniform thickness of 8-9 monolayers (4.0-4.5 nm) in this region. The GaN lattice constant determined from the HRTEM image is 0.457 nm, which is slightly larger than the bulk lattice constant of 0.453 nm for cubic GaN [5]. This observation suggests

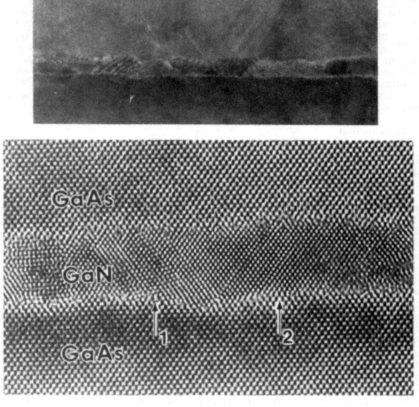

Fig. 2. High resolution electron micrographs of an SIS diode layer. (a) Lower magnification view which shows overall quality of GaN in relation to areas of GaN/GaAs interface roughness. A number of planar defects are also evident in the upper GaAs layer. (b) Arrow 1 indicates region of interface roughness at which numerous planar defects are nucleated. Arrow 2 points to smoother region with excellent GaN crystallinity.

that the GaN is not completely relaxed with the residual tensile strain being caused by the larger GaAs lattice. Most of the GaN defects are planar. Twins and stacking faults propagating along the <111> axes at a 45° angle to the (100) substrate face seem to be the predominant strain relief mechanism. The majority of the defects are nucleated in rough areas of the GaN/GaAs interface (arrow 1), while the quality of the GaN is higher in areas where the interface is fairly smooth (arrow 2).

In order to improve the quality of the heteroepitaxial GaN, as well as to realize high quality GaAs-GaN semiconductor-insulator structures, it is desirable to obtain two dimensional epitaxy as quickly as possible. High energy electron diffraction (HEED) is a non-invasive in-situ technique which is sensitive to the surface roughness of an epitaxial layer. Figures 3 and 4 compare the smoothness of GaN/GaAs heteroepitaxy for GaN grown with continuous and periodically shuttered Ga fluxes, respectively. After 2.7 nm of continuous GaN/GaAs growth, spottiness was evident in Fig. 3 (a) indicating that the GaN surface was rough. After the growth of approximately 10 nm, streaks start to become evident (Fig. 3 (b)) and, by 50 nm of deposition, the surface has reconstructed into a clear (2×2) pattern. When the Ga flux is shuttered for one minute after the deposition of each monolayer, streaks are clearly evident after only 1.4 nm of growth (Fig. 4 (a)) and second order is evident after 2.7 nm of growth (Fig. 4 (b)). The reconstruction into a clear (2×2) pattern develops after 10 nm of GaN/GaAs growth. The HEED data thus indicate that two dimensional GaN/GaAs heteroepitaxy is achieved more quickly by an atomic layer epitaxy approach in which growth is suspended after the deposition of each individual monolayer. Once a smooth GaN surface is attained, two dimensional growth can be maintained under continuous Ga flux.

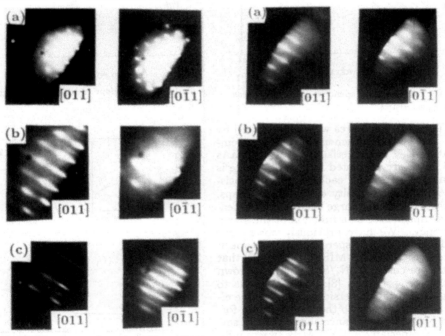

Fig. 3. HEED patterns along the [011] and [0$\bar{1}$1] azimuths of GaN/GaAs grown with continous Ga flux. (a) Spotty after 2.7 nm GaN growth. (b) Streaks appear after 10 nm GaN. (c) Sharp 2×2 reconstruction after 50 nm of GaN.

Fig. 4. HEED patterns of GaN/GaAs grown with shuttered Ga flux. (a) Streaks appear after 1.4 nm GaN growth. (b) Second order evident after 2.7 nm of growth. (c) Sharp 2×2 reconstruction after 10 nm of GaN.

The SIS diode structures were investigated by temperature dependent I-V measurements. The reverse and forward bias I-V characteristics at various temperatures are shown in Figs. 5 and 6 respectively. Rectification is observed. The reverse bias data show a marked temperature dependence which allows an estimate to be made of the GaN-GaAs conduction band offset ΔE_c. A detailed discussion of the calculation of ΔE_c is available elsewhere [4]. Assuming a thermal distribution of electrons, we write $j = Ce^{\frac{-(E_b-\mu)}{k_b T}}$ for the current density where E_b-μ is the energy difference between the barrier top and Fermi level and C is a polynomial in temperature. Plotting $\ln(j)$ vs $1/k_b T$ for the measured temperatures at each voltage gives a straight line with slope $-(E_b-\mu)$ and intercept $\ln(C)$. The interface band bending at each voltage was calculated, and an estimate of 0.9 eV for the conduction band offset was then obtained by finding the best fit to the data for C and E_b at various temperatures and voltages. This value is approximately 50 % of the total difference in the bandgaps of GaN and GaAs.

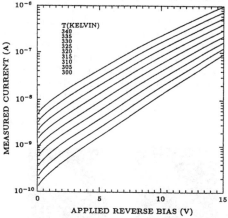

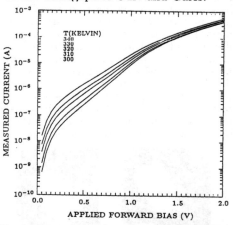

Fig. 5. Temperature dependent SIS reverse bias I-V characteristics.

Fig. 6. Temperature dependent SIS forward bias I-V characteristics.

The MIS structures were investigated by C-V and I-V measurements. The room temperature C-V characteristics of a GaN/GaAs MIS diode are presented in Fig. 7. There is a small hysteresis at each frequency consistent with a low density of bulk GaN traps. However, there is a large change in the measured capacitance as a function of the sweep frequency. We have previously observed a similar frequency dispersion in the capacitance of Si_3N_4/GaAs MIS structures [7] that is not present in Si_3N_4/Si structures grown by the same method [8]. This leads us to surmise that the dispersion is an interface effect and not related to the bulk GaN. In forward bias the capacitance shows signs of saturation, especially in the 1 MHz sweep, indicating the possibility of accumulation. The large leakage current of the diode under forward bias made it impossible to observe complete saturation. The C-V data have the S-shape indicative of an unpinned interface

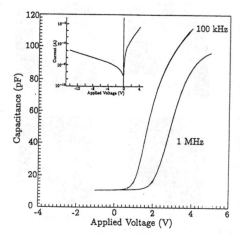

Fig. 7. GaN/GaAs MIS C-V characteristics at 100 kHz and 1MHz. Inset: Room temperature I-V characteristics.

Fermi level being swept from inversion possibly to accumulation. Following the methods of Ref. 9, by comparing the amount of experimentally observed stretch-out to a theoretically generated ideal C-V curve, we estimate an interface state density of approximately $5 \times 10^{11}/\text{cm}^2 \cdot \text{eV}$ for our GaN/GaAs MIS structure.

Room temperature I-V measurements (Fig. 7 inset) on a GaN/GaAs MIS device show the rectifying properties of the diode. However, the current in both forward and reverse bias is fairly large which indicates that the GaN insulator is leaky. Whether the leakiness stems from structural defects or from n-type conductivity in the GaN is under investigation. The insulator breakdown occurred at 22 V reverse bias. A breakdown field of 4×10^6 V/cm for the GaN was calculated by estimating a 50 nm GaN thickness from the dielectric constant of wurtzite GaN ($\epsilon_0 = 9.5$ [10]) and the measured capacitance of the MIS diode.

IV. Summary

We have investigated the structural and insulating properties of cubic GaN with a view towards applications in GaAs/GaN/GaAs SIS and GaN/GaAs MIS device structures. High resolution electron micrographs reveal a chemically abrupt GaN/GaAs interface, but with a roughness of 4-5 monolayers with most of the GaN defects being nucleated at regions of maximum roughness. HEED investigations indicate that two dimensional GaN/GaAs growth is obtained more quickly when the Ga flux is interrupted after the deposition of each monolayer. Rectification is observed in the I-V characteristics of both GaN-GaAs SIS and MIS diodes. An analysis of the temperature dependent I-V characteristics of an SIS structure allows an estimate of 0.9 eV for the conduction band offset between GaN-GaAs to be made. Room temperature C-V measurements on GaN-GaAs MIS structures show that the device can be swept from inversion to a possible accumulation state, indicating that the interface Fermi level is unpinned. An estimate of the interface state density of $5 \times 10^{11}/\text{cm}^2 \cdot \text{eV}$ was made.

Acknowledgements

This research is supported by the Office for Naval Research contract No. N00014-89-J-1780 and monitored by M. Yoder. The authors would like to thank G. Demaggio formerly of Wavemat Inc. and B. Bowdish of the University of Illinois Epicenter for their tireless technical support, and Dr. C. W. Litton of Wright Patterson AFB for his interest. One of us (S. S.) wishes to acknowledge the support of an AFOSR Graduate Fellowship and an NSF Graduate Fellowship during different stages of this research.

References

[1] S. Strite, J. Ruan, Z. Li, N. Manning, A. Salvador, H. Chen, David J. Smith, W. J. Choyke, and H. Morkoç, J. Vac. Sci. Technol. B 9 (1991) 1924.

[2] A. Ourmazd and J. Beck, in The Physics and Chemistry of SiO₂ and the Si-SiO₂ Interface (Plenum Press, New York, 1988).

[3] T. Hariu, T. Usuba, H. Adachi, and Y. Shibata, Appl. Phys. Lett. 32 (1978) 252.

[4] G. Martin, S. Strite, J. Thornton, and H. Morkoç, Appl. Phys. Lett. 58 (1991) 2375.

[5] R. C. Powell, G. A. Tomasch, Y.-W. Kim, J. A. Thornton, and J. E. Greene, in Diamond, Boron Nitride, Silicon Carbide and Related Wide Band Gap Semiconductors (Material Research Society, Pittsburgh, 1989).

[6] M. J. Paisley, Z. Sitar, J. B. Posthill, and R. F. Davis, J. Vac. Sci. Technol. A 7 (1989) 701.

[7] D. S. L. Mui, H. Liaw, A. L. Demirel, S. Strite, and H. Morkoç, Appl. Phys. Lett. (to be published).

[8] D. S. L. Mui, S. F. Fang, and H. Morkoç, Appl. Phys. Lett. (to be published).

[9] E. H. Nicollian and J. R. Brews, in MOS Physics and Technology (Wiley, New York, 1982).

[10] A. S. Barker and M. Ilegems, Phys. Rev. B 7 (1973) 743.

Inst. Phys. Conf. Ser. No 120: Chapter 2
Paper presented at Int. Symp. GaAs and Related Compounds, Seattle, 1991

95

Transmission electron microscopy study of intermetallic compound $Fe_3Al_xSi_{1-x}$ epitaxially grown on GaAs

Y.-F. Hsieh[*], M. Hong, J. Kwo, A. R. Kortan, H. S. Chen, and J. P. Mannaerts

AT&T Bell Laboratories, Murray Hill, New Jersey

ABSTRACT

Epitaxial growth of single crystal thin films of $Fe_3Al_xSi_{1-x}$ intermetallic compounds on GaAs (001) substrates is carried out in a multi-chamber molecular beam epitaxy (MBE) system. Uniform thickness of the metal films and abrupt interfaces of the metal/GaAs are obtained. Interfacial misfit dislocations are observed in the as-grown Fe_3Al(15nm)/GaAs with the substrate temperature held at 250°C. The dislocation density resulting from strain relaxation is found to increase with film thickness. In contrast, a fully strained Fe_3Si layer 60nm thick has been obtained at a growth temperature as high as 500°C. Strain relaxation of Fe_3Si/GaAs interface is only observed in the samples grown at temperatures higher than 500°C or post-annealed at 500°C for an extensive period of time. High resolution X-ray diffraction and TEM observations show that a high quality ternary film, $Fe_3Al_{0.13}Si_{0.87}$, was grown on GaAs with a 0% lattice mismatch.

I. INTRODUCTION

Epitaxial metal-semiconductor heterostructures continue to receive much attention due to interests in the studies of fundamental physics such as mechanisms of Schottky barrier formation, interfacial phenomena, and electronic transport properties. Potential applications of the heterostructural systems include monolithic vertical integration in microelectronics, hot electron high speed devices, and metallic quantum well devices.

Metallic elements of Al[1], Ag[2], and Fe[3] have been shown to grow epitaxially on GaAs. These elements have either large lattice mismatches or dissimilar crystal structures with GaAs substrates. Moreover, their interfaces with GaAs are thermodynamically unstable. Growth or post annealing at modest temperatures, 200-500°C, has led to interdiffusion and formation of metal-Ga or metal-As compounds. The shortcomings of large lattice mismatches and unstable interfaces may be circumvented by the employment of intermetallic compounds.

Previously, two groups of intermetallic compounds were found to form stable interfaces with GaAs or its related III-V compounds. The first group, including NiAl, FeAl, CoGa, etc [4], has a cubic CsCl (B2) structure. The lattice parameters are 1-3% larger than one-half of that of GaAs. The second group, rare earth monopnictides (a NaCl (B1) structure), consists of ErAs, LuAs, YbAs, Er(P,As), and (Sc,Er)As [5]. The lattice constants of the latter compounds are close to that of GaAs. Interesting crystal growth and electrical transport properties were reported from the studies of both types of materials. However, the crystal quality of the heteroepitaxial growth might be further improved by selecting compounds with a similar crystallographic structure as that of GaAs substrate and/or pursuing a closer lattice match by the approach of ternary compounds.

---

* Currently with ERSO/ITRI, Chutung, Hsinchu, Taiwan, R.O.C.

More recently, we have proposed and demonstrated a new intermetallic compound, $Fe_3 Al_x Si_{1-x}$, epitaxially grown on GaAs(001) by MBE.[6] This compound has a $DO_3(BiF_3)$ structure. The crystal lattice can be regarded as a derivative from the BCC lattice of Fe by stacking 8 unit cells of Fe and replacing the atoms on the face-center sites with Al or Si atoms. Thus, the lattice constant of $Fe_3 Al_x Si_{1-x}$ is about two times that of Fe and very close to that of GaAs. This system also fulfills both the necessary structural and thermodynamical requirements for the growth on GaAs by having (1) the same crystallography, FCC, with GaAs, (2) small lattice mismatches, Fe_3Al (+1.54%), Fe_3Si (-0.23%), and $Fe_3 Al_{.13} Si_{.87}$ (0%), (3) thermodynamical stability with GaAs at a typical growth temperature of ~ 400-500°C, and (4) high melting point around 1500°C. In this paper, we have used transmission electron microscopy (TEM) to investigate the film morphology, the interfacial properties, and the film relaxation as a function of growth and annealing conditions.

II. EXPERIMENTAL

The crystal growth was carried out in a multi-chamber MBE system which includes a solid source III-V (GaAs, AlGaAs, InGaAs) MBE chamber and a metal deposition chamber. The metal chamber equipped with two electron guns and four effusion cells is capable of evaporating high melting point materials. No As material has ever been evaporated in this chamber. This would eliminate the possibility of As contamination during the metal deposition process, which is always a major concern for metal films grown in the same chamber which contains As-based III-V materials.

The crystal growth started with a standard GaAs buffer layer 0.5 μm thick on GaAs(001) substrates. After the buffer layer growth, the samples were cooled down to 250-300°C under an As overpressure in the III-V chamber and were then transferred to the second growth chamber through several connecting chambers over a period of 5-10 min. The vacuum of these connecting chambers was below 1×10^{-10} torr. As soon as the samples were introduced into the second growth chamber, they were heated up to ~ 500°C to obtain a Ga-stabilized or an As-stabilized surface. Fe and Si were evaporated from electron beam sources, and Al from an effusion cell. The substrate temperature was kept at 200-500°C during the growth of $Fe_3 Al_x Si_{1-x}$. The co-deposition rate was controlled at 0.05-0.2 nm/sec.

In-situ reflection high energy electron diffraction (RHEED) was used to examine the crystal quality throughout the growth. The composition and thickness of the metal films were studied by Rutherford backscattering spectrometry (RBS). In addition, high resolution X-ray diffraction was used for the structural characterizations. Plan-view and cross-sectional specimens for TEM studies were prepared first using chemical etching on the backside of the wafers and later by a combination of mechanical thinning and ion milling using 4 keV Ar^+. The microstructural analysis is performed using a JEOL-2000FX microscope.

III. RESULTS AND DISCUSSION

Epitaxial single crystal films of $Fe_3 Al_x Si_{1-x}$ on GaAs (001) were obtained regardless of the x values. For one monolayer (ML) growth of $Fe_3 Al_x Si_{1-x}$, we found elongated sharp streaky RHEED patterns as shown in Fig. 1(a). It indicates the attainment of an atomically smooth surface even for films as thin as 1 ML. The RHEED patterns continue to sharpen up when the metal films grow thicker. For an $Fe_3 Al_{0.13} Si_{0.87}$ film 30nm thick, the presence of Kikuchi arcs indicates a high-quality and well ordered single crystal surface (Fig. 1(b)). By examining the RHEED patterns, we found that the crystal quality of $Fe_3 Si$ and $Fe_3 Al_{0.13} Si_{0.87}$ is better than that of $Fe_3 Al$. This may be attributed

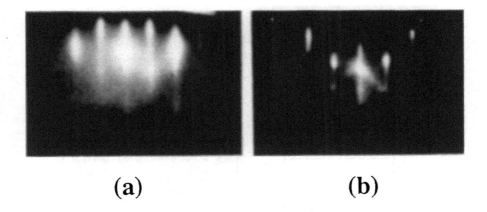

(a)   **(b)**

Fig.1 RHEED patterns recorded along [110] direction during the MBE growth of $Fe_3 Al_x Si_{1-x}$ on GaAs (001). Fig.1(a) is the pattern after 1 ML growth of $Fe_3 Al$. Fig.1(b) is the one after 30nm growth of $Fe_3 Al_{.13} Si_{.87}$.

to smaller lattice mismatches in $Fe_3 Si$ and $Fe_3 Al_{0.13} Si_{0.87}$.

The epitaxial relationship of $Fe_3 Al_x Si_{1-x}$/GaAs was identified by RHEED, X-ray diffraction, and TEM to be: (This relationship holds for any value of x.)

$$[1\bar{1}0]Fe_3 Al_x Si_{1-x} \,// \,[1\bar{1}0]GaAs \text{ and } (001)Fe_3 Al_x Si_{1-x} \,// \,(001)GaAs.$$

Figure 2(a) is a transmission electron diffraction pattern taken from a region consisting of both $Fe_3 Al$ and GaAs. The cross section is shown in Fig. 2(b) for the $Fe_3 Al$ film 30nm thick. The diffraction pattern analysis is shown in the indexed pattern. Splitting of the (004) spots reveals that the lattice mismatch along the direction perpendicular to the interface is estimated to be about 2%. This is consistent with the earlier X-ray θ-2θ scan in which a tetragonal lattice distortion of 3.4% was also observed. However, overlapping of (220) spots indicates a very small lattice mismatch along the in-plane directions on the interface, which is beyond the detection limit of TEM.

A cross-section bright field (BF) micrograph, as shown in Fig. 2(b), exhibits that $Fe_3 Al$ was grown with a uniform thickness and an abrupt interface. Interfacial dislocations resulting from lattice mismatch were observed in the plan-view samples, which have been imaged by g, 3g weak beam diffraction conditions. Comparing the inter-dislocation spacings in films 15 nm and 30 nm thick (shown in Figs. 2(c) & (d)), we observe that the dislocation density is increased with the strain relaxation in the thicker film. However, the antiphase domain boundary, which is generally observed in bulk $Fe_3 Al$ single crystals, was not found in our thin film samples.

Owing to a very small lattice mismatch (-0.23%), fully strained layers of $Fe_3 Si$ were grown on GaAs(001). No misfit dislocations were observed in as-grown samples even for films as thick as 60 nm. This is in strong contrast with the case in $Fe_3 Al$. The diffraction pattern taken from the cross-sectional area as shown in Fig. 3(a) does not reveal the splitting of the (004) spots. This is again expected due to the small lattice mismatch. From Figs. 3(b) and (c), an abrupt interface and a featureless morphology in both the plan-view and cross-section images have demonstrated the growth capability of a thermally stable, strained layer even for substrate temperatures as high as 500°C during the growth.

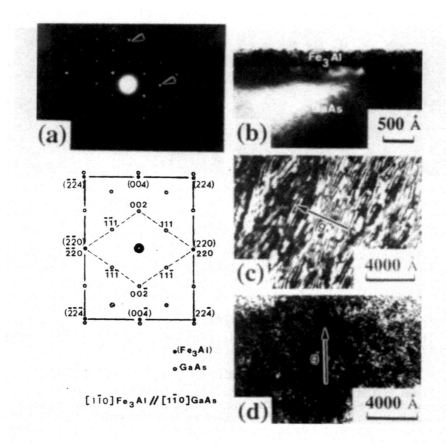

Fig.2 (a) diffraction pattern and (b) bright field (BF) (g=[004]) of an as-grown Fe$_3$Al(30nm)/GaAs(001) sample viewing along the cross-sectional direction close to (110) pole (g=[004]), (c) plan-view BF image (g=[220]) of a film 30nm thick, and (d) plan-view BF image (g=[220]) of a film 60 nm thick.

Strain relaxation of Fe$_3$Si films was observed in samples subjected to post-annealing at 500°C for 24 hours in the metal growth chamber inside which the vacuum was maintained at 10^{-11} torr. The relaxation of the films was studied by TEM on a plan-view sample. As shown in Fig. 4, some of the dislocations along <110> direction are conventional misfit dislocations gliding on {111} plane. However, most of the dislocations are along <100> and <010> directions which are found to be of the edge type with the Burgers vector of 1/2[110]. This suggests a slip system operating on the {011} planes instead of the {111} planes. A previous study of In$_x$Ga$_{1-x}$As/GaAs heterostructures [7] reported that the <100> and <010> dislocation-dominated structure was observed only in samples with higher In composition (x≥0.4). It was proposed that the operation of this secondary slip system is due to the extremely high stresses in the films. Extensive studies of the strain relaxation process in Fe$_3$Si/GaAs films will be published elsewhere [8].

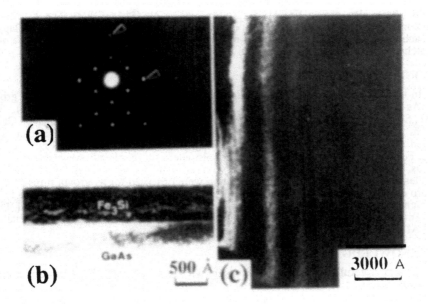

Fig.3 (a) diffraction pattern and (b) BF (g=[004]) of an Fe$_3$Si(60nm)/GaAs(001) sample viewing along the cross-sectional direction close to (110) pole (g=[004]), and (c) plan-view BF image (g=[220]) of an as-grown sample.

Fig.4 plan-view BF image (g=[220]) of an Fe$_3$Si(60nm)/GaAs(001) sample post-annealed at 500°C for 24 hours in the metal MBE chamber. Strain relaxation of this structure results in a dislocation network with most of the dislocation segments running along <100> or <010> direction.

We have also grown $Fe_3Al_xSi_{1-x}$ for x=0.13 on GaAs (001). The diffraction peaks coming from both $Fe_3Al_{0.13}Si_{0.87}$ (004) and GaAs substrates (004) overlapped with each other in a high resolution X-ray diffraction θ-2θ scan. It shows that a zero mismatched metal film is obtained by fine tunning the composition. [8]

IV. CONCLUSION

In this work, we have studied single crystal $Fe_3Al_xSi_{1-x}$, films epitaxially grown on GaAs (001) for x=1, 0, and 0.13 by transmission electron microscopy (TEM). The $Fe_3Al_xSi_{1-x}$ films have a DO_3 structure and lattice constants similar to that of GaAs. Uniform thickness and abrupt interfaces over an entire wafer 5 cm in diameter were obtained for all the $Fe_3Al_xSi_{1-x}$/GaAs films grown at substrate temperatures of 200-500°C. For Fe_3Al/GaAs with a 1.54% lattice mismatch, strain relaxation was observed for films as thin as 15nm. For Fe_3Si/GaAs with a smaller lattice mismatch of -0.23%, featureless morphology and no strain relaxation were observed for as-grown films as thick as 60nm even at substrate temperatures of 500°C. However, after 24 hrs annealing in the ultra-high vacuum, the films also relaxed. High resolution X-ray diffraction and TEM observations show that a perfect lattice matched metal film grown on GaAs was obtained in $Fe_3Al_{0.13}Si_{0.87}$.

V. ACKNOWLEDGMENTS

We are grateful for the useful discussion with R. Hull and C. H. Chen on some of the TEM results.

REFERENCES

[1]   A. Y. Cho and P. D. Derniere, J. Appl. Phys. *49*, 3328 (1978).

[2]   J. Massies and N. T. Linh, J. Crystal Growth, *56*, 25 (1982).

[3]   G. A. Prinz and J. J. Krebs, Appl. Phys. Lett., *39*, 397 (1981).

[4]   J. P. Harbison, T. Sands, N. Tabatabaie, W. K. Chan, L. T. Florez, and V. G. Keramidas, Appl. Phys. Lett., *53*, 2608 (1988).

[5]   C. J. Palmatrom, N. Tabatabaie, and S. J. Allen, Jr., Appl. Phys. Lett., *53*, 2608 (1988).

[6]   M. Hong, H. S. Chen, J. Kwo, A. R. Kortan, J. P. Mannaerts, B. E. Weir, and L. C. Feldman, J. Crystal Growth *111*, 984(1991).

[7]   J. M. Bonar, R. Hull, R. J. Malik, R. W. Ryan, and J. F. Walker, Mat. Res. Soc. Symp. Proc., *160*, 117 (1990).

[8]   M. Hong, A. R. Kortan, J. Kwo, H. S. Chen, Y.-F. Hsieh, and J. P. Mannaerts, to be published.

Inst. Phys. Conf. Ser. No 120: Chapter 2
Paper presented at Int. Symp. GaAs and Related Compounds, Seattle, 1991

101

Magneto-optic and Schottky barrier properties of MnAl/AlAs/GaAs heterostructures

T.L. Cheeks, R.E. Nahory, T. Sands, J.P. Harbison, M.J.S.P. Brasil. H.L. Gilchrist, S.A. Schwarz, M.A.A. Pudensi, S.J. Allen Jr., L.T. Florez and V.G. Keramidas, Bellcore, Red Bank, N.J.

Abstract

Advances in the epitaxial growth of magnetic and magneto-optic materials on compound semiconductors foster progress towards the integration of magnetic devices with optoelectronics. In this study we have investigated the magneto-optic properties and Schottky diode characteristics of epitaxial MBE grown τ MnAl/AlAs/GaAs heterostructures. The ferromagnetic τ phase of MnAl was observed and perpendicular anisotropy was confirmed by the nearly square hysteresis loops and magneto-optic polar Kerr effect. Schottky diode characteristics revealed good quality interfaces and effective barrier heights of 1.03 eV (ideality factor n=1.1) and 0.55 eV (n=1.09) for n and p-type GaAs, respectively. The potential of this material for devices such as photodetectors is discussed.

I. Introduction

The integration of magnetic materials with compound semiconductors can create new possibilities for devices which exploit the properties of both materials. Integration of magnetic devices with compound semiconductor electronic and photonic devices has become more feasible with advances in the epitaxial growth of magnetic and magneto-optic materials on compound semiconductor substrates. The growth of epitaxial magnetic metal films on GaAs and ZnSe, using molecular beam epitaxy (MBE) was first demonstrated by Prinz et al (1981,1986). Prinz describes the potential of these materials for magnetic memory elements using magnetic sandwiches and spin injection devices.(Prinz 1990) Recently, Sands et. al. (1990) reported successful MBE growth of epitaxial ferromagnetic τ MnAl films on {100} GaAs. τ MnAl, a metastable ferromagnetic phase, can be formed in the bulk by rapid cooling from the high temperature hexagonal phase. Other techniques, such as evaporation or magnetron sputtering can also be used to deposit MnAl films that contain the ferromagnetic τ phase, but the films were generally polycrystalline (Shen 1989, Morisako 1987). Growth by MBE under certain conditions (Sands 1991), results in epitaxial τ MnAl films with the magnetically easy c-axis of the tetragonal unit cell oriented normal to the (100) GaAs surface. These epitaxial τ MnAl films show rectangular shaped hysteresis loops and 100 % remanence as measured by magnetotransport, indicating perpendicular magnetization. The favorable properties of τ MnAl, such as perpendicular anisotropy and predictions of large Kerr rotations (Morisako 1987) compounded with the opto-electronic capabilities of III-V semiconductors suggests potential for magneto-optic and information storage applications.

Progress toward new devices will require an understanding of the magnetic properties of the metal, its effect on electrical and optical properties of the underlying compound semiconductor and the interfaces that couple them. In this paper, we describe the magneto-optic properties, such as perpendicular magnetization and polar Kerr rotation, and the Schottky diode characteristics of MnAl/AlAs/n and p-type GaAs heterostructures. We discuss the potential of this material system for devices such as photodetectors.

II. Materials Growth and Characterization

Growth of epitaxial τ MnAl films on GaAs by MBE was described in detail in previous publications (Sands 1990, Harbison 1991). Heterostructures containing GaAs/AlAs were grown using standard III-V growth conditions. The metal layers were grown using a template technique (Sands 1990, Harbison 1991) involving a maximum temperature of about 400 C. The presence of the τ phase of MnAl was verified using in-situ reflection high energy electron diffraction (RHEED) and ex-situ x-ray diffraction.

The heterostructures grown for the present study consisted of an n^+ GaAs substrate with a 100 nm n^+ GaAs buffer layer followed by 1-2 microns of n-type, Si-doped, 2-4×10^{16} cm^{-3} or p-type, Be doped, 2-4×10^{16} cm^{-3} GaAs. A thin layer (2 nm) of AlAs was then deposited followed by a 20 nm metal layer. The magneto-optic properties were measured on samples prepared in a similar way except that the substrate was semi-insulating, the metal layer was thinner (<15 nm) and capped with 20 nm GaAs. Electrical characterization was performed on diodes fabricated using standard photolithography and wet etching.

III. Magneto-optic Properties:

Two important properties for magneto-optic recording are perpendicular anisotropy and the magneto-optic Kerr rotation. Reading information using the magneto-optic Kerr effect requires that the magnetization within the film be parallel to the incident laser path. Perpendicular anisotropy was investigated in the τ MnAl films using extraordinary Hall effect measurements (Leadbeater 1991). The nearly square hysteresis loops, shown in Figure 1, indeed verify the perpendicular anisotropy. A coercivity of about 4.3 kOe is obtained from the hysteresis loop which is similar to the values obtained for bulk MnAl.

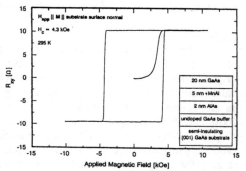

Figure 1. Extraordinary Hall effect measurement at 295 K with the applied magnetic field normal to the film plane.

The spectra in Figure 2 show the polar Kerr rotation of a MnAl/AlAs/GaAs structure as a function of wavelength over a wide range from IR to UV. The spectra were measured after the samples were poled either parallel or anti-parallel to the film normal by exposure to a 0.5 Tesla magnetic field. The shape of the spectra as a function of wavelength were similar for both applied magnetic field directions and were sensitive to the direction of the field. A Kerr rotation of 0.08 degrees was measured for a 10 nm thick MnAl film with little variation throughout the 0.82 to 0.22 micron wavelength range. This near constancy is important, particularly for high density storage applications which require shorter wavelength lasers for reading and writing.

The 0.08 degree value is modest compared to the 0.2-0.4 degree reported for materials such as TbFeCo currently used in the industry. Based on previous predictions for MnAl, and the work on MnBi (Shen 1989), a higher value should be obtained. Thicker films of MnAl are expected to give higher Kerr rotations.

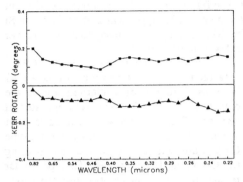

Figure 2. Magneto-optic Kerr rotation as a function of wavelength.

IV. Schottky Diode Characterization

Table I. Schottky diode characterization of MnAl/(Al,Ga)As heterostructures.

Material	n	IV	CV	IPE
MnAl/5 nm AlAs/n-type GaAs	1.1	1.03 eV	0.87 eV	1.2 eV
MnAl/5 nm AlAs/p-type GaAs	1.09	0.55 eV	0.59 eV	----

Schottky diode characteristics were measured using current-voltage (I-V), capacitance-voltage (C-V) and internal photoemission (I P E) measurements for MnAl/(Al,Ga)As/GaAs heterostructures. The GaAs was either n or p-type. Table I shows that the effective barrier heights deduced from I-V measurements were 1.03 (ideality factor n=1.1) for MnAl/AlAs/n-type GaAs and 0.55 eV (n=1.09) for MnAl/AlAs/p-type GaAs diodes. The effective barrier heights deduced for MnAl/AlAs/p-type GaAs were similar to the reported values for other metals on p-type GaAs (Wilmsen 1985). For MnAl/AlAs/n-type GaAs diodes, the effective barrier height values were substantially higher than the values typically reported for metals on GaAs (0.7-0.8 eV) (Wilmsen 1985), but were similar to other transition metal aluminide/GaAs heterostructures (Cheeks 1990). In general, the effective barrier heights in heterostructures containing thin AlAs layers were dependent on the thickness of the AlAs and could be enhanced by as much as 200 mV. The low ideality factors measured for both n and p-type diodes indicated good quality interfaces. The barrier heights deduced from C-V and IPE measurements were also consistent with other transition metal aluminide/GaAs heterostructures (Cheeks 1991).

Mn diffusion in the temperature range from 20 to 400 C should be negligible based on diffusion coefficients obtained for Mn in n and p-type GaAs (Skoryatina 1986). However, there have been reports of enhanced Mn diffusion in GaAs when Si, acting as an n-type dopant, was present. The proposed diffusion mechanism was Si site conversion from Ga to As (Sasaki 1985). Mn, normally an acceptor dopant on a Ga site, has a deep binding energy of about 0.11 eV (Yu 1979). We have investigated whether Mn resides in the MnAl/AlAs/n-type GaAs (Si doped) epilayers using secondary ion mass spectrometry (SIMS) and low temperature (5K) photoluminescence.

An Atomika 3000-30 ion microprobe was employed with 12 keV O_2^+ bombardment at normal incidence. Figure 3 shows the SIMS depth profile of a MnAl/AlAs/n-type GaAs structure with the MnAl and AlAs removed prior to analysis with a wet etch. The Mn and Si concentration were calibrated from a Mn implanted standard and the Si

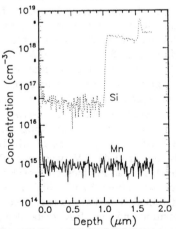

Figure 3. Secondary Ion Mass Spectroscopy profile of MnAl/AlAs/n-type GaAs.

concentration within the substrate, respectively. The figure shows Si concentrations of about 3×10^{16} cm^{-3} throughout the GaAs epilayer. As the n$^+$ buffer layer was approached, the Si concentration increased to about 10^{18} cm^{-3}. The Mn concentration throughout the epilayer and buffer layer was below the SIMS detection level of 10^{15} cm^{-3}.

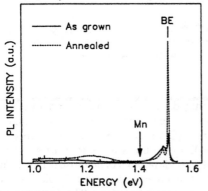

Figure 4. A comparison of photoluminescence spectra of MnAl/AlAs/n-type GaAs with and without ex-situ heating to 450 C. The solid and dashed line represent the heated and unheated sample, respectively.

Photoluminescence (PL) measurements were performed using an Ar laser as a source of excitation and a Ge detector. The 20 mW laser was operated at 488 nm. Figure 4 shows a comparison of spectra obtained from a MnAl/AlAs/n-type GaAs wafer with and without ex-situ heating to 450 C in forming gas. This particular temperature was chosen because the diode fabrication process requires a maximum temperature of 410 C. The luminescence consists of peaks related to donor bound excitons at 1.516 eV and features corresponding to shallow Si and C acceptors in both samples. The absence of a 1.41 eV emission, a signature for Mn (Yu 1979), indicates that Mn was not optically active in these samples. After heating, a Mn signal was still not present but a stronger signal from the donor bound exciton was observed.

The results reported above show excellent Schottky barrier characteristics, good quality interfaces and low levels ($<10^{15}$ cm^{-3}) of Mn in the underlying GaAs epilayers. However, we have observed poor Schottky diode characteristics with barrier heights of 0.814 (n=2.6) when Mn was unintentionally introduced as a dopant in the GaAs. For the n-type GaAs containing Mn, SIMS verified the presence of Mn throughout the GaAs epilayer at concentrations of about 10^{16} cm^{-3}. Also, PL measurements showed a luminescence peak at 1.41 eV, indicating the presence of Mn in the n-type samples. For the p-type GaAs containing Mn, SIMS also verified the presence of Mn but light emission at 1.41 eV was not observed.

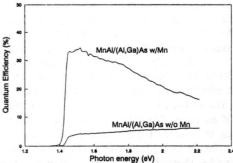

Figure 5. A comparison of quantum efficiency vs. photon energy for MnAl/AlAs/n-type GaAs photodiodes with and without Mn.

Instead, a broad luminescence peak of as yet undetermined origin was observed at 1.464 eV which was about 50 mV higher in energy than Mn. For these latter samples, Mn cross contamination was suspected, since the MBE growth chamber was used for both the III-V deposition and the MnAl metal deposition. Separate chambers for metallization and GaAs growth may be necessary to ensure better quality heterostructures.

V. Optical Characterization

Schottky photodetectors fabricated using MnAl/AlAs/GaAs heterostructures have several advantages namely, clean interfaces due to in-situ growth, an enhanced barrier height due to AlAs incorporation, the potential for integration and a low cost simple structure. Stimulated by preliminary measurements which showed a higher photocurrent (factor of up to 30) for wafers containing Mn, the photoresponse of GaAs photodiodes with and without Mn were compared. Figure 5 shows the quantum efficiency for MnAl/AlAs/GaAs with and without Mn in the GaAs. The spectra shape for MnAl/(Al,Ga)As w/Mn photodiodes was markedly different from that of the MnAl/(Al,Ga)As w/o Mn photodiodes. The photoresponse of photodiodes with Mn in the GaAs decreased with increasing photon energy while that of photodiodes without Mn in the GaAs remained fairly constant. This was attributed to a stronger surface recombination in photodiodes with Mn in the surface region of GaAs. Without optimizing the structure or adding an antireflection coating the quantum efficiency for photodetectors with Mn in the GaAs was about 21 % at 632 nm and zero bias. The quantum efficiency for photodiodes without Mn in the GaAs under identical conditions was only about 7 %. The enhancement in quantum efficiency appears to be related to the presence of Mn in the GaAs. One possible explanation is that the incorporation of Mn may lower the Fermi level and thereby alter the trap activity, affecting the photosensitivity.

VI. Conclusion

We have shown that MBE grown MnAl/AlAs/GaAs structures exhibit very good magneto-optic properties i.e. a square loop and a reasonable Kerr rotation, and excellent Schottky diode behavior for both n and p-type GaAs. The Schottky diode characteristics indicated good quality interfaces. SIMS and PL measurements show that the underlying GaAs was free of diffused Mn. Schottky photodetectors were successfully fabricated with good photoresponse and a quantum efficiency as high as 21 %. Based on these results we believe that this material structure shows potential for integrating magnetic with optical and electronic devices.

Acknowledgements:

We would like to acknowledge D. Aspnes for his assistance and the use of his equipment for Kerr effect measurements.

REFERENCES:
Cheeks T L, Sands T, Nahory R E, Harbison J P, Tabatabaie N, Gilchrist H L, Wilkens B J and Keramidas V G 1990 Appl. Phys. Lett. **56** 1043
Cheeks T L, Sands T, Nahory R E, Harbison J P, Gilchrist H L and Keramidas V G 1991 J. Electronic Mat. in press.
Harbison J P, Sands T, Ramesh R, Florez L T, Wilkens B J and Keramidas V G 1991 J. Crystal Growth **111** 978
Leadbeater M L, Allen S J, DeRosa F, Harbison J P, Sands T, Ramesh R, Florez L T and Keramidas V G 1991 J. Appl. Phys in press
Morisako A, Matsumoto M and Naoe M 1987 J. Appl. Phys. **61** 4281
Prinz G A and Krebs J J 1981 Appl. Phys. Lett. **39** 397
Prinz G A, Zonker B T, Krebs J J, Ferrari J M and Kovanic F 1986 Appl. Phys. Lett. **48** 1756
Prinz G A 1991 Science **250** 1092
Sands T, Harbison J P, Leadbeater M L, Allen S J, Hull G W, Ramesh R and Keramidas V G 1990 Appl. Phys. Lett. **57** 2609

Sands T, Harbison J P, Allen S J, Leadbeater M L, Cheeks T L, Brasil M J S P, Chang C C, Ramesh R, Florez L T, DeRosa F and Keramidas V G 1991 Proc. Spring Meeting of MRS, Symposium S, in press.
Sasaki Y, Sato T, Matsushita K, Hariu T and Shibata Y 1985 J. Appl. Phys 57 1109
Shen J, Kirby R D and Sellmyer D J 1989 J. Magn. Mat. 81 107
Skoryatina E A 1986 Sov. Phys. Semiconductors 20 1177
Wilmsen C W 1985 Physics and Chemistry of III-V Compound Semiconductor Interfaces (Plenum, New York)
Yu P W and Park Y S 1979 J. Appl. Phys. 50 1097

MBE growth optimization and thermal stability of strained $In_{0.25}Ga_{0.75}As$ layers in MODFET layer structures

H. Nickel, R. Lösch, W. Schlapp, H. Kräutle, A. Kieslich[+], A. Forchel[+]

DBP Telekom Forschungsinstitut beim FTZ, P.O.Box 10 00 03, D-6100 Darmstadt, FRG
[+]Technische Physik, Universität Würzburg, D-8700 Würzburg, FRG

ABSTRACT   We have grown by MBE a series of MODFET structures containing $In_{.25}Ga_{.75}As$ channel layers. The channel layer thickness and MBE growth conditions (growth temperature and V/III ratio) have been optimized in order to achieve mobilities as high as possible. The structures were characterized by Hall and low temperature photoluminescence measurements. Non optimized growth conditions cause a degradation of the carrier mobility and give rise to an additional peak in the PL spectrum in parts of the wafer. We have performed rapid thermal annealing experiments and have studied the dependence of the PL spectrum on the annealing temperature.

1. INTRODUCTION

Strained (pseudomorphic) $In_xGa_{1-x}As$ layers are increasingly used in high speed semi-conductor devices such as MODFETs. However, the thickness and indium content of strained layers are restricted by the pseudomorphic limit, which defines the onset of strain relaxation by the formation of dislocations. The double kink dislocation mechanism of Matthews et al. (1974) has proven to predict the critical thickness very well for indium contents up to about 20%. It has been shown by Nickel et al. (1990) that the thermal stability of such layers is comparable to that of GaAs. This no longer holds when the indium content is about 25% or above (Guha et al. 1990). MODFETs incorporating $In_xGa_{1-x}As$ layers with x $\gtrsim$ 0.25 and widths not far below the critical limit, according to the double kink dislocation model, show drastic degradation in carrier mobility, depending on the growth temperature. This is explained either by a change of growth mode (even below the critical thickness) or by relaxation (Berger et al. 1990, Ballingal et al. 1990). In this paper we present Hall measurements and low temperature (2K and 10K) photoluminescence (PL) measurements on MODFET structures containing $In_xGa_{1-x}As$ channel layers grown by MBE at different growth conditions. We demonstrate, that the degradation of the carrier mobility is correlated with the appearence of additional lines in the PL spectrum and we study the influence of rapid thermal annealing (RTA) on the PL spectrum.

2. EXPERIMENTAL

All of the samples were grown in a Varian Gen II MBE machine on two inch D-shaped (Bridgeman type) semiinsulating GaAs (100) wafers which were fixed to molybdenum holders by indium. The layer structure and growth temperature sequence are depicted in Figure 1. The structure consists of an undoped 1.3 μm GaAs buffer layer, an undoped 0.3 μm $Al_{.2}Ga_{.8}As/GaAs$ superlattice, an undoped 1.9 μm buffer layer, an undoped

In $_{.25}$Ga$_{.75}$As channel layer of variable thickness, an undoped 2.2 nm Al$_{.2}$Ga$_{.8}$As spacer layer, a doped 20 nm Al$_{.2}$Ga$_{.8}$As electron supplying layer, a doped Al$_x$Ga$_{1-x}$As grading layer $(x = .2 \rightarrow 0)$ and (in most structures) a doped GaAs cap layer. A graded AlGaAs layer and a doped cap layer were chosen to improve ohmic contacts and hence to obtain good high frequency performance. Two sequences of growth temperatures were used to investigate whether the growth of the upper layers at elevated temperatures serves to anneal, causing strain relief. The growth temperatures were controlled by a pyrometer.

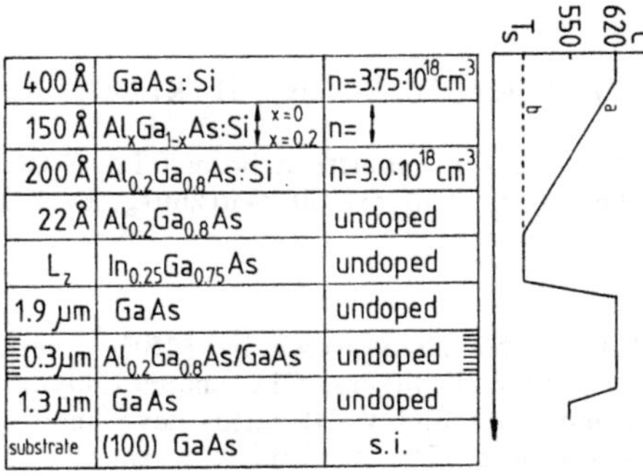

Figure 1

Schematic cross section of the
pseudomorphic In$_{.25}$Ga$_{.75}$As
MODFET structures and the
growth temperature profile.
The temperature profile would
be symmetric at the sides of
the InGaAs channel, if the
vertical scale were linear.

400 Å	GaAs:Si	$n=3.75 \cdot 10^{18}$cm^{-3}
150 Å	Al$_x$Ga$_{1-x}$As:Si $\begin{smallmatrix}x=0\\x=0.2\end{smallmatrix}$	$n=$ ↓
200 Å	Al$_{0.2}$Ga$_{0.8}$As:Si	$n=3.0 \cdot 10^{18}$cm^{-3}
22 Å	Al$_{0.2}$Ga$_{0.8}$As	undoped
L$_z$	In$_{0.25}$Ga$_{0.75}$As	undoped
1.9 µm	GaAs	undoped
0.3µm	Al$_{0.2}$Ga$_{0.8}$As/GaAs	undoped
1.3 µm	GaAs	undoped
substrate	(100) GaAs	s.i.

The critical thickness d_c according to the double kink dislocation mechanism is determined to be about 12 nm for an indium content $x_{In} = 0.25$ when using the formula

$$d_c \text{ (nm)} = 0.676 \times \{ \ln (d_c/0.4) + 1 \}/x_{In} \tag{1}$$

Therefore the channel thickness L_z was initially chosen to be 12 nm. In subsequent structures it was reduced to 9 nm as shown in table 1. As the electron supply layer and the cap layer are significantly doped, there is a conducting bypass in the electron supply layer parallel to the channel layer. We have determined the 2D channel electron mobility using a channel carrier density of 2×10^{12}/cm^2 and a mobility of 1000 cm^2/Vs in the parallel conductance path. The value chosen for the channel carrier density was deduced from Shubnikov de Haas measurements on similar structures and Hall measurements on structures with undoped cap layers. The slight uncertainty of that value will not affect the subsequent conclusions concerning the influence of growth conditions on channel carrier mobility. Photoluminescence measurements have been done either in a closed cycle refrigerator at 10K or in a pumped helium cryostat at 2K. An argon laser (514.5nm) was used as a PL excitation source. For rapid thermal annealing the samples were placed upside down on a silicon or GaAs wafer and covered by GaAs pieces. The annealing time was 1 minute.

3. RESULTS AND DISCUSSION

Table 1 shows growth temperature, V/III beam equivalent pressure (b.e.p.) ratio, and type of growth temperature profile for different wafers. For a determination of the absolute V/III ratios, we should mention that we observe the change from Ga-controlled to As-controlled MBE growth at a V/III b.e.p. ratio of 14. Layer structure parameters such as the channel thickness L_z and the type of doping of the cap layer are also depicted in Table 1. In addition the table shows the 2D channel mobility at liquid nitrogen temperature (calculated from the Hall data), PL intensity, occurence of additional structures in the PL spectrum (which we will assign to relaxation below) and the effective electron saturation velocity determined from MODFETs made from that wafer. Extrinsic values of the saturation velocity v_s were calculated using $v_s = 2\pi \times L_g \times f_T$ where L_g is the gate length and f_T is the current gain cutoff frequency. The intrinsic value was determined according to Dickmann et al. (1989).

MBE No	$T_s(^oC)$	L_z (nm)	V/III b.e.p.	type	$\mu_{2D}(77K)$ (cm^2/Vs)	PL max./relax.	$v_s(10^7 cm/s)$ extr. / intr.
1963	480	12	16.5	a,d	8, 200	50	yes	
1964	460	12	18	a,d	13, 300	10	no	1.62/1.89 [c]
1965/68	480	9	18	a,d	13, 600	40	no	1.63/—— [d]
1967	520	9	18	a,d	8, 500	200	yes	
1998	520	9	37.5	a,d	13, 900	220	yes	
3004	540	9	41	a,d	14, 700	240	?	1.86/—— [e]
3098	520	9	37.5	b,u	13, 510	180	?	

type a,b: cf. fig. 1   type d/u: doped/undoped cap layer
c: cf. Dickmann (1991)   d: cf. Kraus et al. (1990)
e: J. Kraus, private communication

Table 1   Comparison of MBE growth conditions, Hall and PL measurement results, and device results of pseudomorphic $In_{.25}Ga_{.75}As$ MODFET structures.

MODFETs with 12 nm $In_{.25}Ga_{.75}As$ channel layers grown at 520^oC with a V/III b.e.p. ratio of 18 (similar to that used for GaAs) are degraded and show electron mobilities at 77K of less than 2000 cm^2/Vs. Reducing the growth temperature will improve it. At a growth temperature of 460^oC the mobility rises drastically and the PL spectrum is of the type shown in Figure 2, but the PL intensity stays rather low. Such structures are well suited for device fabrication, as shown by the v_s values. The PL spectrum in Figure 2 shows a broad e_1hh_1 transition line at ~980 nm and a high energetic shoulder often ending in a sharp peak at ~ 930 nm associated with the e_2hh_1 transition.

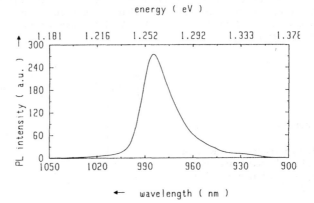

Figure 2

PL spectrum at 10 K of a "normal" $In_{.25}Ga_{.75}As$ MODFET structure

Raising the growth temperature to 480^oC and reducing the width of the strained layer to 9 nm yields slightly higher mobilities and PL intensities. Maintaining the channel width at 9 nm and further increasing the growth temperature to 520^oC will improve the PL intensity but degrade the mobility drastically. The drastic degradation of the mobility is accompanied by the appearance of a new transition line in the PL spectrum at ~1000 nm to ~ 1020 nm. That additional line is more pronounced in rim parts of the wafer and will be discussed in more detail below. Nevertheless, very good mobilities can be achieved at high growth temperatures when the V/III ratio is increased by a factor of about 2. The best mobilities and PL intensities have been reached at a growth temperature as high as 540°C and a V/III b.e.p. ratio of 41. Excellent MODFET devices have been made from that wafer, even though the PL spectrum showed slight indications of the additional features, which we will explain below as the onset of relaxation. We have seen no difference in the PL spectrum when growing the upper layers at lower temperature. So we believe that the appearence of the additional line, as shown in Figure 3, is solely caused by the growth of the strained InGaAs layer.

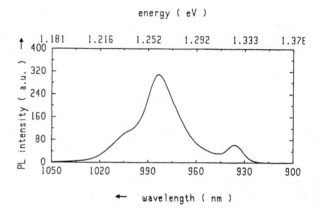

Figure 3

PL spectrum at 10 K of a "relaxed" In$_{.25}$Ga$_{.75}$As MODFET structure

The degradation of the mobility might be explained in two ways: An increase in electron scattering in the InGaAs MODFET channel can be caused either by lattice dislocations related to relaxation or by a roughening of the InGaAs/AlGaAs interface inferred from three-dimensional growth. But, while the appearance of an additional low energy PL emission line can be explained by the decrease of the bandgap by relaxation in parts of the area, it is much less probable that it is caused by an increase in channel width in the case of three-dimensional growth. A final assessment can be made through annealing experiments.

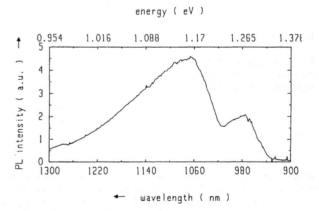

Figure 4

PL spectrum at 10 K of a "relaxed" structure after rapid thermal annealing at 725°C

The PL emission spectra of "relaxed" wafers, where an additional line is already present after growth, show drastic changes after annealing at rather low temperatures $(650^\circ$ to $700^\circ C)$. The intensity of all three lines decreases by a factor of ~100 and simulaneously new broad lines appear at ~1070 nm (~1.15 eV) and ~1160 nm, as shown in Figure 4. That transition must happen in a very small temperature interval, as we have not observed intermediate states. A further increase of the annealing temperature causes only minor changes: the low energy lines will broaden and slightly decrease in intensity. The e_1hh_1 line of unrelaxed areas shifts to higher energies, merges with the e_2hh_1 line and regains intensity.

The additional line at ~1010 nm in wafers as grown rather seems to disappear by annealing than to move to ~1070 nm. Therefore, we tend to explain that emission by defects, caused by relaxation (Wang 1991), rather than by partially relaxed areas. The difference between the energy gaps of strained and relaxed $In_{.25}Ga_{.75}As$ is 100 meV (see e.g. Nickel 1990). Therefore, we think, that the new emission peak appearing at ~1070 nm after annealing is the e_1hh_1 emission from fully relaxed areas. The PL emission at wavelengths beyond 1100 nm after annealing must also be related to defects induced by lattice dislocations.

The Pl spectrum of "normal" wafers, when grown at $500^\circ C$ or above, changes quite similarly on annealing: The intensity decreases, when annealed above $750^\circ C$. A broad emission band appears at lower energy (~1100 nm), when annealed above $850^\circ C$. The e_1hh_1 line of the strained InGaAs layer shifts to higher energies, when annealed above $700^\circ C$, and regains intensity, when the annealing temperature exceeds $900^\circ C$.

The shift of the original e_1hh_1 line to higher energies after annealing is due to a reduction in well width (at the lowest energy levels) and in indium content by atomic interdiffusion (Nickel 1990). The line, originally located at ~1.25 eV (990 nm), is shifted to ~1.34 eV after annealing at $950^\circ C$.

Our results indicate, that $In_{.25}Ga_{.75}As$ layers, when grown at $500^\circ C$ or above, show signs of relaxation, when annealed at a sufficiently high temperature. The relaxation starts earlier, when the additional line is already present after growth. When relaxing, a new emission line appears at an energy corresponding to the transition in a relaxed $In_{.25}Ga_{.75}As$ layer, and the emission intensity decreases drastically. In cases, where after growth the separation between the additional PL emission line and the main e_1hh_1 line is less than 30 meV, annealing at temperatures below $800^\circ C$ does not cause relaxation. In that cases we have inserted question marks in Table 1.

Similar to Elman et al. (1990), we have observed the disappearance of the additional line, when present, and the blue-shift of the original e_1hh_1 emission line after annealing above $800^\circ C$, too. But that is, to our understanding, not a reversal of the relaxation by annealing, but a consequence of the interdiffusion of the type III elements. Atomic interdiffusion, as mentioned above, reduces the width and the indium content of the overcritically strained layer. Thus it will reduce the strain and will inhibit the strained layer from total relaxation. But the low energy emission from relaxed areas, that is already present, does not disappear on further annealing.

The additional emission line already present in nonannealed wafers, and usually located at 1000 nm to 1020 nm is more pronounced in rim parts of the wafers. The most probable explanation is, that the higher defect density at the rim of D-shaped, semiinsulating wafers from MCP Electronic Materials Ltd. favours relaxation.

4. CONCLUSION

We have grown by MBE a series of MODFET layer structures containing an $In_{.25}Ga_{.75}As$ channel layer. The width of the channel layer is kept close to the critical limit described by the double kink dislocation model. When using MBE growth temperatures higher than 460°C and low V/III ratios we observe a degradation of the channel electron mobility. When the degradation is severe, it is accompanied by the appearance of an additional emission line in the low temperature PL spectrum. Annealing studies have revealed that the additional PL emission line originates from relaxation in the InGaAs layer.

5. REFERENCES

Ballingal J M, Pin Ho, Martin P A, Tessmer G J and Yu T H 1990
   J. Electron. Mat. **19** 509
Berger P R, Chen Y C, Singh J and Bhattacharya P K 1990
   Inst. Phys. Conf. Ser. No. **106** 183
Dickmann J, Heedt C H and Daembkes H 1989 IEEE Trans. Electron. Devices **36** 2315
Dickmann J 1991 submitted as Dissertation (in German) University of Aachen, FRG
Elman B, Koteles E S, Melman P, Jagannath C and Dugger D 1990
   Inst. Phys. Conf. Ser. No. **106** 171
Guha S, Madhukar A and Rajkumar K C 1990 Appl. Phys. Lett. **57** 2110
Kraus J, Meschede H, Brockerhoff W, Prost W, Nickel H, Lösch R and Schlapp W 1990
   ITG-Fachbericht **112** "Heterostruktur-Bauelemente" PD 29 (in German)
Matthews J W and Blakeslee A E 1974 J. Cryst. Growth **27** 118
Nickel H, Lösch R, Schlapp W, Leier H and Forchel A 1990
   Surface Science **228** 340
Wang Shu-Min, Andersson T G, Kulakovskii D and Yao Ji-Yong 1991
   Superlattices and Microstructures **9** 123

Thermal annealing effects on the defect and stress reduction in undercut GaAs on Si

Shiro Sakai, Naoki Wada* and Chun Lin Shao**

*Department of Electrical and Electronic Engineering, Tokushima University, Minami-josanjima, Tokushima 770, Japan, * Matsushita Kotobuki Electronics Ltd., Japan, **Nan-tong Textile Engineering Institute, China*

Abstract This paper describes the thermal annealing effects of UCGAS (undercut GaAs on Si) in which the GaAs layer is partially separated from the Si substrate by the post-growth lateral etching. A drastic defect reduction is found in UCGAS after annealing at 800 ˚C, while no significant EPD reduction is obtained in the planar region. The photoluminescence intensity ratio of planar GaAs on Si and UCGAS {PL(UCGAS)/PL(planar)} is about 3 and more than 10 before and after the annealing at 800 ˚C, respectively. A model which assumes the GaAs/Si interface to be a dislocation source predicts that the high temperature annealing is effective in reducing dislocation density in UCGAS.

1. Introduction

GaAs layers grown epitaxially on Si substrates still have an unacceptably high defect density and residual stress. We proposed a UCGAS (undercut GaAs on Si) to reduce both stress and dislocation density (Sakai *et al* 1990 a b, Kawasaki *et al* 1990). The UCGAS has a GaAs layer partly separated from the Si substrate as shown in Fig.1. Stress is released, since the UCGAS is free to move. The dislocation density is also decreased by annealing the UCGAS, because GaAs/Si interface where high density of defect is confined is eliminated in UCGAS. This paper describes the annealing effects on the stress and defect distributions in UCGAS and planar GaAs on Si and discusses the dislocation reduction mechanism.

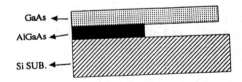

Fig.1. Cross-sectional view of UCGAS

2. Experimental procedure and results

The fabrication procedure of UCGAS was reported previously (Kawasaki *et al* 1990). UCGAS with various shapes were fabricated. Some of the fabricated structures are shown schematically in Fig.3. The fabricated structures were then annealed at 400 ˚C, 600 ˚C and 800 ˚C for 10 min. The annealing ambience was nitrogen at 400 °C and mixture of hydrogen and arsine at 600 °C and 800 °C, respectively.

The defect was characterized by etching in molten KOH and by cross-sectional TEM (transmission electron microscope). EPD (etch pit density) of the sample before the annealing was about 1×10^8 cm^{-2}. A great difficulty was encountered in etching UCGAS in KOH, since Si strongly reacts with molten KOH and the structure very often breaks. EPD reductions by annealing both in UCGAS and in planar GaAs on Si were found, but that in UCGAS was much more drastic than that in the planar GaAs on Si as already reported (Sakai *et al* 1991). A significant dislocation reduction in UCGAS by the annealing was also verified by TEM observation. No dislocation was found in UCGAS annealed at 800 °C, while planar GaAs on Si always had threading dislocation. The TEM characterization is now in progress, and the results will be published in the future.

The stress on the GaAs epilayer is evaluated by PL (photoluminescence) at 43 K. Typical PL spectra are shown in Fig.2. Both UCGAS and planar GaAs/Si samples are annealed at 800 °C. Homoepitaxial GaAs has two main peaks of exciton-related peak Ex and free electron to neutral carbon acceptor e C_A^0. The biaxial stress in planar GaAs on Si splits the valence band degeneracy of $m_J = \pm 1/2, \pm 3/2$. Accordingly, the exciton-related peak splits into two peaks, and at a same time, the hydro-static pressure component of the biaxial stress shifts the peaks to longer wavelength (Zemon *et al* 1986 Freundlich *et al* 1988). The stress is evaluated by the peak shift of $m_J = \pm 1/2$ with respect to that of homoepitaxial GaAs.

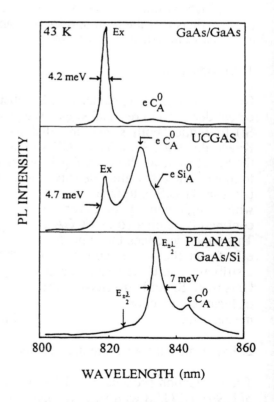

Fig.2. PL spectra at 43 K

The peak shift by the stress is -10.4×10^{-9} meV cm^2/dyn (Zemon et al 1986). Therefore, the stress in the planar GaAs on Si shown in Fig.2 is calculated to be 2.6×10^9 dyn/cm^2. Contrarily, the spectrum of UCGAS shows neither red shift of the spectrum nor valence band splitting indicating that the stress is released. The half width of the exciton-related peak of UCGAS, 4.7 meV, is almost the same as that of homoepitaxial GaAs and kT=3.7 meV at 43 K. Since the peak position reproducibility in our experiment is about 1 meV, the accuracy and the minimum detectable stress value is about 1×10^8 dyn/cm^2.

The stress distribution in UCGAS can be measured by mapping the PL peak wavelength of $E_{\pm 1/2}$. Figure 3 shows the stress distribution in UCGAS having three different geometries annealed at different temperatures. In all structures, the stress at the center is higher than that in the UCGAS part, and the stress redistributes by the annealing even at 400 °C. Especially, the annealing divides the stress in the rectangle into three clearly separated regions as schematically shown in the insert of Fig.4. In each of the three regions, the stress is almost uniform. Comparing the stress distribution after the annealing in circle, rectangle and triangle, the stress reduction at the sharper corner is found to be larger.

Figure 4 shows photoluminescence intensities of $E_{\pm 1/2}$ peak in UCGAS and in the wide area planar GaAs on Si, as a function of the annealing temperature. The intensity from UCGAS is about 3 times stronger than that of the planar GaAs on Si even before the annealing due to the reflection at air/Si and air/GaAs interfaces. The PL intensity is increased by the annealing, and the annealing effect is larger in UCGAS than that in the planar GaAs on Si. The intensity from the corner becomes about one and 3 orders of magnitude stronger by the annealing at 400 °C and 800 °C, respectively. In consequence of this, the PL intensity ratio of UCGAS and planar GaAs/Si increases from 3 before the annealing to more than 10 after the annealing at 800 °C. The intensity from the center where GaAs layer is not undercut is also increased by the annealing. This results is similar to the results of Chand *et al* (1991) who reported that the mesa-etched and annealed GaAs on Si has less PL dark spot. The intensities from the planar GaAs on Si and from the center of the rectangle exponentially depends on.1/T with the activation energy of 0.27 eV.

3. Discussion

The experimental results shown above indicate that the crystal quality is improved by annealing, and the effect of the annealing is larger for UCGAS than that for the planar GaAs on Si. EPD is known to corresponds to the dislocation density (Yamaguchi *et al* 1988). On the other hand, the PL intensity, if the surface conditions are the same, represents the density of the

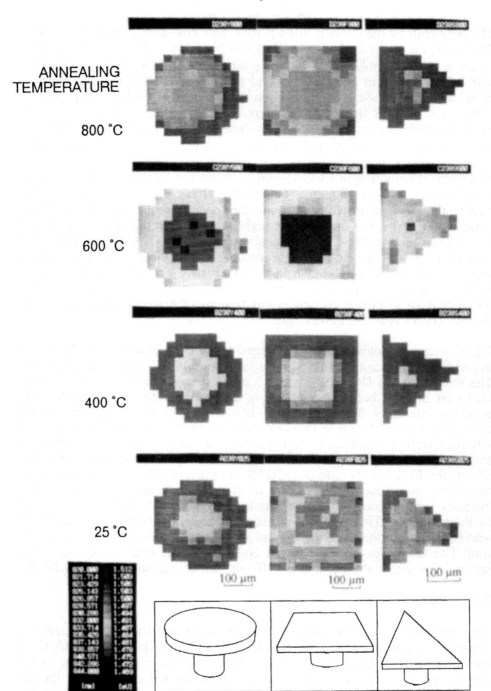

Fig.3 The PL peak wavelength mapping of UCGAS having three different geometries annealed at three different temperature. (Original is color)

non-radiative recombination center which includes various kinds of point defects, micro-twins and dislocations. Since the dislocation motion at 400 °C is almost negligible (Yamaguchi *et al* 1988), the dislocation is not responsible for PL improvement at 400 °C. Therefore, we suspect that the main reason for PL improvement by 400 °C anneal is the decrease of point defect. Actually, the deep level concentration detected by DLTS is changed by the low temperature annealing (Kim *et al* 1991). On the other hand, the high temperature annealing activates dislocation motion.

The UCGAS has two features; the stress is smaller and the GaAs/Si interface is eliminated. Although the UCGAS has much less stress compared to the planar GaAs/Si, stress less than 10^8 dyn/cm^2, which is still high enough to activate dislocation motion, exists in the layer. This is confirmed by the fact that a GaAs layer totally removed from the substrate is bent. The GaAs layer must be plastically deformed, and the residual stress is not released even after the removal of the substrate. The existence of the residual stress in the layer was also verified by the T E M observation (Fang *et al* 1990). Therefore, the main difference of the UCGAS and the planar GaAs/Si is the absence of the GaAs/Si

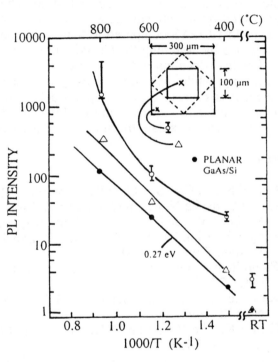

Fig.4 PL intensity versus annealing temperature

interface in UCGAS. Some of the dislocations existing near the GaAs/Si interface are released into the layer by the annealing. This process can be modeled by modifying the treatment of Yamaguchi *et al*(1988). Although we do not describe the details here, we can show that the dislocation density in UCGAS (planar GaAs on Si) decreases (decreases and saturates) with increasing annealing temperature. Although the stress in planar GaAs on Si can be applied by the thermal cycle annealing, UCGAS does not have this opportunity. Therefore, the high temperature annealing rather than the thermal cycle annealing is more effective in UCGAS. Once the residual stress in UCGAS is relaxed, no more dislocation reduction can be expected.

4. Summary

We have investigated and compared the annealing effects of UCGAS and planar GaAs on Si. An EPD reduction in UCGAS is obtained by high temperature annealing, while no drastic reduction is found in planar GaAs on Si. The stress measured by the PL peak wavelength mapping is found to redistribute in UCGAS even at 400 °C, and the sharper corner in UCGAS has more significant effect of the annealing. The PL intensity is also drastically improved by the annealing. All the results show that the annealing effect is larger in UCGAS compared to that in the planar GaAs on Si. The results are explained by assuming the GaAs/Si interface to be a source of the dislocation. A drastic dislocation reduction in UCGAS is expected by the high temperature annealing, while dislocation density saturate at some level in the planar GaAs on Si. It is suggested that the point defect in the layer is annealed out at low temperature and improves the PL intensity. However, some more work may be necessary to explain the annealing effects on PL intensity.

References

Chand N and Chu S N G, 1991 Appl. Phys. Lett., **58**, 74
Choi C, Otsuka N, Munn G, Houdre R, Morkoc H, Zhang S L, Levi D and Klein M V, 1987 Appl. Phys. Lett., **50**, 992
Fang S F, Adomi K, Iyer S, Morkoc H, Zabel H, Choi C and Otsuka N, 1990 J. Appl. Phys., **68**, R31
Freundlich A, Grenet J C, Neu G, Leycuras A and Verie C, 1988 Appl. Phys. Lett., **52**, 1976
Kawasaki K, Sakai S, Wada N and Shintani Y, 1990 Inst. Phys. Conf. Ser. No.112, 269
Kim E K, Cho Y C, Kim Y, Kim H S, Kim M S and Min S, 1991 Appl. Phys. Lett., **58**, 2405
Lee J W, Shichijo H, Tsai H L and Matyi R J, 1987 Appl. Phys. Lett., **50**, 31
Sakai S, Kawasaki K and Wada N, 1990a Jpn. J. Appl. Phys. **29**, L853
Sakai S, Kawasaki K and Wada N, 1990b Jpn. J. Appl. Phys. **29**, 2077
Sakai S, Kawasaki K, Okada M, Wada N and Shintani Y, 1991 Electron. Lett., **27**, 1371
Yamaguchi M, Yamamoto A, Tachikawa M, Itoh Y and Sugo M, 1988 Appl. Phys. Lett., **53**, 2293
Zemon S, Shastry S K, Norris P, Jagannath C and Lambert G, 1986 Solid State Communi., **58**, 457

Inst. Phys. Conf. Ser. No 120: Chapter 3
Paper presented at Int. Symp. GaAs and Related Compounds, Seattle, 1991

119

Extensive study on the effect of undoped GaAs layers on MESFET channels and its application for Ku-band extra high output power devices

Naotaka IWATA, Hiroshi MIZUTANI*, Satoshi ICHIKAWA*,
Akira MOCHIZUKI**, and Hiromitsu HIRAYAMA

Kansai Electronics Research Laboratory, NEC Corporation
9-1, Seiran 2-Chome, Otsu, Shiga 520, Japan
*System LSI Development Division, NEC Corporation
1753 Simonumabe, Nakahara-ku, Kawasaki, Kanagawa 213, Japan
**VLSI Development Division, NEC Corporation
9-1, Seiran 2-Chome, Otsu, Shiga 520, Japan

ABSTRACT: GaAs MESFETs have been successfully fabricated with undoped GaAs layers up to 2000Å thick on the n-GaAs channels. Gate contacts are put on the channels, by boring through the undoped layers. It was found that the undoped layer eliminates a surface depletion layer for the channel, as well as suppresses channel reduction due to the GaAs surface trapping effect, even under high-voltage operations. Consequently, the undoped layer acts as an ideal passivation layer for the channel. The fabricated FET with 1000Å thick undoped GaAs layer and 0.69μm long and 16.128mm wide gate contact has shown 2dB gain compression output power as high as 39.6dBm(9.1W) at 12.575GHz. This FET has great potentials for use in extra high output power devices.

1. INTRODUCTION

High power and high efficiency devices are essential for solid-state amplifier applications. For high output power GaAs MESFETs, which have been developed and used widely for micro-wave applications, premature power saturation is still a serious problem. It has been reported that this phenomenon would be caused by a channel reduction due to carrier trapping effect of interface states at the n-GaAs channel and a passivation film. Consequently, a drastic output power increase has been hampered by lack of good passivation technology for the GaAs surface.

To overcome this GaAs surface problem, new device design approaches, less sensitive to the surface effect, have been presented utilizing an undoped GaAs layer or undoped AlGaAs layer on the n-GaAs channel by Sriram *et al*(1989) and by Takikawa *et al*(1986), respectively. These studies showed the merit of the new device structure using a small size MESFET. However, detailed undoped layer effects on the channels and effects on output power characteristics in practical use have not been clarified.

This paper shows results of an extensive study on the undoped layer effect and its application for Ku-band power GaAs MESFETs, with as high as about 10W output at the Ku-band by one chip.

2. EXPERIMENTAL

Epitaxial wafers used in this study were grown on semi-insulating (001) oriented GaAs wafers by MBE. A wafer series has the same structure regarding an 8000Å thick undoped GaAs buffer layer, a 2000Å thick undoped $Al_{0.3}GaAs$ buffer layer, the 1200Å thick n-GaAs channel layer with donor concentration of $3.5 \times 10^{17} cm^{-3}$, the undoped GaAs layer and a 1000Å thick n^+-GaAs cap layer, while changing the undoped GaAs layer thickness up to 2000Å, for example, type I for 0Å, type II for 1000Å and type III for 2000Å, as shown in Fig.1. Another type of wafer(Type IV), which has the 4000Å thick undoped layer on the channels without the n^+-GaAs cap layers, was also employed to study the undoped layer effect.

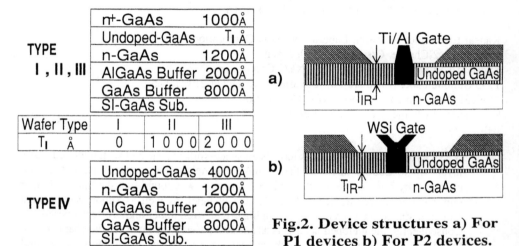

Fig.1. Epitaxial wafer structures.

Fig.2. Device structures a) For P1 devices b) For P2 devices.

Two kinds of FETs have been fabricated, using two kinds of fabrication processes. One fabrication process is the so-called Ti/Al lift-of gate process with the oxide/resist combination by using a single lithography step(the P1 process), reported by Sriram *et al*(1989). This process, with successive chemical etching steps of GaAs and oxide followed by oxide RIE, results in a gate recess narrower than the gate, as shown in Fig.2 a). For the undoped layer thickness of the recessed regions:T_{IR}(i.e. T_{IR} is the undoped GaAs layer thickness on the channel for both source-to-gate and gate-to-drain regions), T_{IR} was changed with T_I. The relations between T_{IR} and T_I are as follow, $T_{IR}=T_I$(Type 1), $T_{IR}=0.5T_I$(Type 2) and $T_{IR}=0.25T_I$(Type 3). Epitaxial wafers used in this P1 process are Types I, II and III, as shown in Fig.1. Each device was named after its fabrication process, wafer type and T_{IR}, for example, P1-2-2 denotes the device with 500Å T_{IR}, utilizing Type II wafer(i.e. T_I=1000Å), and fabricated through the P1 process. The P1 devices, which include several kinds of FETs and Hall devices with gate contacts, were characterized mainly regarding the undoped layer effect on the channels.

The other type of fabrication process(the P2 process) has two chemical etching steps in the recess forming process which are for a wide recess and the recess narrower than the gate, as can be seen in Fig.2 b). Gate contacts for these FETs(P2s) were WSi, and the devices were passivated with plasma enhanced CVD silicon dioxide. Figure 3 shows a cross sectional SEM image of the P2 device with the 1000Å undoped layer(i.e. P2-2-1). Plated heat sink, via-holes and air-bridge structures were employed for the P2 devices. Several types of wafers were used for the P2 device fabrication. For the Type IV wafer, that is the P2-4 devices, Si ion implantation technique has been employed to make source and drain contacts. Then, The P2 devices were characterized with RF power characteristics for practical use. In either case, device fabrications were carried out with good control of recess dimensions, and the gate contacts were put on the n-GaAs channel with a gate length(L_G) of about 0.7μm.

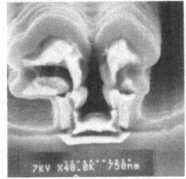

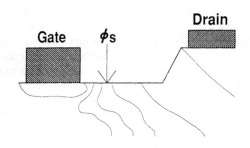

Fig.3. Cross sectional SEM image of P2-2-1.

Fig.4. Potential distribution diagram for the MESFET gate-to-drain region.

3. RESULTS AND DISCUSSION

To study the undoped layer effect on the channels, Hall devices with gate contacts for the P1s were characterized. This experiment corresponds to the characterization of the sheet carrier concentration(n_s) of some portion between gate and drain, as shown in Fig.4. That is, when the surface potential(Φ_s) is changed by electron capture of surface states under high-voltage operations, n_s would be changed with the device geometry.

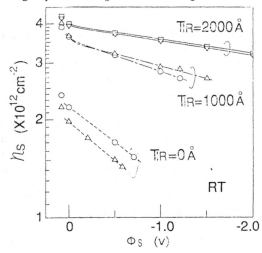

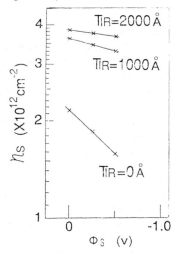

Fig.5. Φ_S dependence on n_S for P1 devices.

Fig.6. Calculation results for Φ_S dependence on n_S.

Figure 5 shows Φ_S dependence on n_S for the Hall devices with various undoped layer thickness T_{IR} values. It was found that, with increasing T_{IR}, n_S tended to be less sensitive to Φ_S. Figure 6 shows calculation results for Φ_S dependence on n_S for various undoped layer thickness T_{IR} values by one dimensional potential calculation. Experimental results, shown in Fig.5, coincide with theoretical one shown in Fig.6. These very closely coinciding results also revealed that, even if Φ_S was biased highly due to the high reverse gate bias(V_G) and the capture of electrons at the high density interface states, a small change in n_S was recognized, in case of more than 1000Å undoped layer thickness. This is due to consumption of the surface depletion layer of channels, because most of Φ_S(and also changes in Φ_S) was consumed in the thick undoped layers. It was found that the undoped layer acts as an ideal passivation layer for the channel, even under high-voltage operations. Since the gate fabrication process, by boring through the undoped layer, tends to be difficult, and resistances both between the channel and source and between the channel and drain increase with increasing in T_I, optimum T_I and T_{IR} values exist. The authors have concluded that 1000Å T_I and T_{IR} would be optimum, with good enough Φ_S suppression effect, as mentioned above.

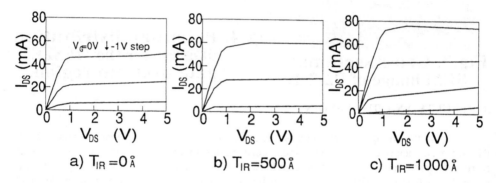

Fig.7. DC characteristics a) For P1-1 b) For P1-2-2 c) For P1-2-1.

Figure 7 shows DC characteristics for the P1 devices up to 1000Å T_{IR} with 200μm gate width(W_G). It was found that transconductances(g_m) and drain currents at $V_G=0V$(I_{DSS}) were increased with increasing T_{IR}. Since the undoped layer eliminates a surface depletion layer in the channel by the Φ_S suppression effect to the channel, as discussed above using Figures 5 and 6, increases in both g_m and I_{DSS} would be due to increases in carrier concentrations beside the gate contact.

Scattering parameter measurements were also carried out for the frequency ranging from 0.5 to 26.5GHz, using on-wafer RF probes and an automatic network analyzer. The current cut off frequency(F_t) and the maximum oscillation frequency(F_{max}) were derived from extrapolations of the current gain and the maximum unilateral gain, respectively, based on -6dB/oct. decay characteristics. The maximum F_t and F_{max} values, 14GHz and 40GHz, were obtained with about 0.7μm L_G through the P1s. Scarcely no difference in F_t was recognized. For F_{max}, very small decreases in F_{max} were recognized with an increase in T_{IR}. Since the gate contact for this FET structure was piled in the undoped layer, this decrease in F_{max} may be due to an increase in parasitic capacitance of the gate. However, since n_S, beside the gate, tend to increase with an increase in T_{IR}, resulting decreases in both a source resistance and a drain resistance should be pronounced. Consequently, the increase in the parasitic capacitance of the gate and the decreases in the resistances beside the gate, with increasing in T_{IR}, would be a trade-off in F_{max}.

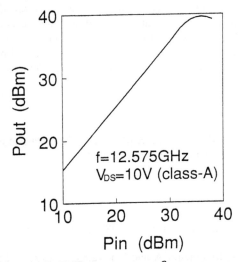

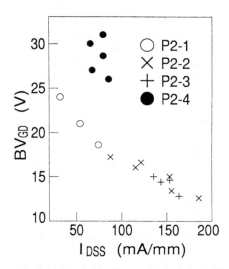

Fig.8. RF power performance for P2-2-1.

Fig.9. Relations between BV_{GD} and I_{DSS} for the P2 devices.

Through the P2s characterization, it was found that the new FET has an RF power performance with excellent gain linearity and high saturation output power. Figure 8 shows the RF power performance for P2-2-1 with 16.128mm W_G under the "class-A"(1/2 I_{DSS}) mode operation at 12.575GHz. P2-2-1 delivered 39.6dBm(9.1W) output power with 5dB gain and a 30% power added efficiency(PAE), at 2dB gain compression. This power handling capability per unit W_G is twice as large as that for conventional GaAs power MESFETs. In addition, the new device has excellent linearity regarding input power vs. output power, as can be seen in Fig.8. This is also due to the undoped layer effect for eliminating the surface depletion layer in the channels. The reason for slightly low gain and PAE would be due to a relatively long L_G, 0.69μm, for P2-2-1.

For P2-4s, though I_{DSS} for P2-4s were a little bit small, due to carrier depletion through the thermal process of the ion implantation technique, high gate-drain breakdown voltages(BV_{GD}) were obtained, as can be seen in fig.9. To make still further good contact to the channel through the undoped layer, Si ion implantation would be a powerful technique, reducing a field strength concentration between the cap layer and the channel.

The latest power GaAs MESFETs, just fabricated, with 0.55μm L_G and 1000Å T_{IR} have 9.2dB gain and 40.1% PAE at 12GHz, remaining its high power handling capability unchanged. This result with calculations for the MESFET with the undoped layer as a function of T_{IR}, using a two-dimensional device simulator, will be reported by Takahashi *et al*(1991). Consequently, this new MESFET structure, with the undoped GaAs layer on the channel, brings high output power as well as high PAE.

4. CONCLUSION

GaAs MESFETs have been successfully fabricated with undoped GaAs layers up to 2000Å thick on the n-GaAs channels. Gate contacts are put on the channels, by boring through the undoped layers. It was found that the undoped layer eliminates a surface depletion layer for the channel, as well as suppresses a channel reduction due to the GaAs surface trapping effect, even under high-voltage operations. Consequently, the undoped layer acts as an ideal passivation layer for the channel. The MESFET with 1000Å T_{IR}, 0.69μm L_G and 16.128mm W_G, has shown 2dB gain compression output power of as high as 39.6dBm(9.1W) at 12.575GHz. This FET has great potentials for use in extra high output power devices.

ACKNOWLEDGEMENTS

The authors would like to thank H.Takahashi, K.Asano and K.Matsunaga for discussions. They also wish to thank M.Ohta, H.Abe and T.Nozaki for encouragement.

REFERENCE

Sriram S, Clarke R C, Messham R L, Smith T J and Driver M C 1989 *IEEE Cornell University Conference* pp 218-227

Takahashi H, Asano K, Matsunaga K, Iwata N, Mochizuki A and Hirayama H 1991 *to be published in the 1991 International Electron Device Meeting Technical Digest*

Takikawa M, Kasai K, Ozeki M, Hirachi Y and Shibatomi A 1986 *Semi-Insulating III-V Materials*(Tokyo: OHMSHA) pp 603-608

Inst. Phys. Conf. Ser. No 120: Chapter 3
Paper presented at Int. Symp. GaAs and Related Compounds, Seattle, 1991

125

Characterization of anomalous frequency dispersion in GaAs BP-MESFETs by direct large-signal $I-V$ measurement

Y.Arai, M.Kasashima, N.Kobayashi, H.I.Fujishiro, H.Nakamura, S.Nishi

Semiconductor Technology Laboratory, Research and Development Group,
Oki Electric Industry Co.,Ltd., 550-5, Higashiasakawa, Hachioji, Tokyo 193, Japan

ABSTRACT: GaAs FETs' anomalous frequency dispersion can be observed clearly with large-signal direct I-V measurement, which can emphasize rate-dependent and/or field-dependent phenomena causing such dispersion. GaAs MESFETs showed behavior specific to structures, such as drain current decrease and drain conductance increase at high frequencies. Characterization and consideration of mechanism for buried p-layer GaAs MESFETs and that with low-temperature buffer layer are presented. Frequency dispersion was apparent in the device with low-temperature buffer. As there proved to be shifts in threshold voltage, such frequency dispersion may be attributed to phenomena related to electron traps in channel-buffer interface.

1. INTRODUCTION

GaAs FETs have been known to exhibit anomalous low-frequency dispersion of drain conductance (gd) and/or transconductance (gm) (Asai *et al.* 1973, Larson *et al.* 1985, Ladbrooke and Blight 1988, Golio *et al.* 1990, Ng and Pavlidis 1991), which even bring about limitations in their applications such as in wide-band analog circuits or dc-coupled circuits. This dispersion has been attributed to the presence of deep-levels originating from semi-insulating GaAs wafer, at the channel/substrate interface and to surface-states, so it is intrinsic for the layer structures of the devices. Compared to small-signal characterization studies, however, much less has been reported on large-signal frequency dispersion behavior in spite of the significance (Makram-Ebeid and Minondo 1985, Canfield *et al.* 1990). By applying large-signal voltage swings to device, not only rate-dependent phenomena but also field-dependent phenomena can affect the behavior.

This time, we applied cyclic voltage changes to drain with measuring corresponding drain current at various frequencies and various swings, i.e., directly measured large-signal I-V characteristics at various operating condition. This characterization is comprehensive because it can clearly reveal the point (bias condition and frequency) where anomalies take place and trend of the behavior relevant to conditions may be also observable, but the papers on it are few (Makram-Ebeid and Minondo 1985) and not so much attention has been paid. This measurement is of course versatile in application phase: for example, the deviation of I-V characteristics from dc to RF, which often brings about unpredictable deviation of operating condition at FET circuit design, can be estimated.

Many papers and our preliminary study verified that the introduction of buried p-GaAs thin layer into undoped GaAs buffer in GaAs MESFETs improves carrier confinement in the channel and is effective for suppressing small-signal frequency dispersion (Anderson *et al.* 1982).

In this study, we characterized and compared GaAs MESFET with buried p-layer and that with additional low-temperature buffer (Itoh *et al.* 1986) to suppress sidegating (Smith *et al.* 1988, Delaney *et al.* 1989) by above-mentioned direct large-signal I-V measurement.

2. MEASUREMENTS AND MEASUREMENT TECHNIQUES

The measurements here were carried out using the configuration in Figure 1, including an emitter-follower circuit through which drain bias is applied and drain voltage/current are monitored. Coaxial transmission lines were used for bias and signal paths to avoid oscillation. Microwave probe heads with 50 Ω impedance were utilized to enable on-wafer device measurement. Gate static bias (Vg) was applied with power supply unit and drain AC bias (Vd) with pulse-generator. Drain AC bias as a triangular wave was applied so that actual drain voltage swung from 0 to a certain positive value.

Output signals (Vx and Vy) measured with digitizing oscilloscope were then transferred to computer for data processing. Drain voltage is calculated from the value of Vx, whereas drain current is calculated from Vy with correcting the current in Vx path.

Frequency of drain voltage swing was ranging from 10Hz to 100kHz.

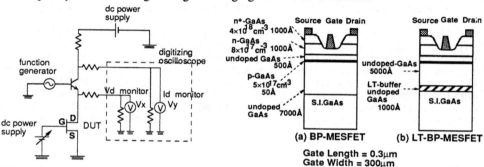

Figure 1.   Measurement configuration for large-signal I-V.

Figure 2.   Layer structure of the devices.

3. DEVICE STRUCTURE AND RESULT OF SIDEGATING

Devices reported here were fabricated on undoped LEC semi-insulating GaAs wafers with the epitaxial layers MBE-grown. Layer structures are shown in Figure 2: the structure (a), conventional BP-MESFET as we call simply "BP-MESFET" from now on, and (b) with additional 300°C-grown 1000Å low-temperature buffer layer ("LT-BP-MESFET").

The Ti/Al gate has the length of 0.3μm and the width of 300μm for both devices.

Brief fabrication procedures are: 1)O[16]-ion-implantation for device isolation, 2)AuGe/Ni/Au ohmic electrode metallization, 3)gate recess formation with wet etchant and gate metallization, 4)formation of Ti/Pt/Au interconnection, and 5)silicon-nitride passivation film deposition.

Results on sidegating characteristics with the gate width of 10μm are shown in Figure 3 (note that drain voltage sweep is dc). Sidegate voltage was applied from the terminal 20μm from the devices. With LT-BP-MESFET, sidegating is almost eliminated for sidegate voltages from 0V to -20V, compared that there exists sidegating to a certain degree with BP-MESFET.

In this I-V characteristics in dc mode, kink is apparent at Vd=1.8V in LT-BP-MESFET,

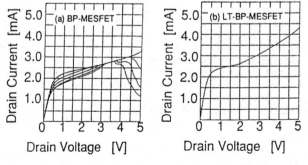

sidegating voltage : 0V to -20V, -5V step

Figure 3.   Sidegating effect for the 10μm-width devices.

whereas it is not observed in BP-MESFET. According to several papers, kink effect is attributed to hole generation due to impact ionization (Fujishiro *et al.* 1988, Zeghbroeck *et al.* 1987). However in this case of LT-BP-MESFET, the kink may not be the result of impact ionization, for our other data (measurement of the hole current to sidegate terminal) suggests that impact ionization takes place at drain voltage larger than 3.5V for LT-BP-MESFET.

4. RESULTS

Figure 4 shows the large-signal I-V characteristics measured at frequency 10Hz and 10kHz for the samples. In LT-BP-MESFET, remarkable anomalies appeared, i.e., decrease of Id and increase of gd were observed for higher frequencies at lower field around Vd<1.8V. Such frequency dispersion is not so apparent in BP-MESFET. Figure 5 shows the frequency dependence of drain conductance at gate bias 0V. Compared to BP-MESFET, LT-BP-MESFET has severe dependence on frequency with a sudden change between 100Hz and 1kHz.

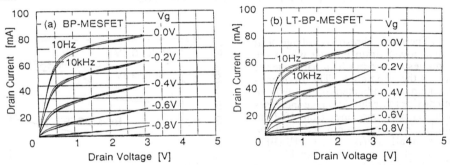

Figure 4. I-V curvrs for the devices at 10Hz and 10kHz.

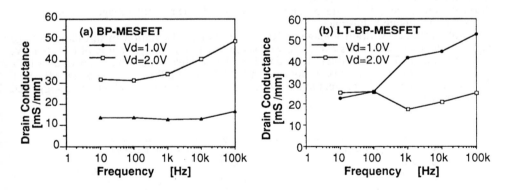

Figure 5. Frequency dependence of drain conductance.

In the region where the Id decrease occurs, there proved to be shifts in threshold voltage (Vth) as plotted in Figure 6. The relation between gm and Vg has practically no dependence on frequency when such threshold voltage shift is taken into consideration. So at least with these operating conditions where only the drain bias is modulated, the Id decreasing anomalies is not the result of the change of gm. Expressing in another words, the dispersion may be dominantly affected by certain modulation of the channel thickness from the back side of the channel, rather than phenomena relating to surface-state.

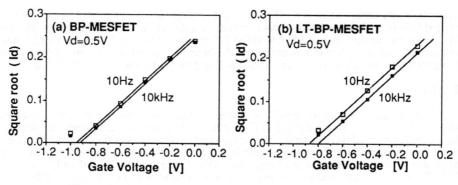

Figure 6.   Square root (Id) vs. gate voltage plots.

Therefore, threshold voltage shift at such lower drain voltage where the anomalies occur, can be a parameter to express the anomalies. The frequency dependence is indicated in Figure 7. In both devices threshold voltage increases gradually with frequency from 10Hz to 100kHz, however, compared with the rather small shifts in BP-MESFET (0.05V), it is almost twice in LT-BP-MESFET (0.09V). Especially the sudden increase is observed in LT-BP-MESFET from 100Hz to 1kHz. The tendency is the same with that of the drain conductance (cf. Figure 5) and drain current. Figure 8 illustrates

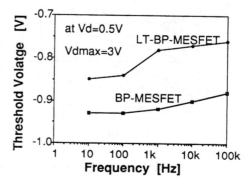

Figure 7.   Threshold voltage vs. frequency.

threshold voltage vs. drain voltage at 10Hz and 10kHz. Deviation in threshold voltage between the two frequencies exclusively exists in low drain voltage region of 0<Vd<1.8V. Furthermore, in LT-BP-MESFET, 10Hz data itself has a "kink" around the drain voltage 1.8V as I-V kink, however, this "kink" is not clear in LT-BP-MESFET 10kHz data or BP-MESFET for both frequencies. So the relation between the anomalies and dc kink effect is suggested from this fact.

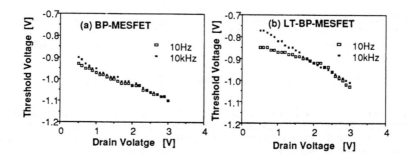

Figure 8.   Relation between threshold voltage and drain voltage
at 10Hz and 10kHz.

Maximum applied field also affected the magnitude of anomalies. Drain current difference between 10Hz and 10kHz depends on the maximum drain voltage applied (Vdmax), as shown in Figure 9. The vertical axis represents the decrease of drain current such as found in Figure 4. When Vdmax is 1V, that is, drain voltage is swept within low-field, or sufficiently below the kink point, decrease of drain current is almost negligible in both devices. The dependence on Vdmax is more strong with LT-BP-MESFET. Such dependence on Vdmax is already evident for Vd>1.5V, that means slightly below the kink point. The shift gets smaller for Vdmax above 1.5V or 2V for both devices.

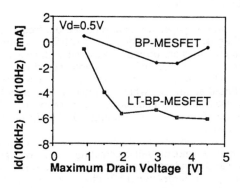

Figure 9.   Drain current change vs. maximum applied drain voltage.

5. DISCUSSION

By large-signal I-V measurement, frequency-dependent anomalous decrease in Id was revealed with LT-BP-MESFET. Here, we summarize some of the specific feature of such anomalies relevant to the applied field: 1) kink is apparent not only in dc I-V also in large-signal I-V characteristics with lower sweep frequency (10Hz to 1kHz) , 2) kink voltage for large-signal I-V is the same as that of dc I-V characteristics, 3) Id change takes place exclusively in the low-field region, below the kink point and 4) magnitude of the Id change is dependent on the maximum applied voltage. From these results, it may be concluded that such anomalies of LT-BP-MESFET are related to field-induced phenomena. As mentioned above, hole current is not observed at that kink voltage, so the field-ionization of electron trap may play a role for the anomalies. The existence of the electron traps as deep-level donors near the channel/buffer interface has been suggested by many papers.

Here, we discuss the I-V anomalies in LT-BP-MESFET as the combination of two phenomena: one is "drain current change" in the low-field region, the other is the field-dependent phenomena with small dependence on frequency, which is apparent in high-field region. When Vd is applied up to a higher voltage, the electron traps cause serious frequency-dependent phenomena at low-field and low frequencies. It is explained as follows: at lower field and lower applied frequency, rate-dependent phenomena is emphasized, i.e., the ratio of the occupied traps changes with the magnitude and rate of the change of applied field. The time constant for emission is so large compared to capture (Canfield *et al.* 1990) that once electrons are trapped due to field-enhanced electron injection/capture, the emission cannot respond to decreasing field if the voltage swing is carried out rapidly. At lower field (0<Vd<1.8V) with much higher frequency, therefore, the quantity of electron emission (injected by application of high field) remains small with the increasing frequency, which alters Id less. This frequency dependent phenomena of capture/emission is known as "self-backgating"(Wager and McCant 1987, Lee and Forbes 1990). At higher field (Vd>1.8V), much electron traps are field-ionized (Kuang *et al.* 1988, Brown *et al.* 1989) and less traps take part in emission/capture phenomena, so frequency-dependence gets smaller. This is why the difference of Id at different frequency takes place only in the low-field region. It is consistent with the result that kink voltage for "kink" in Vth vs. Vd plot and I-V kink voltage are almost the same, because both the frequency-dependent behavior and the kink in I-V may be related to field-ionization. It is noticeable that I-V characteristics for LT-BP-MESFET in higher frequencies is without any kink but not free from the influence of much traps because field-ionization takes place for the whole voltage swing range.

6. CONCLUSIONS

In this paper we have reported the frequency-dependent large-signal characteristics in MBE-grown buried-p layer GaAs MESFETs with/without low-temperature buffer layer. I-V characteristics are directly measured with sweeping the drain voltage from 0 to maximum at various frequencies and various voltage swings.

The sample with low-temperature buffer, with sidegating effectively suppressed, exhibited anomalies when operated to higher-field conditions: decrease of drain current and increase of drain conductance take place exclusively in the drain voltage region below the kink point, at frequencies exceeding 100Hz. In the region of I-V anomalies, threshold voltage shifts were observed.

These anomalies can be explained by field-induced "field-ionization" of the electron traps and rate-dependent "self-backgating". Due to large-signal operation, field-induced phenomena and rate-dependent phenomena take place simultaneously. As we employed the measurement technique of large-signal I-V characteristics, trap-related anomalies of the device with low-temperature buffer can be revealed clearly. This fact implies the significance and versatility of large-signal characterization.

REFERENCES

Anderson W T Jr., Simons M King E E, Dietrich H B and Lambert R J
    1982 IEEE Nucl. Sci. NS-**29** 1533
Asai S, Ishioka S, Kurono H, Takahashi S and Kodera H
    1973 Jpn. J. Appl. Phys. **42** 71
Brown A S, Mishra U K, Chou C S, Hooper C E, Melendes M A, Thompson M, Larson L E,
    Rosenbaum S E and Delaney M J 1989 IEEE EDL-**10** 565
Canfield P C, Lam S C F and Allstot D J 1990 IEEE Solid-State Circuit **25** 299
Delaney M J, Brown A S, Mishra U K, Larson L E, Nguyen L and Jensen J
    1989 GaAs and Related Compounds, Inst. Phys. Conf. Ser. **106** 189
Fujishiro H I, Saito T, Nishi S and Sano Y 1988 Jpn. J. Appl. Phys. **27** 1742
Golio J M, Miller M G, Maracas G N and Jhonson D A
    1990 IEEE ED-**37** 1217
Itoh T, Matsuura H, Nakamura H, Nishi S, Akiyama M and Kaminishi K
    1985 Semi-Insulating III-V Materials 503
Kuang J B, Tasker P J, Wang G W, Chen Y K, Eastman L F, Aina O A Hier H
    and Fathimulla A 1988 IEEE EDL-**9** 630
Ladbrooke P H and Blight S R, 1988 IEEE ED-**35** 257
Larson L E, Jensen J F, levy H M, Greiling P T and Temes G C
    1985 GaAs IC Symp. Dig. 19
Lee M and Forbes L 1990 IEEE ED-**37** 2148
Makram-Ebeid S and Minondo P 1985 IEEE ED-**32** 632
Ng G I and Pavlidis D, 1991 IEEE ED-**38** 862
Smith F W, Calawa A R, Chen C L, Manfra M J and Mahoney L J
    1988 IEEE EDL-**9** 77
Wager J F and McCant A J, 1987 IEEE ED-**38** 1001
Zeghbroeck B J V, Patrick W, Meier H and Vettinger P
    1987 IEEE EDL-**8** 188

Inst. Phys. Conf. Ser. No 120: Chapter 3
Paper presented at Int. Symp. GaAs and Related Compounds, Seattle, 1991

High quality and very thin active layer formation for ion implanted GaAs
MESFETS

Suehiro Sugitani, Kiyomitsu Onodera, Kazumi Nishimura, Fumiaki Hyuga,
and Kazuyoshi Asai

NTT LSI Laboratories
3-1, Morinosato Wakamiya, Atsugi-shi, Kanagawa, 243-01, Japan

Abstract   High quality and very thin active layer formation for Si ion implanted
GaAs MESFETs is investigated by capacitance-voltage and photoluminescence (PL)
measurements. In lower temperature region, carrier concentration increases as the PL
peak intensity at wavelength of about 1.3 μm decreases, which is thought to be associated with a Ga-vacancy (V_{Ga}) -related deep acceptor. In the higher temperature region,
the decrease in carrier concentration corresponds to the Si_{As} acceptor generation. Very
thin active layers with few acceptors due to V_{Ga} and Si_{As}, are applied to MESFETs
with Au/WSiN gates. The MESFETs with 0.35-μm gate-length exhibit a maximum
transconductance of 660 mS/mm and a cutoff frequency of 76 GHz, which are better
than the performance of GaAs HEMTs with the same gate-length.

1. INTRODUCTION

GaAs MESFETs with active layer formed by ion implantation are promising devices
for ultrahigh-speed LSIs because of their simplicity, uniformity, and reproducibility. The
performance of implanted GaAs MESFETs has steadily improved and their potential has
been pointed out by a number of authors (Yamane et al. 1988, Wang et al. 1989). We have
been developing ion implanted n$^+$ layer self-aligned GaAs MESFETs with Au/WSiN refractory metal gates and have achieved good device performance (Tokumitsu et al. 1988).
To devise MESFETs with higher performance, however, a high quality and very thin active
layer is indispensable. To form this active layer, it is necessary to clarify the activation
mechanism and to optimize the annealing condition.

This paper discusses the optimum annealing condition necessary to obtain a high
quality active layer, and demonstrates that the very thin active layer formed using this condition is better than the active layer of a GaAs HEMT with the same gate-length.

2. HIGH QUALITY ACTIVE LAYER FORMATION

2.1 Experiment

The substrates used in this work were undoped semi-insulating (100) GaAs wafers
grown by the liquid-encapsulated Czochralski (LEC) method. Silicon ions were implanted
directly into bare wafers with three combinations of energies and doses (30 keV, 1.6×10^{12}
cm^{-2}; 60 keV, 3.2×10^{12} cm^{-2}; 160 keV, 1.52×10^{13} cm^{-2}) so as to produce a uniform con-

centration profile of about 1×10^{18} cm^{-3} to a depth of 0.15 μm from the surface. An incident ion beam was directed 7° off the <100> crystal axis to minimize ion-channeling effects. After implantation, the wafers were encapsulated with a 0.1-μm thick SiN$_x$ film by plasma-enhanced chemical vapor deposition (PECVD). For comparison, some wafers were also encapsulated with a 0.1-μm thick SiO$_2$ film by PECVD. All wafers were then annealed at temperatures ranging from 600 to 1000°C for periods ranging from 0.1 s to 60 min in an N$_2$ atmosphere. Short period annealing of 0.1 s ~ 60 s was carried out using the rapid thermal annealing (RTA) system. The conventional furnace annealing system was used in long period annealing of 10 min ~ 60 min.

The carrier concentration depth profiles were measured by the electrochemical capacitance-voltage (C-V) profiler. The photoluminescence (PL) spectra were measured at 4.2 K between wavelengths of 0.8 μm to 2.4 μm. Illumination was provided by Ar-ion laser at a wavelength of 0.5145 μm with an intensity of 100 mW. The PL light was dispersed in a 0.64-m monochromator and detected with a cooled S-1 type photomultiplier between wavelengths of 0.8 μm to 0.9 μm. Between wavelengths of 0.9 μm to 2.4 μm, PL light was dispersed in a 0.32-m monochromator and detected with a PbS detector cooled with dry ice.

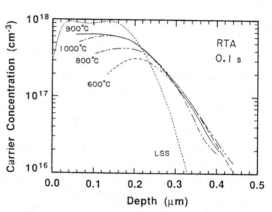

Fig. 1. Carrier concentration depth profiles for the samples annealed using SiN$_x$ film at different temperatures for 0.1 s.

2.2 Results and Discussion

Figure 1 shows the carrier concentration depth profiles for the samples annealed using SiN$_x$ film at different temperatures for 0.1 s. The most noticeable feature in this figure is that the carrier concentration in the shallow region (<~0.2 μm) changes with annealing temperature. In the annealing at 600 ~ 900°C, the carrier concentration increases with the annealing temperature, and a maximum carrier concentration of 6.4×10^{17} cm^{-3} is obtained at 900 °C. However, it should be noted that the carrier concentration is reduced at 1000°C.

To clarify the cause of these carrier concentration changes, PL measurements were carried out at 4.2 K for these samples. The variation of

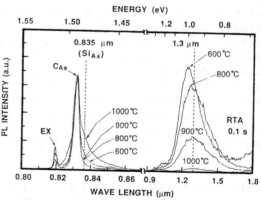

Fig. 2. Dependence of photoluminescence spectra on the annealing temperature for the samples annealed for 0.1 s using SiN$_x$ film.

PL spectra with the annealing temperature is shown in Fig. 2. The peaks at 0.819 μm and 0.830 μm are attributed to the bound exciton (EX) and carbon on As site (C_{As}) acceptor (Ashen et al. 1975), respectively. These two peaks are also observed in the as-grown wafers. In lower temperature region (≤900°C), the noticeable change is that the PL peak intensity at wavelength of about 1.3 μm is reduced as the annealing temperature increases. The ratio of the 1.3-μm peak to the EX peak (I[1.3]/I[EX]) decreases from 0.64 at 600°C to 0.14 at 900°C. In this region, carrier concentration increases with the annealing temperature as mentioned above. These results suggest that the increase in carrier concentration corresponds to the decrease in 1.3-μm PL peak intensity.

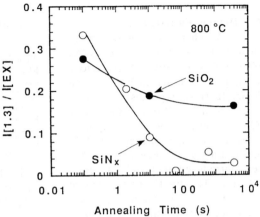

Fig. 3. Dependence of I[1.3]/I[EX] on the annealing period for the samples annealed at 800°C using SiN_x and SiO_2 film.

On the other hand, in the higher temperature region (>900°C), although the 1.3-μm PL peak intensity becomes very weak (I[1.3]/I[EX]= 0.02), the carrier concentration decreases. Figure 2 shows that the PL peak intensity at wavelength of about 0.835 μm clearly appears at 1000°C. It is well known that the 0.835-μm PL peak corresponds to Si on the As site (Si_{As}) acceptor (Ashen et al. 1975). Therefore, this carrier concentration decrease is due to the generation of the Si_{As} acceptor.

The annealing time dependences of carrier concentration and PL spectra were also investigated for the samples annealed at 800°C using SiN_x film. In the shorter period region (≤10 s), the carrier concentration increased as the 1.3-μm PL peak intensity was reduced. In the longer period region (>10 s), the carrier concentration decreased as the Si_{As} peak intensity increased. These results were the same as those for the temperature dependence.

Although the 1.3-μm PL peak has not been defined yet, it was reported that this peak is associated with a deep acceptor state (Kung et al. 1974). Moreover, it was reported that the 1.3-μm peak appears in undoped GaAs epi-layers grown under As-rich condition (Shinohara 1987). The 1.3-μm peak intensity increased in the phosphorus ion implanted samples (Hyuga et al. 1989), but this was not observed in the unimplanted samples with and without annealing. Therefore, this peak is thought not to be related to Si, but rather to defects generated by ion implantation.

In the annealing at 800°C for 60 min, the 1.3-μm peak intensity for the sample annealed using SiN_x film becomes very weak (I[1.3]/I[EX]=0.03), but the intensity of the sample annealed using SiO_2 film remains strong (I[1.3]/I[EX]=0.16), as shown in Fig. 3. It is well known that SiO_2 film enhances Ga outdiffusion from GaAs substrates during annealing (Kuzuhara et al. 1989). These results suggest that the 1.3-μm peak is associated with Ga vacancy (V_{Ga})-related deep acceptor which is generated by ion implantation.

Therefore, the optimum annealing condition for realizing a high quality active layer was determined, where the peak intensities due to V_{Ga}-related deep acceptor and the Si_{As} acceptor are very weak.

3. APPLICATION TO MESFET

3.1 Device Fabrication

A schematic cross-sectional view of the n+ self-aligned GaAs MESFET with Au/WSiN refractory metal gate is shown in Fig. 4. The device fabrication process have been described in detail in an other paper (Onodera et al. 1991). To suppress substrate leakage current, a lightly doped drain and source structure has been introduced along with a buried p-layer. The lightly doped source also obtains a low gate-to-source capacitance, which improves a cutoff frequency. The buried p-layer was formed by Be ion implantation at 50 keV. An n'-layer and an n+-layer were implanted with Si ion at an energy of 40 keV and 80 keV, respectively.

To obtain a high quality and very thin channel layer, Si ion implantation at 10 keV with a dose of 4.1×10^{13} cm^{-2} and RTA was carried out. SIMS analysis revealed that the effective thickness (Sugitani et al. 1987) of this implanted layer was 0.038 μm. RTA is useful in obtaining the active layer because it minimizes the diffusion of implanted ions and activates them with high activation efficiency.

The annealing condition was determined using the above mentioned results. PL measurements were carried out for the samples implanted at 10 keV and annealed using WSiN

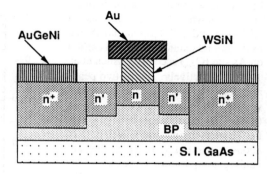

Fig. 4. Schematic cross-section view of n+ layer self-aligned GaAs MESFET with Au/WSiN refractory metal gate.

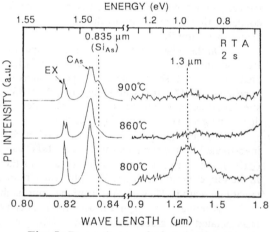

Fig. 5. Dependence of photoluminescence spectra on the annealing temperature for the samples implanted at 10 keV and annealed for 2 s using WSiN film.

gate metal. Figure 5 shows the PL spectra for the samples at different temperatures for 2 s. In the annealing at 860°C, the peaks at both 0.835 μm and 1.3 μm are very weak. The optimum annealing condition of 860°C for 2 s was determined from these results. In this active layer, the effective thickness of carrier concentration profile was estimated to be 0.046 μm from C-V measurements, and a standard deviation of sheet carrier concentration was 5.6% across the 3-inch wafer and 6.4% among the wafers. Consequently, we have succeeded in fabricating a high quality and very thin active layer for the MESFETs.

3.2 MESFET performance

Figure 6 exhibits the measured I-V characteristics of a MESFET with a gate length of 0.35 μm and a gate width of 10 μm. In this device, threshold voltage is -0.2 V and maximum transconductance is about 660 mS/mm, which is comparable to the transconductance of the 0.22-μm gate length epitaxial-channel GaAs MESFET (Jackson et al. 1990) reported recently. The intrinsic transconductance is estimated to be 980 mS/mm from source resistance of 0.5 Ω·mm.

The S-parameter measurements were made from 0.5 to 25.5 GHz using a network analyzer. The current gain was calculated from the measured S-parameters and the current gain cutoff frequency (f_T) was determined from the extrapolation of the current gain to unity using a -6 dB/octave decay in the higher frequency region. A maximum f_T of 76 GHz was obtained on the 0.35-μm MESFET as shown in Fig. 7, where the bias conditions are V_{gs} = 0.5 V and V_{ds} = 2.0 V.

Figure 8 shows the dependence of f_T on the gate length (Lg) at room temperature. Reported GaAs HEMT results (Abe et al. 1990) are included for comparison. Our results are better than the results for a GaAs HEMT with the same gate length.

The channel layer applied to these MESFETs is thought to have high electrical activation efficiency because these channel layers have few acceptors due to V_{Ga} and Si_{As}. High electrical activation efficiency decreases the effective thickness of channel and the excellent performance of MESFETs is thought to be achieved.

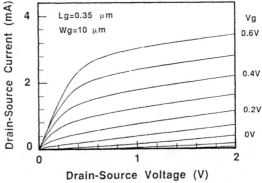

Fig. 6. I-V characteristics of the MESFET with a gate length of 0.35 μm and a gate width of 10 μm. Gate voltage ranged from 0.6 to -0.8 V in 0.1 V step.

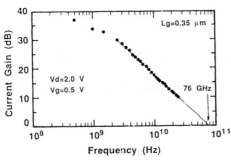

Fig. 7. Current gain as a function of frequency for a current gain cutoff frequency value of 76 GHz.

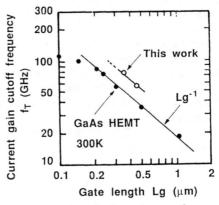

Fig. 8. Gate-length dependence of current gain cutoff frequency.

4. SUMMARY

The characteristics of Si ion implanted and annealed GaAs active layers was investigated by C-V and PL measurements. In the lower temperature region, the increase in the carrier concentration corresponded to the decrease in the 1.3-μm PL peak intensity, which is thought to be due to a V_{Ga}-related deep acceptor. In the higher temperature region, the carrier concentration decrease corresponded to Si_{As} acceptor generation. From these findings, the optimum annealing condition for devising a high quality active layer was determined, i.e., both acceptor peak intensities are very weak. The very thin active layer formed using this condition was applied to MESFETs with Au/WSiN gates. The MESFETs with the gate length of 0.35 μm exhibited a maximum transconductance of 660 mS/mm and a cutoff frequency of 76 GHz, which are better than the performance of GaAs HEMTs with the same gate-length. This high quality and very thin active layer is useful for deep submicron-gate-length MESFETs which are applicable to ultrahigh-speed LSIs.

Acknowledgments

The authors wish to thank K. Hirata for his helpful discussions and continuous encouragement.

References

Abe M and Mimura T 1990 Tech. Dig. IEEE GaAs IC Symp. pp 127-130.

Ashen D J, Dean P J, Hurle D T, Mullin J B, White A M, and Greene PD 1975 J. Phys. Chem. Solids **36** 1041

Hyuga F, Yamazaki H, Ishida S, and Kato N 1989 J. Appl. Phys. **66** 2719.

Jackson T N, Pepper G, DeGelormo J F, and Kuech T F 1990 IEEE International Devices Meeting Technical Digest pp 507-510

Kung J K and Spitzer W G 1974 J. Appl. Phys. **45** 4477.

Kuzuhara M, Nozaki T, and Kamejima T 1989 J. Appl. Phys. **66** 5833.

Onodera K, Tokumitsu M, Tomizawa M, and Asai K 1991 IEEE Trans. Electron Devices **38** 429.

Shinohara M 1987 J. Appl. Phys. **61** 365.

Sugitani S, Yamasaki K, and Yamazaki H 1987 Appl. Phys. Lett. **51** 806.

Tokumitsu M, Onodera K, and Asai K 1988 Extended Abstract of 46th Device Research Conference VA-2 (Boulder, CO).

Wang G W, Feng M, Lau C L, Ito C, and Lepkowski T R 1989 IEEE Electron Device Letters **10** 95.

Yamane Y, Enoki T, Sugitani S, and Hirayama M 1988 IEEE International Devices Meeting Technical Digest pp 894-895

Inst. Phys. Conf. Ser. No 120: Chapter 3
Paper presented at Int. Symp. GaAs and Related Compounds, Seattle, 1991

137

Novel carbon-doped p-channel GaAs MESFET grown by MOMBE

F. Ren, C. R. Abernathy and S. J. Pearton

Abstract

A high performance GaAs p-MESFET using carbon as the p-dopant is demonstrated for the first time. The channel and contact layers were grown by metal organic molecular beam epitaxy (MOMBE). The cap contact layer was highly doped with carbon (5×10^{20} cm^{-3}) in order to minimize the parasitic resistance in the FET structure. The sheet resistivity and transfer resistance of the contacts were 220 ohm/$\square$ and 0.2 ohm-mm respectively. These are comparable to values achieved on n-type GaAs. The room temperature extrinsic transconductance and K-factor values were 50 mS/mm and 165 mS/V·mm with 1 μm gate length and 3.5 μm source-to drain spacings. There are the highest room temperature values ever demonstrated for p-GaAs MESFET.

Introduction

To develop GaAs-based high-speed and low power LSI digital circuits, a complementary technology (n and p-channel devices) is required. Although some ring oscillators have been demonstrated, the room temperature performance was limited by the low hole mobility in the p-channel FET. Many attempts have been made with modulation doped AlGaAs/GaAs FETs,[1] p-channel InGaAs strained FETs,[2] and AlGaAs/GaAs MISFETs[3] to enhance the hole mobility, leading to a somewhat improved low temperature (77 K) extrinsic transconductance. The room temperature performance was still quite poor due to the high parasitic resistances and low hole mobility. To reduce the parasitic resistance, the cap contact layer should be heavily doped. However, the doping levels for conventional p-type dopants, Be and Zn, are restricted by their relatively high diffusion coefficients,[4,5] and solubilities generally less than 10^{20} cm^{-3} for the usual epitaxial growth temperatures. On the other hand, carbon as a p-type dopant has a several order of magnitude lower diffusion coefficient than Be or Zn, and a very high solubility under appropriate conditions.[6] The use of carbon for p-type doping has been demonstrated in AlGaAs/GaAs heterojunction bipolar transistors[7] and quantum well lasers[8] with both MOMBE and MOCVD growth. The unique growth chemistry of MOMBE has been used to achieve electrically active carbon concentrations in GaAs up to the level of 10^{21} cm^{-3} [6] using trimethylgallium (TMGa) as the growth III precursor.

In this work, we report the highest room temperature performance ever achieved in a p-GaAs MESFET. This is possible by using MOMBE carbon doping to produce low resistance ohmic contacts to the FET structure.

Layer Structure and Device Fabrication

The device structure, Fig. 1, consists of a 4000 Å GaAs buffer layer, a 800 Å C-doped ($p = 3 \times 10^{17}$ cm^{-3}) channel layer, a 50 Å undoped $Al_x Ga_{1-x} As$ ($x = 0.3$) dry etch-stop layer, and a 500 Å C-doped GaAs ($p = 5 \times 10^{20}$ cm^{-3}) cap layer. The epilayers were grown in a INTEVAC GaAs Source Gen II MOMBE system at a substrate temperature of 500°C. The carbon dopant was introduced through the use of TMGa, and the doping level controlled by mixing TMGa with triethylgallium (TEGa).[6] The undoped GaAs and AlGaAs were grown with TEGa. Trimethylamine alane (TMAA) was used as the Al source. All of the layers were grown with AsH$_3$ which was cracked in a low pressure cracker. The background doping level for the GaAs buffer layer is $\leq 10^{14}$ cm^{-3}.

500 Å	GaAs	c-doped	p = 5E20
50 Å	Al$_{0.3}$GaAs$_{0.7}$As	Undoped	
800 Å	GaAs	c-doped	p = 3E17
4000 Å	GaAs	Undoped	
	GaAs	S.I. Substrate	

Fig. 1.   Layer structure for the p-GaAs MESFET.

The device isolation was done by argon ion milling, followed by a wet chemical etch to remove the ion bombardment damage. Au-Be based metallization was used for the p-ohmic contact. The gate was recessed by selectively reactive ion etching the heavily doped GaAs cap layer using a CCl$_2$F$_2$/O$_2$ based plasma, and stopping on the thin AlGaAs layer. The selectivity between AlGaAs and GaAs was greater than 200. Ti/Pt/Au was used as the gate metal. A cross-sectional schematic view of the device is shown in Fig. 2.

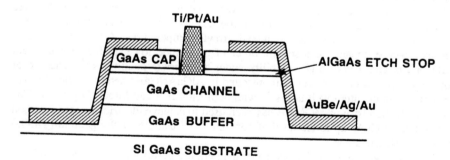

Fig. 2.   A cross-sectional schematic view of the device.

Results and Discussion

Carbon-doped MESFETs are expected to demonstrate a much higher thermal stability than devices utilizing Be- or Zn-doping. This is due to the much lower diffusivity of C at elevated temperature (6×10^{-15} cm/s at 900°C) compared to other acceptor dopants, Be, Mg, Zn and Cd, for which the diffusivities are at least several orders of magnitude higher. The excellent thermal stability of the carbon-doped GaAs allows for a flexible processing sequence, and high temperature anneals to activate implanted n-type dopants are possible for complementary applications without degrading the p-type MESFET. Hall measurements on the p-MESFET structure in this work showed no significant change in sheet carrier density or mobility as a result of a 900 C, 30 sec anneal. The initial sheet carrier density of 1.5×10^{15} cm^{-2} decreased to 1.4×10^{15} cm^{-2} while the 300 K hole mobility decreased from 37 cm^2/V·sec to 34 cm^2/V·sec after annealing. These changes are within the measurement error. By contrast it is generally observed that annealing of highly Be or Zn-doped structures above 800 C causes substantial redistribution of the dopants and loss of dopant to the surface. SIMS data shows that the use of carbon eliminates these problems.

Besides the high thermal stability for the C-doped MESFET, the high doping level capability of this structure is another advantage over the conventional Zn or Be dopants. A heavily doped contact layer reduces not only the contact resistance but also the material sheet resistivity, which are the key factors limiting the device performance. Both contact resistance and sheet resistivity are the major components of the source resistance. High source resistance is extremely detrimental to the device because the negative feedback causes the extrinsic transconductance to be reduced from the intrinsic value.[9,10] From transmission line method (TLM) measurements,[11] transfer resistance and sheet resistivity were determined to be 0.2 ohm-mm and 220 ohm/□, respectively. With these extremely low value of the parasitic resistances, the transistor source resistance which was estimated by forward biasing the gate-to-source diode and measuring the drain voltage as a function of the injection current, was only 2.3 ohm-mm. Based on these values, as shown in Fig. 3, the values for the extrinsic transconductance and K value were 50 mS/mm and 165 mS/V·mm. These are the highest room temperature numbers ever demonstrated for a p-GaAs MESFET. Figure 4 shows the typical drain current-voltage characteristics for 1 μm gate length and 30 μm gate width. The device exhibited excellent pinch-off characteristics. The gate leakage current was also reduced by contacting the gate on the high energy bandgap AlGaAs etch-stop layer.

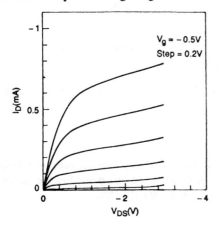

Fig. 3.   Drain I-V characteristics of p-GaAs MESFET.

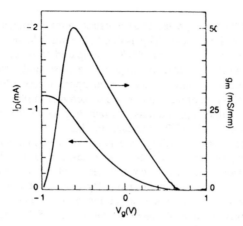

Fig. 4.   Extrinsic transconductance and drain current as a function gate bias for a 30 μm wide p-GaAs MESFET at room temperature.

Similarly, heavily carbon-doped cap layers can be applied to p-channel MODFETs to take advantage of the higher hole mobility in these structures. In this work, the calculated mobility for the carbon doped channel layer was $110 \text{ cm}^2/\text{V·sec}$. A doubling of the hole mobility with p-channel MODFETs could be achieved, as shown in Fig. 5, so that even better device performance can be expected by coupling these techniques.

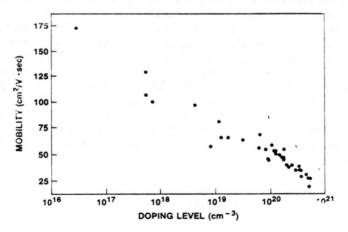

Fig. 5.   The hole mobilities of C-doped GaAs as the function of carrier concentration.

Conclusion

A state-of-the-art p-channel GaAs MESFET with extremely low parasitic resistance has been demonstrated. Even superior results can be expected by combining the present low contact resistance technology into a p-channel MODFET structure.

References

1. R. A. Kiehl and A. C. Gossard, IEEE Electron Dev. Lett., **EDL-5**, p. 521, 1984.

2. T. J. Drummond, T. E. Zipperian, I. J. Fritz, J. E. Schirbe and I. A. Plut, Appl. Phys. Lett., **49**, 461, 1986.

3. S. Fujita and I. Mizutani, IEEE Trans. Electron Devices, **ED-34**, 1889, 1987.

4. D. L. Miller and P. M. Asbeck, J. Appl. Phys., **57**, p. 1816, 1985.

5. P. M. Enquist, J. A. Hutchby and T. J. Delyon, J. Appl. Phys., **83**, p. 4485, 1989.

6. C. R. Abernathy, S. J. Pearton, N. Caruso, F. Ren and J. Kovalchick, Appl. Phys. Lett., **55**, p. 1750, 1989.

7. F. Ren, C. R. Abernathy, S. J. Pearton, I. R. Fullowan, J. Lothian and A. S. Jordan, Electron. Lett., **26**, p. 724, 1990.

8. L. J. Guido, G. S. Jackson, D. C. Hall, W. E. Plano and N. Holonyak, Jr., Appl. Phys. Lett., **52**, p. 522, 1988.

9. N. C. Cirillo, Jr., M. S. Shur, P. J. Void, J. K. Abrokwah and O. N. Tufte, IEEE Electron Device Lett., **EDL-6**, p. 645, 1985.

10. C. P. Lee, H. I. Wang, G. J. Sullivan and D. L. Miller, IEEE Electron Device Lett. **EDL-8**, p. 85, 1987.

11. H. H. Berger, J. Electrochem. Soc., **119**, 509 (1972).

Pseudomorphic GaAs/GaInAs pulse-doped MESFETs grown by organometallic vapor phase epitaxy

N.KUWATA, S.NAKAJIMA, T.KATSUYAMA, K.OTOBE, K.MATSUZAKI, T.SEKIGUCHI, N. SHIGA, and H. HAYASHI.

Optoelectronics R&D Laboratories,SUMITOMO ELECTRIC INDUSTRIES, LTD.
1, Taya-cho, Sakae-ku, Yokohama, 244 Japan

Abstract: Pseudomorphic pulse-doped (PPD) MESFETs with a Si-doped GaInAs active layer have been successfully developed for low noise application. The devices with 0.35μm gate-length exhibited as good linearity as pulse-doped MESFETs with a Si-doped GaAs active layer. The maximum transconductance (gm_{max}) and the transconductance (gm) at low drain current (≈ 0.2 Idss) were 380mS/mm and 250mS/mm, respectively. These values are 10~20% higher than those of the pulse-doped MESFETs. The minimum noise figure (Fmin) of the PPD-MESFETs is 0.74dB at 12GHz for 0.35μm devices, which is 0.1dB lower than that of the pulse-doped MESFETs.

1.INTRODUCTION

Over the past several years, high electron mobility transistors (HEMTs) have gained widespread application in low noise microwave and millimeter-wave amplifiers (Mishra U K et al. 1988, Chao P C et al. 1989). It has been thought that HEMTs give performance superior to MESFETs due to their high electron mobility.

Recent theoretical analysis, however, indicates that the parameter most affecting gm_{max} is the gate-channel spacing rather than mobility(Hasegawa F 1985, 1987). In reality, if the ion implantation profile is optimized, MESFETs exhibit gm_{max} and f_T equal to those achieved with HEMTs (Feng M et al. 1990).

A Fmin of 0.7dB at 12GHz was achieved for 0.3μm GaAs pulse-doped MESFETs with a thin Si-doped GaAs active layer in spite of a low mobility of 1500cm^2/V·s (Nakajima S et al. 1988,1990a). An X-band monolithic four-stage low noise amplifier with 0.5 μm-gate pulse-doped MESFETs showed a Fmin of 1.67dB at 12GHz (Shiga N et al. 1990). These noise characteristics, which are due to a saturation velocity of pulse-doped structure higher than uniform Si-doped GaAs (Nakajima S et al. 1991), compare favorably with those of HEMTs. These facts suggest that not only HEMTs but also pulse-doped MESFETs are promising for low noise application. The device performance of pulse-doped MESFETs is much affected by the following factors :

(1)electron transport property in a high field rather than in a low field
(2)good carrier confinement due to V-shape potential at doping interface
    (Nakajima S et al. 1990b)

With these factors in mind, a GaInAs active layer was applied to PPD-MESFETs. GaInAs shows an electron transport property superior to GaAs. Band offset at GaAs/GaInAs interfaces contributes to better carrier confinement.

2.GROWTH CONDITIONS

OMVPE growth conditions for GaInAs/GaAs strained single quantum well (SSQW) were optimized to obtain high quality GaInAs channel material prior to device fabrication. The OMVPE system has a conventional vertical reactor with a vent-run system for rapid gas switching. Triethylgallium (TEG), trimethylindium (TMI) and arsine (AsH_3, 10% in H_2) were used for GaInAs. Trimethylgallium(TMG) and AsH_3 were used for GaAs. The In content, x, in $Ga_{1-x}In_xAs$ was determined by the lattice constant of bulk $Ga_{1-x}In_xAs$ ($>1\mu m$) measured by x-ray diffraction. The well widths of GaAs/$Ga_{0.82}In_{0.18}As$ SSQWs ranged from 12Å to 150Å below the critical layer thickness ($\approx$170Å) (Matthews J W and Blakeslee A E 1974). Optical properties of the grown SSQWs were characterized by a photo-luminescence (PL) measurement at 4.2K using an Ar+ ion laser ($\lambda = 5145$Å).

For OMVPE growth conditions, growth pressure was fixed as 60torr. Three parameters - growth temperture, V/III ratio, and misorientation of substrate, were varied.

Figure 1 shows PL spectra of the samples shown in the inset grown at 600°C, 625°C and 650°C. The peak intensity and linewidth of the PL spectra in Fig.1 are plotted as a function of growth temperature (Fig.2). The SSQWs grown at 625°C exhibited the highest intensity and narrowest linewidth of 1.5meV.

The V/III ratio was optimized in a similar way. The V/III ratio for the GaInAs layer ranged from 43 to 256, and the optical property of SSQWs was evaluated by PL measurements. A V/III ratio of around 171 is found to be the best condition.

The dependence of SSQW optical properties on misorientation of substrate was also investigated. Figure 3 shows PL spectra of GaAs/GaInAs SSQWs grown on GaAs substrates oriented exactly on (100) and misoriented by 2° from (100) towards <110> at 625°C with a V/III ratio of 171. Four peaks corresponding to the transition in the individual well shown in the inset of Fig.3 are observed. The PL linewidth of SSQWs grown on a 2° misoriented substrate was narrower than that of SSQWs on a nominal (100) substrate. The origin of this difference in

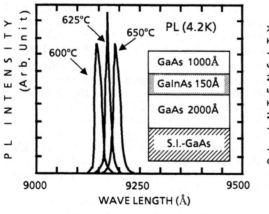

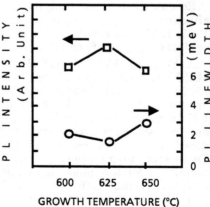

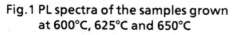

Fig.1 PL spectra of the samples grown at 600°C, 625°C and 650°C

Fig.2 Peak intensity and linewidth of PL spectra as a function of Tg

results is as follows. In a GaAs/GaInAs strained layer structure, the epitaxial layer on a misoriented substrate is two-dimensional, while on a (100) surface a three-dimensional island growth is more probable as it is energetically favorable (Morris et al. 1988). The 12Å-GaAs/$Ga_{0.72}In_{0.18}As$ SSQW exhibited the narrowest PL linewidth of 1.3meV. This is comparable to the best reported value of 0.9 meV for 11Å-GaAs/$Ga_{0.88}In_{0.12}As$ SSQW at 2.5K(Bertolet D C et al. 1988).

Figure 4 shows the epitaxial layer structure of PPD-MESFETs. An undoped p-GaAs buffer layer (10,000Å), a Si-doped GaInAs layer ($4 \times 10^{18}cm^{-3}$, 100Å) and an undoped n- GaAs layer (400Å) were successively grown under the above mentioned optimum OMVPE growth conditions. Background carrier concentrations of the p- buffer layer and the n- cap layer were controlled by a V/III ratio using TMG and AsH_3. Disilane (Si_2H_6) was used as a dopant for the Si-GaInAs layer. The uniformity of sheet carrier density (Ns) and mobility μ are shown in Fig.5. The variation of Ns is less than 2% over a 3-inch wafer.

3.DEVICE FABRICATION

3-1. FABRICATION PROCEDURE

A structure of a pseudomorphic pulse-doped MESFET is shown in Fig.6. After OMVPE epitaxial growth, the device is isolated by mesa etching. SiNx film (800Å) was deposited by p-CVD for passivation and annealing cap. Submicron gates were defined by means of

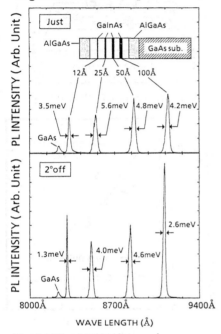

Fig.3 PL spectra dependence on misorientation of substrate

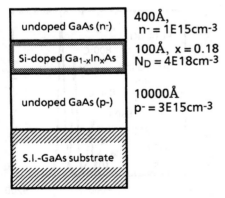

Fig.4 Epitaxial layer structure of PPD-MESFETs.

undoped GaAs (n-) 400Å, $n^- = 1E15cm^{-3}$

Si-doped $Ga_{1-x}In_xAs$ 100Å, x = 0.18 $N_D = 4E18cm^{-3}$

undoped GaAs (p-) 10000Å $p^- = 3E15cm^{-3}$

S.I.-GaAs substrate

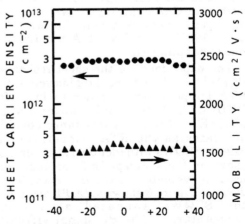

DISTANCE FROM CENTER OF WAFER (mm)

Fig.5 Uniformity of PPD Structure

conventional photolithography based on T-shaped dummy gate techniques. Self-aligned Si ion implantation was performed to reduce source resistance using this T-shaped resist mask with a 0.2μm undercut. After Si ion implantation, the T-shaped resist mask is substituted for the SiO$_2$ film window and then rapid thermal annealing (RTA) is adopted for activation of the implanted Si. The AuGe/Ni ohmic metal is alloyed at 450°C. The Ti/Pt/Au gate metal is overlapped on the SiO$_2$ film window for the formation of a mushroom electrode to reduce gate resistance. The gate length is 0.35μm and the length of the top of mushroom gate is 1.0μm. A first metal interconnection consisting of Ti/Pt/Au is used to provide low resistance interconnects. Air bridges are used to interconnect the three gate feeds of the device.

3-2. RTA CONDITIONS

During device fabrication, RTA is one of the most important processes affecting device performance. A higher RTA temperature is preferable to activate ion-implanted Si in the n+ region for the reduction of source resistance. On the other hand, diffusion of Si and In atoms from the Si-GaInAs channel is enhanced as the RTA temperature is raised. This degrades carrier confinement, which leads to a short channel effect. Furthermore, it was reported that GaAs/GaInAs SSQWs with thin GaAs cap layers were greatly damaged by heat treatment (Zipperian T E et al. 1988). Therefore, RTA temperature had to be optimized.

The dependence of the sheet resistance (R_{sh}) of the n+ implanted region versus the RTA temperature is shown in Fig.7 for PPD-MESFETs and pulse-doped MESFETs. The sheet resistance steadily decreases as the RTA temperature is raised. SSQWs with identical structures as shown in Fig.4, except for Si doping, were used to evaluate the thermal stability of devices. Figure 8 shows 4.2K PL spectra of the SSQWs annealed at various temperatures. The PL signal virtually disappears for the sample annealed at 900°C.

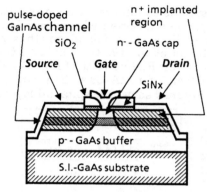

Fig.6   Structure of PPD MESFETs

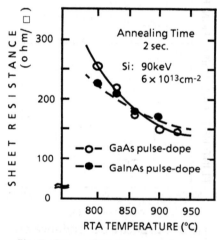

Fig.7 R_{sh} vs RTA Temperature

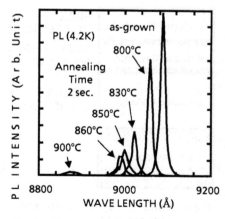

Fig.8   Thermal stability of SSQWs

In accord with these experimental results, 860°C RTA was adopted for the device fabrication in this report, but this condition is not necessarily optimum in our view.

4. DEVICE PERFORMANCE

The typical characteristics of 0.35×20-μm devices are shown in Fig.9. The maximum transconductance (gm_{max}) and the transconductance(gm) at low drain current ($\simeq 0.2$ Idss) were 380mS/mm and 250mS/mm, respectively, which were 20% higher than those of the conventional pulse-doped MESFETs. The pseudomorphic pulse-doped (PPD) MESFETs exhibited as good linearity as the pulse-doped (PD) MESFETs.

Figure 10 shows distribution of threshold voltage (V_{th}) for the 0.35 μm devices, over a 3-inch wafer. The standard deviation of V_{th} was 54 mV (7% of the average value), which was comparable to that of the conventional pulse-doped MESFETs and fully ion implanted MESFETs. This excellent uniformity is due to the good uniformity of the OMVPE grown epitaxial structure as shown in Fig.5 and the n + self-aligned implantation technique.

Typical noise and RF characteristics for 0.35×280-μm devices of PPD-MESFETs and PD-MESFETs are shown in Table 1. Fmin of the PPD-MESFETs is 0.74dB with associated gain (Ga) of 9.3 dB at 12GHz. Fmin of the PPD-MESFETs is 0.1dB lower than that of the PD-MESFETs.

Measured Fmin and Ga dependence on drain current (I_{DS}) are shown in Fig.11. Fmin is relatively independent of drain current, which is very convenient for designing low noise amplifiers because of a large margin for bias conditions.

These improved performances were due to better carrier confinement arising from band offset at GaAs/GaInAs interfaces and the superior electron transport properties of GaInAs, especially higher drift velocity in a high field (Park D H et al. 1989).

5. CONCLUSION

Pseudomorphic GaAs/GaInAs pulse-doped MESFETs with a Si-doped GaInAs active layer have been successfully developed. The minimum noise figure (Fmin)

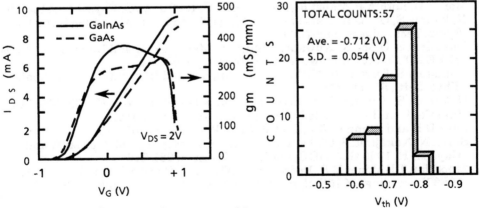

Fig.9 Measured DC gm values vs. gate bias   Fig.10 Distributions of V_{th}

@36%Idss	PPD-MESFETs	PD-MESFETs
Fmin(dB)	0.74	0.83
Ga(dB)	9.3	8.6
F_T(GHz)	32	30
gm(mS)	80	73
Cgs(fF)	310	300
Cgd(fF)	90	90
Cds(fF)	90	100
Rs(Ω)	1.7	1.7
Rd(Ω)	3.3	3.3
Rg(Ω)	1.2	1.3
Rds(Ω)	144	140

Table1   Noise and RF Performance   Fig.11   Measured Fmin and Ga vs. I_D/I_{DSS}

is 0.74 dB at 12GHz for 0.35μm devices, which is 0.1dB lower than that of GaAs pulse-doped MESFETs. A Fmin is relatively independent of drain current. These results indicate that pseudomorphic GaAs/GaInAs pulse-doped MESFETs have higher potential for low noise application than GaAs pulse-doped MESFETs.

ACKNOWLEDGEMENTS

The authors are grateful to K.Yoshida, M. Koyama, H. Kotani, A. Ishida, and K. Koe for their encouragement and support.

REFERENCES

Bertolet D C, Hsu J, Jones S H and Lau K M, 1988 Appl. Phys. Lett., 52, pp293-5
Chao P C, Shur M S, Tiberio R C, Duh K H G, Smith P M, Ballingall J M, Ho P, Jabra A A, 1989 IEEE Trans. Electron Devices, 36, pp461-473
Feng M, Lau C L, Eu V, and Ito C, 1990 Appl. Phys. Lett., 57, pp1233-5
Hasegawa F, 1985 IEEE Trans. Electron Devices, 32, pp2528
Hasegawa F, 1987 Extended Abstracts of the 19th Conf. on SSDM, p387-90
Matthews J W and Blakeslee A E, 1974 J. Crystal Growth, 27, pp118-125
Mishra U K, Brown A S, Rosenbaum S E, Hooper C E, Pierce M W, Delaney M J, Vaughn S and White K, 1988 IEEE Electron Device Lett., 9, pp647-9
Morris D, Roth A P, Maust R A and Lacelle C, 1988 J. Appl. Phys. 64, p4135-40
Nakajima S, Otobe K, Katsuyama T, Shiga N and Hayashi H, 1988 GaAs IC Symp. Tech. Digest, pp297-300
Nakajima S, Otobe K, Kuwata N, Shiga N, Matsuzaki K and Hayashi H, 1990a IEEE MTT-s Int'l Microwave Symp. Digest, pp1081-84
Nakajima S, Kuwata N, Nishiyama N, Shiga N and Hayashi H, 1990b Appl. Phys. Lett., 57, pp1316-7
Nakajima S, Kuwata N, Nishiyama N, Shiga N and Hayashi H, 1991 The 7th Int'l Conf. on Hot Carriers in Semiconductors Digest, pp151
Park D H and Brennan K F, 1989 J. Appl. Phys., 65, pp1615-20
Shiga N, Nakajima S, Otobe K, Sekiguchi T, Kuwayta N, Matsuzaki K and Hayashi H, 1990 GaAs IC Symp. Tech. Digest, pp237-240
Zipperian T E, Jones E D, Dodson B W, Kiem J F, Gourley P L and Plut T A, 1988 Proceedings of 15th Symp. GaAs and Related Compounds, pp365-70

Inst. Phys. Conf. Ser. No 120: Chapter 3
Paper presented at Int. Symp. GaAs and Related Compounds, Seattle, 1991

149

Two-dimensional electron gas analysis on pseudomorphic heterojunction field-effect transistor structures by photoluminescence

H. Brugger,[1] H. Müssig,[1,2] C. Wölk,[1] F.J. Berlec,[1] R. Sauer,[2] K. Kern,[3] and D. Heitmann[3]

[1] *Daimler Benz Research Center, P.O. Box 2360, D-7900 Ulm (Germany)*
[2] *Universität Ulm, Abteilung Halbleiterphysik, D-7900 Ulm (Germany)*
[3] *Max-Planck-Institut für Festkörperforschung, D-7000 Stuttgart 80 (Germany)*

ABSTRACT: The photoluminescence (PL) of a high density two-dimensional electron gas (2DEG) in pseudomorphic AlGaAs/InGaAs/GaAs hetero-field-effect transistor structures is investigated. The PL is dominated by a strong band from the 2DEG in the InGaAs quantum well (QW). The band width increases linearly with applied gate voltage, which directly reflects the two-dimensional density of states (2DDOS) below the Fermi-level. Using the effective electron mass from cyclotron resonance experiments to evaluate the 2DDOS, we determine directly the 2DEG density n_s from the PL spectral width. The n_s values obtained are in excellent agreement with n_s values from Shubnikov-de Haas measurements. The QWs have In-contentrations between 14% and 27%, widths between 7nm and 17nm, and n_s up to $2.6 \times 10^{12} \text{cm}^{-2}$. PL-mapping experiments on wafers show excellent lateral homogeneity of the material composition.

1. INTRODUCTION

Hetero-field-effect transisitor (HFET) structures provide the best performance of all three-terminal semiconductor devices for high speed and low noise amplifiers [1]. GaAs-based HFETs have shown good reliability and long term stability, and offer a high potential for modern microwave and millimeterwave communication and radar systems. Improved device performance of AlGaAs/GaAs HFETs is achieved by incorporation of a pseudomorphic (PM) InGaAs quantum well (QW) [2, 3], which is mainly due to the higher achievable two-dimensional electron gas (2DEG) carrier density (n_s). Information about n_s on a device wafer is beneficial in many respects.

In the present work we report about a photoluminescence (PL) analysis on PM HFET structures. The observed PL signal gives information about the energy spectrum of the 2DEG. The spectral width depends linearly on n_s. This allows a non-destructive determination of n_s on wafers for device fabrication. The method is useful for an in-line control of wafers between MBE-growth and device processing. Independently, we have carried out Shubnikov-de Haas (SdH) measurements on the same samples to substantiate our results.

2. LAYERED STRUCTURE, GROWTH AND HALL RESULTS

The epitaxial layers of the PM HFET structures are shown in Fig. 1. They are grown by molecular beam epitaxy (MBE) on semi-insulating (s.i.) GaAs substrates with an AlAs/GaAs superlattice (SL) buffer layer. Various elastically strained $In_xGa_{1-x}As$ QWs are incorporated, with different In-concentrations from nominally 14% up to 27%, as determined by RHEED oscillations on test samples prior to the growth of HFET structures. The well width, L_z, is changed from 7 nm to 17 nm and is kept below the critical thickness

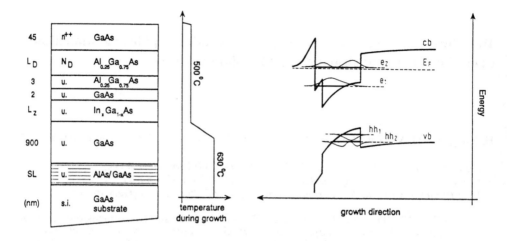

Fig. 1. MBE layer sequence, substrate temperature profile during growth and calculated band diagram of a PM HFET structure with x = 0.18, L_z = 10 nm and n_s = 2*10^{12}cm^{-2}.

of the strained layer. The growth temperature was lowered from 630°C to 500°C before the growth of $In_xGa_{1-x}As$ to avoid a significant desorption of In [4]. The electron supplying layer consists of highly doped 30nm to 42nm thick $Al_{0.25}Ga_{0.75}As$, with Si doping levels between 1.3 and 4x10^{18}cm^{-3}, incorporated in a homogeneously or planar-doped fashion.

For the Schottky barrier depletion PL experiments a 16nm thick Ti/Au semitransparent gate contact and a standard alloyed AuGe ohmic 2DEG ground contact were fabricated. This allows a tuning of n_s and the electron/hole wavefunction overlap via the applied gate voltage. Information about the DC transport behavior was obtained from Hall and SdH measurements performed on patterned samples under perpendicular magnetic field $B_\perp$.

The measurement of n_s and the mobility (μ) of the 2DEG by the Hall method on as-grown samples is complicated by the presence of simultaneous parallel conduction in the doped GaAs cap and the AlGaAs layer. Therefore, a simple Hall measurement on a HFET structure yields parameters that are an average over the entire structure. Information about the 2DEG can be extracted by repeated etch removal (stripping Hall) and surface depletion. Examples are shown in Fig. 2 for a typical PM HFET structure and for a standard AlGaAs/GaAs HFET structure (x = 0). We have achieved maximum n_s-values of 1.0*10^{12}cm^{-2} for standard structures (x = 0) and 2.6*10^{12}cm^{-2} for PM HFET structures (x = 27%, L_z = 12 nm). The results also demonstrate that n_s in PM HFET structures can be increased by more than a factor of 2, in comparison to standard HFET structures, which is also reflected in an improved device performance as discussed below. Measured Hall mobilities on PM HFET structures are between 6000 cm^2/Vs and 7000 cm^2/Vs at 300 K, values similar to those of our standard structures (x = 0).

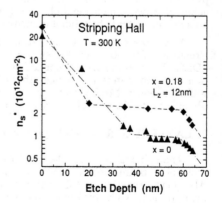

Fig. 2. Hall sheet carrier density as a function of etch depth.

3. PHOTOLUMINESCENCE BEHAVIOUR

Fig. 3 exhibits typical low temperature PL spectra of as-grown PM HFET structures with one (b) and two (a) populated electron subbands, taken with a Fourier Transform spectrometer and a liquid nitrogen cooled Ge-detector under low power excitation with the 488nm line of an Ar^+-laser. They consist of one or two spectral bands and are due to radiative recombinations of electrons from the occupied $n=1$ (e_1), and $n=2$ (e_2) subbands, respectively, with holes in the $n=1$, heavy hole (hh_1) subband. A schematic of the band diagram is shown in Fig. 1. Due to the strong band bending potential the e_2-hh_1 wavefunction overlap integral is several times larger than that of the e_1-hh_1 one, which greatly enhances the E_2 transition, which is parity-forbidden in undoped QWs (see Fig. 3a).

The assignment of the observed transitions is also supported by a calculation of the confining potential, electron and hole subband energies and envelope wavefunctions in a self-consistent way by solving simultaneously the Poisson and Schrödinger equation. Non-parabolicity and strain effects are taken into account. The x-values are fitting parameters. The arrows mark the theoretically expected subband transitions E_1 and E_2 and the n_s-dependent Fermi-energy E_F [5].

The PL intensity increases on the low-energy side within 5 meV and 10 meV (half FWHM) to peak maximum (see Fig. 3). On the high-energy side the spectra broaden significantly, up to a maximum energy which is indicative of E_F and mainly due to band filling effects of the high density 2DEG. For comparison the PL spectrum of a depleted PM HFET structure with negative applied gate voltage on a sample with a semi-transparent Schottky-gate contact is shown in Fig. 3c, which is similar to a PL-signal from an undoped QW. In that case the linewidth is expected to be dominated by inhomogeneous broadening mechanisms due to well width and alloy fluctuations in the ternary material.

The PL transition energies E_1 and E_2 are very sensitive to the In-concentration x and the QW width. We used this dependence to investigate the lateral homogeneity of our two-inch and three-inch PM HFET wafers in PL mapping experiments. The results of a line scan across a typical device wafer with

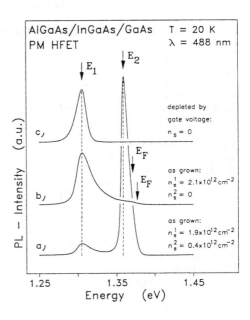

Fig. 3. PL spectra from PM HFETs with a) two and b) one populated subbands and from a depleted PM HFET with applied gate voltage. Curve b) and c) are shifted in energy to match the E_1-transition. N_s-values are from SdH.

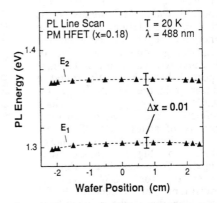

Fig. 4. Energy variation of PL peaks (E_1^*, E_2^*) across PM HFET wafer. The bars indicate expected ΔE for $\Delta x = 0.01$.

x = 0.18 and $L_z = 12$ nm are shown in Fig. 4. From the calculations we obtain: $\Delta E_1(meV) \approx -190 \cdot \Delta x/x$, $\Delta E_2(meV) \approx -175 \cdot \Delta x/x$, $\Delta E_1(meV) \approx -74 \cdot \Delta L_z/L_z$ and $\Delta E_2(meV) \approx -120 \cdot \Delta L_z/L_z$ for x = 0.18 and $L_z = 12$ nm. Therefore, the transition energies in Fig. 4 are mainly influenced by fluctuations in the In-concentration, x. These results demonstrate the excellent homogeneity of the grown material, which is necessary for high-yield device fabrication and large-area microwave and millimeter wave integrated circuits.

4. OPTICAL DETERMINATION OF CARRIER DENSITY

PL spectroscopy has been used successfully to investigate the plasma behaviour and many-body effects of a highly degenerate 2DEG in AlGaAs/GaAs [6-8], AlGaAs/InGaAs [9-11], InP/InGaAs [12, 13], AlInAs/-GaInAs [14, 15] heterostructures and delta-doped GaAs [16]. In the AlGaAs/InGaAs/GaAs PM HFET structure both the 2DEG and the photogenerated holes are confined in the InGaAs QW (see Fig. 1), which greatly enhances the PL efficiency. Additionally, the lower energy gap of InGaAs shifts the transition energies to lower values in comparison to the PL-signals from GaAs and AlGaAs material. This allows a convenient light detection well separated from PL light of the buffer layer and highly doped cap layers.

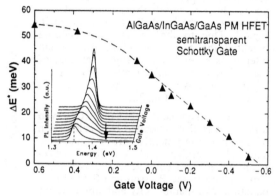

Fig. 5. PL linewidth ΔE^* on gated PM HFET.

In Fig. 3 we have shown that the high-density 2DEG contributes significantly to the linewidth of the PL response. We used samples with a semitransparent Schottky contact to investigate the linewidth dependence on applied gate voltage V_g. In Fig. 5 the measured spectral width ΔE^* versus V_g is shown. We define ΔE^* as the energetic separation between the high-energy cutoff, indicative of E_F, and the maximum of the first spectral band E_1^*. We found a linear relationship $\Delta E^*(V_g)$ for $0.2\ V > V_g > V_{threshold}$, which reflects the two-dimensional density of states (2DDOS) below E_F. Additionally, the E_1^* energy shifts to higher values with negative front gate bias, which is expected for QW subband transitions in depleted HFET structures due to the reduced band bending.

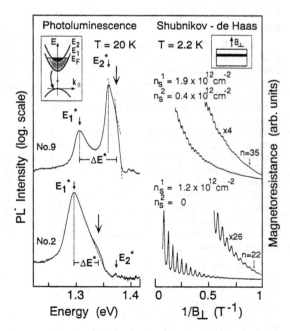

Fig. 6. PL-spectra and SdH-curves of PM HFET.

In Fig. 6 typical PL-spectra from as-grown samples with one (sample No. 9) and two (sample No. 2) occupied electron subbands are drawn on a logarithmic intensity scale, which clearly shows the high-energy cutoffs marked by arrows. The SdH results of the same samples are also drawn in Fig. 6. The measurements were performed at 2.2 K under illumination with a red light-emitting diode. A superposition of two SdH oscillations is observed on sample No. 9. From a fan chart of the oscillation period versus $1/B_\perp$ we can accurately determine the carrier density in each subband. In all samples investigated we found a linear correlation between ΔE^* from PL and n_s from SdH as shown in Fig. 7. This relationship allows a determination of n_s on PM HFET structures from PL in a non-destructive way [5].

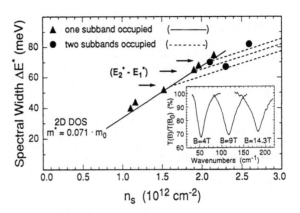

Fig. 7. PL width ΔE^* versus measured 2DEG n_s from SdH from PM HFET samples with one and two occupied electron subbands. The lines are 2DDOS-curves derived from CR. The inset shows CR transmission curves at different magnetic fields.

We have also performed far-infrared transmission experiments on the same samples. Well pronounced dips in the transmission indicate the excitation of cyclotron resonances (CR) (see inset of Fig. 7). From the amplitude and linewidth of the CR we can also determine, with a slightly smaller accuracy as compared to SdH measurements, the total carrier density n_s and found again agreement with the n_s-values determined by PL and SdH. Furthermore, from the resonance position ω_c, the effective cyclotron mass $m^* = eB_\perp/\omega_c$ is determined. Extrapolated to $B_\perp = 0$, a value of $m^* = 0.071 m_0$ is found for the high density PM HFET structures. We use m^* from CR to evaluate the 2DDOS, which is $4\pi m^*/h^2$ in a two-dimensional system. From the PL spectral width we can thus directly determine n_s via 2DDOS. The 2DDOS curves are also drawn in Fig. 7 by dashed lines. The onset of the population of the second subband depends on the energetic separation of the $n = 1$ and $n = 2$ subbands, and therefore leads to individual 2DDOS curves for samples with two occupied subbands due to the different QW composition (x, L_z). The energetic position of the second subband is taken from the E_2^* PL intensity maximum, which is marked by arrows in Fig. 7 for three different samples. The PL linewidth data and the corresponding SdH n_s-values from samples with one (triangles) and two (dots) occupied subbands are drawn in Fig. 7, together with the 2DDOS-curves. The n_s-values deduced from PL-results and 2DDOS curves are in good agreement with n_s-values from SdH experiments within a range of 10%.

5. DEVICE RESULTS

PM HFETs are fabricated from the material mentioned above. Single-gate PM HFETs with x = 0.18, L_z = 12 nm and a gate length of L_g = 0.25 μm show excellent DC, RF and noise performance at room temperature. An extrinsic transconductance g_{mext} = 700 mS/mm and a maximum drain current I_{DSmax} = 600 mA/mm are obtained. The unity current gain frequency (f_T) is 100 GHz, the maximum frequency of oscillation (f_{max}) is 230 GHz. A very low noise figure of 0.6 dB and an associated gain (G_{ass}) of 14 dB are measured at 12 GHz [17]. Dual-gate PM HFETs exhibit even higher gain values. At 12 GHz a maximum stable gain exceeding 22 dB is achieved [18].

For comparision, the following data on AlGaAs/GaAs $(x = 0)$ HFETs $(L_g = 0.25 \ \mu m)$ are obtained: $g_{mext} = 360 \ mS/mm$, $I_{DSmax} = 300 \ mA/mm$, $f_T = 55 \ GHz$ and $f_{max} = 120 \ GHz$ [17]. A minimum noise figure of $0.5 \ dB$ is measured at $G_{ass} = 11 \ dB$ and $12 \ GHz$. The improvement of transconductance and maximum current in PM HFETs is about a factor of 2, which we attribute mainly to the higher achievable n_s in the PM HFET structure.

ACKNOWLEDGEMENT

We would like to thank P. Narozny and J. Wenger for device results. The continuous support of this work by H. Dämbkes is gratefully acknowledged. The work was partly supported by the Bundesministerium für Forschung und Technologie (Bonn, Germany) under contract number NT 2754 2 and ESPRIT Project 5032 AIMS.

REFERENCES

1. For an overview, e.g. Modulation-Doped Field-Effect Transistors, Applications and Circuits, ed. H. Dämbkes, IEEE Press, New York 1991
2. L.D. Nguyen, D.C. Radulescu, M.C. Foisy, P.J. Tasker and L.J. Eastman, IEEE Trans. on Electron Devices 36, 833 (1989)
3. M.Y. Kao, P.M. Smith, P.Ho, P.C. Chao, K.H.G. Duh, A.A. Jabra, and J.M. Ballingall, IEEE Electron Device Lett. 10, 580 (1989)
4. J.P. Reithmaier, H. Riechert, H. Schlötterer, and G. Weimann, Journal of Crystal Growth 111, 407 (1991)
5. H. Brugger, H. Müssig, C. Wölk, K. Kern, and D. Heitmann, accepted for publication in Applied Physics Letters
6. A. Pinczuk, J. Shah, R.C. Miller, A.C. Gossard, and W. Wiegmann, Solid State Commun. 50, 735 (1984)
7. C. Delalande, G. Bastard, J. Orgonasi, J.A. Brum, H.W. Liu, M. Voos, G. Weimann, and W. Schlapp, Phys. Rev. Lett. 59, 2690 (1987)
8. I.V. Kukushkin, K.v. Klitzing, and K. Ploog, Phys. Rev. B37, 8509 (1988)
9. C. Colvard, N. Nouri, H. Lee, and D. Ackley, Phys. Rev. B39, 8033 (1989)
10. W. Chen, M. Fritze, A.V. Nurmikko, D. Ackley, C. Colvard and H. Lee, Phys. Rev. Lett. 64, 2434 (1990)
11. M.S. Skolnick, K.J. Whittaker, P.E. Simmonds, T.A. Fisher, M.K. Saker, J.M. Rorison, R.S. Smith, P.B. Kirby, and C.R.H. White, Phys. Rev. B 43, 7354 (1991)
12. A.F.S. Penna, J. Shah, A. Pinczuk, D. Sivco, and A.Y. Cho, Appl. Phys. Lett. 46, 184 (1985)
13. M.S. Skolnick, J.M. Rorison, K.J. Nash, D.J. Mowbray, P.R. Tapster, S.J. Bass, and A.D. Pitt, Phys. Rev. Lett. 58, 2130 (1987)
14. Y.H. Zhang, D.S. Jiang, R. Cingolani, and K. Ploog, Appl. Phys. Lett. 56, 2195 (1990)
15. Y.H. Zhang, N.N. Ledentsov, and K. Ploog, appears in Phys. Rev. B. (1991)
16. J. Wagner, A. Fischer, and K. Ploog, Phys. Rev. B42, 7280 (1990)
17. H. Dämbkes, P. Narozny, J. Wenger, C. Wölk, B. Adelseck, J. Schroth, K. Schmegner, J. Splettstößer, C. Werres, and A. Colquhoun, Proc. 3rd International Symposium on Recent Advances in Microwave Technology, Reno, Nevada, Aug. 1991, in press
18. J. Wenger, P. Narozny, H. Dämbkes, J. Splettstößer, and C. Werres, Proc. 21st European Microwave Conference, Stuttgart, Germany, Sept. 1991, in press

Inst. Phys. Conf. Ser. No 120: Chapter 3
Paper presented at Int. Symp. GaAs and Related Compounds, Seattle, 1991

155

Investigation of pseudomorphic InGaAs HEMT interfaces

S. Mottet, P.Audren[+], J.M.Dumas, C.Vuchener, J.Paugam and M.P.Favennec[+]

Centre National d'Etudes des Télécommunications, 22300 Lannion, France
[+]Institut Universitaire de Technologie, 22300 Lannion, France

ABSTRACT : Pseudomorphic InGaAs high mobility transistors (PM.HEMTs) which exhibit the best performances are thermodynamically metastable. Investigations on both GaAs/InGaAs and AlGaAs/InGaAs interface vicinities have been performed through deep level characterizations and lifetesting experiments. Results indicate that interfaces of InGaAs strained quantum well HEMTs do not suffer particular degradation after 3000 hours of biased aging tests.

1. Introduction

The development of strained-layer structures greatly increased the material available for microwave device applications and consequently considerable progress is now reported on pseudomorphic InGaAs HEMTs [1]. However, structures which exhibit the best performances are thermodynamically metastable and, in this contribution, the structural stability of such devices during aging is examined for the first time.

2. Device fabrication

The molecular beam epitaxy (MBE)-grown structure of tested PM.HEMTs is illustrated in Fig. 1. Because InGaAs has a lower bandgap than AlGaAs (the bandgaps difference is $\simeq$ 0.55 eV in our case), an AlGaAs/InGaAs quantum well is formed

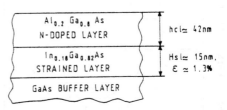

Fig. 1 *MBE-grown structure of tested PM.HEMTs.*

with a conduction band offset $\simeq$ 0.30 eV. Due to the lattice mismatch between InGaAs and GaAs ($\epsilon \simeq$ 1.3 %), the 15 nm thick buried InGaAs layer (hsl) is strained. Consequently, we obtain an AlGaAs/InGaAs strained quantum well (SQW) device.

Based on previously published material studies, this structure is thermodynamically metastable : the equivalent strained overlayer thickness,

15 + 42 = 57 nm , is higher than the critical equivalent overlayer thickness calculated to be 39 nm [2],[3],[4]. Thus, dislocation propagation along the lower interface (single-kink mechanism) or both interfaces (double-kink mechanism) may occur, inducing SQW degradation and drifts in the PM.HEMT performances.

The devices are of a commercially available X-band type fabricated with technological standards : 0.3×200 µm recessed NiAl gate, AuGeNi alloy for ohmic contacts, TiPtAu metallization for overlayers ; finally an Si_3N_4 layer insures the surface passivation. Chips are mounted in hermetically sealed microwave packages.

Typical Noise Figure (NF) and Associated Gain (Ga) of 0.8 and 11 dB are measured repectively for an operating frequency (F) of 11 GHz.

3. Deep level characterizations

The isothermal relaxation technique is used to identify deep levels. The drain current is measured under gate switching conditions. The near pinch-off drain current transient is governed by the free carrier emission and/or capture processes from deep levels located in the reverse bias gate space charge regions, after a zero gate bias filling pulse has been applied (Fig. 2). The full transient curve (1000 discritized values) is measured (Fig. 3), which gives better accuracy and multiexponential discriminating capabilities than the Deep Level Transient Spectroscopy technique (DLTS) which is based only on the measurement of the difference between two discrete points of the curve. Since simultaneous exponential processes may occur, a numerical method discriminates the contribution of each process to the measured transient [5]. Analysis of the Arrhenius plot of the emission coefficients as a function of the measurement temperature yields the deep level characteristics such as energy and capture cross section.

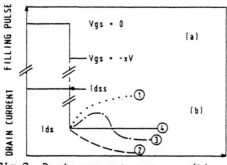

Fig.2 *Drain current response (b) to a gate filling pulse (a) for a device operating in the saturation region. Several processes can be measured : (1) electron emission, (2) electron capture (hole like) or (3) combined processes. (4) is the ideal response.*

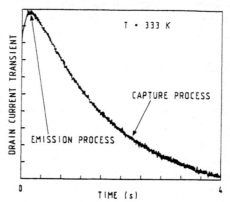

Fig. 3 *Typical transient obtained on a tested PM.HEMT for V_{GS} switching from 0 to -0.7 V at V_{DS}= 2 V.*

Two different gate reverse bias voltages have been used for the
isothermal drain current relaxation measurements. Due the width of the
space charge region, V_{GS}= - 0.5 V (I_{DS}= 10% of I_{DSS}) permits us to
characterize the deep levels in the vicinity of the AlGaAs/InGaAs
interface, while V_{GS}= -0.7 V (pinch-off voltage I_{DS}= 3% of I_{DSS}) permits us
to reach the InGaAs/GaAs interface. In both cases two simultaneous
processes are observed (Fig. 3) : a trapped electron emission from deep
centers and an electron capture by deep centers. The latter mechanism is
often named "hole-like" behavior, but in fact may not involve any hole at
all. The corresponding signatures are deduced from the Arrhenius plots
shown in Fig. 4 and 5. Those figures also show the deep level signature
evolutions during aging.

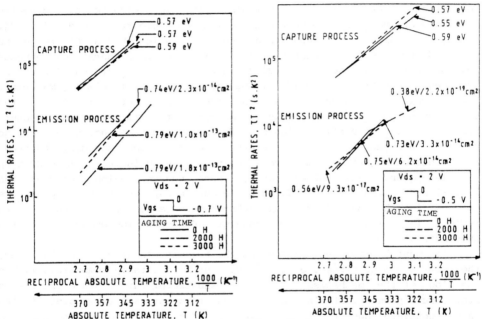

Fig. 4 *Emission and capture rates versus reciprocal temperature as a function of aging time, for V_{GS} switching from 0 to -0.7 V.*

Fig. 5 *Emission and capture rates versus reciprocal temperature as a function of aging time, for V_{GS} switching from 0 to -0.5 V.*

An electron capture process is observed for both gate biases. This
capture process is not observable at lower gate voltage. This would
indicate the location of the mechanism to be in the InGaAs layer vicinity.
The process time constant is temperature dependent and exhibits a related
activation energy E_a= 0.55 ~ 0.59 eV. One can mention that a similar
behavior has been found in the vicinity of the AlGaAs/GaAs interface in a
conventional HEMT [6],[7] by DLTS where the measured activation energies
are respectively 0.49 eV and 0.55 eV . An electron emission process is
simultaneously measured. Its signature parameters are : the energy from the

conduction band $E_a = 0.73 \sim 0.79$ eV and the capture cross-section in the range of $3 \sim 10 \times 10^{-14}$ cm^2. This signature is obviously E_{L2} when located in GaAs [8].

The gate bias and filling pulse avoid the presence of holes during transient measurements. So hole capture or hole emission cannot be explained. In the under gate space charge region, the electron capture process cannot occur since this region is electron depleted during measurement in order to measure the electron emissions. Thus, explaining the "hole-like" signature in HEMTs is impossible when only considering the active layers.

Our previous numerical simulations of the conventional HEMT, including the semi-insulating substrate equilibrium together with electrical conduction under gate bias switching conditions highlighted a dramatic influence of the substrate on the two dimensional electron gas density [9]. First, the simulations confirmed that no "hole-like" signature can be observed from a deep center located in the space charge region under reverse gate bias. Then, we studied the behavior of structures without any deep center excepted E_{L2} (0.72 eV) located in the semi-insulating material. The Fermi-level energy close to the deep center level, in the semi-insulating material, makes comparable the emission and capture processes : E_{L2} is partially ionized. The simulations demonstrated that the electron and the "hole-like" signatures could be simultaneously observed. The electron signature comes from electron emission in the semi-insulating material (and GaAs buffer layer) space charge region. Whereas the "hole-like" signature is induced by electron capture on E_{L2} deep centers located in the semi-insulating substrate outside of the space charge region. This latter mechanism changes the conduction properties of the substrate when biased.

It must be pointed out that in the present case for an InGaAs layer, the same type of signature is found, which is totally compatible with a location of the mechanism in the GaAs substrate. This capture process is stable during aging. It also suggests that this emission takes place in the GaAs substrate and/or buffer layer as for the abovementioned capture process.

4. Lifetesting experiments

In order to assess the stability of the InGaAs buried strained layer during operation, lifetesting experiments have been carried out. The investigation of interface vicinities has been also performed, every 1000 hours, through deep level characterizations.

An homogeneous set of 10 devices issued from the same wafer was biased at the quiescent point for the optimum low noise microwave operation ($V_{DS} = 2$ V, $I_{DS} = 10$ mA). In this contribution, results are presented for an ambient temperature Ta = 175°C leading to a channel temperature Tch $\simeq$ 180°C. Thus, metallurgy induced failure mechanisms which appear for

Tch $\gtrsim$ 230°C are not expected. A set of DC and RF parameters has been measured at 0, 100, 200, 500, 1000 and every 1000 hours.

The measurement conditions of the electrical and microwave parameters related to material properties are reported in table 1.

parameters	measurement conditions
I_{DSS} = 25 mA	V_{DS} = 2 V ; V_{GS} = 0
gm1 = 60 ms	V_{DS} = 2 V ; V_{GS} = 0
gm2 = 45 ms	V_{DS} = 2 V ; I_{DS} = 1 mA
V_{GS}off= 0.55 V	V_{DS} = 2 V ; I_{DS} = 1 mA
I_{GS} < 2 μA	V_{DS} = 2 V ; V_{GS}off
NF = 0.8 dB Ga = 11 dB	V_{DS} = 2 V I_{DS} = 10 mA F = 11 GHz

Table 1 : *Electrical DC and RF parameter measurement conditions.*

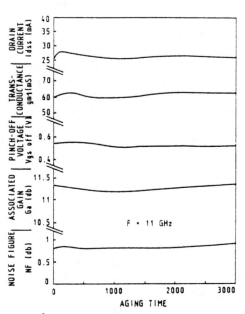

Fig. 6. *Aging characteristics averaged on the set of devices*

The measured parameters remain pratically unchanged over the whole aging experiment, as illustrated in Fig. 6. No metallurgical (Shottky gate contact, ohmic contacts and overlayers) nor surface induced degradation mechanisms have been identified. These results indicate that PM.HEMTs are reliable according to the actual "state of art" [10].

During aging, the E_{L2} signature does not change for V_{GS}= -0.7 V, that is to say close to the GaAs buffer layer. However for V_{GS}= -0.5 V, a change is observed in the measured deep center emission after 3000 hours. The new deep center characteristics are : the energy from conduction band is E_a= 0.56 eV the capture cross-section is 9.3×10^{-17}cm^2 and the concentration is $\simeq 5 \times 10^{14}$cm^{-3} . The same emission signature is still observable when decreasing the gate voltage. This clearly indicates that this deep center appeared in the n-AlGaAs strained layer.

From previous results such a defect has been identified in the first 10 to 20 nm of n-AlGaAs grown on GaAs [11]. According to the authors, this defect may be due to the nonoptimum growth conditions. This point is still under investigation. Other authors have reported structural change in PM.HEMT after rapid thermal annealing [12]. They observed element interdiffusions for InGaAs layer thicknesses greater than the critical thickness. This mechanism is enhanced by misfit dislocations. On the presently studied device, a similar mechanism might be postulated for a lower temperature but longer time (3000 hours) in the presence of an electric field (bias voltage). Therefore there is no significant effect on the electrical parameters since the defects remain in a low concentration.

A new signature appears after 3000 h into aging : $E_a = 0.38$ eV (Fig. 5). The amplitude of the signal is at the limit of detection indicating a very low concentration. There is no indentification of the nature of this deep level center.

5. Conclusion

InGaAs PM.HEMTs are now available, however, structures which exhibit the best performances are thermodynamically metastable. The stability of both InGaAs/GaAs and AlGaAs/InGaAs interface vicinities have been investigated through deep level characterizations and lifetesting experiments. After 3000 hours, a deep level center appears in the n-AlGaAs layer. The defect density remains low enough and does not affect the device performances.

References

1. K.L. TAN, R.M. DIA, D.W. STREIT, T. LIN, T.Q. TRINH, A.C. HAN, P.H. LIU, P-M. D. CHOW and H.C. YEN, IEEE Elect. Dev. Lett. 11 (1990) 585.
2. J.V. MATTHEWS and A.E. BLAKESLEE, J. Cryst. Growth 27 (1974) 118.
3. J.Y. TSAO and B.W. DODSON, Appl. Phys. Lett. 53 (10) (1988) 848.
4. P.S. PEERCY, B.W. DODSON, J.Y. TSAO, E.D. JONES, D.R. MYERS, T.E. ZIPPERIAN, L.R. DAWSON, R.M. BIEFELD, J.F. KLEM and R. HILLS, IEEE Elect. Dev. Lett. 9 (1988) 621.
5. S. MOTTET, P. BELLAN and R. SCHUTTLER, J. of Electostat. 3 (1977) 105
6. T. TAKIKAWA, J. J. Appl. Phys. 26 (1987) 2026.
7. J. GOOSTRAY, H. THOMAS, D.V. MORGAN, E. KOHN, A. CHRISTOU and S. MOTTET, Elect. Lett. 26 (1990) 159.
8. G.M. MARTIN, A. MITONNEAU and A. MIRCEA, Elect. Lett. 13 (1977) 191.
9. S. MOTTET and J.M. DUMAS, CNET contribution to the Commission of European Communities ESPRIT project N°1270, Advanced processing technology for FETs and HEMTs, fifth 6th-month report (1989) 73.
10. S. MOTTET and J.M. DUMAS, in Semiconductor Device Reliability (Kluwer Academic Plublishers, Dordrecht, The Netherlands, 1990) 439.
11. D.J. AS, P.W. EPPERLEIN and P.M. MOONEY, J. of Appl. Phys. 64 (1988) 2408.
12. A. OKUBORA, K. TANAKA, M. OGAWA, J. KASAHARA, T. HAGA and Y. ABE, Inst. Phys. Conf. Ser. N°112 (1990) 447.

Inst. Phys. Conf. Ser. No 120: Chapter 3
Paper presented at Int. Symp. GaAs and Related Compounds, Seattle, 1991

Investigation of transport phenomena in pseudomorphic MODFETS

J. Braunstein, P. Tasker, K. Köhler, T. Schweizer,
A. Hülsmann, M. Schlechtweg, and G. Kaufel
Fraunhofer Institute for Applied Solid State Physics
Tullastrasse 72, 7800 Freiburg, Germany
Phone (+49)-761-5159-534

Abstract

A systematic experimental investigation of the influence of charge modulation on transport phenomena in pseudomorphic $Al_xGa_{1-x}As/In_yGa_{1-y}As/GaAs$ field effect transistors as a function of In mole fraction (y) in the channel is presented. There is evidence that improved charge modulation can totally account for increases in both effective electron velocity and low-field drift mobility with increasing y. Transistor performance was observed to improve by 30% for y=0.25, in comparison to the conventional $Al_{0.3}Ga_{0.7}As/GaAs$ HEMT. A state-of-the-art measured f_T of 120 GHz was achieved with an f_{max} of 200 GHz for y=0.25 with a 0.18 μm gate length.

Introduction

In the past few years the high frequency performance of pseudomorphic MODFETs continued to improve with increasing In mole fraction in the channel. Hence, pseudomorphic MODFETs are an attractive candidate for millimeter-wave, low-noise and power applications. The possible advantages of the pseudomorphic system are i) better charge modulation [1] because of the enlarged conduction band energy offset $(\Delta E_c(y))$ which, with respect to $Al_{0.3}Ga_{0.7}As$, is approximately given by $(247.5 + 800 \cdot y)meV$ for infinite well width and ii) increased electron velocity. These result in improved carrier transport and higher carrier density in the pseudomorphic channel.

Experimental

To allow a systematic study, modulation doped $Al_{0.3}Ga_{0.7}As/In_yGa_{1-y}As/GaAs$ high electron mobility transistors (HEMT), with various In mole fractions and identical supply layer design were fabricated. The layer design is given in figure 1. A 50 Å wide spacer separates a Si-delta-doping of $5 \cdot 10^{12}$ cm^{-2} from the channel. In this investigation the channel width (w) was chosen to be smaller than the critical layer thickness, as determined by Hall mobility measurements [2]. The selected values are given in the following table.

In mole fraction y	0	0.10	0.20	0.25	0.30
Channel width w [Å]	∞	150	130	100	80

Highly reproducible dry recess etching to an AlGaAs etch-stop was utilized to provide a constant gate-to-channel separation, and therefore transistors with identical aspect ratios for all the wafers [3]. This approach eliminates all the geometrical effects on transistor performance, as for example C_{gs} is independent of In mole fraction. For the 0.5 μm gates we measure 1.228 pF/mm with a standard deviation of 0.077 pF/mm for all of the wafers.

Material properties and device results

Solving the Poisson equation for structures with $y > 0$ and various well widths w, and comparing with the case for y=0 gives the following expression for the change in V_{th} :

$$V_{th}(In_y Ga_{1-y} As) - V_{th}(GaAs) = -\frac{\Delta E_c'(y)}{q} + \frac{\Delta E_0'\big(w(y)\big)}{q} = -\frac{\Delta E_{c,eff}'(y,w)}{q}$$

where $\Delta E_c'(y)$ and $\Delta E_0'\big(w(y)\big)$ are the changes in conduction band and ground state energy levels with respect to GaAs channels. The threshold voltage determined by extrapolation of the linear region of DC channel conductance measurements thus reflects the change in effective conduction band energy offset $\Delta E_{c,\text{eff}}'(y)$.

V_{th} becomes lower with increasing In mole fraction, with the same slope as $-\Delta E_c(y)$ ($\to$ figure 2), as long as the channel width can be regarded to be infinite. However, due to the effect of quantization in narrower channels the ground state is rising and therefore reducing the effective conduction band energy offset. This behaviour is observed for a In mole fraction of 0.3, where V_{th} follows no longer the slope since the well is only 80 Å wide.

Two dimensional electron gas (2 DEG) carrier concentration n_{sm}, the maximum 2 DEG sheet density available before the gate begins to significantly modulate parasitic AlGaAs charge, was determined by both Hall measurements on adequate test structures and CV characteristics of MODFETs. n_{sm} was found to increase from $1.2 \cdot 10^{12}$ cm^{-2} for y=0 to $1.9 \cdot 10^{12}$ cm^{-2} for y=0.3, because of increasing conduction band energy offset ($\to$ figure 3).

Differential drift mobility (μ_{diff}) as a function of 2 DEG density was extracted from CV and channel conductance measurements. Maximum μ_{diff} was found to first increase with increasing In mole fraction, then to reach a maximum ($\to$ figure 4), and to decrease for high y; this maximum value being a function of In mole fraction increased from 4200cm^2/Vs for y=0 to 6200cm^2/Vs for 0.2<y<0.25. This is contradictory to Hall mobility which shows a rolloff with increasing y from 7500cm^2/Vs for y=0 to 6200cm^2/Vs for y=0.3. This variation in μ_{Hall} has been explained by increased scattering and increased electron effective mass in channels with higher In[4].

Hall mobility

$$\mu_{Hall} = \frac{\mu_{2DEG}^2 \cdot n_{2DEG} + \mu_{AlGaAs}^2 \cdot n_{AlGaAS}}{\mu_{2DEG} \cdot n_{2DEG} + \mu_{AlGaAs} \cdot n_{AlGaAs}}$$

basically gives the mobility of the 2 DEG because of squaring the high μ_{2DEG}, whereas the drift mobility as determined from CV mesurements

$$\mu_{diff} = \frac{\mu_{2DEG} \cdot n_{2DEG} + \mu_{AlGaAs} \cdot n_{AlGaAS}}{n_{2DEG} + n_{AlGaAs}}$$

is much more sensitive to the AlGaAs, and thus gives an equally weighted average of all different charges. For y=0.1 the two above equations deliver $\mu_{diff} = 5559$ cm^2/Vs and $\mu_{Hall} = 7158$ cm^2/Vs with typical values of $1.2 \cdot 10^{12}$ cm^{-2} for n_s in the 2 DEG and $0.5 \cdot 10^{12}$ cm^{-2} for n_s in the AlGaAs with associated mobilities of 7500 cm^2/Vs in the 2 DEG and 1000 cm^2/Vs in the AlGaAs repectively. These values can be found on the curves of figure 4. Hence, the measured drift mobility is also effected by charge modulation being reduced by parasitic charge modulation in the AlGaAs. As a consequence, when n_{sm} increases ($\rightarrow$ that means increasing the ratio of n_{2DEG}/n_{AlGaAs}) μ_{diff} increases.

Comparing the modulation efficiencies of devices with 0 and 0.15 In mole fraction shows [1] no significant difference in the low current region ($\rightarrow$ low n_s), as there is no significant parasitic charge in the AlGaAs. Plotting μ_{diff} versus n_s for MODFETs with 0 and 0.25 In mole fraction ($\rightarrow$ figure 5) shows identical behaviour for n_s between 0 and $0.5 \cdot 10^{12}$ cm^{-2} where the modulation efficiency is the same. μ_{diff} increases rapidly with inreasing n_s in this low carrier concentration region. This can possibly be explained by increased electron screening reducing the dominant effect of remote ionized impurity scattering for low n_s [4]. In both cases the maximum μ_{diff} is found for $n_s = n_{sm}$ with the above given values. With the higher In mole fraction, the maximum μ_{diff} occurs at substantially higher n_{sm} levels, thus allowing the device to be used for applications requiring a wider range of drain currents.

From the variation of intrinsic f_T with L_g an effective velocity of $1.3 \cdot 10^7$ cm/s and $1.7 \cdot 10^7$ cm/s was extracted for y=0 and 0.25, respectively ($\rightarrow$ figure 6). This improvement is as predicted by the charge modulation theory [1], which assumes identical transport properties but accounts for the effect of reduced parasitic charge modulation in the AlGaAs with higher In mole fraction in the channel.

Transconductance versus current also shows improved g_m and higher saturation current for higher y only when n_{sm} increases. The behaviour of f_T is identical. The improvement of g_m versus drain current is given by figure 7 and the 30% improvement of f_T from 54 GHz to 70 GHz for y=0.25 versus In mole fraction is given by figure 8.

By extrapolating current Gain with a slope of -6 dB per octave of a MODFET with 0.18 μm gate length and y=0.25 an extrinsic f_T of 120 GHz was achieved with an extrinsic f_{max} of 200 GHz.

Conclusion

Investigation of a series of MODFETs with varying In mole fraction in the channel has shown the influence of modulation efficiency on electronic transport properties. A 30% increase of v_{eff} and f_T, and a 50% increase of maximum low-field differential drift mobility was observed with improved modulation efficiency for 0.25 In mole fraction in the channel.

With the present structure, for a reduced gate length of 0.18 μm, a peak f_T of 120 GHz was achieved with 0.25 In mole fraction in the channel. This result was obtained even though the layer structure was not thinned down to provide a fixed aspect ratio and so minimize parasitic capacitance effects.

Acknowledgements

The authors gratefully acknowledge M.-L. Andres, F. Becker, S. Emminger, P. Ganser, K.-H. Glorer, R. Haddad, J. Hornung, P. Hofmann, T. Jakobus, M. Krieg, D.

Luick, T. Norz, E. Olander, B. Raynor, and B. Weismann, for processing of the wafers, and R. Bosch, W. Haydl, J. Hornung, W. Reinert, and J. Rosenzweig for characterizing the wafers and discussions, and H.-S. Rupprecht for encouraging this work and project management.

References

[1] Foisy, M.C., Tasker, P.J., Hughes, B., and Eastman, L.F. " The Role of Inefficient Charge Modulation in Limiting the Current-Gain Cutoff Frequency of the MODFET", IEEE Trans. Electron Devices, **ED-35**, 871-878, July 1988.

[2] Schweizer, T., Köhler, K., Ganser, P., Hülsmann, A., and Tasker, P.J. "Materials and Device Characteristics of Single and Double Sided δ-doped Pseudomorphic $In_y Ga_{1-y} As/Al_{0.3} Ga_{0.7} As/GaAs$ HEMTs $(0 < y < 0.5)$", MRS Fall Meeting 1990, Extended Abstracts, **EA-21**, 305-308.

[3] Hülsmann, A., Kaufel, G., Köhler, K., Schweizer, T., Braunstein, J., Schlechtweg, M., Tasker, P.J., and Jakobus, T. "Mushroom Shaped Gates defined by E-Beam Lithography down to 80 nm Gatelength and Fabrication of Pseudomorphic HEMTs with a Dry-Etched Gate Recess", Proceedings SPIE, Electron Beam, X-Ray, and Ion Beam Technology, San Jose 1991.

[4] Lou, J.-K., Ohno, H., Matsuzaki, K., and Hasegawa, H. "Low Field Transport Properties of Two-Dimensional Electron Gas in Selectively Doped N-AlGaAs/GaInAs-/GaAs Pseudomorphic Structures", Japanese Journal of Applyed Physics, Vol. 27, No. 10, October, 1988, pp. 1831-1840.

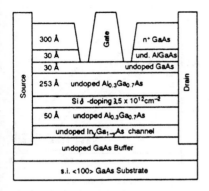

Fig.1: HEMT cross section

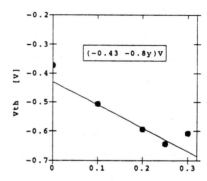

Fig.2: V_{th} versus In mole fraction

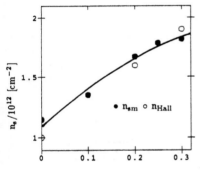

Fig.3: n_{sm} versus In mole fraction

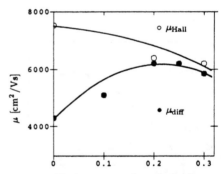

Fig.4: μ versus In mole fraction

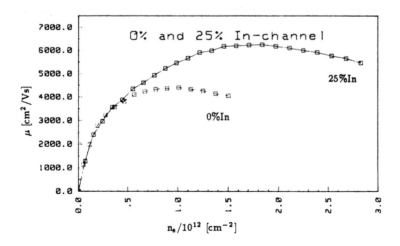

Fig.5: μ_{diff} versus n_s

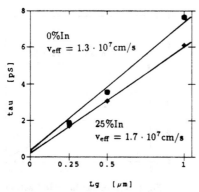

Fig.6: Transit time for HEMTs with
y=0 and y=0.25

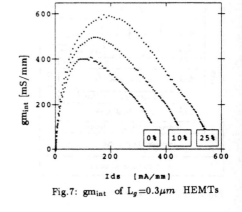

Fig.7: gm_{int} of $L_g=0.3\mu m$ HEMTs

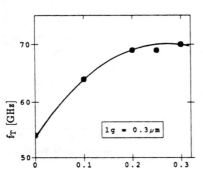

Fig.8: f_T versus In mole fraction

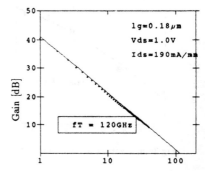

Fig.9: f_T extrapolated from current Gain

Influence of the doping position on the performance of high speed AlGaAs/InGaAs HFETs with doped channels

J.Dickmann, C.Woelk, A.Schurr, H.Daembkes
Daimler-Benz Research Center Ulm
Wilhelm-Runge Straße 11, D-7900 Ulm
H.Nickel, R.Lösch, W.Schlapp
DBP Telekom Research Center, D-6100 Darmstadt

Abstract

This paper presents a theoretical and experimental investigation on the influence of the doping position on the high speed performance of doped channel HFET's. As the result of the investigation we propose a new concept for Doped Channel HFETs: the Backside Pulse Doped Channel HFET. Depending on the doping position we measured for 0.35μm gate length transconductances between 400 and 625mS/mm, saturation currents between 720 and 1100mA/mm and power gain cut-off frequencies between 90GHz and 195GHz.

Introduction

AlGaAs/InGaAs based pseudomorphic MODFET devices have shown to be capable of excellent microwave performance with power gain cut-off frequencies (f_{max}) of 350GHz for 0.1μm gate length [1] and minimum noise figures of 1.9dB at 60GHz [2]. However, MODFET devices with single 2DEG interfaces suffer from only moderate current levels which limits their use in microwave power FET application. Therefore in the past, many attempts have been made to increase the sheet charge concentration in the device channel by introducing new doping techniques. An interesting effective way to further increase the saturation current is intentionally doping the channel. This approach has been several times successfully applied to AlGaAs/InGaAs MODFET devices [3], [4], [5]. The highest saturation current has been reported in [3] for a two channel layer structure having homogeneously doped channels with an equivalent doping sheet carrier concentration of $3x10^{12}cm^{-2}$. For this layer structure, a maximum saturation current of 1A/mm with a transconductance of 530mS/mm were reported for 0.2μm gate length. The cut-off frequencies given are f_T=55GHz and f_{max}=155GHz. For a symmetrically pulse doped channel device with $1.1x10^{12}cm^{-2}$ doping sheet carrier concentration a transconductance of 625mS/mm and a saturation current of 600mA/mm have been reported. For a 0.25μm gate length the current gain cut-off frequency was f_T=90GHz, [5]. This strong difference in device performance raises the question, "What is the best way to introduce the doping in the channel that allows for both high speed and high power operation".

In this paper we report about the theoretical and experimental investigation on the influence of channel doping on the device performance of AlGaAs/InGaAs MODFET's.

Structural Aspects

In order to optimize the device performance it is important to understand the role and importance of the parameters of the device layer structure. Fig.1 shows a schematical drawing of the layer structure used in the investigation.

The model uses a self consistent solution of the Poisson equation and the Schrödinger equation and effective-mass matching techniques. The parameter d_1 is the total distance of the doping pulse from the upper AlGaAs-InGaAs interface. d_2 is the width of the doping pulse and n_{DCH} is the equivalent doping sheet carrier concentration of the doping pulse. The width of the quantum well L_z is kept constant to 12nm. The model used to predict the influence of the channel parameters in Fig.1 on the device characteristics is extensively described in [6].

Since intentionally doping the channel of a MODFET device is in contradiction to the HEMT principle the doped channel FET is therefore in this paper called heterostructure field effect transistor (HFET).

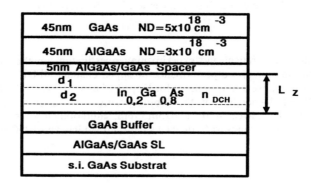

Fig.1:   Layer structure for the doped channel HFET's used in this investigation.

The intention of the HEMT principle is to spatially separate the electrons and their parent donors to reduce Coulomb scattering. In addition, Coulomb scattering by remote parent donors is virtually suppressed by incorporating a thin undoped spacer layer with a typical thickness of 2nm. In a doped channel device, however, the channel is intentionally doped and free electrons and donors interact with each other via the Coulomb potentials.

The effect of the position of the doping pulse can effectively be analysed by profiling the conduction band. Fig. 2 shows conduction band diagrams as a function of applied gate voltage for a channel structure, where the doping pulse is introduced as a 2nm thin pulse (d_2) into the channel at the backside interface of the channel and the buffer layer (d_1=10nm).

In order to reduce the coupling length between the free electrons and the donors in the doping pulse, the width was choosen to be very small, d_2=2nm. The potential of this new approach will become obvious if one considers the position of the electrons in the channel as a function of applied gate voltage. For negative gate bias, the maximum of the electron distribution coincides with the position of the doping pulse (Fig.2a). For this bias point, the electrons of the 2DEG share the same space with the donors in the doping pulse and Coulomb scattering is maximum. But, by increasing the gate voltage towards positive values, the distribution changes entirely (Fig.2b). For positive voltage electrons move towards the upper interface thus generating a continuous increase in spatial separation from the donors in the channel. At V_G=+0.5V this separation is in the order of 6nm. This distance is equal to three times the thickness of a normal spacer layer currently introduced in HFET's at the upper interface to suppress Coulomb scattering. As a result, the electrons in the channel can be expected to have "HEMT-like" transport properties for this gate bias. This means, that the transport properties of the electrons in the channel can be modulated by the gate voltage between "MESFET-like" or "HEMT-like".

The next point that should be clarified is the influence of d_1 and d_2 on the total sheet carrier concentration of the 2DEG. Fig.3a,b, show the total 2DEG concentration as a function of d_1 and d_2, respectively. From Fig.3 one common effect can be deduced. Shifting the doping pulse towards the backside interface increases the 2DEG concentration from approximately $2.5 \times 10^{12} \text{cm}^{-2}$ up to $3.2 \times 10^{12} \text{cm}^{-2}$. The calculated concentration for the layer structure without channel doping is given in the figure for comparative purpose. The explanation of this effect is as follows. From band diagram profiling we found that the backside part of the quantum well flattens and the energy levels of the subband energies drop, allowing for higher electron population.

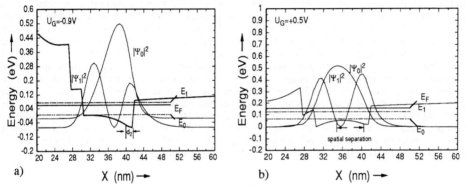

Fig.2:   Conduction band, subbandenergies and square amplitude of electron wave function for different applied gate voltage. a)V_G=-0.9V, b)V_G=+0,5V

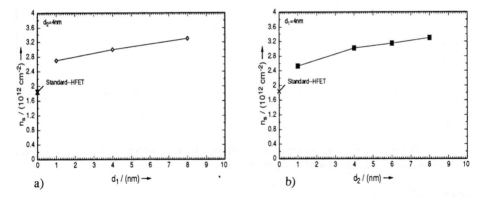

Fig 3:   Total 2DEG concentration as a function of a) distance d_1 of the doping pulse from the upper interface and of b) the width of the doping pulse d_2.

For these calculations either the width of the doping pulse was kept constant to d_2=4nm or the distance d_1 was kept constant to 4nm. The sheet carrier concentration n_{DCH} was kept to 1.6×10^{12}cm^{-2}. For high speed and low noise devices it is desirable to have only the ground state occupied, because electrons in higher states exhibit lower mobilities and possible occupation of higher states would lead to intersubband scattering. Another effect is increased probability of carrier injection into the buffer layer resulting in an undesired increase of the output conductance. For this reason it is important to determine the influence of d_1 and d_2 on the onset of the population of the second subband E_1. We considered the sheet carrier concentration of the 2DEG at which the occupation of E_1 occurs as the figure of merit and defined it n_{sc}.

Fig.4 a and Fig.4b show the dependence of n_{sc} on the distance d_1 of the doping pulse from the upper interface and the width of the doping pulse d_2, respectively. In the calculations either d_2 or d_1 was kept constant to 4nm. The pulse doping concentration was set to 1.6×10^{12}cm^{-2}. Fig.4 clearly demonstrates, that population of the second subband occurs earlier with shifting the channel doping towards the backside interface and as a result intersubband scattering becomes more pronounced. The next information that can be obtained is, that occupation of the second subband is expected for each doped channel device, but the onset of the population can be controlled by d_1 and d_2. In high speed and low noise devices intersubband scattering should be avoided or at least be postboned at a gate bias point that is far beside the optimum bias point of operation. From Fig.4 this requires the introduction of the doping pulse in the vicinity of the upper interface. This requirement is unfortunately in contradiction to the demand for a large value of d_1 to reduce Coulomb scattering and to obtain a high n_s. We consider intersubband scattering of minor priority compared to the demand for low Coulomb scattering and high n_s. Therefore the channel parameter d_1=10nm, d_2=2nm were choosen for device fabrication.

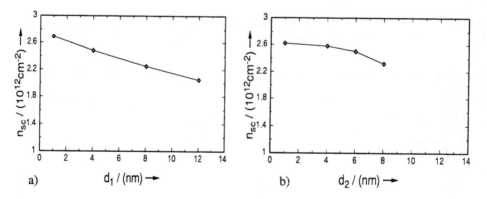

Fig 4:   The sheet carrier concentration nsc of the 2DEG at which the occupation of E_1 occurs as a function of a) d_1 and b) d_2.

Experimental results:
Fig.5 shows typical IV-characteristics of a centrically and backside pulse doped channel device with $L_G \times W_G$=0.35μm$\times$60μm and d_1=d_2=4nm, n_{DCH}=2$\times$10^{12}cm^{-2} and d_1=10nm, d_2=2nm, n_{DCH}=1$\times$10^{12}cm^{-2}. The transconductance of the centrically doped device is 400mS/mm and the maximum saturation current is 1.1A/mm at V_{GS}=+0.8V and V_{DS}=2.5V. This saturation current is higher than the best values reported so far [5]. For the backside pulse doped device, we measured a maximum transconductance of 625mS/mm and a maximum saturation current of I_{DSS}=720mA/mm at V_{GS}=+0.8V. These device results demonstrate the superiority of the backside doped channel approach in high speed

performance. The saturation current is lower due to the reduced doping concentration used in the channel, but considerably higher than for conventional HFET's.

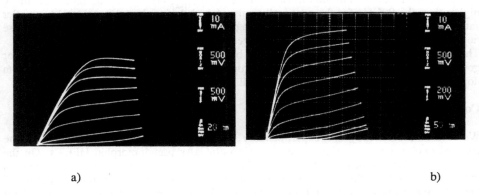

a)   b)

Fig.5:   Typical IV-characteristic of a 0.35µmx60µm doped channel device with
           a)d_1=d_2=4nm, n_{DCH}=2x10^{12}cm^{-2} and b) d_1=10nm, d_2=2nm, n_{DCH}=1x10^{12}cm^{-2}.

Microwave characterisation was performed from 0.45GHz to 26.5GHz using an on-wafer Cascade Microtech probe station and an HP8510 network analyzer. The devices measured were conditionally stable (i.e. stability factor k<1) over the entire measurement frequency range. Therefore we used the maximum unilateral gain (MUG) to determine the power gain cut-off frequency f_{max} with a -20dB/dek roll off. The device S-parameters and the saturation current were measured at different bias conditions V_{GS}, V_{DS} and f_{max} was determined as described above. Fig.6 shows f_{max} determined as a function of the measured saturation current I_{DS}.

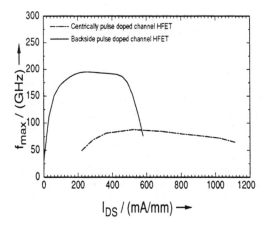

Fig.6: f_{max} for centrically and backside doped channel devices.

It can clearly been seen that the centrically pulse doped device is able to operate at a very high saturation current in excess of 1A/mm, but at the expense of high speed performance. The value of f_{max}=90GHz is very low for a 0.35µm gate length device. Devices with a lower channel doping density resulted in slightly higher f_{max}, but considerably smaller I_{DSS}. Drastically improved high speed performance can be observed from the backside pulse doped channel device. For this device a maximum f_{max}=195GHz was measured much higher than the 90GHz for the centrically pulse doped device, but at considerably smaller saturation current. The current gain cut-off frequency for the centrically and backside pulse doped device are f_T=40GHz and f_T=85GHz, respectively.

These results underline the existing trade off between either very high saturation current or good high speed performance. Further improvements are expected for either using delta doping or/and by using a thicker quantum well thickness L_z.

Conclusion

We studied theoretically and experimentally the influence of the doping position of the channel doping on the performance of AlGaAs/InGaAs pseudomorphic HFET's. On the basis of conduction band profiling and charge control modelling, a layer structure is proposed, that allows the gate voltage controlled modulation of the transport properties of the electrons between "MESFET like" and "HEMT like". For a centrically pulse doped channel HFET a record saturation current of 1.1A/mm was measured.

This work was supported by the German Ministery for Research and Technology under Contract NT 2754-2.

References

[1]   L.F.Lester, P.M.Smith, P.Ho, P.C.Chao, R.C.Tiberio, K.H.G.Duh, E.D.Wolf, "0.15µm Gate-Length Double Recess Pseudomorphic HEMT with Fmax of 350GHz", IEDM Tech.Dig., pp.172-175, 1988

[2]   P.C.Chao, M.S.Shur, R.C.Tiberio, K.H.Duh, P.M.Smith, J.M.Ballingall, P.Ho, A.A.Jabra , "DC and Microwave Characteristics of Sub-0.1µm Gate-Length Planar-Doped Pseudomorphic HEMT's ", IEEE Trans. Electron Devices, vol.36, No.3, pp.461-471, 1989

[3]   P.Saunier, H.Q.Tserng, "AlGaAs/InGaAs Heterostructures with Doped Channels for Discrete Devices and Monolithic Amplifiers", IEEE Trans.Electron Dev., vol.ED-36, No.10, pp.2231-2234, 1989

[4]   P.Saunier, R.J.Matyi,K.Bradshaw, "A Double--Heterojunction Doped--Channel Pseudomorphic Power HEMT with a Power Density of 0.85W/mm at 55GHz", IEEE Electron Device Letters, vol.EDL-9, No.8, pp.397-398, 1988

[5]   P.M.Smith, L.F.Lester, P.C.Chao, P.Ho, R.P.Smith, J.M.Ballingall, M.Y.Kao, "A 0.25- m Gate-Length Pseudomorphic HFET with 32mW Output Power at 94GHz", IEEE Electron Device Letters., vol.EDL-10, No.10, pp.437-439, 1989

[6]   P.Roblin, H.Rohdin, C.J.Hung, S.W.Chiu, "Capacitance-Voltage and Current Modelling of pulse doped MODFET's", IEEE Trans.Electron Dev., vol.ED-36, No.11, pp.2394-2404, 1989

Inst. Phys. Conf. Ser. No 120: Chapter 3
Paper presented at Int. Symp. GaAs and Related Compounds, Seattle, 1991

173

A p-channel GaSb heterojunction field-effect transistor based on a vertically integrated complementary circuit structure

Kanji Yoh, Kazumasa Kiyomi, Hiroaki Taniguchi, Mitsuaki Yano, and Masataka Inoue

Department of Electrical Engineering, Osaka Institute of Technology

5-16-1 Omiya, Asahi-ku, Osaka 535, JAPAN

abstract

We report on the fabrication and characterization of GaSb p-channel HFETs (Heterojunction Field-Effect Transistors) based on a vertically integrated $(Al_{.5}Ga_{.5})Sb/$ $InAs/(Al_{.5}Ga_{.5})Sb/$ $GaSb/(Al_{.5}Ga_{.5})Sb$ double quantum well heterostructure grown by molecular beam epitaxy (MBE). The paper mainly deals with the structural design and fabrication considerations of the p-channel HFETs buried under the n-channel HFETs. The operation of the integrated p- and n-channel HFETs has been demonstrated for the first time showing decent I-V characteristics. Maximum transconductances of the buried p-HFETs ranged 22mS/mm to 50mS/mm at 77K. A p-HFET operation of enhancement mode has been confirmed in the stratified structure, which is an important step toward the realization of the real complementary circuits based on antimonides.

I.INTRODUCTION

Complementary HFET circuits based on the (AlGa)As/GaAs heterostructures or related compounds (Yoh 1987, Ruden 1989) have been shown to have advantages over the silicon counterpart by a factor of 3 to 4 in the speed-power product (Yoh 1988) at 77K operation. Unfortunately, the superiority of a facor of 3 to 4 does not seem to have been enough to stimulate the efforts to develop the complementary HFET (CHFET) technology which could replace silicon CMOS devices. Heterostructures of antimonides are intersting system in both physics (Sai-Halasz 1978, Chang1980) and bandgap engineering point of view. InAs/(AlGa)Sb heterostructures have advantages of high low-field mobility (Tuttle 1989), high-lying satellite valleys, deep quantum well and are suitable for high performance InAs channel field-effect transistors (Tuttle 1987, Luo 1989, Yoh 1990a). GaSb/(AlGa)Sb heterostructures, on the other hand, is suitable for high performance GaSb channel MODFETs (Luo 1990). Both types of high performance HFETs are based on antimonide heterostructures with lattice mismatch of 1.3% maximum. Vertical combination of these high performance p- and n-HFETs would lead to a CHFET circuitry (Longenbach 1990) which would outperform complementary AlGaAs/(InGa)As HFETs. We have investigated the structural design and fabrication method of GaSb p-channel HFETs based on vertically integrated complementary circuit structures. In this paper, the electrical characteristics dependences on the structural and process scheme of the p-channel HFETs buried under the InAs n-channel HFETs is reported.

II. STRUCTURAL AND PROCESS DESIGN

The basic strategy is to integrate standard heterostructures of n- and p-channel antimonide-based HFETs, i.e., GaSb/(AlGa)Sb/InAs/(AlGa)Sb and GaSb/(AlGa)Sb/GaSb/ (AlGa)Sb heterostructures respectively as shown in Figures 1(a) and (b). The basic

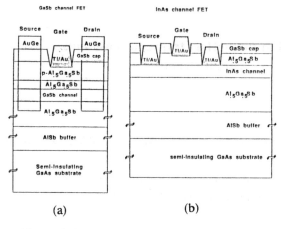

(a)   (b)

Figure 1. Schematic heterostructure diagram of "standard" antimonide-based HFETs: (a) a GaSb p-channel HFET structure, (b) an InAs n-channel HFET structure.

design issues are (i) the order of stratifying n- and p-channel, (ii) how to improve the ohmic contact without GaSb cap layer of the buried p-HFET, (iii) whether or not to maintain even the barrier layer between two channels, and if we do have to keep the barrier layer, (iv) choice of thickness and doping type and concentration of the barrier material. All these issues are closely related with the process design and material limitations. In addition, the following boundary conditions also have to be taken into account : (i) the not-intentionally-doped donors in (AlGa)Sb layer near the (AlGa)Sb/InAs interface (Tuttle 1989), (ii) the only one dopant species in our system, silicon, acts as a donor in InAs layer and as an acceptor in GaSb (Rossi 1990) and (AlGa)Sb layers.

For the better process control and uniformity it is desirable to minimize the thickness of the barrier layer between the channels and to use InAs channel as an etch stop for we can use the high selectivity (Yoh 1990b) in the etching rate of AlGaSb over InAs. Secondly, we do need to maintain the (AlGa)Sb barrier layer between the two channels to avoid the carrier transfer. These requirements force one to investigate vertically integrated structure with InAs well on top of the GaSb well as shown in Figure 2. The proposed structure has an additional

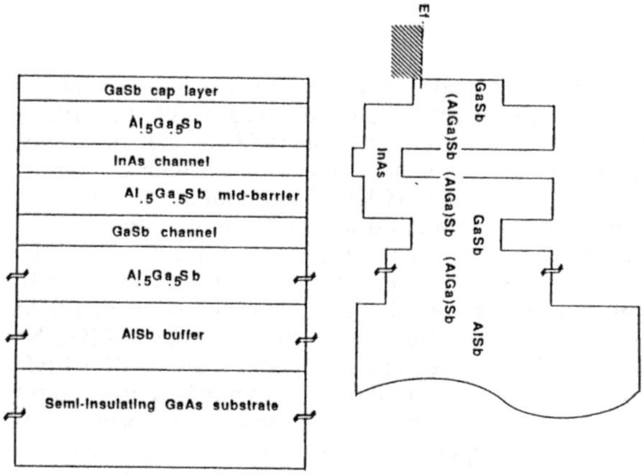

(a)   (b)

Figure 2. Schematic diagram of vertically integrated InAs channel HFET and GaSb channel HFET structure. Heterostructure is shown in (a) and energyband diagram in flatband condition without doping is shown in (b).

advantage that the non-alloyed ohmic contact to the n-channel can be used as an alternative technology to alloyed ohmic contact which may result in the formation of parasitic p-HFET under the n-HFET. Based on the initial experimental results of the standard GaSb p-channel HFETs, which will be discussed later, it turned out that the single Au/Ge alloy process can be used for ohmic contacts of both p- and n-HFETs simultaneously probably because of the amphoteric nature of germanium atoms just as silicon in III-V materials.

First, we will investigate the standard p-FET (structure "A") which would serve as a reference to buried p-HFETs fabricated based on various CHFET structures in evaluating the electrical characteristics. Then, we will investigate p-HFETs in a vertically integrated CHFET structure with acceptors doped in the mid-barrier as a natural combination of "standard" p- and n-HFETs (structure "B"). Finally, we will investigate a CHFET structure without intentional doping in the mid-barrier (structure "C"). In the following two sections, these three types of GaSb p-HFETs based on the three basic structures "A","B", and "C" , as shown in Figure 3 , are described.

III. FABRICATION

The structure "A" consists of 2000Å of GaAs buffer layer grown on undoped GaAs substrate, 1μm of AlSb buffer layer, 2000Å of $(Al_{.5}Ga_{.5})Sb$ bottom barrier layer, 150Å of GaSb channel, 100 Å of $(Al_{.5}Ga_{.5})Sb$ spacer layer, 350 Å of p-$(Al_{.5}Ga_{.5})Sb$ layer, and 100 Å of GaSb cap layer. Silicon atoms with the concentration of $1.0 \times 10^{18} cm^{-3}$ were doped in the $(Al_{.5}Ga_{.5})Sb$ barrier layer in order to provide holes to the GaSb channel. This structure represent the standard GaSb/AlGaSb p-HFET structure designed from the analogy of GaAs/AlGaAs p-HFET. The present structure differs from the work of Luo *et al* (1990) in its barrier material and is basically the same exploratory structure as the vertically integrated complementary circuit structures "B" and "C" when the surface n-channel layers are removed. Ohmic contacts to the p-channel FETs were done by deposition and alloying of Au/Ge/Ni/Au. Then, gate recess etching was carefully done in order to etch-off the GaSb cap layer.

The structure "B" consists of 2000Å of GaAs buffer layer grown on undoped GaAs substrate, 1μm of AlSb buffer layer, 2000Å of $(Al_{.5}Ga_{.5})Sb$ bottom barrier layer, 150Å of GaSb, 30Å of $(Al_{.5}Ga_{.5})Sb$ spacer layer, 150Å of p-$(Al_{.5}Ga_{.5})Sb$ layer, 150Å of $(Al_{.5}Ga_{.5})Sb$ layer, 150 Å of InAs, 150 Å of $(Al_{.5}Ga_{.5})Sb$, and 100 Å of GaSb cap layer. Silicon atoms with the concentration of $5.0 \times 10^{18} cm^{-3}$ were doped in the middle of the $(Al_{.5}Ga_{.5})Sb$ layer which separates InAs and GaSb channels. This doping was intended to serve for three purposes: (i) to provide holes

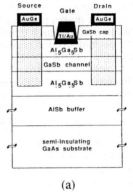

(a)

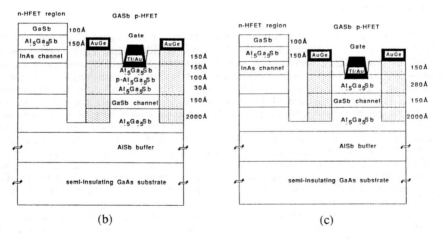

(b)   (c)

Figure 3. Three basic structures of GaSb p-HFETs: Structure "A" represent the standard GaSb p-FET, structure "B" represents p-HFETs in a vertically integrated CHFET structure with acceptors doped in the mid-barrier as a natural combination of "standard" p- and n-HFETs, structure "C" represents p-HFETs in a CHFET structure without intentional doping in the mid-barrier.

to the GaSb channel, (ii) to ameliorate the ohmic contact to the p-channel, (iii) and to increase the threshold voltage of the n-HFET by compensating the not-intentionary-doped donors near the InAs/AlGaSb interface. The 150Å of undoped AlGaSb layer inserted between the InAs channel and the p-AlGaSb charge supplying layer is intended to help preventing silicon outdiffusion to the InAs channel during the growth. Initially, all the surface layers of the p-channel FET region were selectively etched down to the upper interface of the InAs layer. Ohmic contacts to the p-channel FETs were done by deposition and alloying of Au/Ge/Ni/Au multi-layer metals. Surface InAs layer helps to obtain low contact resistance to the GaSb channel. Then, gate recess etching was carefully done in order to etch-off not only InAs layer but also the $(Al_{.5}Ga_{.5})Sb$ surface layer which contains large amount of not-intentionally-doped donors near the InAs interface. Finally, Ti/Au is deposited as a Schottky gate metal.

The structure "C" consists of 2000Å of GaAs buffer layer grown on undoped GaAs substrate, 1μm of AlSb buffer layer, 2000Å of $(Al_{.5}Ga_{.5})Sb$ bottom barrier layer, 150Å of GaSb, 280Å of $(Al_{.5}Ga_{.5})Sb$ layer, 150 Å of InAs, 150 Å of $(Al_{.5}Ga_{.5})Sb$, and 100 Å of GaSb cap layer. Unlike structure B, present structure contains no intentional doping. This structure makes p-HFET to operate in enhancement mode and allows wider process margin because of the small threshold voltage dependence on the gate recess depth. Surface layers of the p-channel FET region were selectively etched down to the upper interface of the InAs channel. Ohmic contacts to the p-channel HFETs were done by deposition and alloying of Au/GeNi/Au multi-layer metals. Surface InAs layer helps to obtain low contact resistance to the GaSb channel. Gate recess etching and the following Shottky metal deposition are done in the same way as the structure "B".

Selective etching of the GaSb and (AlGa)Sb layers is the key technique for both precise control of the gate recess etch and the non-alloyed ohmic contacts (Yoh 1990b) to the InAs thin layer for the n-chnnel FETs. In the present structure, the InAs layer serves not only as the channel layer of the n-channel FET but as the cap layer of the p-channel FETs which improves ohmic contact to the GaSb channel layer as will be discussed in detail in the next section. Sructures "B" and "C" allows the minimul thickness of the AlGaSb layer which separates p- and n-channels, which in turn makes the vertical integration practical.

IV. RESULTS AND DISCUSSIONS

The hole mobility and sheet carrier concentrations of structure "A" by van der Pauw method were 177 $cm^2/Vsec$ and $4.37 \times 10^{12} cm^{-2}$ at 300K and 1,526 $cm^2/Vsec$ and $2.47 \times 10^{12} cm^{-2}$ at 77K, respectively. A typical current voltage characteristics is shown in Figure 4. Maximum transconductance of a p-HFET with the gate length of 1.7μm was 50mS/mm at 77 K. Contact resistance of the p-channel FET with InAs cap were 0.2 Ωmm. It has been shown that germanium alloy can be used as a ohmic contact material for the GaSb based p-channel HFET devices as effective as in InAs n-channel HFETs. This opens up the way to the simplified process technology for the antimonide-based CHFETs. The FETs operated in depletion mode with threshold voltage of 0.08V. Theoretically, the threshold voltage of each device can be adjusted by controlling the depth of gate recess etching, but it should be noted that the controllability over many devices with a variety of gate lengths becomes very tight. Reasonably promising results were obtained in the "standard" GaSb p-HFET and the validity of the AuGe alloy as an ohmic metal to the p-HFET as well as n-HFET has been confirmed.

The hole mobility and sheet carrier concentrations of structure "B" by van der Pauw method were 277 $cm^2/Vsec$ and $2.16 \times 10^{13} cm^{-2}$ at 300K and 744 $cm^2/Vsec$ and $1.30 \times 10^{13} cm^{-2}$ at 77K, respectively. Reduced hole mobility at 77K may be caused by the outdiffusion of the silicon atoms from the heavily doped mid-barrier. The heavy doping was intended to increase the supplied carriers in the channel. The possible outdiffusion to the upper InAs channel was also suspected for the reduces electron mobility at 77K in the InAs channel layer. A typical current voltage characteristics is shown in Figure 5. Maximum transconductance of a p-HFET with the gate length of 2.3 μm was 26 mS/mm at 77 K. The FETs operated in depletion mode with threshold voltage of 0.37V. Contact resistance of the

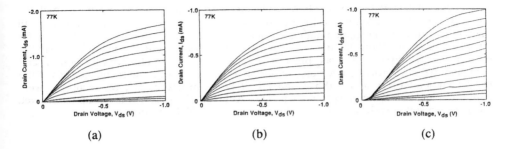

(a)   (b)   (c)

Figure 4. I_{ds}-V_{ds} characteristics of the various p-channel HFETs. Figures of (a), (b) and (c) correspond to structures "A", "B", and "C". Measured gate length of each structure were 1.7μm, 2.3μm, and 1.0μm respectively. V_{gs} of all devices was changed by -0.1V step starting from 0.2V, 0.2V and 0V respectively.

p-channel FET with InAs cap were also 0.2 Ωmm but it should noted that this value represent the contact resistance to the surface InAs layer as well as GaSb channel layer. By this experiment of structure "B", it has been varified that the operation of both InAs n-HFET and GaSb p-HFET vertical integratied on a single wafer. We believe that this indicates one more step advance toward the realization of the antimonide-based CHFETs.

The ungated Hall-effect measurements of holes in the structure "C" was impossible because the carriers were depleted as expected. The Hall-effect results by van der Pauw method for electrons on the top InAs channel were 7900 cm²/Vsec and 2.53x10¹²cm⁻² at 300K and 10,300 cm²/Vsec and 2.04x10¹²cm⁻² at 77K, respectively. We believe the non-idealistic numbers do not represent the the effect of the underlying p-channel layers, but simply an indication of present status of our MBE machine. A typical current voltage characteristics of the p-HFETs is shown in Figure 4. Maximum transconductance of a p-HFET with the gate length of 1.0μm was 22mS/mm at 77 K. The p-FET of structure "C" operated completely in enhancement mode with threshold voltage of -0.22V. Contact resistance to the GaSb channel cannot be separated from the contact resistance to the InAs cap layer as was the case with structure "B". By this experiment of structure "C", it has been varified that not only the operation of both InAs n-HFET and GaSb p-HFET vertically integratied on a single wafer, but also enhancement mode p-HFET buried under InAs n-channel HFET is feasible. The final task regards to the threshold voltage adjustment of InAs n-channel HFET, which is beyond the scope of this paper.

In the vertically integrated structure (structure "B" and "C"), ohmic contact was made small enough by maintaining the InAs layer as a cap layer for the p-HFETs. The experimental results, which is not shown here, that both drain current and transconductance of the "InAs-stripped" p-HFET were greately reduced compared with the "InAs-capped" p-HFET was probably caused by the degraded contact resistance due to the high barrier height of the AlGaSb Schottky gate. In Figure 5, the effectively reduced barrier height of the "InAs-capped" p-HFETs is schematically depicted and compared with the "InAs-stripped" structure.

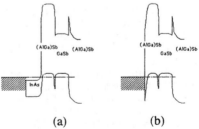

(a)   (b)

Figure 5. Schematic energyband diagram of ohmic contact region in (a) "InAs-capped" p-HFETs and (b) "InAs-stripped" p-HFETs.

The I_{ds}-V_{gs} characteristics and the transconductances of the p-HFETs in the three structures "A", "B", and "C" are shown in Figures 6(a) and (b), respectively. We notice that both transconductance and maximum drain current of the stratified structures are smaller than the "standard" structure. Presently, it is not clear that it comes from the structural difference or the material quality difference. Good pinch-off characteristics were observed for both n- and p-channel HFETs vertically integrated on the same wafer. Typical transconductances of the InAs channel HFETs and GaSb channel HFETs in structure "C", for example, were 88mS/mm and 22mS/mm, respectively at 77K. Although the structure is not optimized yet, the present results clearly show the feasibility of the CHFETs based on antimonides.

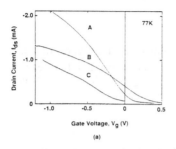
(a)

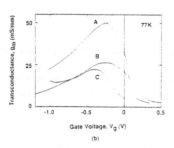
(b)

Figure 6. I_{ds}-V_{gs} characteristics (a) and transconductances (b) of the p-HFETs in the three structures, "A", "B", and "C".

V. SUMMARY

The structural design and fabrication process of the GaSb p-HFETs vertically integrated with InAs n-HFETs has been investigated. Maximum transconductances of the buried p-HFETs ranged from 22mS/mm to 50mS/mm at 77K. A p-HFET operation of enhancement mode as well as the n-channel HFETs fabricated on the same wafer has been confirmed, which is an important step toward the realization of the real complementary circuits based on antimonides.

VI. ACKNOWLEDGMENTS

The authors are grateful to H.Kawahara, T.Mizuguchi for the technical assistance of the device fabrication. This work was supported in part by Grant-in-Aid for Scientific Research on Priority Area from The Ministry of Education, Science and Culture.

REFERENCES

Chang L L, Kawai N, Sai-Halasz G A, Ludeke R, and Esaki L 1980, *Appl.Phys.Lett.* **35**, 939
Luo L F, Beresford R, Wang W I, and Munekata H 1989, *Appl.Phys.Lett.* **55**, 789 (10)
Luo L F, Longenbach K F, and Wang W I 1990, *IEDM Technical Digest*, San Fransisco (IEEE, New York,1990) pp.515-8
Longenbach K F, Beresford R, and Wang W I 1990, IEEE Trans. Electron Devices **37**, No.10, 2265
Rossi T M, Collins D A, ChowD H, and McGill T C 1990, *Appl.Phys.Lett.* **57**, 2256
Ruden P P, Akiwande A I, Narum D, Grider D E, and Nohava J 1989, *IEDM Technical Digest*, Washington, D.C.(IEEE, New York,1990), pp.117-120
Sai-Halasz G A, Esaki L and Harrison W A 1978, Phys. Rev. B **18**, 2812
Tuttle G and Kroemer H 1987, *IEEE Trans. Electron. Devices* , **ED-34**, p2358
Tuttle G, Kroemer H, and English J H 1989, *J. Appl. Phys.* **15**, 5239
Yoh K, Harris J S 1987, *IEDM Technical Digest*, Washington, D.C. (IEEE, New York, 1987), pp.892-4
Yoh K 1988, Ph.D Thesis, Stanford University
Yoh K, Moriuchi T, and Inoue M 1990a, *IEEE Electron Device Lett.* **11**, no.11, pp 526-8
Yoh K, Moriuchi T, and Inoue M 1990b, *Jpn.J.Appl.Phys.* **29**, No.12, L2445

Inst. Phys. Conf. Ser. No 120: Chapter 4
Paper presented at Int. Symp. GaAs and Related Compounds, Seattle, 1991

179

Thermally stable In-based ohmic contacts to p-type GaAs

P.-E. Hallali[a], Masanori Murakami[b], W. H. Price, and M. H. Norcott

IBM T. J. Watson Research Center, Yorktown Heights, New York 10598, USA
a) Present address: IBM France, Corbeil-Essonnes, Cedex, France
b) Present address: Department of Metal Science and Technology, Kyoto University, Sakyo-ku, Kyoto 606, Japan

ABSTRACT: Thermally stable, low resistance p-type ohmic contacts have been developed by depositing NiInW alloys on GaAs substrates in which Be and F were co-implanted. The contacts provided resistances of about $0.6\,\Omega$-mm after annealing at temperatures in the range of 300 to 800 °C for short times. The electrical properties did not deteriorate after annealing at 400 °C for more than 100 hrs, which far exceeds the requirements for current GaAs device fabrication. The present study demonstrates, for the first time, that thermally stable, low resistance ohmic contacts to both n- and p-type GaAs can be fabricated using the same metallurgy.

1. INTRODUCTION

Typical commercial p-type ohmic contact materials contain Zn, Mn, Be, or Mg. These contacts provide low contact resistances. However, most contacts have poor thermal stability after contact formation, caused by a large diffusion of acceptors in the GaAs substrates and formation of low melting point compounds at the metal/GaAs interfaces. The purpose of the present study was to prepare thermally stable, low resistance NiInW ohmic contacts to p-type GaAs. The reason we selected this contact metal systems is that it provided low contact resistance to n-type GaAs and excellent thermal stability during isothermal annealing at 400 °C after contact formation (Murakami et al, 1987). This low resistance is believed to be due to separation of the high barrier at the metal/n-GaAs contact into two low barriers at the metal/In$_x$Ga$_{1-x}$As and the In$_x$Ga$_{1-x}$As/GaAs contacts (Muralami et al, 1988). The excellent thermal stability was due to a lack of low melting point compound formation at the metal/GaAs interfaces. Therefore, thermally stable, low resistance contacts to p-type GaAs are expected for NiInW systems.

2. EXPERIMENTAL PROCEDURES

The Be implant to GaAs substrates was carried out at 20 keV with a dose of 10^{15} cm^{-2}. For the wafers denoted as "Be+F" the F$^+$ was subsequently co-implanted at 10 keV with a dose of 1x10^{15} cm^{-2} (Hallali, et al, 1990).   The activation of the implanted Be was performed by annealing at 850℃ for 5 sec in an atmosphere of flowing argon using a rapid thermal annealer (RTA), which provided a conduction resistance of $\sim$220 Ω/□.   The wafers were chemically cleaned with a HCl solution just before loading them in the vacuum chamber. Ni/Ni-In/Ni/W ohmic contacts were prepared by depositing sequentially Ni(5nm), Ni-In(10nm), Ni(5nm), and W(40nm) on the p-GaAs substrates where a slash "/" between two metals indicates the sequential deposition and a hyphen "-" indicates the co-deposition.   The NiInW contacts were coated with a 50 nm thick Si$_x$N$_4$ cap layer.   These samples were then annealed for $\sim$ 1 sec at various temperatures by RTA.   The total heating and cooling time is about 15 sec.

3. EXPERIMENTAL RESULTS

The electrical properties of NiInW contacts which were prepared with the GaAs substrates with a Be single-implant or Be+F co-implants were measured after annealing at 300℃ for 30 min which was required for the Si$_x$N$_4$ cap layer coating.   Contacts showed ohmic behavior without additional annealing which indicated that the carrier concentrations at the metal/GaAs interfaces were high.   The contact resistance (R$_c$) of the contact prepared using the Be single-implant substrate was measured by the transmission line method to be 2.5 Ω-mm.   About a factor of two reduction in the R$_c$ value (1.4 Ω-mm) was obtained for the contacts prepared with the Be + F co-implanted substrates.   These contacts were annealed at various temperatures and the R$_c$ values are shown in Fig. 1.   For the contact prepared with the Be single-implant substrate, the R$_c$ value increased with increasing annealing temperature.   For the contact prepared with the Be+F co-implanted substrate, the R$_c$ value of 1.4 Ω-mm remained almost constant throughout the entire annealing temperature range.   This result indicates that the Be atoms in the GaAs substrate with single-implant moved away from the metal/GaAs interface at temperatures lower than the implant/activation temperature of 850℃ and that this Be movement was significantly reduced by the F co-implantation.

Further reduction of the R$_c$ values is expected when the metal contacts the GaAs interface at a depth close to the Be concentration peak position.   The peak position was observed by SIMS analysis to be about 150 nm away from the GaAs surface for the substrate in which Be+F were co-implanted and activated at 850℃ (Hallali et al 1990).   The substrates were chemically etched at various depths prior to the metal deposition so that the Be concentration at which the metal contacts to GaAs increases.   The R$_c$ values of the NiInW contacts annealed at 850℃ for 5 sec are shown in Fig. 2. Note that the R$_c$ values decrease with increasing etched depth.   The lowest R$_c$ value of $\sim$0.6 Ω-mm was obtained by etching the GaAs substrate at the depth of 45 nm at which the Be concentration is close to 2×10^{19} cm^{-3}. This R_c value is sufficiently low for 1μ-HFET(Heterostructural Field-Effect Transistor) devices.

The thermal stability of the NiInW contact prepared with the substrate etched to a depth of 45 nm was examined by annealing isothermally at 400℃ after ohmic contact annealing at 800℃ for 2 sec. The R_c values of these contacts are shown in Fig. 3 where no deterioration of the contact resistance is observed for more than 30 hrs. This is believed to be the most stable p-type ohmic contacts prepared by the conventional implant/activation method.

4. DISCUSSION

In order to explain the constant R_c values of In-based contacts annealed at various temperatures as shown in Fig.2, the barrier heights (ϕ_a) at the metal/p-In$_x$Ga$_{1-x}$As contacts with various In concentrations (x) were estimated. The barrier height at the metal/p-GaAs is close to 0.5 eV (Spicer et al, 1980). However, the barrier heights at the metal/p-In$_x$Ga$_{1-x}$As were not measured except x=0.53 and thus these values were calculated assuming $E_g \sim \phi_b(n) + \phi_b(p)$, where E_g is the energy gap of In$_x$Ga$_{1-x}$As and $\phi_b(n)$ and $\phi_b(p)$ are the barrier heights at the metal/n-In$_x$Ga$_{1-x}$As and metal/p-In$_x$Ga$_{1-x}$As, respectively. In this calculation, the E_g and $\phi_b(n)$ values measured by Kajiyama et al (1973) were used. The calculated $\phi_b(p)$ values at the metal/p-In$_x$Ga$_{1-x}$As are almost constant (0.5 ± 0.05 eV) for x values in the range of 0 to 1. This barrier height calculation is consistent with the contact resistance measurement of Fig. 2 where the R_c values were observed not to change in the annealing temperature range of 300-800℃, although the In$_x$Ga$_{1-x}$As layers covered about 60% of the GaAs surface after annealing at 800℃. Thus, the present result indicates that the reduction of the contact resistances is not achieved by barrier height reduction for the In-based ohmic contacts.

5. SUMMARY

P-type ohmic contacts have been successfully fabricated by depositing NiInW metals on the GaAs substrates which were co-implanted by Be and F. The R_c value of $\sim 1.4 \Omega$-mm (which corresponds to the specific resistance of $\sim 8 \times 10^{-5} \Omega$-cm^2) was obtained over a wide annealing temperature range of 300-800℃. The independence of the R_c values on the annealing temperatures, albeit the interfacial microstructure change after annealing at these temperatures, was explained to be due to similar barrier heights at the metal/p-GaAs and the metal/p-In$_x$Ga$_{1-x}$As contacts.
Further reduction of the contact resistance of the NiInW contacts was achieved by increasing the doping level in GaAs close to the metal/GaAs interface and by etching the GaAs surface at depth of ~ 50 nm. The resistance of $\sim 0.6 \Omega$-

surface at depth of ∼50 nm. The resistance of ∼0.6 Ω-mm(∼1x10^{-5} Ω-cm^2) was obtained by these methods and this value is sufficiently low to use these contacts in 1 μ-HFET devices. The present experiment demonstrated the feasibility of using the same NiInW contact metallurgy for both n and p-type ohmic contacts.

REFERENCES

Hallali P.-E., Baratte H., Cardone F., Norcott M., Legoues F., and Sadana D. K., 1990, Appl. Phys. Lett. 57, 569 .

Kajiyama K., Mizushima Y., and Sakata S., 1973, Appl. Phys. Lett. 23, 458.

Murakami M. and Price W. H., 1987, Appl. Phys. Lett. 51, 664.

Murakami M., Shih Y. C., Price W. H., Wilkie E. L., Childs K. D., and Parks C. C., 1988, J. Appl. Phys. 64, 1974.

Spicer W. E., Lindau I., Skeath P., Su C. Y., and Chye P., 1980, Phys. Rev. Lett. 44, 420.

Veteran J. L., Mullin D. P., and Elder D. I., 1982, Thin Solid Films 97, 187.

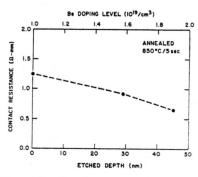

Fig. 1. Contact resistances of NiInW contacts prepared with the GaAs substrates with Be single-implant or Be + F co-implant.

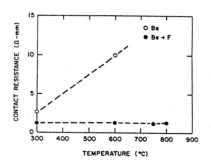

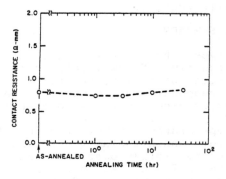

Fig. 2. Contact resistances of NiInW contacts prepared with the GaAs substrate which were chemically etched at various depths.

Fig. 3. Contact resistance of NiInW during isothermal annealing at 400 ℃.

Inst. Phys. Conf. Ser. No 120: Chapter 4
Paper presented at Int. Symp. GaAs and Related Compounds, Seattle, 1991

183

Mushroom shaped gates in a dry etched recessed gate process

G.Kaufel, A.Hülsmann, B.Raynor, K.Köhler, P.Hofmann, Jo.Schneider,
J.Hornung, M.Berroth, and T.Jakobus

Fraunhofer-Institute for Applied Solid State Physics
Tullastraße 72, D-7800 Freiburg, FRG

ABSTRACT: We have developed a recess gate process to fabricate high electron mobility transistors (HEMTs) of enhancement and depletion type on two inch GaAs wafers. The vertical HEMT structure is grown by molecular beam epitaxy (MBE). The gates are written by e-beam lithography and are recessed by dry etching. Mushroom shaped gates of dimension of 0.4 µm for L_g = 0.1 µm are used to reduce the gate resistance. The average threshold voltage for D- and E-type FETs is -500 mV and 100 mV respectively with a standard deviation of $\leq$ 20 mV over a 2" Wafer with L_g = 0.3 µm.

1. Introduction

In order to increase the speed of integrated circuits, the gate-lengths must be decreased in the range from 0.5 µm to 0.1 µm. Mushroom shaped gates have been fabricated using a three layer resist technique and a dry etched gate recess process. The T-gate technique decreases the input resistance and the dry etching enhances the reproducibility and decreases the source resistance of the device.

2. Device Fabrication

First an AlGaAs/GaAs superlattice buffer is grown by MBE on 2" s.i. GaAs wafers. The 2DEG channel is formed by an AlGaAs/GaAs/AlGaAs single quantum well (SQW) [1]. Two 3 nm AlGaAs etch stops are included into an undoped GaAs layer to control precisely the threshold voltage of D- and E-FETs. On top is a highly doped GaAs cap layer to reduce source and drain resistances. The first lithography step defines the ohmic contacts and the alignment marks for e-beam direct write of the gates. Device isolation is done by oxygen implantation at 80 keV. E-beam direct-write lithography is then carried out followed by dry etching onto the first and second AlGaAs etch stops, for D-FETs and E-FETs, respectively. First level metallization is evaporated and for second level metallization gold-plated airbridges are used.

3. E-beam

Resist System:
The resist system for e-beam lithography involves a three layer structure, consisting of a crosslinked P(MMA/MAA) (copolymer of methyl methacrylate and methacrylic acid) prebaked at 170°C and two PMMA

layers with molecular weights of 50,000 and 500,000 with nominal thicknesses of 260 nm. The crosslinked P(MMA/MAA) is used as the bottom layer to define the small footprint of the gate. The solid content of this resist is 9 %, diluted in EGMEA (ethyl glycol monoethyl ether acetate). At low prebake temperatures the resist is highly sensitive but shows low contrast when developed in xylene. Fig. 1 shows the contrast curves of P(MMA/MAA) prebaked at different temperatures and developed in xylene.

In addition crosslinked P(MMA/MAA) shows very high etch selectivity to GaAs during dry-etching of the gate recess. An intermediate layer of PMMA 50k prebaked at 170°C is used as a high sensitive layer to achieve an undercut of the resist profile for better lift-off.

The top layer of the three layer resist system is 6 % PMMA 500K diluted in EGMEA. This resist defines the wide resist opening for the mushroom head.

The gate is positioned between two AuGe/Ni/Au alloyed ohmic contacts with a nominal thickness of 300 nm which are recessed in a 50 nm SiN PECVD protection layer. Oxygen implantation isolation avoids mesa topography steps. We studied the bottom layer resist planarisation between various source to drain distances [2]. Details describing the mechanism of resist planarisation are explained by La Vergne et. al. [3]. The T-gates for our HEMT designs are aligned into a 1.1 to 1.5 µm source to drain gap in which case full planarisation at the bottom layer resist occurs. Thus the bottom layer resist thickness of the crosslinked P(MMA/MAA) is 300 nm.

Exposure:

The exposure is carried out in a Philips EPBG-3 at 50 keV. The spot size is set to 50 nm. Depending on the conditions of the LaB_6 cathode the beam current is between 2 and 5 nA. The design of the 0.1 µm gate consists of three patterns, one in the center with a size of 50 nm and two 100 nm side patterns separated by 100 nm. Because the beam step size is 50 nm the center pattern is defined by a line exposure at a high dose of 600-800 µC/cm². This line defines the foot print of the T-gate and thus the gate length. The side patterns with equal sizes define the mushroom shaped head of the gate. The side patterns are separated by

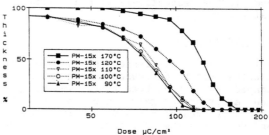

Fig. 1 Contrast curves of P(MMA/MAA) prebaked at different temperatures

Fig. 2 Micrograph of an E-HEMT with 100 nm gatelength in T-gate technology

a space of 100 nm from the center line. A FSDD (feature size dependent dose) software used for postprocessing the layout data gives the side patterns a dose of 15% of the center line.

For gate lengths greater than 0.2 µm the side beam exposure technique is omitted because the cross section for these types of gates is sufficient for low gate line resistance. **Fig. 2 is a micrograph of a** 0.1 μm T-gate.

4. Dry Etching

All dry etch processes are carried out in a commercial 13.56 MHz parallel plate reactor (MPE 3003) from Leybold, which are described in detail in [4] and [5]. The vertical HEMT structure, as far as the etching is concerned, is composed of a highly doped GaAs cap layer (30 nm), two 3 nm thick AlGaAs layers which act as etch barriers for the dry

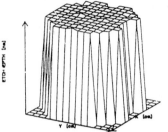

Fig. 3 Distribution of etch depth on a 2" GaAs wafer for 60 μbar, 55 W, 2 min, mean = 656 nm, σ = 6.5 nm

etch process, and a 7.5 nm thick GaAs layer between the two etch stops. The etching conditions used for this work were 60 μbar, 55 V und 90 s and Freon 12 as etch gas. The uniformity of the etch process without

etchstop is shown in Fig. 3. The etch depth is 650 nm with a standard deviation of ≤ 1% across a 2" Wafer.

Fig. 4 shows the etch rates for various pressures. From the curves it can be seen that the etch delay time decreases with increasing pressure.

CF_2Cl_2 can be used for the reactive ion etching of GaAs and AlGaAs. A plasma power of ≥ 90 Watt is necessary to etch AlGaAs. For etching conditions of 60 μbar, 90 W and 150 V, the selectivity of GaAs to AlGaAs is 300:1.

To study radiation damage caused by the plasma, we etched a HEMT structure using dichloro difluoro metane at 100 μbar, 55 W and 55 V for one, two and three minutes. Measurements of the resistance, mobility and carrier concentration showed

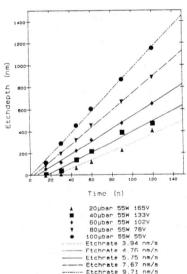

	20μbar 55W 165V
■	40μbar 55W 133V
♦	60μbar 55W 102V
▼	80μbar 55W 78V
●	100μbar 55W 55V
....	Etchrate 3.94 nm/s
....	Etchrate 4.76 nm/s
———	Etchrate 5.75 nm/s
————	Etchrate 7.67 nm/s
————	Etchrate 9.71 nm/s

Fig. 4 Etchrate of GaAs at different pressures

only minimal deviations compared to unetched samples, see reference [6].

5. Results

We have developed a process to produce enhancement and depletion GaAs/AlGaAs SQW-HEMTs for DCFL circuits in a T-gate technology on 2" wafers. The MBE structure results in a 2-DEG carrier concentration of

1.8 x 10^{12} cm^{-2} and a mobility of 7000 cm^{-2}/Vs at 300 K. Fig. 5 shows I_{ds} versus V_{ds} for T-gate SQW-HEMTs in enhancement and depletion mode. The normalized transconductances are 600 mS/mm and 450 mS/mm respectively.

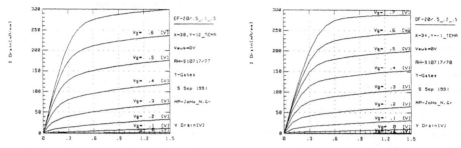

Fig. 5 I_{ds} versus V_{ds} for T-gate SWQ-HEMTs in enhancement (left) and depletion mode (right). The gatelength is 0.1 and 0.2 μm respectively.

6. Acknowledgement

We would like to thank P.Ganser, J.Schaub and T.Schweizer for MBE growth, M.L.Andres, P.Roman, and B.Weismann for optical lithography, E.Olander for optical lithography and dry etching, K.H.Glorer and M.Krieg for e-beam direct write, F.Becker for mask making, D.Luick for the metallization, R.Haddad for alloying, implantation and galvanic, N.Grün, J.Windscheif, and W.Benz for DC-measurements, R.Bosch and W.Reinert for AC-measurements, V.Hurm, M.Lang, and P.Wennekers for circuit design and mask layout, and H.S.Rupprecht for project management.

References

1. K.Köhler, P.Ganser, M.Maier, J.Hornung, A.Hülsmann, Inst. of Physics Conf. Ser. No.112, Ch.7, 521 (1990)
2. A.Hülsmann, G.Kaufel, K.Köhler, B.Raynor, J.Schneider and T.Jakobus Jpn. J. Appl. Phys.29, 2317 (1990)
3. D.LaVergne and D.Hofer, SPIE Vol.539 Advances in Resist Technology and Processing II, 1985, p.115
4. Gudrun Kaufel and Elfriede Olander, Mat. Res. Soc. Symp. Proc. Vol.158 pp.401-406 (1989)
5. A.Hülsmann, G.Kaufel, K.Köhler, B.Raynor, K.H.Glorer, E.Olander, B.Weismann, J.Schneider, T.Jakobus, Inst. of Physics Conf. Ser. No.112, Ch.7, 429 (1990)
6. H.P.Zappe, G.Kaufel Appl. Surface Sci., accepted for publication

Inst. Phys. Conf. Ser. No 120: Chapter 4
Paper presented at Int. Symp. GaAs and Related Compounds, Seattle, 1991

187

Epitaxial Al on δ-doped GaAs: a reproducible and very thermally stable low resistance non-alloyed ohmic contact to GaAs

M. Missous

Department of Electrical Engineering and Electronics
and Centre for Electronic Materials
University of Manchester Institute of Science and Technology
Po Box 88, Manchester M60 1QD, England, UK

ABSTRACT: A new approach to fabricating very reproducible and extremely stable non alloyed ohmic contacts to n-GaAs is reported. The technique relies on the use of surface δ-doping and epitaxial Al metallisation. The advantages of this new type of ohmic contact are :
(i)   Very low specific contact resistivity ($< 10^{-6}$ Ωcm^{-2}).
(ii) Thermal stability up to 550 °C.
(iii) Abrupt interfaces with GaAs with virtually no chemical reactions or interdiffusion leading to penetration depths of less than 90 Å (compared with 2000 Å for the Au-Ge contact).
(iv) Al metallisation allowing submicron patterns to be achieved.
(v)   Simplicity of fabrication and reproducibility.
This approach open up the way to hybrid configurations in MESFETs were the epitaxial Al can be used as both the gate and source and drain metallisations.

I. INTRODUCTION :

The fabrication of thermally stable, low resistance ohmic contacts to GaAs has been a prime objective for decades and whereas spectacular progress has been made on the metal-semiconductor Schottky barrier front [1], the ohmic contact realm has largely stayed an art. The dominant reason is that ohmic contact metallisation is an incredibly complicated process. The most widely used contact [2], namely Au-Ge-Ni, while having excellent contact resistance suffers from thermal instabilities at temperatures much higher than 400 °C and has poor lithographic profiles [3] and high penetration depths [4], all of which are incompatible with advanced GaAs integated circuits (ICs). However as GaAs ICs are becoming a reality, the need for a well controlled and above all well characterised contact metallurgy is urgently felt.

In the last few years, several schemes of non-alloyed contacts have emerged [4-6], designed primarily to circumvent some of the drawbacks of the classical Au-Ge system. The emphasis of all these schemes is to abondon melting, the root of the alloyed contact problems. These contacts include the use of heavily doped graded InGaAs epitaxially grown on GaAs by Molecular Beam Epitaxy (MBE) [5] , growth on n^{++} Ge layers on GaAs [6] and the use of compound conductors such as NiSb [7].

However none of the above fulfils the simultaneous requirement of low contact resistance, thermal stability (at least up to 500 °C), mimimum chemical interaction, uniformity on an atomic scale at the contact-GaAs interface and above all simplicity of fabrication to ensure reproducibility.

In this paper we have investigated the coupling of two ideas concerning the ohmic interface. One is the use of δ-doping just beneath the surface, an approach used successfully by Schubert et al [9], the other is the extremely high thermal stability and uniformity of epitaxial Al on GaAs as reported by Missous et al [10]. It will be shown that this combination respond to all the requirement of modern GaAs ICs. The motivation to use epitaxial Al, besides its thermal stability, was also prompted by the renewed interest in Al metallisation in the GaAs industry recently [11]. A similar approach to ours has also been recently reported by Goossen et al [16].

II. EXPERIMENTAL :

All the layers were grown by MBE in an all- solid- source system. The structures consisted of 1 μm degenerately doped layers of GaAs deposited epitaxially on (100) substrates (2-5 x 10^{18} cm^{-3}) followed by growth interruption of the Ga flux and simultaneous deposition of Si δ-doped layers having a Si atom density ranging from 1.4 to 3.5 x 10^{13} cm^{-2} and finally growth of a 20 Å thick undoped GaAs, the whole structure being grown at 580 °C. The growth of the δ-layer was deliberately carried out at 580 °C to ensure thermal stability at least up to that temperature. After growth termination, all the cells in the MBE chamber were shuttered and the vacuum allowed to reach ~ 2 x 10^{-10} Torr before (100) oriented epitaxial Al 2000 Å thick was deposited at ~ 50 °C as reported in detail in [10].

After removal from the MBE system, samples were prepared both for Hall measurements after etching away the single crystal Al film and for electrical measurements using the transmission line method (TLM) where a four point probe arrangement was used to minimise the effects of probe resistance [11].Annealing experiments were performed using a Rapid Thermal Annealer (RTA) in a nitrogen ambient , at temperatures up to 550 °C, the purpose of which was to study the thermal stability of the electrical properties of the contacts, any chemical reaction or interdiffusion at the interface using Auger Depth Profiling (ADP), and Scanning Electron Microscopy (SEM) and Optical microscopy to monitor surface morphology.

III. RESULTS AND DISCUSSIONS :

1. Hall effect measurements :

The electrically active donor concentrations were determined by Hall measurements using a Van Der Pauw geometry.Furthermore to study the effect of the surface doped δ-layer, a slow etchant was used to remove ~300 Å of the GaAs surface. Two groups of samples were thus obtained: those without a δ-layer (δ-OFF samples) and those still incorporating the δ (δ-ON samples).The Hall measurements for a typical sample, having a Si atom density of 1.4×10^{13} cm^{-2} ,are shown in Table I.It can clearly be seen that the effect of the δ-layer has resulted in an increase in the free electron concentration (n) and a decrease of the Hall mobility (μ). These changes, 12 % for μ and 22 % for n , are outside any experimental errors which we estimate at better than $\pm$ 2%. While it is true that the layer which was etched was thinner by some 300 Å (ie 3% less than the standard 1 μm) that alone cannot explain a reduction of 22 % in n; also the mobility calculations are independent of sample thickness and are seen to decrease by 12 %. SIMS and C-V measurements on similar δ-doped layers, situated 1000-2000 Å below the surface, reveal a Full Width at Half Maximum of 60 Å ,this figure can be taken as an upper limit on the width of the surface δ-layer.The structure can therefore be considered as a strip ~ 60 Å thin in series with a much thicker 1μm , 4.5×10^{18} cm^{-3} layer. Note that this value of doping concentration is very near to the maximum that can be attained using Si at normal growth temperatures of 580 °C [12]. The depletion layer width at this concentration is ~ 140 Å, whereas for the δ-doped region one can extract a 3 dimensional doping density greater than 3×10^{19} cm^{-3} (corresponding to a depletion width of 60 Å) since the δ-layer is not fully depleted. This free electron concentration is higher than any 3 dimensional bulk doping of GaAs using Si and is a factor 2 to 3 higher than the maximum n values obtained at low growth temperatures (<450 °C) [13].One should , of course, be weary about using classical arguments when dealing with quantum mechanical systems such as δ-doped layers, however the example above serves to illustrates the improvements in doping efficiency achievable by δ-doping. And whereas it is true that tunnelling across the δ-layer is the desirable feature for ohmic contact fabrication, very heavy, electrically active, doping is necessary to ensure maximum steepness in the band bending and therefore minimum tunnelling width.

TABLE I

Sample	δ-ON		δ-OFF	
	n (10^{18} cm^{-3})	μ (cm^2/V.s)	n (10^{18} cm^{-3})	μ (cm^2/V.s)
A	5.5	1190	4.5	1352

2. Electrical characteristics :

(a) As deposited contacts :

The electrical properties of the contacts were assessed by current-voltage measurements using an HP 4145 A parameter analyser and by the transmission line method to determine the specific contact resistance (R$_c$). Figure 1 shows the I-V plots of a sample with a doping concentration of 4.5×10^{18} cm^{-3} both with and without a δ-layer. It can clearly be seen that for the case where no surface δ-layer exists, and despite the heavy doping, the I-V curves are not linear but have the characteristics of a two back-to-back schottky contacts. By contrast, the inclusion of a δ-doped layer leads to a strictly linear characteristics between $\pm$ 30 mA.The measured specific contact resistance is 9.5×10^{-7} Ωcm , comparable to what is achievable in state of the art Au-Ge-Ni contacts.Since it has been shown that δ-doping above ~ 1.2×10^{13} cm^{-2} leads to saturation of the free carrier concentration, the excess Si being electrically inactive [14], a sample was prepared having a doping of 2.5 x 10^{18} cm^{-3} and a surface δ-layer of 3.5 x 10^{13} cm^{-2}. Figure 2 shows the corresponding I-V plots and as before the incorporation of the δ-layer leads to a linear contact. However the measured specific contact resistance is now 8.8 x 10^{-6} Ωcm , almost an order of magnitude higher than that of the previous sample. The slightly lower bulk doping of 2.5 x 10^{18} cm^{-3} is of no importance in the determination of R$_c$ since we have shown elsewhere [11] that for degenerately doped GaAs (n > 5 x 10^{17} cm^{-3}), only the δ-layer controls the current flow across the structure.

It is interesting to compare the value 8.8 x 10^{-6} Ωcm obtained for a Si atom density of 3.5 x 10^{13} cm^{-2} with that obtained by Schubert et al [8] who measured a value of 6.3 x 10^{-6} Ωcm for a bulk layer of 10^{18} cm^{-3} having a surface δ of 5 x 10^{13} cm^{-2}. Therefore the excess Si atoms are detrimental to the contact as they lower R$_c$.

(b) The effect of heat treatment :

Thermal annealing of the contacts was performed in an RTA system at 550 °C for 3 minutes. Note that for these studies, the purpose of the annealing is not to promote any dopant diffusion, as in the case of the Au-Ge contacts, but to study the thermal stability of the structures. The linearity of the I-V curves was not affected by the heat treatment whatsoever and the only effect was to increase R$_c$ by a factor of ~ 1.6 for the low δ-doped sample and by a factor of ~ 3 for the highly δ-doped layer. These changes can be understood by the fact that annealing will spread the δ-layer in width and height , both of which would tend to impede the tunnelling current. Note that the final value of R$_c$ for the low δ-doped sample (1.6 x 10^{-6} Ωcm^{-6}) is still an excellent one and would not compromise device performances.

The surface morphology of the contacts was examined using an SEM system both before and after annealing at 550 °C. Figure 3(a) shows a typical micrograph of the epitaxial contact which depicts a featureless surface indicative of a faily flat and smooth thin film. Figure 3(b) shows the same contact after annealing at 550 °C. The epitaxial contact now shows some surface roughness in the form of

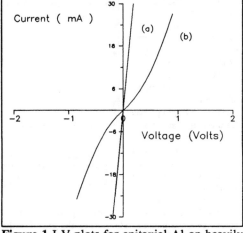

Figure 1 I-V plots for epitaxial Al on heavily doped GaAs (a) with and (b) without a surface δ-doped layer of 1.4 x 10^{13} cm^{-2}.

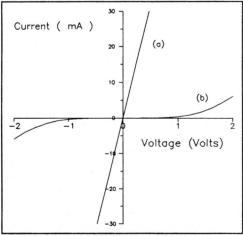

Figure 2 I-V plots for epitaxial Al on heavily doped GaAs (a) with and (b) without a surface δ-doped layer of 3.5 x 10^{13} cm^{-2}.

depressions extending into the Al film, however this effect is very small and has no effect whatsoever either on the lithographic profiles or the morphology of the contact.

The integrity of the contacts is further demonstrated in ADP profile studies. The epitaxial Al on GaAs had abrupt interfaces both at room temperature and after annealing at 500 C for one hour because of the reduced grain boundary diffusion [10]. The transition width was ~ 90 Å almost certainly limited by the resolution of the ADP method. This penetration depth is orders of magnitude lower than that of the Au-Ge-Ni system which usually extends uo to ~ 2000 Å [4]. This is of considerable importance for modern submicron devices with very thin channels.

IV. CONCLUSIONS :

In conclusion, a novel approach to synthesising very reproducible and extremely stable ohmic contacts has been reported. The coupling of epitaxial Al and surface δ-doping has resulted in an almost ideal ohmic contact for use in high speed devices. This new ohmic contact scheme relies on a process based purely on band structure engineering using GaAs and its dopant Si and is therefore readily adaptable to AlGaAs and to a range of other semiconductors. This is not the case for the alloyed Au-Ge based contact which has to be re-examined in the presence of Al [15] and usually needs temperature in excess of 500 °C, exacerbating the surface morphology problem.

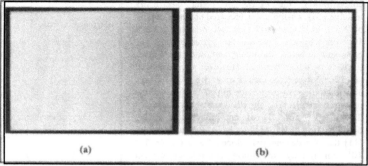

Figure 3 Surface morphology of epitaxial Al on GaAs (a) as deposited and (b) after annealing at 550 °C.

The contacting epitaxial Al is not only ideal for small scale geometries because of its excellent surface morphology , its minimal interaction with GaAs even at temperatures as high as 500 °C results in penetration depths of less than 90 Å compared with > 2000 Å with the Au-Ge contact. This is of tremendous importance for advanced very high speed GaAs ICs using extremely thin channel layers.

ACKNOWLEDGEMENT :

I would like to acknowledge the support and help of many peoples who made this studies possible and in particular Dr. T. Taskin who made the specific contact resistance measurements, Professor A.R. Peaker for many useful discussions and Professor K.E. Singer for permission to use the V90H MBE system. This work was supported by the Science and Engineering Research Council.

REFERENCES :

1. E.H. Rhoderick ands R.H. Williams, Metal semicondutor contacts, 2nd edition, Clarendon, Oxford (1988)
2. N. Braslau, J.Vac.Sci.Technol. 19, 803 (1981)
3. Y.C. Shih, M. Murakami and A. Callegari, J.Appl.Phys. 62,582 (1987)
4. M. Murakami, Y. Shih, W.H. Price, N. Braslau, K.D. Childs and C.C. Parks, Inst.Phys.Conf.Series No 91, 55 (1987)
5. T. Nitton, H. Ito, O. Nakajima and T. Ishibashi, Japan.J.Appl.Phys. 25(10), L865 (1986)
6. R. Stall, C.E.C. Wood, K. Board and L.F. Eastman, Electron.Lett. 15,800 (1979)
7. R. Dutta, A. Lahav, M. Robins and V.G. Lambrecht, J.Appl.Phys.67,3136 (1990)
8. R. Cates, IEEE SPECTRUM 27 (4), 25 (1990)
9. E.F. Schubert, J.E. Cunningham, W.T. Tsang and T.H. Chiu, Appl.Phys.Lett. 49, 292 (1987)
10. M. Missous, E.H. Rhoderick and K.E. Singer, J.Appl.Phys 59(9), 3189 (1986)
11. M. Missous and T. Taskin, to be published
12. M. Missous, K.E. Singer and D.J. Nicholas, J.Cryst.Growth 81, 314 (1986)
13 M. Ogawa and T. Baba, Japan.J.Appl.Phys 24(8), L572 (1985)
14. A. Zrenner F. Koch and K. Ploog, Inst.Phys.Conf.Series No 91,171 (1987)
15. P. Zwicknagi, S.D. Mukherjee, P.M. Capani, H. Lee, H.T. Griem, L. Rathburn, J.D. Berry, W.L. Jones and L.F. Eastman, J.Vac.Sci.Technol. B4(2), 476 (1986)
16. K.W. Goossen, J.E. Cunningham, T.H. Chiu, D.A. Miller and D.S. Chemla, Intern.Electr. Devices Meet. (IEDM) (1989)

Inst. Phys. Conf. Ser. No 120: Chapter 4
Paper presented at Int. Symp. GaAs and Related Compounds, Seattle, 1991

191

Characteristics of dry-etched GaAs p−n junctions grown by MOMBE

C. R. Abernathy, S. J. Pearton, F. Ren, T. R. Fullowan and J. R. Lothian
AT&T Bell Laboratories, Murray Hill, NJ 07974

ABSTRACT

We demonstrate that MOMBE using all gas-sources (AsH_3, TMGa, TEGa and TESn) is capable of producing p-n junctions suitable for high performance GaAs/AlGaAs HBTs. Both Sn and C profiles measured by SIMS showed sharp turn-on and turn-off of the respective dopants, and n-on-p and p-on-n diodes had 4 decade linear regions with ideality factors of 1.55. The absence of cross-contamination of C in the n-GaAs and Sn in the p-GaAs produced diodes with improved characteristics over those grown with solid As reported earlier. HBTs fabricated on our MOMBE grown layers exhibit unilateral current gains and maximum frequencies of oscillation above 60 GHz. To simulate the ion bombardment during plasma etching with either CCl_2F_2/O_2 or CH_4/H_2 mixtures, we exposed our samples to 1 mTorr O_2 or H_2 plasmas for periods of 1-20 min with DC biases of -25 to -400V on the cathode. For O_2 ion bombardment the collector resistance showed only minor ($\leq 10\%$) increases for biases up to -200 V. The base resistance displayed only a minor increase over the pre-exposure value, even for O^+ energies of 375 eV, due to the very high doping in the base. More significant increases in both R_c and R_B were observed for H^+ ion bombardment due to hydrogen passivation effects.

Introduction

Much recent attention has been focussed on the growth and processing of GaAs/AlGaAs HBTs. Structures utilizing Be as the base dopant often display a decrease in gain during bias-stress annealing of the completed devices.[1] This is presumably a result of motion of charged Be interstitials under the action of the high current density present in the junction. By sharp contrast, HBTs with carbon-doped base layers are stable during similar bias-stress cycles. While carbon eliminates the redistribution and reliability problems associated with Ga-site acceptors, there have been reports of unacceptably high levels of carbon contamination in nominally n-type layers grown adjacent to p^+ carbon-doped layers.[2] Moreover, both p-on-n and n-on-p GaAs diodes grown using TMGa and As_2 in a gas-source MBE system showed ideality factors close to 2 for small biases. We show that MOMBE using all gas sources is capable of producing high quality p-n junctions. The dopants exhibit sharp profiles with no cross-contamination into adjacent layers. We also examined the effects of simultaneous exposure of both the n- and p-type layers to O_2 or H_2 discharges as a function of both the DC bias on the sample during the plasma exposure and of the exposure time. Significant changes in the ideality factor of an n-on-p junction are only observed for O^+ and H^+ ion energies above 200 eV. Hydrogen passivation effects can be removed by annealing at 400°C, but high energy (≥ 300 eV) O^+ ion bombardment causes irreversible damage to the diode.

Experimental

The samples were grown in a Varian Gas Source Gen II at a growth temperature of $500°C$ as measured by the substrate thermocouple.[3] Carbon-doped layers were grown using TMG while n-type layers were grown using triethylgallium (TEG). Tetraethyltin (TESn) was used as the n-type dopant source. The n^+np^+ structures consisted of $2100Å$ of carbon-doped GaAs ($p = 4 - 7 \times 10^{19}$ cm^{-3}), $7000Å$ of Sn-doped GaAs ($n = 1-5 \times 10^{16}$ cm^{-3}) and $500Å$ of Sn-doped GaAs ($n = 1.5 \times 10^{19}$ cm^{-3}), while the p^+n n^+ structure was comprised of $700Å$ of Sn-doped GaAs ($n = 2 \times 10^{18}$ cm^{-3}), $4000Å$ of Sn-doped GaAs ($n = 5 \times 10^{16}$ cm^{-3}), and $7000Å$ of carbon-doped GaAs ($p = 4 \times 10^{19}$ cm^{-3}). Secondary ion mass spectrometry (SIMS) profiles showed that both the Sn and C profiles had sharp turn-on and turn-off of the respective dopants, and the carbon concentration was below the SIMS detection limit in the Sn-doped layers. Individual layers of the same Sn doping levels showed n-type conductivity with a 1:1 correspondence between carrier concentration and Sn concentration. This indicates there is no significant compensation of the Sn by background carbon contamination. The mesa diode TLM patterns were defined with $H_3PO_4:H_2O_2:H_2O$ based wet chemical etching by using photoresist as the etch mask (Figure 1). E-beam evaporated AuBe/Au and AuGe/Ni/Au metallization provided the ohmic contacts for p and n layers, respectively.

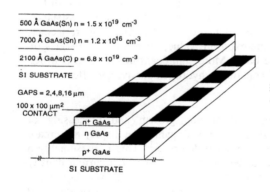

Fig. 1.   Schematic of transmission line patterns on the mesa diode structures used for plasma exposure experiments.

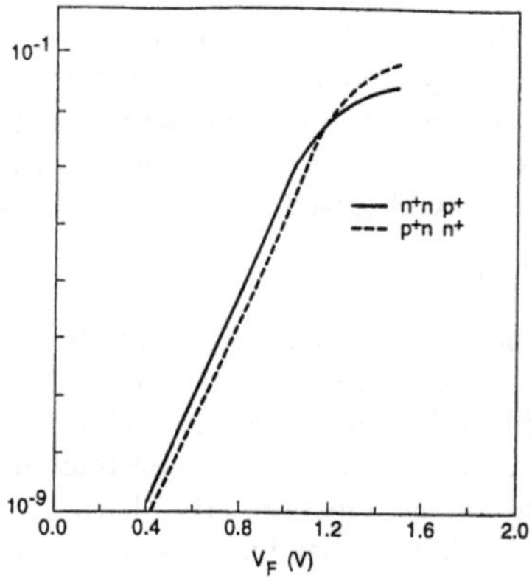

Fig. 2.   Forward I-V characteristics of GaAs n-on-p or p-on-n diodes.

Results and Discussion

Figure 2 shows the forward I-V characteristics of the diodes. Both n-on-p and p-on-n diodes have at least 4-decade linear regions with ideality factors of 1.55. The differences in V_{bi} are caused by the different doping levels in the n layers. Due to the high quality of the epilayers and the absence of cross-contamination of C in the n-GaAs, or Sn in the p-GaAs, the characteristics of both diodes showed drastic improvement as compared with similar work with C and Si for p and n layers respectively.[2] Thus it is not surprising that GaAs/AlGaAs npn HBTs using these C and Sn doped layers have achieved unilateral current gains and maximum frequencies of oscillations above 60 GHz.

We exposed our samples to 1 mTorr O_2 or H_2 plasmas for periods of 1-20 min with DC biases of -25 to -400V on the cathode. In these cases there is ion bombardment, but not etching. Figure 3 shows the variation of collector and base sheet resistance with DC bias during the 5 min plasma exposure. For O_2 ion bombardment the collector resistance shows only minor ($\leq 10\%$) increases for biases up to -200 V, and more rapid increases thereafter. This resistance is mostly determined by the contribution from the lightly doped n-region rather than the n^+ contact layer, and indicates that bombardment-induced point defects penetrate at least $500 \text{\AA}$ of GaAs for oxygen ion energies of ≥ 200 eV. The base resistance displays only a minor increase ($\sim 10\%$) over the pre-exposure value even for oxygen ion energies of 375 eV. This is a result of the very high doping in the base-as for ion implant isolation, it is difficult to introduce a density of deep levels high enough to produce compensation in these heavily-doped layers. More significant increases in both collector and base resistances were observed for hydrogen ion bombardment. Figure 3 shows that the collector resistance increases by a factor of approximately seven over the range of DC biases investigated (-256 to -400 V). We ascribe this greater degradation of the resistance relative to oxygen ion bombardment to hydrogen passivation of the Sn donors in both the collector and contact layers. This phenomenon is well established for reactive ion etching or plasma exposure of GaAs using hydrogen-containing discharges. Exposure of the base to hydrogen ions also produces a substantial ($\geq 30\%$) increase in layer sheet resistance. Acceptor passivation is usually less efficient than donor passivation in GaAs, and once again with the very high p-type doping level it is difficult to fully passivate the carbon in the base.

We also examined the dependence of collector and base sheet resistance on the plasma exposure time for DC bias values of -200 V (Fig. 4). At this bias there were only minor but monotonic increases in both base ($\sim 15\%$) and collector ($\sim 10\%$) sheet resistances for oxygen plasma exposure times in the range 1-20 min. In contrast, hydrogen plasma exposure produced more substantial increases ($\sim 50\%$) in both resistances within the first 5 min, and essentially a saturation behaviour thereafter. This may be a result of the retardation of hydrogen permeation through the formation of hydrogen aggregates or platelets in the uppermost part of each exposed layer. Annealing of the plasma-exposed samples was performed at 350-450°C for 5 min in a He ambient. The hydrogen bombarded samples showed near-complete recovery of the pre-exposed base and collector sheet resistances at 400°C for H^+ ion energies of ≤ 300 V. This recovery is predominantly a result of reactivation of the formerly hydrogen passivated dopants. For -400 V bias bombardment, annealing at 400°C produced base and collector resistance values approximately 20% higher than in the unexposed samples, indicating the presence of residual damage. Oxygen plasma damage was more resistant to annealing than that produced by hydrogen discharges. For oxygen ion energies of ≥ 300 eV, annealing at 450°C for 5 min actually produced a slight increase ($\sim 15 - 20\%$) in both base and collector resistances, possibly as a result of further defect migration. It is clear that such high energy ion bombardment can cause irreversible damage to GaAs.

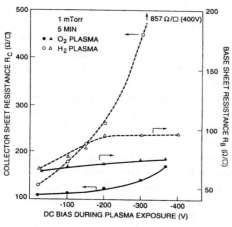

Fig. 3. Variation of base (p^+ GaAs) and collector (n-GaAs) sheet resistances as a function of the DC bias on the sample during exposure to either an O_2 or H_2 discharge.

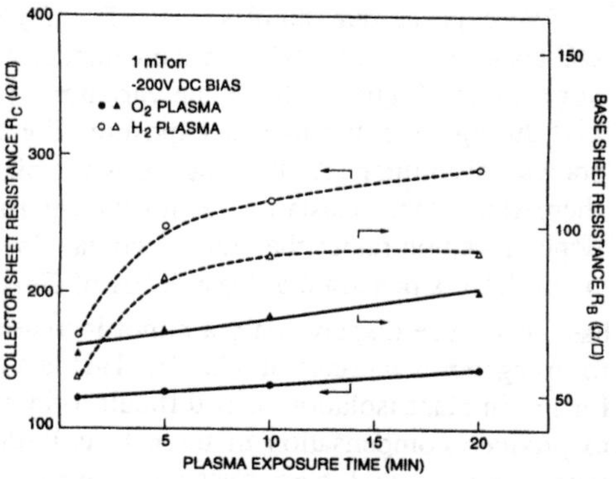

Fig. 4. Variation of base and collector sheet resistances as a function of exposure time to -200 V DC O_2 or H_2 discharges.

References

1. M. E. Kim, B. Bayraktaroglu and A. Gupta, in HEMTs and HBTs: Devices, Fabrication and Circuits, ed. F. Ali and A. Gupta (Artech House, Boston 1991).

2. S. Nozaki, R. Miyake, T. Yamada, M. Konagai and K. Takahashi, Jap. J. Appl. Phys. *29* L1731 (1990).

3. C. R. Abernathy, S. J. Pearton, R. Caruso, F. Ren and J. Kovalichuh, Appl. Phys. Lett. *55* 1750 (1989).

Inst. Phys. Conf. Ser. No 120: Chapter 4
Paper presented at Int. Symp. GaAs and Related Compounds, Seattle, 1991

195

Incorporation of hydrogen into III–V semiconductors during growth and processing

S. J. Pearton[+], C. R. Abernathy[+], W. S. Hobson[+], F. Ren[+], T. R. Fullowan[+], J. Lopata[+], U. K. Chakrabarti[+], D. M. Kozuch* and M. Stavola*

[+]AT&T Bell Laboratories, Murray Hill, NJ 07974 USA
*Lehigh University, Bethlehem, PA 18015 USA

ABSTRACT

Atomic hydrogen is shown to be unintentionally incorporated into GaAs, AlGaAs and InP during gas-source (MOMBE, MOCVD) growth and subsequent processing steps. For example, hydrogen injection into p-type GaAs during boiling in water or etching in $H_2SO_4:H_2O_2:H_2O$ has been detected by the resultant acceptor passivation and more directly by SIMS after processing with deuterated chemicals. Dry etching of these materials in C_2H_6/H_2 or CH_3Cl plasmas leads to substantial dopant passivation, and this can occur even when hydrogen is not a specific part of the plasma chemistry because of the presence of water vapor, photoresist mask erosion and so on. Hydrogen also readily permeates into III-V materials during annealing at moderate temperatures ($\sim 500\,^\circ C$) in H_2-containing ambients.

Introduction

There is an increasing trend toward lower temperature growth and processing of III-V epitaxial layers. One of the problems encountered as a result has been the incorporation of significant amounts of hydrogen in the films, leading to passivation of the electrical activity of both donor and acceptor dopants. The amount of hydrogen incorporated is directly dependent on the initial doping density of the material, with the hydrogen being present in neutral donor-hydrogen and acceptor-hydrogen complexes, and in diatomic or larger clusters.

In this paper we give examples of the introduction of hydrogen into GaAs, AlGaAs, InP and related materials during growth and subsequent processing. Some of the fabrication steps during which hydrogen may be incorporated into the semiconductor include wet chemical cleaning or etching, dry etching, plasma-enhanced chemical vapor deposition and annealing in H_2-containing ambients.

Experimental

Epitaxial layers of C-doped ($p = 6 \times 10^{18} - 4 \times 10^{20}$ cm^{-3}) GaAs were grown on undoped GaAs substrates by either Metal Organic Molecular Beam Epitaxy (MO-MBE)[1] or Metal Organic Chemical Vapor Deposition (MOCVD)[2] and the hydrogen content of the films measured by Cs$^+$ ion beam secondary ion mass spectrometry (SIMS). Doped (n- and p-type) GaAs and InP substrates were boiled in D_2O for periods of 1-6 hours or chemically etched in $D_2SO_4:H_2O_2:D_2O$ mixtures at 25°C for up to 4h. Deuterium incorporation into these samples was also measured by SIMS. Similar samples were dry etched in either C_2H_6/D_2 discharges at 4 mTorr pressure with self-biases of 100-350V on the cathode. The carrier profiles after the dry etching treatment were obtained by Hg-probe, capacitance-voltage (C-V) profiling. The carbon-doped epitaxial samples were also heated at 450-500°C in sealed ampoules containing either D_2 or H_2/D_2 ambients, and the formation of carbon-hydrogen complexes monitored by high-resolution infra-red spectroscopy.

Results and Discussion

A SIMS profile of the hydrogen concentration in an epitaxial GaAs (C) layer grown by MOMBE is shown in Fig. 1. The initial free carrier concentration was 10^{19} cm^{-3} and the layer thickness was 0.42 μm. The hydrogen is present at a concentration of $5 - 10 \times 10^{17}$ cm^{-3} in the epi-layer. In this case the source chemicals were AsH_3, trimethylgallium (TMGa) and triethylgallium (TEGa).[3] A series of different experiments showed that the hydrogen can be introduced from any of these precursors, since, for example, significant concentrations were still detected in layers grown with solid As rather than AsH_3.[4] Similar results were observed with growth of the layers by MOCVD. In these cases the hydrogen concentration is not high enough to lead to significant reductions in hole density, but such a situation is possible under appropriate growth conditions. The passivation of Zn acceptors in InP-based heterostructures during cool-down under AsH_3 following growth has previously been reported.[5] These effects are likely to be of more importance at lower growth temperatures.

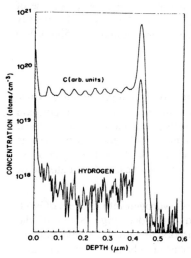

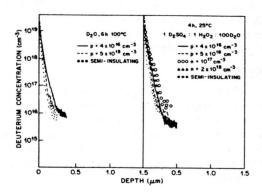

Figure 2. SIMS profiles of deuterium in GaAs of various doping densities boiled in D_2O for 6h (left) or etched in D_2SO_4 : H_2O_2 : 100 D_2O for 4h at 25 °C.

Figure 1. SIMS profile of a MOMBE grown GaAs(C) epitaxial layer.

SIMS profiles of deuterium in water-boiled or wet-etched GaAs of different doping densities are shown in Fig. 2. We found that the samples had to be boiled in the dark for measurable injection of the hydrogen. For both water-boiling and wet etching treatments the incorporation depth of the deuterium is relatively small, but shows a reproducible trend with the doping density in the GaAs. In each case an increase in either the p- or n-type doping level in the material lead to a decrease in the permeation depth of the deuterium. There was also an excellent correlation between the electrical passivation depth and the distance to which the deuterium was incorporated.

Carrier profiles in n-type (10^{17} cm^{-3}) GaAs or AlGaAs samples exposed to $C_2H_6/H_2/Ar$ plasmas are shown in Fig. 3 as a function of post-RIE annealing temperature. The evolution of the carrier profiles with annealing is due to two effects-damage-related deep levels and passivation of Si donors by atomic hydrogen. Upon annealing at 200°C the carrier reduction moves to greater depths, which is due to motion of the hydrogen, a fact confirmed by SIMS measurements. The reduction in carrier concentration is much greater with H_2-based discharges than in the case of etching under the same conditions with a CCl_2F_2/O_2 mixture, which produces compensation by damage but not hydrogen passivation. Similar results are obtained with InP and InGaAs or with etching in CH_3Cl/H_2. The deposition of SiN_x dielectric films by plasma-enhanced (CVD) using silane can also lead to hydrogen incorporation into III-V materials,[6] with associated passivation of near-surface dopants.

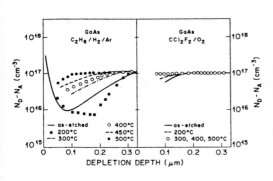

Figure 3. Carrier profiles in n-type (10^{17} cm^{-3}) GaAs etched in either a 19:1 $CCl_2F_2:O_2$ or 1:10:3 $C_2H_6:H_2:Ar$ (4 mTorr, 0.85 W · cm^{-2}), as a function of post- RIE annealing temperatures.

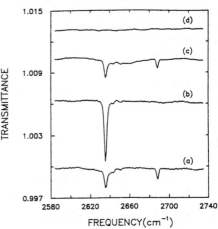

Figure 4. IR spectra from GaAs (C) MOMBE-grown epitaxial layers (a) as-grown (b) after annealing at 450°C for 20 min in H_2 (c) after 5 min, 600°C anneal in N_2/H_2 and (d) after 5 min, 600°C anneal in He.

There are a variety of different heat treatments of III-V's such as ohmic contact alloying at ~400°C, implant isolation annealing at ~500°C and implant activation annealing at ~800°C, which are generally performed in H_2-containing ambients. To study whether or not hydrogen might be introduced during this kind of annealing, the C-H absorption signal from GaAs (C) epi layers was monitored in as-grown and heat-treated samples. Figure 4 shows IR absorption spectra from as-grown material, with two features at 2635.2 cm^{-1} (C-H) and 2688.2 cm^{-1} (suggested to be due to a multiple carbon-hydrogen complex[3]). The presence of these two features confirms there is hydrogen in the as-grown material. After a thermal anneal for 20 min at 450°C in H_2 there is an increase in the strength of the C-H line, and entry of hydrogen was confirmed by SIMS. Subsequent annealing for 5 mins at 600°C in a 90% N_2 : 10% H_2 ambient with the wafer face-down on another GaAs substrate reduced the intensity of the C-H lines, but they are still comparable to the as-grown sample. Similar annealing in a He ambient removed the IR signals to the sensitivity of the apparatus and this again emphasizes the influence of the annealing ambient on the indiffusion and outdiffuson of hydrogen.

Conclusion

We have demonstrated a number of different growth and processing steps during which atomic hydrogen is incorporated into III-V semiconductors. The major effect of this phenomenon is a reduction of the carrier concentration in the near-surface (≤ 1 μm) region, a problem in most device structures where one wishes to maintain the initial doping level.

Acknowledgment

The work by M. Stavola and D. M. Kozuch was supported by NSF Grant No. DMR-9023419.

References

1. C. R. Abernathy, S. J. Pearton, F. A. Baiocchi, T. Ambrose, A. S. Jordan, D. A. Bohling and G. J. Muhr, J. Cryst. Growth *110* 457 (1991).

2. W. S. Hobson, T. D. Harris, C. R. Abernathy and S. J. Pearton, Appl. Phys. Lett. *58* 77 (1991).

3. D. M. Kozuch, M. Stavola, S. J. Pearton, C. R. Abernathy and J. Lopata, Appl. Phys. Lett. *57* 2161 (1990).

4. D. M. Kozuch, M. Stavola, C. R. Abernathy, W. S. Hobson, J. Lopata and S. J. Pearton, to be published.

5. G. R. Antell, A. Briggs, B. Butler, S. Kitching, J. Stagg, A. Chen and D. Sykes, Appl. Phys. Lett. *53* 1 758 (1988).

6. J. P. DeSouza, D. K. Sadana, H. Bacatte and F. Cardone, Appl. Phys. Lett. *57* 1129 (1990).

Inst. Phys. Conf. Ser. No 120: Chapter 4
Paper presented at Int. Symp. GaAs and Related Compounds, Seattle, 1991

199

Improvement of breakdown voltage characteristics of GaAs junction by damage-creation of ion-implantation

Yukiharu Shimamoto, Tsuyoshi Tanaka, Keijiro Itakura, and Daisuke Ueda
Matsushita Electronics Corp., Takatsuki Osaka 569, JAPAN

ABSTRACT

We have developed a technique which improves breakdown voltage of GaAs junction (Schottky or pn junction) by damage-creation of ion-implantation around the junction edge followed by appropriate annealing. Owing to the high-resistive region formed at the surface by ion-implantation, depletion layer is widened laterally in the region, so that the field intensity at the junction-edge is weakened. The technique allows us to obtain ideal characteristics of breakdown voltage. This paper reports the characterization of thus obtained diodes and the investigation of the ion-implantation effect.

1.INTRODUCTION

Increasing attention has been paid to improve the breakdown voltages of HBT or Schottky Barrier Diode (SBD) aiming at the applications to power devices. Mesa etching or field plate structure has been widely used for that purpose (M.F.Chang et al.1990, D.A.Grant et al 1989). However, the ideality factor (simply defined as; measured breakdown voltage / ideal breakdown voltage) of those conventional structures are still around 80 to 90 %. This is due to the electric field at the junction edge that is not weakened enough to suppress the avalanche breakdown.

To solve this problem, we developed a simple edge-termination technique that can achieve an ideal breakdown voltage. The new technique employs the ion-implantation at the periphery of the junction, so that the carrier concentration at the surface is reduced by the carrier compensation of the implantation-damage. Owing to this damaged region formed around the junction, depletion layer is much more widened laterally at the surface, so that the field intensity at the junction-edge is weakened enough to be below the critical electric field where the avalanche breakdown occurs. This paper describes the breakdown characteristics of these diodes and the investigation of the ion-implantation effect.

2.EXPERIMENTAL

The used wafers were n^- / n^+ GaAs epitaxial wafers grown by vapor phase epitaxy. The n^- epi-layer has the concentration of $5 \times 10^{14} cm^{-3}$ and the thickness of 20 μm. AuGeNi/Au (1500/2500Å) was deposited on the backside of the wafers and alloyed at 500°C for 5 min to make Ohmic contacts. The circular shaped Schottky electrode (Ti /Al : 500/20000Å) was formed on n^- layer using an e-beam evaporator. The diameter of the electrode was 300 μm. Ion-implantation was carried out into the uncovered region of the electrode using B^+ or H^+ ion. The acceleration voltages used in the ion-implantation were 50 or 100kV. The condition of the dose is ranging from 5×10^{10} to 5×10^{13} cm^{-2}. Low temperature annealing was also performed at 300 and 500 °C for 5min in a furnace.

In order to compare the breakdown performance of this new technique with conventional ones, two different mesa structure were fabricated using the same wafer. One is conventional tapered (positive-type) mesa obtained by wet etching, the other is vertical mesa obtained by dry etching. The former mesa was formed by the solution of H_3PO_4 :H_2O_2:H_2O = 4:1:45. The height of the resultant mesa is about 3000Å. The latter mesa structure was formed by reactive ion etching using a gas mixture of CCl_2F_2 and He. The latter one was formed to the depth of 20000Å, which reached n$^+$ substrate. This vertical mesa structure is expected to show ideal breakdown performance because there is no junction curvature inside the device.

3.RESULTS & DISCUSSION

Obtained breakdown voltage as a function of dose is shown in Figure 1. The reverse voltages of these samples were measured at the current of 500 μ A. The figure indicates that there is an optimum condition of the dose giving the maximum breakdown voltage. The reason why the sample with the dose more than $5x10^{13}cm^{-2}$ shows poor breakdown characteristic is that the leakage current increases as the increase of the dose. To identify the leakage current under the reverse bias condition, the current was measured varying the temperature of the device under test. We plotted this leakage current on the arrhenius chart shown in Figure 2 (a). It is observed that there are three regions that has different activation energy in the temperature range.

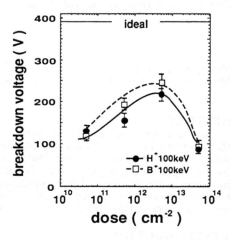

Figure 1 Obtained breakdown voltages as a function of doses as implanted.

Namely, the ion-implantation damage caused three different states in the energy gap. The activation energies are 0.8, 0.4, 0.02eV from valence band.

In order to reduce the leakage current to the allowable range, low temperature annealings were performed at the temperature of 300 and 500 °C . The highest slope (0.8eV) diminished after annealing as shown in Figure 2 (b).

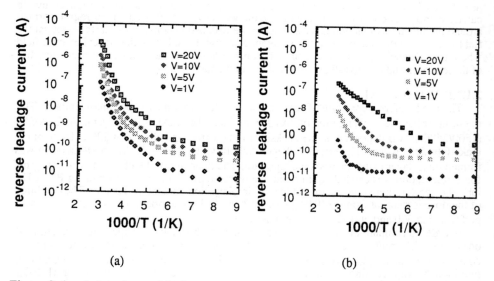

(a)   (b)

Figure.2 Arrehnius chart of leakage current (a) as implanted and (b) after annealing, respectively.

This results in the drastic reduction of leakage current that is based on SRH statistics. This means that the damage with large activation energy is easily annealed out below the temperature of 500 °C (K.Itakura et al to be published). It is noted that no change of barrier height was observed by forward conduction characteristics before and after annealing.

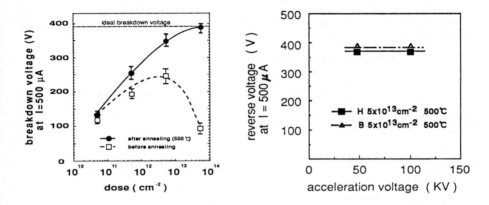

Figure 3 Annealing effects on breakdown voltage.   Figure 4 Obtained breakdown voltages as function of accelerate voltage.

As a result, the breakdown voltage after annealing reached the ideal one calculated for the present epi-layer, as shown in Figure 3. Note that the ideal breakdown voltage for the present epi wafer is 390V (S. M.Sze 1981). Although the damage that is relevant to the leakage current was cured, residual defects keep compensating the carrier at the surface.

These samples have shown the same ideal breakdown voltage regardless of the acceleration voltages of ion-implantation or the kind of ions as shown in Figure 4. It is noted that the breakdown voltage is unchanged as long as the sufficient amount of dose (5×10^{13}) are provided. Consequently, a certain minimum damage is essential to obtain the ideal breakdown voltage. This characteristics give us a wide range of tolerance in the processing.

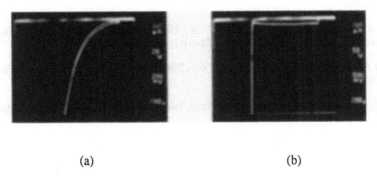

(a)   (b)

Figure 5 Breakdown characteristics of SBD (a) as implanted and (b) after annealing, respectively.

The sample after annealing shows a quite hard breakdown characteristic in contrast with the one before annealing as shown in Figure 5.

The comparative results with other device structures are summarized in table 1. Present edge-termination structure achieved the highest breakdown voltage among all the structures compared. It is noted that the vertical mesa structure did not achieved the ideal breakdown voltage. The reason would be the surface contamination at the vertical sidewall caused local increase of the electric field. On the contrary, the ion-implanted devices are quite stable since the surface region was turned into high resistive. So, the electric field in the semiconductor is less affected by the contaminants.

Table 1 A comparison of the ideality factors of SBDs.

	breakdown voltage	η	Fabrication technique
positive mesa structure	330V	87%	Wet etching based on phosphoric acid
vertical mesa structure	350V	90%	Reactive ion etching
ion-implanted structure	390V	100%	Ion implantation

4.SUMMARY

We found that the ideal breakdown voltage of GaAs junction (Schottky or pn junction) is obtainable by the damage-creation of ion-implantation around the junction-edge followed by appropriate annealing. This newly developed technique is quite simple and applicable to various GaAs power devices with pn junctions as well as with Schottky ones.

5.ACKNOWLEDGMENTS

The authors would like to thank Dr. I. Teramoto, Dr.G. Kano, Mr.M. Kazumura, and Dr.M. Takeshima for their encouragement in this work.

REFERENCES

Chang M F, Wang N and Asbeck P M, 1990 Extended Abstracts of the 22nd Conference on Solid State Devices and Materials (Sendai) pp 47

Grant D A and Gowar J, 1989 power MOSFETs Theory and Applications, (New York: John Wiley & Sons) pp147

Sze S M, Physics of Semiconductor Devices 2nd edition 1981, (New York: John Willy & Sons) pp103

Itakura K, Shimamoto Y and Ueda D , J.Appl.Phys., to be published

Inst. Phys. Conf. Ser. No 120: Chapter 4
Paper presented at Int. Symp. GaAs and Related Compounds, Seattle, 1991

Small size collector-up AlGaAs/GaAs HBTs fabricated using H+ implantation

Takashi Hirose, Kaoru Inoue and Masanori Inada

Semiconductor Research Center, Matsushita Electric Industrial Co., Ltd.
3-15 Yagumo-nakamachi, Moriguchi, Osaka 570, Japan

Abstract : Small size collector-up AlGaAs/GaAs HBTs have been developed making the entire emitter layer under the external base highly resistive by oblique H+ implantation from both sides of the collector normal mesa . A collector cap layer of In0.5Ga0.5As and a high dose base layer doped with Be of 4×10^{19} cm^{-3} are formed in a multilayer structure for the HBTs. A current gain cutoff frequency f_T of 31GHz and a maximum oscillating frequency f_{max} of 50 GHz have been obtained in the HBT with a collector size of 1×10 μm^2 .

1. Introduction

A collector-up heterojunction bipolar transistor (C-up HBT) has such merits [1,2] as the ability to improve high-frequency properties by decreasing the base/collector capacitance, and the ability to make the emitter-common IC, for example the integrated injection logic (I^{2}L). However, the C-up HBT necessitates the suppression of leakage current from the emitter to the external base. To solve this problem, it has been reported with regard to a large size HBT that H+ implantation through the external base is effective [3]. However, in order to improve the high-frequency properties of the HBT, it is imperative not only to minimize each part of the HBT device but also to make the overall emitter area under the external base highly resistive. This is necessary because the influence of leakage current from an intrinsic emitter region to the external base region becomes noticeable as the size of HBT becomes smaller.

In this work, we present small size C-up HBTs with good high-frequency properties by a very simple method using oblique H+ implantation from both sides of the collector normal mesa and a pattern inversion self-alignment process.

2. Experimental

A multilayer structure for the C-up HBT used in this experiment is shown in Table I, which was grown by molecular beam epitaxy (MBE) on a semi-insulating (100) GaAs substrate. An abrupt base/emitter junction was adopted, which consists of an N-Al0.3Ga0.7As / u-GaAs / p+-GaAs layer with an undoped GaAs (u-GaAs) layer of 2.5 nm as a spacer layer. A collector cap layer of n+-In0.5Ga0.5As was formed via composition-graded n+-InyGa1-yAs (y=0-0.5) layer.

A cross section of the C-up HBT

Table I. A multilayer structure for the C-up HBT grown by MBE.

Layer type and composition	Doping (cm^{-3})	Thickness (nm)
n+-In0.5Ga0.5As	2×10^{19}	50
n+-InyGa1-yAs (y=0.5–0.0)	2×10^{19} ~ 5×10^{16}	50
n -GaAs	5×10^{16}	300
p+-GaAs	4×10^{19}	100
u -GaAs	—	2.5
N -Al0.3Ga0.7As	5×10^{17}	500
N -AlxGa1-xAs (x=0.3–0.0)	5×10^{17} ~ 5×10^{18}	30
n+-GaAs	5×10^{18}	600
SI - GaAs substrate		

fabricated is shown in Fig. 1. The fabrication process of the HBT is shown in Figs. 2(a)-2(e).

(a) The device isolation was carried out by H+ implantation using a mask consisting of resist/Au.

(b) A collector normal mesa was made by wet etching using a SiO2 dummy collector elongated to the [0$\bar{1}$1] direction as a mask. A solution with a volume ratio of H2SO4(96%) : H2O2(30%) : H2O = 1 : 8 : 160 was used for the wet etching [5]. The emitter layer under the external base was made highly resistive by implanting H+ at 80 keV and 5 x 10^{13} cm^{-2} from both sides of the collector normal mesa with a tilt angle of 7°.

(c) The dummy collector was replaced with the collector electrode by means of a pattern inversion method [4], and emitter electrodes were formed by a conventional lift-off method.

(d) The base electrodes were formed close to the collector normal mesa by self-alignment [4] using the collector electrode as a mask.

(e) The C-up HBT was completed by carrying out wiring over a SiO2 insulating film.

Metals of AuGe / Ni / Ti / Au, Cr / AuZn / Au and Ti / Au were used for n-type electrodes, p-type electrodes and wiring, respectively.

DC characteristics of the fabricated C-up HBT were measured by a curve tracer, and on-wafer measurements of its high-frequency properties were performed from 1 GHz to 26.5 GHz by a network analyzer.

In order to analyze the high-frequency properties thus obtained, the contact resistivities of respective electrodes were measured with a transmission line method (TLM) having an electrode size of 100 x 100 μm^2 and the distance between the electrodes of 5, 10, 20 and 40 μm in the same wafer as the fabricated HBT. The capacitances between the electrodes of the HBT were measured by a C-V meter at 1 MHz with zero bias.

3.Results and Discussion

DC characteristics for the small size C-up HBT with a collector size of 1 x 10 μm^2 and a base size of 4 x 10 μm^2 are

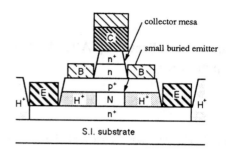

Fig. 1. Cross section of a C-up HBT fabricated by a one-mask multiple self-alignment process.

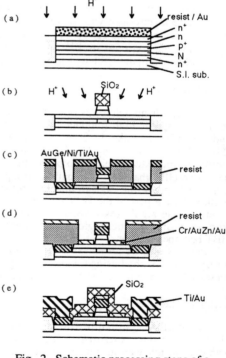

Fig. 2 . Schematic processing steps of a C-up HBT.

(a) Device isolation.

(b) Fabrication of a small buried emitter by oblique H+ implantation using a dummy collector as a mask.

(c) Simultaneous fabrication of emitter and collector electrodes.

(d) Fabrication of base electrodes.

(e) Final structure of a C-up HBT.

shown in Fig. 3. The DC current gain of about 5 is obtained in this small size HBT,which indicates that the leakage current from the emitter layer to the external base is considerably suppressed by using our process technology in which the base/emitter junction area was effectively made smaller than the base/collector junction area using oblique H^+ implantation from both sides of the collector normal mesa. This is also clear from the fact that no DC current gain could be obtained in the C-up HBT, which was made using the collector invert mesa and ordinary H^+ implantation, because the emitter/base junction area is larger than the collector/base area and hence the leakage current from the emitter to the base electrodes is quite large 6).

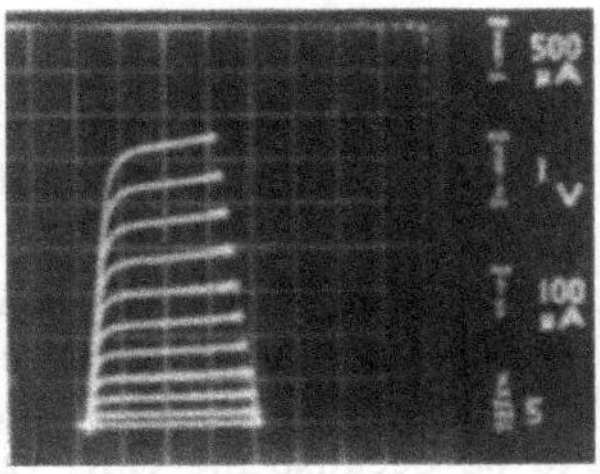

In the fabricated C-up HBT, a small size collector of 1×10 μm^2 and minimization of the distance between the collector normal mesa and the base electrode are realized by using the pattern inversion method 4) and the wet etching technique 5) which makes it possible to form the bottom of the collector normal mesa very close to the base electrode .

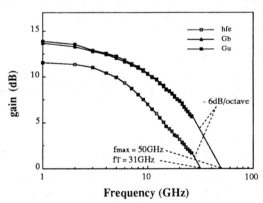

Fig. 3. DC characteristics of a C-up HBT of a collector size of 1x10 μm^2 with a small buried emitter fabricated by H^+ implantation.

Figure 4 shows the high-frequency current gain (hfe), the bilateral power gain (Gb) and the unilateral power gain (Gu) of the same HBT as that in Fig. 3 at the bias condition of Vce = 2.9 V and Ic = 3.7 mA. The bilateral power gain (Gb) was in a state of the most stable gain (MSG) in a measuring range from 1 GHz to 26.5 GHz. The current gain cut-off frequency (fT) of 31 GHz and the maximum oscillating frequency (fmax) of 50 GHz were obtained by extrapolating the values of hfe and Gu at 26.5 GHz using the -6 dB/oct. lines, respectively.

Next, the high-frequency properties were analyzed by using the experimentally obtained parameters of the emitter resistance (RE), the collector resistance (RC), the base/emitter capacitance (CBE) and the base/collector capacitance (CBC), and by applying the saturation velocity model in the collector depletion layer and the diffusion model in the base. RE = 7.3 Ω and RC = 5.9 Ω were obtained by calculating from the contact resistivities obtained by TLM and the size of each part of the HBT. CBE = 190 fF and CBC = 20 fF were obtained by calculating those values at the above-mentioned bias condition from the C-V measurement values at zero bias.

Using these values, the emitter charging time (τE) of 0.68 ps and the collector charging time (τCC) of 0.20

Fig. 4 High-frequency properties of a C-up HBT of a collector size of 1x10 μm^2. Vce = 2.9 V, Ic = 3.7mA

ps were obtained. The collector depletion layer transit time (τC) of 1.93 ps was obtained for the depletion layer width (WC) of 270 nm and the electron saturation velocity (vsat) of 7.0 x 10^6 cm/s. The base transit time (τB) of 1.6 ps was obtained by assuming the electron mobility (μe) of 1200 cm^2/(V·s). The calculated fT value was 36 GHz, which is in good agreement with the measured fT value of 31 GHz.

The base resistance (RB) in high-frequency operation was roughly estimated using the values of fT = 31GHz, fmax = 50GHz and CBC = 20fF, and the following relation (eq.(1)), because the double oblique H$^+$ implantation caused a significantly different damaged state between the p$^+$-GaAs layer for the TLM and the p$^+$-GaAs external base layer. Hence, RB at DC of the HBT device could not be calculated accurately from the TLM values.

$$f_{max} = \left(\frac{f_T}{8\pi\, R_B\, C_{BC}} \right)^{1/2} \qquad \text{--- (1)}$$

The obtained RB of 24.7 Ω was nearly 4 times larger than that of the emitter-up (E-up) HBT [7] with nearly the same geometry as that of the C-up HBT. On the other hand, the CBC value of the fabricated C-up HBT was almost half that of the E-up HBT [7]. This is considered to be because the CBC value of the C-up HBT does not contain parasitic capacitance. The increase in RB is considered to result from the damage to the external base by the H$^+$ implantation. Accordingly, the effect of the reduction in CBC by the C-up HBT is canceled by the increase in RB. To strengthen the effect of the reduction in CBC, a reduction in RB will be necessary. It is considered that the reduction in the dosage of H$^+$ implantation by using an emitter layer with much lower doping level and the use of a much highly doped base layer will further improve the fmax value.

4. Summary

A small size C-up AlGaAs/GaAs HBT with a collector size of 1 x 10 μm^2 was developed by making the entire emitter layer under the external base highly resistive by oblique H$^+$ implantation from both sides along the elongated direction of the collector normal mesa and using the pattern inversion self-alignment process. The HBT showed high-frequency properties with fT = 31 GHz and fmax = 50 GHz. However, the effect of the reduction in CBC was canceled by the increase in RB incurred by the H$^+$ implantation. By optimizing the condition of the H$^+$ implantation and the multilayer structure, much a higher fmax value will be possible.

5.Acknowledgements

The authors are grateful to M. Yanagihara for his cooperation with the ion implantation, and H. Ikeda for his cooperation in performing the high-frequency measurement.

References

1) H. Kroemer: Proc. IEEE 70 (1982) 13.
2) H. T. Yuan, H. D. Shih, J. Delaney and C. Fuller: IEEE Trans. Electron Devices ED-36 (1989) 2083.
3) M. Yanagihara, Y. Ota, A. Ryoji and M. Inada: Jpn. J. Appl. Phys. 29 (1990) L2174.
4) M. Inada, Y. Ota, A. Nakagawa, M. Yanagihara, T. Hirose and K. Eda: IEEE Trans. Electron Devices ED-34 (1987) 2405.
5) D. W. Shaw: J. Electrochem. Soc. 128 (1981) 875.
6) M. Yanagihara : private communication.
7) Y. Ota, T. Hirose, A. Ryoji and M. Inada: Electron. Lett. 26 (1990) 203.

Inst. Phys. Conf. Ser. No 120: Chapter 4
Paper presented at Int. Symp. GaAs and Related Compounds, Seattle, 1991

207

Elimination of mesa-sidewall gate-leakage in InAlAs/InGaAs HFETs by selective sidewall recessing

Sandeep R. Bahl and Jesús A. del Alamo

Massachusetts Institute of Technology, Cambridge, MA 02139

Abstract: Conventional mesa isolation in InAlAs/InGaAs HFET's results in the gate contacting the exposed channel at the mesa sidewall, forming a parasitic gate-leakage path. We successfully propose and demonstrate a novel and simple method of recessing the channel edge into the mesa sidewall using a succinic acid based selective etchant for InGaAs over InAlAs. SEM photographs confirm the recessing of the channel along the sidewall. Measurements on HFET's and on specially designed test diodes fabricated with and without the sidewall isolation step confirm the complete elimination of sidewall-leakage.

1. Introduction

InAlAs/InGaAs Heterostructure Field Effect Transistors (HFET's) on InP have recently emerged as an optimum choice for a variety of microwave and photonics applications (Smith *et al.* 1990). Fabrication of these HFET's by conventional mesa isolation, however, results in sidewalls where the InGaAs channel is exposed and comes in contact with the gate metallization running up the mesa (fig. 1). The low Schottky barrier height of metals on InGaAs (Wieder 1981) potentially results in a sidewall leak-

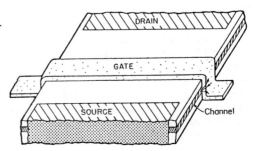

Fig. 1: 3-D Perspective of the HFET showing the mesa-sidewall gate-leakage path.

age path from the gate to the channel. Brown *et al.* (1989) have associated sidewall-leakage in InGaAs HFET's with excess gate leakage current and a reduced breakdown voltage. Bahl and del Alamo (1990) found that this worsens with a high doping and increased x ($x > 0.53$) in the $In_xGa_{1-x}As$ channel. This is particularly consequential, since HFET's with InAs enriched channels have shown excellent electron transport (Ng *et al.* 1989, Bahl and del Alamo 1990).

In this work, we have selectively recessed the the exposed InGaAs channel into the sidewall, forming an air cavity. This isolates the gate from the channel at the sidewall.

2. Experimental

The MBE grown HFET, lattice-matched to S.I. InP, consists of (bottom to top) a 1000Å undoped $In_{0.52}Al_{0.48}As$ buffer layer, a 100Å undoped $In_{0.53}Ga_{0.47}As$ subchannel, a 80Å heavily Si doped ($N_D=6 \times 10^{18}$ cm^{-3}) $In_{0.53}Ga_{0.47}As$ channel, a 300Å undoped $In_{0.52}Al_{0.48}As$ gate insulator layer, and an undoped 50Å $In_{0.53}Ga_{0.47}As$ cap. The heavy doping in the channel makes this heterostructure particularly sensitive for studying sidewall-leakage.

Devices were fabricated by first chemically etching a mesa down to the InP substrate using a $H_2SO_4:H_2O_2:H_2O$ 1:10:220 etch. Then, before removing the mesa-level photoresist mask, the wafer was dipped for 45 sec. into a $SA:H_2O_2$ 6:1 solution, developed by Broekaert and Fonstad (1991), to selectively etch the exposed portion of the $In_{0.53}Ga_{0.47}As$ channel. The SA solution was prepared by adding 1 liter H_2O to 200 g. succinic acid with the addition of ammonium hydroxide until the pH was 5.5. A planar selectivity of 23:1 was measured, with the

$In_{0.52}Al_{0.48}As$ etching at 25 Å/min. For reference, a portion of the wafer was masked during this etch by coating with photoresist using a paintbrush. In this portion, mesa etching was performed, but no selective etching was carried out. For the ohmic contacts, 2000Å of AuGe and 600Å of Ni were evaporated, lifted off, and RTA alloyed at 360°C. For the gate and pad, 300Å of Ti, 300 Å of Pt, and 2000 Å of Au were e-beam evaporated and lifted off.

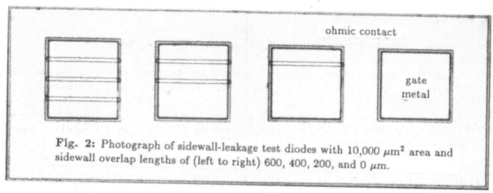

Fig. 2: Photograph of sidewall-leakage test diodes with 10,000 μm^2 area and sidewall overlap lengths of (left to right) 600, 400, 200, and 0 μm.

To study sidewall-leakage, we fabricated special-purpose heterojunction diodes with a mesa area of 10,000 μm^2. Gate/mesa sidewall overlaps were created by etching grooves through the mesa, while keeping the unetched mesa area constant. Twelve diodes were fabricated with sidewall-overlap lengths of 0, 200, 400, and 600 μm, running in each of [011], [001], and [0$\bar{1}$1] crystallographic directions. Fig. 2 shows a photograph of the diode test structure with various sidewall-overlap lengths. HFET's were fabricated with nominal gate-length, L_G=1, 2, and 5 μm and gate-width, W_G=30 μm with mesa sidewalls running in the [011], [001], and [0$\bar{1}$1] crystallographic directions.

3. Results and discussion

Fig. 3 shows an SEM photograph and a sketch showing a HFET structure with a 420 Å $In_{0.53}Ga_{0.47}As$ channel processed with the other device samples. The thicker channel was chosen for clarity in SEM imaging. The presence of a sidewall cavity is clearly revealed. This

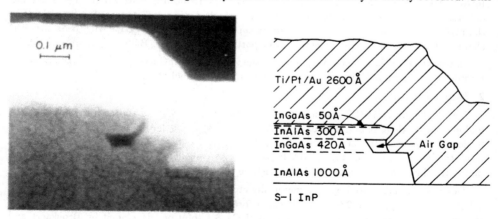

Fig. 3: SEM photograph and explanatory sketch of the HFET sidewall, showing an air gap isolating the gate metal from the channel.

confirms the successful action of the selective etchant. The overhang formed by the InAlAs insulator prevents the gate from contacting the channel edge.

Fig. 4 shows the I-V characteristics of typical heterojunction diodes from the non-isolated portion of the wafer with grooves along the [0$\bar{1}$1] direction. Both forward and reverse currents increase with sidewall-overlap length, as expected from the increase in contact area. The figure shows that sidewall-leakage plays a major role from forward-bias till threshold (-0.82 V), when the channel is depleted. We also find (not shown) that sidewall-leakage decreases with crystallographic orientation as [0$\bar{1}$1] > [001] > [011]. This is consistent with a V profile along the [0$\bar{1}$1] oriented sidewall (Shaw 1981) making a better contact with the gate than an inverted V on the [011] sidewall.

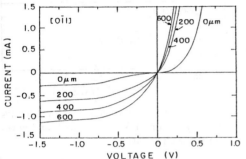

Fig. 4: I-V characteristics of test diodes without sidewall isolation as a function of sidewall overlap length along [0$\bar{1}$1].

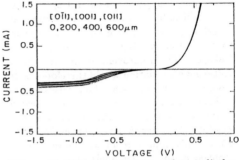

Fig. 5: I-V characteristics of test diodes with sidewall isolation for different orientations and sidewall overlap lengths.

The elimination of sidewall-leakage results in the disappearance of both sidewall length and orientation dependence. Fig. 5 shows the I-V characteristics of twelve diodes (one of each stripe and orientation) from the sidewall isolated portion of the wafer. All reverse characteristics are tightly clustered around the baseline area leakage characteristic observed for zero overlap length. The absence of orientation and overlap length dependence confirms that sidewall-leakage has been completely eliminated.

To detect the presence of, and to confirm the elimination of sidewall-leakage in the HFET's we measured the subthreshold current, $I_{D(SUB)}$ at $V_{DS}=4$ V and V_{GS} approximately 0.5 V below threshold. Fig. 6 is a statistical (averaged over 5 HFET's) plot showing the dependence of $I_{D(SUB)}$ on crystallographic orientation for $L_G=1$, 2, and 5 μm on HFET's with and without sidewall isolation. For the non-isolated HFET's, $I_{D(SUB)}$ increases with L_G (sidewall-overlap

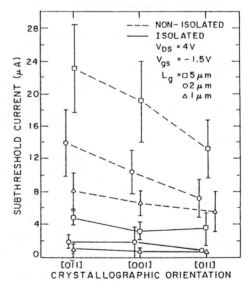

Fig. 6: Subthreshold current vs. crystallographic orientation for HFET's with $L_g=1$, 2, and 5 μm and $W_g=30$ μm with and without sidewall isolation.

length) and shows a strong dependence on crystallographic orientation. Both these dependences are consistent with our findings from the diodes. Sidewall isolation, however, results in a reduction of $I_{D(SUB)}$ between 3 and 9 times for these L_G's, and the disappearance of orientation dependence. The residual dependence on L_G is due to the increased gate area. These results confirm that our sidewall isolation scheme has successfully eliminated sidewall-leakage in HFET's for all crystallographic orientations.

Fig. 7 shows the gate characteristics of typical HFET's with $L_G=1$ μm fabricated with and without sidewall isolation. In forward-bias, sidewall-isolation results in a much larger turn-on voltage. For reverse-bias, both characteristics saturate at threshold. Beyond threshold, they appear to be shifted by the amount of sidewall-leakage. The improvement in breakdown voltage with isolation would therefore result from the reduced pre-threshold reverse gate-leakage current. In these devices, sidewall isolation has increased the breakdown voltage (at -50 μA over 5 devices) from 18.9 V to 20 V. This improvement will be more substantial in thicker channels and in channels with higher InAs mole fractions (Bahl and del Alamo, 1990).

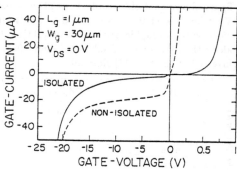

Fig. 7: Gate characteristics of HFET's with $L_g=1$ μm at $V_{ds}=0$ V, with and without sidewall isolation.

4. Conclusion

A simple self-aligned technique for eliminating mesa-sidewall gate-leakage has been developed. This technique uses selective etching to etch the exposed part of the InGaAs channel into the mesa-sidewall, creating a cavity to isolate the channel from the gate. Measurements on specially designed diodes and HFET's have shown complete elimination of sidewall-leakage.

Acknowledgements

We thank Prof. C. G. Fonstad for the use of his MBE, T. P. E. Broekaert for valuable help with the etch recipe, and W. Y. Choi for suggesting this method of sidewall isolation. This work has been funded by the Joint Services Electronic Program under the Research Laboratory of Electronics (DAAL-03-89-C-0001) and the C. S. Draper Laboratory (DL-H-418488).

References

Bahl S R and del Alamo J A 1990, *Proc. 2nd Int. Conf. on InP and Related Mat.*, p. 100.
Broekaert T P E and Fonstad C G 1991, *IEEE Trans. Elec. Dev.*, Accepted for publication.
Brown A S, Chou C S, Delaney M J, Hooper C E, Jensen J F, Larson L E, Mishra U K, Nguyen L D, and Thompson M S 1989 *Proc. IEEE GaAs IC Symp.*, 143.
Ng G-I, Pavlidis D, Jaffe M, Singh J, and Chau H-F 1989, *IEEE Trans. Elec. Dev.*, **36** 2249.
Shaw D W 1981, *J. Electrochem. Soc.* **128** 874.
Smith P M, Chao P C, Ho P, Duh K H, Kao M Y, Ballingall J M, Allen S T, and Tessmer A 1990 *Proc. 2nd Int. Conf. on InP and Related Materials*, Denver, CO. 39.
Wieder H H 1981, *Appl. Phys. Lett.*, **38** 170

Inst. Phys. Conf. Ser. No 120: Chapter 4
Paper presented at Int. Symp. GaAs and Related Compounds, Seattle, 1991

211

Electrical properties of n-type and p-type $Al_{0.48}In_{0.52}As$ Schottky barriers

L.P. Sadwick[*] C.W. Kim[*], K.L. Tan[†] and D.C. Streit[†]
[*]Electrical Engineering Department ,The University of Utah, Salt Lake City, Utah 84112
[†]TRW, Inc., Redondo Beach, California 92072

ABSTRACT

We report electrical measurements on four different metal contacts which formed Schottky barriers to lightly doped complementary n- and p-type $Al_{0.48}In_{0.52}As$ epitaxial material grown by molecular beam epitaxy on semi-insulating InP substrates. The Schottky contact metals studied were Au, Al, Pt and the tri-layer Ti/Pt/Au. Schottky barrier heights were determined by current versus voltage and capacitance versus voltage measurements. The Schottky barrier heights obtained from current versus voltage measurements varied from 0.560 eV for Al on n-type $Al_{0.48}In_{0.52}As$ to 0.905 eV for Al on p-type $Al_{0.48}In_{0.52}As$, with intermediate values for the other metals studied. The sum of the n- and p-type Schottky barrier heights for each metal contact ranged from 1.440 to 1.465 eV, in good agreement with the accepted $Al_{0.48}In_{0.52}As$ band-gap value of 1.45 eV . The agreement between the sum of the n- and p-type Schottky barrier heights obtained from capacitance versus voltage measurements and the accepted $Al_{0.48}In_{0.52}As$ band-gap value were not as good as those obtained from current versus voltage measurements. An anomalously large enhanced Schottky barrier height was observed for the Al on p-type $Al_{0.48}In_{0.52}As$ system from capacitance versus voltage measurements.

I. INTRODUCTION

Lattice-matched AlInAs/InGaAs/InP devices are very attractive for high speed, high frequency transistor applications. As such, the characteristics of Schottky barriers to $Al_xIn_{1-x}As$ (x = 0.48) are of fundamental interest for device applications. AlInAs epitaxial layers have been the subject of numerous investigations in the last decade. These studies have included AlInAs energy band-gap dependence on alloy composition by optical methods (Wakefield, et al., 1984); growth and characterization of MBE grown AlInAs (Davies, et al., 1984); photoemisson studies (Hsieh, et al., 1985); AlInAs Schottky barrier height compositional dependence work (Lin, et al., 1986); further investigations of the Schottky barrier height of AlInAs (Chu, et al., 1986), (Chu, et al., 1988), (Chu, et al., 1988); and barrier height studies (Ohno, et al., 1981)-(Ohno, et al., 1980).

Previous investigations of the barrier height of n-type $Al_{0.48}In_{0.52}As$ have been based on CV and internal photoemission measurements (Wakefield, et al., 1984), (Davies, et al. 1984), (Hsieh, et al. 1985) (Lin, et al., 1986), (Chu, et al., 1986), (Chu, et al., 1988), (Chu, et al., 1988),(Landolt-Bornstein, 1987). Chu *et al.* (Chu, et al., 1988) investigated the barrier height of n-type $Al_xIn_{1-x}As$ grown by MBE at a substrate temperature of 580°C over an In mole fraction range of 0.435 < 1-x < 0.620 or, equivalently, 0.360 < x < 0.565. They determined, within experimental error, that the barrier height was essentially independent of composition and Schottky metal contact (Au, Al, or Hg) and given by $\phi_n = 0.621 \pm 0.05$ eV. They also observed that an anomalous composition-dependent barrier height for samples grown at a substrate temperature of 500°C which they believed was possibly due to the formation of AlAs clusters in the $Al_xIn_{1-x}As$ epilayer. Ohno *et al.* (Ohno, et al., 1981)-

(Ohno, et al., 1980) determined a barrier height for n-type $Al_{0.48}In_{0.52}As$ of 0.8 eV primarily from CV measurements on 10-mil-diameter Al dots. Hsieh *et al.* (Hsieh, et al., 1985) have determined a value for the barrier height of n-type lattice matched AlInAs of 0.64 eV from internal photoemission studies of Au/AlInAs diodes.

II. EXPERIMENTAL

In this work we present experimental data on the fundamental electronic properties of the Schottky barrier height of complementary p- and n-type $Al_{0.48}In_{0.52}As$ Schottky barriers grown by molecular beam epitaxy. The n-type and p-type $Al_{0.48}In_{0.52}As$ samples were grown on semi-insulating InP and consisted of a 3000 Å heavily-doped 1×10^{19} cm^{-3} buried ohmic contact layer upon which approximately 3000 to 4000 Å of a lightly-doped Schottky surface layer was then grown. The group III fluxes were calibrated using RHEED intensity oscillations and the layers were determined to be within ±1% of the lattice-matched composition using double-crystal X-ray diffraction. The uniform, near-surface, low doping concentrations for both the p- and n-type $Al_{0.48}In_{0.52}As$ samples were chosen to be roughly 2×10^{16} cm^{-3}; were determined by capacitance versus voltage (CV) measurements to be approximately 1.4×10^{16} cm^{-3} ±3% and 2.4×10^{16} cm^{-3} ±2% for the p- and n-type devices, respectively. 4.15-mil-diameter (1 mil = 0.001 inch) metal dots were evaporated following a 7-mil-diameter mesa etching to expose the buried heavily-doped layer. One benefit of studying complementary p- and n-type Schottky devices is that the sum of the Schottky barrier heights, ϕ_p and ϕ_n, should be equal to the energy band-gap, E_g, of the semiconductor material (Mead, 1966) and (Sze, 1981). We will show that the barrier height summation rule is well satisfied for $Al_{0.48}In_{0.52}As$ using the four different metal contacts discussed below.

The four metal contacts which have been investigated in this work are Au, Al, Pt, and TiPtAu. The Schottky barrier height was extracted from current versus voltage (IV) and capacitance versus voltage (CV) measurements. For each Schottky contact, the value of the barrier height for p- and n-type AlInAs was determined from an average of a minimum of five individual device values. We used an effective Richardson constant of $0.15A^*$ for n-type $Al_{0.48}In_{0.52}As$ and $0.5A^*$ for p-type $Al_{0.48}In_{0.52}As$, based on measured data (Chu, et al., 1988) and (Landolt-Bornstein, 1987) and a linear approximation of the effective masses, m^*_n and m^*_p, of InAs and AlAs (Mead, 1966), and (Landolt-Bornstein, 1987). The value of A^* is 120 $Acm^{-2}K^{-2}$ (Sze, 1981). For the CV Schottky barrier height results we assumed a room temperature intrinsic carrier concentration of 1.8×10^6 cm^{-3}. Sensitivity and error analysis showed that a ±25% error in the respective effective Richardson constants and the intrinsic carrier concentration resulted in, at most, a 2% error in the barrier heights and band-gap.

III. DISCUSSION

The IV results obtained in this work could be accurately described by the ideal Schottky diode equation with an ideality factor, n, very close to unity (Sze, 1981). Table I summarizes the most pertinent properties obtained from IV measurements during our investigations. The summation of the p- and n-type Schottky barrier heights for a given contact is within ±1% of the accepted room temperature energy band-gap for $Al_{0.48}In_{0.52}As$ of 1.45 eV (Davies, et al., 1984) as determined by optical measurements. Specifically, the relative errors with respect to the accepted value of 1.45 eV are Au: 0.7%; Al: 0.7%; Pt: 0%; and Ti/Pt/Au: -0.7%. The value determined for the Au/n-type $Al_{0.48}In_{0.52}As$ is in good agreement with the value of 0.64 eV obtained by the internal photoemission technique (Hsieh, et al., 1985). It is also in fair agreement with the value determined by Chu *et al.* (Chu, et al., 1988) of 0.62 eV. However, our values are not in good agreement with the value of 0.8 eV obtained by Ohno et al. (Ohno, et al., 1981).

The C-V results obtained on the metal/ $Al_{0.48}In_{0.52}As$ samples were accurately described and modelled by the depletion approximation equation found in standard textbooks (Sze, 1981). The values of the doping concentrations obtained from the slopes of the individual $1/C^2$ vs. V curves were in excellent agreement with the average values of the doping concentration calculated over the corresponding depth range of 1600 to 3500 Å. The CV Schottky barrier heights were determined by the following equations:

$$|\phi_{pCV}| = |V_{pin}| + (E_{F_p} - E_V)/q - \Delta\phi_{peff} \tag{1}$$

$$\phi_{nCV} = V_{nin} + (E_C - E_{F_n})/q - \Delta\phi_{neff} \tag{2}$$

Where ϕ_{pCV} and ϕ_{nCV} are the p-type and n-type Schottky barrier heights, respectively, determined from CV measurements; V_{pin} and V_{nin} are the intercept voltages of the $1/C^2$ vs. V curves for the p-type and n-type samples, respectively; E_C is the conduction band energy of $Al_{0.48}In_{0.52}As$; E_V is the valence band energy of $Al_{0.48}In_{0.52}As$; E_{F_p} and E_{F_n} are the Fermi Energies for the p-type and n-type samples, respectively; q is the normalized elementary charge; and $\Delta\phi_{peff}$ and $\Delta\phi_{neff}$ are the effective Schottky barrier height (also known as the image-force-induced) lowering terms for the p-type and n-type samples, respectively. It should be noted that the intercept voltage of the $1/C^2$ vs. V curve, V_{in}, is equivalent to the built-in potential, V_{bi}, and that terms on the order of the room-temperature thermal voltage, kT/q, (where k is the Boltzmann constant) have been absorbed into the effective image-force-induced Schottky barrier height term, $\Delta\phi_{eff}$. The image-force-induced lowering term, $\Delta\phi_{eff}$, for a zero-bias built-in electric field of approximately 2.8×10^5 V/cm, is on the order of 18 meV, which is similar in magnitude to the thermal voltage.

Table II lists the most pertinent information obtained from the CV measurements on the p-type and n-type $Al_{0.48}In_{0.52}As$ Schottky barriers. Due to the complexities associated with equations (1) and (2), the Schottky barrier heights obtained from the CV data are more prone to systematic errors and are usually less accurate (by approximately ± kT/q) than those obtained from IV data. The differences between the Schottky barrier heights, $\phi_{CV} - \phi_{IV}$, determined by the CV and IV techniques, respectively, are listed in Table III. As can be seen from Table III, the differences in the respective Schottky barrier heights are typically within or near kT/q with the exceptions of the Al and Pt barriers to n-type $Al_{0.48}In_{0.52}As$ (approximately 50 meV) and the Al barrier to p-type $Al_{0.48}In_{0.52}As$ (85 meV). The summation rule for the Schottky barrier heights obtained from the CV technique did not produce values in close agreement with the accepted $Al_{0.48}In_{0.52}As$ band-gap of 1.45 eV. Most notably, the values determined form the $Al/Al_{0.48}In_{0.52}As$ Schottky diodes was 157 meV greater than the accepted $Al_{0.48}In_{0.52}As$ band-gap with the majority of the band-gap overshoot being contributed by the value of the Al/p-type $Al_{0.48}In_{0.52}As$ Schottky barrier height. The band-gap error for the $Au/Al_{0.48}In_{0.52}As$ and $Pt/Al_{0.48}In_{0.52}As$ systems were approximately 30 and 60 meV, respectively. The band-gap error for the $Ti/Pt/Au/Al_{0.48}In_{0.52}As$ system was less than 1%, well within the experimental error margin.

IV. CONCLUSIONS

In conclusion, we have performed a detailed study on the electrical properties of p-and n-type $Al_{0.48}In_{0.52}As$ for four different metallization schemes using the IV and CV techniques. For the IV technique, the sum of the respective p- and n-type barrier heights always equaled the accepted value of the band-gap to within experimental error. The summation rule for the Schottky barrier heights obtained from the CV technique did not produce band-gap values in close agreement with the accepted $Al_{0.48}In_{0.52}As$ band-gap. With the exception of the Al/p-type $Al_{0.48}In_{0.52}As$ samples, the difference between the

Schottky barrier heights determined from CV and IV measurements were within ± 50 meV of each other. An anomalously large, enhanced Schottky barrier height was observed for the Al/p-type $Al_{0.48}In_{0.52}As$ samples. Further work is underway to investigate and understand the origins of this enhanced Schottky barrier height.

Table I. Pertinent IV parameters of $Al_{0.48}In_{0.52}As$.

Contact	n-type			p-type			
Metal	J_S (A/cm^2)	ϕ_n (eV)	n	J_S (A/cm^2)	ϕ_p (eV)	n	$\phi_n+\phi_p$ (eV)
Au	2.82×10^{-6}	0.699	1.07	7.33×10^{-7}	0.764	1.06	1.46
Al	5.49×10^{-4}	0.560	1.14	3.07×10^{-9}	0.905	1.18	1.46
Pt	1.00×10^{-6}	0.725	1.07	3.47×10^{-6}	0.724	1.04	1.45
Ti/Pt/Au	1.54×10^{-5}	0.655	1.05	3.86×10^{-7}	0.781	1.08	1.44

Table II. Pertinent CV parameters of $Al_{0.48}In_{0.52}As$.

Contact	n-type			p-type			
Metal	Vin (V)	ϕ_n (eV)	N_d (x 10^{16}cm^{-3})	Vin (V)	ϕ_p (eV)	N_a (x10^{16}cm^{-3})	$\phi_n+\phi_p$(eV)
Au	0.605	0.730	2.40	0.621	0.760	1.40	1.49
Al	0.492	0.617	2.35	0.852	0.990	1.30	1.61
Pt	0.650	0.775	2.40	0.598	0.734	1.70	1.51
Ti/Pt/Au	0.560	0.685	2.60	0.618	0.757	1.44	1.44

Table III. Comparison between CV and IV Schottky Barrier Heights of $Al_{0.48}In_{0.52}As$.

Contact	n-type	p-type
Metal	$\phi_{n_{CV}} - \phi_{n_{IV}}$ (eV)	$\phi_{p_{CV}} - \phi_{p_{IV}}$ (eV)
Au	0.031	-0.004
Al	0.057	0.085
Pt	0.051	0.013
Ti/Pt/Au	0.030	-0.024

REFERENCES

Chu P, Lin C L, and Wieder H H, "Schottky barrier height of $In_{0.43}Al_{0.57}As$," *Electron. Lett.*, Vol. 22, pp. 890-92, 1986.

Chu P, Lin, C L and Wieder H H, "Schottky barrier height of $In_xAl_{1-x}As$ epitaxial and strained layers," *Appl. Phys. Lett.*, Vol.. 53, pp. 2423-25, 1988.

Chu P and Wieder H H, "Properties of strained layer $In_xAl_{1-x}As/InP$," *J. Vac. Sci. Technol. B*, Vol. 6, pp. 1369-1372, 1988.

Davies G J, Kerr T Tuppen C G, Wakefield B, and Andrews D A, "The Growth and Characterization of Nominally Undoped $Al_{1-x}In_xAs$," *J. Vac. Sci. Technol. B*, Vol.. 2, pp. 219-223, 1984.

Hsieh K H, Wicks G, Calawa A R, and Eastman L F, "Internal Photoemission studies of (GaIn)As,(AlIn)As Schottky diodes and (GaIn)As/(AlIn)As heterojunction grown by molecular beam epitaxy," *J. Vac. Sci. Technol. B*, Vol. 3, pp. 700-702,1985.

Landolt-Bornstein, Numerical Data and Functional Relationships in Science and Technology, (Springer, Berlin, 1987) Vol. 22, pp. 63,117,140,261,299,304,335f, and 350.

Lin C L, Chu P, Kellner A L, Wieder H H, and Rezek Edward A, "Composition Dependence of $Au/In_xAl_{1-x}As$ Schottky Barrier Heights," *Appl. Phys. Lett.*, Vol.. 49, pp. 1593-95, 1986.

Mead C A, "Metal-semiconductor surface barriers," *Solid State Electron.*, Vol.. 9, pp. 1023-1033, 1966.

Ohno H, Barnard J, Wood C E C, and Eastman L F, "Double heterostructure $Ga_{0.47}In_{0.53}As$ MESFET's by MBE," *Electron. Dev. Lett.*, EDL-1, no. 1, pp. 154-155, 1980.

Ohno H, Wood C E C, Rathburn L, Morgan D V, Wicks G W, and Eastman L F, "GaInAs-AlInAs structures grown by molecular beam epitaxy," *J. Appl. Phys.*, Vol.. 52, pp. 4033-4037, 1981.

Sze S M, Physics of Semiconductor Devices, 2nd ed., (Wiley, New York, 1981), pp. 245-293, and 849-851.

Wakefield B, Halliwell M A G, Kerr T, Andrews D A, Davies G J, and Wood D R, "Direct energy gap of $Al_{1-x}In_xAs$ lattice matched to InP," *Appl. Phys. Lett.*, Vol. 44, pp. 341-343, 1985.

Inst. Phys. Conf. Ser. No 120: Chapter 4
Paper presented at Int. Symp. GaAs and Related Compounds, Seattle, 1991

215

Selective area epitaxial growth and fabrication of GaAs MESFETs for monolithic microwave circuits

H. Kanber*, M. Sokolich*, S.X. Bar*, M.I. Herman*, P. Norris[+], C. Beckham[+], and D. Walker[+]

*Hughes Aircraft Company, Microwave Products Division, Torrance, CA 90509
[+]EMCORE Corporation, Somerset, New Jersey 08873

ABSTRACT: GaAs MESFET device structures have been grown on silicon nitride or silicon dioxide masked GaAs substrates by low pressure metalorganic chemical vapor deposition. Very smooth, featureless morphology and 100 percent selectivity of GaAs have been achieved over a range of growth conditions. Size limitations and electrical activity of these islands have been investigated. FET dc characteristics of fabricated devices on the islands show excellent uniformity of 4 percent over the entire 50 mm diameter wafers, indicating good control of doping-thickness product in the FET channels.

1. INTRODUCTION

Selective area epitaxial growth offers the advantages of monolithic integration of microwave devices with optoelectronic devices for close packing density and vertical integration for future systems applications. Metalorganic chemical vapor deposition has been the growth technique of choice in the earlier work of Gale et al and Hollis et al with much of the emphasis being placed upon mask materials, overgrowth over the mask, its aspect ratio, areal resolution and to less extents upon electrical properties. More recently the finer details such as small scale resolution of growth, edge uniformities and electrical properties of both the primary area and perimeter are receiving attention. Recently, Azoulay et al achieved square areas with edge lengths of 17 μm. Elongated structures with (111) facet limited sides were grown with cross sectional base widths of 7 μm.using additions of $AsCl_3$ to prohibit mask nucleation. An alternative to area restriction by masking is restricted growth by focused laser assisted MOCVD growth. Donnelly et al used laser assisted growth to epitaxially grow areas with edge dimensions down to 70 μm but estimated the ultimate limits would be about 0.4 μm. Liu et al have demonstrated the restricted growth of a MESFET structure with a 0.16 μm thick channel with lateral dimensions of 1.5 μm by 75 μm. Molecular beam epitaxy with SiO_2/Si_3N_4 has also been recently used by Metze el al to make selective epi MESFETs with gate dimensions of 0.5 μm by 1 mm. Both dc and RF characteristics were encouraging.

Our results in this paper demonstrate progress in selective area GaAs epitaxy using low pressure (15-76 torr) MOCVD. Growth limits on area resolution, boundary uniformity, doping and mask nucleation were investigated. Essentially 100 percent selectivity has been achieved over a range of growth conditions. At low growth pressure (15-30 torr) smooth, regular morphology was observed on GaAs islands as small as 5 x 6 μm. These islands were found to be electrically active after device fabrication. Device characteristics of fabricated MESFET structures are presented.

2. EXPERIMENTAL

Two inch diameter semi-insulating GaAs LEC substrates masked with SiO_2 or Si_3N_4 were used for the low pressure MOCVD growth. All growth runs were performed in an EMCORE stainless steel high speed rotating disc reactor where the rotation speed of the suspector was 800 rpm. The substrate temperature was 710 °C with a total hydrogen flow rate of 8 slm/minute. One hundred percent AsH_3 was introduced at the beginning of the high temperature bake period and the V/III ratio of 25:1 was used for the p-GaAs buffer layer and a higher V/III ratio of 50:1 was used for the doped channel layer. The growth rate for both the buffer layer and the channel layer was 435 Å/minute. A 3×10^{17} cm^{-3} flat Si doped profile, 0.2 μm thick, was grown for the low noise channel layer profile with a 0.2 μm buffer layer. The GaAs buffer layer was measured by Hall effect to determine its type and the channel layer carrier concentration was calibrated by Polaron profiling. The surface morphology was characterized by Nomarski interference contrast optical microscopy and scanning electron microscopy. Three different device masks were used to cover island dimensions from 5 x 6 μm, 2 x 200 μm to a few several hundred micron size rectangles. FET fabrication was standard processing as described by Wang et al.

3. RESULTS AND DISCUSSION

All growth runs performed showed 100 percent selectivity and had very smooth epilayer surfaces after optimizing the growth conditions. Figure 1 shows an SEM micrograph of the 10 μm square GaAs islands while Figure 2 shows a tilted crossectional view of an array of these islands. No GaAs deposition was observed on any of the mask areas even over dimensions of several hundred microns away from islands of epitaxial GaAs. The smooth morphology seen in Figure 1 was carried down in size to 5 by 6 μm islands, where the (110) facets on the sides of the rectangle and the (111) facets at the corners of the rectangle became more pronounced. Cross-sections of the islands also showed a sharp rectangular profile, with no lip at the edges as observed by other researchers. Size limitations of how close the islands be grown together were investigated and it was observed from cross-sectional SEM measurements that the islands started merging together when the gap between them was 1 μm or less.

Electrical evaluation of the size limitations of the selectively grown GaAs islands was done by FET fabrication. Growth parallel to and perpendicular to the major flat was evaluated by conductivity and FET channel current I_{sat} measurements as a function of FET dimensions ranging from 1 x 10 μm to 5 x 100 μm. Figure 3 shows a linear relationship between the inverse resistance and the nominal line width for these thin long islands varying in width between 1 and 5 μm. Ideally the line should intersect the x-axis at 0 μm. From this we deduce that the selective growth mesas grown parallel to the major flat spread about 1 μm on each side resulting in a total width change of 2 μm. Similar measurements were done for growth perpendicular to the major flat. Choosing the 5 μm long FET

Figure 1   SEM micrographs of 10 μm square GaAs islands grown by selective area low pressure MOCVD.

Figure 2   Edge view of an array of selective area grown GaAs islands with excellent morphology.

channel, the FET channel current was measured for widths ranging from 10 μm to 100 μm. Figure 4 shows a linear relationship between the I_{sat} for the 5 μm channel and the channel width. Again, ideally the line should intersect the x-axis at 0 μm, instead it does so at negative 6 μm. From this data we deduce that selective growth mesas spread about 3 μm on each side for growth perpendicular to the major flat resulting in a total width change of 6 μm.

A flat doped low noise profile was grown for the SAE GaAs islands. Initial measurements on 50 mm wafers showed the average I_{sat} for the structure was 1.4 A/mm with about 10 percent deviation across the 50 mm diameter for 50 μm FETs. The low field sheet resistance of 320 Ω/square indicates a mobility of roughly 3000 cm^2/V-sec, a very reasonable value comparable to values obtained in high quality epitaxial growth at similar doping levels. The calculated saturation velocity is 1.6 x 10^7 cm/sec, again a reasonable value. These initial low noise FET profile wafers showed poor pinch-off characteristics because of the n-type buffer. SAE growth was optimized further by decreasing the V/III ratio to convert from an n- buffer to a p- buffer. Hall measurements were used to determine the type conversion.

An evaluation of the improved low noise profile was characterized by FET fabrication and electrical measurements. Ohmic contact resistance, sheet resistance, gateless saturation current I_{sat} and leakage was measured on 5 by 100 μm FETs. Figure 5 shows the sheet resistance map on a 50 mm diameter SAE wafer (3 SA 608-62) with an average sheet resistance of 293 Ω per square and a standard deviation of 26 Ω per square, indicating a uniformity of 9 percent across the wafer. The gateless saturation current on the same FETs had an average value of I_{sat} = 2.57 mA/μm with a standard deviation of 0.09 mA/μm, indicating an excellent uniformity of 4 percent across the wafer. I_{sat} can be directly related to N_d x t product where N_d is the dopant density and t is the thickness of the active layer. Control of N_d x t product to the tune of 4 percent across the 50 mm diameter indicates very good control of growth parameters during selective area growth. The improved buffer breakdown strength was also evaluated further by etching only the FET channel after ohmic metallization to where the I_{sat} dropped to zero and then etching away half of the buffer layer. The leakage current was measured between source and drain contacts at a constant bias of 2.5 volts. The average leakage on this wafer was 0.1 μA/μm at 0.5 V/μm, indicating an order of magnitude improvement over previous wafers with n- type buffers.

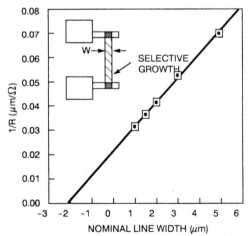

EDGE EFFECT ON SELECTIVELY GROWN GaAs

Selective Growth Mesas Spread about 1 μm on each side resulting in a total width change of 2 μm.

Figure 3   Electrical characterization of SAE GaAs for growth parallel to the major flat.

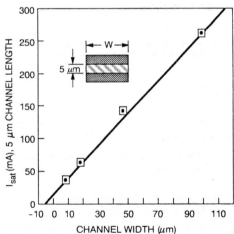

EDGE EFFECT ON SELECTIVELY GROWN MESAS

Selective Growth Mesas Spread about 3 μm on each side resulting in a total width change of 6 μm.

Figure 4   Electrical characterization of SAE GaAs for growth perpendicular to the major flat.

At present, the low noise MESFET profile devices with selective area grown GaAs are employed in an X-band (10 GHz) microwave oscillator as the test vehicle. Two types of gate geometries are being studied while the SAE device doping profiles are the same. For oscillator applications we strive to achieve a linear transconductance to improve phase noise characteristics. FET evaluation presented above and ongoing promises SAE devices as viable components for monolithic integration.

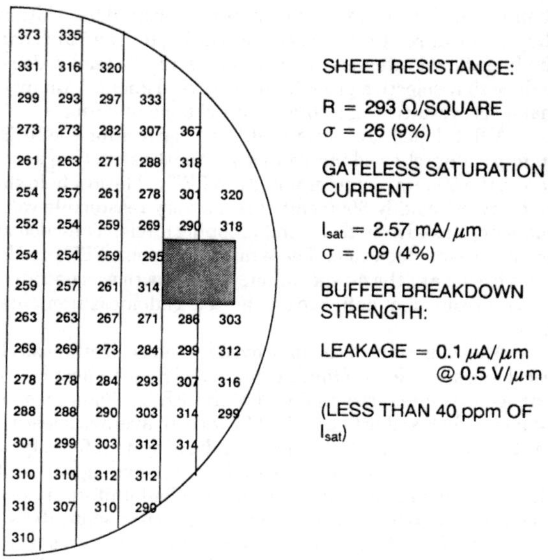

4. SUMMARY

Selective area growth by low pressure MOCVD has been used to grow GaAs MESFET islands on masked GaAs substrates. Smooth, featureless morphology and control of N x t product on these islands are demonstrated. FET electrical evaluation of different size islands indicates comparable dc characteristics to nonselective epi material and very good uniformity over the wafers.

SHEET RESISTANCE:

R = 293 Ω/SQUARE
σ = 26 (9%)

GATELESS SATURATION CURRENT

I_{sat} = 2.57 mA/μm
σ = .09 (4%)

BUFFER BREAKDOWN STRENGTH:

LEAKAGE = 0.1 μA/μm
@ 0.5 V/μm

(LESS THAN 40 ppm OF I_{sat})

Figure 5   FET mapping of SAE grown wafer 3 SA 608-62 showing excellent uniformity across the 50 mm diameter.

5. ACKNOWLEDGMENT

This work was partially supported by Department of the Air Force, Electronic Technology Laboratory, Wright Patterson Air Force Base under Contract Number F33615-89-C-1147. We would like to thank Mr. Mark Pacer for his encouragement and support.

REFERENCES

Azoulay R and Dugrand L 1991, Appl. Phys. Lett. **58**, 129

Donnelly VM and McCaulley JA 1989, Appl. Phys. Lett. **54**, 2458

Gale RP, McClelland RW, Fan JCC and Bozler CO, in International Symposium on GaAs and Related Compounds, Albuquerque 1982, Inst. Phys. Con. Ser. No. 65, 1983, Institute of Physics, Bristol UK, 101

Hollis MA, Nichols KB, Murphy RA, and Bozler CO 1987, in Advanced processing of Semiconductor Devices, SPIE Vol. **797**, 335

Lui H, Robert JC, Ramdani J and Bedair SM 1991, Appl. Phys. Lett. **58**, 1659

Metze GM, Cornfeld AB, Laux PE, HO TC and Pande KP 1990, Appl Phys. Lett. **57**, 2576

Wang KG and Wang SK 1987, IEEE Trans. Electron Devices **ED-34**, 2610

Inst. Phys. Conf. Ser. No 120: Chapter 4
Paper presented at Int. Symp. GaAs and Related Compounds, Seattle, 1991

219

Mn diffusion in GaAs and its effect on impurity-induced layer disordering in GaAs–AlGaAs superlattices

K.C. Hsieh, C.H. Wu, G.E. Höfler, N. EL-Zein and N. Holonyak, Jr..
Department of Electrical & Computer Engineering, University of Illinois.

I. Introduction

The development of using Mn as a p-type dopant in GaAs- and/or InP-based compound semiconductors grown by molecular beam epitaxy (MBE) has been hindered by the inability to obtain a smooth surface with a high doping. Unlike doping with Be, ripple-like features parallel to [-110] are observed in MBE-grown GaAs with Mn doping higher than $1\times10^{17}/cm^3$. The origin of the ripple structure has been suggested to be associated with the surface segregation of the doping species during crystal growth.[1] DeSimone et al.[2] later reported that the GaAs surface started to degrade severely when the Mn concentration was higher than $2\times10^{18}/cm^3$, and attributed the degradation to Mn desorption and accumulation on the growing surface in competition with incorporation. Very limited data on diffusing Mn into GaAs are available in the literature, and a maximum surface concentration of only $2\times10^{18}/cm^3$ accompanied by a nonuniform diffusion profile has been reported by Seltzer, who has done the diffusion by coating a layer of radioactive Mn^{54} film on the GaAs surface.[3] Recently, there is an increasing interest in forming Mn-based ohmic contacts to p-type GaAs.[4] A more detailed study of Mn diffusion in GaAs will be beneficial. In the present work we clarify that $2\times10^{18}/cm^3$ should not be the Mn solid solubility in GaAs and report that very high carrier concentrations ($\sim 10^{20}/cm^3$) from Mn diffusion can be realized with smooth surface morphology. Our results indicate that using different Mn-containing sources and As overpressures for the diffusion of Mn in GaAs can have a profound influence on the ultimate surface carrier concentration, surface morphology, and as a result, on the effectiveness of impurity-induced layer disordering (IILD) in AlGaAs-GaAs superlattices.[5]

II. Experiment

Four different Mn-containing sources have been used to introduce Mn into GaAs, including separate solid sources of Mn, MnAs, or Mn_3As (granules) enclosed in the quartz diffusion ampoules, as well as thin Mn films that are deposited directly onto the substrates by electron beam evaporation. The substrates for Mn diffusion are semi-insulating GaAs. The undoped AlGaAs-GaAs superlattices investigated here for IILD have been grown by low-pressure metalorganic chemical vapor deposition in an Emcore GS 3100 reactor. The superlattice structure consists of 20 periods of $Al_xGa_{1-x}As$ (x~0.4) and GaAs, with both wells and barriers approxiamtely 180Å thick.

Sample preparation for Mn diffusion consists of the usual surface cleaning procedures, followed by an NH_4OH etch. The samples were then loaded into degreased and etched quartz ampoules with the Mn sources, and are evacuated to $\sim 2 \times 10^{-6}$ Torr and sealed. The diffusions are performed between 700^0C to 900^0C for times less than 12 h. Nomarski optical microscopy and transmission electron microscopy (TEM) are used to investigate the surface morphology and annealing surface reactions. Slant-angle lapping technique has also been used for the disordering experiments. Mn diffusion profiles were obtained by either secondary ion mass spectroscopy (SIMS) or capacitance-voltage (C-V) electrochemical etch profiler.

III. Results

Mn and Mn/GaAs

The diffusion runs of Mn in GaAs using separate Mn granules in the sealed ampoules resemble to MBE incorporation of Mn doping under As-deficient conditions, while the diffusion runs using contact Mn films on the GaAs substrates are actually identical to Seltzer's diffusion experiments. Nomarski optical micrographs show that both GaAs surfaces apparently degrade from their initial shiny and flat conditions after Mn-diffusion at 800^{0}C (2 h). For the former, several attempts of adding extra As in the ampoules have been tried unsuccessfully to keep off the degradation, but longitudinal features parallel to one of the <110> directions can be clearly observed on the GaAs surface. These results seem to be consistent with what has been observed in Ilegems' MBE work on Mn-doping. In the case of diffusion with a 500Å-Mn film evaporated onto the GaAs surface, cross sectional TEM indicates that a surface chemical reaction has actually taken place resulting in a layer of minute (Ga,Mn) crystals. The particle size is typically around 500 Å or less, and the depth of reaction, though nonuniform across the sample, depends on the length of diffusion time. Diffraction and energy dispersive x-ray microanalysis suggest that these particles are cubic in structure with a lattice constant close to 8.4Å and a composition about MnGa$_2$, a structure which does not quite fit with any existing data of Mn-Ga alloys.

To further study the effect of this reaction on the amount of Mn diffusing across the reacted layer and into the substrate, SIMS has been performed on the Mn-diffused heterostructure superlattice samples. The known dimension of each layer in the heterostructure provides the calibrated "yardstick" for the Mn diffusion depth profile. Unfortunately, the profiles vary drastically depending on the position as well as the dimension of the measurement area. Fig. 1 shows the Mn ion images at two different depths beyond the reacted layer. The area of view is about 150 μm in diameter. An area-dependence of the Mn concentration can be clearly seen. In Seltzer's work, varying from run to run under the same diffusion conditions, two different profiles have been reported. A single complementary error function can be used to fit nicely for one type of his experimental diffusion profiles. For the other, the anomalous type, two functions are needed for the fitting.[3] We believe that the inconsistency in this earlier work may relate to the chemical reaction and nonuniform diffusion observed in our work.

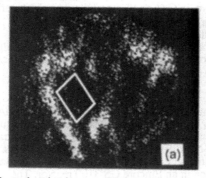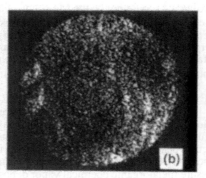

Fig. 1. Secondary ion images of Mn obtained at two different depths from a GaAs sample coated with 500Å Mn and annealed at 800^{0}C (2h) in a sealed quartz ampoule, (a) at about 1200Å (b) at about 1700Å from the initial surface.

MnAs and Mn$_3$As

The surface of the GaAs substrate after Mn diffusion using MnAs as the source remains smooth and shiny, and further TEM analysis shows no surface reaction and no apparent dislocation generation. A typical carrier depth profile of Mn-diffused GaAs at 800^{0}C (6h) measured by the electrochemical C-V technique is shown in Fig. 2(a). It shows a double-kink shape similar to Zn diffusion profiles obtained at high temperatures, suggesting that a substitutional-interstitial mechanism may also be the principal diffusion mechanism for Mn in GaAs, as in the case for Zn.[6] In general, the surface concentration remains the same but the concentration and depth of the plateau region increases with the diffusion time. From 700^0C to 900^0C, the surface carrier concentrations (typically of $5\times10^{19}/cm^3$) do not show a strong dependence on the diffusion temperature. They can, however, be increased to more than $1\times10^{20}/cm^3$ by adding extra As crystals in the quartz ampoules during annealing. Accompanied with the increase in the ultimate surface concentration, a significant reduction in the depth of the plateau region is observed in Fig. 2(b) and 2(c) as the As overpressure increases. Possible reasons for these increase and reduction lie on the distribution of various point defects present in the sample. Bear in mind that a higher As overpressure induces a higher concentration of Ga vacancies near the surface, which in turn trap more in-diffusion Mn atoms, to become the slower substitutional atoms. As a result, less interstitial Mn atoms can diffuse quickly and deeply into the substrate, and a shorter plateau region is formed. It is worth comparing the data of Fig. 2 with Seltzer's work[3]. Since Mn is a deep acceptor having an ionization energy of about 95 meV,[7] the carrier concentration obtained by C-V techniques underestimates the actual atomic Mn concentration obtainable by SIMS. The maximum surface concentration of Seltzer's work, 2×10^{18} atoms/cm^3, is about two orders of magnitude smaller than that for the present work.

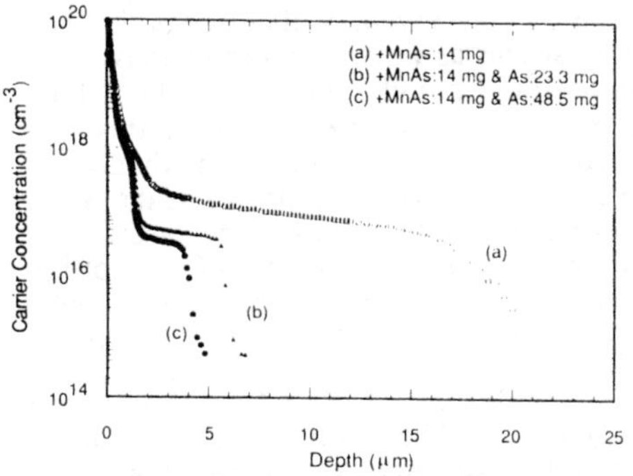

Fig. 2. C-V electrochemical profiles showing the hole concentration profiles in Mn-diffused GaAs. The Mn diffusion into the GaAs (with 14 mg MnAs in the sealed quartz ampoules) occurs at 800^0C (6h) with (a) no additional As, (b) 23.3 mg As, and (c) 48.5 mg As.

The surface morphology of GaAs is surprisingly poor after diffusion using Mn_3As as the source. Since Mn_3As is a Mn-bearing as well as As-bearing material, its high temperature vapor should contain some As vapor necessary for the suppression of surface degradation, though the exact amount of As pressure is not known. Several more diffusion runs have been performed using extra As crystals, in addition to the Mn_3As source, to determine whether Mn_3As can ever be effectively used for Mn diffusion as far as the smooth surface is concerned. Data show that with a high enough As overpressure a smooth surface can indeed be preserved and diffusion profiles as well as the surface concentration similar to those diffused with MnAs are also obtained. However, the amount of As overpressure seems dependent on the diffusion temperature. Such a strong dependence of surface morphology on the choice of Mn-containing diffusion source is in sharp contrast to Zn diffusion. It is well known that Zn can be diffused very fast into GaAs with smooth surface morphologies, irrespective of the use of various Zn-containing sources, e.g., Zn,[8] $ZnAs_2$,[5] and Zn_3As_2;[9] but in the case of Mn, the choice of Mn-containing source is critical. It is not well understood yet as to why there is such a big difference.

Mn Induced Layer Disordering

Since MnAs seems to be the only Mn source which can easily perform the diffusion, it is used in most of our diffusion and layer disordering experiments. To separate pure thermally induced interdiffusion from impurity-induced interdiffusion, unintentionally doped AlGaAs-GaAs superlattice samples have been first thermally annealed at 800^0C (< 12h), and TEM results indicate that that there is negligible interdiffusion near the AlGaAs-GaAs interfaces. With additional MnAs in the sealed quartz ampoule during annealing, IILD is indeed realized. Depending on the Mn diffusion time, partial or complete loss of layering can be easily seen by TEM. The enhanced Al and Ga interdiffusion, together with the fast/slow interstitial/substitutional diffusion mechanism, suggest that point defects of charged interstitial Ga ions rather than Ga vacancies may also be responsible for the layer disordering, as in the case for Zn.[10] From the processing point of view, selective area diffusion is very often needed, and SiO_2 or Si_3N_4 is commonly used as the masking dielectrics. Care needs to be taken when SiO_2 is used as the diffusion barrier for Mn. A slant-angle lapping technique has been used to study the masking ability of SiO_2 and Si_3N_4 in several disordering experiments.

As shown in the optical micrographs in Fig. 3, the superlattice layers under the Si_3N_4 stripes are preserved, while those under the SiO_2 stripes as well as exposed barely to the vapor from MnAs have been lost after a high temperature anneal at 800^0C (12h). It is also noted that the preservation of superlattice layers under the SiO_2 improves as the As overpressure in the ampoule increases.

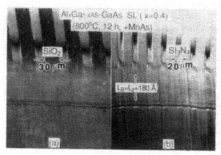

Comparison of electrochemical C-V profiles from broad area SiO_2-masked GaAs samples diffused with the same amount of MnAs, but different As overpressure, indicates that significantly less Mn can diffuse through the SiO_2 layer and be incorporated in the GaAs substrate in the high-As-overpressure environment.

Fig. 3. Optical micrographs of baveled cross sections of an unintentionally doped $Al_{0.4}Ga_{0.6}As$-GaAs superlattice patterned with (a) SiO_2 and (b) Si_3N_4 stripes and Mn-diffused at 800^0C (12h) showing that SiO_2 is not an effective diffusion barrier for layer disordering.

IV. Summary

Using MnAs as the diffusion source, very high concentrations of Mn ($\sim10^{20}/cm^3$) can be diffused into GaAs without inducing surface degradation. Other sources, however, such as Mn, Mn_3As, or evaporated Mn thin film contacting the substrate, may induce surface reactions or even etching at high temperatures. Similar to Zn in GaAs, Mn diffuses interstitial-substitutionally, can be impeded better by Si_3N_4 than SiO_2, and can otherwise be used effectively also to induce disordering in AlGaAs-GaAs superlattices.

V. Acknowledgments

The authors would like to acknowledge the Microanalysis Center at the University of Illinois for the use of these facilities. This work has been supported by the National Science Foundation (ECD-89-43166 and DMR-89-20538), and Army Research Office contract No. DAAL-03-89-K-0008.

REFERENCES
1. M. Ilegems, J. Appl. Phys. **48**, 1278 (1977).
2. D. DeSimone and C.E.C. Wood, J. Appl. Phys. **53**, 4938 (1982).
3. M.S. Seltzer, J. Phys. Chem. Solids **26**, 243 (1965).
4. C. Dubon-Chevalliar, M. Gauneau, J.F. Bresse, A. Israel and D. Ankri, J. Appl. Phys. **59,** 3783 (1986).
5. W.D. Laidig, N. Holonyak, Jr., M.D. Camras, K. Hess, J.J. Coleman, P.D. Dapkus, and J. Bardeen, Appl. Phys. Lett. **38**, 776 (1981).
6. R.L. Longini, Solid-State Electron. **5**, 127 (1962).
7. A.G. Milnes, *Deep Impurities in Semiconductors* (John Wiley & Sons, New York, 1973).
8. K.Y. Hsieh, Y.L. Hwang, J.H. Lee and R.M. Kolbas, J. Electron. Mater. **19**, 1417 (1990).
9. K. Meehan, J.M. Brown, M.D. Camras, N. Holonyak, Jr., R.D. Burnham, T.L. Paoli and W. Streifer, Appl.Phys. Lett. **44**, 428 (1984).
10.T.Y. Tan and U. Gösele, Appl. Phys. Lett. **52**, 1240 (1988).

Inst. Phys. Conf. Ser. No 120: Chapter 5
Paper presented at Int. Symp. GaAs and Related Compounds, Seattle, 1991

223

The effect of defects and interfacial stress on InGaAs/GaAs heterojunction FET reliability

A. Christou and J. M. Hu
CALCE Center for Electronics
Microelectronics Devices Laboratory
University of Maryland College Park, MD

Abstract

Reliability testing of pseudomorphic high electron mobility transistors has identified a possible failure mechanism related to traps and dislocations. The doped channel pseudomorphic HEMTs have a 2.5 times lower mean time to failure compared to single channel conventional pseudomorphic HEMTs.

1. Introduction

A reliability investigation of pseudomorphic HEMTs (PMHEMTs) has been carried out by investigating the effect on reliability of the active layer dislocation density, interfacial defects and interfacial trap centers. The experiments carried out were accelerated life testing and transmission electron microscopy, deep level transient spectroscopy (DLTS) and microwave characterizations. The presence of thin (less that 200 Å) lattice mismatched pseudomorphic structure of InGaAs/A1GaAs and InGaAs/GaAs may result in mechanical stress problems which would limit the transistor's reliability (Anderson and Christou, 1990). The present investigation reports on the reliability results of an optimized doped channel heterojunction field effect transistor (DCHEMT) with a delta doped InGaAs channel of 1.5×10^{12} cm^{-2} (Huang, et al. 1989; Mishra, et al. 1985; Smith, et al. 1987). The transistors were fabricated with a T-gate process and with 0.50μm x 50μm TiPtAu gates (Christou, et al. 1991).

Both the DCHEMT and the PMHEMT degraded devices showed the presence of a minority carrier trap in excess of 10^{16}cm^{-3}. Since the minority carrier trap identified was similar to the distribution of traps present when the InGaAs layer creates dislocations, its presence after accelerated life testing has been attributed to dislocations present at the InGaAs/GaAs interface.

2. Experimental

Standard deep ultra-violet photolithographic techniques were used for gate-lengths of 0.5μm. The processing of such device structures has been reported previously (Christou, et al. 1991). Figure 1 shows the cross-sections for both the DC pseudomorphic HEMT (DCHEMT) and for the Single Channel pseudomorphic HEMT (SHEMT). The DC accelerated life tests were conducted at a base plate temperature of 180°C, 195°C, 210°C and 240°C until fifty percent of the devices tested (eight) degraded. The following DC characteristics were measured before and after accelerated stress testing: I_{DS} - V_{GS} behavior, transconductance (g_m), output conductance $(g_o)_T$, drain-gate breakdown characteristics (BV_{DG}), drain-source breakdown characteristics (BV_{DS}), threshold voltage, and saturated drain current (I_{DSS}). The failure criteria selected was a twenty percent decrease in saturated drain current.

The degradation was analyzed by deep level transient spectroscopy (DLTS) techniques and by cross-sectional transmission electron microscopy (TEM). DLTS techniques have been applied previously to HEMTs (Anderson and Christou, 1990; Christou, et al. 1991) and have successfully reported on the presence of both minority and majority carrier traps due to accelerated stress testing. In addition, source resistance (R_s), drain resistance (R_d), channel resistance (R_{ch}), barrier height (ϕ_b), and ideality factor (η) were measured before and after accelerated stress tests. Typical data showing characteristic changes as a result of the DC bias life tests at 200°C and 110 hours is shown in Table I, for the DCHEMT.

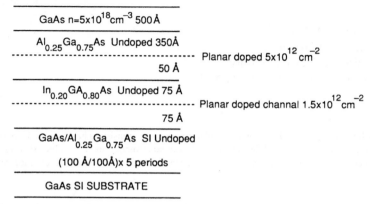

Figure 1   Cross-section of Pseudomorphic InGaAs DCHEMT showing the placement of the planar doped layer in the InGaAs channel.

Table I.   Parameter changes due to DC bias life testing at 200°C and 110 hours.

PARAMETER	INITIAL	AFTER (200°c/110HRS)	Δ
R_s(ohm)	4.3	12.5	8.2
R_d(ohm)	7.8	20.0	12.2
R_{ch}(ohm)	19.5	22.7	3.2
ϕ_b(EV)	0.75	0.75	0
η	1.70	1.75	0.05

In the above case, the R_s and R_d changes may be attributed to ohmic contact interdiffusion, while the R_{ch} change may be dominated by traps in the channel.

3. Experimental Results

A. Failure Distributions
        The failure distribution for the DCHEMTs shows a log-normal distribution for channel temperatures of 180°C, 195°C, and 210°C. The failure criteria was previously selected as a 20 percent decrease in I_{DSS}. A straight line fit of the data has been obtained. It indicates the validity of the log-normal distribution function and the utilization of a single activation energy in explaining the results. The fifty percent cummulative failure then determines the mean time to failure (MTF). The variation of MTF with temperature allows the activation energy to be determined. A similar distribution plot was obtained for the SHEMTs. The activation energy calculated from the data for the SHEMTs is 1.05eV. The extrapolated median time to failure at typical bias temperatures of 110°C is 10 x 10^4 hours for SHEMTs and 4 x 10^4 hours for DCHEMTs each determined at a 95 percent confidence level. The reliability data suggests that the higher doping in the channel results in a shorter life for the DCHEMT. The factor of 2.5 is significant in indicating a difference in reliability levels between the two structures.

B. Failure Mode Analysis

The DCHEMT and SHEMTs were analyzed by DLTS techniques after accelerated life testing at 210°C. Besides the techniques based on capacitance and current DLTS, both the transistor transconductance and low frequency 1/f noise are considered as two parameters sensitive to deep levels, dislocations and interface states. Figure 2 shows typical DLTS for the SHEMTs. Levels at 470 meV and 550 meV from the conduction band were detected and are similar to minority carrier traps due to dislocations present in PMHEMT layers where the InGaAs thickness exceeds the critical layer thickness as reported previously (Christou, et al, 1991). Since no minority carrier traps were present in the DCHEMTs or SHEMTs prior to accelerated life testing we conclude that such traps in the InGaAs layer may have been generated as a result of the reliability tests. The large g_m variation with frequency was also measured which suggests the presence of frequency instabilities due to traps at approximately 0.5 eV (Christou, et al. 1991).

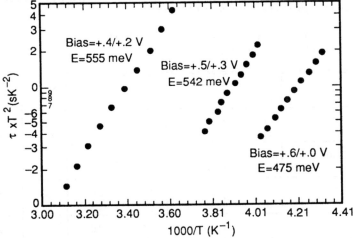

Figure 2   Current DLTS and 1/f spectra for DCHEMTS and SHEMTS tested at 210° C.

In order to attain a better correlation between the minority carrier traps and dislocations in the channel, DCHEMTs were analyzed by cross section TEM and planar TEM techniques. The InGaAs planar TEM image of the DCHEMT indicates the presence of edge dislocations in the cross sectional image. The dislocation density after accelerated life testing has been measured in planar TEM images to be in excess of 10^9 cm^{-2}, which represents a significant increase from 10^7 cm^{-2} measured on TEM control structures prior to testing.

C. Noise Parameters

The minority carrier traps in the InGaAs were further investigated by measuring the four noise parameters: T_{min} minimum noise temperature, $Z_{g,opt}$, optimum source impedance, and g_n, the noise conductance. The noise parameters were measured in a single-stage amplifier configuration at room temperature and up to 150°C. The source impedance measurement allows us to probe changes in both gate resistance and channel resistance introduced by accelerated stress testing where the source impedance is given by: $Z_{g,opt} = (R_g + R_{ch})_{opt} + jX_{g,opt}$. The room temperature noise parameters of both the DCHEMT and SHEMT are shown in Table II for before and after accelerated stress testing at 210°C for 100 hours. For comparison, the results of the SHEMT noise characteristics are also given. As shown in Table II, an increase in the minimum noise temperature and $(R_g + R_{ch})$ occurred in both the DCHEMT and SHEMTs in comparison with a typical noise temperature of 82-90K measured prior to reliability testing. The increases in T_{min} observed further corroborates the observation that traps in the InGaAs channel may introduce a hole emission/capture process which is still present at room temperature after reliability testing. The observed degradation has also affected the dependance of the transconductance g_n on drain current as shown in Table II.

These experiments suggest that the mechanism responsible for the changes observed during accelerated stress testing may be related to the introduction of traps and is also a source of additional noise even at frequencies as high as 8.0 GHz. Trap occupancy, however, is determined among other factors by the bias conditions of the DCHEMT and SHEMT after cooling to room temperature.

4. Summary

The present investigation suggests that pseudomorphic HEMTs degrade from traps in the InGaAs channel probably due to the generation of dislocation in the strained, lattice mismatched layer. The minority carrier traps investigated by DLTS are identical to those created in layers which exceed the strain accommodation thickness (150).   Finally it has been shown that although doped channel PMHEMTs have improved dc and rf performance their reliability level is approximately 2.5 times less compared to single channel conventional PMHEMTs.

Table II   Noise Parameters of DCHEMTs and SHEMTs taken at room temperature, before and after degradation.

	I_{ds} mA	V_{ds} V	T_{min} K	$R_{g,ch}$ Ω	X_{gopt} Ω	g_n mS
SHEMT 200°C/100 hrs	5	1.5	119	19.9	47	8.4
	10	1.5	115	22.0	46	8.0
DCHEMT 200°C/100°C	5	2.5	124	26.0	40	10.3
	10	2.5	119	24.0	39	8.0
DCHEMT BEFORE	5	3.7	95	12.0	35	9.5
	10	3.2	92	12.5	34	9.7
SHEMT BEFORE	5	3.5	90	12.0	36	9.2
	10	3.0	83	13.2	37	9.9

5. Acknowledgements

This investigation has been partially supported by the National Science Foundation as part of the IUCRC program, by the State of Maryland through the Engineering Research Center.

References

Anderson W.T., Christou A. 1990 Semiconductor Device Reliability ed A. Christou and B.A. Unger, (Amsterdam: Kluwer), pp 423-37.
Christou A., Hu J.M. and Anderson W.T. 1991, *Proc. Rel. Phys. Symp.* (New York: IEEE) pp. 200-206.
Huang J.C. and Zaithir M., 1989, *IEEE Election Dev. Lett.* **10**, pp 511.
Mishra U.K., Palmateer S.C., Chao P.C., Smith P.M. and Hwang J.C.M. 1985, *IEEE Electron Dev. Lett.* **6**, pp 142.
Smith P.M., Chao P.C. and Duh K. H., 1987, *IEEE MTT-5* **5** pp 749.

Inst. Phys. Conf. Ser. No 120: Chapter 5
Paper presented at Int. Symp. GaAs and Related Compounds, Seattle, 1991

227

Magneto-quantum tunneling phenomena in AlGaAs/GaAs resonant tunneling diodes

H.M. Yoo, S.M. Goodnick[*], T.G. Stoebe, and J.R. Arthur[*]

Department of Materials Science and Engineering, University of Washington, Seattle, WA 98195

[*] Center for Advanced Materials Research, Department of Electrical and Computer Engineering, Oregon State University, Corvallis, OR 97331

ABSTRACT: $Al_{.65}Ga_{.35}As$/GaAs resonant tunneling diodes (RTDs) with asymmetric spacer layers and barrier configurations have been studied via electric and magnetic field measurements. RTDs with an asymmetric barrier configuration exhibit a higher asymmetry in their I-V characteristics than these with asymmetric spacer layers. However, Shubnikov-de Haas (SdH) oscillations obtained from the former are comparable for both bias directions while those from the latter exhibit very high asymmetries. The asymmetric spacer layer RTDs display SdH oscillations only when the diodes are biased in such a way that the thick spacer layer is in the leading edge of the diodes in contrast to the RTD with asymmetric barriers.

Resonant tunneling diodes (RTDs) grown by molecular beam epitaxy (MBE) with asymmetric barriers and spacer layer configurations have been investigated via electric and magnetic field measurements. The RTD structure with asymmetric barriers consists of the following layers in order of growth from the silicon doped ($n = 1\times10^{18}$ cm^{-3}) <100> GaAs substrate: 1.0 μm of Si 1×10^{18} cm^{-3} doped GaAs buffer layer, 500 Å thick undoped spacer layer, 85 Å undoped $Al_{.35}Ga_{.65}As$ barrier, 50 Å undoped GaAs well, 50 Å undoped $Al_{.35}Ga_{.65}As$ barrier, 500 Å undoped spacer layer and finally a Si 1×10^{18} cm^{-3} doped 4000 Å top contact layer. This RTD is referred to as the 50/85 RTD. The RTDs with asymmetric spacer layer configurations have the same structural parameters with identical 85 Å thick symmetric $Al_{.35}Ga_{.65}As$ barriers and asymmetric spacer layers: in one case the RTD has 50 and 500 Å thick asymmetric spacer layers on the top (emitter) and substrate side, respectively while in the second case the RTD has its spacer layers order reversed. The first and second cases discussed above are called the 50/500 and 500/50 RTD, respectively. The RTDstructures were fabricated with lateral dimensions of 50 x 50 μm^2 and isolated by wet chemical mesa etching utilizing Au/Au-Ge/Ni ohmic contact metallurgy as an etch mask.

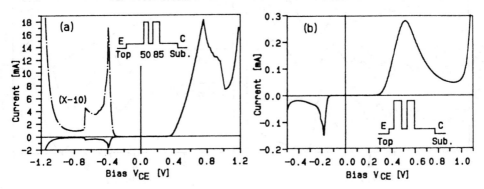

Fig. 1. I-V characteristics of the RTDs measured at 1.8 K. (a) 50/85 RTD. (b) 50/500 RTD. Inside figures represent a metallurgical junctions of the RTD.

Figs. 1 (a) and (b), respectively, show the current-voltage (I-V) characteristics of the 50/85 and 50/500 RTDs measured at 1.8 K. Both RTDs show similar behavior: the RTDs exhibit a poor peak to valley current ratio (PVCR) but a high peak current (Ip) in forward bias, where the tunneling electron first encounters the thinner of the two barriers or spacer layers, while they exhibit better PVCRs but a smaller Ip in reverse bias. The measured PVCRs of the 50/85 and 50/500 RTD at 1.8 K are, respectively, 2.5:1 and 6:1 in forward bias, while those in reverse bias are 18:1 and 8:1. This asymmetric nature of the I-V characteristics have been predicted and observed from the RTDs with asymmetric layer configurations and can be explained in terms of symmetry breaking of the tunneling structures, depending upon the bias direction [1,2,3].

The current (I) through an RTD at a fixed bias, as a function of magnetic field (B) applied normal to the barrier (B//I), shows oscillatory behavior. The oscillation period in 1/B should become periodic with a frequency of Bf, where $1/B_f = \Delta(1/B)$ [4]. Figs. 2 (a) and (b), respectively, show Shubnikov-de Haas (SdH) oscillations obtained from the 50/85 and 50/500 RTDs under different bias conditions. The numbers in the figure represent the applied bias, V_{CE}, to the RTD. As shown in the figure, the 50/85 RTD has comparable SdH oscillations in both bias direction whereas the 50/500 RTD displays SdH oscillations in only one bias direction. The other asymmetric spacer layer diode (500/50 RTD) also shows a similar behavior to the 50/500 RTD: the oscillations are observed only when the diode is biased in such a way that the tunneling electron first encounters the thicker of the two spacer layers. It is interesting to note that the 50/500 RTD shows SdH oscillations in the bias direction where it exhibits a better PVCR, but the 50/85 RTD displays the oscillations regardless of the PVCRs. It is worth mentioning again that the 50/85 and the 50/500 RTDs have PVCRs of 2.5 and 6:1 in forward bias and 18 and 8:1 in reverse bias, respectively. The former displays very strong magneto-quantum oscillations on both bias direction even though it exhibits a poorer PVCR (2.5:1) in forward bias. However, the latter does not show SdH oscillations in forward bias even though it has a PVCR of 6:1, which is much higher than the 2.5:1 observed from the 50/85 RTD for the same bias configuration. This result is in contrast to the linear relationship expected between the strength of the magneto-quantum oscillation and device quality, especially in the PVCR, observed from the RTDs with symmetric layer configurations [5]. These magneto-quantum oscillations may originate from the bulk, from the accumulation layer, or from a quasi-bound state in the well [4,6]. Oscillations originating from the bulk, which are not considered here due to the bias direction dependence of the oscillations, can also be distinguished from the others by the angular dependence of the magnetic field. The latter oscillations are either due to the Fermi energy of the accumulation layer passing through the Landau levels in that region followed by an

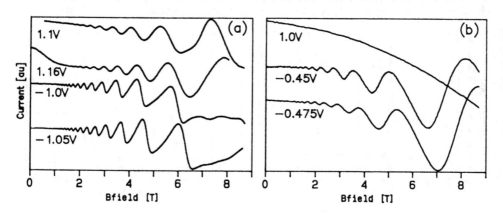

Fig. 2. Shubnikov-de Haas oscillations obtained from (a) 50/85 RTD and (b) 50/500 RTD. The numbers inside figures represent an applied bias, V_{CE}.

elastic tunneling through the double barrier structure or due to an inelastic tunneling of electrons into a Landau level in the well with the emission of an acoustic or longitudinal optical (LO) phonon [7].

The SdH oscillations originating from an accumulation layer are the result of a two step conduction process for an electron from the emitter contact to reach the collector contact [3,8]. The first step is transport into the emitter accumulation layer, followed by tunneling through the rest of the double barrier structure as the second step. An electron in the accumulation layer has a finite life-time and experiences scattering during its residency. Scattering in an accumulation layer broadens the magnetic field induced Landau levels and results in undetectable magneto-quantum features. Electrons in the accumulation layer formed by a thinner spacer layer suffer from a higher scattering rate than that of a thicker spacer layer. The 50/85 RTD, which comprises 500 Å thick spacer layer on both sides, allows for a relatively low scattering rate due to importance in the accumulation layer in both bias directions whereas the 50/500 RTD does so only in a forward bias where the accumulation layer is formed by a 500 Å thick spacer layer. Thus, the 50/85 RTD shows comparable SdH oscillations in both bias direction while the 50/500 RTD shows the oscillations only in a bias direction where the tunneling electron resides in an accumulation layer formed by a thicker spacer layer. The 50/500 RTD does not show any SdH oscillations when the tunneling electron resides in an accumulation layer formed by a thin spacer layer, due to higher scattering rate there. Thus we believe the SdH oscillations observed from the RTDs with asymmetric layer configurations studied here are due to the Fermi energy of the accumulation layer passing through the Landau levels in that region. Assuming that the energy separation between the Landau ladder is greater than the thermal energy of the electron (kT), the SdH oscillations related to the accumulation layer are expected to improve as the spacer layer becomes thicker as it should reduce scattering there.

Fig. 3 shows the SdH oscillation period of the RTDs in 1/B as a function of applied bias. The periodicity of magneto-quantum oscillation is directly related to the carrier density $n_{2d} = 2eB_f/h$, where n_{2d} is a two dimensional carrier density [4] in an accumulation layer which is shown in the right ordinate of Figs. 3. The 50/85 RTD shows a comparable n_{2d} in both bias direction. However, I-V characteristics of the 50/85 RTD shown in Fig. 1 (a) exhibit 10 times more current through the diode in forward bias than that in reverse bias while n_{2d}, displayed in Fig. 3 (a), shows comparable values in both bias. This signifies that the huge difference of current through the 50/85 RTD is not due to a difference in n_{2d} but mainly due to a difference of tunneling probability through a double barrier structure. A slightly higher n_{2d} for this diode in reverse bias is probably due to the effect of silicon impurity out-diffusion during molecular beam epitaxial growth. The silicon dopant atoms incorporated in the MBE process show a surface segregation due to Fermi level pinning at

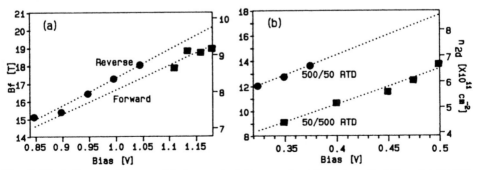

Fig. 3. Shubnikov-de Haas oscillation period, Bf, as a function of applied bias. (a) 50/500 RTD. (b) 50/500 and 500/50 RTDs. The dotted lines in the figures are a result of least squares fit of the measured data. The right ordinate in the figures denote n_{2d}.

the growth surface [9]. Thus the collector spacer layer becomes virtually thinner than the emitter spacer layer due to dopant out-diffusion and shows a higher n_{2d} in reverse bias as a result. The influence of dopant atom out-diffusion can also be seen more clearly from the 500/50 and 50/500 RTDs shown in Fig. 3 (b). These two diodes have the same structural parameters except for an order of the spacer layers (diffusion barriers). A higher n_{2d} of the 500/50 RTD compared to the 50/500 RTD at the same bias configuration can be explained by simply considering the dopant atom out-diffusion length. The separation between the heavily doped collector contact layer and the emitter barrier of the 500/50 RTD is 270 Å (=50+85+50+85) and that of the 50/500 RTD is 720 Å (=500+85+50+85). Thus the active region of the 500/50 RTD has effectively higher doping concentration than the 50/500 RTD and as a result the former shows a higher n_{2d}.

In summary, we have shown the origin of the SdH oscillations from RTDs fabricated with asymmetric layer configurations. The magneto-quantum oscillations originating from the accumulation layer depend only on the scattering rate in that region and their strength improves as the undoped spacer layer becomes thicker, since ionized impurity scattering will be reduced.

Acknowledgements: The authors would like to thank Mr. L. Ungier for discussions related to the work and Mr. S.B. Kim for deposition of plasma enhanced chemical deposition oxide. The work at OSU was partially supported by Office of Naval Research Contract No. N00014-89-J-1894. The work at UW was partially supported by the Washington Technology Center.

References

1. E.O. Kane, "Basic concepts of tunneling", in Tunneling phenomena in solids, pp.1-11, edited by Elias Burstein and Stig Lundqvist, Plenum Press, New York, 1969
2. B. Rico and M. Ya. Azbel, Phys. Rev. B39, 1970, 1984
3. M.L. Leadbeater, L. Eaves, M. Henini, O.H. Hughes, G. Hill and M.A. Pate, Solid-State Electronics, Vol. 32, No. 12, pp. 1467, 1989
4. C.A. Payling et al, Superlattice and Microstrutures 6, 193, 1989
5. H.M. Yoo, PhD dissertation, Oregon State University, 1990
6. H.M. Yoo, S.M. Goodnick, J.R. Arthur, and M.A. Leed, J. Vac. Sci. Technol., B8, 370, 1990
7. V.J. Goldman, D.C. Tsui and J.E. Cunningham, Phys. Rev. B36, 7635, 1987
8. H.M. Yoo, S.M. Goodnick and J.R. Arthur, Appl. Phys. Lett. 56, 86, 1990
9. E.F. Schubert, J. Vac. Sci. Technol. A8, 2980, 1990

Study of a two dimensional electron gas by a new approach in modulation spectroscopy

Ali Badakhshan, and M. Sydor
Department of Physics, University of Minnesota-Duluth, Duluth, MN 55812

ABSTRACT

A new approach in the electric field modulation spectroscopy is presented which effectively extracts a selective photoreflectance (PR) response from a certain layer within a thin multilayer structure. This technique, which we call differential photoreflectance (DPR) spectroscopy, is based on two oppositely phased modulations from two lasers with different wavelengths, so that modulation has a gradient with respect to the depth in the sample. The application of this technique is demonstrated for a two dimensional electron gas in a high electron mobility transistor (HEMT) structure.

INTRODUCTION

In recent years photoreflectance (PR) has emerged as a promising characterization technique which yields sensitive response and detailed information even at room temperature. PR proved to be very successful in accurate determination of semiconductor properties, such as the energy of critical point transitions, the magnitude of the surface electric field, interface quality, and determination of excitonic and some impurity transitions[1-4]. Nevertheless, there are many difficulties associated with the interpretation of PR response at the GaAs bandgap region from multilayer structures[5,6].

Differential photoreflectance (DPR) spectroscopy can be applied effectively to suppress the PR from an intervening layer of most multilayer structures. This technique works on the principle that PR from any given interface is insensitive to the pump wavelength but the laser penetration depth does depend on laser wavelength[7,8]. The layout of the DPR setup is shown in Fig 1. The DPR technique is based on consecutive modulation of the sample by two laser pumps with different energies and half a cycle respective phase difference. The photon fluxes can be made equal "near" a certain interface (see Fig. 2), therefore, the interface is practically unmodulated, while other regions of the sample experience various degrees of modulation in phase with one or the other pump beam. Successive matching of photon fluxes at various interfaces provides a clear understanding of the origin of a certain feature observed in ordinary PR and identify its relative depth.

An additional feature of the DPR apparatus is the use of two identical detectors in and out of the plane of incidence of the probe beam. In Fig. 1 detectors A and B, which are looking at the sample from two different angles, are matched in their background (mostly PL) ac signal. The lock-in amplifier in $A - B$ mode of operation provides PR response with highly reduced PL, as well as some degree of noise reduction.

RESULT AND DISCUSSION

We applied DPR technique to study the PR signal from a 2DEG in a HEMT. Detection of the 2DEG response by PR, provides an important assessment of the electronic quality of HEMT and δ-doped structures. Usually the 2DEG signal is rather weak and its true shape and energy is somewhat ambiguous.[2,9] The PR from 2DEG in modulation doped heterojunctions and from δ-doped wells reported by Glembocki et al.[2] and Bernussi et al.[10] show very short period oscillatory signals confined to a 30 meV region beyond the GaAs band edge. However HEMT structures usually contain undoped GaAs layers which could give rise to a short period FKO signal. These oscillations have periods quite comparable with the signal from a 2DEG observed in PR studies.[2,6,9,11]

Results of this study are merely described in figures and their captions. Comparison between ordinary PR and DPR from HEMT with and without a cap is used to identify and isolate the PR signatures from various layers of the sample. Our results demonstrate that unwanted PR signal from protective caps and room temperature PL can be eliminated more effectively by the use of DPR than by etching off the protective caps.

ACKNOWLEDGEMENTS

We thank Drs. W. Mitchel and M.O. Manasreh of Wright Laboratory, Wright Patterson Air Force Base, Ohio for their valuable help during the course of this study.

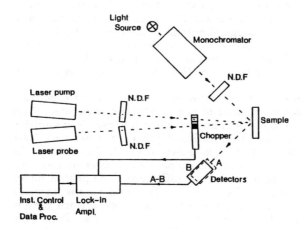

Fig. 1 Differential Photoreflectance apparatus. The laser pumps have different wavelengths and their chopping cycle is shifted by exactly a half a cycle. The two detectors are in and out of the plane of incidence of the monochromator. Thus, the net (A-B) signal minimizes the response to scattered light, photoluminescence, and temporal laser fluctuations.

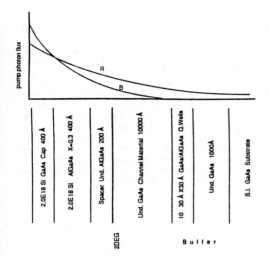

Fig. 2 Sample structure and approximate penetration depths for red (633 nm) and blue (variable 445-545 nm) laser pumps with equal photon fluxes at the (n-GaAs/n-AlGaAs) interface. Matching of laser fluxes at this interface eliminates the broad cap/n-AlGaAs Photoreflectance signal, as will be shown in subsequent figures.

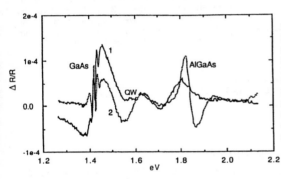

Fig. 3 Curve 1 shows the ordinary Photoreflectance signal (PR) for the sample structure in Fig. 2. The PR was taken using only the 545 nm pump. Curve 2 shows the PR for the same sample after the protective n-GaAs cap was etched off. Notice the change in AlGaAs signal when the cap is removed. Curve 2 still shows a diminished broad PR as though the effect of the cap has not been completely removed by etching.

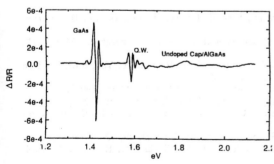

Fig. 4 PR for a HEMT structure similar to that of Fig.2 but having an <u>undoped</u> protective GaAs cap. Here, the PR is quite devoid of the broad signal at 1.4- 1.5 eV which dominated the entire PR spectrum in Fig. 3. The quantum well signals above 1.58 eV can be fitted very closely by derivatives of Gaussian lines. However, even in this sample which is relatively free of the unwanted PR form the cap/AlGaAs, the GaAs band-edge signal from the channel (1.42 eV) still overwhelms a small signal from the two-dimensional electron gas (2DEG). The 2DEG is just barely observable at the tail end of the GaAs band-edge signal (as was first reported by Glembocki et al. in reference 2).

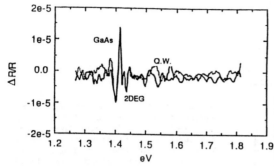

Fig. 5 DPR from etched (curve 1), and unetched (curve 2) sample whose PR was shown in Fig. 3. Photoreflectance signals from cap/n-AlGaAs, GaAs channel layer, 2DEG, and Quantum Wells, all of which were distorted in Fig. 3, can be individually isolated or suppressed in DPR by a proper choice of laser pump intensities and wavelengths. The broad PR signal which was so dominant in Fig. 3 is suppressed here by matching the laser pump intensities at the front surface of n-AlGaAs . Note that here the AlGaAs signal is almost eliminated in both instances and that certain small features are inverted.

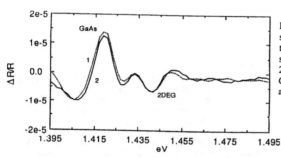

Fig. 6 Details of the bandgap and the 2DEG signal from DPR shown in Fig. 5. Note that the overall signal here is a factor of 10 smaller than the the PR in Fig. 3. Here, the 2DEG signal is well separated from the GaAs band-edge signature, showing the advantage of DPR over PR study.

REFERENCES

1- J.L. Shay, Phys. Rev. B 2, 803 (1970); Shay and R.E. Nahory, Solid State Commun. 7, 945 (1969).

2- O.J. Glembocki, B.V. Shanabrook, N. Bottka, W.T. Beard, and J. Commas, SPIE 524, 86(1985).

3- R.N. Battacharya, H. Shen, P. Parayanthal, Fred H. Pollak. T. Coutts, and H. Aharoni Phys. Rev. B 37, 4044 (1988).

4- N. Bottka, D.E. Gaskill, R. S. Sillmon, R. Henry and R. Glosser, J. Electron. Mater. 17, 161 (1988).

5- M. Sydor, N. Jahran, W.C. Mitchel, W.V. Lampert, T.W. Hass, M.Y. Yen, S.M. Mudare and D.H. Tomich, J. Appl. Phys. 67 (12), 7423 (1990).

6- N. Pan, X.L. Zheng, H. Hendriks and J. Carter, J. Appl. Phys. 68, 2355 (1990).

7- M. Sydor, Ali Badakhshan, and J.R. Engholm Appl. Phys. Lett. 58, 949 (1991).

8- M. Sydor, J.R. Engholm, M.O. Manasreh, C.E. Stutz, L. Liou and K.R. Evans, Appl. Phys. Lett., 56, 1769(1990).

9- E.E. Snow, O.J. Glembocki, and B.V. Shanabrook, Phys. Rev. B, 38, 12484 (1988).

10- A.A. Bernussi, Ikawa F.,Motisuke P., P. Basmaji, Li. M. Siu and O. Hipolito, J. Appl. Phys. 67, 4149 (1990).

11- M. Sydor, Ali Badakhshan, J.R. Engholm, Appl. Phys. Lett., 59 (1991).

12- J.R. Engholm, M. Sydor, F. Szmulowicz, and T. Vaughon (unpublished).

Temperature dependence of photoluminescence decay time in tunneling bi-quantum-well structures

Yoshihiro Sugiyama, Atsushi Tackeuchi, Tsuguo Inata, and Sunichi Muto
Fujitsu Laboratories Ltd.
10-1 Morinosato-Wakamiya, Atsugi, Kanagawa 243-01, Japan

ABSTRACT: We studied the nonresonant electron tunneling time in tunneling bi-quantum-well structures with a pump-probe photoluminescence (PL) system using optical mixing. The PL decay time of narrow wells for an $Al_xGa_{1-x}As$ (x=0.51) barrier of 4.0 nm, determined by tunneling, decreases from 40 ps to 24 ps as the temperature increases from 6K to 132K. We attribute this to electrons in an exciton state thermalizing into free electrons having a faster tunneling rate.

1. INTRODUCTION

Carrier dynamics in heterostructures with tunneling from two-dimensional (2D) states have been intensively studied using time-resolved photoluminescence (PL) measurement with ultrashort pulse lasers in pico or femto second regimes. These include resonant and nonresonant tunneling between 2D states in quantum wells (Matsusue *et al.*, Oberli *et al.*, Sawaki *et al.*, 1989, Deveaud *et al.*, Nido *et al.* 1990) and tunneling escape from the 2D state to a three-dimensional state in resonant tunneling barrier (RTB) structures (Tsuchiya *et al.* 1987, Jackson *et al.* 1989). In these experiments, results were interpreted in terms of free electron (or hole) tunneling and no attempt has apparently been made to describe tunneling dynamics in terms of excitons, although photogenerated carriers tend to form excitons especially at low temperatures. Recently, Tackeuchi *et al.* 1988 proposed a Tunneling Bi-Quantum-well (TBQ) which is a kind of asymmetric coupled quantum well and demonstrated its ultrafast exciton absorption recovery due to electron tunneling between nonresonant 2D states. So far as the contribution of exciton nature to 2D electrons is concerned, TBQs are not different from previously studied resonant and nonresonant tunneling structures. However, this aspect could not be clarified because the measurements (Tackeuchi *et al.*) were done only at room temperature (RT). In this letter, we study the temperature dependence of electron tunneling time between 2D nonresonant states of TBQs using time-resolved PL.

2. EXPERIMENTAL

Figure 1 shows the band diagram of the TBQs we studied. They consist of 60 periods of two alternated quantum wells, 4.5 nm GaAs for narrow wells and 9.0 nm GaAs for wide wells, combined with a potential barrier of $Al_xGa_{1-x}As$ (x=0.51). The energy difference between the lowest quantized levels in the conduction band is estimated to be 60 meV, larger

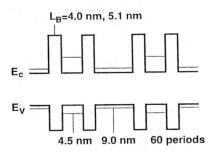

Fig. 1. The band diagram of TBQ.

than the LO phonon energy in GaAs. Barrier thicknesses (L$_B$) were 4.0 nm and 5.1 nm. As a reference, we also measured a conventional MQW consisting of 120 periods of 4.5-nm GaAs well and 4.0-nm barrier Al$_x$Ga$_{1-x}$As (x=0.51). Samples were grown by MBE. We measured the PL decay time of the narrow well of TBQs using a system with optical mixing (Shah 1988). A dye laser (Rh6G, 2.06 eV, 2.5 ps) synchronously pumped by a cw mode-locked Nd:YAG laser with a repetition rate of 82 MHz was separated into pump and probe beams by polarized beam splitter. The average pump power and beam spot on the sample were 9.5 mW and 100 μm$^\phi$. After optical mixing, the sum frequency light generated was put through a 0.32-m monochromator and detected by a photon-counting system with energy resolution less than 30 meV. The temperature of samples was varied from 6K to 132K, in which the Γ band energy of the Al$_x$Ga$_{1-x}$As (x=0.51) barrier was higher (Casey and Panish 1978) than that of the dye laser.

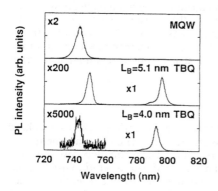

Fig. 2. Photoluminescence (PL) of MQW, TBQ with L$_B$=4.0 nm and L$_B$=5.1 nm at 77K.

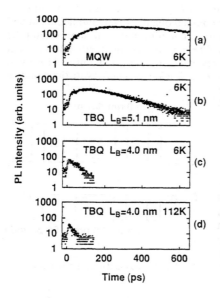

Fig. 3. Time evolution of PL from (a) MQW and from the narrow well of TBQ with (b) L$_B$=5.1 nm (c) L$_B$=4.0 nm at 6K and (d) L$_B$=4.0 nm at 112K.

3. RESULTS

Figure 2 shows the cw PL spectra measured at 77K with Kr ion laser and 1-m monochromator. Figure 3 shows the time evolution of PL from MQW and the narrow well of TBQs at 6K and 112K. A decay time of 350 ps in MQW was obtained from the temporal region between 250 ps and 600 ps. The PL from the narrow well of TBQ with L$_B$=4.0 nm at 6K decays exponentially with a time constant of about 40 ps, about nine times smaller than that of the MQW. This indicates that the PL decays dominantly by tunneling from the lowest quantized level of the narrow well to that of the wide well. Increasing L$_B$ to 5.1 nm increases the decay time to 131 ps, which is regarded as strongly influenced by radiative recombination (350 ps) within the narrow well. The decay time of TBQ with L$_B$=4.0 nm increases as temperature increases. Figure 4 shows the temperature dependence of the PL decay time. It stays almost constant up to 100K. Above 100K it falls, as temperature increases, towards the value obtained by time-resolved absorption measurement (Tackeuchi *et al.*) at RT.

4. DISCUSSION

The origin of the temperature dependence of decay time is not quite clear at the moment. We considered three temperature factors related to electron tunneling in accounting for this dependence and neglected hole tunneling which has a much smaller tunneling rate than the observed decay time. First, the temperature dependence of the barrier tunneling probability is considered to be negligible because the peak current density of a resonant tunneling barrier, proportional to the barrier tunneling probability,

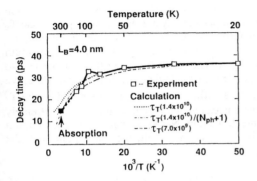

Fig. 4. Temperature dependence of PL decay time for TBQ with $L_B=4.0$ nm. Also shown are the calculated tunneling times for $N_0=1.4 \times 10^{10}$ cm^{-2} and for $N_0=7.0 \times 10^9$ cm^{-2}.

shows little temperature dependence (Sugiyama *et al.* 1988). Second, the non-resonant tunneling between 2D states can be regarded as a kind of intersubband scattering (Sawaki 1989), which should be temperature-dependent. However, the dominant scattering mechanism of electrons between ground states in narrow and wide quantum wells is regarded (Muto *et al.* 1991) to be LO phonon emission (LOE) even at RT. LOE has only a slight temperature dependence, $N_{ph}+1$ ($N_{ph}= 1/[\exp(\hbar\omega_{LO}/k_BT)-1]$), too weak to account for the temperature dependence in experiments. We concentrated on the third factor that is the population ratio of electron states between excitons and free electrons.

A calculation based on variational approximation of excitons indicates (Tackeuchi *et al.*) that the electron tunneling probability is smaller for electrons in the exciton state than that for free electrons because of the attractive force from the holes left in the narrow well. As the temperature rises, more electrons populate the free electron state which has a faster tunneling rate. Therefore, the tendency is consistent with observed decay times. To be more specific, we can express the total tunneling time τ as follows;

$$\frac{1}{\tau} = \frac{\dfrac{N_e}{\tau_e} + \dfrac{N_x}{\tau_x}}{N_e + N_x} \tag{1}$$

where N_e (N_x) is the sheet density of free electrons (excitons), and τ_e (τ_x) the tunneling time of free electrons (electrons in an exciton state). The relationship of $\tau_x = 3\tau_e$ for $L_B=4.0$ nm was derived by Tackeuchi *et al*. To calculate N_e and N_x, we approximated the density of electrons in the exciton state by the density of electrons trapped by imaginary donors of sheet density $N_e +N_x$ and with an energy level equal to the exciton binding energy. Since carrier thermalization time (300 fs at RT and 10 ps at 15K, Knox *et al.* 1985) is faster than decay time in our experiments, we can describe electron distribution by electron temperature T_e and electron quasi-Fermi level E_F. N_e and N_x are given by

$$N_e = \frac{m_e^*}{\pi \hbar^2} \int \frac{dE}{1 + \exp((E - E_F)/k_B T_e)} \tag{2}$$

$$N_e = \frac{N_x + N_h}{1 + g \exp((E_F - (E_0 - B)/k_B T_e)} \tag{3}$$

$$N_0 = N_x + N_h \tag{4}$$

$$N_e = N_h \tag{5}$$

where m_e^* is electron effective mass, g degeneracy factor 2 for the exciton state, N_0 the total number of electron-hole pairs generated by a pump pulse per narrow well, N_h the sheet density of free holes per narrow well, E_0 the lowest electron quantized level of the narrow well, and B the binding energy of free excitons in the narrow well (assumed 10 meV).

Note that eqs. (2) and (3) can be reduced, in the limit of Boltzmann statics, to a 2D version (Chemla 1984) of the well-known Saha equation in plasma physics. Here, we approximate T_e to the sample temperature T. However, the electron temperature may still be slightly higher than T. N_0 is estimated to be 1.4×10^{10} cm^{-2} per narrow well assuming 100% internal quantum efficiency. The calculated result is shown in Fig. 4. The result for $N_0 = 7.0 \times 10^9$ cm^{-2} is also shown in Fig. 4 to indicate a rather weak dependence of calculated results on N_0. Here, we used the free electron tunneling time τ_e as only a fitting parameter, not essential to the T-dependence argument. The observed decay time and calculated tunneling time agree extremely well. Also shown in Fig. 4 is the tunneling time divided by $N_{ph}+1$ to include the contribution from LOE scattering process. Although this correction is small, the agreement with the experiment becomes even better.

Although further study using the determination of the electron temperature at each delayed time is required to confirm the above identification, the temperature dependence of tunneling time in TBQ is regarded to be the temperature dependence of the population ratio of electrons in exciton states to free electrons in the narrow quantum well.

5. SUMMARY

We studied the PL decay time of narrow quantum well in TBQs with a pump-probe PL system using optical mixing. The PL decay time of the narrow well for $Al_xGa_{1-x}As$ (x=0.51) barrier of 4.0 nm, determined dominantly by tunneling process, decreases from 40 ps to 24 ps as the temperature increases from 6K to 132K tending to the decay time at 300K obtained by absorption measurements. This is attributed to the reduction of electron tunneling probability of exciton states compared to that of free electron states and the temperature dependence of population ratio of electrons in the two states.

6. ACKNOWLEDGMENTS

We thank Drs. T. Fujii, O. Ohtsuki, T. Yamaoka, H. Ishikawa, and T. Misugi for their encouragement during this work. We also thank T. Kameda and T. Kurihara for their help in MBE growth.

REFERENCES

Casey, Jr H C and Panish M B *Heterostructure Lasers* partB, p16 (Academic Press, New York, 1978)
Chemla D S, Miller D A B, Smith P W, Gossard A C and Wiegmann W 1984 IEEE J Quantum.Electron. **QE20** 265
Deveaud B, Clerot F, Chomette A, Regreny A, Ferreira R, Bastard G and Sermage B 1990 Europhys. Lett. **11** 367
Jackson M K, Johnson M B, Chow D H, McGill T C and Nieh C W 1989 Appl. Phys. Lett. **54** 552
Knox W H, Fork R L, Downer M C, Miller D A B, Chemla D S, Shank C V, Gossard A C and Wiegmann W 1985 Phys. Rev. Lett. **54** 1306
Matsusue T, Tsuchiya M and Sakaki H *Quantum Wells for Optics and Optoelectronics* Technical Digest, Vol. 10 226 (Opt. Soc. America, Salt Lake City, 1989)
Muto S, Inata T, Tackeuchi A, Sugiyama Y and Fujii T Appl. Phys.Lett. 1991 **58** 2393
Nido M, Alexander M G, Rühle R, Schweizer T and Köhler K 1990 Appl. Phys. Lett. **56** 355
Oberli D Y, Shah J, Damen T C, Tu C W, Chang T Y, Miller D A B, Henry J E, Kopf R F, Sauer N and DiGiovanni A E 1989 Phys. Rev. **B40** 3028
Sawaki N, Höpfel R A, Gornik E and Kano H 1989 Appl. Phys. Lett. **55** 1996
Shah J 1988 IEEE J. Quantum Electron. **QE24** 276
Sugiyama Y, Inata T, Muto S, Nakata Y and Hiyamizu S 1988 Appl. Phys. Lett. **52** 314
Tackeuchi A, Muto S, Inata T and Fujii T 1988 Jpn. J. Appl. Phys. **28** L1098
Tsuchiya M, Matsusue T and Sakaki H 1987 Phys. Rev. Lett. **59** 2356

Inst. Phys. Conf. Ser. No 120: Chapter 5
Paper presented at Int. Symp. GaAs and Related Compounds, Seattle, 1991

239

The substrate current by impact ionization in GaAs MESFETs

Takahiro Yokoyama and Akiyoshi Tamura

Semiconductor Research Center, Matsushita Electric Industrial Co., Ltd.,
3-15,Yagumo-Nakamachi, Moriguchi, Osaka 570, Japan

ABSTRACT :   An effective method was presented to study impact ionization in GaAs MESFET's. It was found that the substrate current (I_{sub}) starts flowing at the onset of breakdown on the drain current (I_{DS}). I_{sub} is attributed to hole injection into a semi-insulating GaAs substrate by impact ionization in the channel. To characterize the substrate current and impact ionization theoretically, a simplified model of lateral field distribution in the channel was used. As a result, the ratio of I_{sub} to I_{DS} (I_{sub}/I_{DS}) can be shown to be proportional to an exponential of $-1/\sqrt{V_{DS}}$.

1.Introduction

As the device dimensions become smaller, the increased electric field in an active region induces the hot carrier effects such as breakdown phenomenon. The hot carrier effect has been investigated intensively in Si MOSFET's. However, the breakdown phenomenon has not been fully underdstood in GaAs MESFET's. This breakdown phenomenon is explained to be occured by impact ionization in the channel (Van Zeghbroeck et al. 1987). In order to reduce the internal electric field of the channel, MESFET's with lightly-doped drain (LDD) structure are widely used as compared with conventional MESFET's. However, if carrier concentration of LDD region is high, low breakdown voltages are observed in LDD-MESFET's, even in offset gate MESFET's. Therefore, the study of impact ionization in GaAs MESFET's is very important to design the FET structure. In this work, we have found that substrate current (I_{sub}) starts flowing at the onset of impact ionization in the channel and the measurement of I_{sub} gives a useful method to understand impact ionization in GaAs MESFET's.

2.Experimental

GaAs MESFET's with LDD structure were fabricated on undoped semi-insulating LEC GaAs substrate. Si^{29} ions were implanted at 25keV for the active layer and at 50keV with a dose of 6.0×10^{12} cm^{-2} for LDD region. The n$^+$ region was formed by Si^{28} implantation at 150keV with a dose of 5×10^{13} cm^{-2} through a 200-nm SiO$_2$ film. WSiN was used for the gate metal. The gate length and width of FET were 1.0μm and 50μm, respectively. The typical threshold voltage of FET was -0.4V. In order to measure I_{sub}, the substrate electrode composed of an ohmic contact was located at 20μm from the edge of the source n$^+$ region, just like a side gate electrode. The I-V characteristics of FET and I_{sub} were measured under the substrate voltage (V_{sub}) of 0V.

3.Results and Discussions

Figures 1(a) and 1(b) show gate leakage current (I_G) versus drain voltage (V_{DS})

and I_{sub} versus V_{DS}, respectively. Typical I-V characteristics of MESFET are also shown in Figure 1(c) and soft breakdown behavior (kink) was observed on the drain current (I_{DS}) at around 4V of V_{DS}.

It was found that I_{sub} starts flowing from FET to the substrate electrode at the onset of breakdown (at the kink), as shown in Figs.1(b) and 1(c). On the other hand, in Fig.1(a), there was a little change in the gate leakage current at the onset of breakdown. I_{sub} is explained by hole injection into a semi-insulating GaAs substrate by impact ionization. If small positive voltage ($\simeq 1V$) was applied to the substrate electrode, there was no change of I_{sub} at the kink and I_{sub} was very small (comparable to the measurement lower limit of current). If small negative voltage ($\simeq -1V$) was applied to the substrate electrode, I_{sub} starts flowing at around the kink points. Therefore, it can be concluded that I_{sub} is hole flowing from FET to the substrate electrode. Once impact ionization occurs in the channel, electron and hole pairs are generated. Holes are easily injected into the substrate and swept to the substrate electrode by the electric field. From the above results, it was found that I_{sub}, rather than I_G, is a good monitor of impact ionization.

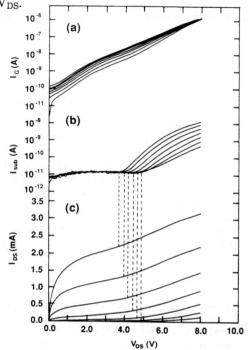

Fig.1 (a) Gate leakage current I_G versus drain voltage V_{DS} for different gate voltage V_{GS}. (b) Substrate current I_{sub} versus V_{DS} for different V_{GS}. (c) Drain current I_{DS} versus V_{DS} for different V_{GS}. V_{GS} varies from -0.6V to 0.0V in steps of 0.1V, respectively. The curves for $V_{GS} = -0.6V$ are top curve in (a) and bottom curves in (b) and (c).

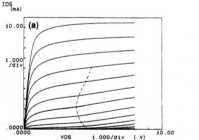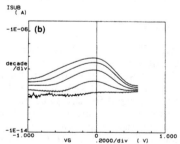

Fig.2 (a) I_{DS} versus drain voltage V_{DS} for different gate voltage. V_{GS} varies from -0.6V (bottom curve) to 0.6V in steps of 0.1V. (b) I_{sub} versus V_{GS} for different V_{DS}. V_{DS} varies from 4V (bottom curve) to 8V in steps of 1V.

Figure 2 shows I-V characteristics of MESFET and I_{sub} versus V_{GS} characteristics for different V_{DS}. The gate voltage dependence of the onset of breakdown is clearly understood from Fig.2(b). If $V_{DS} = 4V$, the breakdown did not occur, so I_{sub} was very small and there was no dependence on the gate voltage. But, $V_{DS} \geq 5V$, I_{sub} flew and had a peak value (I_{submax}) at $V_{GS} \approx 0V$. As mentioned above, I_{sub} was closely related to impact ionization, therefore, impact ionization occurred most intensively at $V_{GS} \approx 0V$ and the onset of breakdown appeared at the lowest V_{DS}, as shown a dashed line in Fig.2(a). Theoretical background of the dependence of I_{sub} on V_{GS} is explained by equation (3) and this is discussed later.

To characterize I_{sub}, it is essential to know an analytical model of electric field in the channel, especially an analytical expression of the maximum electric field (E_m). It is well known that when V_{DS} is relatively high, the high field region is generated at the drain side of the gate edge or the gate side of the drain edge and that is affected by gate bias and the surface conditions (Wroblewski et al. 1983). Because of these complications, such an analytical model is not still available. One way to solve this problem was presented by Hui et al.(1990). They assumed the expression of E_m as the same one as in the case of Si MOSFET's. However, it is not clear that such assumption is exactly correct for GaAs MESFET's. In this case, for theoretical calculation of I_{sub}, a simplified model of lateral electric field distribution in GaAs MESFET, as shown in Figure 3, was used on the assumption that high electric field region is generated at the drain side of the gate edge and electron impact ionization coefficient α_n in GaAs was assumed to be expressed as a following formula

$$\alpha_n = \alpha_i \exp\left\{-\left(\frac{\beta_i}{\varepsilon}\right)\right\} \quad (1) \quad \text{here } \varepsilon : \text{electric field}$$

According to reference (Shur 1978), the maximum electric field is expressed as

$$E_m \propto \sqrt{\frac{(V_{DS} - V_{DSAT})}{\Delta L_2}} \quad (2) \quad \text{here } \Delta L_2 : \text{length of high field domain}$$

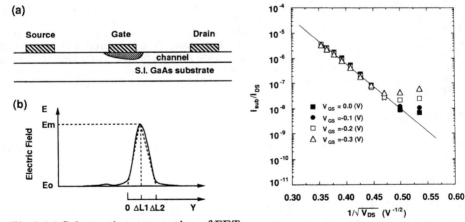

Fig.3 (a) Schematic cross section of FET
(b) Electric field distribution

Fig.4 $\ln I_{sub}/I_{DS}$ versus $1/\sqrt{V_{DS}}$ for different V_{GS}

The outline of the calculation was the same as that in the case of Si IGFET's (El-Mansy et al. 1977). If V_{DS} is applied, high field region is generated at the drain side of the gate edge, and this region was simplified to the triangular shape, as shown dashed lines in Fig.3(b). Here, surface depletion layer was neglected. According to reference (El-Mansy et al. 1977), I_{sub} is expressed as

$$I_{sub} = (M-1)I_{DS} \quad (3) \quad M : \text{avalanche multiplication factor}$$

Here M is a decreasing function of V_{GS} and I_{DS} is an increasing function of V_{GS}, therefore, the product of M and I_{DS} can have a peak value, as shown in Fig.2(b). Using α_n, M is expressed as

$$M - 1 = \int_0^{\Delta L_2} \alpha_n dy$$

$$\simeq \frac{\Delta L_2 \alpha_i}{E_m - E_0}\left\{ E_m \exp\left(-\frac{\beta_i}{E_m}\right) - E_0 \exp\left(-\frac{\beta_i}{E_0}\right)\right\} \simeq \alpha_i \Delta L_2 \exp\left(-\frac{\beta_i}{E_m}\right) \quad (4)$$

putting Eq.(4) into Eq.(3), we get

$$I_{sub} \propto \alpha_i \Delta L_2 \exp\left(-\frac{\beta_i}{E_m}\right) I_{DS} \quad (5)$$

From equations (2) and (5), we can obtain that the ratio of I_{sub} to I_{DS} (I_{sub}/I_{DS}) is proportinal to an exponential of $-1/\sqrt{V_{DS}}$.

Figure 4 shows I_{sub}/I_{DS} versus $1/\sqrt{V_{DS}}$ characteristics and these data give a straight line. On the basis of this results, it can be concluded that this calculation is a good approximation of substrate current by impact ionization, in spite of using several assumptions in this calculation. In Fig.4, when $1/\sqrt{V_{DS}}$ is larger than 0.46, where V_{DS} is lower than about 4.5V, there are discrepancies from theoretical calculation and it may be because impact ionization is so weak around this region that I_{sub} is very small and contains some background current.

4.Conclusions

I_{sub} starts flowing from FET to the substrate electrode at the onset of breakdown. I_{sub} is explained by hole injection into the substrate due to impact ionization in the channel. The gate voltage dependence of the onset of breakdown is clearly understood from I_{sub} versus V_{GS} characteristics. A simplified electric field distribution model was used for theoretical calculation of I_{sub}. As a result, it was found that I_{sub}/I_{DS} is proportional to an exponential of $-1/\sqrt{V_{DS}}$ and that measuring I_{sub} gives a useful method to study impact ionization in GaAs MESFET's.

5.Acknowledgements

We would like to thank Dr. T.Takemoto, T.Onuma and Dr. M.Inada for their encouragement throughout this work.

6.References

Van Zeghbroeck B J et al. 1987 IEEE Electron Device Lett., vol. EDL-8, 188
Wroblewski R et al. 1983 IEEE Trans. Electron Device, vol. ED-30, 154
Hui K et al. 1990 IEEE Electron Device Lett., vol. EDL-11, 113
Shur M S 1978 IEEE Trans. Electron Device, vol. ED-25, 612
El-Mansy Y A et al. 1977 IEEE Trans. Electron Device, vol. ED-24, 1148

Inst. Phys. Conf. Ser. No 120: Chapter 5
Paper presented at Int. Symp. GaAs and Related Compounds, Seattle, 1991

243

RHEED analysis for stoichiometric GaAs growth of migration enhanced epitaxy

Mitsuaki Yano, Kou Masahara, Kanji Yoh, Yoshio Iwai, and Masataka Inoue

New Material Research Center, Osaka Institute of Technology
Asahi-ku Ohmiya, Osaka 535, Japan

abstract
We have studied the mechanism of RHEED oscillation during MEE growth of (100)-GaAs. Contrary to the case of MBE, the mechanism can be understood by the alternating changes in surface composition due to the beam-switching. This analysis revealed that excess As adatoms can exist on the GaAs surface up to 530 °C.

Migration of Ga adatoms can be enhanced when the GaAs surface is free from As_4 molecules. This characteristic behavior has been used for migration enhanced epitaxy (MEE) in which the Ga- and As-beam are alternatively supplied to grow GaAs even at low temperatures (Horikoshi et al. 1988 and 1989a). In order to obtain high-quality GaAs, however, each deposition period of Ga- and As-beam must be optimized to the stoichiometric condition during growth (Horikoshi et al. 1989b, Iwai et al. 1991). The optimum condition of Ga can be easily set independently of the growth temperature due to the low vapor pressure. On the other hand, the control of highly volatile As_4 is very sensitive to the growth temperature (Iwai et al. 1991).

In this paper, we propose an in-situ monitoring of the surface composition by using reflection high energy electron diffraction (RHEED) to control precisely the As deposition during MEE growth. We analyzed the oscillation waveform of RHEED intensity to understand the growth mechanism under various MEE conditions. In order to interpret the waveform, the electron-beam reflectivity was examined for the surfaces with different As coverage.

The growth was done using an ANELVA-620 MBE apparatus containing metallic Ga and As sources. In this study, the epitaxial layer was deposited on (100)-oriented GaAs. An incident electron-beam of 25 KeV was introduced from the [011] and [0$\bar{1}$1] directions. The Ga deposition period, for example 2.0 s, was adjusted to grow just one atomic plane of the cation site, N_s on the surface. The following As deposition period was controlled within the accuracy of a 0.1 s switching interval. The typical As-beam intensity was 8×10^{-7} Torr in the present experiment, and the background pressure was kept below 7×10^{-10} Torr. The growth temperature was measured by a pyrometer calibrated assuming the native Ga-oxide desorption temperature to be 580 °C.

Figure 1(a) and (b) show oscillation patterns of the RHEED specular-beam intensity observed in the [011] and [0$\bar{1}$1] directions, respectively. The growth temperature decreased continuously from 600 °C to 400 °C. The total supply of As_4 molecules per pulse, that is $N_{As} =$ (flux intensity) x (pulse width), was kept at the constant value of $N_{As} = N_{Ga} = N_s$. Between each period of Ga and As deposition, the growth was interrupted for 1.0 s to enhance the migration of adatoms. Solid and dashed lines in Figure 1 correspond to the switch-off timings of Ga- and As-beam, respectively.

As can be seen in Figure 1, the amplitude of RHEED oscillation increased initially and then decreased with decreasing growth temperature. Near the maximum amplitude at 550 °C, the temperature becomes optimum to grow stoichiometric GaAs (Iwai et al. 1991). Under this optimum condition, reconstructions just after As and Ga deposition were (2x4)As and (4x2)Ga, respectively. With decreasing growth temperature, the reconstruction of Ga-stabilized surface clearly changed, although the As-stabilized surface continued as (2x4)As. Between 530 °C and 450 °C, the (4x2)Ga surface was replaced by a (4x8)Ga. This replacement is due to a reaction of impinging Ga atoms with excess As adatoms which have been remained on the surface before Ga deposition. During the following growth interruption of 1.0 s, the (4x8)Ga reconstruction quickly changed into (4x6)Ga, and then gradually turned to (2x4)As through a blurred pattern. The observed change during growth interruption is presumably due to As incorporation from background atmosphere. Below 450 °C, the predominant reconstruction became to be (2x4)As even for the surface just after Ga deposition. As the result, the reconstruction below 420 °C was completely dominated by the (2x4)As independently of the Ga deposition. These experimental data indicate that the surface after Ga deposition increases its As coverage with decreasing growth

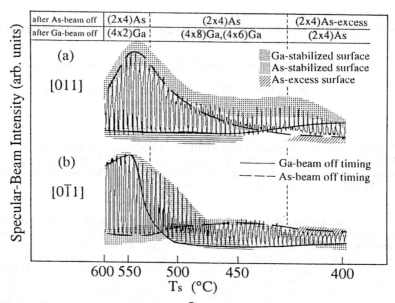

Fig.1   Change of the [011] and [0$\bar{1}$1] RHEED oscillation waveforms with decreasing growth temperature.

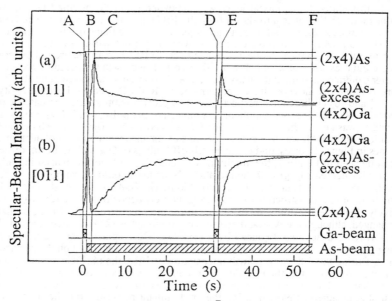

Fig.2   Typical variation of the [011] and [0$\bar{1}$1] RHEED intensities associated with Ga and As deposition at 400 °C.

temperature. According to the recent report by Deparis et al. (1991), the degrees of As coverage are 3, 10, 22 and more than 75% for the surfaces with (4x2)Ga, (4x8)Ga, (4x6)Ga and (2x4)As reconstructions, respectively.

As well as excess As on the growth surface after Ga deposition, an increase of As adatoms with decreasing growth temperature is expected for the surface after As deposition although the effect was not clearly reflected in the RHEED pattern. In order to study the effect of As adatoms, we analyzed the variation of specular-beam intensity induced by As adsorption at a fixed low growth temperature. Figure 2 shows a typical RHEED intensity variation associated with Ga and As deposition, in which (a) and (b) are signals from [011] and [0$\bar{1}$1] directions, respectively. At 400 °C, Ga atoms (in an amount N_s) were first deposited on the (2x4)As surface from A to B to produce the (4x2)Ga reconstruction at B. After this Ga deposition, the first As deposition, B to D, followed by the second Ga deposition, D to E, was introduced to the surface. The second As deposition was started at E. During the course of deposition, the (4x2)Ga reconstruction at B changed into (2x4)As at C, and continued to F independently to the Ga deposition between D and E. The clearest (2x4)As reconstruction, however, was observed at the times of C and E.

One can understand the different RHEED signal response to the first and second Ga depositions as follows. The first Ga deposition between A and B must have been performed on a (2x4)As surface free from excess As adatoms because the deposition of Ga atoms in an amount of N_s has established the (4x2)Ga surface of which As coverage is negligibly small. The clear (2x4)As reconstruction observed at C also corresponds to a surface free from excess As adatoms. Further As deposition between C and D accumulates excess As adatoms on the surface to saturate at D, which is responsible for the appearance of a less clear (2x4)As reconstruction at D. (Hereafter, we express the less clear (2x4)As reconstruction to be (2x4)As-excess.) The second Ga deposition from D to E must have been done on the As-excess surface. The amount of excess As adatoms at D must be estimated as N_s because the same amount of Ga atoms have been consumed to establish the clear (2x4)As surface at E. The following As deposition also accumulates the same amount of excess As adatoms on the surface at F. The data of Figure 2 should indicate that the degree of As-excess, E_{As}, increase from zero ($E_{As}=0\%$ i.e., (2x4)As) up to N_s ($E_{As}=100\%$) with the progression of As adsorption.

Here we come back to Figure 1 to understand the temperature-depended RHEED oscillation in relation to the results in Figure 2. Typical waveforms of Figure 1 are magnified in Figure 3. The optimum oscillations at around 550 °C, which is shown in Figure 3(a), corresponds to the alternation of the points A and B in Figure 2, i.e., the waveform of Figure 3(a) can be understood as the alternative change of (2x4)As reconstruction with (4x2)Ga. On the other hand, the oscillation under As-excess condition below 420 °C can be explained by the alternation of the points D and E in Figure 2, i.e., the waveform of Figure 3(e) corresponds to the alternative change of (2x4)As-excess reconstruction ($E_{As}=100\%$) with (2x4)As ($E_{As}=0\%$). Notice that the opposite response of RHEED intensity to As and Ga deposition (phase of the waveform) is the characteristic of these

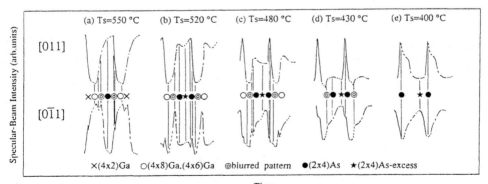

Fig.3   Magnified waveforms of Figure 1. Shutter-timings are specified by the line drawings; ——— Ga, —·—As, and - - - - - growth interruption.

two conditions, the optimum and the completely As-excess conditions. As well, the oscillation between 530 °C and 420 °C can be understood as the intermediate case of these two representative conditions mentioned above, i.e., each one cycle of the oscillations in Figure 3(b), (c) and (d) can be understood by the alternative change of (2x4)As-excess reconstruction (0% < E_{As} < 100%) with (4x8)Ga and/or (4x6)Ga. The following sequences were observed for specular-beam intensities of the stably reconstructed states; (2x As) > (2x As-excess) > (4x Ga) in the [011] direction; and (x2 Ga) ≈ (the blurred pattern intermediate to (2x4)As and (4x6)Ga) > (x4 As-excess) > (x4 As) and (x6 Ga) > (x8 Ga) in the [01̄1] direction; respectively. The sequence between (x4 As) and (x6 Ga) is not clear at present. Although not identified in this experiment, the observed blurred pattern might correspond to the (3x1)Ga reconstruction which has been known as an intermediate structure stabilized between (2x4)As and (4x6)Ga (Deparis et al. 1991).

In order to estimate the degree of As-excess through the oscillation waveform, we simulated the [011] waveform by using a numerical calculation based on a two-dimensional semi-empirical model which reflects the effect of excess As adatoms. In the simulation, we calculated the occupancy of Ga-, As- and excess As-stabilized surfaces on the two-dimensional growth front at every impinging times of Ga and As atoms of which sticking and migration probabilities were considered. The RHEED intensity at every moment was determined as the total reflection from above mentioned three different surfaces, i.e., the waveform was calculated by the time-depended occupancy of Ga-, As- and excess As-stabilized surfaces of which [011] reflectivities have been assigned from the experiment of Figure 2. Typical calculated results are shown in Figure 4 of which (a), (b), (c), (d) and (e) correspond to the experimental waveforms in Figure 3. As we can see, calculated results in Figure 4 agrees with the observed waveforms in Figure 3. The degree of As-excess, E_{As} can be evaluated as shown in Figure 4 by the calculation. We can understand the change of [011] waveform as a function of E_{As} which increases with decreasing growth temperature. Note that E_{As} = 15% calculated for Figure 4(b) is close to the experimental result of Figure 3(b) of which E_{As} is estimated to be 10% to 20% since the observed reconstruction after Ga deposition is (4x8)Ga and/or (4x6)Ga.

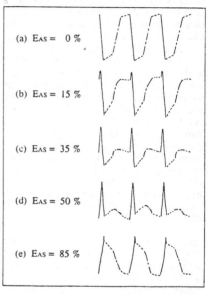

(a) E_{As} = 0 %

(b) E_{As} = 15 %

(c) E_{As} = 35 %

(d) E_{As} = 50 %

(e) E_{As} = 85 %

Fig.4 Simulated RHEED oscillation waveforms as a function of E_{As}. Shutter timings are specified by the same symboles as Fig.3.

The proposed mechanism of MEE oscillation is completely different from the rough-smooth alternation model (Harris et al. 1981) responsible for the case of conventional MBE, although the increase of surface roughness must decrease the MEE oscillation amplitude. This computer simulation, of which detail will be discussed elsewhere (Yano et al. to be submitted), enables us to estimate the degree of excess As adatoms by monitoring the RHEED oscillation waveform.

This work was supported in part by a Grant-in-Aid for Scientific Research on Priority Area "Electron Wave Interference Effect in Mesoscopic Structures" from The Ministry of Education, Science and Culture.

References
Deparis C and Massies J 1991 *J.Cryst. Growth* **108** 157
Harris J J, Joyce B A and Dobson P J 1981 *Surface Sci.* **103** L90
Horikoshi Y, Kawashima M and Yamaguchi H 1988 *Jpn.J.Appl.Phys.* **27** 169
Horikoshi Y and Kawashima M 1989a *J.Cryst.Growth* **95** 17
Horikoshi Y, Kawashima M and Yamaguchi H 1989b *Proc.Symp.Heteroepitaxial Approaches Semicond. Lattice Mismatch 1st Conseq.USA* pp 358-68
Iwai Y, Yano M, Hagiwara R and Inoue M 1991 *Semicond.Sci.Technol.* in press
Yano M, Masahara K and Inoue M to be submitted elsewhere

Inst. Phys. Conf. Ser. No 120: Chapter 5
Paper presented at Int. Symp. GaAs and Related Compounds, Seattle, 1991

247

Problems in the use of epitaxial AlAs layers as calibration standards for the Al content of AlGaAs/GaAs layers

I.C.Bassignana, D.A.Macquistan and A.J.SpringThorpe
Bell-Northern Research Ltd., Advanced Technology Laboratory, Ottawa, Canada.

Introduction

Accurate determination of the Al content of AlGaAs/GaAs epitaxial layers is of great practical interest. Usually the Al content of epitaxial layers is calculated from a measurement of the difference in lattice parameters between the epitaxial AlGaAs layer and the GaAs substrate. Measured lattice parameter differences are converted into compositions by using the Poisson's ratio and end-member lattice parameters (Bassignana et al 1989). However, these numbers are not well known which makes the determination of Al concentration problematic. An alternative approach has been to use epitaxial layers of the end-member AlAs as a calibration standard; there has recently been renewed interest in this method. Interestingly, several groups have reported significantly different values for the double crystal x-ray diffraction (DCD) rocking curve peak separation ($\Delta\Theta$) of AlAs and GaAs, ranging from 365 arc-sec to 386 arc-sec. This is disturbing since this quantity is the 'internal standard'. Two factors could give rise to this discrepancy in the measurements; either the absolute step size calibration of the diffractometers varies, or alternatively, AlAs grown under different conditions can vary significantly. In this work we investigate the reproducibility of the measurements and the effect of some growth parameters.

Verification of Instrument Calibration

One approach to instrument calibration which depends only on substrate materials is performed in the following way. A standard double crystal diffractometer is configured to use different first and second crystals; in principle, any two crystals can be used. The resulting rocking curve traces out $K\alpha_1$ and $K\alpha_2$. The separation of these two peaks depends only on the reflection plane of the material used and the internal step size calibration of the instrument. The former can be calculated; the latter is the difference between theoretical and experimental step size. When GaAs is used as the first crystal and InP as the second crystal using CuKα radiation, the calculated separation is 16.9 arc sec for the (+,-) setting and 644 arc-sec for the (+,+) setting. The experimental peak separation of 610 arc-sec is ~5% smaller than the calculated value, consequently the diffractometer step size is 5% greater than true. All of the values in this report have been corrected for this calibration. We suspect that all instruments of similar vintage may require this type of calibration.

Reproducibility of the Measurements

Epitaxial layers of AlAs/GaAs provide a convenient standard for DCD measurements because AlAs is an end-member compound the composition of which cannot vary from point to point on an epitaxial layer. Six AlAs/GaAs layers ranging in thickness from 0.6 to 1.3μm. were grown by MBE. Rocking curves were recorded on two Bede 6" double crystal diffractometers aligned in the parallel (+,-) setting, using CuKα radiation. The quantity of interest is $\Delta\Theta$, the angular separation between the AlAs and GaAs peaks. Three factors must be considered: 1.) for a given instrument, the reproducibility of a $\Delta\Theta$ measurement made repeatedly at a given point on a sample, 2.) for a given instrument, the reproducibility of a $\Delta\Theta$

measurement made at different but nearby points on a given sample and, 3.) diffractometer to diffractometer variation in the measurement of $\Delta\Theta$ made at a given area on a sample.

Rocking curves recorded consecutively, under identical conditions, at a given point on a sample showed that on a given diffractometer the reproducibility of the measurement is approx. 1% i.e. within 1-5 arc-sec in ~365 arc-sec. Measurements of $\Delta\Theta$ made at nearby but different points on a given AlAs/GaAs layer were usually within 5-10 arc-sec, but in some cases differences in $\Delta\Theta$ measured at different points could be as large as 20 arc-sec. It is difficult to attribute these large differences to instrument variation since we have shown above that rocking curves measured repeatedly at a given point are more reproducible. Since very significant differences can be found on one wafer, comparisons between different wafers are questionable and possibly valid only if averages are compared. Further investigation using topography showed that these layers are highly dislocated, and possibly the differences can be explained in terms of variations in the layers themselves, as discussed below.

Another approach to Al calibration was to grow a layer of AlAs as part of the structure so that it could be used as a true internal standard. Rocking curves for several such stacks were measured. All the layers in the stacks were about 1μm thick and many were of high Al content, for example, one stack consisted of four layers, ~1μm $Al_{.4}Ga_{.6}As$, ~1.7 μm $Al_{.7}Ga_{.3}As$, ~2.2μm $Al_{.88}Ga_{.22}As$ and a ~2.4 μm layer of AlAs. The rocking curve data for the AlAs layer in these stacks was no different from single AlAs layers. This experiment suffers much from the same problem as that cited above for single layers: depending on the exact growth conditions the AlAs layer can be relaxed to different extents. The AlAs grown in the same stack does not guarantee that the relaxation takes place to the same extent in all the layers.

Limit for Critical Thickness
Epitaxial layers below the pseudomorphic limit can accommodate lattice mismatch with the substrate elastically; above the pseudomorphic limit the epitaxial layer forms a semi-coherent interface with the substrate and the layer is dislocated to accommodate the strain. A number of theories exist for estimating the critical thickness and strain relaxation in heteroepitaxial layers. Experimentally relaxation can be detected by the observation of misfit dislocations which appear as crosshatching in x-ray topographs. Asymmetric crystal topography (ACT) is ideally suited to this observation because of its extreme orientation sensitivity (Boettinger et al 1976).

The Matthew-Blakeslee (MB) estimate (Matthews et al 1974) of the minimum critical thickness for this system is approx. 0.2 μm. ACT topography was used to characterize all the in-house MBE grown layers of AlAs/GaAs. All these layers, the thinnest of which had a thickness of 0.6μm, showed crosshatching. These observations are in agreement with the theory and indicate relaxation in the layers, making them a poor choice for calibration standards. All ACT topographs of these layers are very similar in showing clusters of dislocations on a macroscopic scale; the layer does not relax uniformly and the clusters of dislocations appear in the topograph as lighter and darker areas.

We noted above that, for AlAs layers, considerable differences in $\Delta\Theta$ can be measured. Since the topographs show clusters of dislocations, it is possible that the x-ray probe beam of the DCD, which has a diameter of approx. 500μm, is sampling areas with varying densities of dislocations. A larger probe beam would average the differences. Since relaxation is a chaotic phenomenon, it is very likely not to be uniform and the topographs confirm this. Rocking curves recorded at different points sample more or less dislocated areas giving rise to different $\Delta\Theta$s. Diffractometers with much larger sampling areas can average this effect so that it is not always observed. Alternatively, layers well below the critical thickness should show little difference in $\Delta\Theta$ measured at different points. A preliminary review of the data we have collected on many layers grown below the critical thickness shows that this is indeed the case.

Calculating the Degree of Relaxation

The (004) or any surface symmetric reflection measures the lattice parameter along the axis normal to the substrate. The mismatch ($\Delta d/d$) between the epitaxial layer and the substrate is calculated from the rocking curve peak separation ($\Delta\Theta$) using the differentiated form of Bragg's law, $\Delta d/d = m^* \cot\Theta \, \Delta\Theta$. This equation is corrected to calculate the 'relaxed' or actual mismatch by multiplying by m^*, the correction factor calculated from the Poisson's ratios (Bartels et al 1978); the inherent assumption is that the epitaxial layer is completely coherent. The difference in the perpendicular lattice parameter between a coherent epitaxial AlAs layer and a completely relaxed AlAs layer is ~0.07%. The $\Delta\Theta$ for a coherent AlAs layer (~365 arc-sec) differs from that for a relaxed layer ($\Delta\Theta$ ~ 183 arc-sec). The measured $\Delta\Theta$ between these two rocking curves differs by a factor of ~2, i.e. the Poisson's correction. In the most commonly encountered situations, the relaxation of the AlAs layer is not complete so that the measured rocking curve will be somewhere between these two values. The partially relaxed layer will have a $\Delta\Theta$ which is reduced by $\Delta\Theta$(coherent layer)/(2 X 100) arc-sec per percent of relaxation. For AlAs/GaAs the change in $\Delta\Theta$ is about 1.8 arc-sec per percent of relaxation. Possibly the differences in $\Delta\Theta$ for a given layer can be related to different degrees of relaxation across a wafer, as is suggested by the topographs. A difference of 20 arc-sec between rocking curves represents a relaxation of about 11%, a plausible number for this system.

Comparison of Epitaxial Layers from Different Sources

The table below compares rocking curves and topographs for samples from different laboratories. Two comparisons are important: (1) the comparison between the $\Delta\Theta$s recorded for these samples on our diffractometers and those reported by the other investigators, and (2) the comparison between the various AlAs epitaxial layers grown by different groups.

The variation from instrument to instrument is difficult to gauge but one approach is to compare the average of all measurements made on each instrument. All of the in-house measured $\Delta\Theta$s on in-house samples are averages recorded on two diffractometers at several places on each of many samples. The in-house measurements of $\Delta\Theta$ for samples A,B and C are also averages for several measurements made at a few points on the samples. Within experimental error there seems to be no difference between our measurements of $\Delta\Theta$ and those obtained elsewhere. However it is not known how the rocking curves for these other measurements were recorded and it is assumed that the numbers also are averages; otherwise, all of the difficulties discussed above are to be expected.

Sample	$\Delta\Theta$ in-house	$\Delta\Theta$ other	AlAs (μm)	Growth method	Topograph crosshatch
in-house	365	n/a	1	MBE	yes
Sample A	382	372	2	MOCVD	no
Sample B	381	386	1	MBE	yes
Sample C	369	368	1	MBE	no

The in-house samples showed somewhat smaller $\Delta\Theta$s (~15 arc-sec) than those grown elsewhere indicating a greater degree of relaxation in these samples. However sample B, which has a larger $\Delta\Theta$ than measured by us also showed the presence of some misfit dislocations in the ACT topographs indicating that it also is relaxed.

All the layers analyzed are beyond the MB limit for the critical thickness; however, the coherency of a layer depends heavily on the details of the growth conditions. Layers which should be relaxed in an equilibrium state can exist in a coherent metastable state. These data indicate that the quality of the layer depends heavily on the details of the growth since topographs of the AlAs layers of approximately the same thickness from different sources do not all show relaxation. This may be due to a variety of factors, including the growth temperature, the cooling rate, or even the particulate count on the surface prior to growth, since all of these can initiate the nucleation of dislocations. In order to address the question of growth temperature, MBE AlAs layers were grown between 569 C and 720 C, but showed no significant difference; all $\Delta\Theta$ values are approximately 365 arc-sec and topographs of all these layers show crosshatch. Other samples where the layers were grown using As_4 instead of As_2 showed increased relaxation and $\Delta\Theta$ ~341 arc-sec. These results are preliminary and more work is in progress (Bassignana et al in preparation).

Conclusions

DCD measurements are highly stable and reproducible, internal calibration of the step size on both our diffractometers shows that the absolute step size are about 5% larger than expected. In-situ AlAs standards grown as part of a calibration stack have no proven advantage over external standards. AlAs epitaxial layers grown beyond the critical thickness cannot be used as a standard for calibration since layers which are thick enough to give reliable $\Delta\Theta$ measurements have also been found to be partially relaxed. Topographs of AlAs (>0.6μm) layers do not always show relaxation. 1μm thick AlAs layers grown at temperatures between 569 and 725 C all show relaxation. A round robin of AlAs layers from 4 laboratories showed that AlAs layers of similar thickness can have different degrees of relaxation which may depend on growth conditions.

References

Bartels W J and Nijman W 1978 *J. Cryst Growth* **44** 518-525
Bassignana I C and Tan C C 1989 *J. Appl. Cryst.* **22** 269-276
Bassignana I C and Macquistan D A *in preparation*
Boettinger W J, Burdette H E and Kuriyama M 1976 *Rev. Sci. Inst.* **47**(8) 906
Matthews J W and Blakeslee A E 1974 *J. Cryst. Grow.* **27** 118-125

Inst. Phys. Conf. Ser. No 120: Chapter 5
Paper presented at Int. Symp. GaAs and Related Compounds, Seattle, 1991

251

Hot-electron-acceptor luminescence in quantum wells: a quantitative measurement of the hole dispersion curves

M. Zachau, J.A. Kash, and W.T. Masselink

IBM Research Division

T.J. Watson Research Center, Yorktown Heights, New York 10598, USA

Abstract: — We have quantitatively measured the dispersion curves of holes in GaAs/Al$_x$Ga$_{1-x}$As quantum wells using hot-electron-acceptor luminescence in combination with band-edge luminescence. The directionally averaged heavy-hole and light-hole energies are determined for in-plane wave vectors between 3% and 7% of the Brillouin zone in quantum wells of different well width.

I. Introduction

The in-plane hole dispersion in AlGaAs/GaAs heterostructures and quantum wells (QWs) has been investigated theoretically in some detail [1-4]. Although a profound qualitative understanding has been achieved, the uncertainties in the Luttinger parameters [5] and the problem of matching the wave functions at the interfaces [6, 7] result in a significant uncertainty in the calculations. Hayden et al. studied the hole dispersion in GaAs/AlGaAs QWs experimentally [8], but only a qualitative comparison with theory was possible. Here we present the first quantitative measurements of the hole dispersion curves in QWs, using the technique of hot-electron-acceptor luminescence. [9-12]

II. Experimental Technique

Our QWs are p-type doped with an acceptor level A^0 separated from the valence band edge by the acceptor binding energy E$_A$. In the experiment an incoming photon of energy $\hbar\omega_L$ excites an electron from the lowest heavy hole subband H$_1$ to the lowest electron subband E$_1$, creating a heavy hole and an electron both with wave vector k. The hot electron will most likely relax quickly by emission of LO phonons. With a small probability of $\simeq 10^{-5}$, however, the electron radiatively recombines with a hole at a neutral acceptor before it can emit phonons. From the corresponding transition energy (e$_H$,A^0) we determine one point on the heavy-hole dispersion curve as follows (Fig. 1):

$$H_1(k) = \hbar\omega_L - (e_H,A^0) - E_A \tag{1}$$

$$E_1(k) = (e_H,A^0) - E_g^{QW} + E_A \tag{2}$$

The QW band gap E$_g^{QW}$ and the acceptor binding energy E$_A$ are measured using band-edge luminescence as discussed below. Then Eq. (1) directly gives the heavy-hole energy H$_1$(k). Eq. (2) gives the electron energy E$_1$(k), from which we will obtain the wave vector k. Other points on the dispersion curve with different wave vector and energy are obtained by varying the laser energy $\hbar\omega_L$. For each $\hbar\omega_L$ one observes luminescence not only from (e$_H$,A^0) but also from a second transition (e$_L$,A^0), as indicated in Fig. 1. From (e$_L$,A^0) the dispersion of the light-hole subband L$_1$ can be determined in the same way as the heavy-hole dispersion is determined from (e$_H$,A^0).

The three samples used in this study are GaAs/Al$_x$Ga$_{1-x}$As multiple QWs grown by molecular beam epitaxy. L$_z$, the barrier thickness, and x were precisely determined from x-ray rocking curves. The alloy composition x was confirmed by electron microprobe measurements. The samples have a well width L$_z$ = 54 (75, 98) Å, 88 (142, 102) Å thick Al$_x$Ga$_{1-x}$As barriers with x = 0.25 (0.32, 0.39), 100 (30, 100) periods, and are 3×10^{17} (2×10^{18}, 3×10^{17}) Be-doped in the central 30 (10, 35) Å of the wells. The luminescence was detected in cross-polarization using a triple monochromator and an imaging photomultiplier. All measurements are made at a temperature of 2K.

III. Results and Discussion

Fig. 2a shows a typical hot-electron-acceptor luminescence spectrum of the 54Å wide QW. The peaks (e_H, A^0), $(e_H, A^0) - \hbar\omega_{LO}$, and (e_L, A^0) can be easily identified. Fig. 2b shows the energies of the transitions (e_H, A^0) and (e_L, A^0) as a function of laser energy $\hbar\omega_L$. We also include the observed acceptor transitions from electrons which have emitted one or two LO phonons before recombining. For the determination of the hole dispersion, however, we only need to consider the zero-phonon transitions. We note that the hot-electron-acceptor luminescence can generally only be observed in a limited range of $\hbar\omega_L$. The band-edge luminescence, which exponentially increases with decreasing detection energy, determines the low-energy limit, which is 1.64 eV in case of the 54Å wide QW. The high-energy limit comes from excitation of higher electronic transitions resulting in a strong luminescence background.

In order to obtain the hole dispersion from Eqs. (1) and (2) and the data of Fig. 2b, we need to know the QW band gap E_g^{QW}, the acceptor binding energy E_A, and the dispersion $E_1(k)$ of the electron subband. E_g^{QW} can be obtained by adding the exciton binding energy to the transition energy of the free exciton, which we measure by PLE. For the exciton binding energy we use the recent calculation

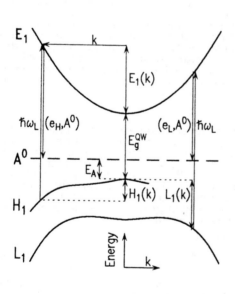

Fig. 1. Schematic diagram of the energy levels versus wave vector in a QW.

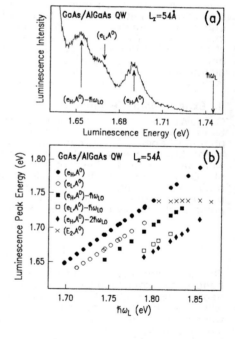

Fig. 2. Hot-electron-acceptor luminescence for the 54Å-wide QW. (a) spectrum for $\hbar\omega_L = 1.745$ eV. (b) Peak energies versus $\hbar\omega_L$

by Andreani et al. [13] We note that this calculation gives exactly the same energy of the 2s state of the free exciton as we measure by PLE. E_A is obtained as the difference between E_g^{QW} and the transition energy (E_1, A^0) of cold electrons at the bottom of E_1 to the acceptor, which we measure by luminescence. The so-obtained E_A agrees within 1 meV with the calculation of Fraizzoli et al. [14].

Let us now discuss how we obtain the wave vector k from the energy $E_1(k)$. The task is to account for the deviation from a parabolic dispersion relation as precisely as possible. We do not know of any measurements of the in-plane electron dispersion in a QW for the energy range 50 meV $\lesssim E_1(k) \lesssim$ 250 meV. Therefore we determine the electron dispersion in the QW by a calculation. We choose the approach of Lassnig [15], because it is not restricted to small energies $E_1(k)$, and adjust the k • p parameters so that the calculation reproduces the bulk dispersion measured by Ruf et al. [16] This fitting procedure eliminates the most important source of error in the calculation, namely the uncertainty in the k • p parameters [17]. For $Al_xGa_{1-x}As$ we use $E_0 = (1.519 + 1.445 x)$ eV as fundamental band gap [18], $m_0^*/m_0 = 0.0665 + 0.084 x$ as band-edge effective mass [19], and the same k • p parameters as in GaAs [20]. We assume that 65% of the band-gap difference occurs in the conduction band [21]. The calculated electron dispersion can be expressed as $k^2[10^{-2} \times 4\pi^2/a^2] = c_1E + c_2E^2 + c_3E^3$ with the lattice constant $a = 5.65$Å of GaAs. The expansion coefficients for the 54 (75, 98) Å wide QW are found to be $c_1 = 1.54$ (1.51, 1.48) eV^{-1}, $c_2 = 1.05$ (1.03, 1.04) eV^{-2}, and $c_3 = 0.137$ (0.173, 0.09) eV^{-3} respectively. As expected, the nonparabolicity becomes stronger with decreasing well width L_z.

We are now in a position to calculate the hole dispersion curves using Eqs. (1) and (2). The results are shown in Figs. 3 and 4. Because of the limited accessible range of $\hbar\omega_L$, we cannot provide quantitative data close to k=0 at this time. Comparing the hole dispersions of the three different QWs, we find that the splitting between H_1 and L_1 increases with decreasing L_z, in agreement with calculations [22, 23]. We note that the hot-electron-acceptor luminescence comes from all directions in k in the QW plane. Therefore the band structures shown in Figs. 3 and 4 represent an appropriately weighted average over all in-plane directions. At present it is difficult to compare our experimentally determined hole dispersion with any calculated dispersion from the literature, because L_z and the

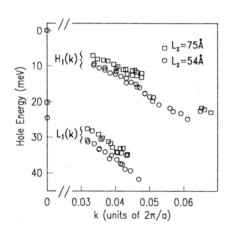

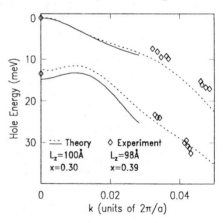

Fig. 3. In-plane dispersion of H_1 and L_1 for the two narrower QWs. $L_1(0)$ is determined from PLE assuming that the exciton binding energy is 2.8 meV larger for the light-hole exciton than for the heavy-hole exciton [13].

Fig. 4. Same as Fig. 3, but for our widest QW. The gap in the data corresponds to a laser energy close to resonant excitation from H_3 to E_1. Calculated dispersions are shown as full [22] and dotted [23] lines.

alloy composition x of the barriers used in most calculations are not the same as in our QWs. To initiate the discussion we compare in Fig. 4 our experimental data for the widest QW with two calculations using the same well width of $\sim 100\text{Å}$ but a smaller Al-content in the barriers ($x = 0.3$). Taking into account the random and systematic errors in the experiment which we estimate as ± 1.5 meV and ± 2 meV respectively, our measured hole energies agree with the calculation of Batty et al. [23], as far as absolute energies are concerned. The heavy-hole light-hole splitting, however, for which the systematic error can be neglected, is significantly larger than predicted by the calculation. This may be partly due to the different values of x used. The earlier calculation of Bastard et al. [22] only applies to wave vectors smaller than those of our data Nevertheless it appears that this calculation predicts larger hole energies than we have measured.

IV. Conclusions

We have measured the in-plane dispersion of the lowest heavy-hole and light-hole subbands in the wave vector range $0.06\pi/a \lesssim k \lesssim 0.14\pi/a$ for GaAs/AlGaAs QWs with $50\text{Å} \lesssim L_z \lesssim 100\text{Å}$. Our data are accurate enough to serve as an experimental reference to future calculations that use the same material parameters as our structures.

We thank Marc Goorsky for the x-ray data and useful discussions.

[1] D. A. Broido and L. J. Sham, Phys. Rev. B **31**, 888, (1985).
[2] U. Ekenberg and M. Altarelli, Phys. Rev. B **32**, 3712, (1985).
[3] T. Ando, J. Phys. Soc. Jpn. **54**, 1528, (1985).
[4] L. C. Andreani, A. Pasquarello, and F. Bassani, Phys. Rev. B **36**, 5887, (1987).
[5] G. Bastard "Wave Mechanics Applied to Semiconductor Heterostructures" (Les Editions de Physique, Paris, 1988).
[6] D. L. Smith and C. Mailhiot, Phys. Rev. B **33**, 8345, (1986).
[7] R. F. Taylor and M. J. Burt, Semicond. Sci. Technol. **2**, 485, (1987).
[8] R. K. Hayden, D. K. Maude, L. Eaves, E. C. Valadares, M. Henini, F. W. Sheard, O. H. Hughes, J. C. Portal, and L. Cury, Phys. Rev. Lett. **66**, 1749, (1991).
[9] B. P. Zakharchenya, D. N. Mirlin, V. I. Perel, and I. I. Reshina, Sov. Phys. Usp. **25**, 143, (1982).
[10] R. G. Ulbrich, J. A. Kash, and J. C. Tsang, Phys. Rev. Lett. **62**, 949, (1989).
[11] J. A. Kash, Phys. Rev. B **40**, 3455, (1989).
[12] G. Fasol, W. Hackenberg, H. P. Hughes, K. Ploog, E. Bauser, and H. Kano, Phys. Rev. B **41**, 1461, (1990).
[13] L. C. Andreani and A. Pasquarello, Phys. Rev. B **42**, 8928, (1990).
[14] S. Fraizzoli and A. Pasquarello, Phys. Rev. B **42**, 5349, (1990).
[15] R. Lassnig, Phys. Rev. B **31**, 8076, (1985).
[16] T. Ruf and M. Cardona, Phys. Rev. B **41**, 10747, (1990).
[17] The so obtained k•p parameters ($P_\xi^2 = 27.8$ eV,$P_\xi^2 = 1.9$ eV,and $C = -2.49$) are only meaningful together with the Hamiltonian used by Lassnig.
[18] T. F. Kuech, D. J. Wolford, R. Potemski, J. A. Bradley, K. H. Kelleher, D. Yan, J. P. Farrell, P. M. Lesser, and F. H. Pollak, Appl. Phys. Lett. **51**, 505, (1987).
[19] M. Zachau, F. Koch, G. Weimann, and W. Schlapp, Phys. Rev. B **33**, 8564, (1986).
[20] The k•p parameters of the AlGaAs barriers do not affect the results significantly.
[21] D. Arnold, A. Ketterson, T. Henderson, J. Klem, and H. Morkoç, J. Appl. Phys. **57**, 2880, (1985).
[22] G. Bastard and J. A. Brum, IEEE J.Quantum Electron. **22**, 1625, (1986).
[23] W. Batty, U. Ekenberg, A. Ghiti, and E. P. O'Reilly, Semicond. Sci. Technol. **4**, 904, (1989).

Inst. Phys. Conf. Ser. No 120: Chapter 5
Paper presented at Int. Symp. GaAs and Related Compounds, Seattle, 1991

255

Novel frequency dispersive transconductance measurement technique for interface states in FETs

Wonshik Lee, S. A. Tabatabaei, and A. A. Iliadis

Joint Program for Advanced Electronic Materials
Electrical Engineering Department, University of Maryland, College Park, MD 20742, USA

ABSTRACT: We report the development of an improved frequency dispersive transconductance technique capable of measuring the density and frequency distribution of interface states at the gates and source-drain access regions of MESFET's, and its application to study the state of the interface in SiO_2 insulated gate and enhanced barrier height gate n-channel InP FET's.

1. Introduction

The interface states at the gates and access regions of FET's play a critical role to the DC/AC performance and operational stability of these devices. However, due to the lack of appropriate measurement techniques, these interfaces have not yet been well characterized. Traditional capacitance techniques require large area metallizations for accurate interface states measurements, which make them unsuitable for micron and submicron gate devices. Conductance techniques[1,2] have been applied with success on insulated gate FET's[3] but met with limited success in the case of metal-semiconductor FET's (MESFET's) where the difficulty is increased and fitting parameters were needed to derive interface states densities[4]. In this work we report the development of a frequency dispersive transconductance technique which, in our case, takes into account both the gate interface and the source-drain access regions interfaces to measure the density and frequency distribution of interface states in MESFET's. The density and frequency distribution of the states is derived from measured parameters of the frequency dispersion relationship of the extrinsic transconductance without the need for numerical fitting parameters. We used the technique to study the interfaces of PECVD SiO_2 insulated gate and enhanced barrier height gate n-channel InP FET's.

2. The Measurement Technique

The technique is based on modeling the interfaces both at the gates and the source and drain access regions to incorporate the effects of all interface states. At the gate, interface states introduce electron trapping that results in charge sharing between the states and the depletion layer. The equivalent circuit model for such a system is shown in Fig.1(a). The access regions are modeled using the distributed element model shown in Fig.1(b), which takes into account the varying surface potential between source and gate.

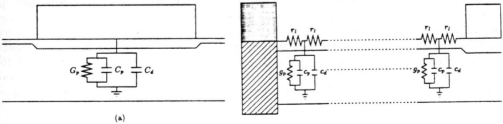

(a)

(b)

Fig. 1. Equivalent Circuit Model

The interface states contribute to the frequency dispersion of the extrinsic transconductance of the device, given in this case by:

$$g_m(w) = \frac{g_{mo}(w)}{1 + g_{do}(w)[R_s(w) + R_D(w)] + g_{mo}(w)R_s(w)} \tag{1}$$

where g_{mo} is the intrinsic transconductance, g_{do} is the intrinsic channel conductance, R_s, R_L the parasitic source-drain resistances (we omit contact resistances for simplicity) and w the angular frequency. Based on the model these parameters are shown to be frequency dependent and are given by:

$$g_{mo}(w) = -q\mu_n N_D \frac{W}{L} V_{DS} \frac{\Delta x_d(w)}{v_g} \tag{2}$$

$$g_{do} = q\mu_n N_D \frac{W}{L}[t - (x_d(w) + \Delta x_d(w))] \tag{3}$$

$$R_{s,k}(w) = (q\mu_n N_D)^{-1} \frac{L_s/n_s}{W[t - x_{ds,s}(w, y_{s,k})]} \tag{4}$$

where L_s is the length of the source access region, $x_{ds,s}(w, y_{s,k})$ is the source access region depletion width at frequency w at point $y_{s,k}$ where

$$y_{s,k} = [(k - 1/2)/n_s]L_s$$

is the distance of the k-th segment from the edge of the gate. v_g is the gate voltage and:

$$\Delta x_d(w) = -\frac{G_p(w) + jw[C_p(w) + C_d(w)]}{jwC_{do}} \frac{v_g}{2\phi_b} x_{do} \tag{5}$$

where ϕ_b is the barrier height, $C_d(w)$ the depletion layer capacitance and subscript o refer to equilibrium values. The rest of the symbols have their usual meaning. The measurement involves the simultaneous measurement of the real and imaginary part of the transconductance with frequency. Since $g_m(w)$ is linearly related to the voltage gain A_v, a measurement of the magnitude and phase of $A_v(w)$ is equivalent to the measurement of the transconductance, then the values of $G_p(w)$ and $C_p(w)$ are determined and the density and frequency distribution of the interface states are derived by plotting $G_p(w)/w$ versus frequency.

3. Experimental Results and Discussion

FET's with enhanced barrier height (EBH) gates were fabricated on n-type InP layers with $N_D = 2 \times 10^{17} cm^{-3}$ epitaxially grown by MOCVD on SI(100) InP substrates (the barrier enhancement is achieved by using a UV light assisted surface passivation technique described previously[5]). The DC characteristics of the devices were obtained and the measurement technique was employed to examine the interface. Figure 2(a) and (b) show the I-V characteristics and the interface states distribution of two EBH FET's measured at $v_g = 0V$ and $V_{DS} = 100mV$. As can be seen both interfaces exhibit similar interface states distributions with a low frequency peak at 50 Hz and a higher frequency peak coming up at 10^7 Hz. The interface states density for the low frequency peak are $N_s = 9.6 \times 10^{11} cm^{-2}$ and $N_s = 1.56 \times 10^{12} cm^{-2}$ for FET(a) and (b) respectively. The significantly higher density of interface states for FET(b) correlates well with the observed larger looping in the I-V characteristics of this device, thus establishing for the first time a link between the low frequency interface peak and the DC performance of these devices.

These FET's were compared with SiO_2 insulated gate FET's fabricated on the same InP wafer. The SiO_2 was deposited by plasma enhanced chemical vapor deposition (PECVD) at relatively low substrate temperatures (250 C). One group of samples, group (a), was passivated prior to SiO_2 deposition using the UV light assisted passivation technique that enhances the barrier[5], while the second group, group(b), was not passivated prior to SiO_2 deposition. The measurement of $G_p(w)/w$ *vs* frequency of these devices is shown in Figure 3(a) and (b). The interface states distribution is now dramatically different demonstrating distinct peaks the main one of which is at 50 KHz with a density of $N_s = 1.69 \times 10^{13} cm^{-2}$ in the case of the SiO_2/InP passivated interface. The SiO_2/InP interface without the passivation shows a significant reduction in the low frequency peak at 50 KHz while the main peak now is at the higher frequency of 1.2 MHz with a density of $N_s = 5.2 \times 10^{12} cm^{-2}$. Significantly higher looping is observed in the I-V characteristics of the device with the SiO_2/InP passivated interface which may be attributed to the higher density of this low frequency peak (50 KHz) that dominates the distribution at the interface. Substantially higher transconductance values were routinely obtained from the enhanced barrier (80-100 mS/mm) rather than the SiO_2 insulated gate FET's (20-25 mS/mm), which correlates with the lower densities of states obtained from these devices. This indicates that damage due to the deposition of SiO_2 may be one of the reasons for the degradation of the interface in these FET's.

4. Conclusions

We have developed a frequency dispersive transconductance technique capable of monitoring the state of the interface at the gates and access regions of MESFET's. The technique is modeled by taking into account both the gate interface and the source-drain access regions to measure the frequency dispersion of the transconductance and produce the frequency distribution and density of interface states without the need of numerical fitting parameters. We used the technique to establish for the first time a link between the looping observed in the I-V characteristics of n-channel EBH InP FET's and a low frequency peak of interface states. Comparison between SiO_2 insulated gate InP FET's and the EBH MESFET's showed significantly different interface states structure. The lower quality of the SiO_2/InP interface is attributed to the damage induced on the InP surface by the PECVD deposition of the SiO_2. It is concluded that the technique is a powerful tool for characterizing micron and submicron gate devices.

References

1. E.H. Nicolian, and A. Goetzberger, Bell Syst. Tech. J., 46, 1055(1967)

2. T. Elewa, H. Haddara, and S. Cristoloveanu, Proc. ESSDERC 87, 599(1988)

3. M. Ozeki, K. Kodawa, M. Takikawa, and A. Shibatomi, J. Vac. Sci. Technol., 21, 438(1982)

4. J. Graffeuil, Z. Hadjoub, J.P. Fortea, and M. Pouysegur, Solid State Electron., 29, 1087(1986)

5. A.A. Iliadis, W. Lee, and O.A. Aina, IEEE Electron Dev. Lett., 10,370(1989)

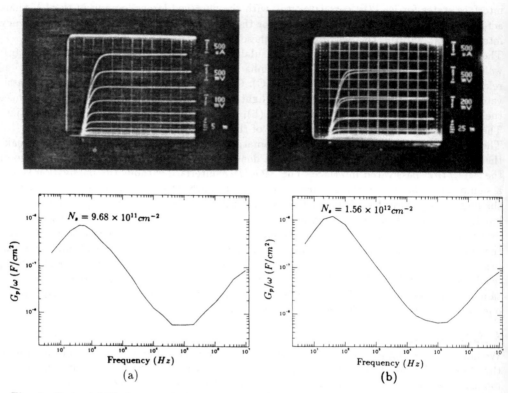

Fig. 2. Output I-V characteristics and interface states distribution of $2\mu m$ gate enhanced
barrier height InP FET's

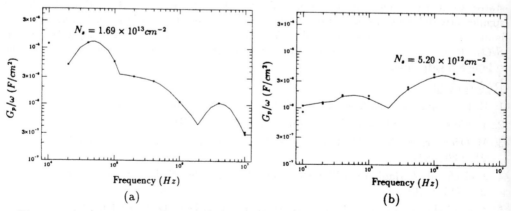

Fig. 3. Interface states distributions of $2\mu m$ gate SiO_2 insulated gate InP FET's:
(a) Passivated prior to SiO_2 deposition, (b) non-passivated

Inst. Phys. Conf. Ser. No 120: Chapter 5
Paper presented at Int. Symp. GaAs and Related Compounds, Seattle, 1991

259

Semi-insulating InP characterized by photoreflectance

D. K. Gaskill, N. Bottka, G. Stauf,[1] and Alok K. Berry[2]

Naval Research Laboratory, Washington, DC 20375
[1]Advanced Technology Materials, Danbury, CT 06810
[2]George Mason University, Fairfax, VA 22030

ABSTRACT: Photoreflectance (PR) has been employed to characterize semi-insulating InP substrates from various vendors. The PR spectra taken at 82 K near the band gap were found to be explainable by the exciton ionization model. The spectra were used to estimate the surface electric field and the surface Fermi level of the substrates at room temperature. The phase of the PR line shape was found to depend on the bulk resistivity of the substrate. A simple physical model is used to explain this dependence, and a method for quantitatively determining resistivity was used to determine resistivity homogeneity throughout a boule and on a substrate. The PR line shape exhibited sensitivity to surface polishing damage and the damage depth was quantified.

1. INTRODUCTION

Ideally, methods for assessing the quality of devices on the semiconductor mass production processing line should be non-destructive in nature. Further, these methods should be: sensitive to critical structural elements that device performance depends on, automatable, and yield data easily comparable to standards for the device. Photoreflectance (PR), a contactless form of modulation spectroscopy, has recently shown the potential for meeting these assessment requirements.

PR measures the change in reflectance, ΔR, of a sample being perturbed by an above band gap pump source of light. It has been shown that the resulting (normalized) PR spectrum depends on the nature of the modulation mechanism and geometrical factors,

$$\frac{\Delta R}{R} = \alpha \Delta \varepsilon_1 + \beta \Delta \varepsilon_2 , \qquad (1)$$

where $\Delta \varepsilon_{1,2}$ are the change in the real and imaginary parts of the dielectric constants in the structure due to the modulation mechanism and α and β are geometrical factors that are determined by the thickness and composition of the constituent layers (Seraphin and Bottka 1965). Some typical modulation mechanisms for device structures arise from the electric field distribution in the structure (the Franz-Keldysh effect (Aspnes and Studna 1973)) or ionization of excitons due to electric fields. The important point to keep in mind is that samples with identical thicknesses, compositions, and electric field distributions will have identical PR spectra. This concept has recently been used by several laboratories to characterize GaAs/AlGaAs heterojunction bipolar transistors (Bottka *et al.* 1991, Yin *et al.* 1990), a subject of ongoing research at many other laboratories.

This laboratory is extending PR research onto InGaAs/InP device structures. Previous PR studies of InGaAs nearly lattice matched to InP have been reported (Gaskill *et al.* 1990a). This work presents a summary of the PR results on semi-insulating (SI) InP substrates since, as described above, the substrate upon which a device structure is grown can contribute to the PR spectrum. The mechanism of the PR signal is identified, the line shape of the signal is shown to be a measure of the resistivity of the substrate, and the effect polishing damage has on the signal is discussed.

2. EXPERIMENTAL

The PR data acquisition system was the same as used by Bottka *et al.* (1988). Measurements were made near the band gap of InP using an Ar⁺ laser operating at 514 nm as the modulation pump. Data were acquired digitally and stored on computer for subsequent analysis.

SI InP wafers, obtained from various vendors, were cleaved, cleaned using an organic solvent, and etched in 0.5% Br methanol at 300 K (Berry *et al.* 1991). No cleaning or etching was done for the initial portion of the polishing damage studies. They were then mounted immediately after etching on a Joule-Thompson refrigerator with temperature control range 82-300 K to minimize changes in surface chemistry due to exposure to the ambient. Room temperature Hall measurements were made on the samples, or adjacent areas, using the Van der Pauw technique. The resistivity of the samples ranged from 2.3 to 230 MΩ cm.

3. RESULTS AND DISCUSSION

The PR signal of an etched sample, having a 92 MΩ cm resistivity, as a function of temperature is displayed in Figure 1. The line shape of the signal exhibits a simple "up-down" behavior at low temperatures, whereas at higher temperatures (>225 K) the simple structure has rotated, becoming "down-up" and an additional high energy structure is evident. The additional structure is probably a Franz-Keldysh oscillation since it lies above the band gap. The behavior of the simpler portion of the spectra is best explained by an exciton ionization model, first proposed by Evangelisti *et al.* (1972). The exciton ionization model is best understood by starting from the line shape function (Aspnes 1980),

$$\frac{\Delta R}{R} = Ce^{i\theta}(E - E_X + i\Gamma)^{-m} ,$$

(2)

where C, θ, E_X, Γ, and m are the amplitude, phase angle, exciton energy, broadening and transition type parameter, which for excitons equals 2, respectively. In the model, θ is modified by an optical path length introduced through the following situation. The volume near the surface is broken up into two regions. In the first region, beginning next to the surface where $x=0$, the electric field is intense enough to ionize excitons. In the second region, beginning at $x=L$, the electric field intensity is insufficient to ionize excitons. Thus the optical response of the media, near E_X, suffers from a discontinuity at $x=L$. This means that the probe light will interfere with itself as different rays are reflected from the regions bounded by $x=0$ and L. Therefore, θ will consist of both an optical path length component, $2\pi L/\lambda_m$, where λ_m is the wavelength in the medium, and an additive component which is the natural phase angle of the intrinsic exciton line shape function. A full rotation corresponds to $\theta=2\pi$. The line shape rotation, or change in θ, shown in Fig. 1 implies an increase in L as the sample's temperature is raised from 82 K.

The rotation of the exciton line shape in Fig. 1 from 82 to 300 K hve been used to calculate the surface electric field of the sample (Gaskill *et al.* 1990b). The 300 K electric fields for samples with resistivities between 72 and 230 MΩ cm were about 17 kV cm^{-1}. Additionally, the position of the surface Fermi level was found to lie 0.1 eV below the conduction band minimum, comparing reasonable well with the 0.16 eV measurement by Spicer *et al.* (1979).

Figure 2 exhibits the 82 K PR spectra for a set of samples, shown in descending resistivity. The spectra also shows rotation of the line shape, similar to Fig. 1. The dependence of the line

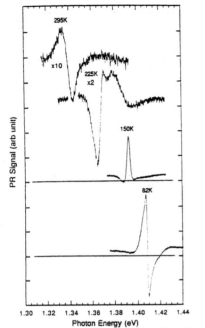

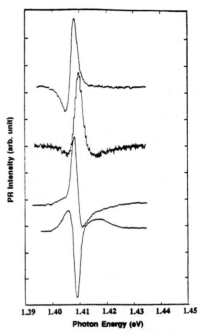

Fig. 1 PR lineshapes as a function of temperature for 92 MΩ cm InP.

Fig. 2 PR lineshapes of InP at 82 K in descending order of resistivity: 230, 140, 92, and 2.3 MΩ cm.

shape on resistivity leads to the postulate that, in this case, θ, consists in part of an optical length component that depends on resistivity. Samples with higher resistivity would have a higher surface electric field, hence a larger value of L, and thus larger θ. A plot of θ, derived from fits using the Aspnes three point method (Aspnes 1973), are plotted as a function of resistivity in Fig. 3. The solid line drawn in Fig. 3 is used as a guide to the eye and has numerical form

$$\rho = 5.6 + 4.0 \times 10^{-3}\theta + 4.6 \times 10^{-3}\theta^2 , \qquad (3)$$

where ρ is in MΩ cm and θ is in degrees.

Figure 4 shows the easily identifiable change in line shape due to resistivity variations of three substrates taken from the tail, middle, and seed end of a boule. PR measurements have also been made on an individual substrate, confirming the manufacturer's claim on the magnitude and uniform homogeneity of the resistivity. (A slight resistivity imhomogeniety can be observed for the 92 MΩ cm spectra shown in Fig. 1 and 2 which were taken from the edge and center of a substrate, respectively.) Since the resistivity of a substrate can be measured by PR without recourse to cleaving or the formation of contacts, a direct impact on optimizing device performance can be obtained. This is because microwave losses for strip lines made on InP substrates are minimized using high resistivity substrates (Aina 1991). Additionally, the performance of complex high frequency devices often depends on the resistivity of the substrate.

Because PR is also sensitive to lattice damage, the extent of substrate polishing damage could be measured by comparing PR spectra before and after controlled etching. When the PR signal stabilized after successive etches, the damaged layer was assumed to be removed. All samples used in this study showed some effects from polishing damage, with widely varying results. In some cases, only small changes in the spectra were observed after etching and in others,

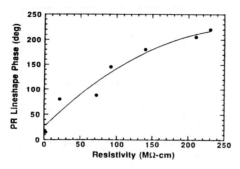

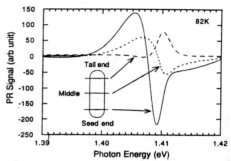

Fig. 3 82 K PR lineshape phase angle, θ, of InP as a function of 300 K resistivity. The solid line is drawn as a guide to the eye and is given as Eq. (3) in the text.

Fig. 4 82 K PR lineshape variations as a function of position in an InP boule. The resistivity can be deduced from the lineshape phase using Fig. 3 or Eq. (3).

completely different spectra were obtained. For those device structures that may depend critically on the substrate interface, etching is highly recommended. Etching 2-4 μm eliminated polishing damage in all cases.

4. SUMMARY

The mechanism of the PR signal has been identified to arise from an exciton interference effect. The phase rotation of the spectra with temperature was used to estimate the magnitude of the surface electric field and the surface Fermi Level. Samples from different vendors having different resistivities also exhibited phase rotation at 82 K. The phase of the PR line shape was demonstrated to be a measure of substrate resistivity. All substrates studied showed polishing damage extending 2-4 μm from the surface. These results demonstrate PR to be a useful characterization tool for the growth of devices on SI InP.

5. ACKNOWLEDGEMENTS

We gratefully acknowledge Drs. H. H. Wieder, C. Hansen, A. Overs, R. Henry, and L. Aina.

6. REFERENCES

Aina L private communication, 1991
Aspnes DE 1973 Surf. Sci. **37** 418
Aspnes DE 1980 Handbook on Semiconductors, Vol. 2 on Optical Properties of Solids, ed. M
Aspnes DE and Studna AA 1973 Phys. Rev. B7 4605
Balkanski (New York: North-Holland)
Berry Alok K, Gaskill DK, Stauf GT, and Bottka N 1991 Appl. Phys. Lett. **58** 2824
Bottka N, Gaskill DK, Sillmon RS, Henry R, and Glosser R 1988 J. Electron. Mater. **17** 161
Bottka N, Gaskill DK, Wright PD, Kaliski RW, and Williams DA 1991 J. Cryst. Gr. **107** 893
Evangelisti F, Frova A, and Fischbach JU 1972 Phys. Rev. Lett. **29** 1001
Gaskill DK, Bottka N, Aina L, and Mattingly M 1990a Appl. Phys. Lett. **56** 1269
Gaskill DK, Bottka N, and Berry Alok K 1990b Conf. Semi-insulating III-V mater, Toronto
(Bristol: Adam Hilger) pp 329-334
Seraphin BO and Bottka N 1965 Phys. Rev. **139** A560
Spicer WE, Chye PW, Skeath PR, Su CY, and Lindau I 1979 J. Vac. Sci. Technol. **16** 1422
Yin X, Pollak FH, Pawlowicz L, O'Neill T, and Hafizi M 1990 Appl. Phys. Lett **56** 1278

Inst. Phys. Conf. Ser. No 120: Chapter 5
Paper presented at Int. Symp. GaAs and Related Compounds, Seattle, 1991

263

Confined states in InGaAs/InAlAs single quantum wells studied by room temperature phototransmitance and electrotransmitance at high electric fields

A. Dimoulas, A. Georgakilas, G. Halkias, C. Zekentes, C. Michelakis and A. Christou.*
Foundation for Research and Technology-Hellas, Heraklion 71110, Crete, Greece
* University of Maryland, College of Engineering, College Park, MD 20742, U.S.A.

ABSTRACT : Room temperature Phototransmitance and Electrotransmitance have been applied to high quality 50 Å and 250 Å $In_xGa_{1-x}As/In_{0.52}Al_{0.48}As$ single quantum wells grown by MBE for two nominal In compositions x=0.53 and x=0.6. Several interband excitonic transitions between confined carrier states have been clearly resolved. An observed red shift of the transition energies when an DC electric field is applied perpendicular to the layers, may be attributed to the Quantum Confined Stark Effect.

INTRODUCTION

Electric field modulation of reflectance, absorption or transmitance, has been extensively used during the last few years for the characterization of Single (SQW) and Multiple (MQW) Quantum Well structures at room temperature . Early works on Electroreflectance (ER) have shown the use of this technique in studying the quantum confined carrier states in GaAs/AlGaAs SQWs and MQWs (Mendez et al. 1981). Other QW structures (InGaAs/AlGaAs, $In_{0.53}Ga_{0.47}As/In_{0.52}Al_{0.48}As$) have also been studied by ER. (Fritz et al. 1990; Satzke et al. 1991). A contactless version of electric field modulation involves the alteration of the surface built-in electric field by using a laser light (Photoreflectance (PR), Phototransmitance (PT)). It was first reported, by Glembocki et al.(1985), that Photoreflectance at 300K exibits sharp, well resolved spectra which correspond to interband transitions between quantized electron and hole states in GaAs/AlGaAs QWs and modulation doped heterojunctions. Since then, PR has been extensively used in AlGaAs/GaAs, InGaAs/GaAs QWs and several other multilayer structures (Ksendsov et al. 1990; Shen et al. 1987). There is little work done on electromodulation in the InGaAs/InAlAs quantum well system. Only very recently (Satzke et al. 1991), the lattice matched $In_{0.53}Ga_{0.47}As/In_{0.52}Al_{0.48}As$ has been measured by electroabsorption at 2K. Photomodulation experiments on the above systems has not yet been reported.

In the present work we show that the interband transitions in lattice matched and strained InGaAs/InAlAs SQWs can be investigated by Photoreflectance and Phototransmitance measurements at room temperature. We also report the response of optical transitions to externally applied electric fields. A red shift of the optical transitions has been observed and was associated to the Quantum Confined Stark Effect (QCSE) which has been previously reported for the InGaAs/InAlAs system (Nojima et al 1988; Miller et al. 1984). The sensitivity of the QW structure to electric fields may be exploited for the fabrication of high quality optical modulators and optical bistable devices (Chemla et al.1987).

EXPERIMENTAL

The SQWs were grown by MBE on n^+ InP(001) substrates having the following structure : 0.3 μm $In_{0.52}Al_{0.48}As/L_w$ $In_xGa_{1-x}As/1μm$ $In_{0.52}Al_{0.48}As/n^+InP(001)$. The $In_{0.52}Al_{0.48}As$ is the barrier layer while the $In_xGa_{1-x}As$ layer is the QW region. Two nominal In compostion of x=0.53 and x=0.6 were grown with thicknesses L_w=50 Å and 250 Å. (See Table I). All layers were undoped and they were grown at 500 ^{0}C. The QW structures were of high quality. The Photoluninescence FWHM, at 12 K was 9.5 meV for the 50 Å QWs. This FWHM value is comparable to the best ever reported in the literature (Welsh et al. 1985; Gupta et al. 1991) and indicates smooth interfaces. The reflectivity and transmitivity was modulated by a 10mW He-Ne laser, chopped at a frequency of 220 Hz. The modulation occurs as a result of the light induced changes in the built-in electric field at the QW region. The AC part of the reflected (or transmited) light is detected by using conventional Lock-in techniques.

In the experiment which involves the application of an external electric field, a semitransparent Shottky contact, made of 100 Å Ti, with a diameter of 500 μm has been deposited on the top $In_{0.52}Al_{0.48}As$ layer. The ohmic contact was made to the n^+ InP substrates.

RESULTS AND DISCUSSION.

In the 250 Å thick samples S1 (x=0.53) and S2 (x=0.6), three and four transitions respectively were clearly resolved in the room temperature Phototransmitance spectra as shown in Fig.1. The resonances, are excitonic transitions of the type ne-nhh between the heavy hole and electron confined states of the InGaAs quantum well, having the same principal quantum number. In the strained QW S2, additional peaks appear as low

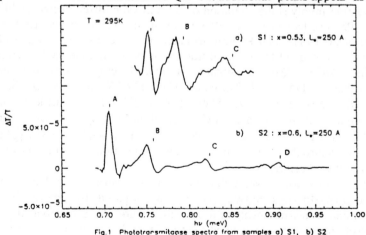

Fig.1 Phototransmitanse spectra from samples a) S1, b) S2

energy shoulders to the main peaks. They are attributed to normally parity - forbitten transitions ne-n'hh, between states with different quantum numbers n, n', which become allowed probably due to valence band mixing effects. In the thin (50 Å) SQW S3 (x=0.53), only the ground state transition 1e-1hh at 0.907eV, is confined. No clear evidence for light hole related transitions was obtained, in agreement with previous observations on this system.

The transition energies have been determined as adjustable parameters from a fit to the PT experimental curve. The absorption coefficient α, was assumed to have a Gaussian functional form as a result of the combined effects of thermal, alloy and interface disorder. The data was then fitted by a first derivative of the absorption coefficient α, with respect to energy, lifetime and intensity of the transition. Results on the energy positions of the transitions are summarized in Table I. In the strained, thin sample S4, as many as 3 transitions were resolved. We have not been able to identify them unambiguously at the present time.

Sample	x_{In}	$L_w(A)$	$E_A(eV)$	$E_B(eV)$	$E_C(eV)$	$E_D(eV)$
S1	0.53	250	0.753 1e-1hh	0.79 2e-2hh	0.843 3e-3hh	
S2	0.6	250	0.705 1e-1hh	0.75 2e-2hh	0.822 3e-3hh	0.91 4e-4hh
S3	0.53	50	0.907 1e-1hh			
S4	0.6	50	0.71	0.805	0.935	

TABLE 1

The response of confined states to electric fields was examined by Electrotransmittance by applying, in addition to the modulating voltage V_m = 1 Volt peak-to-peak, a reverse bias V_b . Data obtained from sample S1, showed that there is a red shift to all of the transitions . The shift starts from a lower value of -1.5 meV at V_b=- 2 Volts and increases, by increasing reverse bias, up to -15 meV at V_b= - 6 Volts as shown in Figure 2. Similar behaviour was observed previously for the ground state transition 1e-1hh in the InGaAs/ InAlAs system by using Photocurrent and Electroabsorption spectroscopies(Satzke et al. 1991; Nojima et al. 1988). They associated, the red shift to the QCSE, first reported in the GaAs/AlGaAs MQWs They also found, that the higher lying transitions were insensitive to electric fields ; a situation which is expected also from theory. Their results, however, were obtained in QWs of indermediate thicknesses of 100 Å or less. More experimental work has to be done for QWs of larger width (L_w > 250 Å) where, according to our observations (Fig. 2), the higher lying transitions B, C, shift to lower energies by the same amount as the 1e-1hh ground state.

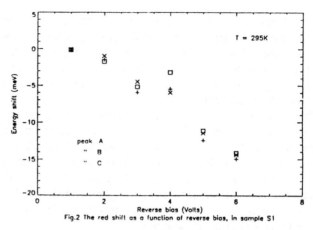

Fig.2 The red shift as a function of reverse bias, in sample S1

CONCLUSIONS

We have investigated high quality strained (x=0.6) and lattice matched (x=0.53) $In_xGa_{1-x}As/In_{0.52}Al_{0.48}As$ single quantum wells of thicknesses 50Å and 250 Å, grown by MBE. By using Phototransmitance and Photoreflectance at 300 K we have determined the excitonic transitions between quantized valence and conduction bstates.

An electric field, applied through a semitransparent Shottky contact, produced a red shift to the transition energies. The red shift was increased by increasing the reverse bias and reached a value of -15 meV at - 6 Volts bias voltage. This behaviour may be attriduted to the QCSE. The excitonic resonances persist up to -9 Volts although they are very weak. This makes InGaAs/InAlAs system a good candidate for optical modulators and other electro-optic devices. However the linewidth is rather broad (of the same magnitude as the red shift) and this is a drawback for the fabrication of efficient electrooptic devices. More insight has to be gained on the factors, such as interface roughness, which affect the excitonic line broadening. The growth of high quality interfaces is of vital importance for the high performance of optoelectronic devices.

ACKNOWLEDGEMENTS : The present work has been supported by the European ESPRIT project N^0 3086. One of us (A.D) wish to thank Dr. O.J.Glembocki, and Prof. G. Guillot for useful discussions.

REFERENCES

1. D.S. Chemla, D.A.B. Miller, P.W Smith, Semiconductors and Semimetals, Vol. 24p. 279, Eds Williardson and Beer, 1987.
2. O.J. Glembocki, B.V. Shanabrook, N. Bottka,W.T. Beard, and J.Comas, Appl. Phys. Lett. 46, 970, (1985).
3. S. Gupta, P.K. Bhattacharya, J. Pamulapati, and G. Mairou, J. Appl. Phys. 69 3219 (1991)
4. I.J. Fritz, T.M. Brennan, J.R.Wendt, and D.S. Ginley, Appl. Phys. Lett. 57, 1245 (1990).
5. A. Ksendzov, H. Shen, F.H. Pollak,and D. P. Bour, Sd. St. Commun. 73, 11 (1990).
6. E. Mendez, LL. Chang, A. Landgren R. Ludeke, L. Esaki, and F.H. Pollak, Phys, Rev. Lett. 46, 1230 (1981).
7. D.A. Miller, D.S Ghemla, T.C. Damen,A.C. Grossard, W. Wiegmann, T.H. Wood and C.A. Burrus, Phys. Rev. Lett. 53, 2173 (1984)
8. S. Nojima, Y. Kawamura, K. Wakita, and O. Mikanni, J. Appl. Phys. 64, 2795 (1988)
9. K. Satzke, G. Weiser, W. Stolz, and K. Ploog, Phys. Rev. B43, 2263 (1991).
10. H. Shen, S.H. Pan, F.H. Pollak, M. Dutta and T.R. AuCoin, Phys. Rev. B36, 9384 (1987)
11.D.F. Welch, G. W. Wicks and L. F. Eastman, Appl. Phys. Lett. 46, 991 (1985).

Inst. Phys. Conf. Ser. No 120: Chapter 5
Paper presented at Int. Symp. GaAs and Related Compounds, Seattle, 1991

267

Impact ionization phenomena in GaAs MESFETs: experimental results and simulations[1]

A. Neviani, C. Tedesco, and E. Zanoni
Dipartimento di Elettronica e Informatica,
Via Gradenigo 6a, Padova, Italy - Phone +39-49-8287663

C.U.Naldi and M. Pirola
Politecnico di Torino, Dipartimento di Elettronica,
Corso Duca degli Abruzzi 24, Torino, Italy - Phone +39-11-5644101

1. Introduction

The trend toward the reduction of device dimensions has found a severe limitation in impact ionization phenomena. The high fields occurring in the region under the drain edge of the gate of submicron FET devices cause carrier multiplication. Most of the generated holes are collected at the gate electrode, accounting for the sudden increase in the gate current that can be observed as the drain potential overcomes a certain threshold. This pre-avalanche behaviour imposes severe limitations on the range of operating biases, particularly in power applications.

Hui *et al* (1990) and Canali *et al* (1991) measured gate current I_g versus drain to source potential V_{ds} at constant gate to source potential V_{gs} in ion implanted MESFETs manufactured in $\langle 110 \rangle$ direction and succeded in extending the previously available data for the ionization coefficient α_n versus electric field (Pearsall *et al* 1978) by over five orders of magnitude in the low field regime. No investigations on the I_g dependence on V_{gs} at constant V_{ds} were carried out in that work, while the substrate current I_b vs V_{gs} curves in Si MOSFETs were published by Chan *et al* (1984) for different gate lengths and drain polarizations, showing a characteristic bell shape. This behaviour was qualitatively explained ascribing the initial rise to the increase of the drain current and the subsequent fall to the drop of the electric field in the open channel.

This work reports measured I_g vs V_{gs} curves for GaAs MESFETs showing the same bell shape found in Si MOSFET samples for I_b. In order to check the validity of the above considerations, numerical simulations were performed to yield the longitudinal electric field profile in the device channel. Recessed gate MESFETs were simulated using a two-dimensional Drift-Diffusion program. The value of the channel field was picked up for fixed bias condition and two different recess depths and its behaviour versus drain and gate potential is reported. The corresponding ionization coefficient α_n is evaluated according to Lee *et al* (1980), and an estimate of the ionization current is attempted following Chan *et al* (1985). The results are shown together with experimental data for I_g current.

[1]Work partially supported by C.N.R.

2. Samples and Electrical Measurements

All experimental data reported in this work refer to GaAs MESFETs manufactured on a semi-insulating GaAs substrate epitaxially grown in $\langle 110 \rangle$ direction, implanted with a Si^+ dose of 5×10^{12} cm^{-2} at 100 KeV (channel implant) and 1×10^{13} cm^{-2} at 40 KeV (n^+ shallow implant for the ohmic contact regions). The channel implant under the gate has a peak concentration of 2×10^{18} cm^{-3}. Device layout was defined with 0.5×300 μm^2 recessed gate, a gate to source spacing $L_{gs} = 1$ μm and a gate to drain spacing $L_{gd} = 3.5$ μm. The Al gate contact was placed in the middle of the recessed region, the distance from the trench walls being 0.25 μm in both source and drain direction.

Measurements of gate and drain current versus V_{ds} at constant V_{gs} are shown in Fig.1. The sudden increase of I_g occurs when the drain potential is raised over a certain threshold. This pre-avalanche behaviour cannot be ascribed to the breakdown of the gate-drain Schottky diode since I_g initially increase as V_g rises from the pinch-off value, see Fig.2, thus reducing the inverse polarization of the Schottky diode. Moreover, the negative exponential dependence of the increase of I_g on V_{ds}^{-1} is an evidence that impact ionization takes place. In Fig.3 we report a semi-logarithmic plot of I_g vs V_{ds} at constant V_{gs}, showing in detail the two contributions to gate current and their relative importance: at low drain voltages the electric field is not sufficiently strong to heat carriers and produce impact ionization, thus I_g is dominated by reverse current, while increasing V_{ds} the contribution of holes generated by impact ionization becomes more and more important. The electrons generated in this process are collected at the drain contact, but they do not affect I_d since the device is kept in pre-avalanche regime. The discussion on these phenomena was carried out by Hui *et al* (1990). We focused our attention on the $I_g - V_{gs}$ relation, shown

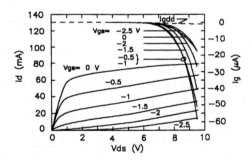

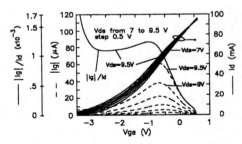

Fig.1 Experimental gate and drain currents measured on a # C device. I_{gd0} is the characteristic of the gate-drain diode with source kept floating.

Fig.2 Drain current (continuous lines), gate current (dashed lines) and I_g/I_d ratio (dash-dotted line) vs gate bias at fixed drain potential.

in Fig.2, together with I_d and the I_g/I_d ratio, obtained with V_{ds} fixed in the ionization regime. I_g shows the same bell shape already observed in Si MOSFETs. At high negative V_{gs} the ratio I_g/I_d is determined by the presence of strong reverse and negligible value of I_d due to the pinch-off condition. Moving V_{gs} from pinch-off toward positive values, as an increasing number of hot electrons becomes available for impact ionization, I_g becomes proportional to I_d. When V_{gs} is further increased, the longitudinal electric field drops due to the opening of the channel, causing a decrease of carrier heating, while I_d continues to increase. As the impact ionization rate is exponentially dependent on the electric field, the net result is a strong decrease of hole current.

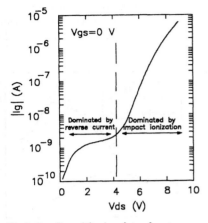

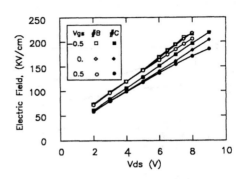

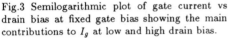

Fig.3 Semilogarithmic plot of gate current vs drain bias at fixed gate bias showing the main contributions to I_g at low and high drain bias.

Fig.4 Maximum longitudinal electric field vs drain potential for different gate biases in device # B (empty marks) and # C (filled marks.)

3. Drift-diffusion simulation results

The numerical simulations were performed using a code implementing a standard drift-diffusion model, based on the coupled solution of Poisson and carrier continuity equations. The simulation domain represents a two dimensional cross section of the device. The contact dimensions and spacings used in the simulation were the same reported in section 2, while recess depths of 400Å (#C device) and 600Å (#B device) were considered.

The simulation output consisted of longitudinal electric field profiles along the channel of the device. Particular care was taken to evaluate the field in the conductive channel and not in the space charge region, which is characterized by the highest fields but is completely depleted of carriers, thus giving a negligible contribution to impact ionization current. Moving from source to drain, the field profile in the channel reaches a peak value E_{max} under the drain edge of the gate.

The resulting E_{max} vs V_{ds} curves for different gate biases are shown in Fig.4 for both #B and #C devices. Considering the approximately linear behaviour, we proceed as Hui *et al* (1990) and define an effective length l_{eff} for the ionization process as the inverse of the slope of the curves, to be used later in the calculation of the gate current. E_{max} vs V_{gs} is reported in Fig.5 for three different drain biases: as expected, when V_{gs} moves from pinch-off toward positive values the electric field in the channel drops. In order to support the previous assertion, we have calculated the ionization coefficient from the formula:

$$\alpha_n = \alpha_\infty e^{-E_0/E_{max}}$$

with $\alpha_\infty = 2.19 \times 10^7$ cm^{-1} and $E_0 = 2.95 \times 10^6$ V/cm, reported by Lee *et al* (1980), and then estimated I_g following Hui *et al* (1990):

$$I_{g,sim} = I_d l_{eff} \alpha_n$$

were I_d is taken from experiment. Notice that here l_{eff} is not a fitting parameter but is derived from the simulation (see above). The results reported in Fig.6 show a sufficiently good agreement with experimental data. The value of the electric field inside the channel varies slightly also when moving in a direction perpendicular to the device surface. Anyway, the different I_g vs V_{gs} curves that can be drawn according to the vertical position in which the field is evaluated, are identical to those shown in Fig.6, except for a slight

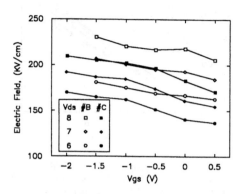

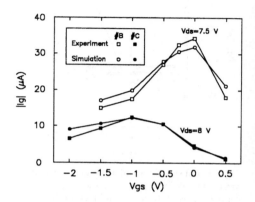

Fig.5 Maximum longitudinal electric field vs gate potential for three different drain polarizations in device # B (empty marks) and # C (filled marks).

Fig.6 Experimental (square marks) and calculated (triangular marks) gate current vs gate bias at fixed drain potential in device # B (empty marks) and # C (filled marks).

shift in value. For a better match a more accurate device physical model of the avalanche phenomena is needed. We also notice that the two values of the recess depth in the simulated devices account for the different experimental data of ionization current in device #B and #C. Anyway, a strong influence of surface density of states in determining the different behaviour cannot be excluded.

4. Summary

Measurements of gate current performed on GaAs MESFETs, where at sufficently high drain bias the main contribution to I_g has been shown to come from impact-ionization-generated holes collected at the gate electrode, show a bell-shape dependence on gate potential. Two-dimensional numerical simulations were performed to obtain the peak value of the electric field in the device channel, showing that the decrease of I_g at positive gate bias can be ascribed to the drop in the electric field. Fittings of I_g experimental data are also reported.

5. Acknowledgments

The authors wish to thank A. Cetronio and C. Lanzieri, ALENIA R&D Lab..

<u>References</u>:

K Hui, C Hu, P George and P K Ko, 1990 *IEEE EDL* **17** 113

T P Pearsall, F Capasso, R E Nahory, M A Pollack and J. R. Chelikowsky, 1978 *Solid State Electronics* **21** 297

T Y Chan, P K Ko and C Hu, 1984 *IEEE EDL* **5** 505

M H Lee and S M Sze, 1980 *Solid State Electronics* **23** 1007

T Y Chan, P K Ko and C Hu, 1985 *IEEE EDL* **6** 551

C Canali, A Paccagnella, E Zanoni, C Lanzieri and A Cetronio, 1991 *IEEE EDL* **12** 80

Inst. Phys. Conf. Ser. No 120: Chapter 5
Paper presented at Int. Symp. GaAs and Related Compounds, Seattle, 1991

271

Optically determined low-temperature, high-mobility transport in "interface-free" GaAs heterostructures

D.J. Wolford, G.D. Gilliland, T.F. Kuech,[+] and J.A. Bradley
IBM Research Division, T.J. Watson Research Center
Yorktown Heights, NY 10598

H.P. Hjalmarson
Sandia National Laboratories
Albuquerque, NM 87185

At low temperatures free electrons and holes condense into free excitons, which are electrically neutral, and are thus unobservable in conventional, electrical transport measurements. To study the physical limitations of low-temperature, high-mobility transport we have resorted to a new all-optical, time-resolved photoluminescence (PL) confocal imaging technique which is sensitive to both charged and neutral-particle transport. We have thus measured spatial transport of spectrally-resolved free-excitonic species at low-temperatures (1.8 - 50 K) in a series of high-quality MOCVD-prepared $GaAs/Al_{0.3}Ga_{0.7}As$ double heterostructures, with varying GaAs layer thicknesses (10 μm to 50 Å). We find such constants to be sensitive to, and monotonic in, laser power — with 1.8-K peak diffusion constants of > 1000 cm^2/s and minimums of ≈ 1 cm^2/s. We find that this anomalous transport results from the joint diffusion of free excitons and free carriers coupled through temperature-dependent capture and ionization. We conclude that the high-mobilities for joint neutral-particle (excitonic) and free carrier (electron and/or hole) transport are the result of diminished charged-center scattering and the simulation of modulation-doping which photoexcitation creates in the screening of charged centers — and are thus limited by only by intrinsic lattice (deformation potential) scattering.

1. Introduction.

Transport of free carriers in GaAs structures at moderate temperatures (300 - 50 K) has been studied extensively, and is relatively well understood, with mobilities limited by various intrinsic and extrinsic temperature-dependent scattering mechanisms.[1-4] However, at low temperatures (< 50 K) photoexcited free electrons and holes may condense into free excitons, which are electrically neutral, and thus unobservable with conventional electrical transport measurement techniques. Also, at these temperatures, conventional phonon scattering ceases to be the dominant scattering mechanism as impurity (both immediate and remote) and piezoelectric scattering become increasingly important. Moreover, the effects of degeneracy may also become important. Thus, while some of these inherent physical limitations may be studied through electrical transport measurements, these complications to the low-temperature transport are difficult to address. In order to study these issues we have resorted to an all-optical, time-resolved photoluminescence (PL) imaging technique, which is sensitive to both charged and neutral particle transport.[5-7]

Conventional transport techniques attempt to study the absolute physical mobility limitations through growth of special heterostructures, modulation-doped structures. These structures largely remove the effects of Coulomb scattering by ionized impurities, and have yielded structures with electron mobilities exceeding 10,000,000 cm^2/Vs.[3] We find that our all-optical measurement technique allows similar studies of these high mobilities, in simpler structures, in a contactless fashion. We find that optical excitation results in similar transport properties as does modulation-doping, and allows for studies of exciton transport. We find that the observed diffusion results from the joint, coupled diffusion of free carriers and free excitons.

We have used this technique to study minority-carrier transport in both GaAs/ $Al_{0.3}Ga_{0.7}As$ double heterostructures and $n^+/n^-/n^+$ GaAs homostructures at temperatures from 50 - 300 K.[6,7] We find <u>identical</u> minority-hole mobilities in these n-type structures

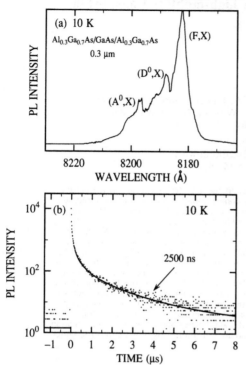

versus temperature, and find minority-hole mobilities in quantitative agreement with majority-hole mobilities determined electrically. In addition, we observe the effects of p-type modulation doping in the heterostructures being due to the p-type $Al_{0.3}Ga_{0.7}As$, and may thus measure minority-electron mobilities in some of these structures. All of these results are well-understood, and demonstrate the power, accuracy, and utility of this new, all-optical technique.

2. Experiment/Materials.

Our samples were MOCVD-prepared, nominally undoped $Al_{0.3}Ga_{0.7}As/GaAs/$ $Al_{0.3}Ga_{0.7}As$ heterostructures with GaAs thicknesses ranging from 10 μm to 50 Å and $Al_{0.3}Ga_{0.7}As$ thicknesses of 0.5 μm. Samples were grown at 750°C. High sample purity is confirmed by the low-temperature PL spectra shown in Fig. 1a, dominated by free-exciton emission. In addition, the observed long lifetimes of Fig. 1b enable our time-resolved study of minority-carrier and exciton transport. These lifetimes have been studied in detail and are reported elsewhere.[8] Our all-optical transport measurements rely on the time-resolved imaging of the PL induced by laser excitation. High spatial resolution is

Figure 1. (a) PL spectrum of 0.3 μm heterostructure at 10 K. (b) Free exciton decay kinetics at 10 K.

obtained by tightly focussing the laser beam onto the sample, to the diffraction limit. Spectral resolution is obtained by imaging through a spectrometer, and temporal resolution of < 1 ns is obtained through time-correlated single photon counting.

3. Results and Discussion.

We find free exciton PL distributions which are well-fit by expanding Gaussians versus time, shown in Fig. 2. For truly diffusive motion, PL spatial full-widths-at-half-maximums (FWHM) squared should vary linearly versus time. We find transport which is clearly <u>diffusive,</u> with diffusion constants, obtained from the slope, increasing with decreasing temperature. We have previously observed diffusive minority-carrier transport at temperatures above 50 K, with diffusion constants independent of laser excitation power. Below 50 K, however, we find transport to be <u>laser power-dependent,</u> with PL distributions still accurately fit with Gaussian distributions. Figure 3 shows an example of our anomalous power-dependent transport results at 1.8 K, where the diffusion constant increases with and is monotonic in laser power. This power-dependent diffusion is most prominent at 1.8 K and decreases with increasing temperature to 50 K. We find peak 1.8-K diffusion constants of ~ 300 cm²/s and 1300 cm²/s for 9.82- and 0.30-μm thick structures, respectively. These diffusion constants vary with laser power over <u>3 orders-of-magnitude</u> at 1.8 K.

In an attempt to further understand the dynamics of these anomalous transport results we have measured spatially- and time-resolved PL spectra at both 1.8 K and 10.0 K. We find the PL spectra dominated by free excitons at all times, except just after the laser pulse,

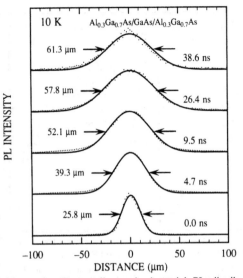

Figure 2. Temporally resolved spatial PL distributions.

for high laser powers. Since our laser excitation is above gap, we initially create free electrons and holes, which rapidly condense into free excitons. We have conducted an extensive study of free-exciton and free-carrier decay kinetics, and find that even at the lowest temperatures, free excitons are influenced by the presence of even a small number of free carriers.[8] Thus, even though our experiments spectrally resolve only free excitons, our results may be influenced by free carriers. In order to determine the dominant transport species we compare mobilities obtained from our diffusion constants with electrical mobilities using the Einstein relation, $\mu = De/kT\alpha$, where α is a correction factor which depends on the carrier temperature, density, and type (fermions or bosons). For fermions (bosons) at low temperatures and high-densities this correction serves to reduce (enhance) the power-dependent diffusivities. Since we are measuring the <u>joint diffusion</u> of free excitons and free carriers, and the corresponding correction factors may cancel, we have used the nondegenerate ($\alpha = 1$) Einstein relation to derive mobilities. The observed long lifetimes (approaching 14 μs at 25 K in some structures) suggest that carriers may thermalize during their lifetime, and we have thus used the lattice temperature in the Einstein relation. Figure 4 shows

mobilities for the two structures obtained in this manner. We obtain peak 1.8-K joint mobilities of ~ 1,500,000 cm²/Vs and ~ 7,000,000 cm²/Vs for our 9.82- and 0.30-μm-thick heterostructures, respectively. These exceedingly high exciton diffusion constants and mobilities are not completely unheard of in semiconductors. Recently, Trauernicht et al.[9] have reported 1.8-K exciton diffusion constants of ~ 1000 cm²/s in Cu_2O.

We have performed a two-dimensional PL imaging experiment in order to assess possible "phonon-wind" driven transport, and find no evidence for this type of transport (shell-shaped PL distributions). However, one possible mechanism for our power-dependent transport is similar to modulation-doping. Our low-power mobilities at each temperature are similar to

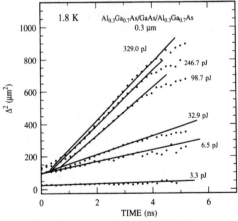

Figure 3. Power dependence of squared PL spatial distribution FWHM versus time.

classically obtained mobilities dominated by ionized-impurity scattering, and are very nearly identical in both samples. At higher laser powers the increase in mobility may result from both an increase in carrier density and the charge neutrality of excitons. Thus, for example, excitons are insensitive to Coulomb scattering, and may thus have large mobilities. Further, the photoexcitation of additional free carriers may overwhelm, or screen out, the residual impurities in the GaAs, thus yielding larger mobilities. These effects are similar to that observed in modulation-doped structures, whereby Coulomb scattering is significantly reduced. In electrical transport measurements, for every charge carrier there is a charged impurity. The advantage of modulation-doping is that this charged impurity is spatially removed from the charged carrier conduction channel. In our transport technique, photoexcitation may generate many times more carriers than there are residual impurities, thus simulating modulation doping, and allowing studies of absolute mobility limits.

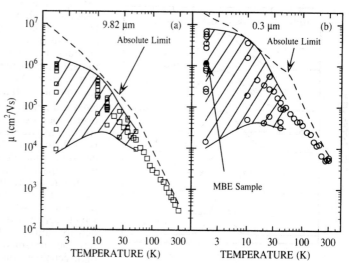

Figure 4. Mobilities derived from measured diffusion constants for (a) 9.82 μm and (b) 0.30 μm thick heterostructures. Solid point represents corresponding results obtained for a high-purity MBE heterostructure (non-modulation doped). Absolute limits are for holes and electrons, respectively, and were taken from Refs. 10 and 2.

4. Conclusions.

We conclude that the high-mobilities for joint neutral-particle (excitonic) and free carrier (electron and/or hole) transport are the result of diminished charged-center scattering and the simulation of modulation-doping which photoexcitation creates in the screening out of charged centers — thus resultant mobilities are limited only by intrinsic lattice (deformation potential) scattering. These results are the subject of further study in an effort to additionally elucidate the observed dynamics.

We thank L.M. Smith, G.A. Northrop, and J. Martinsen for useful discussions. Supported in part by the ONR under contract N00014-85-C-0868 and N00014-90-C-0077 and DOE under contract DE-AC04-756DP00789. †Present Address: Univ. of Wisconsin, Dept. of Chem. Eng., 1415 Johnson Drive, Madison WI 53706.

REFERENCES

1. C.M. Wolfe, G.E. Stillman, and W.T. Lindley, J. Appl. Phys. **41**, 3088 (1970).
2. W. Walukiewicz, H.E. Ruda, J. Lagowski, and H.C. Gatos, Phys. Rev. B **30**, 4571 (1984).
3. L. Pfeiffer, K.W. West, H.L. Stormer, and K.W. Baldwin, Appl. Phys. Lett. **55**, 1888 (1989).
4. E.E. Mendez and W.I. Wang, Appl. Phys. Lett. **46**, 1159 (1985).
5. A. Olsson, D.J. erskine, Z.Y. Xu, A. Schremer, and C.L. Tang, Appl. Phys. Lett. **41**, 659 (1982).
6. G.D. Gilliland, D.J. Wolford, T.F. Kuech, and J.A. Bradley, Appl. Phys. Lett. **59**, 216 (1991).
7. L.M. Smith, D.J. Wolford, J. Martinsen, R. Venkatasubramanian, and S.K. Ghandhi, J. Vac. Sci. Technol. B **8**, 787 (1990).
8. D.J. Wolford, G.D. Gilliland, T.F. Kuech, L.M. Smith, J. Martinsen, J.A. Bradley, C.F. Tsang, R. Venkatasubramanian, S.K. Ghandhi, and H.P. Hjalmarson, J. Vac. Sci. Technol. B. **9**, 2369 (1991).
9. D.P. Trauernicht and J.P. Wolfe, Phys. Rev. B **33**, 8506 (1986).
10. W. Walukiewicz, J. Appl. Phys. **59**, 3577 (1986).

Inst. Phys. Conf. Ser. No 120: Chapter 6
Paper presented at Int. Symp. GaAs and Related Compounds, Seattle, 1991

275

$In_{0.52}Al_{0.48}As/In_{0.53}Ga_{0.47}As$ HIGFETs using novel 0.2 μm self-aligned T-gate technology

Yi-Jen Chan, Dimitris Pavlidis and Tim Brock
Center for High Frequency Microelectronics
Department of Electrical Engineering and Computer Science
The University of Michigan, Ann Arbor, MI 48109-2122, USA

Abstract:
0.2μm self-aligned T-gate InAlAs/InGaAs HIGFET's have been fabricated and investigated experimentally. The submicron gate technology is based on lift-off of sputtered WSi_x. Low gate leakage current and high gate/drain breakdown were obtained with the help of a T-gate structure which controls the lateral spreading of the source/drain implants. The peak extrinsic g_m is 650 mS/mm and g_{ds} shows a low value of 35 mS/mm using a p-buffer design. The DC gain ratio (g_m/g_{ds}) is much higher (~18) with a p-buffer than with an undoped buffer approach where the ratio is only 4. The short-channel effects associated with substrate injection are also substantially reduced using the p-doped buffer design.

I. INTRODUCTION:

InAlAs/InGaAs Heterostructure Insulated-Gate Field-Effect-Transistors (HIGFET's) have recently demonstrated very promising characteristics for high speed digital applications. The two-dimensional electron or hole gas (2DEG or 2DHG) can be induced in the channel of HIGFET's by applying a forward gate bias. Since no dopant exists in the wide bandgap material used for the insulator, as for example in HEMT's, the induced carriers are less susceptible to impurity scattering, and good carrier transport properties are expected. Good InAlAs/InGaAs HIGFET characteristics with a f_T of 63 GHz for 0.3μm long gates have been presented (Feuer et. al. 1989). Lattice-matched and also strained designs with this material were reported by the authors (Chan et. al. 1991a), and electrical performance enhancement was shown with up to x=0.65 In content. The use of selective ion implantation in HIGFET's also allowed the demonstration of complementary (Swirhun et. al. 1991) and enhancement/depletion (E/D) mode logic (Feuer et. al. 1991), (Chan et. al. 1991b).

To futher explore the ultimate performance of HIGFET's, it is essential to investigate devices with submicron gates. However, the fabrication of submicron gate HIGFET's imposes a number of difficulties compared to standard HEMT technology. This is due to the incorporation of self-aligned ion implantation and high temperature annealing processes in the HIGFET's. As a result, lateral diffusion of implant species during high temperature annealing needs to be well controlled particularly when the gate length is small; lateral spreading under the gate can degrade the diode characteristics and increase the leakage. To avoid this problem a novel submicron T gate technology based on lift-off was developed and is reported in this paper. The T-shape avoids the dielectric sidewall protection which has been previously reported (Mishra et. al. 1989) to cause device degradation. Furthermore, it allows the source/drain implants to be separated from the footprint of the T-gate by a spacer resulting in increased gate/drain breakdown.

This work is supported by U.S. Army Research Office (Contract No. DAAL03-87-K-0007).

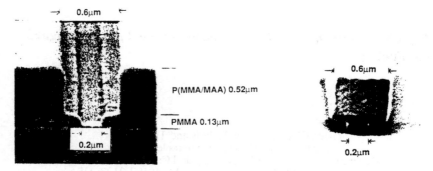

Fig.1a P(MMA-MAA)/ PMMA resist profile after MIBK/ IPA development.

Fig.1b 0.2µm-long self-aligned WSi$_x$ gate after lift-off.

Short-channel effects caused by high-field substrate injection can be another limitation in submicron HIGFET operation and result in reduced transconductance (g_m) and increased output conductance (g_{ds}). To reduce short-channel effects and provide a better carrier confinement, a slightly p-doped InAlAs buffer layer was used under the InGaAs channel and device characteristics obtained with this approach are presented.

II. 0.2µm WSi$_x$ T-GATES BY LIFT-OFF AND DEVICE FABRICATION:

The 0.2µm long T-gates were defined by the JEOL 5DIIF electron beam system with an energy of 50KeV and a dose of 6 nC/cm. A bi-layer, P(MMA-MAA)(0.52 µm)/ PMMA(0.13 µm), photoresist was used to provide the necessary high/ low sensitivity respectively. The low sensitivity PMMA provides a high resolution for the gate footprint to have at least 0.2 µm length, and the high sensitivity P(MMA-MAA) results in a wide opening of 0.6 µm at the top as is necessary for the T-profile. Sidelobe e-beam scanning was used to obtain appropriate T-shape resist profiles with high uniformity over the wafer. Fig 1(a) shows the resist profile after development in the MIBK/ IPA solution. A uniform 0.2 µm footprint opening was obtained all along the gate. A 2500Å WSi$_x$ refractory gate metal was then sputtered. Its thickness was tailored to the resist profile so that reproducible metal breaking could be achieved at the edge of the T-gate, as necessary for good lift-off. Fig. 1(b) shows a 0.2µm WSi$_x$ gate after lift-off. The aspect ratio of this gate is around 3, and the width of the spacer separating the implanted region from the gate is 0.2 µm.

The HIGFET layers were grown by MBE in a Varian GEN II system at a temperature of 500°C. The device cross-section is shown in Fig. 2. A 150Å quantum-well In$_{0.53}$Ga$_{0.47}$As channel was sandwiched between two undoped InAlAs layers. The top 400Å InAlAs was used as a gate insulator. A 1000Å thick Be-doped InAlAs buffer with a concentration of 2E17 cm^{-3} was incorporated, and was separated from the channel by a 400Å thick undoped InAlAs layer. The latter reduces the possibility of hole conduction in the InGaAs channel. For comparison, a layer with a conventional undoped-InAlAs buffer was also grown.

Following the formation of the 0.2µm WSi$_x$ T-gate, Si was implanted at a double-energy using 30 KeV, 2E13 cm-2, and 50 KeV, 3E13 cm-2 conditions. The WSi$_x$ gate was the only implant mask used in the process. The samples were subsequently Rapid-Thermal-Annealed (RTA) at a temperature of 850°C for 5 secs to activate the implants. Following mesa etching, ohmic contacts were realized using Ge/Au/Ni/Ti/Au, and were RTA treated at 400°C for 10 secs. Finally, a Ti/ Au interconnect layer was deposited by lift-off for microwave measurements.

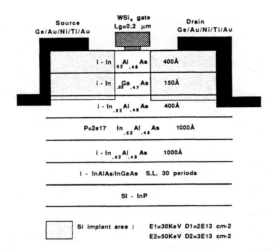

Fig.2 Device cross-section of InAlAs/InGaAs HIGFET's with p-buffer.

III. P-BUFFER DC DEVICE CHARACTERISTICS:

Extremely low gate leakage current (< 100nA) was achieved at -2V with a 0.2 μm x35 μm gate. This suggests that WSi_x is thermally stable after high temperature annealing and that no significant lateral spreading of the ion implants takes place. A comparison to HIGFET's fabricated in our laboratory using square gate technology showed that the latter devices had much higher leakage $I_{g,leak}$= 1.5 $\mu A/\mu m^2$ at V_{gs}=-4V then the 0.2 μm T-gate HIGFET's where $I_{g,leak}$ is 0.2 $\mu A/\mu m^2$ at V_{gs}= -4V. In addition to the low leakage due to reduced lateral spreading a high gate-drain breakdown voltage, typically 3.5V, was also observed.

Fig. 3 shows the drain-source current (I_{ds}) versus drain-source voltage (V_{ds}) characteristics of 0.2 μm InAlAs/InGaAs HIGFET's with a p-doped buffer. A good channel pinch-off was observed. The output-conductance (g_{ds}) at V_{gs}=0.5V was around 6 mS/mm and 35 mS/mm at V_{ds}= 0.75V and V_{ds}=3V respectively. The presence of kink-effect can be seen above V_{ds}=1V. The transconductance (g_m) versus V_{gs} transfer characteristics are shown in Fig. 4. The devices showed a peak extrinsic g_m of 650 mS/mm at V_{ds}=3V and 400 mS/mm at V_{ds}=1V. This represents a substantial improvement compared with 1μm HIGFET's (Chan et. al. 1991a) of our laboratory where g_m~220mS/mm at V_{ds}=3V. The g_m-V_{gs} characteristics showed the same trends independent on whether V_{ds} exceeds the threshold for kink effects. Frequency dispersion characterization from 10Hz to 1MHz showed that the kink effect causes only 3% transconductance drop. This suggests that equally good performance at V_{ds}=3V should be expected from the device at DC, as well as, high frequencies. The DC gain (g_m/g_{ds}) at V_{ds}=3V was of the order of 18 and represents a substantial improvement over previously reported submicron HIGFET's (Feuer et. al. 1989). A high K-value ($\Delta g_m/\Delta V_{gs}$) of the order of 5400 mS/mm-V and 1480 mS/mm-V was evaluated at V_{ds}=3V and V_{ds}=1V respectively, suggesting the device suitability for logic gates operating at very fast switching speeds. The threshold voltage (V_{th}) evaluated from the $\sqrt{I_{ds}}$ -V_{gs} characteristics was 0.27V.

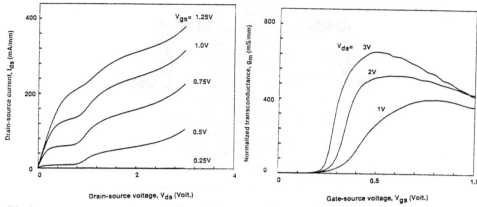

Fig.3 I_{ds}- V_{ds} characteristics of a 0.2 μm InAlAs /InGaAs HIGFET with p-doped buffer design.

Fig.4 g_m-V_{gs} transfer characteristics of a 0.2μm InAlAs/InGaAs HIGFET at V_{ds}= 1, 2, and 3V.

IV. IMPACT OF P-BUFFER ON SHORT-CHANNEL EFFECTS AND MICROWAVE PERFORMANCE:

The impact of the p-buffer on short-channel effects was studied by evaluating the subthreshold current characteristics and the V_{ds} dependence of V_{th}. Fig 5 shows the subthreshold current versus V_{gs} for different V_{ds} values in the case of a 25 μm wide device. I_{ds} remains very low (~ 1E-7A) under subthreshold bias conditions. Compared to similar size devices (Han et. al. 1988) with an undoped buffer, our results demonstrate a reduction in subthreshold current of two orders of magnitude.

Substrate injection in submicron FET's can also cause V_{th} to depend on V_{ds} (Smith et. al. 1986). As V_{ds} increases, the potential barrier between the channel and the buffer becomes smaller and the position at which it is lowest shifts toward the source end of the gate. The barrier height is also expected to be lower when the gate length is reduced, and the substrate injection is in this case enhanced. A linear relation between V_{th} and V_{ds} (Conger et. al. 1988) can provide an indication of the degree of substrate injection. This is expressed as V_{th}= V_{to}-R*V_{ds}. A larger R-value represents a higher dependence of V_{th} on V_{ds}, and suggests enhanced substrate injection. The 0.2 μm devices reported here have a low R-value of 0.035 which is of the same order as 0.026 obtained from 1μm long HIGFET's without p-buffer fabricated in our laboratory. Since a reduction of gate length from 0.7 μm to 0.3 μm in conventional (undoped-buffer) AlGaAs/GaAs HEMT's, is typically resulting in much higher R-changes, of the order of 25 (from 0.01 to 0.25: Han et. al 1988), it seems that substrate injection is indeed limited with the help of the p-buffer.

The gate-length dependence of V_{th} is another measure of short-channel effects. The V_{th} of FATFET's (L_g=50 μm) fabricated on the same p-buffer wafer was 0.35V. Therefore the V_{th} shift in changing L_g from 50 μm to 0.2 μm is extremely small and only of the order of 80 mV. This very low V_{th} shift together with the low subthreshold current and small V_{th} dependence of V_{ds} discussed above supports the benefits drawn out of the p-buffer designs.

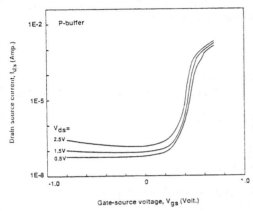

Fig.5 Subthreshold current-V_{gs}-V_{ds} characteristics of a 0.2 μm InAlAs / InGaAs HIGFET with a p-doped buffer.

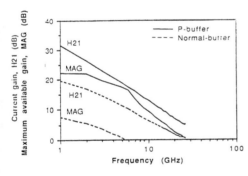

Fig.6 Frequency dependence of current (H21) and power gain (MAG) for both p- and undoped buffer 0.2μm InAlAs/ InGaAs HIGFET's.
V_{gs}=0.7V, V_{ds}=2.5V (p-buffer).
V_{gs}=0.6V, V_{ds}=2.5V (undoped-buffer).

A systematic study of the characteristics of 0.2 μm T-gate InAlAs/InGaAs HIGFET's with conventional undoped InAlAs buffer reveals that the p-buffer design improves indeed the device performance. The peak g_m's at V_{ds}=3V and V_{ds}=0.75V were 247 mS/mm and 178 mS/mm respectively, and g_{ds} was typically 62 mS/mm at V_{ds}=3V. The DC gain (at V_{ds}=3V) drops from 18 for the p-buffer to 4 for the undoped-buffer device. Substrate injection is increased substantially more by short-channel effects in the conventional device and contributes in the resulting small value of DC gain. The large V_{th} shift for the undoped-buffer designs also demonstrates a significant improvement in device characteristics using the p-buffer; the V_{th} shift corresponding to a 50 μm to 0.2 μm gate length reduction was 0.25V (from 0.13V to -0.12V) for undoped-buffer design, and 80mV (from 0.35V to 0.27V) in the p-buffer. Similar improvements by p-buffer designs have also been reported (Yamasaki et. al 1985) in submicron self-aligned GaAs MESFET's.

Microwave characterization of both p- and i-buffer devices was done from 0.5 to 26.5 GHz with an HP 8510B network analyzer and a Cascade prober. Fig. 6 shows the measured current (H_{21}) and maximum available power (MAG) gain for 0.2 μmx150 μm device. A cut-off frequency (f_T) of 45 GHz and a maximum oscillation frequency (f_{max}) of 27 GHz were observed for the p-buffer devices. However, these two values drop to 27 GHz and 6 GHz for the undoped-buffer devices, respectively. The microwave results confirm the performance improvement of the p-buffer designs in 0.2 μm T-gate InAlAs/InGaAs HIGFET's.

V. CONCLUSIONS:

In summary, a noval technology approach is reported for the fabrication of submicron (0.2 μm) HIGFET's based on self-aligned sputtered WSi_x T-gates by lift-off. The 0.2 μm InAlAs/InGaAs HIGFET's demonstrated a low gate leakage (0.2 μA/μm^2 at V_{gs}=-4V) and high gate-drain breakdown (3.5V). Compared to the conventional undoped-buffer designs, the p-buffer devices present better DC (g_m= 650 mS/mm vs. 247 mS/mm, g_{ds}= 35 mS/mm vs. 62 mS/mm) and microwave characteristics (f_T= 45 GHz vs. 27 GHz, f_{max}= 27 GHz vs. 6 GHz). The substrate injection related to short-channel effects can also be dramatically reduced with the help of p-buffer designs.

ACKNOWLEDGMENT: The authors wish to thank Prof. P.K. Bhattacharya, Dr. W.-Q. Li and Mr. Y.C. Chen for stimulating discussions and guidance in MBE growth.

REFERENCES:

Chan Y.J. and Pavlidis D., IEEE ED-38, p.1999 , 1991a.
Chan Y.J. and Pavlidis D., Proceedings of 3rd InP and Related Materials Conf. p. 242, 1991b.
Conger J., Peczakski A., Shur M.S., IEEE EDL-9, p. 128, 1988.
Feuer M.D. et. al., IEEE EDL-10, p.70, 1989.
Feuer M.D. et. al., IEEE EDL-12, p.98, 1991.
Han C.J. et. al., IEDM Tech. Digest, p. 696, 1988.
Mishra U.K. et. al., IEDM Tech. Digest, p.101, 1989.
Smith T., IEEE EDL-7, p.188, 1986.
Swirhun S. et. al., Proceedings of 3rd InP and Related Materials Conf. p. 238, 1991.
Yamasaki K., Kato N., Hirayama M., IEEE ED-33, p. 2420, 1985.

Inst. Phys. Conf. Ser. No 120: Chapter 6
Paper presented at Int. Symp. GaAs and Related Compounds, Seattle, 1991

Reduced silicon movement in GaInAs/AlInAs HEMT structures with low temperature AlInAs spacers

A.S. Brown, L.D. Nguyen, R.A. Metzger, M. Matloubian, A.E. Schmitz, M. Lui, R.G. Wilson, and J.A. Henige

Hughes Research Laboratories
3011 Malibu Canyon Road
Malibu, CA 90265

1. INTRODUCTION

High Electron Mobility Transistors (HEMTs) fabricated from GaInAs/AlInAs modulation-doped structures currently exhibit the highest current gain cut-off frequency [1], highest maximum frequency of oscillation [2] and lowest noise figure [2] of any three terminal device. Inverted (GaInAs on AlInAs) and double-doped device structures have not been widely utilized in this material system because of the poor mobility of the two-dimensional electron gas (2DEG) formed at the inverted modulation-doped interface. The data presented in this paper show that the degraded characteristics of the inverted modulation-doped interface, as compared with the normal (AlInAs on GaInAs), result from the segregation of silicon in AlInAs. When the inverted interface is modulation doped, silicon moves into the GaInAs channel, increasing the ionized impurity scattering of the 2DEG.

Silicon segregation also degrades the efficiency of the modulation doping of normal structures. The silicon movement, which occurs during growth, effectively widens the doped layer and displaces more of the charge away from the heterojunction interface; this, in turn, reduces the transfer efficiency of the structure.

We have reduced the magnitude of these deleterious effects by growing a thin layer of AlInAs immediately following the doped region at significantly reduced substrate temperatures (300 to 350°C). This paper will discuss the degree of silicon segregation that occurs in AlInAs as a function of the substrate temperature, the improvement in the mobility of the 2DEG formed at the inverted interface grown with a low temperature spacer, the improvement in transfer efficiency obtained with the low temperature spacer, and device results from fabricated wafers utilizing these improved growth conditions.

2. EXPERIMENTAL

All the modulation-doped structures were grown in a Riber 2300 MBE system. The substrate temperature was referenced to the (2x4) to (4x2) transition that occurs while heating the InP substrate in an As_4 beam before growth is initiated. For an As_4 beam equivalent pressure of 9×10^{-6} Torr, we assume that this transition occurs at 540°C. Unless otherwise stated, the growth temperature is 500°C and the growth rate is ~100 Å/min. Secondary Ion Mass Spectrometry (SIMS) measurements were performed at Charles Evans and Associates. The device processing sequence has been described elsewhere for both the power device [3] and the inverted device [4].

3.   RESULTS

A.   SIMS Analysis of Silicon Segregation in AlInAs

Secondary Ion Mass Spectrometry (SIMS) analysis was performed on test structures to aid in determining both the magnitude of the effect and the mechanism (either diffusion or surface segregation) that gives rise to it. The structure grown consists of a 2500-Å AlInAs buffer layer followed by five periods of a GaInAs/AlInAs inverted modulation-doped structure. This structure consists of 450 Å of AlInAs, a silicon delta-doped layer consisting of either 1.5×10^{12} cm^{-2} or 3.0×10^{12} cm^{-2}, a 50-Å AlInAs spacer grown at either 500, 425, 350, or 275°C, and finally a 500-Å GaInAs layer. From the substrate the first spacer temperature used was 275°C, followed by 375, 425, 500°C, and finally ending with 275°C again for the last part of the structure nearest the surface of the epitaxial layer. The 275°C layer was grown twice to determine resolution changes with depth of the layer, as well as possible diffusion effects that may occur for the spikes grown earlier in the run and held at an elevated substrate temperature during the completion of the run.

Figure 1 shows the silicon and aluminum yields for the sample grown with intended areal doping densities of 3.0×10^{12} cm^{-2} for each of the spikes. When comparing just the maximum charge obtained for the spikes, we can see that spreading of the charge is enhanced for the spike with the 500°C spacer layer. Figure 2 is a summary of the data obtained for each of the samples. The raw data were analyzed to determine the slope of the silicon profile towards the surface of the layer. Comparing the data for the two different spike concentration levels, we can see that the spreading is enhanced for the higher concentration at all temperatures, with the exception of the spacer grown at 275°C. For both doping levels, a significant increase in the slope (approximately a factor of 2) occurs when the growth temperature of the spacer layer is reduced from 500 to 350°C. No reduction in the spreading is observed by further lowering the growth temperature to 275°C.

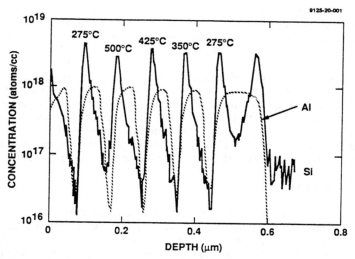

Fig. 1. Aluminum and silicon concentrations versus depth as determined by SIMS analysis. The growth temperature used immediately after silicon deposition is shown above the spike.

RATE OF DECAY OF Si (Å/DECADE)
130
120
110
100
90
80
70
60
50
40
0
250
300
350
400
450
500
550
600
T_s (°C)
3.0 x 10^{12} cm^{-2}
1.5 x 10^{12} cm^{-2}

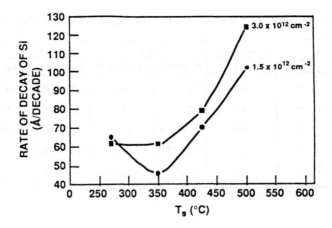

Fig. 2. Rate of decay of silicon versus substrate temperature.

B. Electrical Properties of Inverted Modulation-Doped Structures

Figure 3 is a schematic of a typical inverted GaInAs/AlInAs modulation-doped structure. In these experiments, the active channel thickness was held constant at 200 Å. The spacer thickness X_s was varied from 30 to 100 Å for the bottom heterojunction, and was held at 50 Å to modulation-dope the top heterojunction interface. Doping levels were varied to obtain 2DEG densities at the inverted interface from 1.5×10^{12} cm^{-2} to 3.0×10^{12} cm^{-2}.

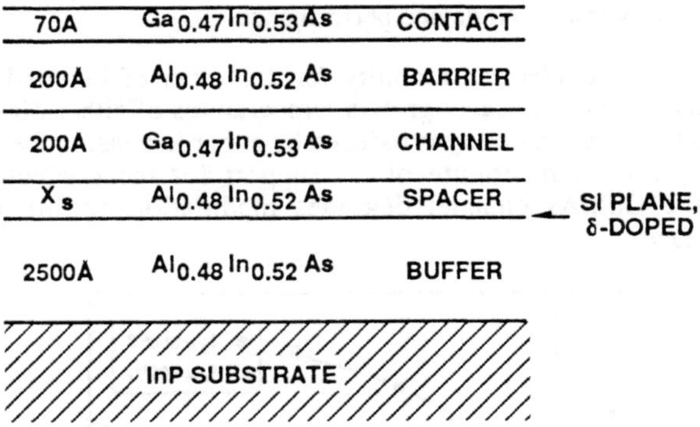

70A
$Ga_{0.47}In_{0.53}As$
CONTACT
200Å
$Al_{0.48}In_{0.52}As$
BARRIER
200Å
$Ga_{0.47}In_{0.53}As$
CHANNEL
X_s
$Al_{0.48}In_{0.52}As$
SPACER
SI PLANE, δ-DOPED
2500Å
$Al_{0.48}In_{0.52}As$
BUFFER
InP SUBSTRATE

Fig. 3. Schematic of inverted GaInAs/AlInAs modulation-doped structure.

Table 1 shows Hall data obtained at 300 and 77°K. The spacer layer thickness, as well as the temperature used for the growth of the spacer layer, is listed after the run number. These data show that a significant increase in the mobility of the 2DEG is obtained when the spacer is grown at 300°C. Samples A1613 and A1616 have identical designs that result in a 2DEG density of 1.5×10^{12} cm^{-2}. The mobility obtained using the low temperature spacer is comparable to that which would be obtained in a normal structure for

Table 1. Electron mobility for inverted and double-doped modulation-doped structures.

RUN #	X_s(A)	T_s (SPACER) (°C)	300 K		77 K	
			n_s(cm^{-2})	μ_s(cm^2/V-s)	n_s(cm^{-2})	μ_s(cm^2/V-s)
A 1613	40	300	1.5×10^{12}	8061	2.0×10^{12}	44,707
A 1616	40	500	1.5×10^{12}	3111	2.4×10^{12}	4,816
A1642	60	300	3.0×10^{12}	9131	3.6×10^{12}	35,250
A 1609	60	500	2.3×10^{12}	4655	2.6×10^{12}	7,662
A 1663 *	50	300	5.5×10^{12}	10,093	5.3×10^{12}	27,773

* (DOUBLE DOPED)

a similar 2DEG density. Significant degradation of the mobility is seen to occur for the sample grown under standard growth conditions for these materials. Other samples grown under these conditions show similar degradation; however, as the spacer layer thickness is increased the mobility improves. Sample A1642 has a higher 2DEG concentration but higher mobility than A1609. A1642 has a 2DEG concentration comparable to that which would be used in a normal optimized HEMT in this materials system. A1663 is a sample designed for power applications. Both the inverted and the normal heterojunctions are modulation doped to achieve a total 2DEG concentration of 5.5×10^{12} cm^{-2} in the 200-Å channel. As with the inverted structures, the spacer below the GaInAs channel is grown at 300°C. As can be seen from both the 300 and 77°K electron mobilities, the transport is not degraded as compared with standard HEMT structures. Because of the higher 2DEG concentration, the conductivity of the 2DEG is increased at 300°K by approximately 30%. This leads to significantly improved device performance.

Figure 4 is a plot of 300°K electron mobility for a number of inverted structures with different spacer thicknesses and spacer growth temperatures of either 500 or 300°C. The fact that the degradation in the mobility is reduced by either increasing the spacer thickness or decreasing the growth temperature offers support for the conjecture that silicon segregation into the GaInAs channel degrades modulation doping at the inverted heterojunction interface.

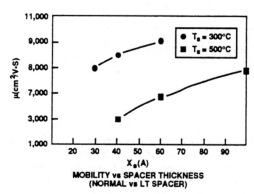

Fig. 4. Electron mobility versus spacer layer thickness for 500 and 300°C spacer layer growth temperature.

C. Electrical Properties of Normal Structures

Normal modulation-doped structures of various designs (variation in doping-thickness product in the AlInAs donor layer) have been grown with a low temperature AlInAs layer (50 Å) immediately following the silicon deposition. In all cases, we observe an increase of the charge by 20 to 50%. No degradation in the mobility was observed.

D. Device Results

By using the low temperature spacer, we have obtained the first high quality inverted AlInAs/GaInAs device, as well as high quality double-doped structures fabricated for power applications. Figure 5 shows the dc I-V characteristics of an inverted GaInAs/AlInAs HEMT. The sheet charge and mobility of this structure were 3×10^{12} cm^{-2} and 9100 cm^2/Vs. Devices with 0.2 μm gates were fabricated and exhibited trans-conductances of 800 mS/mm, f_{TS} of 100 GHz, and f_{max}s of 170 GHz. The dc output conductance of 28 mS/mm is much better than that typically obtained in a normal GaInAs/AlInAs HEMT structure, and no kink effect was observed in the I-V characteristic.

Figure 6 shows the dc I-V characteristic of a 0.2x300-μm-gate power HEMT. This double-doped structure exhibited a sheet charge and mobility of 5.5×10^{12} cm^{-2} and 10020 cm^2/Vs, respectively. A current density of over 900 mA/mm was observed, with a transconductance of 860 mS/mm. The gate-to-drain breakdown voltage was 4.5 V and the drain-to-source breakdown voltage was 8 V. At 12 GHz the device delivered 960 mW/mm with 40% power-added efficiency.

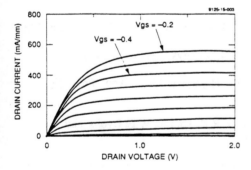 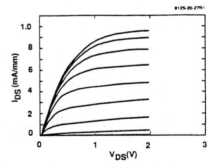

Fig. 5. DC I-V characteristic of inverted HEMT.   Fig. 6. DC I-V characteristic of power HEMT.

4. CONCLUSIONS

The degradation observed in the electrical characteristics of inverted and double-doped modulation-doped structures has been attributed to various phenomena, including poor structural quality of the inverted interface [5], silicon diffusion or surface segregation into the channel region [6], and trapping of ambient impurities [7] at the heterojunction interface. We have shown in this paper that the primary cause of degraded transport properties of inverted modulation-doped structures in the GaInAs/AlInAs heterojunction system is silicon surface segregation into the GaInAs channel region. A technique in which the AlInAs spacer layer is grown at a reduced substrate temperature has been developed that allows for the growth of inverted structures with electron mobilities comparable to those achieved in normal structures. This is different from techniques reported by other workers [6,8] to reduce the effect of silicon movement in AlGaAs/GaAs inverted modulation-doped structures. In these cases, the substrate temperature is reduced throughout the growth of the entire structure or in the doped region. This results in the need to optimize between reduced silicon movement and the reduction in the quality of the material in the active channel. The technique described above has resulted in the ability to grow double-doped structures with the highest conductivity obtained in a 2DEG system. The best inverted HEMTs fabricated to date were grown with the low temperature spacer. The first reported study of InP-based power HEMTs utilized wafers grown with this technique. Also, by reducing the segregation of silicon in normal modulation-doped structures, the transfer efficiency is improved. This should result in devices with reduced gate-to-drain feedback capacitance.

REFERENCES

1. L. Nguyen, L. Jelloian, M. Thompson, and M. Lui, IEDM, San Francisco, CA, 1990.

2. A.J. Tessmer, P.C. Chao, K.H.G. Duh, P. Ho, M.Y. Kao, S.M.J. Lin, P.M. Smith, J.M. Ballingall, A.A. Jabra, and T.Y.H. Yu, 1989 Proc. IEEE/Cornell Conference on Advanced Concepts in High Speed Semiconductor Devices and Circuits, Ithaca, NY, 1989.

3. M. Matloubian, L.D. Nguyen, A.S. Brown, L.E. Larson, M.A. Melendes, and M.A. Thompson, 1991 MTT, Boston, MA.

4. A.E. Schmitz, L.D. Nguyen, A.S. Brown, and R.A. Metzger, 1991 DRC, Boulder, CO.

5. H. Morkoc, T.J. Drummond, and R. Fisher, J. Appl. Phys. 53 (1982), 1030.

6. S. Sasa, J. Saito, K. Nanbu, T. Ishikawa, and S. Hiyamizu, Jpn. J. Appl. Phys. 23 (1984), 573.

7. T.J. Drummond, J. Klem, D. Arnold, R. Fisher, R.E. Thorne, W.E. Lyons, and H. Morkoc, Appl. Phys. Lett. 42 (1983), 615.

8. K. Kohler, P. Ganser, M. Maier, and K. H. Bachem, 6th International MBE Conference, San Diego, CA, 1990.

Inst. Phys. Conf. Ser. No 120: Chapter 6
Paper presented at Int. Symp. GaAs and Related Compounds, Seattle, 1991

287

Effect of n and p channel doping on the $I-V$ characteristics of AlInAs−GaInAs HEMTs

U.K. Mishra*, L.M. Jelloian+, M. Lui+, M. Thompson+, S. E. Rosenbaum+ and K.W. Kim#

*University of California, Santa Barbara, CA. 93106
+Hughes Research Laboratories, Malibu, CA. 90265
#N.C. State University, Raleigh, N.C. 27695

1. Introduction:

Enhancement in the high frequency performance of transistors is dependent both on improved materials properties and smaller transit distances. This has resulted in improved performance as devices have progressed from MESFETs (Wang etal) to GaAs based pseudomorphic HEMTs(Tan etal) to InP based lattice-matched(Mishra etal, Chao etal) and pseudomorphic(Thompson etal) HEMTs. The GaInAs based devices have demonstrated the highest speeds because of a combination of high electron mobility, peak velocity and sheet charge density in the GaInAs channel of an AlInAs-GaInAs HEMT. The reduction in gate length to 50 nm has increased the extrinsic f_T of the HEMT to 292 GHz at room temperature(Thompson etal) but the f_{max} of the HEMT has been restricted by the rapid increase in the output conductance at the small gate length. The output conductance is caused by a combination of substrate injection and channel length modulation. Attempts to reduce the substrate injection by using a combination of p-doped and wide band gap buffer layers have had limited success. In this study, we investigate the effects of doping the channel of an AlInAs-GaInAs modulation doped transistor n and p type and evaluate the effect on the DC and RF characteristics of devices with 0.25 μm gate length. Devices with 1 μm gate length were also studied to determine the gate length dependence. The motivation is to evaluate the effect of the two doping types on the distribution of electric field in the channel and study its effect on output conductance and electron transport.

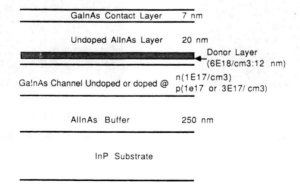

Figure 1: Schematic of the Layer Structures Used in the Study.

2. Material Characteristics:

A schematic of the wafer structure is shown in figure 1. The layer sequence follows that of a conventional AlInAs-GaInAs HEMT with an AlInAs buffer, a GaInAs channel, an undoped AlInAs spacer, followed by an AlInAs donor layer and capped by a sequence of an undoped AlInAs schottky layer and a GaInAs contact layer. The major difference is the doping in the GaInAs channel. Wafer A is a control sample with no doping in the channel region. Wafers B and C have p-type doping in the channel at 1×10^{17}cm^{-3} and 3×10^{17}cm^{-3} respectively. Wafer D is doped n-type in the channel at 1×10^{17}cm^{-3}. The doping in the channel **is in addition** to the modulation doping, resulting in sheet charge densities and Hall mobilities in the wafers as presented in figure 2.

Wafer#	A	B	C	D
channel doping	undoped	p(1×10^{17})	p(3×10^{17})	n(1×10^{17})
μ(cm^2/V-s)	10359	9765	9670	8247
ns($\times10^{12}$cm^{-2})	3.2	3.3	2.94	3.89

Figure 2: Sheet Charge and Hall Mobility of the Device Structures

3. Device Fabrication and Results:

To study the impact of the additional doping in the channel, devices were fabricated with 1μm and 0.25μm gate length. Alloyed AuGeNi metallurgy was used for ohmic contacts and TiPtAu for the gate metal. The distance between the gate and the source metals was 1μm for the long gate length device and 0.5μm for the device with the 0.25μm gate length. The I-V characteristics of representative devices with 1μm gate length are shown in figure 3.

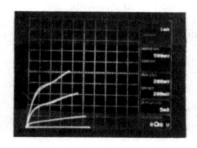

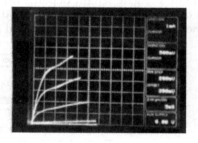

Wafer A

gm = 440mS/mm.

Wafer B

gm = 480mS/mm.

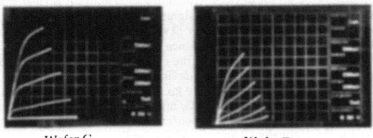

Wafer C
g_m = 520mS/mm.

Wafer D
g_m = 360mS/mm.

Figure 3. I-V Characteristics of Devices with 1μm Gate Length Fabricated on the Various Wafers.

It is apparent from the results that the g_m of the devices increased as the p-type doping in the channel was increased. The additional n-type doping in the channel reduced both the transconductance and the dc output resistance of the transistors. The trends were maintained for the devices with 0.25μm gate length where T-gate structures were employed and the variation in gate length from wafer to wafer and across a wafer was estimated to be less than a maximum 10%. Figure 4 compares the I-V characteristics of the 0.25μm gate length devices from the different wafers.

Wafer A   Wafer B

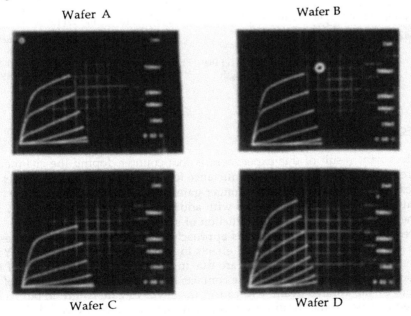

Wafer C   Wafer D

Figure 4. I-V Characteristics of Devices with 0.25μm Gate Length Fabricated on the Various Wafers.

Figure 5 compares the dc g_0 for devices with 1μm gate length . It is clear that the p-type doping in the channel does not significantly affect the output conductance of the devices. However, wafer D shows an extremely high g_0 of 72mS/mm. In contrast, the output conductance of the 0.25μm gate length devices are very sensitive to the doping in the channel as shown in figure 6. The dc and rf output conductance of these devices are very similar and therefore only the rf data are presented. The g_0 decreases from a maximum of 98mS/mm(wafer D) to a minimum of 35mS/mm(wafer C). The f_T of the devices, as determined by S-parameter measurements were 108 GHz(wafer A), 100 GHz(wafer B), 95 GHz(wafer C) and 85 GHz(wafer D).

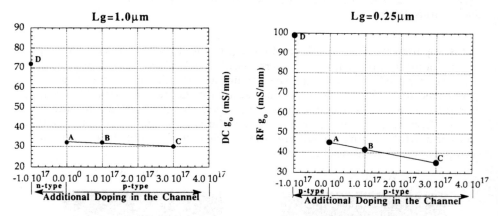

Figure 5. Comparison of the D.C. g_0 of the Various Devices with 1μm Gate Length.

Figure 6. Comparison of the R.F. g_0 of the Various Devices with 0.25μm Gate Length.

4. Discussion:

The dominant result of this experiment is that counter doping the channel p-type results in reduced output conductance for 0.25 μm gate-length devices whereas g_0 was unaffected at the longer gate length. The reduction in the output conductance for the devices with additional p-type doping in the channel could be due to both a reduction of substrate injection and drain induced space charge injection. This approach has been applied with success in the suppression of short channel effects in MOSFETs. As explained by Yau, short channel effects in a MOSFET are due in part to the charge sharing of the bulk space-charge layer between the contacts and the gate. The depletion of the bulk by the applied drain field causes the effective gate length to be reduced and hence the drain (or output) conductance to be increased. The compromise is that, increasing the bulk doping reduces the electron mobility and reduces the breakdown voltage of the drain-substrate n-p junction. Our current **qualitative** understanding of this phenomenon is shown in figures 7a and 7b.

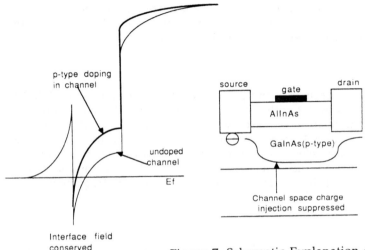

Figure 7. Schematic Explanation of Suppression of g_0 with Enhanced p-type Doping in the Channel.

Assuming that the doping in the AlInAs donor layer is constant, the amount of charge transferred into the GaInAs remains the same independent of the acceptor concentration as it a function only of the component material band structures. The electric field at the interface is therefore constant. In the case of the samples counter-doped p-type, the net negative charge causing the interfacial field is now the sum of the two-dimensional electron charge and the negatively charged acceptors. The latter case results in a steeper confining potential contiguous with the 2DEG. This serves two important purposes. First, the confinement of the electrons to the channel is substantially improved over an undoped channel. Second, the electron concentration in the GaInAs channel further from the gate is rapidly quenched. It is clear from the early work on MOSFETs by Fichtner, that the sub-surface electron concentration strongly influences the magnitude of the space charge injection from the source to the drain. The larger barrier to space charge injection between the source and the drain with the highest p-type doping reduces this parallel current path and therefore reduces the output conductance. The encouraging fact of this approach is that, the dc and rf output conductances are similar and that the p-type doping does not significantly affect the transport properties of the electrons in the channel as borne out by the similar f_T values for wafers A,B and C. The degradation of characteristics for Wafer D we believe is from excess electron concentration in the channel, leading to a reduction in drain field shielding and hence reduced g_m and g_0. Further studies are necessary to understand the combination of doping in the AlInAs and the channel to optimize HEMT characteristics. It is also necessary to perform a more complete two-dimensional simulation of the HEMT to fully understand the nature of the electric fields within the device to minimize drain field penetration beyond pinch-off and optimize device performance. We are currently conducting self-consistent two-dimensional simulations on these structures and will report on the findings shortly.

5. Conclusions

We have studied the effects of n and p-type doping in the channel of AlInAs-GaInAs HEMTs. The devices with high p-type doping($3 \times 10^{17} cm-3$) exhibited reduced output conductance compared with the control sample (undoped channel) and the device with n-type doping ($1 \times 10^{17} cm-3$) in the channel. It is suggested that enhanced electron confinement and reduced source-drain space charge injection are the reason. Two-dimensional transistor analysis is required to fully exploit this approach and enhance the high frequency performance of HEMTs.

6. References

G.W. Wang, R. Kaliski and J.B. Kuang, IEEE EDL, vol. 11, No.9, 394-396 (1990).

K.L. Tan, D.C. Streit, R. M. Dia, S.K. Wang, A.C. Han, P.-M.D Chow, T.Q. Trinh, P.H. Liu, J.R. Velebir and H.C. Yen, IEEE EDL, vol. 12, No. 5, 213-214 (1991).

U.K. Mishra, A.S. Brown, L.M. Jelloian, M. Thompson, S.E. Rosenbaum and L.D. Nguyen, IEDM Tech Digest, (1989).

P.C. Chao, A.J. Tessmer, K.-H.G. Duh, P. Ho, M.-Y. Kao, P.M. Smith, J.M. Ballingall, S.-M.J. Liu and A.A. Jabra, IEEE EDL, Vol. 11, No.1, 59-62 (1990).

M. Thompson, L.M. Jelloian, L.D. Nguyen, A.S. Brown,35th International Symposium on Electron Ion and Photon Beam Technol., Seattle (1991).

L.D. Yau, Solid.St.Electron., 17, 1059 (1974).

W. Fichtner, IEEE Solid St. Circuits and Technol., Workshop onScaling and Microlithography, New York, April (1980).

MOVPE growth, technology and characterization of $Ga_{0.5}In_{0.5}P$/GaAs heterojunction bipolar transistors

K.H. Bachem, Th. Lauterbach, M. Maier, W. Pletschen, and K. Winkler

Fraunhofer-Institut für Angewandte Festkörperphysik, Tullastr.72,

W-7800 Freiburg, Germany

ABSTRACT

From LP-MOVPE grown lattice matched GaInP/GaAs layer structures we fabricated conventional Heterojunction Bipolar Transistors (HBTs) and Bipolar Transistors having only a thin GaInP layer between emitter and base (Tunneling Emitter Bipolar Transistors, TEBTs). The base was doped to above 2×10^{19} cm^{-3} with carbon, which - due to its low diffusivity - allows precise matching of doping transition and heterojunction even at the high doping levels required for a low base sheet resistance. The conventional HBTs exhibit a common emitter current gain of up to 65 at 10^4 A/cm^2 collector current density, whereas TEBTs with a 2 and 5 nm thick GaInP layer show values of 10 and 115, respectively. The devices have nearly ideal output characteristics featuring high Early voltages and small offset voltages.

INTRODUCTION

Heterojunction Bipolar Transistors (HBTs) based on III-V materials have many prospects for digital and microwave applications. Recently a new design, the Tunneling Emitter Bipolar Transistor (TEBT), has been proposed by Xu and Shur [1] in which the solid wide-gap emitter of a conventional HBT is replaced by only a thin layer of wide-gap material between emitter and base. This layer causes potential barriers in both the conduction band and the valence band. While electrons can easily tunnel through the conduction band barrier, the holes are repelled by the valence band barrier because of their larger effective mass. Therefore, hole injection from the base into the emitter is suppressed.

This device has numerous technological advantages compared to the conventional HBT:
(i) all the technology developed for the basic transistor material (e.g. GaAs) can be applied for device fabrication because there is only a thin layer of a different material, (ii) the barrier layer can serve as an etch stop in the base contact formation step, and (iii) the barrier layer provides a passivation of the emitter-base junction and thus improves device performance[2].

For GaAs based systems AlGaAs has widely been used as the wide-gap material. However, Kroemer suggested to consider $Ga_{0.5}In_{0.5}P$/GaAs as early as in 1983 [3] because of its larger valence band offset as compared to AlGaAs/GaAs, which is of course beneficial for npn-HBTs and -TEBTs.

Meanwhile there is sufficient experimental evidence that the difference of band gaps at the GaInP/GaAs heterojunction indeed gives rise to mainly an offset in the valence band [4]. Further advantages of this combination are that (i) contact formation is not troubled by the presence of Aluminium containing layers, (ii) phosphides are easily grown in device quality by MOVPE, and (iii) there is suitable etch selectivity for wet and dry etch processes.

Good RF performance of bipolar transistors is related to a low base sheet resistance and thus to heavy acceptor doping. However, for reproducible device performance no dopant diffusion or segregation from the base to the emitter during growth or subsequent high temperature processing steps may be tolerated. All these requirements can be met by doping the base with carbon. A suitable procedure for carbon doping is the growth of GaAs from Trimethylgallium and Trimethylarsine at deposition temperatures below 600°C [5-7].

In this study we investigated conventional GaInP/GaAs HBTs as well as GaAs TEBTs with 5 and 2 nm thick GaInP layers between emitter and base in order to evaluate the potentials of the different concepts.

EXPERIMENTAL

The intended transistor layer structures shown in Fig. 1 were grown on 2 inch wafers in an LPMOVPE system from TMGa, TMIn, AsH$_3$ and PH$_3$ at a substrate temperature of 585°C. For the carbon doped base layer trimethylarsine was used instead of arsine while silane was used for all n-type layers. Epitaxial layers similar to the ones used for device processing were analized by Secondary Ion Mass Spectrometry in an ATOMICA spectrometer using Cs primary ions.

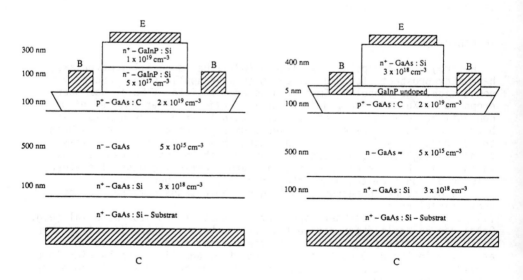

Fig. 1: Schematic cross-sections of conventional HBT (left) and TEBT (right). Intended doping concentrations and layer thicknesses are indicated. The undoped GaInP layer of the TEBTs are 5 or 2 nm thick.

The mesa type device structures were patterned by standard contact photolithography and selective wet etching. All GaAs layers were etched using $H_3PO_4:H_2O_2:H_2O = 3:1:50$ while $H_3PO_4:HCl = 1:3$ was chosen for GaInP. First of all the backside collector contact metallization was performed by depositing NiAuGe. Then the emitter contact was lithographically defined and subsequently metallized in the same manner. The base contact was formed by depositing TiPtAu after etching off all layers above the base layer. All contacts are alloyed at once at 400°C for 5 s. Finally, the mesa was patterned by etching down to the collector layer.

The fabricated structures include transmission line like contact patterns on emitter and base layer as well as large area transistor test structures for easy DC measurements. The DC characteristics were taken with an HP 4142 parameter analyzer on wafer.

RESULTS AND DISCUSSION

MOVPE Growth

Figure 2 shows the SIMS profile of a TEBT structure similar to the one shown in Fig. 1. There are some (intended) differences in the doping levels and layer thicknesses. Nevertheless the most important features of our devices are confirmed by the SIMS profile: the abrupt doping transitions between the heavily doped base and the adjacent layers.

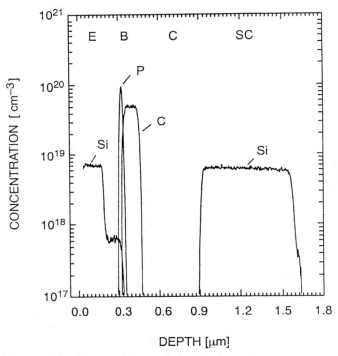

Fig 2: SIMS depth profile of a TEBT layer structure. Profiles refer to silicon (emitter and collector), carbon (base) and phosphorus (barrier layer) as indicated. The phosphorus signal is not to scale.

Based on latest published results [8] we have to assume that the P-N-junction and the compositional transition between GaAs base and GaInP barrier match even more precisely than the chemical profiles because C does not form an acceptor state in GaInP.

From the resistance measurements on emitter and base contact patterns we calculated sheet resistances of about 30 Ohm/Sq and below 400 Ohm/Sq, respectively. These are in good agreement with the intended doping levels and layer thicknesses shown in Fig. 1.

Device Characterization

In Fig. 3 we present the output charcteristics of a TEBT with a 5nm thick GaInP barrier layer with 7.75×10^{-6} cm^2 emitter area. The characteristics of conventional HBTs and of TEBTs with 2nm thick barriers not shown in this paper are not much different from the one of Fig. 3. Common features of all output characteristics are small collector-emitter offset voltages (<100 mV), high breakdown voltages (around 10 V at 10 mA, 7 V at 25 mA collector current) and high Early voltages. These result from the doping profile chosen.

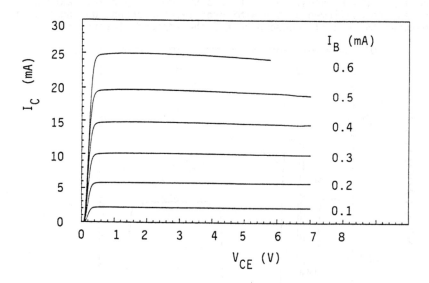

Fig. 3: Output characteristics of a TEBT with 5 nm thick
        GaInP layer.

Concerning the dependence on base-emitter voltage V_{EB}, ideal $\exp(qV_{EB}/kT)$ behavior of collector current I_C is observed for $I_C < 1$ mA, whereas at low current levels the base current is dominated by a leakage current due to the simple fabrication process of the devices. Therefore high current gains up to 65 (conventional HBT) and 115 (TEBT, 5nm) are only observed at collector current densities above 10 A/cm^2, see Fig. 4.

These current gains are clearly higher than would be expected from the elementary bipolar transistor model for a homojunction device with the same doping profile. Obviously the 5nm thick GaInP layer suppresses hole injection from the heavily doped base into the emitter as well as the the solid wide-gap emitter does. The ideal $IC(V_{EB})$ characteristics and the small emitter-collector offset voltage indicate that the electron transport is not strongly affected by a potential barrier in the conduction band.

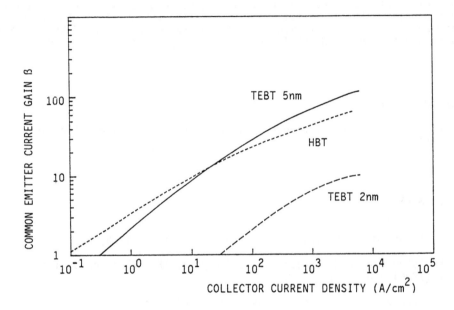

Fig. 4: Common Emitter current gain versus collector current density for a typical HBT and typical TEBTs with 5 and 2 nm thick GaInP layers.

The TEBTs with a 2 nm barrier show a much smaller current gain, which only reaches 10 to 15 at 10^4 A/cm^2. From the Gummel plots we cannot rule out that the lower gain is due to higher leakage currents of these devices caused by technological problems. Still we can conclude that an only 2 nm thick GaInP layer is either not suitable as an etch-stop and for the passivation of the emitter-base junction or does not provide the required suppression of hole injection from the base to the emitter.

SUMMARY AND CONCLUSIONS

Large area GaInP/GaAs HBTs and TEBTs with layer structures suited for the fabrication of RF capable devices have been fabricated from MOVPE grown layers. Carbon doping and the use of GaInP instead of AlGaAs as the wide gap material have been shown to be feasible for transistors with current gains of up to 115 and excellent output characteristics. Even a GaInP barrier as thin as 5 nm was found to effectively reduce the hole injection from the heavily doped base to the emitter. Thus the technological advantages of the TEBT concept combined with the attractive GaAs:C/GaInP material system holds a promising potential for the easy fabrication of high performance microwave transistors using MOVPE for the layer growth and the GaInP layers as etch stops during processing.

ACKNOWLEDGEMENT

We thank K. Winkler, J. Wiegert, and T. Fuchs for excellent technical assistance.

REFERENCES

1   J. Xu and M. Shur, IEEE Electron Device Lett. **7** (1986) 416

2   E. Tokumitsu, A.G. Dentai, and C.H Joyner, IEEE Electron Device Lett. **7** (1986) 416

3   H. Kroemer, J. Vac. Sci. Technol. **B1** (1983) 126

4   M.A. Haase, M.J. Hafich, and G.Y. Robinson, Appl. Phys. Lett. **58** (1991) 616 and references therin

5   T.F. Kuech et al, Appl. Phys. Lett. **53** (1988) 1317

6   R.M. Lum et al, J. Elec. Mat. **17** (1988) 101

7   G. Neumann, Th. Lauterbach, M. Maier, and K.H. Bachem, Inst. Phys. Conf. Series **112** (GaAs and Related Compounds 1990) p. 167

8   J.E. Cunningham et al, Appl. Phys. Lett. **56** (1990) 1760

Accumulation mode GaAlAs/GaAs bipolar transistor with two dimensional hole gas base

K.Matsumoto, M.Ishii, H.Morozumi, S.Imai, K.Sakamoto, Y.Hayashi
Electrotechnical Laboratory, MITI, Tsukuba 305 JAPAN

W.Liu, D.Costa, T.Ma, A.Massengale, and J.S.Harris
Stanford University, Stanford CA 94305 USA

Abstract   The Accumulation Mode Bipolar Transistor is proposed
and investigated for the first time. The transistor uses the two
dimensional hole gas (2DHG) as an ultra thin base layer, which is
accumulated at the AlAs barrier - p$^-$ collector interface.   The
transistor shows the common-emitter current gain of ~30.   The
theoretical results reveal that by adopting the p$^-$ collector
rather than the n$^-$ collector, and with the higher impurity
concentration, the carrier concentration of 2DHG base, the turn-
on emitter-base bias, and the punch-through base-collector bias
are improved drastically.

1, Introduction

A new type of GaAlAs/GaAs bipolar transistor which uses an accumulated
two dimensional hole gas (2DHG) as a base is proposed and investigated.   So
far, we have developed the GaAs Inversion-Base Bipolar Transistor (GaAs
IBT) by Matsumoto et al (1986)(1988a)(1988b) which uses an inversion layer
for the 2DHG base.   Though the GaAs IBT has ceratain good features such as

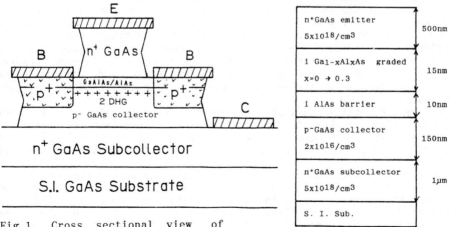

Fig.1, Cross sectional view of
Accumulation   Mode   GaAlAs/GaAs
Bipolar Transistor.

Fig. 2, Layered crystal structure
for Acc BT.

the small emitter size effects, short base transit time, etc., it has some problems to be solved such as high turn-on bias, low punch-through voltage, etc. In order to overcome those problems, the present device is proposed.

2, Experiments
    Figure 1 shows the cross-sectional view of the fully self-aligned Accumulation Mode Bipolar Transistor (Acc BT). The transistor consists of the n^+ GaAs emitter, p^- GaAs collector, and n^+ GaAs subcollector. Between the n^+ GaAs emitter and the p^- GaAs collector, there is not a metallurgical p^+ GaAs base layer but instead there is a semi-insulating barrier layer of GaAlAs/AlAs. The p^+ GaAs external base area is self-aligned to the emitter. Although the present transistor looks like a HBT with a p^- base layer, it operates as a transistor with 2DHG base. That is, the impurity concentration and the thickness of the p^- collector is low and thin enough, respectively. Therefore, the depletion layer extends both from the n^+ emitter and from the n^+ subcollector toward the p^- collector, and the p^- collector becomes completely depleted and there are no hole carriers present at the zero bias condition. By applying the emitter-base bias, the interface between the AlAs barrier and the p^- collector is accumulated and 2DHG which works as a thin base is induced.
    The fabrication process of the present transistor starts with the crystal growth of the following layered structure by MBE ; n^+GaAs emitter (n^+=5x10^{18}/cm^3, 500nm) / undoped Ga$_{1-x}$Al$_x$As graded layer (x=0 $\to$ 0.3, 15nm) / undoped AlAs barrier layer (10nm) / p^- GaAs collector (p^-=2x10^{16}/cm^3, 150nm) / n^+ GaAs subcollector (n^+=5x10^{18}/cm^3, 1µm). The layered crystal structure is shown in Fig. 2.
    The isolation area is formed by an oxgen ion implantation at 400keV with a dose of 1x10^{15}/cm^2 using a two layer mask of Aluminum (3µm) and sputtered SiO2 (1.7µm). After the oxgen ion implantation, the Al/SiO2 mask is removed by HF. The SiO2 layer of 1.4µm is again deposited on the isolated active area by plasma CVD, which is patterned to form the dummy emitter. Using the SiO2 dummy emitter as a mask, the top n^+ GaAs layer is etched down by a chemical etchant of H3PO4:H2O2:H2O to the surface of the graded GaAlAs layer to form the emitter. Using the SiO2 dummy emitter again as a mask, arsenic and beryllium ions are co-implanted at 260keV, 1x10^{14}/cm^2 and 30keV, 5x10^{14}/cm^2, respectively, to form the self-aligned external p$^+$ base area. The sample is then annealed at 750°C for 30 seconds to activate the implanted ions. After the annealing process, the SiO2 dummy emitter is replaced by the emitter ohmic metal of AuGe(130nm) / Mo(50nm) / Au(200nm). An ohmic metal for the external base area of AuBe(130nm) / Mo(50nm) / Au(100nm) is then deposited self-aligned to the emitter metal. Using the base ohmic metal as a mask, the layered crystal structure is etched down to the n^+ GaAs sub-collector, and the collector ohmic metal (the same as the emitter ohmic metal) is deposited self-aligned to the base.
    The emitter area is 1x10µm^2, and the base area is 3x10µm^2. In order to compare with the characteristics of the Acc BT, the GaAs IBT was also fabricated with the n^- collector with a doping of n^-=2x10^{16}/cm^3. The other device parametres of the IBT are all identical to those of the Acc BT.
    Figure 3 shows the common emitter transistor characteristics of (a)IBT and (b)Acc BT, respectively. The base current applied is from 0µA to 45µA with 5µA step. By comparing the characteristics of Acc BT and IBT, the various differences between the two become clear. In both transistors, the current gain of h_{FE}=30 was obtained at V_{CE}=2.5V and I_B=45µA. The output impedance of the Acc BT at V_{CE}=2V and I_c=0.9mA is 5.5 kohm which is two times higher than that of the IBT (2.6 kohm). In other words, the Early

voltage of Acc BT, which is obtained from the intersection of the tangent of the collector current in the active region at V_{CE}=2.25V and I_C=1mA on the voltage axis plus offset voltage, is 5.8V, which is ~6 times higher than that of IBT(~1V, at V_{CE}=2.25V, I_C=0.9mA). In the common-emitter transistor configuration, the direct emitter-collector leakage current at zero base current must be as small as possible. For the Acc BT in Fig. 3 (b), the direct emitter-collector leakage current is 0.05mA at V_{CE}=2V, while that of IBT is two times larger (0.1mA) at the same V_{CE}. The offset emitter-collector bias for both the Acc BT and the IBT, at which collector current begins to flow, is almost the same value of V_{CE}=~0.7V.

Thus, the Acc BT shows several superior characteristics to the IBT. The reasons for these superior characteristics of the Acc BT compared to those of the IBT can be explained as follows: the higher output impedance and the higher Early voltage of the Acc BT over the IBT is attributed to two physical phenomena. One is the higher 2DHG base concentration of the Acc BT over the IBT; the other is the lower direct emitter-collector leakage current of the Acc BT. Owing to the higher built-in potential of the n^+ emitter - p^- collector junction of the Acc BT than that of the n^+ emitter - n^- collector junction of the IBT, the Acc BT can induce a higher 2DHG base concentration than that of the IBT at the same emitter-base bias. This will become more clear in the theoretical analysis in the following section. The Early voltage, V_A, is a function of the base carrier concentration, Q_B, and the base-collector capacitance, C_{BC}, i.e., V_A=Q_B/C_{BC}. Therefore, for higher Q_B, the Early voltage, V_A, becomes higher. Furthermore, the higher built-in potential of the n^+ emiiter - p^- collector junction of Acc BT hinders electrons in the emitter from reaching the n^+ GaAs subcollector at zero base current. Therefore, the direct emitter-collector leakage current of the Acc BT is suppressed compared to that of the IBT.

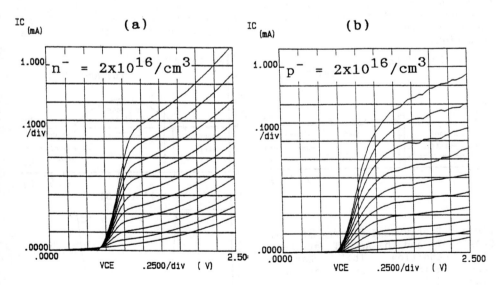

Fig.3, Transistor characteristics of common-emitter mode for (a)IBT and (b)Acc BT.

3. Theory

By using one dimensional simulation based on the classical theory, in which the poisson's equation and the continuity equation of current are solved simultaneously , various merits of the Acc BT over the IBT become clear and the experimental results in the previous section can be understood more deeply. The device structure which was fabricated was selected for the simulation and the doping concentration of the collector in the simulation was varied from $n^-=2\times10^{16}/cm^3$ (IBT) to $p^-=2\times10^{17}/cm^3$ (Acc BT).

Figure 4 shows the energy band diagram of the (a) IBT (n^- collector), and (b) Acc BT (p^- collector) at the zero bias condition. In this figure, the doping concentration of the p^- collector of the Acc BT was set to the highest value of $p^-=2\times10^{17}/cm^3$ to clearly see the effect of the p^- collector. In the IBT, the surface of the n^- collector at the AlAs interface is almost in the flat-band condition, and there are no hole carriers induced at the AlAs barrier / n^- collector interface. In the Acc BT, the surface of the p^- collector at the AlAs interface is drawn upward by the built-in potential of n^+ emitter - p^- collector junction and the energy band diagram of the p^- collector shows a slight peak at its center region. This built-in potential helps to increase the 2DHG base carrier concentration at the same emitter-base bias, and/or to lower the turn-on emitter-base bias as shown in Fig. 6. The higher the doping concentration of the p^- collector, the peak of the energy band diagram becomes higher. The peak of the p^- collector in the energy band diagram helps to reduce the direct emitter-collector current as mentioned in the previous section. As shown in Fig. 4(b), the p^- collector is completely depleted at the zero bias condition, because the valence band of the p^- collector, E_V, is far below the Fermi level, E_F. Figure 4 (b)' shows the carrier concentration of the Acc BT at the zero bias condition. Though the p^- collector is doped to $2\times10^{17}/cm^3$, the depletion layers extend both from the n^+ emitter and from the n^+ subcollector toward the p^- collector, and the peak carrier concentration of the p^- collector (holes) becomes as low as ~ $10^{13}/cm^3$. Therefore, the p^- collector is completely depleted at the zero bias condition.

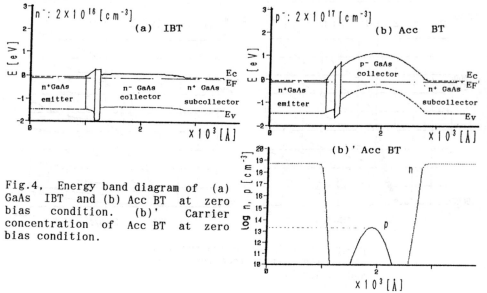

Fig.4, Energy band diagram of (a) GaAs IBT and (b) Acc BT at zero bias condition. (b)' Carrier concentration of Acc BT at zero bias condition.

Figure 5 shows the energy band diagram and the carrier concentration of (a),(a)' IBT and (b),(b)' Acc BT ,respectively, at the normal operation mode of the transistor when the emitter-base and base-collector bias of Veb=2.1V, Vbc=-0.4V, respectively, are applied. In the energy band of the IBT, the n⁻ collector shows the steep slope, and the induced hole carrier at the AlAs/n⁻collector interface shows the sharp triangular profile and forms the 2DHG base. The peak hole carrier concentration reaches almost $10^{19}/cm^3$, and the 2DHG base thickness at the carrier concentration of $10^{17}/cm^3$ is as thin as 65Å. While in the Acc BT, the p⁻ collector has a mild slope as shown in Fig. 5(b), and the induced hole carrier at the AlAs/p⁻ collector interface shows the two dimensional sharp triangular profile at high carrier concentrations, but begins to spread out to the three dimensional profile at lower carrier concentrations. The base thickness at high carrier concentrations of $10^{18}/cm^3$ is as thin as 50Å, and that at lower carrier concentrations of $10^{17}/cm^3$ is 450Å. The higher the emitter-base bias, and/or the higher the doping concentration of the p⁻ collector's, the more three dimensional spreading of the induced hole carrier is promoted. Therefore, in order to keep the induced hole base to be two dimensional, the appropriate emitter-base bias and/or doping concentration of the p⁻ collector must be chosen.

Figure 6 shows the dependence of the carrier concentration of the 2DHG base on the emitter-base bias. The base-collector bias is set to zero volts. The collector type is n or p and the impurity concentration is used as a parameter. By adopting the p⁻ collector rather than n⁻ collector, and using the higher impurity concentrations, the turn-on emitter-base bias which is needed to induce the 2DHG (>$3x10^{10}/cm^2$) decreased drastically from ~ 1.7V to ~ 0.5V. This is because higher impurity concentrations of the p⁻ collector increases the built-in

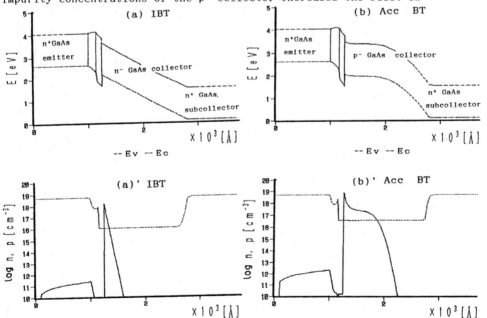

Fig.5, Energy band diagram and carrier concentration of (a),(a)' IBT and (b),(b)' Acc BT, respectively, when Veb=2.1V and Vbc=-0.4V are applied.

potential of the n+ emitter - p⁻
collector junction (see Fig. 4(b)).
Therefore, the surface of the p⁻
collector at the AlAs interface is drawn
further upward and a lower emitter-base
bias can induce the 2DHG, which makes
the emitter-base turn-on bias smaller.
Furthermore, the base carrier
concentration at V_{eb}=2.1V increased more
than 2 times from ~ 2×10^{12}/cm² to ~
4×10^{12}/cm². If the doping
concentration of the n⁻ collector
increases, the surface of the n⁻
collector at the interface of the AlAs
barrier becomes more flat-band (see Fig.
4(a)), and it becomes harder to induce
the 2DHG at the AlAs/n⁻ collector
interface, or in other words, the
induced hole carrier concentration
becomes smaller at the same emitter-base
bias. Similarly, the punch-
through base-collector bias which makes
the 2DHG base vanish ($<3\times10^{10}$/cm²) is
improved from ~4V to ~8V by adopting the
p⁻ collector rather than n⁻ collector
and using a higher impurity
concentration as shown in Fig 7. The
main reason of the improved punch-
through voltage is that for higher
impurity concentrations, the 2DHG
concentration becomes higher (see Fig.
6). Therefore, the punch-through voltage
becomes larger.

4. Conclusion
    The accumulation mode bipolar
transistor is proposed and investigated
for the first time. The experimental
and theoretical results show that by
adopting the p⁻ collector rather than n⁻
collector, the output impedance is
improved by more than two times; the
emitter-base turn-on bias becomes
smaller (~1.7V to ~0.5V); and the
punch-through voltage is improved from
4V to 8V. Thus the Accumulation Mode
Bipolar Transistor overcomes the
problems of the GaAs IBT while still
retaining all the desirable features of
the IBT.

References
Matsumoto K, Hayashi Y, Hashizume N, IEEE Electron Device Lett. Vol.
7,(1986)p.627
1988a Matsumto K, Hayashi Y, Nagata T, and Yoshimoto T, Jpn J. Appl. Phys.
Vol. 27, No. 6, (1988) p. 1154.
1988b Matsumoto K, Hayashi Y, Kojima T, Extended abstracts of 1988
International Conference on Solid State Devices & Materials p.531

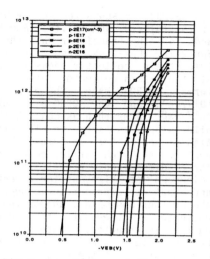

Fig.6, Dependence of the carrier
concentration of the 2DHG on
emitter-base bias. Base-
collector bias is set to zero
volts. Collector is n⁻ type
(IBT) and p⁻ type (Acc BT).

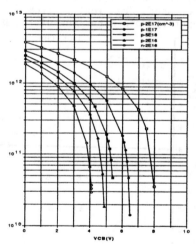

Fig.7, Dependence of the carrier
concentration of the 2DHG on
base-collector bias. Emitter-
base bias is set to V_{eb}=2.1V.
Collector is n⁻ type (IBT) and
p⁻ type (Acc BT).

A two-dimensional electron gas modulated resonant tunneling transistor

K.F. Longenbach, Y.Wang and W.I. Wang

1. Introduction

Resonant tunneling devices have received a great deal of attention because of their potential for high speed electronic devices. Most work has focused on two-terminal devices where the resonant tunneling process has been used to demonstrate frequency performance into the THz range (Sollner *et al* 1983). Three-terminal resonant tunneling devices have also been explored in the hope of exploiting the high speed of the tunneling process and the increased function of a device which exhibits a bi-stable input or output characteristic. Resonant tunneling structures have been incorporated into hot electron transistors (Yokoyama *et al* 1985), bipolar transistors (Capasso *et al* 1986; Reed *et al* 1989), and the gate or source/drain of field-effect transistors (Futatsugi *et al* 1986; Woodward *et al* 1987; Capasso *et al* 1987). In this paper we report the operation of a novel three-terminal resonant tunneling device called a two-dimensional electron gas modulated resonant tunneling transistor (2-DMRTT). Device operation is demonstrated at 80 K as well as room temperature with peak current densities of 420 A/cm^2 and peak-to-valley ratios as high as 5:1.

In the 2-DMRTT, the magnitude of a resonant tunneling characteristic is modulated by controlling the density of carriers in the emitter electrode of a double-barrier structure. Figure 1 shows a schematic cross section of the device as realized in the AlAs/GaAs material system. The structure consists of a wide bandgap AlGaAs modulation-doped layer grown on top of a resonant tunneling device. The top electrode layer of the tunneling structure is undoped so that the two-dimensional electron gas (2-DEG) of the modulation doped structure forms the emitter electrode of the double-barrier structure and the source of this three-terminal device. Resonant tunneling transport from source-to-drain is in the vertical direction beneath the modulation doped layer. A gate-to-source voltage is used to control the density of carriers in the 2-DEG, and thus the number of carriers available for resonant tunneling. Since the transit time from the 2-DEG to the lower electrode of the tunneling structure is quite rapid, high speed operation is primarily limited by how fast the 2-DEG carrier density can be modulated. In an optimally scaled device structure, the RC time constant associated with this process could possibly be minimized to yield cut-off frequency values into the hundred gigahertz range without the need for sub-micron lithography or elaborate device processing. In addition to high speeds, the negative differential resistance(NDR) feature that is present in the output characteristic of this device also makes possible the application of this device to many of the complex analog and digital circuit functions already demonstrated by other resonant tunneling transistors, such as a microwave power oscillator/modulator and a single transistor flip-flop.

The device operation is more clearly illustrated by considering the band diagram in the vertical direction beneath the gate electrode (Figure 2) for various gate-to-source (V_{gs}) and drain-to-source voltages (V_{ds}). As noted above, V_{ds} controls the incoming energy of the tunneling carriers relative to the resonant state in the quantum well. This is illustrated in

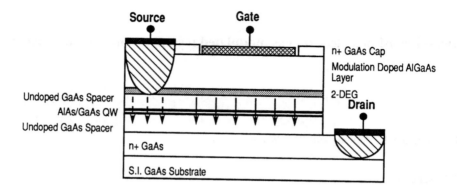

Figure 1 Schematic cross section of the two-dimensional electron gas modulated resonant tunneling transistor. The arrows indicate the current flow due to the modulated (solid) and unmodulated (dashed) portions of the I_{ds}.

Figure 2a, which shows the case of V_{gs}=0 V and V_{ds}=0.3 V. Under this bias condition, a voltage drop appears across the tunneling structure and carriers from the 2-DEG resonantly tunnel from the source to the drain. Since the top electrode of the tunneling structure is undoped, the only carriers available for the tunneling process are supplied by the 2-DEG, whose density can be controlled by V_{gs}. Negative values of V_{gs} will tend to deplete carriers in the 2-DEG, while positive values will increase the carrier density. For sufficiently large negative biases (Figure 2b), all the carriers in the 2-DEG can be depleted, cutting off the source of carriers for tunneling and thus the current transport beneath the gate electrode. Therefore, it is clear that the input characteristics of this device will be similar to a modulation-doped field effect transistor, while the output characteristic will contain a negative differential resistance feature due to the resonant tunneling structure between the source and drain.

2. Device Fabrication

To study this device, samples were prepared by molecular beam epitaxy on semi-insulating GaAs substrates. First a 600 nm GaAs layer doped 3×10^{18} cm^{-3} was grown to form the drain electrode layer. On top of this is a 500 Å undoped GaAs spacer layer followed by a double barrier structure consisting of a 28 Å AlAs barriers and a 70 Å well. The tunneling structure is topped with a 500 Å undoped spacer followed by 40 Å undoped $Al_xGa_{1-x}As$ (x=0.35) spacer layer, a 1×10^{12} cm^{-2} Si planar doped layer, a 350 Å undoped $Al_xGa_{1-x}As$(x=0.35) barrier layer and finally a 400 Å GaAs top layer doped 5×10^{18} cm^{-3}.

Devices were fabricated using a bipolar transistor mask set. First a mesa was defined using a wet chemical etch. The Au/Ge drain (collector) contact was then deposited and rapid thermal annealed. Following this the source contact (emitter) was defined, Au/Ge deposited and rapid thermal annealed at 400 °C for 20 sec. The gate (base) was then defined, a gate recess performed and Au deposited to form the Schottky gate contact. The contact geometry for the devices studied consisted of a 40 µm x 40 µm square source contact with a gate-to-source spacing of 5 µm on all four sides. The gate electrode was a 15 µm wide stripe around three sides of the source with a larger 45 µm x 80 µm pad on one side.

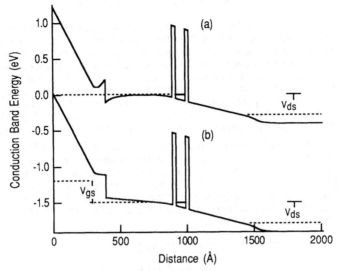

Figure 2 Schematic band diagram in the vertical region below the gate electrode. At (a) $V_{gs}=0$ and $V_{ds}=0.3$ V and at (b) $V_{gs}=-0.3$ V and $V_{ds}=0.3$ V.

3. Results and Discussion

Devices were characterized at both room temperature and 80 K. The room temperature common source output characteristics are shown in Figure 3a for gate voltages from -0.5 V to 0.3 V in 0.2 volt steps. Modulation of the drain current is clearly shown with an NDR feature present that exhibits peak-to-valley ratios in the range of 1.2 and peak source current densities from 140 to 300 A/cm^2. At 80 K(Fig. 3b) for the same gate voltage range, the modulation effect is seen more clearly and the peak-to-valley ratio improves to 5:1. The current density at the peak voltage also improves to 188 A/cm^2 and 420 A/cm^2 at cut-off and saturation respectively and the peak-to-valley ratio remains relatively constant over the entire gate voltage range. In both characteristics below a gate voltage of -0.5 V, further changes in the peak current density are not observed because, although transport beneath the gate is cut-off, conduction beneath the oversized source contact is still present. Above 0.5 V, modulation of the peak current density levels off and gate leakage current starts to become noticeable. Similar peak current densities are also observed for negative drain-to-source voltages indicating that tunneling from the 2-DEG is comparable to tunneling from the 3×10^{18} cm^{-3} doped drain electrode.

These results demonstrate the fundamental operating principles of this device and can be easily improved by using a more optimal mask set in which the gate-to-source spacing is reduced and the source and gate areas are adjusted to maximize modulation beneath the gate and reduce the parasitic current beneath the source. The device performance can also be improved by adjusting its vertical dimensions and incorporating either strained of lattice matched InGaAs/InAlAs layers. Thinner tunneling barriers will increase the resonant tunneling current density, directly improving the transconductance. Utilizing InGaAs for the undoped spacer/channel layers will increase the 2-DEG carrier density and mobility, which should improve the modulation effect and operation speed of the device. Also, further improvement in peak-to-valley ratio and peak current density may also be obtained by optimizing the undoped spacer layer thickness (Yoo *et al* 1990) or utilizing InGaAs layers in the tunneling structure (Broekaert *et al* 1990; Kapre *et al* 1991).

The ultimate operation speed of this device can be estimated to first order by examining its dominant time constants. When the device is biased near the negative differential resistance region, the source to drain transit time is the sum of the tunneling time and the spacer layer transit time. For thin tunneling barriers and spacer layers less than 1000 Å, these transit times will be less than a picosecond and will not limit the performance. Therefore, for an optimal tunneling structure, the high frequency response of this device will be limited by the time required to charge the 2-DEG. This time is given by $R_c C_{gs}$, where R_c is the channel resistance and C_{gs} the gate to source capacitance. For the modulation-doped structure studied here, C_{gs} is about 221 nF/cm^2. Using sheet resistance values commonly achieved in modulation-doped structures (100 Ω/sq.), the channel resistance can be estimated for an interdigitated source/drain layout by using a method similar to that used to estimate base spreading resistance in bipolar transistors. This technique yields a cut-off frequency of 170 GHz for a device with two sources and a total gate-width of 20 μm. In fact, the actual cut-off frequency may be higher since the negative resistance of the tunneling characteristic will feedback current to the 2-DEG, thus effectively reducing the channel resistance.

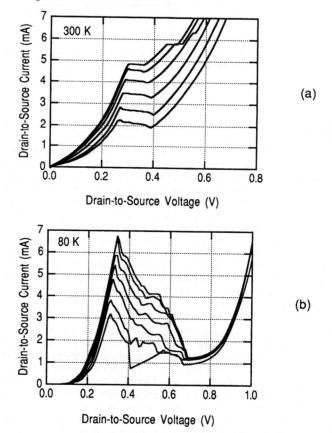

Figure 3 Common source output characteristics at (a) 300 K for V_{gs} = -0.5,-0.3, -0.1, 0.1, 0.3 and 0.5 V. and (b) at 80 K for V_{gs} = -0.5,-0.3, -0.2,-0.1, 0.1, 0.3 and 0.5 V.

4. Conclusion

In conclusion, we have demonstrated the operation of a novel resonant tunneling transistor. A simple analysis of the device operation has shown that operation speeds in into the hundred GHz range are possible. Attainment of these high speeds can be achieved by modifying the contact geometry, optimizing the tunneling structure and utilizing either strained of lattice matched InGaAs/InAlAs layers to reduce the channel resistance. The complex circuit functions realizable with this device as well as its potential for high speed operation make this device an attractive candidate for a wide variety of digital and analog circuit applications.

5.Acknowledgements

This work is supported by the Office of Naval Research (N00014-86-K-0694). Kort F. Longenbach is supported by IBM East Fishkill.

References

Broekaert T.P.E. and Fonstad C.G. 1990 *J. Appl. Phys.* **68** 4310
Capasso F., Sen S., Gossard A.C., Hutchinson A.L.and English J.H. 1986 *IEEE Electron Device Lett.* **EDL-7** 573
Capasso F., Sen S., Beltram F., and Cho A.Y. 1987 *Electron. Lett.* **23** 225
Futatsugi T., Yamaguchi Y., Ishii K., Imamura K., Muto S., Yokoyama N. and Shibatomi A. 1986 *IEDM Tech. Dig.* 286
Kapre R.M., Madhukar A, and Guha S. 1991 *Appl. Phys. Lett.* **58** 2255
Reed M.A., Frensly W.R., Matyi R.J., Randall J.N. and Seabaugh A.C. 1989 *Appl. Phys. Lett.* **54** 1034
Sollner T.C.L.G., Goodhue W.D., Tannenwald P.E., Parker C.D. and Peck D.D. 1983 *Appl. Phys. Lett.* **43** 588
Woodward T.K., McGill T.C., Chung H.F. and Burnham R.D., 1987 *Appl. Phys. Lett.* **51** 1542
Yokoyama N., Imamura K., Muto S., Hiyamizu S. and Nishi H. *Jpn. J. Appl. Phys.*,1985 **24** L853
Yoo H.M., Goodnick S.M. and Arthur J.R. 1990 *Appl. Phys. Lett.* **56** 84

Inst. Phys. Conf. Ser. No 120: Chapter 6
Paper presented at Int. Symp. GaAs and Related Compounds, Seattle, 1991

A two-dimensional electron gas emitter AlGaAs/GaAs heterojunction bipolar transistor with low offset voltage

Y. Wang, Q. Wang, K.F. Longenbach, E.S. Yang and W.I. Wang

1. Introduction

Single-heterojunction bipolar transistors (SHBT) have been studied widely for high performance circuit applications(Chang *et al* 1987) and it is well established that the offset voltage in these devices can be attributed to the difference in turn-on voltages of the emitter and collector junctions (Beneking *et al* 1982; Lee *et al* 1984). Typically the potential spike in the emitter junction of a SHBT produces an emitter-base turn-on voltage higher than that of the collector homojunction. This difference results in a collector offset voltage on the order of 0.2 volts in the output characteristics of the device. This offset voltage is undesirable for digital and microwave applications since it raises the saturation voltage and thus increases the power dissipation of circuits utilizing these devices (Su *et al* 1987). Several methods have been proposed to reduce this offset voltage including; a double heterojunction structure (Beneking *et al* 1982; Su *et al* 1987 Tiwari *et al* 1987), compositionally graded emitter (Hayes *et al* 1983; Mohammed *et al* 1991), and the heterostructure-emitter (Wu *et al* 1990). However, these devices suffer from either lower current gain or increased process complexity.

In this paper, we report a novel 2-DEG emitter HBT which has a low offset voltage and high gain. The concept employed here is similar to a method used in α-Si/Si bipolar transistors(Zhu *et al* 1989) to improve emitter carrier transport. However, the motivation and benefits of using a 2-DEG emitter in an HBT are different. In the conventional SHBT,

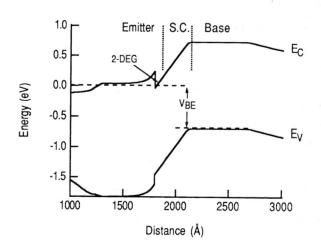

Figure 1 Energy band diagram of 2-DEG emitter HBT at low forward bias.

a moderately doped AlGaAs emitter is grown on top of a heavily doped base. The large valence band offset at the AlGaAs/GaAs heterojunction allows a lightly doped emitter to be used without significantly degrading the emitter injection efficiency. However, a large conduction band discontinuity is also present which causes the formation of a potential spike in the emitter-base space charge region. The presence of this spike increases the turn-on voltage of the emitter-base junction compared to the collector-base homojunction and results in an offset voltage in the output characteristics of the device. In the 2-DEG emitter device, a relatively thick undoped GaAs spacer is added between the AlGaAs emitter and GaAs base of a conventional SHBT. At equilibrium, charge transfer from the n-type AlGaAs results in the formation of a 2-DEG at the AlGaAs/GaAs interface and a lowering of the potential spike compared to the standard SHBT case. At low forward biases, conduction will not be limited by the AlGaAs barrier, hence, the 2-DEG will define the emitter edge of the junction (Figure 1). The results of this is an emitter-base junction with characteristics similar to the GaAs collector homojunction. Using this concept, devices with a 300 Å undoped spacer have been studied and found to have nearly identical emitter and collector turn on voltages. Output characteristics of these devices exhibit offset voltages of 30 mV with current gains greater than 10 at emitter current densities of 3 A/cm^2 and gains up to 600 in the high current density regime.

2. Device Fabrication

The structure of the device studied here is shown in Figure 2 and was prepared by molecular beam epitaxy on a semi-insulating (100) GaAs substrate. The p-type base is doped with Be to 1x10^{19} cm^{-3} and is separated from the n-type AlGaAs portion of the emitter by a 300 Å undoped GaAs spacer layer. On top of the undoped GaAs layer is a 500 Å Al$_x$Ga$_{1-x}$As (x=0.3) layer followed by a 300 Å linearly graded Al$_x$Ga$_{1-x}$As region (x=0.3 to 0) all doped n-type with Si to 2x10^{18} cm^{-3}. Finally, the device is capped by a GaAs layer which is doped with Si to 5x10^{18}cm^{-3}. The collector region consists of 600nm n-type GaAs doped to 3x10^{16} cm^{-3} layer on top of a 600nm 3x10^{18} cm^{-3} GaAs layer.

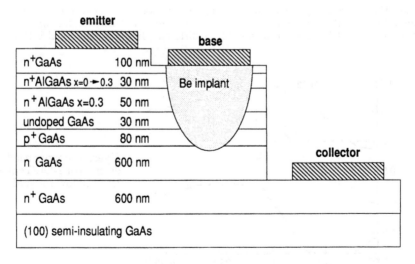

Figure 2 Cross-sectional diagram of 2-DEG emitter HBT device.

Devices were fabricated by etching off 700 Å from the GaAs cap layer in a solution consisting of $NH_4OH:H_2O_2:H_2O$ (3:1:250) to form the emitter mesa. Following this the base reach-through contact region is defined, ion implanted with Be, and annealed for 5 seconds at 850 °C. The base mesa is then defined by an etch down to the heavily doped collector layer. AuGe/Ni and ZnAu are then deposited to form the n- and p-type contacts. The contacts are then sintered in forming gas at 400 °C with the final devices having an effective emitter area of 120x64 μm^2.

3. Results and Discussion

The common-emitter characteristics were characterized on a curve tracer at room temperature with the results for a typical device given in Figure 3. Figure 3a clearly shows the 30 mV offset voltage attained in this device, which to our knowledge is the lowest attained for any abrupt emitter junction HBT. A current gain of 10 at an emitter current density of 3 A/cm^2 is also observed in this low current density portion of the characteristic. At higher current densities (Figure 3b), current gains as high as 400 are observed for this device. By using an alternate fabrication process to reduce the peripheral leakage currents (Lee *et al* 1989), current gains as high 600 have also been observed for these samples. Therefore, it is clear in either case that this device structure produces high current gains even though the doping level in the base is higher than the emitter. These high current gains provide a strong indication that the hole blocking effect of the AlGaAs layer still remains. Measurements of the emitter and collector junctions (Figure 4) show that both junctions have turn-on voltages close to 1 V, making it apparent that the 2-DEG forms the emitter side of the junction and causes the observed reduction in offset voltage. The presence of the 2-DEG was further confirmed by performing Hall Effect measurements on samples with the GaAs cap layer and part of the AlGaAs layer removed. Hall mobilities on these samples improved from 1,200 cm^2/V-s at 300 K to 11,300 cm^2/V-s at 77 K. These results are indicative of the presence of the 2-DEG and in conjunction with the device measurements indicate that in fact the 2-DEG does define the emitter junction edge.

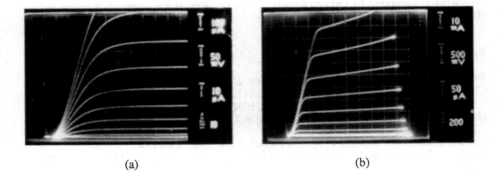

(a)   (b)

Figure 3 Common-emitter characteristics of 2-DEG emitter HBT in the low current regime (a) and in (b) high current regime.

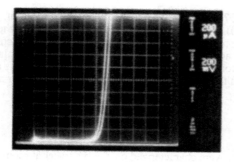

Figure 4 Emitter-base(right) and collector-base (left) forward junction characteristics.

In addition to the low offset voltage, this device may have other advantages over a standard SHBT. Since, the 2-DEG notch forms the emitter side of the junction, the emitter-base capacitance will be the series combination of the depletion layer and spacer layer capacitances. Under forward bias, the depletion layer capacitance will become large compared to the spacer layer capacitance and the emitter-base capacitance will be dominated by the spacer layer capacitance. Thus compared to the standard SHBT structure, the emitter-base capacitance will be reduced and the high frequency performance improved. Furthermore, since large amounts of defects can be present at the AlGaAs/GaAs interface, the removal of this region from the space charge region of the device will reduce the recombination current. As a result, the low current gain of the 2-DEG emitter HBT should be improved(Zhu *et al* 1989).

4. Conclusion

In conclusion an AlGaAs/GaAs 2-DEG emitter HBT structure has been demonstrated for the first time. Devices exhibit an offset voltage of 30 mV and current gains as high as 600. The 2-DEG emitter device structure provides a simple and practical way to achieve an emitter with homojunction characteristics while maintaining the blocking of back injected minority carriers. The high gain and low offset voltage make this device an attractive candidate for high speed circuit applications.

This work is supported by Office of Naval Research. Kort F. Longenbach is supported by IBM East Fishkill.

References

Beneking H. and Su L.M. 1982 *Electron. Lett.* **18** 25
Chang M.F., Asbeck P.M., Wang K.C., Sullivan G.J., Sheng N.H., Higgins J.A. and Miller D.L.1987 *IEEE Elect. Device Lett.* **EDL-8** 303
Hayes J.R., Capasso F., Malik R.J., Gossard A.C. and Wiegmann W. 1983 Appl. Phys. Lett. **43** 949
Lee S.C., Kau J.N. and Lin H.H. 1984 *Appl. Phys. Lett.* **45** 1114
Lee W.S., Ueda D., Ma T., Pao Y.C. and Harris,Jr. J.S. 1989 *Elect. Device Lett.* **10** 200

Mohammad S.N., Chen J., Chyi J.I and Morkoc H. 1991 *J. Appl. Phys.* **69** 1067

Su S.L., Tejayadi O., Drummond T.J., Fischer R. and Morkoc H. 1983 *IEEE Elect. Device Lett.* **EDL-4** 140.

Tiwari S., Wright S.L. and Kleinsasser A.W. 1987 *IEEE Trans. Elect. Device* **ED-34** 185

Wu X., Wang Y., Luo L.F., and Yang E.S.1990 *IEEE Elect. Device. Lett.* **11** 264

Zhu E.J., Zhang S.S., Sheng W.W., Zhao B.Z., Xiong C.K., and Wang Y.S., 1989 *IEEE Elect. Device Lett.* **11** 4.

Inst. Phys. Conf. Ser. No 120: Chapter 6
Paper presented at Int. Symp. GaAs and Related Compounds, Seattle, 1991

317

1/f noise in AlGaAs/GaAs HBTs using ultrasensitive characterization techniques for identifying noise mechanisms

Marcel N. Tutt, Dimitris Pavlidis, and David Pehlke†   Robert Plana and Jacques Graffeuil

Solid-State Electronics Laboratory
Department of Electrical Engineering and Computer Science
University of Michigan
Ann Arbor, MI 48109-2122

Laboratoire d'Automatique et
d'Analyse des Systèmes
Université Paul Sabatier
31077 Toulouse, France

ABSTRACT

The 1/f noise characteristics and mechanisms of GaAs/AlGaAs HBT's are studied using conventional short circuit current measurements ($S_{I_C}(f)$ and $S_{I_B}(f)$) and an indirect highly sensitive technique based on noise figure characterization. The $S_{I_C}(f)$ and $S_{I_B}(f)$ follows a 1/f dependence at low frequencies, for moderate bias conditions. A Lorentz component is found in the spectra. The base resistances are obtained by approximating them to the correlation resistances and range from 17Ω to 63Ω. The intrinsic noise sources are studied using a simple model employing two internal, uncorrelated noise sources. These sources are similar to the $S_{I_C}(f)$ and $S_{I_B}(f)$ at moderate bias conditions. However, the intrinsic collector noise differs from $S_{I_C}(f)$ at low bias conditions.

1. INTRODUCTION

Heterojunction Bipolar Transistors (HBT's) play an important role as sources at microwave and millimeter-wave frequencies. Since these circuits operate in a nonlinear fashion, upconversion of the 1/f noise results in phase noise which introduces limitations in both analog and digital systems. As a result, fundamental device design requires very low 1/f noise characteristics in order to achieve ultimate circuit and system performance. A first study on HBT noise sources has been reported by the authors in the past (Tutt et al (1989)). Full understanding of the HBT noise properties requires; however, accurate determination of both collector and base noise spectra.

Measurement of the collector noise current spectra is straightforward since the output of an HBT is like that of a current source. On the other hand, the base noise current spectra can be more difficult to characterize since the impedance looking into the base is much lower than that of the collector. Hence, the measured spectra will always be somewhat less than the true value and must be corrected for by accounting for the respective input impedances of the device and the receiver. This can be avoided in two ways. First, the device can be scaled down to increase its input impedance. The major drawback here is that for most practical applications envisioned for HBT's, large emitter areas are most desirable. Moreover, the reduced area could enhance periphery effects which may not be as significant in larger area devices. Hence, it is desirable to characterize devices which are truly representative of the HBT's of interest.

A good solution to this problem is through a second, ultrasensitive characterization technique which involves the use of noise theory for two ports. In this approach it is possible to obtain the short circuit noise currents through direct measure of noise on the output port for various input terminations. With an appropriate number of such measurements it is possible to obtain an alternate representation of the noise sources. Then by use of of an exact analytical transformation, the short circuit noise currents can

†This work is supported by: Contract MDA904-90-C-4094, ARO (Contract No. DAAL03-87-U-0007), and NASA (Contract No. NAGW-1334).

be obtained. In this paper, application of this alternate method is shown to provide the correct short circuit currents. The bias dependence of the sources is studied to determine their origin. A simple circuit model, including noise sources, is presented.

2. DEVICE FABRICATION

The GaAs/AlGaAs HBT's were fabricated, at the University of Michigan's Solid State Electronics Laboratory, on MOCVD grown material using a self-aligned technology. An abrupt emitter-base heterojunction structure was used. The emitter was doped (n-type) to $2 \times 10^{17} cm^{-3}$, while the base was doped to $1 \times 10^{19} cm^{-3}$. A thick, lightly doped ($1 \times 10^{16} cm^{-3}$) pre-collector was used for improved breakdown characteristics. The emitter ohmic metal is used as a mask for lift-off of the base ohmic metal. A reactive ion etch step is used to etch to within approximately $1000 \mathring{A}$ of the base layer. Next a wet etch is used to complete the etching to the base and provides for removal of RIE plasma damage as well as a $1000 \mathring{A}$ undercut profile allowing for self-aligned lift-off of the base metal. This $1000 \mathring{A}$ contact separation between base metal and active area yields a significant reduction in base resistance over non-self-aligned alternatives. The base contact is achieved using non-alloyed Pd/Zn/Pd/Au.

3. NOISE MODELING of LINEAR NOISY TWO PORTS

The representation of a noisy two port in terms of short circuit currents is shown in figure 1. This is typically used in low frequency transistor noise studies. i_1 and i_2 are the input and output short circuit noise currents. They are correlated. An alternate representation is also given in figure 1 using the uncorrelated noise sources i and e_n (Rothe et al (1956)). This latter representation is used for the well known noise figure analysis which is used for characterizing circuits at microwave and millimeter-wave frequencies. The noise sources in these two representations can be expressed in terms of each other using the small-signal parameters of the network itself. The short circuit noise currents can therefore be expressed in terms of i and e_n using:

$$i_1 = i(1 - Y_{11}Z_{cor}) - Y_{11}e_n \tag{1}$$

and

$$i_2 = -Y_{21}e_n - iY_{21}Z_{cor} \tag{2}$$

Y_{ij} are the device Y-parameters. i and e_n are the input noise current and noise voltage spectra, respectively. They are uncorrelated. Z_{cor} is the correlation impedance which accounts for the correlation between the quantities that one would measure. i, e_n, and Z_{cor} are determined from the noise parameters F_{min}, R_{opt}, and g_n which are the minimum noise figure, optimum resistance, and noise conductance, respectively. These are in turn obtained from the measurement of the output noise spectra for multiple (>10) source terminations. This approach of determining the low frequency noise characteristics of HBT's is an alternative way to the more conventional method of direct noise measurement at the base and collector terminals. It also allows one to determine the correlation between the noise sources at the input.

4. DEMONSTRATION of the CHARACTERIZATION METHODS

The noise figure charaterization technique was applied to determine the input and output short circuit noise currents. The noise spectra were measured for 12 different source resistances from 0.5Ω to $10k\Omega$. i, e_n, Z_{cor} were evaluated by fitting as described in Section 2. The device small signal parameters, needed to calculate i_1 and i_2 (Eqs.1 and 2), are obtained from gain and impedance measurements using the HP4194A. The gain measurements were made from 100Hz to 100kHz, while the impedance measurements were made from 10kHz to 100kHz. By noting that the device impedance was

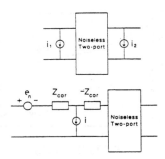

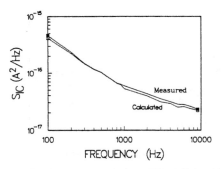

Figure 1: Equivalent Representations of Noisy Two-Ports.

Figure 2: Comparison of $S_{IC}(f)$ from Direct Measurements and $S_{IC}(f)$ Obtained from e_n and i.

constant and real over this range, the impedances were extrapolated down to 100Hz. This permitted $S_{I_C}(f)$ and $S_{I_B}(f)$ to be calculated from the results of the noise figure measurements at the output of the device.

Figure 2 shows the results of direct measurement of $S_{I_C}(f)$ $(S_{I_C}(f) = \overline{i_2^2})$ with that calculated using i and e_n. The HBT had an emitter area$\simeq 200\mu m^2$ and was biased at $I_B = 2.1$mA, and $V_{CE} = 2$V. Good agreement is obtained between the two techniques. Figure 3 shows the results of direct measurement of $S_{I_B}(f)$ $(S_{I_B}(f) = \overline{i_1^2})$ with that calculated using i and e_n. Here the agreement is less satisfactory and the noise level estimated by the direct measurement can be more than 50% lower than the value estimated with the help of the noise modelling technique. Part of this discrepancy is due to the fact that $S_{I_B}(f)$ has to be corrected for the base impedance. However, it is also believed that the indirect method may be sensitive to the number and values of the source impedances. Although a large number of source resistances has been used in the estimates to eliminate this problem, it seems that an even larger number is necessary for higher resolution.

The bias conditions used for the results depicted in Figures 2 and 3 correspond to relatively high gain, $h_{fe,op}$ ($\frac{h_{fe,op}}{h_{fe,max}} = 0.72$) operating conditions. Some limitations may however, exist in the accuracy of the indirect method depending on the gain of the HBT. The device gain is small at low bias (typically, $\frac{h_{fe,op}}{h_{fe,max}} = 0.35$) and as a result, when the source impedance is changed, the output noise variations may be marginal. Since the principle of the indirect technique is based on the evaluation of the noise parameters by examining the output noise variations it follows that when these variations are small, the results may include significant errors.

5. ORIGINS of the NOISE SOURCES

The bias dependence of all of the noise sources was studied in order to determine the apparent origins of $S_{I_B}(f)$, and $S_{I_C}(f)$. The results of this study are practically independent of the characterization technique and either one can be used for this purpose.

The collector current of a HBT (emitter area$\simeq 200\mu m^2$) was varied by a factor of $\simeq 8$ from 1.6mA to 12.5mA, and $S_{I_C}(f)$ was measured. Figure 4 shows the $S_{I_C}(f)$ spectra at the two different collector currents. A noise hump, indicating a Lorentz component, was observed for both $S_{I_C}(f)$, and $S_{I_B}(f)$ at approximately 10kHz. Its significance

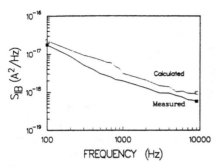

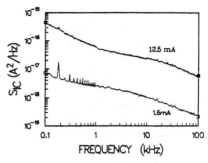

Figure 3: Comparison of $S_{IB}(f)$ from Direct Measurements and $S_{IB}(f)$ Obtained from e_n and i.

Figure 4: $S_{IC}(f)$ for I_C=12.5mA and 1.6mA.

decreased with increasing bias. Moreover, it seemed to be more significant in $S_{I_C}(f)$ than $S_{I_B}(f)$. $S_{I_C}(100Hz)$ is greater than $S_{I_B}(100Hz)$ by a factor of 20, as expected due to the amplification of the base noise at the output of the HBT and the additional direct contribution of noise between the collector and the emitter.

$S_{I_C}(100Hz)$ was found to have an I_C^2 dependence. This may be indicative of generation-recombination effects $(S_I \propto I^2)$. It was determined that $S_{I_B}(100Hz)$ displayed a $I_B^{1.8}$ type of dependence, over the corresponding I_B range. This may be associated with surface recombination $(S_I(f) \propto I^{1.5})$. It was interesting to note that in the i and e_n representation e_n was essentially bias independent, while i had a $I_B^{1.49}$ dependence which is not quite the same as S_{I_B}.

All of the sources $(S_{I_C}, S_{I_B}, i,$ and $e_n)$ followed a $1/f^\alpha$ frequency dependence for all bias conditions. At moderate values of I_C $(I_C$=12.5mA$)$ $\alpha = 1$ for $S_{I_C}(f)$. While for lower values of I_C $(I_C$=1.6mA$)$, the slope decreased to $\alpha = 0.77$ due to the effect of increased Lorentz component, present in the spectrum. α remained at essentially 1 for S_{I_B} for all of the bias conditions. i had α=1 at moderate values of bias, while e_n had a shallower slope of α=0.8. Corner frequencies were estimated to be in the range of 1-20MHz by extrapolating the results of S_{I_C}, at 100kHz, to the estimated shot noise limit assuming a $1/f^1$ dependence.

6. A SIMPLE PHYSICAL MODEL

The measurements, and analyses directly address only the terminal characteristics of the device, so far. In order to obtain a better understanding of what is happening within the device, the internal noise characteristics are required. To accomplish this, a simple 2 source model is used to describe the measured 1/f noise (figure 5). One source is placed in parallel with r_{be}. This source describes the $1/f^\alpha$ noise in the base emitter region. The second source is placed in parallel with r_{ce}. This source is used to account for the $1/f^\alpha$ noise between the collector and emitter. These two sources are uncorrelated. A third source may exist between the base and the collector (van der Ziel et al (1986)); at this time it is assumed to be negligible and it is not included.

At the low frequencies used for characterization (100Hz to 100kHz), the capacitances are assumed to have negligible effect. The equivalent circuit values (figure 5) are obtained from low frequency device measurements made with an HP4194A. This permits rapid device characterization. The value of r_{bb} is obtained by approximating $r_{cor} = r_{bb}$. This

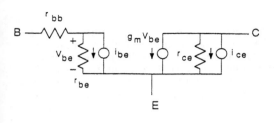

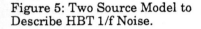

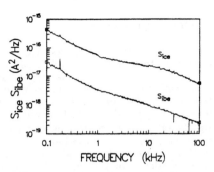

Figure 5: Two Source Model to
Describe HBT 1/f Noise.

Figure 6: Plot of S_{ice} and S_{ibe} at
moderate bias conditions.

implies that the correlation between the measured terminal quantities is due to r_{bb}. This assumption was based on the fact that the correlation impedance did not show any significant frequency dispersion nor bias dependence. Furthermore, DC measurements revealed that the magnitude of r_{bb} is of the same order of magnitude as r_{cor}; r_{cor} in this work ranged from 17Ω to 63Ω. By looking at the value of R_{opt} one finds that R_{opt} is larger by at least 32Ω than r_{bb}. Following (Limsakul (1985)) this indicates that part of the input noise voltage source, is correlated with i via r_{bb}, and another part is uncorrelated with i. Both measurement techniques were used to determine the intrinsic noise sources. As shown in Section 3, the r_{cor} can be obtained from the noise figure technique. The conventional technique was used for obtaining the external short circuit currents since the base impedance was known from the $r_{cor} = r_{bb}$ approximation. The internal noise sources were then obtained from knowledge of the circuit parameters and the short circuit noise currents.

Figure 6 is a plot of the noise spectra for the two internal sources ($S_{ice} = \overline{i_{ce}^2}$ and $S_{ibe} = \overline{i_{be}^2}$) for a five finger ($20\mu m \times 2\mu m$) HBT. The bias condition of I_C=12.5 mA and V_{CE}=2 V, corresponding to a moderate bias condition, was used. For this particular bias condition, S_{ice} is more than one order of magnitude larger than S_{ibe} at 100Hz. In addition, at 100Hz, both S_{ice} and S_{ibe} had essentially $1/f^\alpha$ behavior with α=1. S_{ice} shows a Lorentz component more pronounced than S_{ibe}. At this bias condition, both S_{ice} and S_{ibe} are virtually identical to S_{I_C} and S_{I_B}, respectively.

At a lower bias condition (I_C=1.6 mA and V_{CE}=2 V) the difference between S_{ice} and S_{ibe} was greatly reduced. In fact, at low frequencies S_{ice} and S_{ibe} were found to be virtually equal. S_{ice} still displayed a more significant Lorentz component than did S_{ibe}. The dependence of S_{ice} on I_C is $I_C^{2.4}$, which is even greater than that shown by S_{I_C} ($S_{I_C} \propto I_C^2$). Strong current dependence of measured noise spectra has been reported in the past for S_{I_B} of HBT's and was associated with burst noise ($S_{I_B} \propto I_B^3$) (Zhang et al (1984)). In addition, the dependence of S_{ibe} on I_B is $I_B^{2.2}$ which is greater than that found for S_{I_B} ($S_{I_B} \propto I_B^{1.5}$). Based on the observations, in this section, S_{I_B} and S_{I_C} do not necessarily reflect the internal noise current spectra S_{ibe} and S_{ice}, respectively, as indicated by their different bias dependencies. This indicates that an analysis of device noise characteristics based on terminal noise properties may be highly misleading. Knowledge of the internal sources, as obtained here, is important for complete understanding of the noise sources.

7. CONCLUSIONS

Noise characteristics of GaAs/AlGaAs HBT's were studied using two different characterization techniques. One is based on the direct measurement of short circuit noise currents, while the second is very sensitive and is based on noise figure measurements. The origins of the noise sources were determined to be G-R, and surface recombination for $S_{I_C}(f)$ and $S_{I_B}(f)$, respectively, based on bias dependence. At moderate bias levels, $S_{I_C}(f)$ and $S_{I_B}(f)$ followed a $1/f^\alpha$ trend with $\alpha=1$. At lower values of bias, a significant Lorentz component reduced α to 0.77 for $S_{I_C}(f)$. A simple model was used to describe the $1/f$ noise behavior of the HBT's. Two sources were used: i_{ce} and i_{be}. Both measurement methods were combined to extract the internal noise sources. The I_C dependence of the internal noise source i_{ce} $(Si_{ce} \propto I_C^{2.4})$ is different than that of the terminal noise S_{I_C} $(S_{I_C} \propto I_C^2)$. Similarly, the I_B dependence of the internal noise source i_{be} $(S_{i_{be}} \propto I_B^{2.2})$ is different than that of the terminal noise S_{I_B} $(S_{I_B} \propto I_B^{1.8})$. This clearly indicates the need for evaluating the internal, rather than the external device noise characteristics for a physical understanding of the noise properties of HBT's.

REFERENCES

Limsakul C., Ph.D. Thesis, Institut National des Sciences Appliquées de Toulouse, France, 1985.

Rothe H., and Dahlke W., Proceedings of the IRE 1956, 811.

Tutt M., Pavlidis D., Bayraktaroglu B., 1989 Gallium Arsenide and Related Compounds 97, 701.

van der Ziel A., Zhang X., Pawlikiewicz A. H., IEEE Transactions on Electron Devices, EDL-33, No. 9, 1371.

Zhang X. N., van der Ziel A., Duh K. H., Morkoc H., IEEE Electron Devices Letters EDL-5, No. 7, 1984, 277.

Inst. Phys. Conf. Ser. No 120: Chapter 6
Paper presented at Int. Symp. GaAs and Related Compounds, Seattle, 1991

323

Low-frequency noise characterization of Npn AlGaAs/GaAs heterojunction bipolar transistors

Damian Costa and J.S. Harris, Jr.

226 McCullough Building, Stanford University, Stanford, CA 94305, Tel. 415 723-9484

Abstract: We report low-frequency noise measurements of Npn, $Al_xGa_{1-x}As/GaAs$ heterojunction bipolar transistors (HBTs) as a function of bias current, device geometry, extrinsic-base-surface condition, aluminum mole fraction, and temperature. The 1/f noise is mainly associated with fluctuations in the base-surface recombination velocity. The noise "bump" (observed at intermediate frequencies) is generated by a trap in the AlGaAs, possibly the DX center, near the surface of the emitter-base junction. It is shown that the use of an AlGaAs surface passivation ledge and the reduction of the Al mole fraction from $x = 0.3$ to $x = 0.2$ significantly improves the low-frequency noise performance of our HBTs.

I. INTRODUCTION

The AlGaAs/GaAs heterojunction bipolar transistor (HBT) can extend the operating frequency range of low-phase-noise oscillators beyond that of oscillators incorporating silicon bipolar transistors (Madihian et al 1991). Although the flow of charge in AlGaAs/GaAs HBTs is mostly vertical in nature like in silicon bipolar transistors, the presence of a mesa-type structure (which exposes the base surface) and a higher defect density in the AlGaAs emitter, raises questions about the low-frequency noise (and up-converted phase noise) performance of AlGaAs/GaAs HBTs compared to that of silicon bipolar transistors.

To date, most of the work on the low-frequency noise of AlGaAs/GaAs HBTs (Tutt et al 1988, Jue et al 1989) has focused on the bias and temperature dependence of the noise data. Recent experimental results of AlGaAs/GaAs HBTs (Hayama et al 1990), indicate that a HBT incorporating a thin, depleted AlGaAs layer (AlGaAs surface passivation ledge) over the extrinsic base region shows dramatically lower 1/f noise as compared to a device without the AlGaAs ledge. However, these devices still exhibit a large anomalous noise "bump" over the frequency range of 10 KHz to 1 MHz.

In this work, the low-frequency noise characteristics of Npn, $Al_xGa_{1-x}As/GaAs$ HBTs (grown by MBE) have been studied as a function of bias current, device geometry, extrinsic-base-surface condition, aluminum mole fraction in the emitter, and temperature. With the transistor in the common-emitter configuration, the noise power at the collector was measured from 100 Hz to 10 MHz with a selective level meter (HP 3586). Using a simple low-frequency circuit model for the transistor and the gain of the system , we refer the noise at the output back to the input and define an equivalent input base noise current (Motchenbacher et al 1973). These measurements show the existence of three distinct regions in the noise spectra: a 1/f line-shape at the lower end of the measured frequency range, a Lorenztian spectrum (noise "bump") at intermediate frequencies, and a white noise region at the higher end of the measured frequency range. The origin of the 1/f noise of small-geometry AlGaAs/GaAs HBTs is clearly demonstrated to result primarily from fluctuations in the extrinsic-base surface recombination velocity, as explained by Fonger's model (1956). Our measurements also suggest that the anomalous noise "bump" is generated by a trap, possibly the DX center (Lang et al 1979), in the AlGaAs near the surface of the emitter-base junction. It is shown that the use of an AlGaAs ledge and the reduction of the Al mole fraction (x) from $x = 0.3$ to $x = 0.2$ substantially reduces the low-frequency noise of our HBTs.

II.   SOURCES OF NOISE

The low-frequency noise of bipolar transistors arises from several independent noise sources [van der Ziel et al 1986]. This section describes the more probable noise sources in AlGaAs/GaAs HBTs according to the shapes of the respective noise spectra.

A. 1/f Noise

Many theoretical studies have associated 1/f noise with surface states (McWhorter 1957) and numerous experimental results support this surface noise theory (Sah 1956). Fonger originally attributed 1/f noise to fluctuations in the current recombining at the extrinsic base surface, I_{bs}, resulting from fluctuations in the surface recombination velocity, s. He showed that the spectral density associated with a fluctuation in I_{bs}, $S_{I_{bs}}$, could be expressed as

$$S_{I_{bs}}(f) = \frac{I_{bs}^2}{s^2} S_s(f) \qquad (1) \qquad \text{and} \qquad S_{I_{bs}}(f) = I_c^2 \left(\frac{W_b L_d}{D_n}\right)^2 \left(\frac{P_e}{A_e}\right)^2 S_s(f) \qquad (2)$$

where f is the frequency, $S_s(f)$ is the spectral density associated with a fluctuation in s, I_c is the collector current, W_b is the base width, D_n is the electron diffusivity in the base, P_e is the emitter perimeter, and A_e is the emitter area. $S_s(f)$, which characterizes the noise contribution of the surface, was empirically found to vary as 1/f by Fonger. The quadratic dependence of the spectral density on collector current results because n'(0) can be expressed either in terms of the extrinsic-surface-recombination base current component ($I_{bs} \propto sn'(0)$) or the collector current ($I_c = qA_e D_n n'(0)/W_b$).

B. Burst Noise

Burst noise in bipolar transistors is believed to arise from traps or generation-recombination (g-r) centers in the emitter-base space charge region (Hsu 1970, Jageger 1970). The results of these investigations have shown that the spectral density of burst noise for a single trapping time constant process, takes the form of a Lorenztian

$$S_{I_{burst}}(f) \propto \tau \frac{I_c^m}{1 + (2\pi\tau f)^2} \qquad (3) \qquad \text{and} \qquad \tau = \frac{\tau_0}{T^2} \exp\left(\frac{E_a}{kT}\right) \qquad (4)$$

where τ is a thermally activated trapping time constant, m is a constant in the range of 0.5 to 2, τ_0 is a constant, E_a is the activation energy of the participating trap, inclusive of the energy barrier for capture, T is the temperature, and k is Boltzmann's constant.

C. White Noise

The major part of the white noise of the equivalent input base noise current arises from the shot noise due to the dc base current ($S_{I_{shot}}(f) = 2qI_b$).

III.   RESULTS

A. Current Dependence

The low-frequency noise was measured as a function of bias current. Figure 1 shows the typical equivalent input base noise current spectral densities and the various components of noise current for a 4 um x 10 um emitter $Al_{0.3}Ga_{0.7}As$/GaAs HBT operating in the linear region (V_{ce} = 3V) at three different bias currents. This device does not have an AlGaAs passivation ledge.   The equivalent input base noise current spectral density at 100 Hz, S_{I_b}(100Hz), was found to vary as $I_c^{2.1}$ over the measured collector current range very close to the collector dependence predicted by equation (2). At intermediate frequencies (10^4-10^6 Hz)

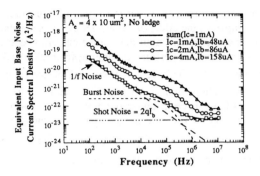

Fig. 1   Equivalent input base noise current spectral densities and the various components of noise current for a 4 um x 10 um emitter $Al_{0.3}Ga_{0.7}As/GaAs$ HBT without the AlGaAs passivation ledge operating at three different bias currents (V_{ce} = 3V).

each noise spectrum exhibits a striking "bump", characteristic of burst noise. This "bump" was fitted to a Lorenztian spectrum of the form given by equation (3). The white noise at the high end of the measured frequency range is roughly equal to the base current shot noise.

B. Device Geometry Dependence

Figure 2 is a comparison of the equivalent input base noise current spectral densities of $Al_{0.3}Ga_{0.7}As/GaAs$ HBTs, without the AlGaAs ledge, for two different values of emitter perimeter-to-area ratio (P_e/A_e) at a fixed collector current of 20 mA. As the P_e/A_e increases from 0.2 to 0.33, S_{I_b}(100Hz) increases by a factor of 2.5. This result agrees closely with the predicted increase of 2.7, calculated from the $(P_e/A_e)^2$ dependence in the 1/f surface theory of equation (2). Khajezadeh et al (1969) also observed a similar dependence of 1/f noise on emitter geometry in silicon bipolar transistors. Figure 2 also shows that the noise "bump", in the frequency range of 10 KHz to 1 MHz, becomes more pronounced and increases in magnitude as the P_e/A_e increases.

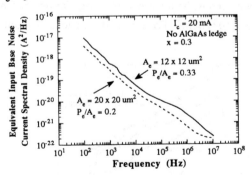

Fig. 2   Comparison of the equivalent input base noise current spectral densities of $Al_{0.3}Ga_{0.7}As/GaAs$ HBTs, without the AlGaAs ledge for two different values of P_e/A_e at a fixed I_c = 20 mA.

The constant collector current condition limits the range of emitter areas, which can be examined, because of the likelihood of appreciable self-heating of the junction as the emitter area is reduced. Because noise-producing processes can be thermally activated (as in eqn 4), variations in the junction temperature can affect the noise measurements. A more useful measurement condition for comparing the effect of device geometry is constant collector current *density*. Figure 3 compares the equivalent input base noise current spectral densities of four $Al_{0.3}Ga_{0.7}As/GaAs$ devices with emitter perimeters ranging from 28 to 80 um (emitter areas ranging from 40 to 400 um^2) without AlGaAs ledges at a fixed the collector current density of 5×10^3 A/cm^2. These curves show that at a constant collector current density, larger-perimeter (larger-area) devices have inferior low-frequency noise performance. These results can be compared to Fonger's theory by substituting $I_c = J_cA_e$ into equation (2). From the data in figure 3 at a fixed collector current density , $S_{I_b}(100Hz)$ was found to vary as $P_e^{2.46}$, which is in reasonable agreement with the predicted quadratic dependence.

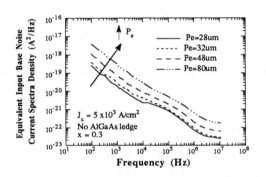

Fig. 3   Comparison of the equivalent input base noise current spectral densities of $Al_{0.3}Ga_{0.7}As/GaAs$ devices without AlGaAs ledges for four emitter perimeters (areas) at a fixed $J_c = 5 \times 10^3$ A/cm^2.

C. *Effect of AlGaAs Ledge*

Figure 4 is a comparison of the equivalent input base noise current spectral densities of $Al_{0.3}Ga_{0.7}As/GaAs$ HBTs with and without the AlGaAs ledge for two different emitter areas at a fixed collector density of 5×10^3 A/cm^2. For emitter dimensions of 4 um x 10 um,

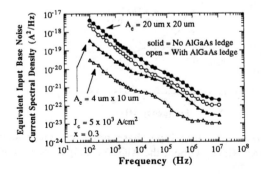

Fig. 4   Comparison of the equivalent input base current spectral densities of $Al_{0.3}Ga_{0.7}As/GaAs$ HBTs with and without the AlGaAs passivation ledge for two emitter sizes at a fixed $J_c = 5x10^3$ A/cm^2.

S_{I_b}(100Hz) of the device with the AlGaAs passivation ledge is approximately a factor of ten (10 dB) lower than that of a similar size HBT without the ledge. For emitter dimensions of 20 um x 20 um, the 1/f noise difference between the devices with and without the ledge is about a factor of 2. These results indicate that the AlGaAs surface passsivation ledge plays an increasingly important role in improving the 1/f noise behavior as emitter dimensions are scaled. It should be pointed out that the magnitude and the shape of the noise "bump" changes with the incorporation of the AlGaAs ledge.

D. Effect of Aluminum Mole Fraction in the Emitter

Figure 5 compares the equivalent input base noise current spectral densities of 4 um x 10 um emitter $Al_{0.3}Ga_{0.7}As$/GaAs HBTs with those of similar size $Al_{0.2}Ga_{0.8}As$/GaAs HBTs at , a collector current density of 5 x 10^3 A/cm^2. Both devices with and without the AlGaAs ledge were compared. While the x = 0.2 and x = 0.3 devices exhibited comparable 1/f noise and white noise levels, the noise "bump" was virtually eliminated for the x = 0.2 devices. For both devices with and without the AlGaAs ledge, S_{I_b} (50 KHz) of the x=0.2 devices was roughly three times lower than that of the x=0.3 devices. These results are very strong evidence that the noise "bump" is associated with traps in the AlGaAs emitter.

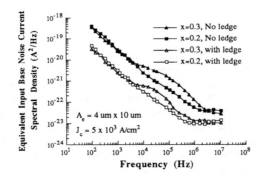

Fig. 5 Comparison of the equivalent input base noise current spectral densities of 4 um x 10 um emitter $Al_{0.3}Ga_{0.7}As$/GaAs HBTs with and without the AlGaAs ledge for two different Al mole fractions (x) at a fixed J_c = 5 x 10^3 A/cm^2.

E. Temperature Dependence

The temperature dependence of the equivalent input base noise current spectral density of a typical 4 um x 10 um $Al_{0.3}Ga_{0.7}As$/GaAs HBT without the AlGaAs ledge was measured at a fixed collector current of 2 mA. Both the 1/f noise and the shot noise (as expected) are nearly invariant with temperature. In contrast, the noise "bump" , which has the form of a slightly rounded Lorenztian spectrum, has a large temperature dependence. For increasing temperature, the magnitude of the low-frequency plateau decreased and the characteristic -3dB frequency (inversely related to the time constant) shifted toward higher frequencies, as predicted by equation (4). The burst-noise "bump" was fitted to a Lorenztian spectrum at several different temperatures. An activation energy of 0.20 eV was estimated from an Arrhenius plot of this time constant.

IV. DISCUSSION

The results in the previous sections clearly demonstrate that the 1/f noise of

AlGaAs/GaAs HBTs primarily originates at the extrinsic base surface. The observed white noise is consistent with the shot noise of the dc base current.

The dependence of the burst noise on collector current (and therefore V_{be}) and Al composition clearly indicate that the noise "bump" is associated with an AlGaAs trap in the emitter-side of the depletion region. The magnitude of the trapping time constants agree with the capture and emission times of the Si DX center. The DX center is then one possible candidate for the noise-"bump"-producing trap. Mooney et al [1990] have shown that the relative occupancy of DX centers decreases by a factor of 2.5 as the Al mole fraction is decreased from x = 0.3 to 0.2. The observed magnitude reduction of the noise "bump" at 50 kHz by a factor of about three as x is decreased from 0.3 to 0.2, is thus consistent with this reduction in the active DX center concentration. Our activation energy of 0.20 eV determined from the noise measurements can be compared to published data on Si-doped $Al_xGa_{1-x}As$. For Al mole fractions in the range of $0.2 < x < 0.4$, capture activation energies of the DX center ranged from 0.2 to 0.4 eV and the emission activation energy of the DX center is about 0.43 eV (Mooney et al 1990).

The observations that relate the noise "bump" to the emitter periphery also provide information about its origin. At a fixed collector current, the noise "bump" is more pronounced as the P_e/A_e increases. The AlGaAs ledge affects the magnitude and shape of the noise "bump". These observations suggest that fluctuations in the occupancy of AlGaAs traps near the surface of the emitter-base junction are the dominant source of the burst-noise "bump".

V. SUMMARY

The dominant source of 1/f noise in our AlGaAs/GaAs HBTs was shown to be defects at the extrinsic base surface. This finding is consistent with Fonger's model which attributes the 1/f noise to fluctuations in the surface recombination velocity. Our measurements suggest that the origin of the anomalous noise "bump" observed at intermediate frequencies is a trap, most likely the DX center, in the AlGaAs emitter near the surface of the emitter-base junction. The results of this study reveal simple modifications in the epitaxial structure and the fabrication process that reduce the low-frequency noise of AlGaAs/GaAs HBTs. We have demonstrated that the use of a protective AlGaAs ledge over the extrinsic base surface and the reduction of the Al mole fraction in the emitter from 0.3 to 0.2 significantly improves the low-frequency-noise performance of our AlGaAs/GaAs HBTs.

Acknowledgements

We wish to thank William Liu for material growth, and Dr. Don D'Avanzo and Noel Fernandez of Hewlett-Packard for assistance in performing the low-frequency noise measurements. This work was supported by JSEP under contract DAAL03-88-C-0011 and DARPA/ONR under contract number N00014-84-K-0077.

References

Fonger W., *Transistors I*, Princeton, N.J. : RCA Labs, p. 239, 1956.

Hayama N. , Tanaka S., and Honjo K.,*Third Asia-Pacific Microwave Conference Proceedings*,p.1039,1990.

Hsu S.T.,Whittier R.J., and Mead C.A., *Solid State Elect.*, vol. 13, p. 1055, 1970.

Jaeger R.C. and Broderson A.J., *IEEE Trans. Electron Devices*, ED-17, p. 128, 1970.

Jue S.C., Day D.J., Margittai A., and Svilans M., *IEEE Trans. Electron Devices*, ED-36, p. 1020, 1989.

Khajezadeh H. and McCaffery T.T., *Proc. IEEE*, vol.57, No.9, p.1518, 1969.

Lang D.V., Logan R.A., and Jaros M., *Phys. Rev. B*, vol. 19, No. 2, p. 1015, 1979.

Madihian M. and Takahishi H., *IEEE Trans. Microwave Theory Tech.*, MTT-39, p.133, 1991.

McWhorter A.L., *Semiconductor Surface Physics*, Philadelphia, p.169, 1957.

Mooney P.M., *J. Appl. Phys.*, vol. 67, No. 3, p. R1, Feb. 1990.

Motchenbacher C.D. and Fitchen F.C., *Low-Noise Electronic Design*, p.66, 1973.

Sah C.T. and Heilscher F.M.,*Phys. Rev. Lett*, vol. 17, p.956, 1966.

Tutt M.H., D. Pavlidis D., and Bayraktaroglu B., *Int. Symp. GaAs and Rel. Compounds*, p. 701, 1988.

van der Ziel A., Zhang X., and Pawlikiewicz A.H., *IEEE Trans. Electron Devices*, ED-33,p. 1371, 1986.

Inst. Phys. Conf. Ser. No 120: Chapter 6
Paper presented at Int. Symp. GaAs and Related Compounds, Seattle, 1991

329

Submicron AlGaAs/GaAs heterojunction bipolar transistor process with high current gain

Won-Seong Lee, Takatomo Enoki, Shoji Yamahata,
Yutaka Matsuoka and Tadao Ishibashi

NTT LSI Laboratories, 3-1, Morinosato Wakamiya,
Atsugi-shi, Kanagawa Pref., 243-01 Japan

ABSTRACT: *A new self-aligned process is developed to obtain sub-micron high performance AlGaAs/GaAs HBTs which can maintain a high current gain for emitter sizes on the order of $1 \mu m^2$. A DC current gain larger than 30 can be typically obtained for HBTs with $0.5x2$ μm^2 emitter-base junction area at sub-mA collector currents. The high current gain at low collector current level and a small device area comparable to silicon bipolar transistors make these devices suitable for LSI applications.*

1. INTRODUCTION

Reducing the operating current of AlGaAs/GaAs HBTs to a sub-mA range is necessary to incorporate a large number of devices without exceeding the limited heat dissipation capability of a package. The reduction of the operating current of bipolar transistors is mostly done by shrinking device dimensions while maintaining optimum operating current density. A high current gain at low collector current level and a small device area comparable to silicon bipolar transistors is necessary to make HBTs suitable for LSI applications.

With current mesa-type self-aligned AlGaAs/GaAs HBTs, however, it is difficult to reduce device size because of the high surface recombination velocity of GaAs and the lack of reliable techniques to remove base metal deposits from the emitter sidewall as well as to make contact to the submicron emitter region. Many self-aligned HBT processes are reported for use in digital circuits. These processes, however, were not suitable for aggressively scaling device dimensions to submicron level for various reasons. Undercut in the emitter is used to separate emitter and base metal while the emitter requires most critical dimension control in submicron region (Chang et al. 1987, Malik et al. 1990); The emitter-base junction periphery is not completely isolated from high recombination velocity regions such as GaAs surface or ion-implanted regions (Hayama et al. 1990). This paper describes a new self-aligned AlGaAs/GaAs HBT process to obtain submicron high-performance AlGaAs/GaAs HBTs, especially at low operating current level.

2. HBT STRUCTURE AND PROCESS

We have developed a new self-
aligned AlGaAs/GaAs HBT process
that can maintain a current gain
larger than 30 for emitter sizes on
the order of 1 μm². While the final
structure is similar to the one
reported earlier (Nagata et al. 1985),
this process incorporates the new
features described below.

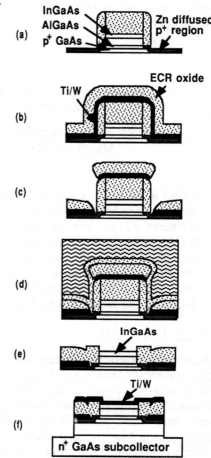

1) Incorporation of an AlGaAs
surface passivation structure around
the entire emitter-base junction
periphery to reduce surface
recombination.

2) Reliable removal of base metal
(Ti/W) deposits from the sidewall.
Electron cyclotron resonance(ECR)
plasma deposited oxide is used to
expose the sidewall region and the
exposed base metal is etched
chemically by ECR plasma.

3) Selective etching of GaAs from
AlGaAs by ECR plasma to obtain a
vertical sidewall and low damage
horizontal surface. This selective
etching step provides a well-defined
thickness for the AlGaAs passivation
layer and maintain a process
uniformity against a thickness
variation.

Fig.1 Process flow: (a) dummy
emitter and extrinsic base formation,
(b) base metal deposition, (c) sidewall
deposit removal, (d) planarization,
(e) etchback, (e) emitter contact
region definition

4) Planarization and etch-back of a
dummy emitter to incorporate high
temperature processes such as
extrinsic base diffusion.

5) Extension of the emitter metal
over the base and self-alignment of
emitter contact area and active device region.

6) Refractory metal (Ti/W) contact for the emitter and extrinsic base.

7) Zinc diffusion for the extrinsic base to reduce base contact resistance and
increase etching tolerance during the base surface exposure.

The HBT fabrication sequence is schematically shown in Fig.1. Device
fabrication starts from SiN_x/SiO_2 deposition and C_2F_6 etching for dummy

emitter formation. The InGaAs and part of the underlying GaAs layer are removed by ECR plasma containing Ar and Cl_2 gases. The gas composition and etching condition is controlled to obtain similar etching rates between InGaAs and GaAs. The remaining GaAs is selectively removed from the AlGaAs layer by ECR plasma etching containing Ar, Cl_2 and NF_3 gases. At a given gas composition and pressure, RF bias applied to the substrate is varied to obtain the desired selectivity and mirror like surface (Suzuki et al. 1985). The maximum potential applied between the plasma and the substrate is less than 100V to minimize the damage in the AlGaAs and underlying p^+ GaAs layer (Knoedler et al. 1988). Then, SiN_x/SiO_2 layers are deposited and etched to form 0.2 μm sidewall and the AlGaAs layer is etched by Ar + Cl_2 ECR plasma. At this step, the AlGaAs surface passivation region is formed. Zn diffusion for extrinsic base is done at 550 °C for 1 min using $ZnAs_2$ as a source (Nakajima et al. 1989). The side diffusion connects the intrinsic and extrinsic base region in case the thin p^+GaAs base layer is over-etched off during the previous ECR etching step (Fig.1, a).

A 200-Å-thick Ti layer is evaporated and a 1000-Å-thick W layer is sputtered on the p^+ GaAs to form a base metal. Subsequently, an oxide layer is deposited by ECR-CVD using O_2 plasma generated from an ECR plasma source and SiH_4 fed into the chamber maintained at 0.5 mTorr during the deposition. An interesting feature of the ECR-CVD oxide is that the oxide deposited on the sidewall region can be removed easily with diluted HF (1% in NH_4F solution) while the one on the horizontal surface shows a negligible etch rate (Ehara et al. 1984) (Fig.1, b).

The exposed base metals on the sidewall region are removed by NF_3 plasma generated by ECR. Maintaining a negligible undercut in the metals (Ti/W) on the horizontal surface is essential to fabricate submicron size devices and to reduce extrinsic base resistances. Fig.2 shows an SEM photograph of a cross section after the sidewall removal step (Fig.1, c).

The small gap formed on p^+ GaAs by the sidewall metal removal step is filled with PECVD oxide. Then the mesa structure is

Fig.2 Cross section of a mesa structure with the Ti/W sidewall deposits removed (after Fig.1 (c) step)

planarized with a photoresist (Fig.1, d). The planarized structure is etched back by O_2, C_2F_6 and SF_6 reactive ion etching (RIE). The incorporation of a dummy pattern and the extensive use of selective etching during the etch back process make it possible to expose InGaAs contact regions reproducibly regardless of the emitter area (Nagy et al. 1991). For instance, 0.5x2 μm² and 100x100 μm² emitter HBTs can be fabricated at the same time (Fig.1, e).

The exposed InGaAs contact region is cleaned before emitter metal deposition to maintain low emitter contact resistances. Ti and W are subsequently deposited on the whole wafer. The emitter and base contact region are defined by photolithography and dry etching (Fig.1, f). A pair of $0.5 \times 2 \ \mu m^2$ emitter size HBTs after the step of Fig.1, (f) is shown in Fig. 3. The white region is the emitter contact area ($2.5 \times 4 \ \mu m^2$) which has the same size as the active base-

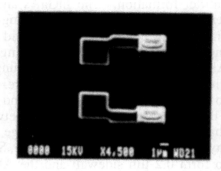

Fig.3   An SEM photograph of 0.5×2 μm^2 emitter HBTs

collector junction. In order to reduce device parasitic, the rest of the regions, including the region underneath the base contact lead, is made semi-insulating by proton implantation except for the active area.

Conventional techniques are used for the subsequent processes. Collector metal of AuGe/Ni/Ti/Au is evaporated and lifted-off and the ohmic contact is formed by annealing at 380 °C for 30 sec. SiO_2 passivation, isolation implantation and Ti/Au interconnect metalization are performed to complete the fabrication.

3. DEVICE PERFORMANCE

Results were obtained using an MBE-grown HBT structure with an 800-Å-thick, 8×10^{18} cm^{-3} Be doped, graded-base and 3000-Å-thick undoped collector layer. Fig. 4 shows the I-V characteristics of an HBT with an emitter size of $0.5 \times 2 \ \mu m^2$. The emitter-size effect of HBTs on the same wafer is characterized at $J_c = 5 \times 10^4$ A/cm^2, which is shown in Fig.5. Although the HBTs still show the emitter-size effect, the effect is reduced to a practically negligible level by

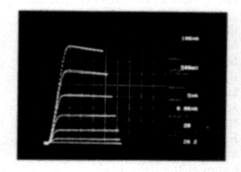

Fig.4   I-V characteristics of the 0.5×2 μm^2 emitter size HBT shown in Fig.3

incorporating a 0.2 μm long AlGaAs surface passivation structure around the emitter periphery (Hayama et al. 1990). With a better quality epitaxial layers, current gain of larger than 70 could be obtained from a $0.5 \times 2 \ \mu m^2$ emitter size HBT at $I_c = 0.3$ mA. The periphery components of the base current calculated from the slope of the graph in Fig.5 is 1.3×10^{-6} A/μm at $J_c = 5 \times 10^4$ A/μm, which is approximately 1/3 of that measured from a control sample without the AlGaAs surface passivation structure. Further reduction of this size effect will be possible by optimizing the length and thickness of the AlGaAs structure as well as the time and temperature of Zn diffusion.

The extrinsic $\Delta I_c/\Delta V_{BE}$ of the device shown in Fig. 4 is $2.5 \times 10^{-3}\ \Omega^{-1}$ at $I_c = 0.5$ mA calculated from gummel plots assuming r_b/h_{FE} contribution is negligible and confirmed by the saturation characteristic of HBTs. This value corresponds to the emitter contact resistivity of $3.5 \times 10^{-6}\ \Omega \cdot cm^2$. Considering that the typical contact resistivity using InGaAs non-alloyed contact is low $10^{-7}\ \Omega \cdot cm^2$ (Nittono et al. 1988), the emitter contact resistance is expected to be reduced.

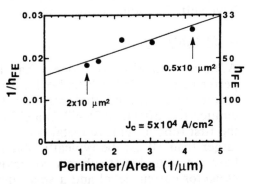

Fig.5 1/hFE versus emitter perimeter/area for HBTs with AlGaAs passivation

In Fig. 6, the current gain cutoff frequency (f_T) versus collector current density is plotted for small size HBTs. The maximum unilateral power gain cutoff frequency (f_{max}) of these devices are approximately 45 GHz and relatively independent of the sizes. For $0.5 \times 2\ \mu m^2$ HBT, f_T of 19 GHz and f_{max} of 28 GHz is obtained at $I_c = 100\ \mu A$. The decrease in f_T with emitter size is mostly due to the increase in emitter resistance with the reduction in emitter size. The

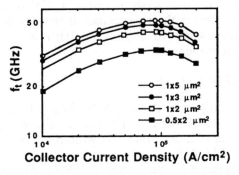

Fig.6 f_T of small geometry HBTs measured at various current densities

effect of emitter resistance on f_T can be found by plotting $1/(2\pi f_T)$ vs g_e^{-1} shown as

$$\frac{1}{2\pi f_T} = (C_e + C_{bc} + C_p) \cdot g_e^{-1} + \tau_b + \tau_c'$$

where C_e is the emitter capacitance, C_{bc} is the base-collector junction capacitance, C_p is the parasitic capacitance, g_e is the input dc conductance, τ_b is the base transit time and τ_c' is the collector signal delay time (Tiwari 1989).

For small size HBTs, the capacitance charging time due to large emitter resistance becomes a major proportion of the electron propagation delay times in $1/(2\pi f_T)$ since the increase rate of emitter resistance is higher than the decrease rate of base-collector and parasitic capacitances as the emitter size is scaled down while keeping a minimum alignment tolerance. The decrease in parasitic capacitances, however, will have large impact in the decrease of the propagation delay time of digital circuits, because the circuit propagation delay components related with emitter resistance is small once it is reduced to have sufficient logic swing at the given bias current level (Tang et al. 1979).

4. SUMMARY

A novel self-aligned process has been developed for sub-micron high performance AlGaAs/GaAs HBTs. An AlGaAs surface passivation structure is incorporated around the entire emitter-base junction periphery to reduce emitter size effect. ECR-CVD and ECR plasma etching technique is used for the first time to remove metal deposits from the sidewall. A DC current gain larger than 30 can be typically obtained for HBTs with 0.5x2 μm^2 emitter-base junction area at sub-mA collector currents. The high current gain at low collector current level and a small device area comparable to silicon bipolar transistors make these devices promising for ultrahigh-speed LSI applications.

ACKNOWLEDGEMENT

The authors wish to thank O. Nakajima, K. Nagata, H. Ito and T. Nittono for providing their expertise to remove all the wrinkles from the process. They also wish to thank Y. Ishii and K. Hirata for their suggestions and encouragement.

REFERENCES

Chang M F, Asbeck P M, Wang K C, Sullivan G J, Sheng N H, Higgins J A and Miller D L 1987 *IEEE Elect. Dev. Lett.* **EDL-8** 303
Ehara K, Morimoto T, Muramoto S and Matsuo S 1984 *J. Electrochem. Soc.* **131** 419
Hayama N and Honjo K 1990 *IEEE Electron Device Lett.* **EDL-11** 388
Knoedler C M, Osterling L and Shtrikman H 1988 *J. Vac. Sci. Technol.* **B6** 1573
Malik R J, Lunardi L M, Ryan R W, Shunk S C and Feuer M D 1989 *Electron. Lett.* **25** 1175
Nagata K, Nakajima O, Yamauchi Y and Ishibashi T 1985 *Inst. Phys. Conf. Ser.* **79** 589
Nagy A and Helbert J 1991 *Solid State Tech.* **34** 53
Nakajima O, Ito H and Ishibashi T 1989 *Inst. Phys. Conf. Ser.* **106** 563
Nittono T, Ito H, Nakajima O and Ishibashi T 1988 *Jpn. J. Appl. Phys.* **27** 1718
Suzuki K, Ninomiya K, Nishimatsu S and Okudaira S 1985 *J. Vac. Sci. Technol.* **B3** 1025
Tang D D and Solomon P M 1979 *IEEE J. Solid-State Circuits* **SC-12** 679
Tiwari S 1989 *IEEE Electron Dev. Lett.* **EDL-10** 574

Inst. Phys. Conf. Ser. No 120: Chapter 7
Paper presented at Int. Symp. GaAs and Related Compounds, Seattle, 1991

335

Photoluminescence analysis of C-doped npn AlGaAs/GaAs heterojunction bipolar transistors

Z.H. Lu[1,2], M.C. Hanna[1], E.G. Oh[1,2], A. Majerfeld[1,2], P.D. Wright[3],and L.W. Yang[4]

[1]Department of Electrical and Computer Engineering and [2]Center for Optoelectronic Computing Systems, University of Colorado, Boulder, Colorado, 80309-0425
[3]Martin Kestrel Company, Colorado Springs, Colorado 80921
[4]Ford Microelectronics, Inc., Colorado Springs, Colorado 80908

ABSTRACT: We report the observation of distinct photoluminescence (PL) emission peaks at 10 K from the collector, base, emitter, and the emitter-base interface of C-doped *npn* heterojunction bipolar transistors (HBT). A calibration of the bandgap reduction, measured from PL versus hole density, was used for a novel determination of the base hole density in the range of 10^{19}-10^{20} cm^{-3}. In addition, we show physically significant correlations between the other PL emission peaks and key electrical properties of the HBT. We also report the application of capacitance-voltage and deep-level transient spectroscopy for quantitative analysis of the AlGaAs emitter.

1. INTRODUCTION

The emitter-base (E-B) junction is known to be a critical region regarding the performance of the heterojunction bipolar transistor (HBT), as the exact configuration of the heterojunction and the *p-n* junction affect greatly the injection efficiency and, thus, the current gain (Tischler *et al* 1989 and Hafizi *et al* 1990). It is generally found (Asbeck *et al* 1989) that the reproducibility of the current gain of the HBT is difficult to control during the growth process by either molecular beam epitaxy (MBE) or organometallic vapor phase epitaxy (OMVPE). Because of the nature of the HBT structure, the critical carrier densities cannot be directly measured, but are generally derived from transfer calibration studies. Therefore, characterization techniques that require only minor wafer processing and that are essentially non-destructive prior to device or circuit processing are highly desirable. Extensive studies were performed using room temperature photoreflectance, from which very valuable information about HBT structures was obtained (Bottka *et al* 1990 and Yin *et al* 1990). We present a study of the characterization of HBT structures by means of low temperature photoluminescence (PL), capacitance-voltage (C-V) and deep-level transient spectroscopy (DLTS). We show that low temperature PL can be used to determine the hole density of the base, the aluminum percentage of the AlGaAs emitter, the doping level of the sub-collector, and that it also provides a measure of carrier recombination in the E-B heterojunction region.

2. EXPERIMENT

The samples used to establish the relationship between the hole density and the energy of the PL emission peak were grown using the OMVPE process (Hanna *et al* 1991). The hole densities were obtained from van der Pauw-Hall measurements at room temperature. PL spectra were taken at 10K with an excitation intensity of 0.02-4 W/cm^2 from an Ar^+ laser. The *npn* AlGaAs/GaAs HBT wafers were grown by low pressure OMVPE (Yang *et al* 1991). The GaAs base was doped with C to $1-10 \times 10^{19}$ cm^{-3}. The $Al_xGa_{1-x}As$ (nominally $x=0.3$) emitter was doped with Se or Si. The n^+-GaAs cap layer was removed by wet chemical etching for PL, C-V and DLTS measurements. For C-V and DLTS measurements, a Au Schottky contact was deposited on the AlGaAs emitter. The DC and the high frequency characteristics of the devices were discussed elsewhere (Yang *et al* 1991).

3. RESULTS

For heavily doped *p*-type GaAs it is well known that the PL spectrum shifts substantially to lower energy due to bandgap renormalization [BGR] at high hole densities (Olego and Cardona 1980). To evaluate the hole density from a PL spectrum, we used the peak energy E_P, shown in Figure 1, as a calibration quantity. Although E_P is not the bandgap energy in heavily doped *p*-type GaAs, it represents the effect of BGR and it can be correlated with the hole density. Therefore, the hole density in heavily doped *p*-type GaAs can be determined from E_P without comprehensive analysis.

Figure 2 shows the PL spectrum in the range of 1.2-2 eV for a HBT structure with a C-doped GaAs base with nominal hole density of 5×10^{19} cm^{-3}. The dashed lines are the line shape fittings for the purpose of extracting peak positions. The peak at 1.458 eV originates from band-to-band recombination in the heavily C-doped base. From the PL calibration curve, a hole density of 5×10^{19} cm^{-3} is determined with an error bar of $\pm 15\%$, in agreement with the specified value for this wafer. The two sharp peaks at 1.512 and 1.49 eV are, respectively, the bound exciton and the free-electron to acceptor transitions from the collector region. The shoulder peak at ~1.55 eV originates from band-to-band transitions from the n^+-GaAs sub-collector. The high energy side of this

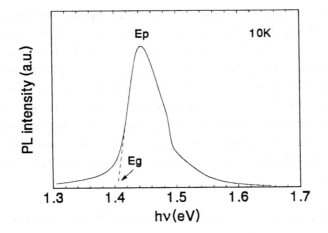

Figure. 1. PL spectrum of heavily C-doped GaAs HP226 at 10 K

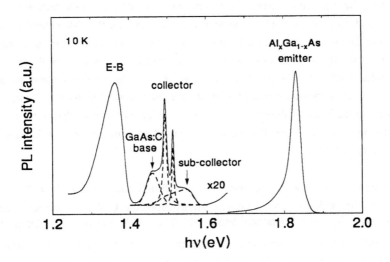

Figure 2. PL spectrum of HBT wafer C2 at 10 K. The dashed lines are line shape fittings to extract peak positions.

emission provides a measure of the degenerate Fermi energy, from which we estimate that the electron density in the sub-collector region is ~2-4x10^{18} cm^{-3} (Szmyd *et al* 1990), which is consistent with the design value of 3x10^{18} cm^{-3} for this wafer. From the temperature dependence of the peak at 1.84 eV, this peak is assigned to band-to-band transitions from the AlGaAs emitter, from which the aluminum percentage x is calculated to be 0.26. We found that the value of x may vary substantially between growth runs and needs to be measured to understand device performance. These results demonstrate that low temperature PL can be used to determine the hole density of the base, the aluminum percentage of the AlGaAs emitter, and to estimate the electron densities of the heavily-doped subcollector.

The PL spectrum also reveals the quality of the E-B heterojunction. The peak between 1.3-1.4 eV appears for all the wafers evaluated. From careful step etching experiments we determined that this peak is related to the E-B heterojunction and not to emission from the neutral region of the AlGaAs emitter. From its excitation dependence, this transition is likely due to the recombination of electrons in the emitter grading region with holes from the base, as suggested by the simulation of Hafizi *et al* (1990). Under forward bias, the injected electrons can recombine with holes from the base through this process --- and possibly assisted by the deep-level donors near the E-B heterojunction --- deteriorating the current gain. Indeed, we have observed a qualitative correlation between the relative intensity of this peak and the DC current gain, the ideality factor and the turn-on voltage of the E-B junction. Because this carrier recombination process is critically dependent on the configuration of the *pn* junction relative to the heterojunction, it causes reproducibility problems from wafer to wafer. We expect that with the aid of this technique, new structure designs can be tested to improve the performance and reproducibility of HBTs.

In addition to low temperature PL, we found that C-V and DLTS provide quantitative information on the AlGaAs emitter. From the C-V measurement, the total doping density N_D can be obtained. From DLTS and C-V analyses, we determine the emission and capture energies, and the thermal ionization energy, from which the chemical identity of the donor is obtained. Furthermore, from an analysis of the charge neutrality equation, we obtain the actual electron density n in the emitter, which is an essential quantity for proper electron injection. In this regard, it is worth pointing out that the actual electron density, because of the large ionization energy of donors in AlGaAs, is about one order of magnitude lower than the doping density for x=0.3 as shown in Table I. These results demonstrate that C-V and DLTS analyses give valuable information about the emitter layer of an HBT wafer.

Table I. C-V and DLTS analyses of the AlGaAs emitter of HBTs

Wafer	x	N_D (cm^{-3})	N_{Se} (cm^{-3})	N_{Si} (cm^{-3})	$N_{0.6}$ (cm^{-3})	$N_{0.8}$ (cm^{-3})	$n_E{}^*$ (cm^{-3})
B1	0.26	1.3×10^{17}	9.0×10^{16}	4.0×10^{16}	-	-	3.8×10^{16}
B2	0.29	9.8×10^{16}	1.2×10^{16}	5.3×10^{16}	4.3×10^{16}	3.8×10^{16}	9.1×10^{15}
C1	0.23	2.1×10^{17}	-	2.1×10^{17}	-	-	$\sim 2 \times 10^{17}$

* n_E is the total electron concentration in Γ, L, and X valleys.

4. CONCLUSIONS

We have demonstrated the application of low temperature PL, C-V and DLTS techniques to characterize HBT structures. The electron density in the emitter can be determined from C-V and DLTS analyses. From low temperature PL, the hole density in the base of an *npn* HBT is obtained. Our initial study indicates that the PL emission from the E-B heterojunction region can be used to estimate the DC performance of HBT wafers prior to processing and may help to design new structures for improved performance and reproducibility.

5. ACKNOWLEDGMENTS

The authors (Lu, Hanna, Oh and Majerfeld) wish to acknowledge the support by the Optoelectronic Computing Systems Center (NSF/ECR grant No. CDR 8622236), the Colorado Advanced Technology Institute through the Advanced Materials Institute, and by Ford Microelectronics Inc.

REFERENCES

Asbeck P M, Chang M-C F, Higgins J A, Sheng N H, Sullivan G J, and Wang K-C, 1989 *IEEE Trans. Electron Device ED*-36 2032

Bottka N, Gaskill D K, Wright P D, Kaliski R W, and Williams D A, 1991 *J. Cryst. Growth* 107 893

Hafizi M E, Crowell C R, Pawlowicz L M, and Kim M E, 1990 *IEEE Trans. Electron Device ED*-37 1779

Hanna M C, Lu Z H, and Majerfeld A, 1991 *Appl. Phys. Lett.* 58 164

Olego D and Cardona M, 1980 *Phys. Rev. B*22 886

Szmyd D M, Porro P, Majerfeld A, and Lagomarsino S, 1990 *J. Appl. Phys.* **68** 2367

Tischler M A, Baratte H, Kuech T F, and Wang P J, 1989 *J. Appl. Phys.* **65** 4928

Yang L W, Wright P D, Brusenback P R, Ko S K, Kaleta A, Lu Z H, and Majerfeld A, 1991 *Electronics Lett.* **27** 1145

Yin X, Pollak F H, Pawlowicz L, O'Neill T, and Hafizi M E, 1990 *Appl. Phys. Lett.* **56** 1278

Inst. Phys. Conf. Ser. No 120: Chapter 7
Paper presented at Int. Symp. GaAs and Related Compounds, Seattle, 1991

341

Photoluminescence investigation of AlGaAs/GaAs heterojunction bipolar transistor layers

H. Tews, R. Neumann, P. Zwicknagl, and U. Schaper
Siemens Research Laboratories
8000 Munich 83, Germany

Abstract:
  Photoluminescence spectroscopy (PL) has been applied as a fast and destruction free method to investigate highly complex AlGaAs/GaAs heterojunction bipolar transistors (HBT) layer sequences. Entire 2" and 3" HBT wafers were characterized at 77K. A large variety of parameters could be obtained, e.g. the emitter material composition and the collector doping. PL gives detailed information on the position of the emitter-base and the base-collector interfaces. Comparisons have been made between electrical measurements and PL.

1. Introduction

Heterojunction bipolar transistors for high-speed devices represent a challenge for both epitaxy and technology. HBT processing requires a large number of highly sophisticated technology steps. A sensitive, destruction free, and quantitative characterization of the total HBT epilayer sequence is very desirable before processing. It has to assure that only high quality wafers enter the time consuming and expensive HBT processing. The epitaxial growth of these HBT layer sequences is complicated both for Metalorganic Vapour Phase Epitaxy (MOVPE) and for Molecular Beam Epitaxy (MBE). Complex n- and p-type doping profiles and material compositions have to be realized. The most difficult task is to grow a very thin ultra-highly doped base with abrupt p^+-n^--junctions on both sides. The choice of the acceptor and the epitaxial process have to guarantee that the base doping profile does not change significantly during technology processing of the HBT-wafer, and that it does not degrade during device operation.

We have applied photoluminescence spectroscopy as a fast and destruction free method to investigate HBT layer sequences. We will demonstrate that a large variety of valuable parameters on the layer properties is obtained.

2. Experimental

The layer sequences studied in this work consist of a 0.5µm thick n^+-GaAs subcollector layer doped to $5x10^{18}cm^{-3}$ followed by 0.5µm GaAs collector doped to $2x10^{16}cm^{-3}$, a thin highly doped 100nm GaAs base or alternatively, a 100nm AlGaAs graded base with a linear grading of the group-III compounds Al and Ga, a similar linear 10-20nm graded AlGaAs at $2x10^{17}cm^{-3}$ between emitter and base, a 100nm AlGaAs emitter with 28% Al-content in the layer and a doping of $2x10^{17}cm^{-3}$, and finally a 200nm GaAs cap layer doped to $5x10^{18}cm^{-3}$. The layers were grown by MOVPE at reduced pressure. Details of the growth are described in Ref. 1. Donors were Si from a Si_2H_6 source. Acceptors were either C from the reaction between trimethylarsenic (TMAs) and trimethylgallium [2], or Mg from bis-cyclopentadienyl- Mg. The hole concentrations were $1x10^{19}cm^{-3}$ for Mg and $5x10^{19}cm^{-3}$ for C. Considering the severe Mg-memory effect in the reactor cell [1] and the extremely low diffusion coefficient for

C [3], carbon would be the preferred dopant for HBT base layers. However the growth of C-doped layers from TMAs requires relatively low temperatures [2]. Therefore the Al containing graded base layers were produced using the Mg dopant. In addition, similar HBT layer sequences grown by molecular beam epitaxy (MBE) with Be acceptors were investigated.

Prior to processing the entire 2-inch or 3-inch HBT wafers were analyzed by photoluminescence spectroscopy at 77K. Excitation was done with 514nm light from an Ar-ion laser. The emission was analyzed in a 1m grating spectrometer and collected in a GaAs photocathode.

3. Photoluminescence Results
3.1 Abrupt Base HBTs

The luminescence spectra of graded and abrupt base HBT layer sequences, respectively, are considerably different. Therefore, they will be discussed here separately. Figure 1 shows the SIMS profil of an MOVPE grown HBT layer sequence with a GaAs:C base. One can clearly observe a high base doping of approximately $7 \times 10^{19} \text{cm}^{-3}$ with sharp transitions on the emitter-base and the base-collector side. The detection limit in our SIMS is $8 \times 10^{17} \text{cm}^{-3}$ for C. The Al-profile is also shown to illustrate the correct position of the C-profile at the emitter-base interface.

When light is directed onto the HBT layer surface, it gets absorbed in the different layers. By assuming a constant absorption coefficient of $7 \times 10^{4} \text{cm}^{-1}$, approximately 75% of the light are absorbed in the 200nm GaAs cap layer, 13% in the AlGaAs-emitter, and some 6% in the base. The emitted light from the layer sequence is composed of emissions from the single layers masked by the reabsorption in the layers closer to the surface. An analysis of the luminescence light thus gives an interesting insight on the layer properties.

We have found that at room temperature only broad luminescence bands are observed. At 77K, however, sharp lines result, see Fig. 2a. Here the luminescence spectra of a complete HBT layer sequence is shown. It consists of three bands: "A" at 670nm, "B" at 790nm, and "C" at 850nm. In order to understand the nature of the luminescence bands, samples were prepared which had single layers successively removed. In Fig. 2b is shown the emission from a layer sequence where all layers down to the top of the subcollector had been removed. One observes two lines: a sharp line "D" at the 77K bandgap position, and a broad band "E" at higher energies. The sharp line is found to be emission from the GaAs substrate. The broad band "E" corresponds to the band-band emission from the n^{+}-GaAs subcollector layer. Compared to undoped GaAs, the recombination is shifted to higher energies as a consequence of the Burstein-Moss-effect. The magnitude of the shift allows a precise determination of the subcollector doping.

Figure 2c shows the luminescence from a

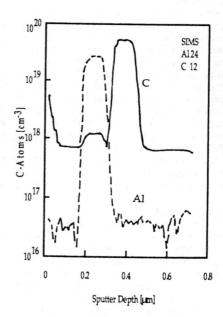

Fig. 1: SIMS profile of C and Al from cap, emitter and base.

sample, where in addition to the subcollector also the low n-doped GaAs collector layer is present. In contrast to Fig. 2b, we now observe only one intense line at the bandgap energy of undoped GaAs. This line "F" is due to recombination in the collector layer. The halfwidth of the line can be used to determine the doping of the collector [4]. In this case, the 12meV halfwidth corresponds to a collector doping level of $4 \times 10^{16} cm^{-3}$. The linewidth of undoped GaAs (i.e. the substrate) is below 6meV at 77K and can thus be distinguished from the collector luminescence. Due to its low luminescence intensity the subcollector luminescence band "E" is not visible in a linear scale. It can be seen, however, if the spectra are plotted in a logarithmic scale.

In Fig. 2d also the base layer is present. The laser light is now absorbed in the base layer to a great extent. Therefore the collector emission is considerably reduced. In Fig. 2d, two emission bands are visible: the collector line "F", plus a new broad band "C" at low energies. This broad band originates from the p^+-GaAs base layer. Highly p-doped GaAs layers are known to emit low-energy luminescence due to a shrinkage of the effective bandgap [5]. Recording the low-energy onset of this line or the peak maximum position, the hole concentration can be determined.

With the information of the luminescence spectra 2b-2d, the complete HBT layer sequence in Fig 2a can now be analyzed. The layers, which contribute to the overall spectrum (Fig. 2a) in addition to the layers in Fig. 2d are the GaAs n^+-cap and the AlGaAs emitter. Therefore line "A" at 670nm can be easily identified as the band-band recombination in the AlGaAs emitter corresponding to an Al-content in the layer of 27.5%. Here the formula of Casey and Panish [6] is used for the calculation of the AlGaAs-bandgap energy. Band "B" is the signal from the n^+-GaAs cap in analogy to the subcollector luminescence in Fig. 2b. Band "C" is the base signal. Due to the high absorption in cap, emitter and base, only little or no information is available on the collector layer.

It is important to note that the intensity

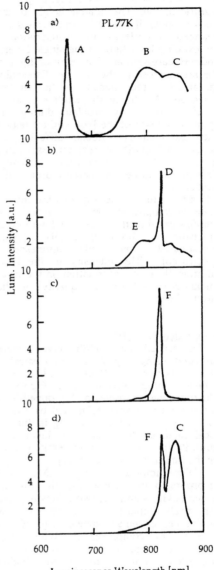

Fig. 2: Luminescence spectra of HBT layers at 77K
a) complete layer sequence, b) only subcollector, c) subcollector and collector, d) subcollector, collector and base

of the luminescence lines in Fig. 2a plays a significant role. The relative height of the lines changes if the layer thicknesses are changed. This effect is due to absorption of the exciting laser light in the single layers, and the reabsorption of the emitted light in the layers close to the surface. Especially important is the intensity of the emitter emission "A". Since the emitter layer is relatively thin, any problem during the epitaxial growth of the base-emitter region such as memory effects, diffusion of dopants or impurity segregation results in a reduction of the luminescence intensity. This point will be considered closer in the next section.

Summarizing we can say, that the emitter material composition, and the cap and base dopings can directly be recorded with great precision in abrupt base HBT layers investigated by photoluminescence spectroscopy. The intensity of the lines additionally gives information on the layer thicknesses and on the quality of the emitter layer.

3.2 Graded Base HBTs

The graded base HBT layers were prepared using Mg as acceptor species. Due to its memory effect it is difficult to achieve abrupt profiles [1]. Figure 3 illustrates two examples of Mg-doped base layers with a) a successful, and b) a non-successful epitaxial growth. Mg is thus an ideal candidate to study the influences of impurity incorporation in the emitter layer on the luminescence spectra [7].

When impurities like Mg are incorporated or diffuse into the low n-doped AlGaAs emitter, two things happen: for small impurity concentrations new luminescence centers appear (e.g. Mg in AlGaAs, or complexes involving Mg), further the emitter layer is partially converted from n to p for higher concentrations. Both effects can be observed. In Fig. 4 are shown the luminescence spectra of the emitter region of the same HBT-layers as in Fig.3. The sample with the abrupt Mg profile exhibits an intense emitter signal as in Fig. 2a, however now two deep centers with binding energies of 54meV and 130meV

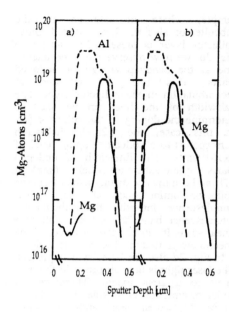

Fig. 3: SIMS spectra of two graded base HBT layer sequences showing the Mg and the Al profile a) successful epitaxy with little Mg diffusion, b) extended Mg diffusion into the emitter

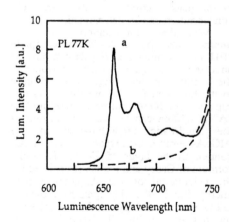

Fig. 4: PL spectra of the emitter luminescence for the two samples of Fig. 3. a) for the abrupt Mg-profile, b) for the broad Mg profile

are observed. We have found from comparisons with Be in MBE HBT layers that very similar emitter luminescence is emitted, therefore these centers are not uniquely related to Mg. The extended Mg-profile in Fig.3 results in a total suppression of the emitter luminescence. Therefore the 77K luminescence of the AlGaAs emitter records very well any base broadening into the emitter layer.

4. Electrical DC-Results

Diode characteristics were measured of the base-emitter heterojunction to get a correlation between the material characteristics determined by PL before the technology processing and the electrical performance of the complete HBT. For several wafers of different lots the base-emitter voltage V_{BE} is measured at a base current $I_B = 1\mu A$ (Fig. 5). V_{BE} is found to depend on diffusion, shifting from 1.1V for Mg-doped HBT bases exhibiting extensive diffusion into the emitter to 0.8V for C-doped HBT bases showing no diffusion of the dopant.

This V_{BE}-shift is caused by a shift of the pN-junction into the AlGaAs emitter layer in the case of broadened Mg-profiles. The advantage of the heterojunction is thereby lost. In the following simulation the effect of broadening of the acceptor profile on the emitter-base diode is investigated.

The base current density J_B is given by the ideal diode equation with V_T as thermal voltage

$$J_B = J_S *(e^{\frac{V_{BE}}{V_T}} - 1).$$

The saturation current density J_S depends on doping N, mobility μ and lifetime τ of the minority carriers, and on the intrinsic carrier concentration n_i of the base (index B) and the emitter (index E) [6]. q is the electrical charge.

$$J_S = q\sqrt{V_T}\left(\sqrt{\frac{\mu_E}{\tau_E}\frac{n_{iE}^2}{N_E}} + \sqrt{\frac{\mu_B}{\tau_B}\frac{n_{iB}^2}{N_B}}\right)$$

The intrinsic carrier concentration n_i is a strong function of the bandgap [6]. An extension of the base dopant into the emitter represents an increase of the base bandgap or, in other words, an increase of the Al-content in an $Al_xGa_{1-x}As$ base. The parameters used in the above formulas are given in Table I. They were used to calculate the ideal diode characteristics for an junction area of 1cm^2. The results are shown in Fig.5. The lifetimes used are τ_E = 5.5ns and τ_B = 20ns [8]. Comparing the influences of mobilities, lifetimes, doping,

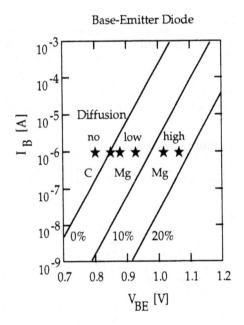

Base-Emitter Diode

Fig 5: Calculated base current I_B as function of the base-emitter voltage V_{BE} and the Al-content [%] of the base for an Al-content in the emitter of 0.3. Shown are also the measured V_{BE}-voltages (*) for 1 μA base current of different wafers.

x	0.0	0.1	0.2
Emitter			
10^{16}	356	318	283
10^{17}	284	254	226
Base			
10^{18}	3160	2977	2794
10^{19}	2089	1986	1847
n_i	$2.1 \cdot 10^6$	$2.4 \cdot 10^5$	$2.3 \cdot 10^4$

Table I: Mobility $\mu[cm^2/Vs]$ and intrinsic carrier concentration $n_i[cm^{-3}]$ as function of doping $N[cm^{-3}]$ in the emitter and base layer and of the Al content x in the base

and the intrinsic carrier concentration, it is interesting to note that n_i has by far the most important influence on the base current density.

The rise of the Al-content x of the base from 0% to 20% which represents the shift of the p-N-junction into the AlGaAs emitter causes a shift of 0.22V to higher V_{BE} voltages. This is consistent with the measured data.

5. Summary

We have applied photoluminescence spectroscopy as a fast and destruction free method to investigate HBT layer sequences. A large variety of valuable parameters can be obtained, e.g. the emitter material composition and the collector doping. PL gives information on the correct position of the emitter-base and the base-collector interfaces. Comparisons have been made between electrical measurements of the emitter-base diode and PL data. We have shown a good agreement between electrical measurements and simulations of the V_{BE}-voltage shift with the material data analysed by PL and SIMS.

6. Acknowledgments

We gratefully acknowledge help from R. Treichler (SIMS), Mrs. U. Yüksegil (PL) and the members of the technology group. The work has been supported by the German Bundesministerium für Forschung und Technologie (NT 2718).

References
1. Tews H, Neumann R, Humer-Hager T and Treichler R 1990
   J. Appl. Phys. **68** 1318
2. Kuech T F, Tischler M A, Wang P J, Scilla G, Potemski R and Cardone F 1988
   Appl. Phys. Lett. **53** 1317
3. Cunningham B T, Guido L J, Baker J E, Major J S, Holoynak N and Stillmann G E
   1989 *Appl. Phys. Lett.* **55** 687
4. Cusano D A 1964 *Solid State Commun.* **2** 353
5. Olego D and Cardona M 1980 *Phys. Rev.* **B22** 886
6. Casey H and Panish M 1978 *Heterostructure Lasers*, Academic Press
7. Humer-Hager T and Tews H 1990 *J. Appl. Phys.* **68** 1310
8. Ho S C M and Pulfrey D L 1989 *IEEE Electron Devices* **ED-36** 2173

Inst. Phys. Conf. Ser. No 120: Chapter 7
Paper presented at Int. Symp. GaAs and Related Compounds, Seattle, 1991

347

Photoluminescence study of GaAs antisite double acceptor in GaAs under hydrostatic pressure

P.W. Yu[*] and A. Kangarlu[+]

[*] University Research Center, Wright State University, Dayton, OH 45435

[+] Physics Department, University of Dayton, Dayton, OH 45469

ABSTRACT: The photoluminescence of Si_{As} and gallium antisite Ga_{As} related emissions of Si–doped p–type GaAs have been measured as a function of hydrostatic pressure using a diamond anvil cell. The Ga_{As} related emissions have energies of ~1.32 and 1.44 eV at 2 K and are due to the donor–acceptor pair transitions. The pressure coefficients of the binding energy of acceptors responsible for the two transitions are 0.2–0.3 meV/kbar in the direct gap $\Gamma_{1c}-\Gamma_{15v}$ region. At pressures larger than the $\Gamma_{1c}-X_{1c}$ crossover, the emission intensity of the 1.44 eV emission decreases drastically and the half width of the emission increases considerably due to the transfer of electron from the Γ_{1c} minimum to the X_{1c} minimum.

1. Introduction

The development of the diamond anvil cell in conjunction with the ruby luminescence calibrating the value of pressure has given new momentum to the study of electronic properties of III–V semiconductors under very high hydrostatic pressure (Barnett *et al* 1973).

Two energy levels at 77–78 and 203–204 meV above the valence bandedge are found in GaAs grown from Ga–rich melts. These levels are commonly attributed to the presence of a gallium antisite double acceptor (Ga_{As}) where the levels of 77–78 and 203–204 meV are assigned by the first (Ga_{As}^{0}/Ga_{As}^{-}) and second (Ga_{As}^{-}/Ga_{As}^{--}) levels of the Ga_{As}. These levels were identified by infrared absorption spectra (Elliot *et al* 1982, Fischer *et al* 1986). Photoluminescence (PL) spectra (Elliot *et al* 1982, Yu *et al* 1982) also show the presence of two emissions at ~1.32 and 1.44 eV at ~4 K. The 1.44 eV emission was identified as arising from the 77–78 meV level, which is due to first level of the Ga_{As}. Another PL emission at 1.32 eV is present together with the 1.44 eV emission in Ga–rich, shallow donor compensated materials. The 1.32 eV emission was interpreted (Elliot 1983) to be due to the second level of Ga_{As}. However, experimental evidence associating the 1.32 eV emission with the second level is not convincing. Nonetheless, we believe that the 1.32 eV emission is associated with Ga_{As} in a certain form. In this paper, we report the PL characteristics of the 1.32 and 1.44 eV emissions using hydrostatic pressure for a better understanding of the two emissions.

2.   Experimental

GaAs crystals grown under Ga—rich condition by the liquid encapsulated Czochralski method were used for the present study.   Two types of crystals, undoped and intentionally Si doped, were used.   Si doping was made to compensate acceptors in p—type crystals.   The room temperature hole concentration of undoped and Si doped crystals was $p = 3 \times 10^{17}$ and 1×10^{16} cm^{-3}, respectively.   PL measurements under atmospheric and elevated hydrostatic pressures were carried out at 2–300 K and 10 K, respectively. The variable temperature PL apparatus dispersed the spectrum through an f=1.29 m Spex spectrometer and a GaAs or an S–1 type photomultiplier tube while the 10 K PL apparatus used an f=0.6 m Spex spectrometer and a GaAs photomultiplier tube.   Elevated hydrostatic measurements were performed in a light Be—Cu diamond anvil cell.   Liquid argon was used as

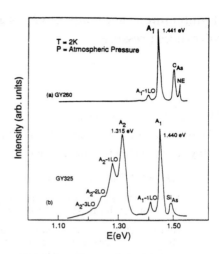

Fig. 1.   Atmospheric pressure PL spectra from two p—type crystals.

pressure transmitting medium and Ruby R_1 and R_2 lines were used for in situ pressure calibration.   The hydrostatic isotrophy and uniformity of the pressure inside the chamber was monitored and found to be within 0.5 kbar.   Samples were thinned to ~30 μm and 50 x 100 μm^2 samples were selected.   An Ar ion laser was used for excitation with a typical power of 100 W/cm^2 for elevated pressure measurements and with a lower power than 100 W/cm^2 for atmospheric measurements.

3.   Results and Discussion

Figure 1 shows PL features at 2 K under atmospheric pressure for two types of p—type crystals we have investigated.   Sample GY260 was not doped intentionally whereas sample GY325 was doped with Si to compensate available acceptor with an intention of driving the Fermi level up from the valence bandedge.   PL of sample GY260 consists of the near bandedge emission NE$_\Gamma$, the C$_{As}$—related emissions of the donor—acceptor $(D_\Gamma^0 - A^0)$ and the free conduction—band electron—to—hole at acceptor $(e-A^0)$ transitions, and A$_1$ emission at 1.441 eV.   PL of sample GY325 is characterized by A$_2$ emission at 1.315 eV and Si$_{As}$—related transitions in addition to A$_1$ emission at 1.440 eV.   Since sample GY325 is doped with Si it can be expected to observe the Si$_{As}$—related emissions rather than the C$_{As}$—related transitions usually present in undoped crystals. Excitation intensity dependence in the range of $I_{ex} = \sim 10^{-3} - 10^2$ W/cm^2 shows that the A$_1$ and A$_2$ emissions are mainly due to the (D^0-A^0) pair transition since the peak position moves to a higher energy with increasing excitation intensity. The peak position also shifts to a higher energy with increasing temperature in the temperature range from 2 to ~50 K. With a further increase of temperature the $(e-A^O)$ transition becomes dominant. Then, the transition energy decreases with increasing temperature at temperatures higher than ~ 50 K. This feature is also the same for the Si_{As}–related transition. The near-bandedge emission NE shows in GY260 whereas the PL intensity of the NE emission in GY325 is very weak compared to those of other emissions. The longitudinal optical (LO) phonon side bands of A_1 and A_2 emission are also shown. We have done hydrostatic measurements on GY260 type crystals. Here we describe our elevated hydrostatic measurements on GY325–type crystals which show A_1 and A_2 emissions.

Figure 2 and 3 show PL spectra under different hydrostatic pressures from atmospheric pressure to 34 kbar and 34 to 47 kbar, respectively. The Si_{As}–related, A_1 and A_2 $(D^O_\Gamma-A^O)$ emissions move to higher energy with increasing pressure following the pressure dependence of the $\Gamma_{1c}-\Gamma_{15v}$ gap. Pressure dependence of the three emissions was fitted with the following equation representing a nonlinear variation of transition energy $E(P)$,

$$E(P) = E(0) + \alpha P + \beta P^2 \qquad (1)$$

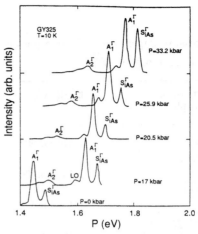

Fig. 2. PL spectra of sample GY325 at pressures from 0 to 33 kbar.

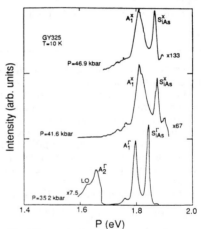

Fig. 3. PL spectra of crystal GY325 at pressures from 35 to 47 kbar.

where $E(0)$ is the transition energy at zero pressure. Table I shows the fitted values of $E(0)$, α and β in the pressure range smaller than that of the $\Gamma_{1c}- X_{1c}$ crossover (P_c). The crossover value is known to be 41.3 kbar (Wolford *et al* 1985). Table I also includes pressure coefficients of the near bandedge emission obtained from GY260. We consider that the pressure coefficient of the near bandedge emission is the same as that of the $\Gamma_{1c}-\Gamma_{15v}$ gap. The nearly identical value of the pressure coefficients of the $\Gamma_{1c}-\Gamma_{15v}$ gap, A_1 and A_2 transitions indicates that the acceptors responsible for A_1 and A_2 emissions are tied to the valence band maximum. The pressure coefficients of the acceptor binding energy responsible for A_1 and A_2 emissions is determined to be 0.2 and 0.3 meV/kbar respectively.

Table I

Pressure dependence of three photoluminescence emissions using the relation $E(P) = E(0) + \alpha P + \beta P^2$

Emission sample	E(0)(eV)	α(meV/kbar)	β(meV/kbar^2)
NE($\Gamma_{1c}-\Gamma_{15v}$) GY260	1.514±0.003	12.2±0.3	−0.04±0.009
Si$_{As}$ GY325	1.486±0.006	11.5±0.66	−0.046±0.017
A$_1$ GY325	1.440±0.005	12.0±0.5	−0.058±0.014
A$_2$(up−stroke) GY325	1.318±0.005	11.9±0.6	−0.044±0.015
A$_2$(down−stroke) GY325	1.321±0.003	10.7±0.4	−0.42±0.014

The small value is consistent with the pressure coefficients of other intermediate deep acceptors: namely, 0.3, 0.5, and 1.1 meV/kbar respectively, for Ag (Pistol *et al* 1988), Cu (Nilsson *et al* 1973) and Mn (Samuelson et al 1988) acceptors, which have the atmospheric pressure binding energies of 240, 150, 110 meV. However, these small values do not agree with calculations by Ren *et al* (1982) which would show the increased pressure coefficient of binding energy with the increasing binding energy. However, this small variation can be, within the effective mass approximation, due to the decrease (Samara 1967) of the static dielectric constant with increasing pressure while the hole mass remains (Neuman *et al* 1973) nearly constant.

At pressures larger than P_c the transition energy of the Si$_{As}$−related and A$_1$ emissions do not increase due to the $\Gamma_{1c}-X_{1c}$ valley mixing and eventual dominance of the $X_{1c}-\Gamma_{15v}$ gap dependence. The $X_{1c}-\Gamma_{15v}$ gap variation is known to range (Wolford *et al* 1985, Mathieu *et al* 1979) from −1.3 to −2.6 meV/kbar. The nature of the A$_1$ and Si$_{As}$ transitions at pressures larger than P_c is due to the $(D_x^0-A^0)$ pair of the deepened donor tied to the X_{1c} valley with Si$_{As}$ and the first level of Ga$_{As}$. Deepening of donor levels up to ~100 meV in the indirect gap compared to those of direct regime can be

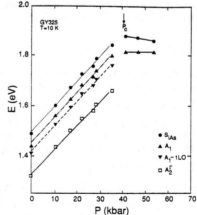

Fig. 4. Transition energy−vs−pressure relation under P=0−60 kbar.

found from the donor binding energies of GaP (Dean *et al* 1968). The transition energy–vs–pressure relationship for the Si_{As}, A_1, A_2, and $A_1^{\Gamma}-1$ LO emissions is shown in Fig. 4 for the direct and indirect regions. A_2 emission was not observed clearly in the $X_{1c}-\Gamma_{15v}$ regions as shown in Fig. 3. Another effect of the pressure induced $\Gamma_{1c}-X_{1c}$ crossover is seen in Fig. 5. The figure shows the relation of the A_1 emission intensity –vs– pressure. We see a drastic decrease of A_1 intensity. The photoexcited electrons relax to the X_{1c} minimum while the population in the Γ_{1c}

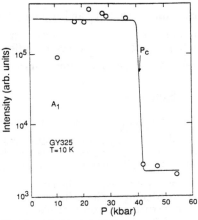

Fig. 5. PL intensity of A_1 emission-vs-pressure.

decreases as $e^{-\Delta E/kT}$, where ΔE is the separation between the Γ_{1c} and X_{1c} minima. The step–like intensity can be fit with the following equation (Olego et al 1980),

$$I = I_0 \left[\left\{1+A \exp (\alpha_\Gamma - \alpha_x) (P - P_c)/kT\right\}^{-1} + \tau_{\Gamma-x}/\tau_{rad}\right] \qquad (2)$$

where I_0 is a constant, P_c the pressure of the $\Gamma_{1c}-X_{1c}$ crossover, α_Γ and α_x the pressure coefficients for the $\Gamma_{1c}-\Gamma_{15v}$ and $X_{1c}-\Gamma_{15v}$ gaps, $\tau_{\Gamma-x}$ an average intervalley scattering time of the electron from the Γ_{1c} to the X_{1c}, and τ_{rad} the direct gap radiative time. The solid line is the fit. We obtain $P_c = 41$ kbar and $\tau_{\Gamma-x}/\tau_{rad} = 7.8 \times 10^{-3}$. We find that the fit is not sensitive to the values of A.

Another effect due to the $\Gamma_{1c}-X_{1c}$ crossover is seen in Fig. 3. We see an abrupt increase of the half width of A_1 emission and small increase of that of Si_{As} emissions around the crossover of the Γ_{1c} to the X_{1c}. In the pressure region of $P < P_c$ electrons can be scattered within the Γ_{1c} minimum by phonons with small wave vectors. For $P > P_c$ phonon–assisted intervalley scattering of electron from Γ_{1c} to X_{1c} becomes possible. So, the emission line broadening is due to the scattering with phonon and can be shown as a function of energy difference between the Γ_{1c} and X_{1c} points.

Table I shows that the pressure coefficient of A_2 emission differs between the two measurements made with increasing (up–stroke) and decreasing (down–stroke) pressures whereas those of the Si_{As} and A_1 emission are the same irrespective of the measurement procedure. The values of α of A_2 emission are 11.9 and 10.7 meV/kbar, respectively, for up– and down– stroke measurements. This observation may indicate that A_1 and A_2 emissions are not involved with the same center. This observation is consistent with our recent PL experiment which shows that the binding energy of A_2 acceptor differs compared to the 203–204 meV of the second level of Ga_{As} obtained from infrared measurements.

4. Summary

Two Ga_{As} related photoluminescence emissions at 1.32 and 1.44 eV were studied under hydrostatic pressures from 0 to 60 kbar at 10 K. The pressure coefficients of the acceptors responsible for the two emissions show the small values of 0.2–0.3 meV/kbar in the direct bandgap region. Effects due to the $\Gamma_{1c} - X_{1c}$ crossover are shown by the step–like decrease of emission intensity and the emission line broadening. The difference of the pressure coefficient of the 1.32 eV emission, for up–stroke and down–stroke measurements, indicates that the center responsible for the 1.32 eV emission may not be a simple substitutional component.

5. Acknowledgments

The work of PWY was performed at Wright Lab, Solid State Electronic Technology Directorate (WL/EL), Wright Patterson Air Force Base under Contract No. F33615–86–C–1062. AK gratefully acknowledges the assistance and expertise of Mr. Fred Pestian and the staff of the University Machine Shop in the construction of the pressure cell.

References

Barnett J D, Block S and Piermarini G J 1973 Rev. Sci. Instrum. 44 1
Dean P J, Frosch C J and Henry C H 1968 J. Appl. Phys. 39 5631
Elliot K R, Holmes D E, Chen R T and Kirkpatrick C G 1982 Appl. Phys. Lett. 40 898
Elliot K R 1983 Appl. Phys. Lett. 42 274
Fischer D W and Yu P W 1986 J. Appl. Phys. 59 1952
Mathieu H, Merle P, Ameziane E L, Archilla B, Camassel J and Poiblaud G 1979 Phys. Rev. B 19 2209
Neumann H, Topol I, Schulze K R and Hess E 1973 Phys. Stat. Sol. 56 K55
Nilsson S and Samuelson L 1978 Solid State Commun. 67 19.
Olego D, Cardona M, and Müller H 1980 Phys. Rev. B 22 894.
Pistol M E, Nilsson S and Samuelson L 1988 Phys. Rev. B 38 8293
Ren S Y, Dow J D and Wolford D J 1982 Phys. Rev. B 25 7661
Samara G A 1967 Phys. Rev. 27 655
Samuelson L and Nilsson S 1988 J. Lum. 40 & 41 127
Wolford D J and Bradley J A 1985 Solid State Commun. 53 1069.
Yu P W, Mitchel W C, Mier M G, Li S S and Wang W L 1982 Appl. Phys. Lett. 41 535

Inst. Phys. Conf. Ser. No 120: Chapter 7
Paper presented at Int. Symp. GaAs and Related Compounds, Seattle, 1991

353

Reflectance modulation studies on laser diode mirrors

P W Epperlein and O J F Martin

IBM Research Division, Zurich Research Laboratory, CH-8803 Rüschlikon

ABSTRACT: The versatility and usefulness of the newly developed reflectance modulation technique for laser mirror characterization will be demonstrated by means of typical examples of application on (110) facets of 10 μm wide ridge waveguide AlGaAs/GaAs quantum well lasers: (1) Typical, maximum temperature increases ΔT derived from the reflectance changes $\Delta R/R$ are $\simeq 1$ K per 1 mW optical diode power for uncoated mirrors and are $\simeq 10\times$ lower for $\lambda/2$-Al$_2$O$_3$ coated ones. ΔT is sensitive to the facet surface treatment prior to coating. (2) Temperature maps show a very localized hot spot within the near-field region and are in accordance with our two-dimensional heat flow calculations. (3) Degradation processes have been monitored in real time. Thus, we determined a critical temperature increase $\Delta T \simeq 120$ K for the occurrence of catastrophic optical mirror damage in the AlGaAs system.

1. INTRODUCTION

The growing importance of the semiconductor laser diode in advanced applications of optical storage, communication and pumping systems concomitantly places higher demands on its reliability. Today's profound knowledge of laser degradation processes, accompanied by essential improvements in laser design and fabrication technologies, has extremely improved the performance, yield, reliability and lengthened the lifetime of laser diodes emitting in the near-infrared regime. However, as the technology proceeds to increase the laser's optical output power from a smaller area, for example in large-scale integrated systems, the stress on materials, especially at the mirrors, becomes a great challenge, because local mirror facet heating is limiting the maximum optical output power P either through output power saturation or catastrophic optical mirror damage (COMD). Therefore, knowing the mirror temperature during laser operation provides useful information for maximizing P by an optimized mirror surface technology and for exploring thermally activated degradation processes. Local mirror temperatures have been measured by micro-Raman spectroscopy using the Stokes/anti-Stokes phonon line intensity ratio (Todoroki 1986, Brugger and Epperlein 1990, Beeck et al 1990). Reflectance modulation (RM) is an alternative, novel method (Epperlein 1990a) for laser mirror characterization. Here the laser is operated with current pulses leading to a temperature modulation of the laser diode which is then detected by changes of the optical surface reflectance on the mirrors. The RM technique offers numerous advantages: it is fast, sensitive and easy to use; it allows the continuous recording of the temperature increase ΔT as a function of the optical output power P; it represents a sensitive local probe for observing the temporal development of degradation processes and it allows the recording of instructive temperature maps of the hot mirror region in active lasers.

In this paper we will demonstrate the maturity of this novel, versatile and powerful technique by means of some typical examples of application on (110) mirrors of ridge waveguide AlGaAs/GaAs single quantum well (SQW) graded-index separate confinement heterostructure (GRIN-SCH) lasers. In particular, we report the sensitivity of ΔT to the mirror surface technology, compare the experimental temperature topographs with numerical simulations and finally report the existence of a critical temperature increase $\Delta T \simeq 120$ K for the onset of COMD.

2. EXPERIMENT

The $Al_xGa_{1-x}As$ lasers were grown by conventional molecular beam epitaxy on n^+-doped, (100) GaAs substrates with an 8 nm GaAs SQW sandwiched between 200 nm graded regions with parabolic Al concentration profiles ($x = 0.2 - 0.36$). Adjacent, $\simeq 1.8$ μm thick $Al_{0.36}Ga_{0.64}As$ cladding layers lead to a spreading of the near-field spot. Ridge waveguides etched into the top p-cladding along (110) directions on the (100) surface produce the lateral optical mode confinement. The ridge widths w are 10 and 15 μm. The facets of the 750 μm long lasers are (110)-oriented and were formed by cleaving in air. They are uncoated or coated with $\lambda/2$-thick Al_2O_3 or Si_3N_4 passivation layers. The laser devices are soldered junction-side-up to a copper-alloy heat sink. Details on laser structure and growth have been described elsewhere (Jaeckel et al 1991).

In the optical modulation techniques the response of the optical constants of a solid to a periodic change of a parameter such as stress, temperature or electric field is measured. Figure 1 illustrates how this principle can be applied to the mirror surface of an active laser diode. The laser is square-pulse power modulated with a low repetition frequency of $f = 220$ Hz and a duty cycle of 50%. This ensures that the reflectance response is maximal during operation and turned off between the pulses. A continuous-wave (cw) Ar^+-laser beam with a wavelength of $\lambda_0 = 457.9$ nm and low intensity of

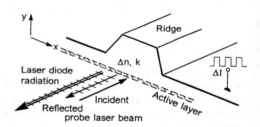

Fig. 1. Schematic diagram of reflectance modulation on the mirror facet of a ridge laser diode operated with current pulses ΔI.

$I_0 < 1$ mW is used for probing the mirror surface within the light penetration depth of 40 nm. The focussed ($\simeq 1.5$ μm diameter) probe beam impinges the mirror at normal incidence. Upon reflection the probe laser beam is intensity-modulated with the frequency f. The modulation depth of the reflected beam $I_0 \Delta R$ is a measure for the strength of the periodic perturbations on the reflectance. The temperature modulation of the optical properties due to the pulsed power operation is mainly responsible for the intensity modulation of the reflected light ($\rightarrow$ thermoreflectance, see Sect. 3).

Both the reflected probe laser light with $\lambda_0 = 457.9$ nm and the laser diode emission with a lasing wavelength of $\lambda \simeq 850$ nm are modulated with the same frequency f. To avoid any artifacts in the reflectance measurements, the 850 nm radiation has been entirely excluded from detection by filtering and dispersion. The spectrally clean 457.9 nm reflectance intensity $I_0 \Delta R$ is measured by a silicon photodiode. Phase-sensitive detection is used. The dc reflectance intensity $I_0 R$ is measured in the same way but with the laser diode turned off and the probe laser beam mechanically chopped with the same frequency f and duty cycle of 50%. The division of $I_0 \Delta R$ by $I_0 R$ is performed and the ratio $\Delta R/R$ is directly proportional to the temperature modulation amplitude ΔT on the mirror surface. A relationship between $\Delta R/R$ and ΔT was derived (Epperlein 1990a) from the normal-incidence reflectance law to be

$$\Delta T \simeq 4 \times 10^3 \times \Delta R/R \qquad [\text{K}]. \qquad (1)$$

Finally, two effects are to be noted which may impact the reflectance results. First, $\lambda/2\text{-}Al_2O_3$ or $\lambda/2\text{-}Si_3N_4$ passivation coatings have optical thicknesses which are almost equal to an integral multiple of half of the probing wavelength $\lambda_0 = 457.9$ nm. In this case the coatings have no influence on the intensity of the reflected radiation. Second, the probe laser energy $\hbar\omega$ is larger than the band gap energy E_g of $Al_{0.36}Ga_{0.64}As$ and thus creates electron-hole pairs with excess energies $\hbar\omega - E_g$. This excess energy is redistributed among the carriers and the lattice via carrier-carrier and carrier-phonon interactions. Reflectivity changes may now also be caused by this photo-induced free carrier plasma (Auston et al 1978). However, this contribution can be neglected in our measurements. Moreover, $\Delta R/R$ signals due

to strong carrier injection into the narrow (8 nm) QW during laser operation are also negligibly small because of the relatively large probe laser spot ($\simeq$ 1.5 μm).

3. RESULTS AND DISCUSSION

3.1. Mirror Temperatures

The dependence of $\Delta R/R$ on P provides useful information on the mirror properties, especially on the thermal behavior. Figure 2 demonstrates a set of curves recorded for 10 μm wide, cleaved GaAs lasers with different mirror treatments and coatings. The focus spot of the probe laser was in the center of the laser diode near-field spot. All curves show the same basic shape. $\Delta R/R$ increases continuously with P. Near the zero-point the curves are usually slightly convex. This signal regime can be ascribed to electroreflectance (ER) caused by a surface potential modulated by the pulsed carrier injection. This interpretation is in agreement with results from electric field-induced Raman scattering (EFIRS) measurements on the band bending of laser mirror surfaces (Beeck et al 1990). The adjacent, approximate-

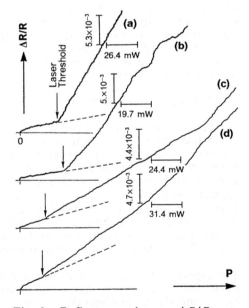

Fig. 2. Reflectance changes $\Delta R/R$ as a function of lasing power P (with reference to laser threshold) of 10 μm wide AlGaAs/GaAs lasers with cleaved mirrors; (a) uncoated, (b) O_2-ashed and $\lambda/2\text{-}Si_3N_4$ coated, (c) O_2-ashed, wet etched and $\lambda/2\text{-}Si_3N_4$ coated and (d) only $\lambda/2\text{-}Al_2O_3$ coated.

Table 1. Reflectance changes $\Delta R/R$ and temperature increases ΔT (according to Eq. (1)) per 1 mW lasing power P (< 50 mW) for the mirrors (a)-(d) in Fig. 2.

Sample	$\dfrac{\delta(\Delta R/R)}{\delta P}$ (lasing) (1/mW)	$\dfrac{\delta(\Delta T)}{\delta P}$ (lasing) (K/mW)
(a)	$\simeq 2.4 \times 10^{-4}$	$\simeq 1.0$
(b)	$\simeq 2.0 \times 10^{-4}$	$\simeq 0.8$
(c)	$\simeq 0.5 \times 10^{-4}$	$\simeq 0.2$
(d)	$\simeq 0.3 \times 10^{-4}$	$\simeq 0.1$

ly linear shape of the curves is due to thermoreflectance (TR) by Joule heating of the diode drive current which is long-ranging outside of the ridge structure due to spreading. At laser threshold an additional TR signal sets in caused by heating of the laser action within the near-field due to strong, nonradiative surface recombination. This signal superimposes on the current-heating TR (dashed lines). The slope of this signal regime with respect to the dashed line is a sensitive parameter for the "heating efficiency" of the mirror surface only under lasing conditions. In Table 1 this parameter is listed for the different mirrors used in Fig. 2. Thus, laser heating is highest with $\simeq 1$ K per 1 mW for uncoated (sample a) and lowest with $\simeq 0.1$ K per 1 mW for $\lambda/2$-Al_2O_3 coated (sample d) mirrors and $P < 50$ mW. The removal of the damage layer formed after an O_2-ashing treatment by a wet etch reduces the heating by a factor of four (compare sample b and c). Measurements on mirrors with the same surface treatment prior to coating showed no significant difference in ΔT between $\lambda/2$-Al_2O_3 and $\lambda/2$-Si_3N_4 passivation layers. The ΔT data from TR measurements are in good agreement with the data obtained from anti-Stokes and Stokes phonon Raman spectra.

3.2. Temperature Topographs

Temperature maps have been obtained by raster scanning the focus spot of the probe laser across the mirror of the lasing diode, and detecting $\Delta R/R$, i.e. ΔT as a function of location. A typical ΔT-map of an as-grown, cleaved and uncoated mirror of a 15 μm wide laser pulse-operated at $\simeq 15$ mW is shown in Fig. 3a. The scanned area was 35 μm $\times$ 15 μm. The first scan (at the top of the 3d-plot) was aligned along the p-n junction. For clarity only scan lines with equidistances of 2 μm are shown. There are two temperature regions, one with a very localized, high ΔT of $\simeq 20$ K within the near-field spot just below the ridge, and the other with a slower decay of temperature in the GaAs substrate according to a thermal conductivity $\simeq 5\times$ larger than in the AlGaAs (Adachi et al 1985).

The experimental ΔT map has been compared with heat flow calculations. From a theoretical point of view, the temperature distribution in the laser can be obtained by solving the three-dimensional heat equation in the device. For simplicity, we have neglected any variation of heat production that may occur along the laser structure. Therefore we can limit our investigations to a cross section of the laser where a solution of the two-dimensional heat equation must be found (Martin et al 1991). Although such an approach neglects particular thermal effects which might occur at the mirror, it gives a good qualitative picture of the temperature distribution in the laser facet. We have used a numerical method to obtain the solution of the two-dimensional heat equation in the cross section of the laser. In our model,

(a) 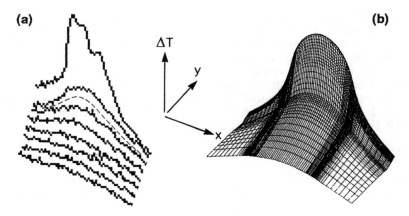 **(b)**

Fig. 3. Topographic temperature image of a cleaved, uncoated mirror of a 15 μm wide ridge AlGaAs laser operated at $\simeq$ 15 mW; from (a) experiment and (b) heat flow calculations. Scanned area 35 μm × 15 μm. Top scan along active layer with highest temperature $\Delta T \simeq$ 20 K. Dashed line = interface AlGaAs layer/GaAs substrate.

we consider that heat is produced in the active layer, in a region that corresponds to the near-field spot of the device. Neglecting radiation and convection losses at the surfaces of the laser, we assume that heat can only flow through the substrate to a heatsink. The temperature at the bottom of the heatsink is held constant (300 K). We define a grid over the cross section of the laser and discretize the heat equation using finite differences. A variable grid is used with a higher refinement in the regions of particular interest (e.g. active layer); the total number of meshes is typically 40,000 over the entire cross section. The discrete problem is then solved using an alternating direction algorithm. Figure 3 shows the good agreement between the calculated and experimental temperature maps.

3.3. Degradation Processes

Degradation processes can be monitored in real time by TR, because laser degradation is accompanied by a temperature increase or output power decrease, or both. In this way, we measured ΔT associated with the time development of dark line defects and with the occurrence of COMD. As a typical example, Fig. 4 exhibits the $\Delta R/R$ vs. P dependence for an ashed/wet etched, $\lambda/2$-Si$_3$N$_4$ coated mirror (sample c) which switched to COMD at a critical temperature $\simeq$120 K above room temperature and at $P \simeq$ 180 mW. The existence of a critical temperature is compatible with the known thermal runaway model for COMD (Henry et al 1979). The heating of the mirror surface, together

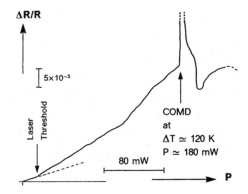

Fig. 4. Reflectance change $\Delta R/R$ vs lasing power P of a $\lambda/2$-Si$_3$N$_4$ coated mirror demonstrating the onset for COMD at a critical temperature $\simeq$120 K above room temperature.

with the rapid exponential increase of the absorption coefficient with temperature (Henry et al 1979) causes the surface to switch into a highly absorbing state for the laser radiation which has an energy not affected by the hot mirrors (Brugger and Epperlein 1990). Band-gap shrinkage and absorption increase form a positive feedback cycle at the critical temperature leading to thermal runaway, i.e. catastrophic self-destruction of the mirror. We have found that the critical temperature is practically independent of surface treatment and coating (samples a-d) and amounts to $\Delta T \simeq 120 \pm 10$ K for the AlGaAs system. This value is confirmed by Raman measurements (Epperlein 1990b, Tang et al 1991).

4. CONCLUSIONS

We have demonstrated that reflectance modulation is a powerful, nondestructive, fast and sensitive new tool for laser mirror characterization. It is especially suited to measure operating temperatures with micrometer spatial resolution, to produce informative temperature maps and to monitor in real time the formation of heat associated with laser degradation processes. Since this new mirror temperature mapping technique can reveal local hot spots, it might prove quite useful for the design of reliable laser structures. Finally, it could also be employed for monitoring the mirror quality by measuring the mirror temperature in a laser production line.

ACKNOWLEDGEMENT

The authors would very much like to acknowledge the excellent technical support of W. Bucher and L. Perriard and the members of the Laser Science and Technology Department for laser fabrication.

REFERENCES

Adachi S 1985 *J. Appl. Phys.* **58** R1
Auston D H, McAfee S, Shank C V, Ippen E P and Teschke O 1978 *Sol. St. Electr.* **21** 147
Beeck S, Egeler T, Abstreiter G, Brugger H, Epperlein P W, Webb D J, Hanke C, Hoyler C and Korte L, 1989 in *ESSDERC'89* Proceedings, Eds Heuberger A, Ryssel H and Lange P, (Springer, Berlin, 1990) 508
Brugger H and Epperlein P W 1990 *Appl. Phys. Lett.* **56** 1049
Epperlein P W 1990a *Inst. Phys. Conf. Ser.* **112** 633
Epperlein P W 1990b unpublished results
Henry C H, Petroff P M, Logan R A and Merritt F R 1979 *J. Appl. Phys.* **50** 3721
Jaeckel H, Bona G L, Buchmann P, Meier H P, Vettiger P, Kozlovsky W J and Lenth W 1991 *IEEE J. Quantum Electron*, Special Issue (June 1991).
Martin O J F, Bona G L and Wolf P 1991 to appear in IEEE J. Quant. Electron.
Tang W C, Rosen H J, Vettiger P and Webb D J 1991 *Appl. Phys. Lett* **58** 557
Todoroki S 1986 *J. Appl. Phys.* **60** 61

Inst. Phys. Conf. Ser. No 120: Chapter 7
Paper presented at Int. Symp. GaAs and Related Compounds, Seattle, 1991

Measurement of minority hole zero-field diffusivity in n⁺-GaAs

M.L. Lovejoy, M.R. Melloch and M.S. Lundstrom
Purdue University, West Lafayette, Indiana, USA 47907

B.M. Keyes and R.K. Ahrenkiel
Solar Energy Research Institute, Golden, Colorado, USA 80401

ABSTRACT: Minority hole diffusivity, or equivalently the hole mobility, was measured in n⁺-GaAs with the zero-field time-of-flight technique. Devices for this study were fabricated on MBE grown films, and effects such as enhanced transport due to photon recycling and circuit effects were minimized with the device design. For a donor doping of 1.8×10^{18} cm⁻³, the minority hole diffusivity was found to be 5.8 cm²/s, which corresponds to a mobility of ~225 cm²/V·s; this is greater than the majority carrier mobility for comparably doped p-GaAs. This result for hole mobility has important implications for the design of devices such as pnp-HBT's.

1. INTRODUCTION

Recent measurements have shown that the minority electron mobility, or equivalently the diffusivity, is about one-half that of majority electrons in correspondingly doped n-GaAs (see e.g. Nathan *et al.* 1988, Furuta *et al.* 1990 and Lovejoy *et al.* 1991a). For bipolar device engineering, there is a great need for the corresponding measurements of minority hole diffusivity. In this paper, the first direct measurement of zero-field minority hole diffusivity in n⁺-GaAs is presented.

For n⁺-GaAs doped to 1.8×10^{18} cm⁻³, the zero-field diffusivity was found to be 5.8 cm²/s, which is a higher diffusivity than that of majority holes in comparably doped p⁺-GaAs. This contrasts with the results for electrons where the minority carrier mobility is observed to be lower than the corresponding majority carrier mobility. These results have important implications for devices, and the zero-field time-of-flight technique (ZFTOF) described in this paper is well-suited for the measurement of minority carrier diffusivity, and its doping dependence, in n⁺-GaAs.

2. EXPERIMENTAL TECHNIQUE

The ZFTOF technique is shown schematically in Figure 1(a). In this technique, devices which resemble thick emitter solar cells are employed. A picosecond laser pulse photoexcites excess minority holes in the quasi-neutral n⁺-GaAs emitter; these carriers diffuse in a zero-field environment to the np-junction where the minority carriers are swept across the high-field depletion region. Such carrier collection effectively charges the np-junction capacitor and gives rise to a transient voltage response that is measured with a Tektronix 7854 sampling oscilloscope. The response characterizes minority carrier transport in the emitter and theoretical fits to the measured response are used to extract the transport parameters. Steady-state internal quantum efficiency *vs.* wavelength (QE) analysis of the film structure corroborates the parameters deduced from the transient study.

A Spectra-Physics mode-locked ion laser synchronously pumping a dye laser with Rhodamine 6G dye was used in this study. The system was tuned for lasing at 600 nm with a 4 MHz repetition rate and 5-7 ps FWHM pulses. QE measurements were performed with measurement equipment from Oriel (halogen light source, monochromator and detection system). Incident and reflected beam fluxes in QE measurements were detected with a calibrated silicon photodetector certified by Newport Corporation.

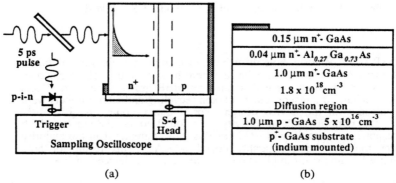

(a)   (b)

Figure 1.   a) Schematic illustration of the measurement apparatus and np-junction test device used for zero-field time-of-flight (ZFTOF) studies of minority hole diffusion in n^+-GaAs.   b) The ZFTOF diode structure for measurement of minority hole diffusivity in n^+-GaAs doped 1.8×10^{18} cm^{-3}.

Devices for this study were grown on (100) p^+-GaAs substrates by molecular beam epitaxy in a Varian GEN-II system. Silicon and beryllium are the n- and p-type dopants, respectively. The ZFTOF device structure used in this study is shown in Figure 1(b). As shown, the exposed surface is passivated with $Al_{0.27}Ga_{0.73}As$ to provide a low surface recombination velocity at the emitter-AlGaAs interface. A thin emitter layer was chosen to minimize photon recycling effects, as demonstrated by Dumke (1957). The emitter thickness was accurately controlled by monitoring RHEED oscillations during film growth. Mesa-isolation by wet etching with NH_4:H_2O_2:DI (10:3.5:500) was used to fabricate the diodes that were 520×540 μm^2 in size. An alloyed AuGe:Ti:Au contact was electron beam evaporated and alloyed at 385°C for 120s. Electrical contact to the substrate was facilitated by the alloyed indium layer formed during substrate mounting for MBE growth.

ZFTOF transients with 10-90% risetimes of ~800 ps dictated that circuit parasitics be minimized to realize accurate measurement of the ZFTOF response. To achieve this, a high-speed package utilizing a Wiltron K Connector® launcher mounted in a small test fixture was used. The device substrate was silver epoxied to the fixture and the emitter was connected directly to the center pin of the launcher. Circuit parasitics were deduced from one-port S-parameter measurements in the spectrum from 45 MHz to 26.5 GHz. Series inductance was found to be below 0.2 nH, and circuit simulation showed that such a low series inductance has a negligible effect on the measurement. In addition, parallel parasitic capacitance was negligible because of the large junction capacitance of the ZFTOF device (~100 pF).

Extraction of the diffusivity was performed by simulating the transport process in the quasi-neutral emitter. Included in the numerical solution of the governing minority carrier diffusion equation (mcde) were effects of finite absorption coefficient and finite interface recombination velocity. The diffusion current at the junction was calculated from the mcde solution. Circuit effects were included by using a lumped element circuit model, as discussed by Lovejoy *et al.* (1991b). In properly designed ZFTOF devices, the derivative of the measured response is closely proportional to the photocurrent, therefore, the derivative of the response is fit with the analogous simulation, which includes circuit effects. Adjustments of the diffusivity (D_p), lifetime (τ_p) and interface recombination velocity (S_f) were made to achieve best fits of the measured data.

3. DISCUSSION

Assessment of the film properties was made by Hall mobility measurements and SIMS analysis. Majority carrier Hall mobility was measured to be 2580 cm^2/V.s, which is above that reported by Ilegems (1985) for a Hall concentration of 1.8×10^{18} cm^{-3}. A Hall factor of 1 was assumed for the Hall measurements. SIMS analysis yielded a total silicon concentration of 2.7×10^{18} cm^{-3}. Silicon is an amphoteric dopant in GaAs, and this data implies a compensation ratio of 0.67. The SIMS analysis also verified that the emitter layer thickness was correct and that the emitter doping profile was uniform.

ZFTOF devices used in this study had 54% of the junction area shadowed by the emitter contact metallization and by a black wax that was applied over the mesa perimeter. The perimeter was covered because we have learned that generation along the perimeter can significantly distort the measured signal. This is a strong effect in GaAs structures with n-epilayers on p-substrates because the minority electrons in the base have a higher diffusivity than the minority holes in the emitter, and generation along the perimeter by the unattenuated beam can create a large enough electron current to distort the signal. Shadowing of the junction can also be a problem since the charge must distribute uniformly across the junction as the junction capacitance is charged. We have used a lumped circuit model presented by Lovejoy *et al.* (1991b) to investigate this effect, and we found that for this sample, the emitter sheet resistance and the junction capacitance are sufficiently low that 54% shadowing does not distort the response.

As shown in Figure 2, the ZFTOF transient experimental data were fit very well. The transport parameters were $D_p = 5.8$ cm^2/s, $\tau_p = 3.5$ ns and $S_f = 0$. This fit was sensitive to variations of ±5% in D_p and was relatively insensitive to longer lifetimes. In addition, the fit was equally good for $S_f \leq 10^4$ cm/s. Such a low S_f is expected for a GaAs-AlGaAs heterojunction with a low mole fraction of aluminum (*0.27*). A diffusion length, ($\sqrt{D_p\tau_p}$), greater than 1.4 μm is implied by these data. Our direct measurement of minority hole mobility agrees with the value of $D_p = 5.6$ cm^2/s which Slater *et al.* (1991) deduced from S-parameter measurements of pnp-heterojunction bipolar transistors with the base doped at 4×10^{18} cm^{-3}.

Additional verification of the transport parameters deduced from ZFTOF transient measurements was obtained from steady-state QE measurements. As shown in Figure 3, the QE data was fit very well by the parameters from the ZFTOF analysis. In terms of solar cell performance, the diode has very low QE due to the thick cap layer that attenuates the light incident on the emitter region; this also is shown in Figure 3 where a simulation with

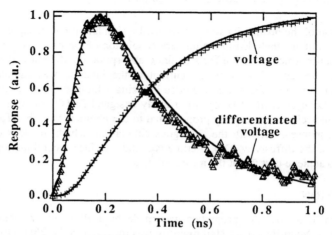

Figure 2. Measured ZFTOF transient response and the differentiated voltage (photocurrent). The solid lines are the best fits of the simulated responses which included circuit modeling.

the cap removed is plotted. The discrepancy near the bandgap energy is believed to result from the doping dependence of the absorption edge, which we did not model.

QE analysis showed that if $S_f > 2 \times 10^4$ cm/s, then the theoretical QE was much lower than the experimental data (see Figure 3). This was true even for an extremely long diffusion length and supports the low S_f that was required to fit the ZFTOF transient data.

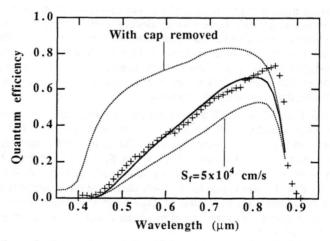

Figure 3. Internal quantum efficiency *vs.* wavelength for the diode structure shown in Figure 2. The solid line is the fit of experimental data with the parameters deduced from the ZFTOF technique, and the dashed lines are variations from this fit as noted.

For $S_f < 2 \times 10^4$ cm/s, QE fits yielded a diffusion length of 1.4-1.7 μm. Combining this with the diffusion coefficient implies a lifetime of 3.5-4.5 ns, which agrees with the value deduced from the transient analysis and with the expected radiative limited lifetime for n$^+$-GaAs doped to 1.8×10^{18} cm^{-3} that was reported by Lundstrom *et al.* (1990).

It is interesting to compare the minority hole diffusivity with that of majority holes in p-GaAs doped 1.8×10^{18} cm^{-3}. The majority diffusivity reported by Ilegems (1985) is 4.2 cm^2/s which is 30% smaller than the minority hole diffusivity that we measured for n$^+$-GaAs. Such a high minority hole diffusivity was not predicted by the theoretical calculations of Lowney and Bennett (1991). They treated impurity scattering with three different methods and found room temperature minority hole diffusivity to be 1.1-2.3 cm^2/s. Clearly, more measurements are needed to understand the important scattering mechanisms in n$^+$-GaAs.

4. SUMMARY

The minority carrier diffusivity in 1.8×10^{18} cm^{-3} silicon doped n$^+$-GaAs was measured to be 5.8 cm^2/s, which corresponds to a room temperature zero-field mobility of 228 cm^2/V.s. This higher than expected mobility has important implications for high performance pnp-heterojunction bipolar transistor design because the cutoff frequency (f_T) is limited by hole transport through the base. The measurement was made with the ZFTOF technique which demonstrates that the technique is suitable for mapping out the mobility doping dependence in degenerately doped n$^+$-GaAs; it is also believed to be well-suited to making temperature dependence measurements of minority carrier mobility. Finally, caveats in implementing the ZFTOF technique, such as circuit effects and two-dimensional effects were discussed.

5. ACKNOWLEDGEMENTS

The work was supported by the National Science Foundation under grant number ECS 8901638. Author Lovejoy acknowledges support from the AT&T Ph.D. Scholarship Program.

6. REFERENCES

Dumke W P 1957 *Phys. Rev.* **105** 139
Furuta T and Tomizawa M 1990 *Appl. Phys. Lett.* **56,** 824
Ilegems M 1985 "Properties of III-V Layers", Chapt. 5 in *The Technology and Physics of Molecular Beam Epitaxy,* Parker E H C, ed., (Plenum Press)
Lovejoy M L, Keyes B M, Klausmeier-Brown M E, Melloch M R, Ahrenkiel R K and Lundstrom M S 1991a *Jap. J. Appl. Phys.* **30** L135
Lovejoy M L, Melloch M R, Ahrenkiel R K and Lundstrom M S 1991b (submitted for publication)
Lowney J R and Bennett H S 1991 *J. Appl. Phys.* **69,** 7102
Lundstrom M S, Klausmeier-Brown M E, Melloch M R, Ahrenkiel R K and Keyes B M 1990 *Solid-State Electron.* **33,** 696
Nathan M I, Dumke W P, Wrenner K, Tiwari S, Wright S L and Jenkins K A 1988 *Appl. Phys. Lett.* **52,** 654
Slater D B Jr., Enquist P M, Najjar F E, Chen M Y, Hutchby J A, Morris A S and Trew R J 1991 *IEEE Electron Device Lett.* **12,** 54

Inst. Phys. Conf. Ser. No 120: Chapter 7
Paper presented at Int. Symp. GaAs and Related Compounds, Seattle, 1991

365

Strained-layer relaxation by partial dislocations

D. M. Hwang, S. A. Schwarz, R. Bhat, and C. Y. Chen
Bellcore, Red Bank, NJ 07701-7040, USA

Abstract: A new strain relief mechanism in strained heteroepitaxial structures of fcc semiconductors is identified. The signature defect of the proposed mechanism is a microtwin along a {111} plane spanning an embedded strained layer. This defect can form when two partial dislocations with antiparallel Burgers vectors of the $<112>/6$ type are generated inside the strained layer and glide to the opposite interfaces, leaving a stacking fault between them. We show that, for typical semiconductor strained layers with misfits $\geq 0.7\%$, the microtwin formation is the lowest-energy strain relaxation channel and poses fundamental limitations for strained layer device structures.

1. Introduction

One of the challenges facing the semiconductor industry today is the integration of materials with different lattice constants. It is possible to grow a thin film epitaxially on a lattice-mismatched substrate with a defect-free interface (Frank et al 1949, van der Merwe 1963). In such a structure, the lattice misfit is accommodated by the distortion of the unit cells and results in a strained pseudomorphic (i.e. coherent) layer which exhibits useful properties for many electronic and optoelectronic applications. However, when a strained layer exceeds a critical thickness, misfit dislocations may be formed at the interface to relieve the strain energy. The presence of misfit dislocations catastrophically degrades the transport and optical properties and thus constrains the design of strained layer devices.

Much research (Matthews and Blakeslee 1974, Matthews 1975, People and Bean 1985, Dodson and Tsao 1987, Tsao and Dodson 1988) has been directed to understanding the strain relaxation mechanism and predicting the critical thickness. The most commonly cited models are the mechanical force balance model of Matthews and Blakeslee (1974) and the energy balance model of People and Bean (1985). Recently, Hu (1991) pointed out that the mechanical force balance approach is fundamentally identical to the free energy minimization approach and that there are some flaws in the model of People and Bean (1985).

For epitaxial growth of face-centered cubic (fcc) systems along [001], the misfit dislocations considered (Matthews and Blakeslee 1974, Matthews 1975, Dodson and Tsao 1987, Hu 1991) for strain relaxation are usually the 60° perfect dislocations. The primitive translation vectors in fcc systems are $<110>/2$. A 60° perfect dislocation has a Burgers vector (e.g. $\vec{b} = [\bar{1}01]/2$) which makes a 60° angle with the dislocation line direction (e.g. [1̄10]). In fcc systems, there also exist partial dislocations with $\vec{b} = <112>/6$ which are always accompanied by stacking faults. The energy per unit length of a partial dislocation itself is only 1/3 that of a perfect dislocation (Hirth and Lothe 1982). Simple projection analysis (Hwang et al 1991a) indicates that a 90° partial dislocation (e.g. $\vec{b} = [11\bar{2}]/6$, which makes a 90° angle with the [1̄10] dislocation line) can relieve a strain 2/3 that of a 60° perfect dislocation. Therefore, in fcc systems with small stacking fault energies, 90° partial dislocation formation is a more effective strain-relief mechanism than the formation of 60° perfect dislocations. Nevertheless, despite the fact that stacking faults and microtwins have been frequently observed in heteroepitaxial systems, partial dislocations are usually considered as growth faults, not the primary strain relaxation products.

Recently, Hwang et al. (1989, 1991a, 1991b, 1991c) demonstrated that partial dislocation formation can be the primary strain-relief mechanism in capped strained layers of fcc structure. The resulting defects are stacking faults and microtwins spanning the strained layers along the {111} glide planes. In this article, examples of this new strain relief mechanism are presented. Using the free energy minimization scheme, we show that the critical thickness for the formation of 90° partial dislocations can be a factor of 2 smaller than that predicted for the 60° perfect dislocations. The new calculation agrees with recent observations.

2. Experimental Results

We first present an example for strain-relaxation during "post-annealing". Fig. 1 is the defect structure for a partially-relaxed strained superlattice (Hwang et al 1989). This is a transmission electron microscopy (TEM) bright-field image obtained with the electron beam along the [1$\bar{1}$0] direction with a [002] two-beam diffraction configuration. This specimen was derived from a superlattice of 40 nm InP and 10 nm $In_{0.53}Ga_{0.47}As$ grown by organometallic chemical vapor deposition. The lattice constant of $In_{0.53}Ga_{0.47}As$ matches that of InP and the as-grown specimen was found by TEM to be free of strain and defects. The specimen of Fig. 1 has been subject to Zn-diffusion in a sealed ampoule at 600°C for 1 hour with a Zn_3As_2 source. Secondary ion mass spectrometry (SIMS) studies (Hwang et al 1989, Schwarz et al 1990) revealed that In and Ga are homogenized in the superlattice region due to the selective enhancement of cation diffusion, while the anions (P and As) remain effectively immobile. The superlattice becomes $In_{0.9}Ga_{0.1}P/In_{0.9}Ga_{0.1}As$ which is no longer lattice matched. The lattice constant of the P layer is now 0.6% smaller than that of the InP substrate while that of the As layer is 2.6% larger. In Fig. 1, we see a high density of microtwins spanning the 10 nm As layer, distinctively displayed due to the diffraction contrast. High resolution lattice images (Hwang et al 1989) indicate that these microtwins are platelets along {111} planes one to three monolayers thick. (In Fig. 1, the microtwins away from the thin edge on the left-hand side are widened due to the artifact of 3-dimensional projection, since the specimen was tilted 1~2° toward [110] to bring out the diffraction contrast.) We have studied (Hwang et al 1991c) similar $In_xGa_{1-x}P/In_xGa_{1-x}As$ superlattices with various As layer thicknesses. We found that, after Zn diffusion, the As layers 10 nm thick or thinner remain defect free, while the As layer 14 nm thick displays many non-overlapping {111} microtwins. For As layers 20 nm thick or thicker, a high density of overlapping faults, including perfect dislocations, are observed.

Fig. 1. Defect structure in a strained $In_{0.9}Ga_{0.1}P/In_{0.9}Ga_{0.1}As$ superlattice. The as-grown superlattice consisting of 40 nm InP and 10 nm $In_{0.53}Ga_{0.47}As$ is lattice-matched and defect-free. The strain is introduced after growth by Zn diffusion which causes cation homogenization.

The next example is for strain-relaxation defects occurring, most probably, during growth. TEM micrographs for a 5 nm AlAs layer (Hwang et al 1991a) sandwiched between InP and $In_{0.53}Ga_{0.47}As$ are shown in Fig. 2. AlAs has a lattice constant 3.5% smaller than that of InP or $In_{0.53}Ga_{0.47}As$. Microtwins spanning the AlAs layer along the {111} planes are observed in Fig. 2a by diffraction contrast. A high-resolution lattice image for one of the defects is shown in Fig. 2b, indicating a monolayer-thick microtwin (i.e. two twinning operations separated by a monolayer) which can also be regarded as an intrinsic stacking fault. The stacking fault is terminated at the two opposite interfaces with a pair of partial dislocations of the $<112>/6$ type. A Burgers circuit drawn around the whole fault yields a null vector, indicating that the Burgers vectors at opposite edges of the stacking fault are anti-parallel. This configuration is different from the usual Shockley partial-dislocation pair which originates from the dissociation of a perfect dislocation, and from the Frank partial dislocation loop which encloses an intrinsic stacking fault resulting from a vacancy disc. The fault shown in Fig. 2b should not exist in crystals which are unstrained, homogeneously strained, or with strain varying monotonically in one direction, since two dislocations with anti-parallel Burgers vectors attract each other and there is no energy barrier to prevent their annihilation. The partial dislocations shown in Fig. 2b are kept apart by the strain field which has a local maximum in the As layer. We have studied (Hwang et al 1991b, 1991c) specimens with AlAs layers of various thicknesses embedded in InP. We found that AlAs layers 2 nm and 3 nm thick are defect free, while those 4 nm and 5 nm thick are partially relaxed with the presence of microtwins only. AlAs layers 10 nm thick are heavily faulted with many microtwins extending into the overlayers.

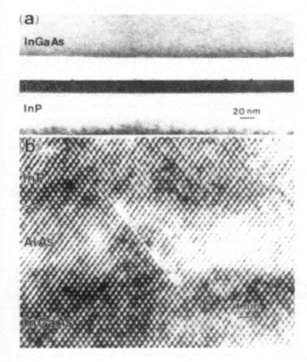

Fig. 2. Defect structure in an as-grown AlAs layer 5 nm thick sandwiched between InP and $In_{0.53}Ga_{0.47}As$. The lattice image (b) shows a stacking fault terminated at both interfaces with partial dislocations.

3. Theory

The critical thickness for strain relaxation in an embedded strained layer can be easily derived following the energy minimization scheme of Matthews and Blakeslee (1974), Matthews

(1975), and Hu (1991). Before relaxation, the strain is equal to the misfit $f = |a - a_0|/a_0$ which yields a stress (van der Merwe 1963) $\sigma = 2fG(1+\nu)/(1-\nu)$, where G is the shear modulus and ν is Poisson's ratio. The amount of strain δu relieved by the formation of a misfit dislocation with Burgers vector $\vec{b}$ is $b\cos\lambda$, where λ is the angle between $\vec{b}$ and the direction of strain relaxation (i.e. the direction in the interfacial plane perpendicular to the dislocation line). The strain energy relieved by the formation of a unit length of misfit dislocation is the product of the stress σ, the strain relief δu, and the strained layer thickness h, i.e.

$$\frac{\delta E_s}{\delta l} = \sigma \cdot \delta u \cdot h = \frac{2(1+\nu)}{1-\nu} fGhb\cos\lambda. \tag{1}$$

The energy required to form a unit length of dislocation is (Hirth and Lothe 1982)

$$\frac{\delta E_d}{\delta l} = \frac{Gb^2}{4\pi} \left[\frac{\sin^2\theta}{1-\nu} + \cos^2\theta \right] \ln\frac{\alpha R}{b}, \tag{2}$$

where θ is the angle between $\vec{b}$ and the dislocation line, R is the range of the dislocation, and α is a material dependent parameter representing the contribution of the core energy. $\alpha = 4$, the value expected for semiconductors (Hirth and Lothe 1982), is used in the following plots. Different choices of α cause a significant numerical deviation only when R/b is small.

At thermodynamic equilibrium, the total free energy should be minimized. For the formation of perfect misfit dislocations in an embedded strain layer at its critical thickness h_c, the strain energy relieved in Eq. (1) should be equal to twice the dislocation energy in Eq. (2) since there are two dislocations on opposite interfaces, i.e. $\delta E_s = 2 \cdot \delta E_d$. We obtain

$$h_c = \frac{b}{4\pi f\cos\lambda} \cdot \frac{1-\nu\cos^2\theta}{1+\nu} \cdot \ln\frac{\alpha h_c}{2b\sin\phi}, \qquad \text{Perfect Dislocations} \quad (3)$$

where we let $R = h_c/(2\sin\phi)$, half the distance between the two dislocations along the glide plane.

For partial dislocations, a stacking fault exists between the two partial dislocation lines which are $h/\sin\phi$ apart $[\phi = \cos^{-1}(1/\sqrt{3}) = 54.7°$ is the angle between the (111) glide plane and the (001) interfacial plane]. The stacking fault has an energy per unit length of partial dislocation

$$\frac{\delta E_f}{\delta l} = \frac{\gamma h}{\sin\phi}, \tag{4}$$

where γ is the stacking fault energy per unit area. The strain energy relieved δE_s must now balance out the additional stacking-fault energy, i.e. $\delta E_s = 2 \cdot \delta E_d + \delta E_f$. We obtain

$$h_c = \frac{\dfrac{b}{4\pi f\cos\lambda} \cdot \dfrac{1-\nu\cos^2\theta}{1+\nu} \cdot \ln\dfrac{\alpha h_c}{2b\sin\phi}}{1 - \dfrac{1}{2f\cos\lambda\sin\phi} \cdot \dfrac{1-\nu}{1+\nu} \cdot \dfrac{\gamma}{bG}}. \qquad \text{Partial Dislocations} \quad (5)$$

For the case of uncapped strained layers, we only need one misfit dislocation $\vec{b}$ to relieve a strain $b\cos\lambda$. Therefore, the right-hand sides of Eqs. 3 and 5 are reduced by a factor of 2. We also need to change the range of dislocation R from $h_c/(2\sin\phi)$ to h_c, the distance from the surface.

4. Discussion

To illustrate the competition of various defect formation mechanisms during strain relaxation, the critical thicknesses h_c for capped strained layers are plotted in Fig. 3 as functions of the lattice misfit f using Eqs. 3 and 5 for various types of misfit dislocations: (a) 60° perfect dislocations with $b = a/\sqrt{2}$ and $\theta = \lambda = 60°$ (e.g. $\vec{b} = [10\bar{1}]/2$, assuming the growth direction [001], the dislocation line along $[1\bar{1}0]$, and the strain relief along [110]); (b) 90° partial dislocations with b= a/√6, θ = 90°, λ = 54.7° (e.g. $\vec{b}$= [$\overline{1}$12]/6), and γ = 50 mJ/m²; (c) same as previous except γ = 0; and (d) 90° perfect dislocations with b= a/√2, θ = 90°, and λ = 0° (e.g. $\vec{b}$= [110]/2). Note that the only material sensitive parameters in Eqs. 3 and 5 are ν and γ/G. We use ν = 0.25 and G= 47 GN/m², the values for AlAs (Madelung 1982) in the plots. For semiconductors, ν is typically (Hirth and Lothe 1982) between 0.2 and 0.3 which would cause a ± 7% variation of the plotted results. γ is pressure and temperature dependent and may even become negative (which would cause an fcc to hcp transition). We plot two cases in Fig. 3 to cover the general range (Hirth and Lothe 1982) of γ.

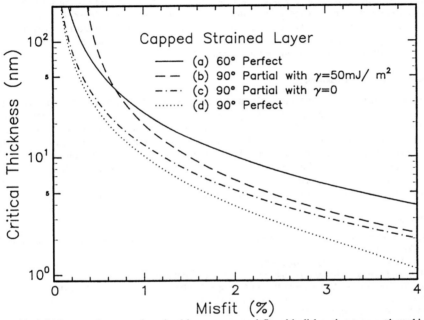

Fig. 3. Critical thicknesses for capped strained layers versus misfit, with dislocation type and stacking fault energy as parameters. Note that the 90° perfect dislocations cannot participate in the initial strain relaxation because they cannot glide into or out of the interface.

Since a dislocation can glide only in the plane defined by its line and its Burgers vector, the 90° perfect dislocations cannot glide into or out of the interfacial plane. Therefore, 90° perfect dislocations cannot participate in the initial relaxation process of a pseudomorphic strained layer. Excluding Curve (d), Fig. 3 indicates that, for semiconductors with γ ≲ 50 mJ/m² and f ≳ 0.7%, 90° partial dislocation formation has the lowest critical thickness and is thus the initial strain relaxation channel. As the misfit decreases, the critical thickness increases and the stacking fault energy (proportional to the layer thickness) may eventually exceed the energy difference between partial and perfect dislocations. Only in such systems is formation of the conventional 60° perfect dislocations the primary relaxation channel. If the stacking fault energy can be neglected, partial dislocation formation is always energetically favored.

Although 90° perfect dislocations cannot participate in the initial strain relaxation, they are the most effective strain relief defects corresponding to the smallest critical thickness shown in Fig. 3. When a strained layer much exceeds the critical thickness, a high density of 90° partial dislocation and 60° perfect dislocations will be formed. These dislocations interact with each other, and finally develop into 90° perfect dislocations – the lowest energy configuration. Previous TEM studies (e.g. Gerthsen et al 1988, Bhat et al 1989) on nearly-fully-relaxed interfaces indeed found that the dominant defects are 90° perfect dislocations.

The critical thickness predicted for the case of 2.6% strain in Fig. 1 is ~4.3 nm. However, we observed defects only when the thickness exceeds ~10 nm. Because the as-grown specimens are defect-free, considerable over-strain may be needed during post-annealing to overcome the activation energy barrier for defect generation.

For the cases illustrated in Fig. 2, the 3.5% strain exists during the growth of AlAs layers. The critical thickness for an uncapped 3.5% strained layer is only ~1.3 nm, about half that for a capped layer. Partial dislocations and stacking faults may have formed during growth when the critical thickness was much exceeded. After capping with a thick overlayer, the defects develop into stacking faults spanning the strained layer if its thickness has exceeded the critical thickness for capped layers; otherwise, the partial dislocations will glide to the free surface during the growth of the overlayer and eliminate the stacking faults.

5. Summary and Conclusions

A new strain relief mechanism in strained heteroepitaxial layers of fcc structure is identified. We demonstrate experimentally and analytically that, for typical semiconductor strained layers with misfits greater than ~0.7%, the formation of 90° partial dislocations and their associated stacking faults is an energetically favored strain relaxation channel as compared to the formation of conventional 60° perfect dislocations. The defects responsible for the initial stage of strain relaxation are thus microtwins spanning the strained layers which catastrophically degrade the transport and optical properties. Therefore, this new mechanism and the proposed critical thickness pose fundamental limitations for strained layer device structures.

References:

Bhat R, Lo Y H, Caneau C, Chang-Hasanin C J, Skromme B J, Hwang D M, Zah C E, Koza M A 1989 *Mat. Res. Symp. Proc.* **145** 367
Dodson B W and Tsao J Y 1987 *Appl. Phys. Lett.* **51** 1325
Frank F C and van der Merwe J H 1949 *Proc. Roy. Soc.* **A198** 216
Gerthsen D, Ponce F A, Anderson G B, and Chung H F 1988 *J. Vac. Sci. Tech.* **B6** 1310
Hirth J P and Lothe J 1982 *Theory of Dislocations* (New York: Wiley) p 91 231 375 and 837
Hu S M 1991 *J. Appl. Phys.* **69** 7901
Hwang D M, Schwarz S A, Mei P, Bhat R, Venkatesan T, Nazar L, and Schwartz C L 1989 *Appl. Phys. Lett.* **54** 1160
Hwang D M, Schwarz S A, Ravi T S, Bhat R, and Chen C Y 1991a *Phys. Rev. Lett.* **66** 739
Hwang D M, Schwarz S A, Ravi T S, Bhat R, and Chen C Y 1991b *Mat. Res. Soc. Proc.* **202** 531
Hwang D M, Schwarz S A, Bhat R, Chen C Y, and Ravi T S 1991c *Optical and Quantum Electronics* **23** S829
Matthews J M and Blakeslee A E 1974 *J. Crys. Gr.* **27** 118
Matthews J M 1975 *J. Vac. Sci. Technol.* **12** 126
Madelung O 1982 *Landolt-Börnstein Numerical Data and Functional Relationship in Science and Technology* (Berlin: Springer-Verlag) **III** 17a 166
People R and Bean J C 1985 *Appl. Phys. Lett.* **47** 322 and **49** 229
Schwarz S A, Hwang D M, Mei P, Schwartz C L, Werner J, Stoffel N G, Bhat R, Chen C Y, Ravi T S, and Koza M 1990 *J. Vac. Sci. Technol.* **A8** 2997
Tsao J Y and Dodson B W 1988 *Appl. Phys. Lett.* **52** 852 and **53** 848
van der Merwe J H 1963 *J. Appl. Phys.* **34** 117 and 123

Inst. Phys. Conf. Ser. No 120: Chapter 8
Paper presented at Int. Symp. GaAs and Related Compounds, Seattle, 1991

371

Local structures in GaInP on GaAs studied by fluorescence-detected EXAFS

Yoshikazu Takeda, Hirotaka Yamaguchi * and Hiroyuki Oyanagi*,

Department of Materials Science and Engineering, Nagoya University, Furocho, Chikusa-ku, Nagoya 464-01, Japan
Electrotechnical Laboratory, Umezono, Tsukuba, Ibaraki 305, Japan

ABSTRACT: Local structures of GaInP grown on GaAs by OMVPE have been studied by fluorescence-detected EXAFS (extended X-ray absorption fine structure) measurements to discuss the anomaly of energy gap in terms of bond length. The growth temperature dependence of the bond length in ordered GaInP was found. Difference in the EXAFS spectrum shape and the bond length between highly ordered GaInP and impurity-induced disordered GaInP both grown at 700°C was very small. The Debye-Waller factor showed no meaningful variation with growth temperature and ordering.

1. INTRODUCTION

Properties of alloys, such as $A_xB_{1-x}C$, have been widely analyzed using the VCA (virtual crystal approximation) and additional disorder effects, where it was assumed that the material parameters vary smoothly and uniquely with alloy composition x and the atoms on each sublattice are randomly distributed (Nordheim 1931, Takeda 1982, Pearsall 1982). Although the energy gap of $A_xB_{1-x}C$ should be primarily determined by x, a band gap variation with growth parameters such as the growth temperature and the input V/III ratio has been observed in $Ga_{1-x}In_xP$ at the fixed composition x=0.51, which is lattice-matched to GaAs, in the growth by organometallic vapor phase epitaxy (OMVPE) (Gomyo 1986, Ohta 1986, Ikeda 1986). The gap width varies with T_g and is a minimum at T_g=650°C.

On the other hand, TEM (transmission electron microscope) observation revealed that this system exhibits long range order in the cation arrangement with the CuPt type structure (Gomyo 1987, Ueda 1987). The fact that the energy gap of the OMVPE-grown GaInP is lower than that of the liquid-phase epitaxy(LPE)-grown GaInP has been attributed to the degree of order. In the same material composed of the same atomic species, the band gap may vary with the bond length, structural disorder and also chemical disorder(compositional disorder). The structural ordering can be observed by the TEM and the crystal structure is determined by the TED (transmission electron diffraction) analysis. However, the bond length (not the average lattice constant) in which even a slight change as small as a few percent causes an observable shift of energy gap of about 0.1eV cannot be determined by TEM or X-ray diffraction.

EXAFS is a useful tool to determine the local structure such as the bond length, structural disorder and chemical disorder. In order to discuss the origin of the band gap anomaly from the microscopic point of view, we performed EXAFS measurements at the Ga K-edge on ordered and disordered $Ga_{0.49}In_{0.51}P$ alloys lattice-matched to GaAs and on disordered alloy powders.

2. EXPERIMENTAL

Ordered samples with Tg=600, 650 and 700°C and a Zn-doped disordered sample with Tg=700°C were grown on GaAs(100) by OMVPE at atmospheric pressure with the fixed V/III ratio of 160 (Kondow1988 a). The GaInP layers were as thick as 1.5μm. Lattice mismatch with the substrate was kept at +0.24±0.04%. The variation of the lattice spacing is far below the resolution of the EXAFS, i.e., 0.5%

As reference samples, GaInP alloy powders were synthesized from GaP and InP powders by solid phase diffusion at 950°C for two weeks in Ar-filled closed ampoules at several alloy compositions.

EXAFS measurements were conducted on beam line BL4C at the Photon Factory with a positron energy of 2.5GeV and a maximum-stored current of 350mA. Data were collected with a fixed-exit monochromator using flat and sagittally bent Si(111) crystals. Higher harmonics were rejected by detuning the parallel setting of the crystals. The fluorescence signal was collected by a NaI scintillation detector array (Oyanagi 1985) with a total reflection technique for the relatively thick (1.5μm) GaInP layers, so that the extra fluorescence signal from Ga atoms comprised in the GaAs substrate is absent. It was confirmed experimentally that the substrate signal is negligible. EXAFS for powder alloys and binary compounds were measured in a transmission mode. The energy resolution of the spectrometer was estimated to be 2eV at 9keV.

3. RESULTS AND DISCUSSION

Figure 1 shows the Ga K-edge absorption spectrum of $Ga_{0.49}In_{0.51}P$ with $T_g=700°C$. Normalized Ga K-EXAFS oscillations multiplied by wave number k of excited photoelectron, i.e., kχ(k), were Fourier transformed over the k range between 3Å^{-1} and 15Å^{-1} with a Hanning window function.

The results are shown in Fig. 2. The prominent nearest-neighbor peak region of the Fourier transform between 1.0Å and 3.0Å is due to P atoms around Ga atoms. This region was extracted by an inverse-Fourier transform into k-space. The resulting Fourier-filtered EXAFS oscillations were curve-fitted with a single-shell model including structural parameters of radial distance r_{Ga-P} and Debye-Waller factor σ. The

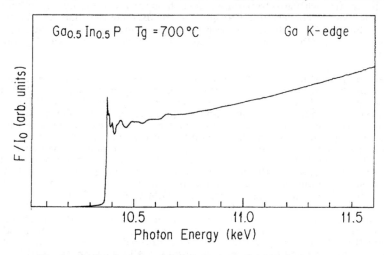

Fig. 1   Ga K-edge absorption spectrum of the ordered $Ga_{0.49}In_{0.51}P$ grown at 700°C.

results of the fitting are listed in Table 1 and plotted in Fig. 3 with circles as a function of In composition x. The triangles are the data for the powder alloys and they are on the straight line connecting the Ga-P bond length in GaP and that in InP at the dilute limit (Oyanagi 1988). Growth-temperature dependences of the energy gap (photoluminescence (PL) peak) and the bond lengths are plotted in Fig 4 left and right, respectively.

The bond length r_{Ga-P} is found to increase with T_g while σ shows no meaningful variation. Compared with the PL peak energy shift shown in Fig. 4 left, an appreciable difference in the bond length between ordered and Zn-doped disordered phases is not

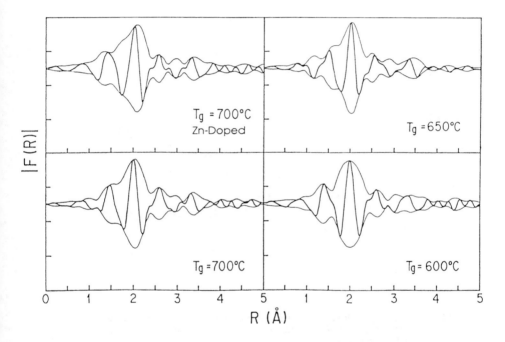

Fig. 2   Imaginary part and magnitude of Fourier transform of the Ga K-edge EXAFS oscillations of $Ga_{0.49}In_{0.51}P$.

Table 1   Ga-P bond lengths and Debye-Waller factors obtained by the present study.

	GaP	GaInP			
Sample		#772	#771	#770	#773 (Zn doped)
T_g (°C)		600	650	700	700
r_{Ga-P} (Å)	2.360	2.368	2.373	2.378	2.384
σ (10^{-3}Å)	6.43	7.44	6.27	7.51	6.90

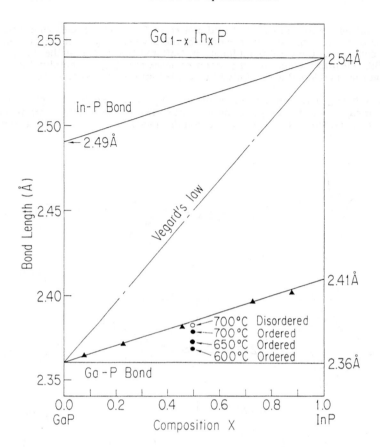

Fig. 3   Ga-P bond length as a function of In composition x. Circles
are for the OMVPE-grown GaInP and triangles are for the
alloy powders.

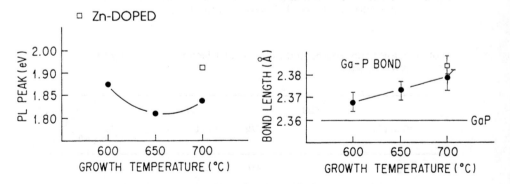

Fig. 4   Photoluminescence peak energy (left) and Ga-P bond length
(right) as functions of growth temperature.

observed.

We also analyzed the EXAFS of the ordered phase by a two-shell model taking the effect of cation ordering in the CuPt-type structure into account, but the results were the same as those of a single-shell model within experimental error.

The main concern in the present study is to investigate the relation between the energy gap and the bond length in GaInP alloys. Two distinct phenomena have been observed; the first is the jump in energy gap at 700°C in ordered and disordered GaInP, and the second is the growth temperature dependence of the energy gap.

Although the energy gap jumps by more than 70meV from ordered (undoped) to disordered (Zn-doped) GaInP, the Ga-P bond length and the Debye-Waller factor are almost the same. The results indicate that as far as the Ga-P bond is concerned, the energy gap jump is due to the chemical disorder in Zn-doped GaInP and the contribution of the bond length variation to the jump is excluded. This result also indicates that the bond length is primarily determined by the nearest neighbor interaction and is hardly dependent on chemical order/disorder of cation as the second nearest neighbors.

The energy gap varies with a downward bowing as a function of growth temperature while the bond length increases almost monotonically with growth temperature. The results suggest that the growth temperature dependence of the energy gap is not simply explained by the bond length only. The energy gap at the Γ–symmetry increases with decreasing lattice spacing which is known as the blue shift of the optical absorption edge at high pressures (Pankove 1972). If this is applicable to the present case, the gradual decrease in the energy gap with growth temperature in Fig. 4 left may be partly explained. On the other hand, the TEM observation shows the increasing degree of ordering and also the size of the domain of the ordered region with growth temperature (Suzuki 1988, Kondow 1988 b) both of which are expected to affect the energy gap. These three factors, bond length, degree of ordering and domain size, so far observed in GaInP, should be competing to determine the downward bowing of the energy gap with growth temperature.

4. SUMMARY

Local structures of ordered and Zn-doped disordered phases in GaInP alloys lattice-matched to GaAs grown by OMVPE at several growth temperatures have been investigated by EXAFS on the Ga K-edge. The Ga-P bond length, which is longer than that in the pure binary compound GaP, increases with T_g above 600°C. From the EXAFS measurements, it was found that two competing factors which govern the energy gap, i.e., the Ga-P bond length and the degree of chemical order(compositional order) are likely to determine the energy gap. The chemical disorder appears to determine the jump in energy gap between ordered and disordered GaInP since the difference in the Ga-P bond lengths between the two phases is very small. The downward bowing of the energy gap with growth temperature may result from competing factors such as the bond length, degree of ordering, and possibly the domain size of the ordered phase.

Acknowledgment
The authors express their thanks to Drs. M. Kondow and S. Minagawa of Hitachi Ltd. for providing GaInP samples lattice-matched to GaAs. This work has been performed as a part of a project (Proposal No. 87-065) approved by the Photon Factory Program Advisory Committee, and was supported in part by the Grant-in-Aid for Scientific Research (C) from the Ministry of Education, Science and Culture.

References
Gomyo A, Kobayashi K, Kawata S, Hino I, Suzuki T and Yuasa T (1986) *J. Cryst. Growth* 77 367
Gomyo A, Suzuki T, Kobayashi K, Kawata S and Hino I (1987) *Appl. Phys. Lett.* 50 673
Ikeda M, Nakano K, Mori Y, Kaneko K and Watanabe N (1986) *J. Cryst. Growth* 77 380

Kondow M, Kakibayashi K and Minagawa S (1988 a) *J. Cryst. Growth* 88 291

Kondow M, Kakibayashi H and Minagawa S (1988 b) *Appl. Phys. Lett.* 53 2053

Nordheim L (1931) *Ann. Phys. (Leipz)* 9 607 and 641

Ohta Y, Ishikawa M, Sugawara H, Yamamoto M and Nakanishi T (1986) *J. Cryst. Growth* 77 374

Oyanagi H, Matsushita T, Tanoue H, Ishiguro T and Kohra K (1985) *Japan. J. Appl. Phys.* 24 610

Oyanagi H, Takeda Y, Matsushita T, Ishiguro T, Yao T and Sasaki A (1988) *Solid State Commun.* 67 453

Pankove J I (1971) *Optical Processes in Semiconductors* (Englewood Cliffs: Prentice-Hall) 24

Pearsall T P (1982) *GaInAsP Alloy Semiconductors* ed Pearsall T P (Chichester: John Wiley & Sons) 295

Suzuki T, Gomyo A, Iijima S, Kobayashi K, Kawata S, Hino I and Yuasa T (1988) *Japan. J. Appl. Phys* 27 2098

Takeda Y (1982) *GaInAsP Alloy Semiconductors* ed Pearsall T P (Chichester: John Wiley & Sons) 213

Ueda O, Takikawa M, Komeno J and Umebu I (1987) *Japan. J. Appl. Phys.* 26 L1824

Inst. Phys. Conf. Ser. No 120: Chapter 8
Paper presented at Int. Symp. GaAs and Related Compounds, Seattle, 1991

Effect of step motion on ordering in GaInP and GaAsP

G. S. Chen and G. B. Stringfellow

Department of Materials Science and Engineering, University of Utah, Salt Lake City, Utah 84112, USA

Abstract. $Ga_{0.5}In_{0.5}P$ and $GaAs_{0.5}P_{0.5}$ layers have been grown on exactly (001) substrates by OMVPE. Only 2 of the 4 possible $L1_1$ variants, $(\bar{1}11)$ and $(1\bar{1}1)$, are observed for both materials. To examine the effects of surface steps on ordering, the substrates were first patterned with [110] oriented grooves ~5 μm wide and from 0.2 to 1 μm deep. TEM observations reveal that the two directions of step motion produce two different variants. For GaInP, one half of the groove is filled with a single domain of the $(\bar{1}11)$ variant while the other half is a single domain of the $(1\bar{1}1)$ variant. For GaAsP, single, large $(\bar{1}11)$ and $(1\bar{1}1)$ domains are also produced. However, large, well-defined regions with absolutely no ordering are also observed. These are shown to be due to the growth on {511} facets inadvertently produced at the edges of the grooves during etching. The {511} planes have a higher growth rate, so disappear during growth of the layers, leaving large domains of ordered material. The domains formed in both materials are the largest ever observed, several square microns in cross section and extending along the entire length of the groove.

1. Introduction

GaInP and GaAsP alloys form the $L1_1$ (or CuPt) ordered structure with alternating monolayers on the {111} planes [1,2]. Only 2 variants of the $L1_1$ structure were observed for epilayers grown by organometallic vapor phase epitaxy (OMVPE) (Bellon et al 1989, Chen et al 1990, Chen et al 1991, Suzuki and Gomyo 1988). Formation of the $L1_1$ structure in III/V alloys was unexpected since total energy calculations indicate that bulk materials with ordering on the {100} and {210} planes are thermodynamically more stable (Dandrea et al 1990, Srivastava et al 1985). Recently, by assuming the formation of distorted group III dimers on the reconstructed, group III-rich (001) surface, total energy calculations (Froyen et al 1991) concluded that formation of the $(\bar{1}11)$ and $(1\bar{1}1)$ variants of the $L1_1$ structure in $Ga_{0.5}In_{0.5}P$ can be completely explained by surface thermodynamics.

On the other hand, indirect evidence indicates that kinetic factors affect ordering, including processes occurring at steps propagating across the surface in the two dimensional growth mode. In this work, $Ga_{0.5}In_{0.5}P$ and $GaAs_{0.5}P_{0.5}$ layers are grown on grooved substrates by OMVPE. The substrates are exactly (001) GaAs for $Ga_{0.5}In_{0.5}P$. For the growth of $GaAs_{0.5}P_{0.5}$, thick (~20 μm) layers of $GaAs_{0.6}P_{0.4}$ grown by hydride VPE to be used for LED applications are used as the substrates to reduce the generation of defects due to lattice parameter mismatch. In order to examine the effects of step motion on ordering, the substrates were patterned with [110] oriented grooves on the surface before growth. The groove is used to provide sources of [110] steps moving in opposite directions from the two edges during epitaxial growth. Transmission electron microscopy (TEM) shows that two large, separate $(\bar{1}11)$ and $(1\bar{1}1)$ domains are formed on each groove for both alloys.

The domains meet at the center of the groove. Thus, the effect of step motion is directly demonstrated to control ordering for alloys with mixing on either the group III (GaInP) or group V (GaAsP) sublattice.

2. Experimental

$Ga_{0.5}In_{0.5}P$ and $GaAs_{0.5}P_{0.5}$ were grown on grooved GaAs and $GaAs_{0.6}P_{0.4}$ substrates, respectively. The grooved GaAs (or $GaAs_{0.6}P_{0.4}$) was prepared by patterning parallel lines with equal width and spacing on the substrate surface using standard photolithographic techniques. The grooves, ~ 5 μm wide on 10 μm centers and oriented in the [$1\bar{1}0$] direction, were obtained by etching the patterned surface for different time periods in a solution of $H_2SO_4:H_2O_2:H_2O=1:1:30$ (Bhat et al 1988) (or 2:1:8 for GaAsP). This results in depths ranging from 0.2 to 1 μm. After removing the photoresist, they were degreased, then etched in H_2SO_4 for 1 min (2 min) followed by etching in $H_2SO_4:H_2O_2:H_2O=1:1:30$ (8:1:1) for another 0.5 min (1 min) to remove residual surface oxides. Finally, the samples were rinsed in deionized water for 3 min, N_2-blown dry, and loaded into the OMVPE reactor (Wang et al 1990). The precursors are trimethylgallium (-10°C), trimethylindium (17°C), phosphine and arsine. For the growth of $Ga_{0.5}In_{0.5}P$, a 0.1 μm thick GaAs buffer layer was first deposited to improve the epilayer quality. The substrates were aligned with the grooves parallel to the gas flow direction during growth. The temperature and growth rate were 670°C and 2 μm/hr, respectively, for both alloys. The V/III ratios were 150 for $Ga_{0.5}In_{0.5}P$ and 160 for $GaAs_{0.5}P_{0.5}$.

The cross-sectional TEM samples were prepared to study the ordered structure by mechanically thinning and then Ar-ion milling at 77°K to electron transparency. The TEM images were obtained using a JEOL 200CX electron microscope operated at 200KV.

3. Results and Discussion

Figure 1 shows the surface morphology observed using a Nomarski interference phase contrast microscope of a $Ga_{0.5}In_{0.5}P$ sample grown on a grooved GaAs substrate. The grooved pattern is clearly seen. The [110] cross-sectional diffraction pattern taken from this sample shows the existence of $1/2(\bar{1}11)$ and $1/2(1\bar{1}1)$ superspots of the $L1_1$ structure, similar to results for epilayers grown on unpatterned substrates. However, the dark field (DF) images, illustrated in figs. 2(a) and 2(b), are dramatically different from those formed on unpatterned substrates where small, intermixed domains of both variants are observed (Morita et al 1988). The residual background contrast in fig. 2 is mainly due to thickness fringes (Edington 1976). Figure 2(c) is a schematic drawing to assist the identification of particular areas of the two dark field images. In fig. 2, two separate ($\bar{1}11$) and ($1\bar{1}1$) domains are formed in each groove (~ 0.3 μm deep). The domains meet at the center of the groove and each fills approximately one half of the groove. This is verified by electron diffraction patterns obtained with the electron beam entirely confined to 1/2

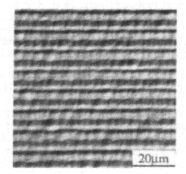

Fig. 1 Differential interference contrast micrograph of $Ga_{0.5}In_{0.5}P$ grown on an exactly (001) GaAs substrate patterned with [$1\bar{1}0$] oriented grooves.

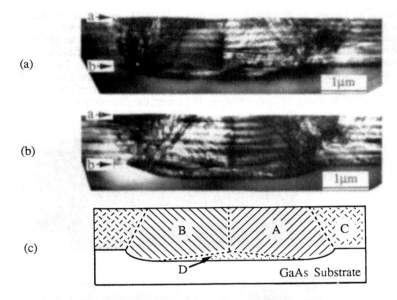

Fig. 2 [110] pole dark field images of $Ga_{0.5}In_{0.5}P$ grown on an exactly (001) GaAs substrate patterned with [110]-oriented grooves: (a) image from the

$1/2(\bar{1}11)$ superspot, (b) image from the $1/2(1\bar{1}1)$ superspot, and (c) schematic drawing of domains. Arrow "a" indicates the top surface of the $Ga_{0.5}In_{0.5}P$. Arrow "b" indicates the $Ga_{0.5}In_{0.5}P$/GaAs interface. The symbol A (or B)

denotes a very large domains of the $1/2(\bar{1}11)$ (or $1/2(1\bar{1}1)$) variant. Symbols C and D denote areas where small domains of both variants coexist.

of the groove. Each half of the groove shows a strong set of superlattice spots for only a single variant. As a result, very large domains, several square microns in cross section running the length of the substrate are formed. It is clear that the [110] oriented steps moving to the right form a single variant (domain B) for the left side of the groove while those moving to the left (on the right side of the groove) produce the other variant (domain A). Similar results were obtained for grooves with depths ranging from 0.2 to 1 µm.

For the part of the epilayer grown on the flat bottom of the groove, the growth occurs initially by two dimensional nucleation and propagation of [110] steps in both the [$\bar{1}$10] and

[$1\bar{1}0$] directions. This results in the the formation of small, intermixed $(\bar{1}11)$ and $(1\bar{1}1)$ domains for the region labelled "D" in fig. 2, the same as for growth on the flat (001) surface between grooves (area labelled C in fig. 2).

These results demonstrate directly that the direction of [110] step motion determines the specific $L1_1$ variant formed for GaInP.

For GaAs$_{0.5}$P$_{0.5}$ grown on grooved GaAs$_{0.6}$P$_{0.4}$ substrates, the effects of step motion on ordering are similar. This is illustrated by the TEM DF images shown in figs. 3(a) and 3(b). Figure 3(c) is a schematic drawing of the various areas. Again, a single, large domain was observed to fill each side of the groove. Comparison of the dark field images in figs.2 and 3 reveals other similarities. The triangular region labelled F in fig. 3 shows even more clearly where small domains of both variants are formed. Similarly the areas labelled C and D in fig. 3, are regions where growth occurs on the flat (001) surface resulting in the formation of small domains of both variants. However, the domain patterns in fig. 3 show

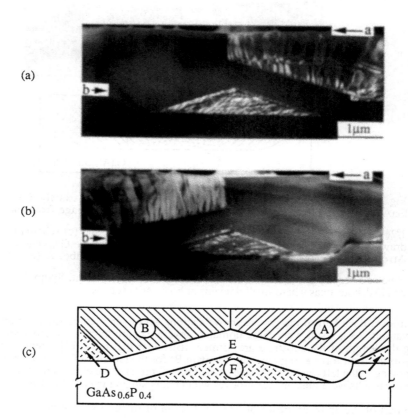

Fig. 3 [110] pole dark field images of GaAs$_{0.5}$P$_{0.5}$ grown on a GaAs$_{0.6}$P$_{0.4}$ substrate patterned with [110]-oriented grooves: (a) image from the 1/2($\bar{1}$11) superspot, (b) image from the 1/2(1$\bar{1}$1) superspot, and (c) schematic drawing of domains. Arrow "a" indicates the top surface of the GaAs$_{0.5}$P$_{0.5}$. Arrow "b" indicates the Ga$_{0.5}$In$_{0.5}$P/GaAs$_{0.6}$P$_{0.4}$ interface. The symbol A (or B) denotes a very large domain of the 1/2($\bar{1}$11) (or 1/2(1$\bar{1}$1)) variant. The symbols C, D and F denote areas where small domains of both variants coexist. The symbol E denotes disordered areas within the groove.

a striking feature not observed for $Ga_{0.5}In_{0.5}P$. The regions labelled E in fig. 3 are completely disordered, i.e., neither variant of the $L1_1$ structure is present.

In order to understand th mechanism for the formation of the domain patterns shown in fig. 3, $GaAs_{0.7}P_{0.3}/GaAs_{0.5}P_{0.5}$ superlattices were grown on the grooved $GaAs_{0.6}P_{0.4}$ substrates. This allows a determination of the surface profile of the epilayers at various times during growth. Figure 4 is a [110] pole dark field image taken from a sample of the superlattice structure. It shows clearly that the disordered region was produced by growth on facets having an angle of approximately 16° with the (001) surface. The angle between

(001) and (511) or ($\bar{5}$11) planes is 15.79°, indicating that the disordered material was grown on the {511} planes. $Ga_{0.5}In_{0.5}P$ grown by OMVPE on a {511} surface has been reported to yield only disordered material (Valster et al 1991). Growth on the {511} facets is more rapid than on the (001)-oriented bottom of the groove, resulting in the formation of the triangular-shaped region labelled F in fig. 3.

Formation of the disordered region is confirmed by the TEM result shown in fig. 4. This photomicrograph was taken using the $1/2(\bar{1}11)$ superspot of the $GaAs_{0.7}P_{0.3}/GaAs_{0.5}P_{0.5}$ superlattice structure. Thus, only the areas containing the ($\bar{1}$11) ordered domains (regions A, C, D, and F in fig. 3) are bright. Similarly, ($1\bar{1}1$) domains were observed in regions B, C, D, and F and not in regions A and E when the $1/2(1\bar{1}1)$ superspot was used to form the DF image (not shown). These observations support the conclusion that the chevron-shaped region labelled E in fig. 3 is completely disordered. In addition, examination of the dark field image in fig. 4 shows that the $GaAs_{0.7}P_{0.3}$ layers of the superlattice structure are also disordered. This is consistent with the previous observations that no $L1_1$ ordering occurs in $GaAs_{1-x}P_x$ for $x \leq 0.35$ (Chen et al 1991).

The mechanism by which ordering occurs is not completely understood. A model based on surface reconstruction has been used to describe the formation of the ($\bar{1}$11) and ($1\bar{1}1$) variants of the $L1_1$ structure for both GaInP and GaAsP (Chen et al 1991). The rows of dimers formed on the group V-rich, reconstructed surface during OMVPE growth are postulated to run in the [110] direction. This yields a single variant of the $L1_1$ structure for motion of a [110] oriented step along the surface. Significantly, the direction of step motion determines which of the 2 possible variants is formed, in agreement with the experimental results presented in this paper.

Fig. 4 [110] pole dark field image taken from the $1/2(\bar{1}11)$ superspot of the $GaAs_{0.7}P_{0.3}/GaAs_{0.5}P_{0.5}$ superlattice structure that was grown on a $GaAs_{0.6}P_{0.4}$ substrate patterned with [110]-oriented grooves. The arrow indicates the interface between the superlattice and $GaAs_{0.6}P_{0.4}$ substrate.

4. Summary

$Ga_{0.5}In_{0.5}P$ and $GaAs_{0.5}P_{0.5}$ have been grown on substrates patterned with an array of [110] oriented grooves ~ 5 μm wide and < 1 μm in depth. The groove is used to examine the effects of step motion on ordering. TEM dark field images and diffraction patterns clearly show that the direction of propagation of the [110] oriented steps determines the specific $L1_1$ variant formed. Thus, the steps moving in the [$\bar{1}$10] direction form the ($\bar{1}$11) variant while those moving in the [1$\bar{1}$0] direction produce the (1$\bar{1}$1) variant. As a consequence, large symmetrical ($\bar{1}$11) and (1$\bar{1}$1) domains, several square microns in cross section running the entire length of the substrate are formed in the groove. The effect of step motion on the selection of the ordered variant formed is essentially the same for $Ga_{0.5}In_{0.5}P$ and $GaAs_{0.5}P_{0.5}$ alloys. These results support a model that the asymmetrical formation of ordered variants of the $L1_1$ structure is due to the selectivity of atomic incorporation at the steps on a reconstructed surface. These results also demonstrate that the domain size can be controlled by regulating the motion of [110] surface steps. This offers the possibility of producing large domains of ordered materials for characterization and perhaps for future device applications.

5. Acknowledgements

The authors wish to thank the Department of Energy for financial support (grant DE-FG02-84ER45061) of the research described in this paper.

References

Bellon P, Chevalier J P, Augarde E, Andre J P, and Martin G P, 1989 J. Appl. Phys. **66**, 2388.
Bhat R, Kapon E, Hwang D M, Koza M A, and Yun C P, 1988 J. Cryst. Growth **93**, 850.
Chen G S, Jaw D H, and Stringfellow G B, 1990 Appl. Phys. Lett. **57**, 2475.
Chen G S, Jaw D H, and Stringfellow G B, 1991 J. Appl. Phys. **69**, 4263.
Dandrea J E, Ferreira L G, Froyen S, Wei S H, and Zunger A, 1990 Appl. Phys. Lett. **56**, 731.
Edington J W, Practical Electron Microscopy (New York, Van Nostrand Reinhold Co. 1976), Section 3.2.3.1.
Froyen S and Zunger A, 1991 Phys. Rev. Lett. **66**, 2132.
Morita E, Ikeda M, Kumagai O, and Kaneko K, 1988 Appl. Phys. Lett. **53**, 216.
Srivastava G P, Martins J L, and Zunger A, 1985 Phys. Rev. **B31**, 2561.
Suzuki T and Gomyo A, 1988 J. Cryst. Growth **93**, 396.
Wang T Y, Jen H R, Chen G S, and Stringfellow G B, 1990 J. Appl. Phys. **67**, 563.
Valster A, Liedenbaum C T, Finke M N, Severens A L, Boermans M J, Vanderhoudt D E, and Lieuwma C W, 1991 J. Cryst. Growth **107**, 403.

Inst. Phys. Conf. Ser. No 120: Chapter 8
Paper presented at Int. Symp. GaAs and Related Compounds, Seattle, 1991

383

Temperature-dependent electron mobility and clustering in GaInP$_2$

D. J. Friedman, A. E. Kibbler, and J. M. Olson
Solar Energy Research Institute, 1617 Cole Blvd., Golden, CO 80401, USA

Abstract. We present temperature-dependent Hall mobility data $\mu(T)$ for GaInP$_2$, an alloy which shows cation-site ordering under certain growth conditions. We find that samples grown to give high ordering have consistently lower mobilities than samples with low ordering, suggesting that the mobility is limited by cluster scattering by ordered domains. By analyzing $\mu(T)$ in terms of a cluster scattering model developed by Harrison, Hauser, and Marsh, we quantify the difference in clustering between the more- and less-ordered states of GaInP$_2$.

1. Introduction

Under the appropriate growth conditions, the pseudobinary semiconductor alloy Ga$_{0.5}$In$_{0.5}$P shows cation-site ordering into a CuPt-type structure (Gomyo *et al* 1988). This ordering, which is believed to be driven by a surface ordering mechanism (Froyen and Zunger 1991, Kurtz *et al* 1990), appears to be correlated with a variation in the band gap from 1.81 eV in the most highly ordered material to 1.89 eV in the least-ordered material . Because of its band gap, its lattice match to GaAs, and its relatively high resistance to oxidation compared to aluminum-containing semiconductors, Ga$_{0.5}$In$_{0.5}$P (hereafter GaInP$_2$) is an important candidate for III-V solar cell (Olson *et al* 1990), laser, and light-emitting diode applications. An understanding of this ordering is thus of technological as well as theoretical interest.

The conduction mobility μ in alloys can be significantly affected by alloy clustering (Harrison and Hauser 1976a,b; Marsh 1982), which has been invoked to explain the temperature dependence of the mobilities of certain arsenic-containing alloys at high temperature (Bhattacharya *et al* 1981,1985; Hong *et al* 1987; Oh *et al* 1990). Here, we present the measured Hall mobility $\mu_H(T)$ in GaInP$_2$ from $T = 300$ to 600 K and analyze it using the Harrison-Hauser/Marsh (HHM) formalism to quantify the difference in clustering between the ordered and disordered states of GaInP$_2$.

2. Experimental

We grew our samples on GaAs substrates by atmospheric-pressure metal-organic chemical vapor deposition, at a growth rate of 100 nm/min. The inlet V/III ratio was 65–80 for most of the samples, and 35 for the remainder. Under these conditions, a growth temperature T_g of 670 °C produces partially-ordered (hereafter "ordered") low-band-gap material, while growth temperatures of 600 °C and 750 °C produce relatively disordered (hereafter "disordered") high-band-gap material (Kurtz *et al* 1990). The samples are nominally undoped, with *n*-type Hall carrier concentrations in the range 10^{15}/cm^3 to 10^{16}/cm^3. The Hall mobility μ_H was measured using the van der Pauw method.

To obtain reliable mobility measurements, considerable care was required to avoid parasitic conduction associated with the substrate, both at room temperature and above 500-600 K. For semi-insulating substrates, mobility measurements are significantly affected by formation of a two-dimensional electron gas (2DEG) (Dingle *et al* 1978) at the GaInP$_2$/GaAs interface (Ikeda and Kaneko 1989). Using *p*-type substrates, or placing a *p*-type GaInP$_2$ buffer layer between the nominally undoped layer and the substrate, proved an adequate solution for measurements below 500-600 K; for higher temperatures it was necessary to remove the substrate.

To verify this, we measured the temperature-dependent Hall mobility μ_H(300K–20K) of two GaInP$_2$ samples grown simultaneously at T_g = 670 °C, one on a semi-insulating substrate and one on a p-type substrate. The sample on the p-type substrate shows a 300-K mobility of 620 cm^2/V·sec, rising to a maximum and then decreasing as the temperature is lowered, consistent with the behavior expected for the mobility of a bulk semiconductor. In contrast, the sample on the semi-insulating substrate shows an apparent room-temperature mobility of 975 cm^2/V·sec, rising to a plateau and then leveling off with decreasing temperature. This latter behavior is characteristic of the presence of a 2DEG, which has artificially inflated the 975 cm^2/V·sec room-temperature value. Furthermore, even with a p-type substrate or a p-type GaInP$_2$ buffer layer between the nominally-undoped layer and the substrate, above 500-600 K we observe a sharp increase in the apparent carrier concentration several hundred degrees below the temperature at which intrinsic conduction in GaInP$_2$ would be significant. This behavior is also due to parasitic conduction.

For this reason, for all of the highest-temperature (500-600 K and above) mobility measurements reported here, the substrate was removed by a combination of mechanical lapping and chemical etching. For lower temperatures, additional measurements were made on samples grown on p-type substrates, or for which a p-type GaInP$_2$ buffer layer was first grown on the substrate. In this lower-temperature regime, the mobilities measured for the samples on p-type substrates or buffer layers agree very well with the mobilities for the substrateless samples. For the samples on substrates, in order to avoid contacting the underlying p-type layer a thin n-doped GaAs contacting layer was grown on the GaInP$_2$, permitting nonalloyed plated gold contacts to the sample (the contacting layer was etched away after plating). For the samples without substrates, annealed indium contacts usually proved more convenient; in this case, the wires to the contacts were mechanically anchored near the contacts, permitting operation above the melting point of indium.

3. Formalism

We shall treat cluster scattering using Marsh's (1982) extension of the Harrison-Hauser (1976a,b) theory of alloy scattering. This model treats the cluster scattering potential as a square well of energy V_0 and cluster radius r_c; formally, the scattering potential $eV(r)$ is given by

$$eV(r) = \begin{cases} V_0: & r \le r_c \\ 0: & r > r_c \end{cases}. \tag{1}$$

We show here the derivation of the momentum-destroying cluster scattering rate $1/\tau_C$, filling in some of the details missing from Marsh (1982). Proceeding from Seeger's (1988) treatment of elastic scattering, the general momentum-destroying scattering rate $1/\tau_m$ is given by

$$1/\tau_m = \frac{2\pi N \hbar k}{m^*} \int_0^\pi \left(\frac{Vm^*}{2\pi \hbar^2}\right)^2 |H_{kk'}|^2 (1 - \cos\theta) \sin\theta \, d\theta, \tag{2}$$

where N is the density of scattering centers, k and k' are the initial and final electron wave vectors, m^* is the conduction band effective mass, V is the crystal volume, and θ is the scattering angle. The matrix element $H_{kk'}$ is given for spherical potentials by

$$H_{kk'} = \frac{4\pi}{V} \int_0^\infty [eV(r)] \frac{r \sin qr}{q} dr, \tag{3}$$

$$q^2 = 2k^2(1 - \cos\theta). \tag{4}$$

Inserting eq. (1) into eq. (3) and integrating, and then inserting the result for $|H_{kk'}|^2$ into eq. (2) and changing the variable of integration from θ to $u = 2qr_c$, we obtain

$$1/\tau_C = \frac{4V_0^2 r_c^3 f m^* k}{3\hbar^3} \left[\frac{9}{4(kr_c)^4} \int_0^{2kr_c} \frac{(\sin u - u \cos u)^2}{u^3} \, du \right], \tag{6}$$

where we have defined $f = 4\pi r_c^3 N/3$, the fraction of the crystal volume occupied by the clusters. In the absence of band nonparabolicity, which will be negligible for GaInP$_2$ at low electron energies, eq. (6) is equivalent to Marsh's eq. (5), but is more convenient. The electron drift mobility μ is given by

$$\mu = e\langle\tau\rangle/m^*, \tag{7}$$

$$\langle\tau\rangle = \frac{4}{3\pi^{1/2}} \int_0^\infty \tau(\varepsilon) \, \varepsilon^{3/2} \exp(-\varepsilon) \, d\varepsilon, \tag{8}$$

where ε is the electron energy in units of k_BT, and Boltzmann statistics apply.

The term in the square brackets of eq. (6) goes to 1 for $kr_c \ll 1$; in this regime, the cluster mobility μ^C goes as $T^{-1/2}$. From eq. (8) the value of $\langle\tau\rangle$ is determined mostly by electrons of energy on the order of k_BT. Therefore the temperature range for which $\mu^C \propto T^{-1/2}$ is given by

$$k_BT \ll \hbar^2/(2m^* r_c^2) ; \tag{9}$$

that is, the temperature at which μ^C starts to deviate from a power law temperature dependence goes as $1/r_c^2$. Because $k = (2m^* k_B T\varepsilon)^{1/2}\hbar$, the small $\{r_oT\}$ limit (as quantified by eq. (9)) of μ^C is

$$\lim_{r_c, T \to 0} \mu^C = (e/m^*) \, \hbar^4 / [m^{*3/2}(2\pi k_BT)^{1/2}V_0^2 r_c^3 f] , \tag{10}$$

in agreement with the Harrison-Hauser model.

The measurements give the Hall mobility $\mu_H = \langle\tau^2\rangle/\langle\tau\rangle^2 \mu$, rather than the the drift mobility μ. Therefore, for comparison with our data we will calculate μ_H, which has qualitatively the same dependence on r_c and T as μ has. Typically, μ_H is within a factor of two of μ. Figure 1 shows the cluster Hall mobility μ_H^C as a function of temperature T for $V_0^2 r_c^3 f = (0.3 \text{ eV})^2 (10 \text{ Å})^3$ and various values of r_c, with m^* appropriate for GaInP$_2$ (see Table I). That is, when r_c is increased $V_0^2 f$ is decreased to keep $V_0^2 r_c^3 f$ constant. As expected, for small r_c and T, $\mu_H^C \propto T^{-1/2}$. However, as r_c is increased, μ_H^C starts to increase with increasing temperature. The temperature dependence of the mobility thus provides a way of studying of r_c.

To include other scattering mechanisms such as acoustic deformation potential scattering in the mobility calculation, we assume that the various types of scattering processes are independent, so that the total scattering rate $1/\tau$ is given by the sum of scattering rates for the individual mechanisms. Because the scattering rates are energy dependent, the reciprocal of the total mobility is not equal to the sum of the reciprocals of the mobilities for the

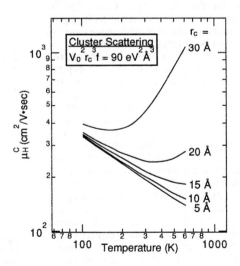

Figure 1. The cluster Hall mobility μ_H^C calculated for GaInP$_2$ conduction band effective mass (see Table I) as a function of temperature for various values of cluster radius r_c, with $V_0^2 r_c^3 f = 90 \text{ eV}^2\text{Å}^3$.

individual scattering mechanisms.

4. Results and Analysis

Figure 2 shows the room temperature Hall mobilities and carrier concentrations for a number of disordered (Tg = 600 °C and 750 °C) and ordered (Tg = 670 °C) samples. The data fall into two clearly defined groups: all of the ordered samples have mobilities μH ~500 cm2/V•sec, while both the Tg = 600 °C and 750 °C disordered samples have μH roughly double this value. There is no significant dependence of μH on the carrier concentration, either between the two groups or within a group.

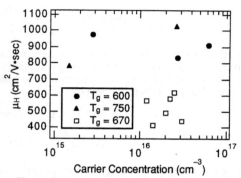

Figure 2. Room-temperature Hall mobilities and carrier concentrations for a number of ordered (growth temperature $T_g = 670$ °C) and disordered ($T_g = 600$ °C, 750 °C) GaInP$_2$ samples.

Therefore, neither ionized-impurity nor space-charge scattering [15](Weisberg 1962; Figure 2 of this paper implies that space-charge scattering is decreasing important with increasing temperature) are significantly affecting the mobility at room temperature. These effects should become even less important with increasing temperature [15]. It is worth noting that Adams et al (1980) [16] have used pressure dependent mobility measurements to show that space-charge scattering is negligible compared to alloy scattering in GaInAsP alloys at room temperature. We

therefore ascribe the difference in mobility between the disordered and ordered samples at room temperature and above to stronger cluster scattering in the ordered samples. While it is possible that Tg = 600 °C and 750 °C disordered samples differ in their degree of ordering, any such differences are too small to be detectable in our data.

Figure 3 shows the measured temperature-dependent μ_H for typical ordered and disordered GaInP$_2$ samples. In contrast to the behavior of $\mu_H(T)$ observed by Bhattacharya *et al* (1981,1985) for certain arsenic-containing alloys, where μ_H was observed to start bending back up at 400-600 K, the mobilities for both the ordered and disordered GaInP$_2$ decrease smoothly with temperature. A comparison with Figure 1 thus suggests that the cluster size r_c must be less than about 20 Å. A more quantitative analysis of the data requires the inclusion of the other scattering mechanisms of importance at room temperature and above: polar optical (PO) and acoustic deformation potential (AD) scattering. For PO scattering, a relaxation time cannot strictly be

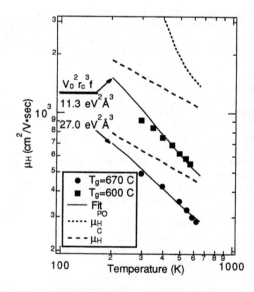

Figure 3. The measured μ_H for typical ordered ($T_g = 670$°C) and disordered ($T_g = 600$°C) GaInP$_2$ samples, and the calculated μ_H including polar optical, acoustic deformation potential, and cluster scattering, from the parameters in Table I. The corresponding individual mobilities μ_H^{PO} and μ_H^{C} are also shown; μ_H^{AD} lies above the upper y-limit of the graph.

defined; we use the relaxation-time approximation of Harrison and Hauser (1976b), which agrees quite well with more sophisticated calculations (Fortini *et al* 1970). Table I shows the material parameters used, from Adachi (1982) and Rode (1975); we neglect the possibility that these parameters might be different for the ordered and disordered states of GaInP$_2$. In the small-r_c ($r_c < \sim 15$ Å) regime of eq. 10, r_c and $V_0^2 f$ cannot be determined independently. Figure 3 shows the fit to the data in this regime. The small-r_c cluster scattering fit parameter $V_0^2 r_c^3 f$ takes the values 27.0 eV^{2}Å^3 and 11.3 eV^{2}Å^3 for the ordered and disordered samples, respectively. If the contribution to the net scattering rate from mechanisms other than cluster scattering has been underestimated in this modeling, the actual difference between the ordered and disordered $V_0^2 r_c^3 f$ values would be greater still.

The value of the acoustic deformation potential E_{AD} is not as precisely determined as that of the other material parameters: for instance, values as high as 18 eV have been given for InP (Nag and Dutta 1978). By adjusting the value of E_{AD} (constraining it to lie within a physically reasonable regime of $E_{AD} \leq 20$ eV), it is possible to fit the data for r_c as large as 20 Å. However, due to the

TABLE I. Material parameters used in the calculation of the Hall mobility for GaInP$_2$. Values are interpolated from the values for GaP and InP from the references, using the interpolation methods of Harrison and Hauser (1976b).

Quantity	Value	Ref.
Acoustic deformation potential E_{AD}	9.9 eV	a
spherically averaged elastic constant c_l	1.437×10^{12} dyn/cm^2	b
Debye optical phonon temperature θ_D	539 K	a
static dielectric constant ε_s	11.74	a
dynamic dielectric constant ε_d	9.33	a
effective mass m^*/m_0	0.109	b

[a]Rode (1975)
[b]Adachi (1982)

behavior of $\mu_H(T)$ for large r_c illustrated in Figure 1, for $r_c > \sim 20$ Å it is not possible to fit a calculated $\mu_H(T)$ to the data even by adjusting the value of E_{AD} within the above constraints. Thus, as argued above, the temperature dependence of μ_H sets an upper limit of about 20 Å on r_c. Because the fractional cluster volume f necessarily satisfies $f \leq 1$, and because at $r_c = 20$ Å the alloy scattering rate τ_C is still decreasing with increasing r_c, the upper limit of 20 Å on r_c provides a lower limit on the well depth V_0. At $r_c = 20$ Å, the data are fit by $V_0 f^{1/2} = 55$ meV and 38 meV for the ordered and disordered samples, respectively. Therefore, the modeling provides a lower limit of a few tens of meV for V_0. Furthermore, the cluster diameter $2r_c$ clearly cannot be significantly less than the 5.65 Å lattice constant. The fit $V_0^2 r_c^3 f = 27.0$ eV^{2}Å^3 (ordered) and 11.3 eV^{2}Å^3 (disordered) thus yields an upper bound on $V_0 f^{1/2}$ of roughly 1 eV (ordered) and 0.4 eV (disordered). These latter bounds are doubtless very conservative.

In a random alloy, the only source of cluster scattering centers would be clusters arising from random statistics. However, in the partially ordered materials under consideration here, clusters of CuPt-like order are present and may be acting as the dominant scattering sites. If we make the reasonable assumption that V_0 is approximately the same for the ordered and disordered materials, either r_c or f (or both) must be greater for the ordered material than for the disordered. There would appear to be two ways this might occur: the scattering centers might be composed of (1) small domains of the ordered material dispersed throughout the surrounding matrix of disordered material, or (2) random clusters in the disordered fraction of the material. However, in case (2) r_c would represent the mean cluster size due to random scattering in the disordered fraction of the material and would be the same in the more- and less-ordered materials. However, f, the fractional volume of scattering cluster centers, would then be greater in the disordered material than in the ordered, so that in case (2) one would expect stronger cluster scattering in the disordered material than in the ordered, contrary to what we observe.

This implies that the mobility-limiting cluster scattering centers are predominantly composed of the ordered fraction of the crystal volume. In this picture, a *perfectly* ordered sample would have a *higher* mobility than any of the samples in Figure 2. Unfortunately, it may not be possible to grow such a sample to verify this prediction. Assuming V_0 and r_c to be the same in both materials, the fit of Figure 3 means that the fractional volumes of ordered cluster scattering sites in the two materials are related by $f_{ordered}/f_{disordered} \approx 2.4$.

5. Summary

In summary, the cluster scattering rate is greater in ordered $GaInP_2$ than in disordered $GaInP_2$. In terms of the HHM model, we obtain an upper limit of ~20 Å on the cluster size for both materials. Fitting to $\mu_H(T)$ permits estimating the relative scattering cluster volumes in the two materials.

Acknowledgments

We thank P. K. Bhattacharya, S. R. Kurtz, and K. A. Bertness for useful discussions.

References

Adachi S 1982 J. Appl. Phys. **53** 8775
Adams A R, Tatham H L, Hayes J R, El-Sabbahy A N and Greene P D 1980 Electron. Lett. **16** 562
Bhattacharya P K and Ku J W 1985 J. Appl. Phys. **58** 1410
Bhattacharya P K, Ku J W, Owen S J T, Olsen G H and Chiao S-H 1981 IEEE J. Quantum Electron. **QE-17** 150
Dingle R, Störmer H L, Gossard A C and Wiegmann W 1978 Appl. Phys. Lett. **33** 665
Fortini A, Diguet D and Lugand J 1970 J. Appl. Phys. **41** 3121
Froyen S and Zunger A 1991 Phys. Rev. Lett. **66** 2132
Gomyo A, Suzuki T and Iijima S 1988 Phys. Rev. Lett. **60** 2645
Harrison J W and Hauser J R 1976a Phys. Rev. B **13** 5347
Harrison J W and Hauser J R 1976b J. Appl. Phys. **47**, 292
Hong W-P, Bhattacharya P K and Singh J 1987 Appl. Phys. Lett. **50** 618
Ikeda M and Kaneko K 1989 J. Appl. Phys. **66** 5285
Kurtz S R, Olson J M and Kibbler A E 1990 Appl. Phys. Lett. **57** 1922
Marsh J H 1982 Appl. Phys. Lett. **41** 732
Nag B R and Dutta G M 1978 J. Phys. C **11** 119
Oh J E, Bhattacharya P K, Chen Y C, Aina O and Mattingly M 1990 J. Electron. Mater. **19** 435
Olson J M, Kurtz S R, Kibbler A E and FaineP 1990 Appl. Phys. Lett. **56** 623
Rode D L 1975, in *Semiconductors and Semimetals*, Vol. 10, edited by R. K. Willardson and A. C. Beer (New York: Academic Press) p 85
Seeger K 1988 *Semiconductor Physics*, 4th ed. (New York: Springer-Verlag)
Weisberg L R 1962 J. Appl. Phys. **33** 1817

Inst. Phys. Conf. Ser. No 120: Chapter 8
Paper presented at Int. Symp. GaAs and Related Compounds, Seattle, 1991

389

Local structures of single-phase and two-phase GaAs$_{1-x}$Sb$_x$ studied by fluorescence-detected EXAFS

Hirotaka Yamaguchi, Yoshikazu Takeda* and Hiroyuki Oyanagi

Electrotechnical Laboratory, Umezono, Tsukuba-shi, Ibaraki 305, Japan
*Department of Materials Science and Engineering, Nagoya University, Furocho, Chikusa-ku, Nagoya 464-01, Japan

ABSTRACT: Local structures of a single-phase GaAs$_{0.5}$Sb$_{0.5}$ alloy grown on InP by MBE and two-phase GaAs$_{1-x}$Sb$_x$ alloys grown from melt were studied by fluorescence-detected EXAFS. We found that in the single-phase GaAs$_{0.5}$Sb$_{0.5}$, the Ga-As bond is stretched from 2.447Å in pure GaAs to 2.469Å and the Ga-Sb bond is contracted from 2.640Å in pure GaSb to 2.619Å. In the phase separated GaAs$_{1-x}$Sb$_x$, the bond lengths were the same as the values in pure binary compounds. These results indicate that, under the lattice-matching constraint, the single-phase GaAs$_{0.5}$Sb$_{0.5}$ is stabilized with the bond length relaxation.

1. INTRODUCTION

The properties of semiconductor alloys of complete miscibility, such as A$_x$B$_{1-x}$C, have been analyzed using the virtual crystal approximation (VCA) where the interatomic distances are assumed to vary linearly from one end to the other with lattice spacing or alloy composition x (Nordheim 1931). However, structural studies by extended absorption fine structure (EXAFS) show that bond lengths in ternary Ga$_x$In$_{1-x}$As(Mikkelsen 1983) and in quaternary Ga$_x$In$_{1-x}$As$_y$P$_{1-y}$(Oyanagi 1985) alloys substantially deviate from the average interatomic distance almost preserving the values in pure binary compounds. These experiments had been done on the semiconductor alloys of complete miscibility. Compared with these miscible alloys not much attention had been paid to GaAs$_{1-x}$Sb$_x$ (GAS) which is a typical example of immiscible alloys (Gratton 1973).

It has been shown that GAS in the immiscible region can be grown by molecular beam epitaxy (MBE) (Chang 1977, Waho 1977, Chiu 1985, McLean 1985, Klem 1987, Nakata 1988) and organometallic vapor phase epitaxy (OMVPE) (Cooper 1982, Bedair 1983, Cherng 1984, 1986). Especially, high quality single-phase GaAs$_{0.5}$Sb$_{0.5}$ lattice-matched with InP is obtained by OMVPE(Cherng 1986) and MBE(Chiu 1985, Klem 1987, Nakata 1988) although it is at the center of miscibility gap.

The reason why immiscible GAS alloy separates into two phases is that the enthalpy of mixing due to the difference between bond lengths (or lattice parameters d$^{(0)}$'s; d$^{(0)}$=r$^{(0)}$4/$\sqrt{3}$) of Ga-As (r$^{(0)}$$_{Ga-As}$=2.447Å) and Ga-Sb (r$^{(0)}$$_{Ga-Sb}$=2.640Å) is too large to keep the single phase and then the alloy separates into a GaAs-rich domain and a GaSb-rich domain to keep the original bond lengths in the pure binary compounds(Osamura 1972 and for review, Zunger 1989 and Ichimura 1989). An EXAFS study on GAS was reported for the miscible regions near both binary ends by Marbeuf et al. (Marbeuf 1986). The bond lengths in the alloys, r$_{Ga-As}$ and r$_{Ga-Sb}$, are found to be almost the same as those in the pure compounds, r$^{(0)}$$_{Ga-As}$ and r$^{(0)}$$_{Ga-Sb}$. In other words, the composition variation of the bond lengths is quite small compared with miscible ternary systems such as $Ga_{1-x}In_xAs$ (Mikkelsen 1983).

In this paper, we report the first measurements of EXAFS for single-phase $GaAs_{0.5}Sb_{0.5}$ grown on InP by MBE to reveal the microstructures of the artificially grown and highly stressed material.

2. EXPERIMENTAL

Single-phase $GaAs_{0.5}Sb_{0.5}$ layers were epitaxially grown on InP (001) surface by MBE at T_g=460, 530 and 560°C (Nakata 1988). The thickness was 2.5µm to avoid a possible effect of the interface region between the substrate and the epitaxial layer for the EXAFS measurements. It was confirmed by X-ray diffraction from the samples grown at 460°C and 530°C that the epitaxial layer was lattice-matched with InP within the accuracy of $3.5\sim4\times10^{-3}$, and half-width of the X-ray rocking curve was 180~190 arcsec.

The Ga K-edge and As K-edge EXAFS spectra for the single phase $GaAs_{0.5}Sb_{0.5}$ layers were taken by the fluorescence-detection technique using a spectrometer on beam line 4C at the Photon Factory in Tsukuba (Oyanagi 1985). The storage ring was operated at 2.5GeV with a typical positron current of 250-100mA. A sagittally bent Si(111) double crystal monochromator was used to monochromatize the white X-ray beam from the storage ring. The harmonic content of the incident beam was reduced by detuning the crystals from a parallel setup.

The phase-separated alloy with compositions of $GaAs_{0.28}Sb_{0.72}$ and $GaAs_{0.98}Sb_{0.02}$ was prepared as a reference sample. The alloy was grown from melt with equal mole fractions of GaAs and GaSb in an Ar-filled quartz ampoule using a similar temperature process as Gratton and Woolley(Gratton 1973). The composition was determined by x-ray diffraction from the powdered alloy. EXAFS for the powdered phase-separated alloy and binary compounds were measured in the transmission mode.

3. RESULTS

The Ga K-EXAFS of the single phase $GaAs_{0.5}Sb_{0.5}$ grown at 460°C is shown in Fig. 1. Figure 2 indicates the Fourier transform of EXAFS on the Ga K-edge. Prominent peaks

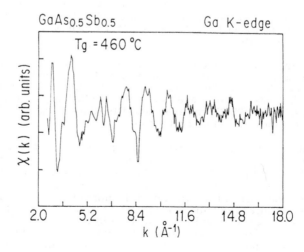

Fig. 1   The Ga K-EXAFS oscillation of single-phase $GaAs_{0.5}Sb_{0.5}$ grown at 460°C.

observed at around 2Å are due to the nearest-neighbor anions around the Ga atom, *i.e.*, As and Sb atoms. In Fig. 2, the first nearest peak apparently consists of two overlapping peaks due to two kinds of anion shells. The prominent peak region between 1.2Å and 3.1Å is Fourier-filtered into k-space and the result is analyzed by a two shell model which is based on the assumption that As and Sb anions are randomly distributed around the Ga atom with average coordination numbers of 2 for each atom. The two shell model is well fitted to the spectra for the samples grown at 460 and 530°C, and the Ga-As bond length $r_{Ga-As}=2.469$Å with the Debye-Waller factor $\sigma_{Ga-As}=0.09$Å and the Ga-Sb bond length $r_{Ga-Sb}=2.619$Å with $\sigma_{Ga-Sb}=0.03$Å are obtained. The same procedure for the sample grown at 560°C, however, shows quite different results from those for the other samples, *i.e.*, the Ga K-EXAFS spectrum cannot be fitted with the two-shell-model.

The As K-EXAFS oscillations are Fourier-transformed and the results are shown in Fig. 3. The nearest-neighbor region between 1.2Å and 3.0Å is analyzed by using a single-

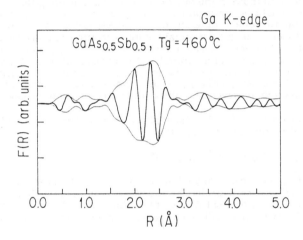

Fig. 2   Fourier transform of the Ga K-EXAFS shown in Fig. 1.

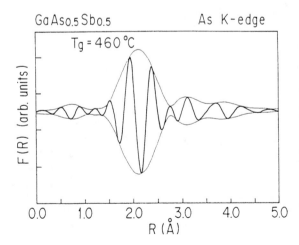

Fig. 3   Fourier transform of the As K-EXAFS of single-phase GaAs$_{0.5}$Sb$_{0.5}$ grown at 460°C.

shell-model. The results for the As-Ga bonds are in agreement with those by Ga K-EXAFS within experimental error (± 0.01Å). The result for the 560°C sample is well fitted to the single-shell-model as well as for other two samples in contrast to the result of the Ga K-edge. The present results of r_{Ga-As} and r_{Ga-Sb} are summarized in Fig. 4 as a function of Sb composition, x, to compare with those in solid solutions in the miscible region near both ends and calculated values by a valence force field (VFF) model by Marbeuf *et al.* (Marbeuf 1986). The bond lengths in the miscible region are almost the same as those in the binary compounds. On the other hand, the bond lengths in the single-phase GaAs$_{0.5}$Sb$_{0.5}$ are found to be slightly relaxed as expected by a VFF model.

4. DISCUSSION

Marbeuf *et al.* reported that the second-nearest-neighbor distance r_{Ga-Ga} in the miscible region was very close to those in the binary compounds, indicating that the lack of bond-bending relaxation might be a reason for the large miscibility gap. Qteish *et al.* (Qteish 1989) showed that the phase diagram of the GaAs$_{1-x}$Sb$_x$ system was successfully reproduced by a theoretical calculation taking into account the relaxation of the second-nearest-neighbors. However, the calculated bond lengths, r_{Ga-As} and r_{Ga-Sb}, were in disagreement with those determined from EXAFS by Marbeuf *et al.* They attributed the disagreement to experimental errors in dilute systems near the end binaries. Our results on a phase-separated GaAs$_{0.28}$Sb$_{0.72}$, whose composition is higher than those which Marbeuf *et al.* measured, however, also shows the deviation from the calculated values.

The relaxation parameter, ε, for the single phase GaAs$_{0.5}$Sb$_{0.5}$ obtained in the present study is calculated by the following equation

$$\varepsilon = (r_{Ga-Sb} - r_{Ga-As})/(r^{(0)}_{Ga-Sb} - r^{(0)}_{Ga-As}) = 0.78$$

where r_{Ga-As} and r_{Ga-Sb} are determined by the present EXAFS analysis. This result is in good agreement with those expected by the VFF. This means that the single-phase GaAs$_{0.5}$Sb$_{0.5}$ is stabilized on the InP substrate with the bond length relaxation given by a simple elastic theory such as VFF which takes only the nearest-neighbor interaction

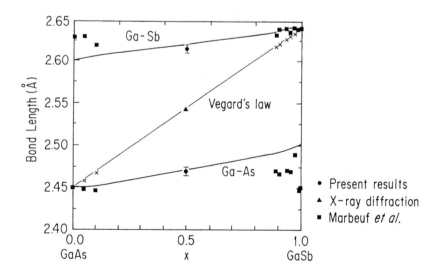

Fig. 4   Interatomic distances r_{Ga-Sb} and r_{Ga-As} in GaAs$_{1-x}$Sb$_x$. Thick solid lines indicate the calculated results from the VFF model after Marbeuf *et al.*

into account. The present EXAFS results confirm the fabrication of the single phase from the microscopic viewpoint. Next, we discuss the substrate temperature dependence of the crystal growth of the single-phase $GaAs_{0.5}Sb_{0.5}$. We observed anomalous Ga K-EXAFS oscillations for the single-phase layer grown at 560°C which cannot be explained by a uniform single phase with the average local structure. Thus the anomaly may originate from the nonuniform distribution of short range order or phase separation. The possibility of phase separation, however, is ruled out on the basis of a careful analysis of EXAFS data measured for the phase-separated sample. Nakata *et al.* (Nakata 1988) reported the optical and the electronic properties of several single-phase $GaAs_{0.5}Sb_{0.5}$ samples grown at the temperatures between 470°C and 560°C. The growth temperature dependence of electron mobility and electron concentration showed the maximum values at 510°C and the samples grown at above 540°C are highly resistive. The crystal structures by X-ray diffraction, however, did not exhibit any critical differences corresponding to these anomalies, although the half-width of the rocking curve, 390arcsec for this sample, is larger than others. The anomalies in the Ga K-EXAFS oscillations for the sample grown at 560°C are, therefore, due to the structural modification not in a long range order but in a short range order involving clustering. The possibility of clustering in this system was indicated by Cherng *et al.* (Cherng 1986).

5. CONCLUSION

The local structures of single-phase $GaAs_{0.5}Sb_{0.5}$ grown on InP by MBE were studied by fluorescence-detected EXAFS. Interatomic distances r_{Ga-As} and r_{Ga-Sb} in the single-phase $GaAs_{0.5}Sb_{0.5}$ are in agreement with the calculated values from the VFF model. In the miscible region, on the other hand, they deviate from a simple elastic calculation and almost the same bond lengths as the pure binary values are obtained. These results indicate that the single-phase $GaAs_{0.5}Sb_{0.5}$ on InP is stabilized by the InP substrate with the bond length relaxation. In the sample with Tg=560°C the possibility of a structural modification in a short range order such as clustering was indicated.

Acknowledgment
The authors express their thanks to Fujitsu Laboratories Limited for providing $GaAs_{0.5}Sb_{0.5}$/InP samples. This work has been performed as a part of a project (Proposal No. 87-065) approved by the Photon Factory Program Advisory Committee, and was supported in part by the Grant-in-Aid for Scientific Research on Priority Areas, "New Functionality Materials-Design, Preparation and Control-" from the Ministry of Education, Science and Culture.

References
Bedair S M, Timmons M L, Chiang P K, Simpson L and Hauser J R (1983) *J. Electron. Mat.* 12 959
Chang C-A, Ludeke R, Chang L L and Esaki L (1977) *Appl. Phys. Lett.* 31 759
Cherng M J, Stringfellow G B and Cohen R M (1984) *Appl. Phys. Lett.* 44 677
Cherng M J, Cherng Y T, Jen H R, Harper P, Cohen R M and Stringfellow G B (1986) *J. Electron. Mat.* 15 79
Chiu T H, Tsang W T, Chu S N G, Shah J and Ditzenberger J A (1985) *Appl. Phys. Lett.* 46 408
Cooper III C B, Saxena R R and Ludowise M J (1982) *J. Electron. Mat.* 11 1001
Gratton M F and Woolley J C (1978) *J. Electrochem. Soc.* 127 657
Ichimura M and Sasaki A (1989) *J. Cryst. Growth* 98 18
Klem J, Huang D, Morkoc H, Ihm Y E and Otsuka N (1987) *Appl. Phys. Lett.* 50 1364
McLean T D, Kerr T M, Westwood D I, Grange J D and Murgatroyd I J (1985) *Proc. of 11th Int'l. Symp. on GaAs and Related Compounds, Biarritz* 1984 145
Marbeuf A, Karouta F, Dexpert H, Lagarde P and Joullie A (1986) *J. Phys. (Paris) Suppl.* 47 369
Mikkelsen Jr. J C and Boyce J B (1983) *Phys. Rev.* B28 7130
Nakata Y, Fujii T, Sandhu A, Sugiyama Y and Miyauchi E (1988) *J. Cryst. Growth* 91 655
Nordheim L (1931) *Ann. Phys. (Leipz)* 9 607 and 641

Osamura K, Nakajima K and Murakami Y (1972) *J. Jpn. Inst. Metals* 36 744

Oyanagi H, Matsushita T, Tanoue H, Ishiguro T and Kohra K (1985) *Jpn. J. Appl. Phys.* 24 610

Oyanagi H, Takeda Y, Matsushita T, Ishiguro T and Sasaki A (1986) *Proc. of 12th Int'l. Symp. on GaAs and Related Compounds, Karuizawa* 1985 295

Qteish A, Motta N and Balzarotti A (1989) *Phys. Rev.* B39 5987

Waho T, Ogawa S and Maruyama S (1977) *Jpn. J. Appl. Phys.* 16 1875

Zunger A and Wood D M (1989) *J. Cryst. Growth* 98 1

Inst. Phys. Conf. Ser. No 120: Chapter 8

395

Paper presented at Int. Symp. GaAs and Related Compounds, Seattle, 1991

Growth and optical properties of natural InAs$_{1-x}$Sb$_x$ strained layer superlattices

Ferguson I T[1], Norman A G[1], Seong T-Y[2], Thomas R H[3], Phillips C C[3], Zhang X M[1], Stradling R A[1,3], Joyce B A[1] and Booker G R[2]

1. IRC for Semiconductor Materials, Blackett Laboratory, Imperial College, Prince Consort Road, London SW7 2BZ
2. Dept. of Materials, University of Oxford, Parks Road, Oxford OX1 3PH
3. Dept. of Physics, Blackett Laboratory, Imperial College, Prince Consort Road, London SW7 2BZ

ABSTRACT: MBE growth of InAs$_{1-x}$Sb$_x$ below 430°C results in its coherent phase separation into tetragonally distorted platelets of two different compositions, that is the formation of a 'natural' strained layer superlattice (n-SLS). The results presented here suggest that the n-SLS is generated at or near to the growth front by a process of surface diffusion. A similar behaviour in GaAs$_{1-x}$Sb$_x$ suggests that it may be associated with a possible miscibility gap in the As-Sb phase diagram rather than that of the ternary alloy. A comparison of the optical properties of n-SLS with those of conventionally grown SLSs is made.

1. INTRODUCTION

The growth of some III-V quaternary and ternary compounds can prove difficult by equilibrium growth techniques, e.g. liquid phase epitaxy (LPE), due to the existence of miscibility gaps within which spinodal decomposition can give rise to either alloy clustering or phase separation. Indeed, plan-view transmission electron microscopy (TEM) of LPE (001) In$_{1-x}$Ga$_x$As$_y$P$_{1-y}$ layers [1-3] has revealed a coarse tweed-like contrast modulation with a periodicity of 1000-2000Å along the orthogonal [100] and [010] directions. Energy dispersive X-ray microanalysis (EDX) of a similar tweed-like structure in Ga$_{0.53}$In$_{0.47}$As showed that it corresponds to a ±5% modulation in composition of gallium and indium [2]. Similar observations have been made for III-V ternary alloys grown by non-equilibrium growth techniques such as MBE [3]. The existence of regions of different composition have also been reported for MBE grown Ga$_x$Al$_{1-x}$As as on (110) orientated substrates but not for (001) substrates [4]. No theoretically predicted equilibrium miscibility gap exists for GaAlAs and this phenomena is not observed for LPE grown material. So phase separation was attributed to a proposed miscibility gap due to an exchange reaction between Al and Ga during MBE growth at the (110) surface. Recent theoretical calculations for quaternary alloys have shown that bulk spinodal decomposition may result in coherent phase separation and consequently the formation of tetragonally strained layers of different composition [5].

During a comprehensive study of the MBE growth of InAs$_{1-x}$Sb$_x$ [6-9], epitaxial layers grown below 430°C were found to have undergone coherent phase separation into tetragonally distorted platelets of material of two different compositions. In particular, the growth of an InAs$_{0.5}$Sb$_{0.5}$ alloy results in an extended platelet structure of approximate compositions InAs$_{0.3}$Sb$_{0.7}$ and InAs$_{0.7}$Sb$_{0.3}$, i.e. the spontaneous formation of a 'natural' strained layer superlattice (n-SLS) [9]. This paper discusses the current understanding of the n-SLS growth mode and a comparison of the optical properties of a n-SLS structure in InAsSb with those of conventionally grown 'artificial' SLS structures is made. Preliminary results of n-SLS formation in GaAsSb are also presented.

2. EXPERIMENTAL DETAILS

InAs$_{1-x}$Sb$_x$ (0≤x≤1) layers were grown by MBE in a VG Semicon V80H system using elemental In, Sb$_4$ and As$_4$ (all 6N purity) as source materials. The layers were grown onto either (001) Cr-doped semi-insulating GaAs substrates or epitaxial layers of InAs deposited on GaAs substrates or InSb substrates of different orientation (001), (110) and (111)B. Reflection high energy electron diffraction (RHEED) intensity oscillations of the specular beam were recorded for an incident beam energy of 12.5 keV at an angle of incidence of ~2° for various azimuths. Layers were examined by TEM using standard two-beam diffraction conditions, and selected area transmission electron diffraction (TED) for the two orthogonal [110] and [$\bar{1}$10] cross-sections in a Philips CM20 electron microscope. Photoresponse spectra were obtained using a Bomem Fourier Transform Infrared Spectrometer on samples of approximate dimensions 3x10 mm² which wer mounted in a CTI-cryogenics cryo-cooler operating at 10 K.

3. RESULTS AND DISCUSSION

3.1 The Growth of InAs$_{1-x}$Sb$_x$ (0≤x≤1)

RHEED oscillations were monitored during the growth to investigate the relative incorporation rates of the three elements, indium, arsenic and antimony. Accurate control of the growth parameters was necessary to reproduce different alloy compositions. Using a technique which has previously been used to calibrate As$_4$ incorporation during the MBE growth of GaAs [10], Sb$_4$-induced RHEED oscillations were observed following the controlled deposition of ~10 monolayers of indium onto a (001) InSb substrate. The period of the oscillation was used to give an accurate incorporation rate for antimony on InSb. A systematic calibration of Sb$_4$ incorporation was completed as a function of the antimony cell temperature at a constant substrate temperature (T_g) of 370 °C, figure 1. T_g was measured relative to $T_t{}^S$, where $T_t{}^S$ is a system measured temperature at which the InSb (001) reconstruction changes from c(4x4) to an asymmetric (1x3), estimated to be 390°C. The apparent activation energy of sublimation for Sb$_4$ was E_{Sb}=1.45 eV but this value was found to depend strongly on the source geometry. InAsSb layers of a specific composition were grown by using calibrated indium and antimony fluxes, corrected for changes in lattice constant, and an incorporation rate of arsenic slightly larger than that required for stoichiometry since antimony tends to incorporate in preference to arsenic [11]. Other groups [12,13] have used very much larger antimony and arsenic fluxes than that required for a particular InAs$_{1-x}$Sb$_x$ (or GaAs$_{1-x}$Sb$_x$) composition, controlling the alloy composition by varying growth temperature and relying on the difference in sublimation and atomisation energies of Sb$_4$ and As$_4$. This has limited InAs$_{1-x}$Sb$_x$ alloy compositions to x≥0.8 for growth temperatures ≤450°C in some instances [11].

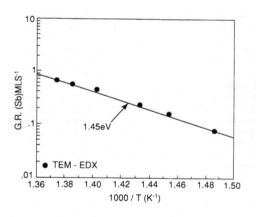

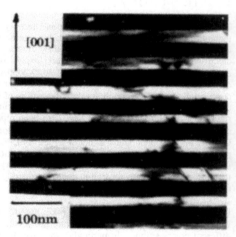

Figure 1. An Arrhenius plot of the apparent activation energy of Sb$_4$ sublimation for a standard Knudsen type cell, 1.45 eV (solid line). TEM-EDX measurements on thick epilayers shows that the InAs$_{1-x}$Sb$_x$ alloy composition follows the apparent activation energy.

Figure 2. g[002] DF TEM [110] cross-section micrograph of a n-SLS of ~285Å InAs$_{0.33}$Sb$_{0.67}$ and ~285Å InAs$_{0.69}$Sb$_{0.31}$ which was grown at 340°C with equal fluxes of antimony and arsenic, and no modulation of the Knudsen cell shutters.

Two distinct growth regimes were found to exist over different temperature ranges. For 430≤T$_g$≤480°C the layers were essentially homogeneous, with some ordering on two sets of {111} planes [6], but below 430°C coherent phase separation occurs (see section 3.2). Ex-situ TEM-EDX measurements of alloy composition on a series of thick layers grown above 430°C confirmed the RHEED oscillation calibration (figure 1). The InAs-like LO phonon energies observed in optical Raman scattering exhibited a linear dependence with composition over the full alloy range [8]. Conventional SLSs were also grown at 430°C where both the layer thickness and the modulation of the alloy composition, the indium and antimony incorporation rates respectively, were calibrated in-situ by RHEED.

3.2 Natural Strained Layer Superlattice

(a) Initial TEM
Using equal and calibrated fluxes of Sb_4 and As_4 to give a nominal composition of $InAs_{0.5}Sb_{0.5}$, an epitaxial layer was grown at 340°C on (001) GaAs. Figure 2 shows the [110] cross-section dark field (DF) micrograph of this sample obtained using the compositionally sensitive (002) reflection. Alternate light and dark platelets of equal thickness are visible corresponding to material of two different alloy compositions. It should be stressed that the Knudsen sources were not modulated. The presence of material of two different compositions was confirmed in <110> pole cross-section TED patterns where the higher order spots were split into two in the [001] growth direction, but not in the <110> directions, indicating the presence of two different lattice constants perpendicular to the (001) layer surface but only one parallel to the layer surface. This indicates that the platelets of different compositions have crystal lattices which are tetragonally distorted, that is coherent phase separation has occurred. For the sample of figure 2 the two tetragonally distorted platelets have lattice constants of 6.11 Å and 6.40 Å perpendicular to the layer surface and 6.28Å parallel to the surface, corresponding to compositions of $InAs_{0.69}Sb_{0.31}$ and $InAs_{0.33}Sb_{0.67}$ respectively (estimated using a procedure reported previously [6]). The lattice constant parallel to the surface corresponds approximately to the composition of $InAs_{0.5}Sb_{0.5}$, indicating that the n-SLS structure has relaxed with respect to the substrate. The interfaces between the platelets are highly regular and run approximately parallel to the (001) surface. The clearly defined change in contrast between the platelets indicates very sharp interfaces which have been revealed to be almost monolayer abrupt by high resolution electron microscopy [7]. A strong anisotropy was apparent in the lateral extent of the platelet structure in the orthogonal <110> directions, with platelets shorter and the morphology less uniform when viewed in the [$\bar{1}$10] direction [7,9].

It is evident from these TEM and TED observations that the characteristic features of coherent phase separation observed in the $InAs_{1-x}Sb_x$ are different to the phase separation reported for MBE growth of $Ga_xAl_{1-x}As$ grown on (110) GaAs substrates [4] or the coarse 'tweed-like' modulation associated with alloy-clustering observed in some alloys [1-3]. Growth of $InAs_{1-x}Sb_x$ below 430°C results in a tetragonally distorted platelet structure with large variations in composition, separated by discrete interfaces along the [001] direction. Hence we refer to this structure as a natural strained layer superlattice (n-SLS) [9].

(b) Variation of n-SLS structure with growth parameters
Neither varying the growth temperature from 295°C to 400°C or composition from x=0.2 to 0.8 at 370°C changed the average platelet compositions significantly from $InAs_{0.3}Sb_{0.7}$ and $InAs_{0.7}Sb_{0.3}$. However, the platelet thickness and its lateral size in the <110> directions exhibited a systematic increase with increasing growth temperature [7,9]. Varying the nominal alloy composition from $InAs_{0.5}Sb_{0.5}$ only changed the relative volumes of the two phases [6]. Figure 3 shows the g[002] DF micrograph of an $InAs_{0.5}Sb_{0.5}$ epilayer grown at 370°C for 1 hour (~1 μm) after which growth was continued at 430°C for another 1 hour (~1 μm). A sharp transition is present between the different temperature regions, clearly demonstrating that n-SLS growth only occurs below a critical temperature for the same growth conditions. Another sample in which the reverse structure was grown at 430°C for 1 hour (~1 μm) and 370°C for another hour (~1 μm) and then annealed produced a similar but inverted micrograph to figure 3 showing that the n-SLS does not form by bulk diffusion [7,9].

Given that the <110> directions are equivalent in the bulk, the lateral anisotropy observed for the platelet structure and its increase with growth temperature suggests the n-SLS may be formed by a process of surface diffusion. A similar anisotropy has been

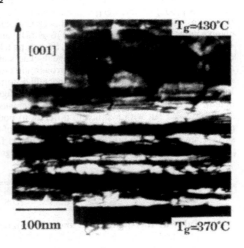

Figure 3. g[002] DF TEM [110] cross-section micrograph showing regions of a layer of composition $InAs_{0.5}Sb_{0.5}$ grown at 370°C and 430°C: 370°C region is phase-separated n-SLS, 430°C region is homogenous.

reported for the MBE growth rate along the orthogonal <110> directions due to a difference in the atomic step structure for GaAs [14]. The low bulk cation/anion diffusivity for III-V compounds also led to the postulation that phase separation occurred by rapid surface diffusion at the liquid-solid interface in LPE [1], and the vapour-solid interface in hydride transport VPE [15], grown $In_{1-x}Ga_xAs_yP_{1-y}$.

(c) n-SLS growth on different orientation substrates

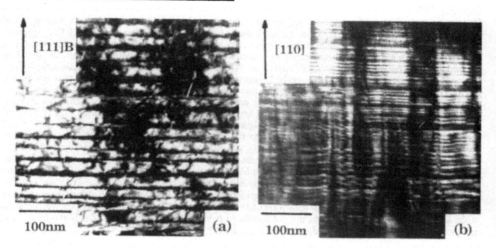

Figure 4. g[002] DF TEM micrographs of a layer of composition $InAs_{0.5}Sb_{0.5}$ grown at 370°C simutaneously onto (a) (111)B orientated InSb substrate and (b) (110) orientated InSb substrate

To investigate further n-SLS formation, layers of nominal composition $InAs_{0.5}Sb_{0.5}$ were grown at 370°C simultaneously on (001), (111)B and (110) orientated InSb substrates. RHEED was performed on the (001) substrate to calibrate the growth parameters enabling the required composition to be grown. The (001) layer exhibited phase separation, producing a n-SLS similar to that described above for other (001) layers (figures 2 and 3) with platelets of average thickness ~30nm and compositions of $InAs_{0.38}Sb_{0.62}$ and $InAs_{0.78}Sb_{0.22}$, table 1. A g[002] DF micrograph taken from a [110] cross-section of the (111)B orientated sample is shown in figure 4(a). Coherent phase separation has again occurred producing a regular platelet structure of two different compositions lying parallel to the (111)B growth surface. The lateral size of the platelets is isotropic in the <110> directions. The darker contrast, Sb-rich platelets have an average thickness of ~5nm whilst the lighter contrast As-rich platelets have an average thickness of ~21nm. The diffraction spots in the [110] TED patterns are split into two spots in the [111] growth direction but not in the <224> direction parallel to the layer surface, indicating the presence of tetragonally distorted material of two different compositions estimated to be $InAs_{0.45}Sb_{0.55}$ and $InAs_{0.77}Sb_{0.23}$, table 1. In figure 4(b) a g[002] DF micrograph taken close to a <100> pole of a <112> cross-section of the (110) orientated sample is shown. Coherent phase separation is once again observed but with a structure different to both the (100) and (111)B samples as very thin extended platelets of average thickness ~3nm (many <1nm) of different compositions lying parallel to the (110) growth surface are present. In the TED patterns obtained from this cross-section at the <112>, <111> and <100> poles the diffraction spots are extended in the [110] growth direction but are not obviously split.

Orientation	a_p	c_1	c_2	a_{r1}	a_{r2}	y_1	y_2
(001)	6.23	6.42	6.08	6.32	6.15	0.38	0.78
(111)B	6.23	6.35	6.08	6.29	6.16	0.45	0.77
(110)	6.25	-	-	-	-	-	-

Table 1: $InAs_{0.5}Sb_{0.5}$ grown on different orientation substrates at 370°C. a_p is the measured lattice parameter parallel to (001) layer surface of phases 1 & 2 (Å); c_1 & c_2 are the measured lattice parameters perpedicular to the layer surface of phases 1 & 2 respectively; a_{r1} & a_{r2} are the relaxed lattice parameters and y_1 & y_2 are the compositions.

Coherent phase separation resulting in n-SLS formation appears to occur regardless of substrate orientation on both polar ((001), (111)B) and non-polar (110) surfaces and along both the elastically soft [001] and hard [111] crystallographic directions. This indicates that the n-SLS growth mode is similar on different orientations, although a drastically different modulation period occurs. This confirms that n-SLS formation is not a bulk equilibrium process but may be associated with the non-equilibrium nature of MBE. This behaviour is again consistent with a surface diffusion process generating the n-SLS at or near to the growth front.

(d) Preliminary results from GaAsSb

(001) MBE GaAs$_{0.5}$Sb$_{0.5}$ layers grown at and below 450°C have also exhibited coherent phase separation producing platelet structures similar to those described above for (001) InAs$_{0.5}$Sb$_{0.5}$ layers except the lateral scale of the platelets in GaAsSb is much smaller and the morphology less regular for the same growth conditions. Similar temperature ranges are found for n-SLS formation in InAsSb and GaAsSb, and the compositions of the resulting platelets are very similar, for example in a layer grown at 450°C the compositions of the platelets are GaAs$_{0.41}$Sb$_{0.59}$ and GaAs$_{0.74}$Sb$_{0.26}$ (c.f. table 1). This suggests that the behaviour may be associated with a possible miscibility gap present in the As-Sb phase diagram [16] at low temperatures rather than miscibility gaps associated with the ternary alloys, InAsSb and GaAsSb, since these alloys are predicted to have significantly different critical temperatures [17].

3.3 A Comparsion of the Optical Properties of Conventional and Natural SLSs

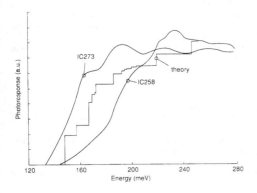
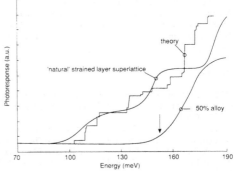

Figure 5. Photoresponse spectra for 10.5nm (IC258) and 27nm (IC273) conventional SLSs of InAs$_{0.2}$Sb$_{0.8}$/InSb. The 'theory' curve is the absorbed optical power in IC258 calculated assuming an unstrained valence band offset ΔE_v=0.06(x_b-x_w)eV.

Figure 6. Measured photoresponse of the natural SLS of figure 2 and 50% alloy control sample. The arrow marks the expected bandedge and the 'theory' curve is the absorbed optical power calculated assuming an unstrained valence band offset ΔE_v=0.06(x_b-x_w)eV.

Strained layer superlattices (SLS) of InAs$_{1-x}$Sb$_x$ which exhibit optical absorption out to wavelengths longer than 10µm (a region of minimum atmospheric absorption) have been grown by MBE [18] and MOCVD [19], but structures reported to date have been mainly limited to antimony rich alloys (0.6<x<1) with comparatively small interlayer strains. Figure 5 shows the photoresponse spectra of two intentionally grown SLS of InAs$_{0.2}$Sb$_{0.8}$/InSb with equal well and barrier thicknesses of 10.5nm (IC258) and 27nm (IC273). The nominal SLS parameters used here and the measured optical absorption edge (at 300cm^{-1}) are similar to those reported previously [20]. A clear quantum confinement effect is apparent, blue shifting the absorption edge of the shorter period SLS IC258 relative to IC273. Also shown is the calculated absorbed optical power spectrum for IC258, modelled by numerically solving the 1-D Ben Daniel-Duke Hamiltonian [21] for the subband energies, using the deformation potential data of [20], the strain data from our TED measurements and the unstrained temperature-dependent alloy bandgap values of [22]. Conduction band non-parabolicity effects were accounted for by self-consistently calculating the electron mass in each layer to be m^*_i=C{m^*_{alloy}(0)+[m^*_{InSb}(0)-m^*_{InSb}(E_i-E_c)]}, where m^*_{alloy}(0) is the band edge effective mass in each layer (linearly extrapolated from the known binary values), m^*_{InSb}(E) is the k.p mass of an electron in InSb at an energy E above the conduction band edge, C is the factor by which the unstrained bandgap is changed by strain and E_i is the energy of the ith electron state. We find the best fit by assuming an unstrained valence band offset of $\approx(x_b-x_w)x0.06\pm0.05eV$ where $X_{b(w)}$ are the barrier (well) antimony concentrations.

The lack of similar peaks in the absorption spectra indicated that the oscillatory structure (with a period of 33.1 meV) in the IC273 spectrum originates not from optical etalon effects, but from LO-phonon mediated oscillatory photoconductivity (OPC) [23], and hence has a period given by $hw_{LO}[1+(m^*_e/m^*_h)]$. Using $hw_{LO}=23.3meV$ for x=0.8 [8] gives $(m^*_e/m^*_h)=0.42$ and using the electron mass described above this implies an in-plane mass of $\sim0.044m_0$ for the lowest energy strain modified hole band. This is in good qualitative agreement with recent cyclotron resonance data from p-type InAsSb SLS samples [24] and confirms the type II band offset. In high quality bulk semiconductors the mobility-lifetime product is typically at a minimum for photo-excited electrons close to $\underline{k}=0$, giving OPC traces which extrapolate back to the band gap energy. For the SLS samples however the carrier lifetime appears to be determined mainly by recombination at defects in the buffer layer, producing a maximum in carrier lifetime for carriers close to $\underline{k}=0$ (with low perpendicular mobilities), and inverting the sign of the oscillations in the OPC spectra. On this basis we use the maxima in the IC273 spectrum to obtain an estimate of 125 meV for the effective bandgap.

Figure 6 shows the 10K photoresponse spectrum for the n-SLS in figure 2 together with that of an x=0.5 homogenous alloy control sample. The arrow marks the x=0.5 alloy band edge expected from recent 10K PL measurements [22]. The n-SLS sample exhibits a long wavelength cutoff of $\approx12.5\mu m$, some 40 meV below this value. The computer modelled absorbed optical power spectrum of the n-SLS again gave a best fit with an unstrained valence band offset of $(x_b-x_w)x0.06\pm0.1eV$, the same value as that obtained for conventional SLSs. This differs significantly from the value of $(x_b-x_w)x0.41\pm0.1eV$ reported in [20] and further experiments are in progress to investigate this discrepancy.

References

[1] P Henoc, A Izrael, M Quillec and H Launois, Appl. Phys. Lett. **40** (1982) 963
[2] F Glas, M M Treacy, M Quillec and H Launois, J.Phys. (Paris), **43**, Colloq. **C5**, Supplement to No. 12 (1985) C5-11
[3] A G Norman and G R Booker Proceedings of the Microscopy of Semiconducting Materials Conference, Oxford, 1985. Published in the Institute of Physics Conference Series **78** (1985) 257
[4] P M Petroff, A Y Cho, F K Reinhart, A C Gossard and W Wiegmann, Phys. Rev. Lett. **48** (1982) 170
[5] I P Ipatova,V. A Shchukin, V.G Malyshkin, A Yu Maslov and E Anastassakis, Solid State Comm. **78** (1991) 19
[6] T Y Seong, A G Norman, G R Booker, R Droopad, R L Williams, S D Parker, P D Wang and R A Stradling, 1990 MRS Fall Meeting USA 1989, Mat. Res. Soc. Symp. Proc. **163** (1990) 907
[7] T Y Seong, A G Norman, J L Hutchison, I T Ferguson, G R Booker, R A Stradling and B A Joyce, Proceedings of the Microscopy of Semiconducting Materials Conference, Oxford, 1991. To be published in the Institute of Physics Conference Series
[8] Y B L Li, S S Dosanjh, I T Ferguson, A G Norman, R A Stradling and R Zallen. Submitted to Semiconductor Science and Technology
[9] I T Ferguson, T Y Seong, R H Thomas, A G Norman, C C Phillips, R A Stradling, G R Booker and B A Joyce, Submitted to Applied Physics Letters
[10] J H Neave, B A Joyce and P J Dobson, Appl. Phys. Lett. **A34** (1984) 179
[11] S Tsukamoto, P Battacharya, Y C Chen and J H Kim, Appl. Phys. **67** (1990) 6819
[12] G S Lee, Y Lo, Y F Lin, S M Bedair and W D Laidig, Appl. Phys. Lett. **47** (1985) 1219
[13] M Yano, M Ashida, A Kawaguchi, Y Iwai and M Inoue, J.Vac. Sci. Technol. **B7** (1989) 199
[14] B A Joyce, J Zhang, C T Foxon, D D Vvedensky, T Shitara and A K Myers-Beaghton, Semicond. Sci. Technol. **5** (1990) 1147
[15] S N Chu, S. Nakahara, K. E. Strege and W D Johnston, J. Appl. Phys. **57** (1985) 4610
[16] M Hansen 'Constitution of Binary Alloys' McGraw-Hill Book Company, New York, Toronto and London. (1958) 158
[17] G B Stringfellow, J. Crystal Growth **58** (1982) 194
[18] G C Osbourn, Semicond. Sci. Technol. **5** (1990) S5
[19] S R Kurtz, R M Biefield and T E Zipperian, Semicond. Sci. Tech. **5** (1990) S24-26
[20] S R Kurtz, G C Osbourne, R M Biefield and S R Lee App. Phys. Lett. **53** (1988) 216
[21] G Bastard and J A Brum J. Quant. Electron. **QE22** (1986) 1625
[22] Z M Fang, K Y Ma, D H Jan, R M Cohen and G B Stringfellow J. App. Phys. **67** (1990) 7034
[23] R W Shaw, Phys. Rev. **B3** (1971) 3283
[24] S Y Lin, D C Tsui, L R Dawson, C P Tigges and J E Schirber App. Phys. Lett. **57** (1990) 1015

Inst. Phys. Conf. Ser. No 120: Chapter 9
Paper presented at Int. Symp. GaAs and Related Compounds, Seattle, 1991

401

Interface-free GaAs structures—from bulk to the quantum limit

D.J. Wolford, G.D. Gilliland, T.F. Kuech,[†] J. Martinsen, J.A. Bradley, and C.F. Tsang
IBM Research Division, T.J. Watson Research Center
Yorktown Heights, NY 10598

R. Venkatasubramanian and S.K. Ghandhi
Rensselaer Polytechnic Institute
Troy, NY 12180

H.P. Hjalmarson and J. Klem
Sandia National Laboratories
Albuquerque, NM 87185

We have studied, through extensive transient PL (1.8 - 300 K) measurements, intrinsic re-combination in "ideal" GaAs structures, passivated by state-of-the-art "surface barriers." Lifetime versus GaAs thickness (10 μm - 0.01 μm) yields the lowest interface recombination velocities yet reported of $\leq$ 40 cm/s for MOCVD-prepared GaAs/Al$_{0.3}$Ga$_{0.7}$As double heterostructures and ~ 60 to 5500 cm/s for MOCVD-prepared all-GaAs n$^+$/n$^-$/n$^+$ homostructures. In comparison, Na$_2$S passivation gives ~ 5500 cm/s, and bare GaAs sur-faces yield $\geq$ 34,000 cm/s. Further, identical measurements made in comparably MBE-prepared GaAs/Al$_{0.3}$Ga$_{0.7}$As double heterostructures suggest corresponding interface recombination velocities of ~ 250 to 5000 cm/s. Thus, we prove — from bulk to quantum wells — that "interface-free" GaAs structures are now achievable, and that intrinsic band-to-band and/or free exciton recombination may dominate in such structures.

1. Introduction.

Despite modern epitaxy, the two "intrinsic" recombination processes in direct-gap semiconductors, such as GaAs — band-to-band and free-exciton recombination — are often dominated by "extrinsic" processes due to defects at surfaces and/or interfaces and bulk defects and, therefore, can not be fully explored.[1-9] The deleterious effects of interfacial and surface recombination may be quantified through such recombination velocities (S), which relate the nonradiative decay rate at both front and back interfaces to the distance between them. These velocities are generally considered to be large (S $\approx$ 10^5 - 10^6 cm/s) ,[3-5] thus making any meaningful analysis of "intrinsic" recombination difficult. Hence, there have been numerous attempts at reducing these nonradiative decays through interfacial passivation with a variety of materials including: Al$_x$Ga$_{1-x}$As,[3,10,11] In$_x$Ga$_{1-x}$P,[5] KOH,[8] Na$_2$S,[9] ZnSe,[12] and P$_2$S$_5$[13] — all designed to confine minority-carriers away from bare-surface states which cause significant, <u>if not entirely dominant,</u> nonradiative decay.

We have thus applied extensive photoluminescence (PL) measurements to the prototypical direct-gap semiconductor, GaAs, using widely differing state-of-the-art "surface-barriers," all aimed at substantially reducing, if not eliminating, interfacial re-combination. We have thus extended the work of Nelson and Sobers[3] in LPE GaAs/Al$_x$Ga$_{1-x}$As structures to both MOCVD- and MBE-prepared structures which, in contrast to LPE-methods, may readily produce truly abrupt interfaces — thus better defining an interface and its properties, while removing possible ambiguity in active-GaAs-layer thickness. We find that "surface-free" structures are now achievable,[14,15] and that the re-sulting, dominant decay is indeed "intrinsic," by free-carriers and/or free-excitons.

2. Experiment.

The MOCVD double homostructures and heterostructures were prepared[15] with re-spective growth conditions of 700°C and 1 atm. with TMG or 680°C and 0.1 atm. with

TEG, and 750°C and 1 atm. with TMG; further, the $Al_{0.3}Ga_{0.7}As$ heterostructure barriers were p-type (~ 10^{16} cm^{-3}) and 0.5-μm thick, while the GaAs layers were n-type ($\leq 10^{15}$ cm^{-3}) and 10 μm to 50 Ås thick. The MBE 0.5-μm/0.5-μm/0.5-μm heterostructures were prepared in a Varian Gen II reactor. Homostructures were silane n$^+$-doped, leading to peak Si doping of 5 – 8x10^{18} cm^{-3}, at $\leq$ 800 Å, bounding the undoped n$^-$ (~ 10^{15}cm^{-3}) regions. Time-resolved photoluminescence (PL) was carried out with 50-ps resolution.

3. Results and Discussion.

Figure 1 shows 300-K cw PL spectra for MOCVD and MBE heterostructures, and an MOCVD homostructure. The high-energy side results in Maxwell-Boltzmann electronic temperatures (T_e) agreeing with sample temperature, thus showing PL is band-to-band of thermalized free carriers. Moreover, the spectra for all samples are identical (same peak energy) — despite wide differences in GaAs-layer thickness — thus confirming that "photon recycling" is unimportant.[16]

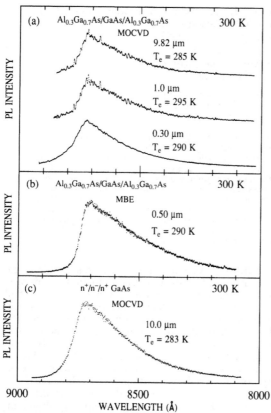

Fig. 1. 300-K PL for (a) MOCVD- and (b) MBE-GaAs/ $Al_{0.3}Ga_{0.7}As$ double heterostructures, and an (c) MOCVD n$^+$/n$^-$/n$^+$ all-GaAs homostructures — each with indicated GaAs active-layer thicknesses. Derived Maxwell-Boltzmann carrier temperatures are shown.

Contrasting to the unanimity of PL in Fig. 1, decay kinetics of Fig. 2 drastically differ. Nonetheless, we find all may be understood using the 1964 band-to-band rate equations of Lasher and Stern,[1] yielding time-dependent minority-carrier density (PL) decay as,

$$p(t) = \frac{Ae^{-t/\tau}}{1 + C[1 - e^{-t/\tau}]}, \quad (1)$$

where $A=p_i$ is the initial photoexcited pair density, C is the "bimolecular" component of decay, and τ is the experimental decay time. Applying this to Fig. 2, we find kinetics for thick heterostructures (> 0.5-μm) to be nonexponential, and easily fit by Eq. 1, with all fits becoming purely exponential in the long-time limit; these long lifetimes (τ), thus derived, represent the 1/e decay constant of the exponential decay tail. These decays may thus be interpreted as being initially dominated by the photoexcited carrier density, but eventually becoming dominated by a small residual nonradiative interfacial recombination term — and thus all decays become asymptotically exponential. Decay kinetics for thinner structures are dominated by the majority <u>hole</u> density — and not the interface recombination term — and are, therefore, strictly exponential at all times. (We may conclude the n-type GaAs becomes *effectively p-type modulation-doped*, from accumulation of holes from the p-type $Al_{0.3}Ga_{0.7}As$, overwhelming the background donor density.) In contrast, decays of all homostructures are rigorously exponential with the differences being governed by the majority <u>electron</u> density, reflecting lower homostructure purity.

Sample preparation plays a key role, as Fig. 3 shows the 300-K decays for identical heterostructures of 0.50-μm-thick GaAs grown by MOCVD at 750°C, and by MBE at from 600 to 740°C. Importantly, the lifetime of the 600°C MBE sample is only marginally better than the 7.5 ns observed for a "bare" GaAs surface (discussed below), while increasing growth temperature apparently results in a reduction of "extrinsic" defect states and, therefore, longer minority lifetimes. Significantly, however, the best MBE sample shows a nearly 4X poorer lifetime than for the identical MOCVD structure.

Lifetimes, thus obtained may then help quantify interface quality through the velocity S (cm/s) by the relation

$$\frac{1}{\tau} = \frac{1}{\tau_{nr}} + \frac{2S}{d} + B(n_0 + p_0),$$ (2)

where B is the "bimolecular" radiative coefficient $(2 \times 10^{-10} cm^3/s)$,[3] p_0 (n_0) is the built-in hole (electron) density, and τ_{nr} is the bulk nonradiative decay lifetime. The n^--GaAs thickness, d, is then varied to directly obtain S, and thus the influence of the GaAs interfaces on the nonradiative decay, and Fig. 4 summarizes both homo- and heterostructure results. As readily noted, two different homostructure sets, prepared under different MOCVD conditions (TMG vs. TEG, in different reactors), have differing surface recombination velocities from ~ 60 to 5500 cm/s. — the lower value being our "typical best" and the higher value being our "typical worst;" the fact a solid fit cannot be drawn smoothly through the TEG data, leading to a unique value for S (1800 cm/s is dashed as a average) suggests each TEG sample may have somewhat different interfacial quality. More is needed to resolve these differences; however, both results lead to lifetimes indicative of good-to-excellent interfacial quality.

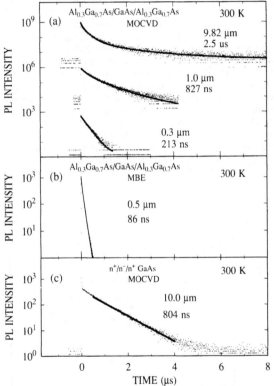

The heterostructures in Fig. 4 are also somewhat complex, in that data does not lie in a smooth line for thinner structures (< 0.3-μm). We have noted above, however, that lifetimes in only the thick samples are dominated by interface recombination. We therefore conclude that S may only be reliably estimated (those for which p-type modulation may be ignored) from structures with d > 0.3-μm, and we find a value for S of ≤ 40 cm/s — among the lowest interface recombination velocities ever demonstrated for any semiconductor structure — and especially in the often problematical GaAs/Al$_x$Ga$_{1-x}$As interfacial system. Indeed, in comparison to the previous most thorough and comprehensive comparison of Nelson and Sobers (Fig.4),[3] our 2.5-μs lifetime in the 9.82-μm structure is effectively 4 times longer than the 1.3-μs lifetime in their 16-μm structure. Furthermore, our confirmed corresponding interface recombination velocity is some 6 times less than their most optimistic fit of 250 cm/s. This comparison is of considerable interest since, in contrast to LPE structures, where interfaces are known to com-

Fig. 2. 300-K PL intensity decays for the corresponding structures in Fig. 1. Lifetimes represent the exponential decay tail derived through least-squares fits to Eq. 1.

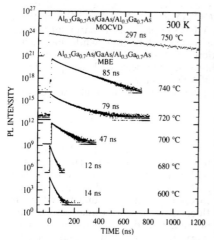

Fig. 3. 300-K PL decays for 750°-prepared MOCVD-GaAs/Al$_{0.3}$Ga$_{0.7}$As double hetero-structure (0.50-μm GaAs layer) compared to identical MBE-GaAs/Al$_{0.3}$Ga$_{0.7}$As structures prepared from 600 to 740°C.

positionally grade over 100's to 1000's of Ås, our heterostructures are abrupt to nearly monolayer scales. Nonetheless, abrupt interfaces are apparently good, but insufficient, since the best MBE-heterostructure lifetime of ~ 85 ns (in the absence of detailed lifetime vs. thickness) suggests an apparent S varying from ~ 250 to 5000 cm/s for the 0.5-μm-thick structures (■) of Fig. 4.

Figure 4 also shows comparison between our double heterostructures, with these same structures etched (leaving a single "bare" GaAs surface), as well as the etched structures repassivated with Na$_2$S. Applying Eq. 2, we obtain S ≥ 34,000 cm/s for bare GaAs, and ~ 5500 cm/s for Na$_2$S-passivated surfaces. In connection with this, Figs. 5(a) and 5(b) show the corresponding PL spectra and time-decays for the 9.82-μm hetero-structure under these various conditions. We find the relative PL efficiencies here scale nearly with the corresponding lifetimes and interface recombination velocities, thus suggesting increased nonradiative interfacial recombination is accompanied by reductions in both PL lifetimes and PL efficiencies. The results contained in Figs. 4 and 5 represent perhaps the most accurate determination of the passivating effects of state-of-the-art surface-barriers for a single interface. Temperature dependent lifetimes for the 10-μm TMG homostructure of Fig. 1 are shown in Fig. 6 (◊). Actual decays are rigorously exponential from 300 K to ~ 40 K, and the derived lifetimes show a decrease with decreasing temperature; below 40 K, however, kinetics become nonexponential and show a lifetime increase with decreasing temperature. Such results may be accurately described by theory for recombining free electron-hole pairs[1] — but only if the temperature-dependent carrier densities are fully accounted for, using the expression (dashed in Fig. 6)

$$\frac{1}{\tau} = B(300/T)^{3/2} n_0(T), \qquad (3)$$

where $n_0(T)$ is the temperature-dependent equilibrium majority-electron density, and $N_A/N_D \approx 0.4$. Corresponding temperature-dependences of two MOCVD heterostructures are also shown in Fig. 6. For thick heterostructures (> 0.5-μm) the decays are nonexponential and "bimolecular" (e.g., Fig. 1) between 300 K and ~ 50 K, with asymptotic lifetimes gotten from fits to Eq. 1; these show the expected $T^{-3/2}$ dependence with decreasing temperature, just as in

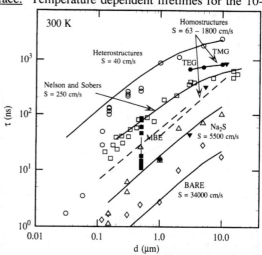

Fig. 4. 300-K lifetimes for MOCVD GaAs/Al$_{0.3}$Ga$_{0.7}$As double heterostructures (o), MOCVD double heterostructures etched and Na$_2$S repassivated (△), "bare" GaAs/Al$_{0.3}$Ga$_{0.7}$As heterostructure (◊), MOCVD n$^+$/n$^-$/n$^+$ all-GaAs homostructures prepared in different reactors using TMG(o) or TEG (▽), MBE GaAs/Al$_{0.3}$Ga$_{0.7}$As double heterostructures (□) prepared at different temperatures (Fig. 3), and Nelson & Sobers' LPE GaAs/Al$_{0.3}$Ga$_{0.7}$As double heterostructure data (Ref. 3) (□) — all versus GaAs Layer thickness. Lines represent least-squares fits yielding interface recombination velocities shown.

the homostructure. The corresponding decays for thin heterostructures (< 0.5-μm) are, in contrast, rigorously exponential between 300 and 50 K (like the homostructure), becoming nonexponential only at low temperatures, with lifetimes which first decrease and then increase with decreasing temperatures — here reaching 14.5 μs near 25 K. Thus, to summarize, higher temperature recombination (~ 30 - 300 K), displays band-to-band decays qualitatively similar for all structures, with majority carriers determining decay in the relatively impure n⁻ homostructure and the thin "p-type, modulation-doped" heterostructure, and by minority carriers in the relatively pure (compared to photo-pumped carrier concentrations) n-type thick heterostructures.

Focussing now on the lowest temperature regime, free-exciton recombination is the only other "intrinsic" radiative process which may occur in GaAs (other than free-carrier), and it must set in for temperatures where kT is comparable to, or less than, the exciton binding energy (~ 4 meV) — hence, below $\approx$ 50 K. Thus, all structures, below ~ 30 K, show lifetimes plunging with temperature, due to temperature-dependent trade-offs between band-to-band and bound state formation, as free excitons begin to form, finally dominating at lowest temperatures; in particular is the continuous and dramatic plunge in lifetime, of over 5 orders-of-magnitude in 20 K, to ~ 200 ps at 1.8 K in the 0.3-μm structure. In comparison, Fig. 7 shows typical 2-K cw PL for the 0.5-μm-thick MOCVD and MBE (79-ns sample) heterostructures of Fig.

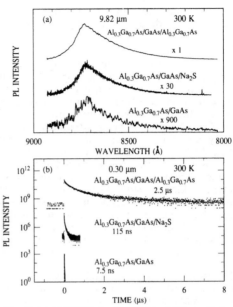

Fig. 5. (a) 300-K PL for MOCVD a 9.82-μm GaAs/Al$_{0.3}$Ga$_{0.7}$As double heterostructure, such a structure etched and Na$_2$S repassivated, or etched leaving "bare-surfaced" GaAs/Al$_{0.3}$Ga$_{0.7}$As. (b) Corresponding 300-K PL decays, with least-squares-fit lifetimes.

3, against the 10-μm homostructure. Importantly, all spectra display a prominent PL peak at the free-excitonic (1.5151 eV) resonance. Note also that, in contrast to the MOCVD heterostructure, which decays at 297-ns in Fig. 3, the MBE-prepared sample shows minimal impurity-excitons and yet its 79-ns lifetime is substantially shorter. Hence, observing significant free-excitonic luminescence at low temperatures appears to be a necessary, but not sufficient condition, for the existence of high interfacial quality.

Furthermore, when any of these heterostructures or homostructures are chemically etched to expose a bare GaAs surface, the clear free-excitonic peaks of Fig. 7 vanish — to be replaced by weak, structured emission (often termed the upper and lower exciton-polariton branches) and strong donor and/or acceptor bound excitons. This suggests that impurity-bound excitons are preferentially seen in low-temperature PL of even the purest GaAs, because of rapid trapping of free excit-

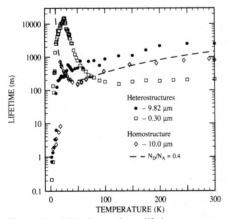

Fig. 6. Minority-carrier lifetimes versus temperature for a thick (9.82 μm) and thin (0.30 μm) MOCVD GaAs/Al$_{0.3}$Ga$_{0.7}$As double heterostructure, and a thick (10 μm) n⁺/n⁻/n⁺ GaAs homostructure ◊ (with dashed fit according to Eq. 3 assuming $N_A/N_D = 0.4$).

ons enroute to imperfect surfaces, where nonradiative decay efficiently quenches their emission.

4. Conclusions.

These results are among the most comprehensive concerning optical study of passivation and intrinsic recombination (band-to-band and free-exciton) in any direct-gap semiconductor. We demonstrate for GaAs/$Al_{0.3}Ga_{0.7}As$ double heterostructures that relative purity and thickness of both the barrier material (e.g., $Al_xGa_{1-x}As$) and the central GaAs layer, as well as the sample growth procedures, are crucial to understanding radiative mechanisms. Moreover, detailed kinetic studies are necessary for an unambiguous understanding of interface recombination, and, thereby allowing for the most accurate determination of S for a single barrier GaAs interface. We find that truly "interface-free" GaAs/$Al_xGa_{1-x}As$ structures are now achievable and are superior to other popular chemical-passivants. Finally, we have shown definitive lifetimes suggesting that, for now, MBE-prepared double heterostructures improve

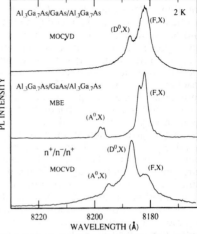

Fig. 7. 2-K PL from MOCVD and MBE heterostructure, and an MOCVD homostructure, with free-excitons (F,X), and donor- (D^0,X), and acceptor-bound excitons (A^0,X) labeled.

with growth temperature, but still cannot equal our best MOCVD homo- or heterostructures. From our results, we also conclude that observation of true free excitonic recombination is a necessary but not sufficient condition for the existence of "high-quality" interfaces.

Acknowledgements

IBM work was supported, in part, by ONR, under contracts N00014-85-C-0868 and N00014-90-C-0077. Sandia work was supported by the U.S. DOE, under contract DE-AC04-76DP00789. RPI work was supported by SERI, Golden, ᴵCO, under contract XL-5-05018-2, and the New York State ERDA, under agreement 970-ERER-ER-87.

REFERENCES

†University of Wisconsin, Dept. of Chemical Engineering, Madison, WI 53706.
1. G. Lasher and F. Stern, Phys. Rev. **133**, 553 (1964).
2. H.C. Casey, Jr. and F. Stern, J. Appl. Phys. **47**, 631 (1976).
3. R.J. Nelson and R.G. Sobers, J. Appl. Phys. **49**, 6103 (1978).
4. E. Yablonovitch, R. Bhat, J.P. Harbison, and R.A. Logan, Appl. Phys. Lett. **50**, 1197 (1987); E. Yablonovitch, C.J. Sandroff, R. Bhat and T. Gmitter, Appl. Phys. Lett. **51**, 439 (1987).
5. J.M. Olsen, R.K. Ahrenkil, D.J. Dunlavy, Brian Keyes, and A.E. Kibbler, Appl. Phys. Lett. **55**, 1208 (1989).
6. L.M. Smith, D.J. Wolford, J. Martinsen, R. Venkatasubramanian, S.K. Ghandhi, Appl. Phys. Lett. **57**, 1572 (1990); L.M. Smith, D.J. Wolford, R. Venkatasubramanian, and and S.K. Ghandhi, J. Vac. Sci. Technol. B **8**, 787 (1990).
7. C.J. Sandroff, R.N. Nottenburg, J.C. Bischoff, and R. Bhat, Appl. Phys. Lett. **51**, 33 (1987).
8. M.G. Mauk, S. Xu, D.J. Arent, R.P. Mertens, and G. Borghs, Appl. Phys. Lett. **54**, 213 (1989).
9. B.J. Skromme, C.J. Sandroff, E. Yablonovitch, and T. Gmitter, Appl. Phys. Lett. **51**, 2022 (1987);
10. L.W. Molenkamp and H.F.J. van't Blik, J. Appl. Phys. **64**, 4253 (1988).
11. P. Dawson and K. Woodbridge, Appl. Phys. Lett. **45**, 1227 (1984).
12. S.K. Ghandhi, S. Tyagi, and R. Venkatasubramanian, Appl. Phys. Lett. **53**, 1308 (1988).
13. K.C. Hwang, and S.S. Li, J. Appl. Phys. **67**, 2162 (1990).
14. G.D. Gilliland, D.J. Wolford, T.F. Kuech, J.A. Bradley, and C.F. Tsang, to be published.
15. D.J. Wolford, G.D. Gilliland, T.F. Kuech, L.M. Smith, J. Martinsen, J.A. Bradley, C.F. Tsang, R. Venkatasubramanian, S.K. Ghandhi, and H.P. Hjalmarson, J. Vac. Sci. Technol. B **9**, 2369 (1991).
16. B. Bensaid, F. Raymond, M. Leroux, C. Verie, and B. Fofana, J. Appl. Phys. **66**, 5542 (1989).

Inst. Phys. Conf. Ser. No 120: Chapter 9
Paper presented at Int. Symp. GaAs and Related Compounds, Seattle, 1991

407

Ground-state in-plane light-holes in GaAs/AlGaAs structures

E. D. Jones, S. K. Lyo, J. F. Klem, and J. E. Schirber,

Sandia National Laboratories, Albuquerque, NM 87185

S. Y. Lin

Electrical Engineering Department, Princeton University, Princeton, NJ 08544

ABSTRACT: We have performed low-temperature magnetoluminescence, far-infrared cyclotron resonance, and magneto-transport measurements on narrow (4.5 nm) GaAs/$Al_{0.25}Ga_{0.75}$As single quantum-wells. The ground-state in-plane valence band was found to be light and nonparabolic, i.e., at zone center, the valence band mass is $m_v \approx 0.1 m_0$ and for Fermi energies $E_f \approx 10$ meV, $m_v \approx 0.25 m_0$.

1. INTRODUCTION

At high electric fields, the requirement for light-hole valence-band masses for high-speed p-type digital electronic structures has been well documented by Osbourn et. al., (1987). The attainment of light-hole masses has been mainly achieved in layered structures by the introduction of compressive biaxial strain in the active quantum layers (Osbourn et. al. 1987, Bastard 1988, Pearsall 1988, and Weisbuch and Winter 1991). This strain is introduced by growing structures with two materials with differing lattice constants such as $In_xGa_{1-x}As$ and GaAs. Recently, the valence-band energy dispersion curves of these kinds of modulation doped strained-single-quantum-well structures were determined from magnetoluminescence measurements (Jones et. al. 1989, 1990). These authors report on the importance of the energy difference ΔE between the heavy and light-hole valence-bands in determining the degree of heavy and light-hole mixing. Large ΔE energy differences (i.e., reduced mixing) give rise to a small ground-state in-plane light-hole mass.

The energy difference ΔE (due to quantum confinement) between the heavy and light-hole valence bands in wide (≈ 10 nm) lattice-matched quantum well structures is small and thus the heavy-hole light-hole mixing is large. For these kinds of p-type structures, the in-plane valence-band ground-state is heavy, e.g., for Fermi energies $E_f > 2$ meV, $m_v \approx 0.3$ (here all effective masses are expressed in units of the free electron mass m_0).

However, by reducing the quantum well width, the energy difference ΔE can be increased. The maximum $\Delta E \approx 30$ meV, occurs near a width of 4.5 nm for GaAs/$Al_{0.25}Ga_{0.75}$As single quantum wells. In this paper we report the measurement of the ground-state valence-band dispersion curves for these narrow quantum-well structures using magnetoluminescence, far-infrared cyclotron resonance, and magneto-transport techniques. We find that for Fermi energies $E_f \approx 5$ meV, the in-plane ground-state in-plane valence-band mass is $m_v \approx 0.15$ and thus is light.

2. EXPERIMENTAL

The GaAs/Al$_{0.25}$Ga$_{0.75}$As 4.5 nm single-quantum well structures were grown by molecular beam epitaxy using a Varian GEN-II MBE machine. A schematic representation of the n-type structure used for the magnetoluminescence measurements is shown in Figure 1. The growth temperature was 600 C and at 4 K, the measured 2D-carrier concentration and mobility were respectively 6.6×10^{11} cm^{-2} and 2.2×10^4 cm^2/Vs. Also, p-type samples were grown for the cyclotron resonance and magneto-transport experiments.

The magnetoluminescence measurements (to 6.5 T) were made by attaching the [100] face of the sample to an Al-coated 100-µm core-diameter optical fiber. The magnetic field direction was parallel to the growth direction, i.e., the Landau orbits are in the plane of the GaAs single quantum well. The sample was illuminated by the 514.5 nm line of an Argon-ion laser through the optical fiber. The luminescence signal, returning back through the fiber was directed to a monochromator and a CAMAC-based data acquisition system (Jones and Wickstrom, 1985).

Cyclotron resonance measurements were made on both n-type and p-type samples. The sample substrate was wedged 5° in order to avoid multiple interference in the cyclotron resonance experiment. These measurements were performed at 4.2 K in magnetic fields (perpendicular to the two-dimensional plane) up to 8.5 T using an optically pumped, linearly polarized, far-infrared molecular gas laser. The transmitted radiation was detected by a Ge bolometer placed below the sample.

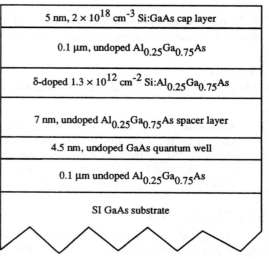

Figure 1. Schematic representation of the 4.5 nm GaAs/Al$_{0.25}$Ga$_{0.75}$As single quantum well structure. Growth temperature was 600 C.

The valence-band effective masses were also obtained from four terminal magneto-transport measurements using the Van der Pauw geometry. Shubnikov-de Haas oscillations in the resistance were measured using both direct current and field modulation techniques, the latter at lower fields where sufficient modulation-field could be achieved. The field (to 10 T) was provided by a rotatable split superconducting coil and sample temperatures from 1.1 to 4 K were obtained by pumping on a ^{4}He bath.

3. RESULTS AND DISCUSSION

For n-type structures and low temperatures, e.g., 4 K (0.34 meV), only the $n_v = 0$ state is populated and hence the magnetoluminescence transitions between the $n_c = 1, 2, 3, \ldots$ and the $n_v = 0$ Landau levels are zeroth-order forbidden (Lyo et. al. 1988). At these temperatures, the conduction-band effective mass m_c can be uniquely determined for the n-type samples by analyzing the field dependence of the magnetoluminescence spectra. The interband luminescence transition energy E is given in terms of the bandgap energy E$_{gap}$

and the conduction and valence-band cyclotron energies $\hbar\omega_{c,v} = (2\mu_\beta H / m_{c,v})$ by

$$E = E_{gap} + (n_c + \tfrac{1}{2})\hbar\omega_c + \tfrac{1}{2}\hbar\omega_v . \qquad (1)$$

At high temperatures, e.g., 77 K (6.7 meV), the $n_v = 0, 1, 2, 3, \ldots$ valence-band Landau-levels are thermally occupied and all transitions obey the $\Delta n \equiv (n_c - n_v) = 0$ selection rule and hence, are *allowed*. For this case, the interband luminescence transition energy E is given by

$$E = E_{gap} + (n + \tfrac{1}{2})(\hbar\omega_c + \hbar\omega_v), \qquad (2)$$

where n = 0, 1, 2, 3, ...

Figure 2 shows two magnetoluminescence spectra for H = 6 T. The upper spectrum is taken at 4 K while for the lower spectrum, T ≈ 76 K. The energy difference δE between the peaks is seen to be about 8 and 10 meV respectively for the 4 and 76 K spectra. This 2 meV

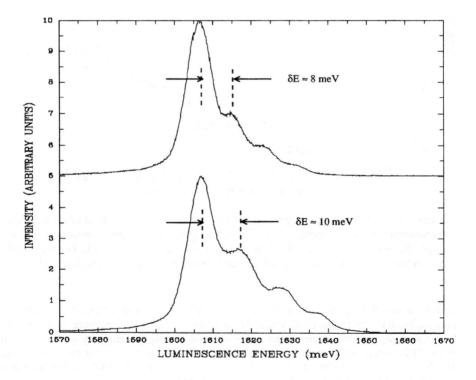

Figure 2. Magnetoluminescence spectra at 6 T at 4 K (upper) and 76 K (lower) for an n-type 4.5 nm GaAs/Al$_{0.25}$Ga$_{0.75}$As single quantum well. The energy separation between the first two Landau transitions is indicated for both spectra.

difference is a direct verification of the light-hole nature of the ground state in-plane valence-band mass, i.e., if m_v were heavy (e.g., 0.45) then the valence-band energy contributions to (1) and (2) can be ignored and the energy difference between the Landau peaks in both spectra would be nearly the same.

The conduction-band dispersion curve can be obtained from the field dependent low temperature magnetoluminescence measurements. For this temperature, the energy difference δE between the $E(n_c + 1)$ and $E(n_c)$ luminescence peaks gives all the necessary infor-

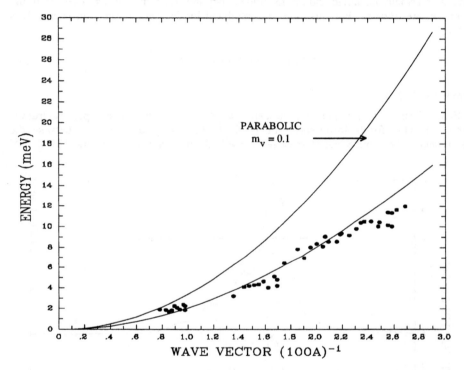

Figure 3. Valence-band dispersion curve determined by magnetoluminescence measurements for an n-type 4.5 nm GaAs/Al$_{0.25}$Ga$_{0.75}$As single quantum well. The upper dispersion curve is parabolic for a valence-band mass $m_v = 0.1$. The solid line drawn through the data points is a result of a $k \cdot p$ calculation (Lyo and Jones 1991).

mation about the conduction band from (1). Within the experimental resolution, the conduction-band is found to be parabolic with $m_c \approx 0.08$. The increase to m_c from the bulk value, $m_c \approx 0.067$, is attributed to the increased bandgap energy by quantum confinement. This is in agreement with previous studies (Singleton, et. al. 1988 and Osório, et. al. 1989) and also corroborated by our cyclotron resonance measurements on the same n-type sample which also gives $m_c \approx 0.08$ at the Fermi energy.

With this knowledge of the conduction-band dispersion curve, the field dependent valence-band energy can be derived from the high temperature 76 K data using (2). Figure 3 shows the valence-band dispersion curve as determined by our measurements. The method and justification used for relating the wavevector k to the magnetic field H has been fully discussed by Lyo and Jones (1990). The upper dispersion curve in Fig. 3 is parabolic with $m_v \approx 0.1$. The dispersion curve drawn through the data is based upon the $k \cdot p$ calculation discussed in this Conference by Lyo and Jones (1991). The maximum value of the wavevector of about 3% of the Brillouin zone is determined by the Fermi energy of the conduction band (Jones et. al. 1989). From Fig. 3, it is very evident that the valence band is highly nonparabolic. For the data at small wavevectors, $m_v \approx 0.15$ and for larger wavevector data (e.g., for $E_v \approx 10$ meV), $m_v \approx 0.3$.

In order to further corroborate the above valence band results, far-infrared cyclotron resonance and magneto-transport measurements were performed on several 4.5 nm p-type samples. The Fermi energies of the samples used for the cyclotron resonance measurements were of the order of 5-7 meV and the measured masses were found to be in the range of 0.20 to 0.25, in good agreement with the values derived by magnetoluminescence.

Magneto-transport values of the effective mass m_v were obtained by fitting the standard Lifshitz and Kosevich expression (Lifshitz and Kosevich 1956),

$$R \propto A(H,T) \sum_{r=1}^{\infty} r^{-\frac{1}{2}} \left(\frac{C_r \sin(2\pi r F/(H+r\phi))}{\sinh(rxTm_v/H)} \right), \qquad (3)$$

to the amplitude of the Shubnikov-de Haas resistance oscillations. Here $x = 2\pi^2 ck/eh$, F is the Shubnikov-de Haas frequency, H is the applied magnetic field, and T is the temperature. This technique depends upon observation of the oscillations at sufficiently low magnetic fields so that only the $r = 1$ term is contributing to (3). The necessity of having a low carrier concentration (low E_f) resulted in rather modest mobilities ($\mu \leq 6000$ cm^2/Vsec) so the data were of necessity taken at higher magnetic fields than desired. The lowest value of the effective mass obtained was $m_v = 0.16$ which should be considered an upper bound. Larger values for m_v were obtained at high fields as expected because of high harmonic contributions in (3), and because of field dependent mixing of the light and heavy-hole valence bands. In spite of these difficulties, there is no doubt that we are observing light holes of mass of the order of 0.16.

4. CONCLUSIONS

Using magnetoluminescence, far-infrared cyclotron resonance, and magneto-transport techniques, we have shown that the ground-state in-plane valence-band for GaAs/Al$_{0.25}$Ga$_{0.75}$As 4.5 nm single-quantum well structures is light. The valence-band mass m_v is energy dependent and varies between 0.15 and 0.3 for nominal Fermi energies. The agreement with $k \cdot p$ calculations is good.

5. ACKNOWLEDGMENTS

Part of this work was performed at Sandia National Laboratories and was supported by the Division of Materials Science, Office of Basic Energy Science, U. S. DOE under Contract No. DE-AC04-76P00789. Part of this work was performed at the Electrical Engineering Department, Princeton University and was supported by the Air Force Office of Scientific Research No. 88-0248.

6. REFERENCES

Bastard G, 1988 *Wave Mechanics Applied to Semiconductor Heterostructures* (New York: Halsted, Academic Press)

Jones E D and Wickstrom G L 1985 *Southwest Conference on Optics*, SPIE **40**, 362

Jones E D, Lyo S K , Fritz I J, Klem J F, Schirber J E, Tigges C P, and Drummond T J 1989 Appl. Phys. Lett. **54** 2227

Jones E D, Biefeld R M, Klem J F, and Lyo S K 1990 *Int. Symp. GaAs and Related Compounds, Karuizawa, Japan, 1989*, Inst. Phys. Conf. Ser. No. **106** 435

Lifshitz I M and Kosevich A M 1956 Soviet Physics, JETP **2** 635

Lyo S K, Jones E D, and Klem J F 1988 Phys. Rev. Lett. **61** 2265

Lyo S K and Jones E D 1990 *Electronic, Optical, and Device Properties of Layered Structures*, Proceedings, Fall Meeting of the Materials Research Society, Boston 271

Lyo S K and Jones E D 1991 *Proc. 18th Int. Symp. GaAs and Related Compounds*, Seattle.

Osbourn G C, Gourley P L , Fritz I J, Biefeld R M, Dawson L R, and Zipperian T E 1987 *Semiconductors and Semimetals, Vol. 24, Applications of Multiquantum Wells, Selective Doping, and Superlattices*, edited by R. Dingle (New York: Academic Press)

Osório F A P, Degani M H, and Hipólito O 1989 Superlattices and Microstructures **6** 107

Pearsall T P 1990 *Semiconductors and Semimetals, 32, Strained-Layer Superlattices: Physics*, edited by T. P. Pearsall, (New York: Academic Press)

Singleton J, Nicholas R J, Rogers D C, and Foxon C T B 1988 Surface Science **196** 29

Weisbuch G and Vinter B 1991 *Quantum Semiconductor Structures* (New York: Academic Press)

Inst. Phys. Conf. Ser. No 120: Chapter 9
Paper presented at Int. Symp. GaAs and Related Compounds, Seattle, 1991

413

Dynamics and transport of excitons confined at high-quality GaAs/Al$_x$Ga$_{1-x}$As interfaces

G.D. Gilliland, D.J. Wolford, G.A. Northrop, T.F. Kuech,[+]
and J.A. Bradley
IBM Research Division, T.J. Watson Research Center
P.O. Box 218
Yorktown Heights, NY 10598

We have studied, through detailed time- and space-resolved spectroscopies and numerical modeling, decay kinetics and spatial transport of free excitons confined near GaAs/Al$_{0.3}$Ga$_{0.7}$As heterointerfaces. We find that free excitons in bulk structures become readily localized by the conduction and valence bandbending at these n-p heterointerfaces. These excitons are thus quasi-two-dimensional and highly polarized in the growth direction, and give rise to a distinct, near-edge PL emission termed the H-band. We find these exciton dynamics to be completely analogous to the Quantum-Confined Stark Effect (QCSE), with excitonic transport which is extremely sensitive to the heterointerface electronic structure. We measure a 1.8-K quasi-2D exciton mobility of ~ 300,000 cm^2/V-s.

1. Introduction.

In 1984, Yuan et al.[1] observed a new photoluminescence (PL) emission in liquid-phase epitaxy (LPE)-prepared GaAs/Al$_x$Ga$_{1-x}$As heterostructures. Their observations showed the extremely complex nature of this emission process. They found the H-band was rapidly quenched with increasing temperature, disappearing altogether at temperatures above ~ 15 K. PL measurements versus cw laser power density revealed the dynamic nature of the emission process — namely the shifting of H-band PL peak to higher energies with increasing laser power. Their most important measurements involved PL measurements after chemically removing a portion of the structure. They found that H-band emission was substantially reduced or eliminated upon removal of the Al$_x$Ga$_{1-x}$As barriers. In addition, they found that H-band emission was not present in structures with intentionally graded interfaces. This proved that the mechanism responsible for H-band emission involved, in some way, the heterointerfaces.

Since the pioneering work of Yuan et al.,[1] there has been substantial interest in H-band emission, which has since has been observed in n-n, n-p, and p-p heterostructures, as well as structures grown by metal organic chemical vapor deposition (MOCVD) and molecular beam epitaxy (MBE).[2-10] In addition, its behavior has been studied in both electric and magnetic fields.[2,3,5] These studies confirm the association with the GaAs/Al$_x$Ga$_{1-x}$As heterointerface. However, there are striking inconsistencies among H-band studies reported thus far. It is therefore not surprising that no single unified picture of H-band emission dynamics has yet emerged.

Three distinct models have been proposed for H-band emission. Yuan et al.[1] suggested, for their n-n structures, that the recombination of free holes and electrons confined in the conduction-band potential notch at the heterointerfaces is responsible for H-band emission, whereas for p-n structures the holes are confined in potential notches and recombine with free electrons. Others have suggested that H-band PL is impurity-induced;[4,7] in these models H-band emission originates from the radiative recombination of donor-acceptor pairs with one of the impurities residing at the heterointerface. Another impurity model suggests, however, that the recombination is of free carriers confined in the heterointerface potential notch, with impurities located within the barrier material, i.e. the Al$_x$Ga$_{1-x}$As.

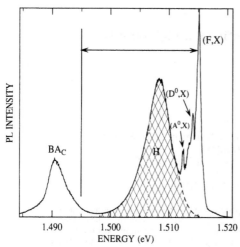

Figure 1. Low temperature (1.8 K) PL spectra of 0.3 μm thick double heterostructure. Hatched area shows H-band emission which may peak at energies shown by the arrow.

Lastly, we[9,10] have advocated a model, based on the high-purity and quality our samples, in which H-band PL results from the recombination of intrinsic <u>excitons</u> confined to the GaAs/$Al_x Ga_{1-x}As$ hetero-interfaces.

We have performed experiments parallel to those of Yuan et al.[1] in our double heterostructure samples, in an effort to confirm that the emission, shown in Fig. 1 (and typical of our samples) is indeed H-band emission. We find excellent agreement with all aspects of the originally reported H-band emission, thus confirming H-band emission in our samples. In order to gain further insight into the H-band dynamics and the emission mechanism, we performed time-resolved PL measurements. We have also numerically modelled our proposed excitonic state, and find excellent agreement with the experimental H-band recombination kinetics. Lastly, we have used a new, novel, all-optical technique to measure the transport properties of these excitons — and find that their transport properties are especially sensitive to the dynamic bandbending at the heterointerfaces, further confirming our excitonic model.

2. Experiment/Materials.

Our GaAs/$Al_{0.3}Ga_{0.7}As$ double heterostructures were grown by MOCVD at 750°C, with growth interruptions (15 s) at each interface to insure abruptness. $Al_{0.3}Ga_{0.7}As$ layers were p-type (1 x $10^{16} cm^{-3}$) and 0.5-μm thick. GaAs layers were n-type (1 x $10^{15} cm^{-3}$) and ranged in thickness from 0.1 to 2.0 μm. Heterointerface quality was assessed through detailed room-temperature lifetime measurements, thus providing an estimate of the interface recombination velocity. We find lifetimes > 2.5 μs and interface recombination velocities of < 40 cm/s.[11,12] These measurements are crucial to an unambiguous understanding of intrinsic dynamics in GaAs structures, since they provide the only means of estimating any possible nonradiative decay at the heterointerfaces which may mask the true, intrinsic decay processes. We find that our structures are wholly dominated by intrinsic processes, as evidenced by the dominance of free-exciton emission over donor and acceptor bound-exciton emission (Fig. 1), and the negligible interface recombination velocities. These low interface recombination velocities also confirm that carriers may reside near an interface without undergoing

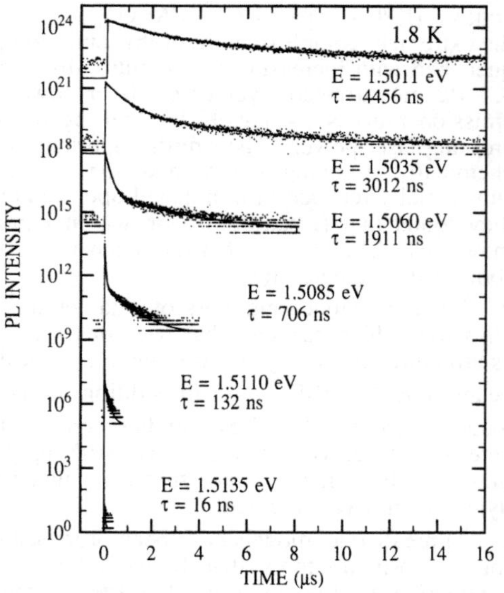

Figure 2. 1.8 K time-decay kinetics obtained at various emission energies within the H-band PL. Results are for a 0.2 μm thick heterostructure, and lifetimes represent the exponential tail of the decay.

nonradiative decay at the possible surface or interface states.

3. Results and Discussion.

Figure 2 shows H-band time decays at 1.8 K and various energies throughout the H-band emission profile, shown with the arrow in Fig. 1. These kinetics were obtained using the time-correlated single-photon counting technique, described elsewhere.[10] The decay kinetics we find are all nonexponential, and become longer for decreasing emission energies. Lifetimes shown in the figure are obtained from the exponential tail of each decay, and are > 50 μs in some samples (> 0.5 μm). We find also that all kinetics, in all samples, are identical on the high energy side of the emission, whereas on the low energy side, we observe lifetime saturation in thin structures (< 0.5 μm); lifetimes, in fact, saturate at lower values for decreasing thicknesses (i.e. 380 ns for 0.1 μm structure). In addition, time-resolved spectra show a time-dependent redshift of the H-band peak energy by up to ~ 25 meV.

Our model for H-band dynamics is based on the band structure of the double heterostructure and its effects on photoexcited carriers, and includes the Coulomb attraction between photoexcited electrons and holes. Figure 3 shows a realistic band structure for a typical double heterostructure. In our experiments, photoexcitation was chosen to be energetically below the $Al_{0.3}Ga_{0.7}As$ bandgap, and thus *only* generated carriers within the GaAs layer. It is clear in this figure, for these types of structures, that holes may become

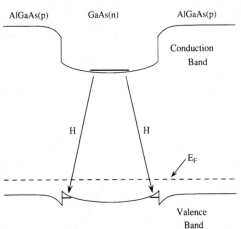

Figure 3. Band structure of a typical heterostructure, showing the potential notches responsible for the confinement of the quasi-2D excitons.

readily bound in the potential-notch at each heterointerface, whereas electrons tend to reside in the center of the layer. This is Yuan et al.'s[1] model for H-band emission and dynamics. However, in high-purity, high-interfacial-quality samples, such as ours, Coulombic effects between electrons and holes must also be included. This is indicated by the strong free-exciton emission typical of our samples, in Fig. 1. We have proposed[9,10] that H-band emission in our samples is from correlated electron-hole pairs, i.e. excitons bound by the Coulomb potential, and we model this by numerically solving Schrödinger's equation for the ground-state envelope wavefunctions of both electrons and holes — including the Coulomb interaction, and the conduction band and valence band potentials. The time-dependent dynamics arise from the time-dependent carrier densities, which produce time-dependent band-bending. These dynamics were thus modeled by parameterizing such potentials. The details of this calculation have been presented elsewhere.[9,10,13] We find that this model accurately predicts the quasi-2D exciton binding energies (and hence H-band temperature dependences), transition energies (H-band peak-energy power dependence and time-dependent PL shifts), structural dependence (lifetime saturation in thin samples), charge densities, and oscillator strengths.

Figure 4 shows the comparison between our model and our experimental results. Here lifetimes obtained for a 0.2-μm thick structure are plotted versus emission energy, together with our model calculation results for this particular structure (inset); the agreement is excellent. Further, we find the saturation in lifetime evident in this figure is not present in thicker structures, also in excellent agreement with experiment. These calculated lifetimes depend crucially on excitonic effects, since without this inclusion the model predicts radically different decay dynamics (smaller relative change in lifetime versus emission energy).

To further elucidate H-band dynamics, we have used a new, all-optical PL-imaging

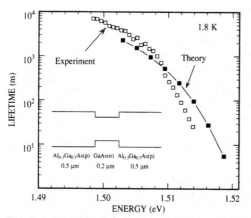

Figure 4. Theoretical results together with summary of lifetime results shown in Fig. 2.

technique,[14] which detects <u>neutral-particle transport,</u> and may prove the intrinsic nature of, and quantify transport of, the quasi-2D excitons responsible for H-band emission in our samples. This technique is powerful - more so than electrical techniques in many ways; thus, it is contactless, has high spectral and spatial resolution, may distinguish between the transport of distinct particles, may distinguish between diffusive and non-diffusive transport, and, most importantly, is sensitive to the transport of electrically neutral species, i.e. excitons. By spatially imaging the luminescence from a tightly focused laser pulse, this technique follows the lateral expansion of carriers with time, with resolutions of ~ 3 μm spatially, < 1 cm^{-1} spectrally, and < 1 ns temporally.

Figure 5 shows <u>time-resolved spatial</u> profiles obtained by spectrally windowing the <u>entire</u> H-band emission region (1.498 - 1.514 eV), and excluding free-exciton emission from the detector. This spectral windowing was necessary since we have shown that H-band emission is not spectrally stationary, and we wish to measure quasi-2D exciton transport from the time of their formation until they recombine. The solid lines in Fig. 5 represent Gaussian fits to the data with full-widths-at-half-maximum (FWHM) at each time after the laser pulse shown. To test for diffusive transport, Fig. 6 shows the squared FWHM versus time obtained from each spatial distribution, and for diffusive transport the data should be linear in time, with the slope proportional to the diffusion constant. Clearly, the data is not linearly distributed; however, we may obtain estimates of possible diffusion constants at both early and late times as shown in the figure — with the result of 3600 cm²/s and 49 cm²/s, respectively. Application of the Einstein relation for a diffusion constant of 3600 cm²/s results in a 1.8-K mobility of ≈ 23,000,000 cm²/V-s — a result much larger than any reported mobility in any GaAs structure ever,[15] and therefore, (in light of the nonlinear results of Fig. 6) might be termed nonphysical. It is reasonable to think that the initial rapid expansion of this exciton distribution results instead from a "driving force." One possible origin of such a force is the topic of the remainder of this paper.

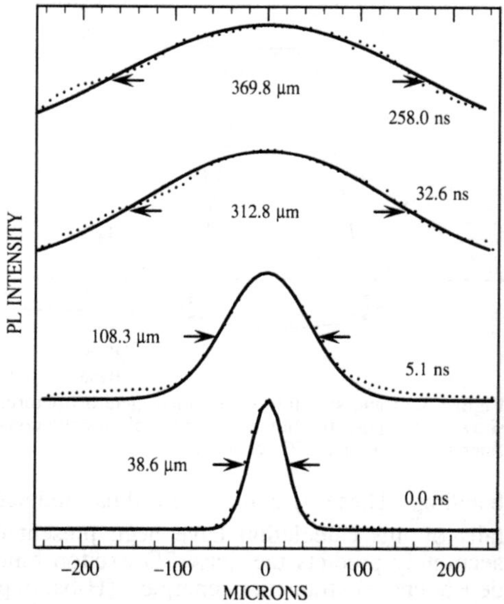

As a means of resolving questions regarding this quasi-2D exciton transport, we have measured both time and spatially resolved PL spectra, obtained by tightly focussing the laser to 3 μm and measuring time-resolved spectra at various distances from the laser spot. The high-density of initially photoexcited carriers may reduce the bandbending in directions

Figure 5. Time-resolved H-band spatial distributions at 1.8 K. Solid lines represent Gaussian fits to the data at various times after the laser pulse, with FWHM shown.

both parallel and perpendicular to the heterointerface. The excitons thus experience "flat" bands and may be viewed initially as spherical 3D excitons. At some distance from the laser spot, along the heterointerface, the bandbending must increase to the equilibrium

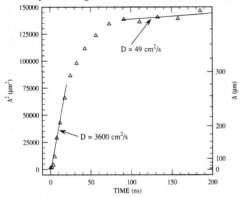

bandbending, since there are no photoexcited carriers away from the laser spot. Thus, 3D excitons are driven to these lower energy states, hence becoming two-dimensional, by the nonuniform bandbending. This force thus originates from the same bandbending which gives rise to the time-dependent shift of H-band PL. Thus, the PL energy at any distance from the excitation spot is a <u>direct measure</u> of this bandbending potential, and the gradient in the PL energy is the driving force for the expansion.

Figure 6. Squared FWHM versus time obtained from Fig. 5. Slope of data is proportional to the diffusion constant. Spatial distribution expands to > 400 μm.

Figure 7 shows the bandbending force, versus time, obtained from the spatial gradient of the H-band PL energy versus time. We find a maximum force of ≈ 1800 meV/cm, which decreases with time after the laser pulse. After > 20 - 40 ns the bandbending is essentially spatially uniform, and the quasi-2D excitons experience no additional force. Therefore, at later times the transport should be purely diffusive and this qualitatively explains our transport results of Fig. 6. At early times the extremely large diffusion constant actually represents the transport of quasi-2D excitons influenced by the bandbending force. At long times the transport is diffusive, and leads to a 2D-exciton mobility of 316,000 $cm^2/V\text{-}s$ — clearly a more physical and reasonable result for normal diffusive processes.

We may model this bandbending force on the quasi-2D excitons through careful consideration of the electrostatics involved. The spatially nonuniform bandbending results from a screening of the heterointerfacial field (which produces the bandbending) by the additional photoexcited carriers, in this case excitons. Thus, the nonuniform spatial distribution of excitons results in a spatially nonuniform bandbending, and, hence, a spatially nonuniform force. In addition, the time-dependent kinetics and transport lead to a time-dependent bandbending force. We have chosen to simplistically model this force through the spatially nonuniform distribution of quasi-2D excitons.

In our p-n structures the quasi-2D exciton center of mass (the hole) is localized next to the heterointerface, whereas the electron primarily exists more toward the center of the structure. Classically, this is a dipole, and all of the quasi-2D exciton dipole moments are aligned parallel to one another, perpendicular to the interface. Assuming a Gaussian distribution of dipoles (because the laser spot is Gaussian), we may thus calculate the peak net force

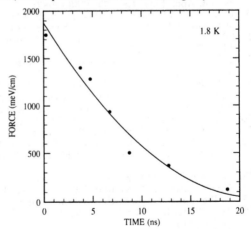

Figure 7. Bandbending force versus time after the laser pulse at 1.8 K.

on the excitons. Assuming a peak sheet density of $10^{10} cm^{-2}$, and a dipole moment of 500 Å, we calculate a peak force of ≈ 2000 meV/cm. Importantly, this is in excellent agreement with the force at t=0 in Fig. 7. Figure 7 also shows that the force decreases dramatically in the first 20 ns, presumably due to the reduction in nonuniformity of the bandbending.

4. Conclusions.

Taken together, these results further confirm that our H-band emission is "intrinsic," arising from highly mobile quantum-confined excitons along the heterointerface — and hence not impurity-induced. We find an intrinsic, quasi-2D exciton mobility of $\approx 300,000$ cm^2/V-s at 1.8 K — a value significantly less than the highest reported 2DEG mobility of $\approx 11,000,000$ cm^2/V-s,[15] but in excellent agreement with the highest 2DHG mobility of $\approx 300,000$ cm^2/V-s.[16] We find that this mobility is almost an order of magnitude lower than measured 3D free-exciton mobilities in these same structures,[17] but this presumably results from differences in Bohr radii and the effects of heterointerface scattering. Moreover, we have measured the time- and carrier density-dependent bandbending at the heterointerfaces, and their effects on the carrier dynamics. We find the screening of the heterointerface field by photoexcited carriers significantly affects the dynamics of excitons confined at the heterointerfaces, with the lateral gradient in excitation density (and hence, bandbending), resulting in an outward driving force. This force qualitatively explains the rapid expansion observed at early times.

We thank J. Martinsen for help in computer-data acquisition and analysis. Supported in part by the ONR under contracts N00014-85-C-0868 and N00014-90-C-0077. [†]Present Address: Univ. of Wisconsin, Dept. of Chem. Eng., 1415 Johnson Drive, Madison WI 53706.

REFERENCES

1. Y.R. Yuan, K. Mohammed, M.A.A. Pudensi, and J.L. Merz, Appl. Phys. Lett. **45**, 739 (1984); Y.R. Yuan, M.A.A. Pudenski, G.A. Vawter, and J.L. Merz, J. Appl. Phys. **58**, 397 (1985); Y.R. Yuan, J.L. Merz, and G.A. Vawter, J. Lumin. **40,41**, 755 (1988).
2. W. Ossau, E. Bangert, and G. Weimann, Solid State Communications **64**, 711 (1987).
3. Zh.I. Alferov, A.M. Vasil'ev, P.S. Kop'ev, V.P. Kochereskho, I.N. Ural'tsev, Ai.L. Efros, and D.R. Yakovlev, JETP Lett. **43**, 570 (1986).
4. L.W. Molenkamp, G.W. 'tHooft, W.A.J.A. van der Poel, and C.T. Foxon, J. de Physique Colloque **C5**, 217 (1987); G.W. 'tHooft, W.A.J.A. van der Poel, L.W. Molenkamp, and C.T. Foxon, Appl. Phys. Lett. **50**, 1388 (1987).
5. Q.X. Zhao, J.P. Bergman, P.O. Holtz, B. Monemar, C. Hallin, M. Sundaram, J.L. Merz, and A.C. Gossard, Semicond. Sci. Technol. **5**, 884 (1990).
6. J.P. Bergman, Q.X. Zhao, P.O. Holtz, B. Monemar, M. Sundaram, J.L. Merz, and A.C. Gossard, Phys. Rev. B **43**, 4771 (1991).
7. P.S. Kop'ev, V.P. Kochereshko, I.N. Uraltsev, Al.L. Efros, and D.R. Yakovlev, J. Lumin. **40,41**, 747 (1988).
8. I. Balslev, Semicond. Sci. Technol. **2**, 437 (1987).
9. G.D. Gilliland, D.J. Wolford, T.F. Kuech, and J.A. Bradley, Proc. of the 20th Int. Conf. on the Phys. of Semicond., Thessaloniki Greece, (World Scientific 1990) p. 1577.
10. G.D. Gilliland, D.J. Wolford, T.F. Kuech, and J.A. Bradley, Phys. Rev. B **43**, 14251 (1991).
11. G.D. Gilliland, D.J. Wolford, T.F. Kuech, C.F. Tsang, and J.A. Bradley, unpublished.
12. D.J. Wolford, G.D. Gilliland, T.F. Kuech, L.M. Smith, J. Martinsen, C.F. Tsang, R. Venkatasubramanian, S.K. Ghandhi, and H.P. Hjalmarson, J. Vac. Sci. Technol. B **9**, 2369 (1991).
13. G.D. Gilliland, D.J. Wolford, T.F. Kuech, and J.A. Bradley, unpublished.
14. G.D. Gilliland, D.J. Wolford, T.F. Kuech, and J.A. Bradley, Appl. Phys. Lett. **59**, 216 (1991).
15. L. Pfeiffer, K.W. West, H.L. Stormer, and K.W. Baldwin, Appl. Phys. Lett. **55**, 1888 (1989).
16. W.I. Wang, E.E. Mendez, Y. Iye, B. Lee, M.H. Kim, and G.E. Stillman, J. Appl. Phys. **60**, 1834 (1986).
17. G.D. Gilliland, D.J. Wolford, H.J. Hjalmarson, T.F. Kuech, and J.A. Bradley, unpublished.

Inst. Phys. Conf. Ser. No 120: Chapter 9
Paper presented at Int. Symp. GaAs and Related Compounds, Seattle, 1991

419

Anisotropy in the interband transitions of (110) oriented quantum wells

D. Gershoni

AT&T Bell Laboratories
Murray Hill, New Jersey 07974 U. S. A., and
Physics Department Technion, Haifa 32000 Israel

I. Brener, G. A. Baraff, S. N. G. Chu, L. N. Pfeiffer, and K. West

AT&T Bell Laboratories
Murray Hill, New Jersey 07974 U. S. A.

Abstract. The reduced symmetry of the quantum confinement direction which is inherent to quantum wells grown on (110) oriented substrates results in in-plane polarization anisotropy of the optical transitions. We present a study of this anisotropy using photoluminescence excitation spectroscopy of (110) and (100) oriented quantum well structures. The spectral features observed in the photoluminescence-excitation spectra of these structures are modeled using an eight band $k \cdot p$-type effective mass theory. We directly determine the valence band anisotropy of the quantum well material from the orientational dependence of the transition energies and oscillator strengths. We found $(\gamma_3 - \gamma_2)/(\gamma_3 + \gamma_2) = 0.18 \pm 0.03$, where $\gamma_{2,3}$ are the GaAs Luttinger parameters.

The spectroscopy of quantum wells (QWs) and superlattices has been intensively investigated since the first observation of quantum confinement by optical means by Dingle Wiegmann and Henry (1974). Most of these investigations have been performed on (100) oriented structures, since high-quality epitaxial growth was traditionally available in this crystallographic orientation only (Miller et al 1984, Nelson et al 1987 and references therein). In order to account for the detailed spectroscopic studies of QW structures, several theoretical models have been developed (See for example Schulman and Chang 1985, Eppenga et al 1987 and references therein.) It became clear from these models, which are far more complicated than the first intuitive "particle in a box" description of Dingle and his coworkers, that the wavefunctions and consequently the optical properties of these quantum confined systems should be sensitive to the crystallographic direction of the epitaxial growth. This is mainly due to the valence band anisotropy of the zinc-blend III-V binaries used for the hetero-epitaxy. Hayakawa et al (1988), Molenkamp et al (1988) and Shanabrook et al (1989) have recently shown experimentally that indeed, both the energies and the oscillator strengths of the interband optical transitions are sensitive to the crystallographic direction of the epitaxial growth.

In this work, we report on detailed spectroscopic studies of single QWs structure formed by high quality epitaxial growth on (110) oriented substrate. We show, both experimentally and theoretically, that the reduced symmetry of the composition induced quantum confinement [(110) here as opposed to (100) or (111) studied previously], results in an in-plane polarization anisotropy of the spectral response of the sample. Our eight band k·p model describes well the optical transition energies as well as their polarization anisotropy. From the observed optical anisotropies, we can thus, quite accurately determine the valence band anisotropy of GaAs - the QW material.

The $Al_{0.26}Ga_{0.74}As$/GaAs multi-quantum well samples were grown by molecular beam epitaxy (MBE) simultaneously on two different GaAs wafers. One was (100) oriented and the other one was (110) oriented, both to within a tenth of a degree. During growth the wafer temperature was $480°C$ as measured by an infrared pyrometer and the As_4 partial pressure was 1.6×10^{-5} Torr. measured at the substrate position. The rate of growth of GaAs under these conditions was $0.5 \mu m$/hour, and a smooth featureless morphology was obtained on both substrates. A more detailed description of the optimal growth conditions for this crystallographic orientation is given by Pfeiffer et al (1990). Modulation doped 250Å wide QWs grown in the same conditions show two dimensional electron gas mobilities as high as $5 \times 10^5 cm^2/V \cdot s$ for the (110) oriented substrate and about a factor of 3 lower for the (100) oriented one. A very thin GaAs buffer layer (~30Å) was deposited prior to the growth of three GaAs QWs of different widths, separated by 150-200Å thick $Al_{0.26}Ga_{0.74}As$ barriers. The samples were not rotated during growth so that thickness gradients of about 7%/cm due to differences in distance from the effusion cells are expected. To circumvent the difficulty of well size determination, we used cross-section transmission electron microscopy (TEM) to determine the dimensions of the QWs in the exact positions where the optical studies had been carried out. The exact Aluminum concentration was then determined from the optical spectroscopy. The value thus determined agreed within experimental uncertainty with the nominal growth value.

For the photoluminescence (PL) and PL excitation (PLE) measurements the samples were oriented using two cleavage planes and placed in a He-flow cryostat. A pyridin-2 dye laser pumped by an Ar^+ ion laser was used as a continuously tunable source of excitation. The laser light was focused at normal incidence onto the sample surface and its polarization was controlled by a polarization plane rotator and a polarizer. The luminescence from the sample was collected at a large solid angle perpendicular to the exciting light direction (See inset to Fig. 1). The PL, collected this way from both samples, was found to be completely unpolarized.

Fig. 1a displays a set of PL spectra from the (110) oriented sample for various polarization angles of the exciting laser light measured from the $(\overline{1}10)$ crystallographic direction. Similarly, Fig. 1b displays a set of PL spectra from the (100) oriented sample. The spectra which were excited at 1.746 eV show a clear polarization angle dependence of the emission intensity from the three (110) oriented QWs. In contrast, the two lowest energy lines, which originated in the GaAs substrate, as well as the emission from the three (100) oriented QWs, do not show this dependence. As can be seen in Fig. 1a, At the excitation energy of 1.746 eV the emissions from the narrowest and from the widest QWs have a similar dependence on the polarization angle. Both have a clear maximum (minimum) when the light is polarized along the (001) $((\overline{1}10))$ crystallographic direction. The emission intensity from the middle size well, shows the opposite dependence.

It reaches maximum (minimum) when the polarization is parallel to the ($\bar{1}$10) ((001)) axis. As mentioned above, it is clear from Fig. 1b that the PL intensity from all the (100) oriented QWs is completely polarization independent.

This unusual behavior of the PL from (110) oriented QWs is described more completely in Fig. 2a. The figure displays the PL spectrum (lowest solid line), the ($\bar{1}$10) polarized PLE spectra (solid lines), and the (001) polarized PLE spectra (dashed lines) from the three (110) oriented QWs. The PLE spectra are vertically displaced for clarity. The energy of excitation for which Fig. 1. was obtained is marked with an arrow in Fig. 2a. Thus, the amplitude of the anisotropy modulation of the PL intensity, exhibited in Fig. 1, also appears as the difference between the two PLE polarizations shown in Fig. 2a. In Fig. 2c we show the PLE spectra from the three QWs in the (100) oriented sample. Since no polarization dependence was observed here, there is only one (solid) line for each PLE spectrum.

The calculated absorption spectra for the (110) ((100)) oriented QWs are shown for comparison in Fig. 2b (Fig. 2d). Our model for calculating absorbtion spectra does not include excitonic effects, therefore the spectra are down shifted by 0.01 eV to roughly account for the two-dimensional exciton binding energy.

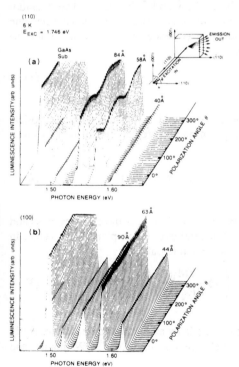

Fig. 1 PL spectra of a) the (110) and b) the (100) oriented MQW structure for various excitation polarization angles θ measured from the ($\bar{1}$10) crystallographic axis (See inset).

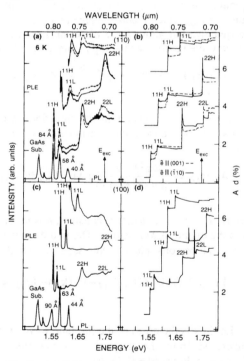

Fig. 2 a) [c)] Measured and b) [d)] calculated PLE spectra of the (110) [(100)] oriented MQW sample. The calculated and measured PLE spectra are vertically displaced above the measured PL spectra (lowest solid lines).

The experimental PLE spectra are dominated by excitonic resonances, thus the similarity between them and the calculated absorption is not immediately evident. However, if those resonances are ignored, the agreement is quite satisfactory. The polarization anisotropy and the peak energies (to within the experimental line widths) are both very well described by our calculations for all six QWs.

Our absorption calculations proceed along the lines described by Baraff and Gershoni (1991). We use a Fourier expansion method to solve the eight-band $k \cdot p$ Kane-Luttinger hamiltonian with bulk parameters for each material region. This converts the coupled differential equations into a matrix eigenvalue problem which is then solved numerically. The discontinuity in the material parameters is taken care of by a procedure discussed by Baraff and Gershoni (1991). The procedure is equivalent to the condition that the normal component of the probability current be continuous across internal interfaces. The method is applied to an arbitrary crystallographic direction of composition induced confinement by a simple rotation of the differential operators.

Once the eigenvalues $E_i(k_{||})$ and the corresponding eigenfunctions $\Psi_i(k_{||},z)$ are found it is straight forward to evaluate the optical matrix element $M_{ij}(k_{||})$ for transitions between the eigen states i and j. Here $\hat{z}$ is defined as the direction perpendicular to the quantum well plane, which is uniform and infinite in the $\hat{x}$ and $\hat{y}$ directions. The quantum number i (j) which labels the eigen values of the system, runs over the discrete energies due to quantization along the $\hat{z}$ direction in the conduction (valence) bands, and $k_{||}$ is the in-plane wave vector (with components k_x and k_y).

$$M_{ij}(k_{||}) = |<\Psi_i(k_{||})|\hat{e} \cdot \frac{\hbar}{i} \nabla |\Psi_j(k_{||})>|^2 \tag{1}$$

where $\hat{e}$ is a unit vector in the direction of the electric field associated with the optical radiation. The absorption coefficient (A) for a given photon frequency (ω) is then given by:

$$A(\hbar\omega) = \frac{4\pi^2\hbar^2\alpha}{ndm^2(\hbar\omega)} \sum_{i,j} \int M_{ij}(k_{||})dk_{||}\delta[E_i(k_{||})-E_j(k_{||})-\hbar\omega] \tag{2}$$

Here α is the fine structure constant, n is the index of refraction, m is the electron mass and d is the quantum well width. The two dimensional integral over the in-plane wave vector in Eq. 2 can be conveniently expressed as a one dimensional integral:

$$\int M_{ij}(k_{||})dk_{||}\delta[E_i(k_{||})-E_j(k_{||})-\hbar\omega] = \int_0^{2\pi}d\phi\frac{kM_{ij}(k,\phi)}{\partial E_{ij}(k,\phi)/\partial k} \tag{3}$$

In Eq. 3, k is the norm of the in-plane wave vector $k_{||}$, ϕ is its azimuthal angle, conveniently measured from the projection of the electric field vector, $\hat{e}$, in the QW plane, and $E_{ij}(k,\phi)=E_i(k,\phi)-E_j(k,\phi)$.

To calculate the integral in the right hand side of Eq. 3 , we evaluate the integrand for evenly spaced values of ϕ, average the results and multiply by 2π. In practice, the process converges well for a very small number of divisions of the interval 2π (For a (110) oriented QW the use of four divisions results in convergence at least as good as the experimental accuracy). Moreover, we have found that for the case of quantum confinement along the principal crystallographic axis (100) and polarization field along a principal crystallographic axis in the QW plane, such as (010), this average is given exactly by evaluating at the single point $\phi = \frac{\pi}{4}$. This yields the well known result that the in-plane absorption of a (100) QW is completely unpolarized. Similarly, we have observed that evaluating at $\phi = \frac{\pi}{4}$ results in a very good approximation for the average over all the azimuthal intervals for the case of the (110) oriented QW with polarization field along the ($\bar{1}$10) or (001) crystallographic directions in the QW plane. Fig. 2b and 2d were calculated by using $\phi = \frac{\pi}{4}$ in the evaluation of the integral in Eq. 3. The material parameters used in these calculations are listed and referenced in Table. 1.

A word of caution is due here: a few of the parameters are not very well determined yet, in particular, those of the AlGaAs ternary barrier. Fortunately, the calculations are not too sensitive to the values of the barrier's band-structure γ_i and P parameters [where γ_i, i=1, 2, 3, are the Luttinger (Luttinger 1956) and P is the conduction-valence band interaction matrix element]. The computed spectra are much more sensitive to the band structure parameters of the QW material. We were able to compute spectra all of which equally resembled the experimental results using different **combinations** of those γ_i and P parameters. We found, however, that the quality of the fit to both the transition energies and polarization anisotropies is a very sensitive function of the valence band anisotropy of the QW material which we define as $\gamma_a = (\gamma_3 - \gamma_2)/(\gamma_3 + \gamma_2)$. Since our goal was to experimentally estimate the valence band anisotropy, we chose accepted parameters (except γ_3) from the literature. We obtained γ_3 from the best simultaneous fits to the six different PLE spectra in Fig. 2a. Thus, comparison between the computed spectra and the experimental measurements provides an all optical method of measuring this otherwise difficult to determine parameter.

Table 1. Parameters used in calculations.

	E_g (eV)	Δ_{so} (eV)	γ_1	γ_2	γ_3	$\frac{2mP^2}{\hbar^2}$ (eV)	m_{el}^* (e.m.u)	Val. band offset (eV)
GaAs	1.519[a]	0.340[a]	6.790[b]	1.924[b]	2.782[c]	28.8[d]	0.0665[a]	0.0
$Al_{0.26}Ga_{0.74}As$	1.845[e]	0.323[f]	6.942[g]	1.658[g]	2.838[c]	27.0[f]	0.089[f]	-0.131[h]

a)   Madelung, Schulz and Weiss (1982).

b)   Nelson Miller and Kleinman (1987).

c)   Best fit to our optical data.

d)   Hermann and Weisbuch (1977)

e)   We use AlAs direct bandgap of 3.04 eV and bowing parameter of 0.37 eV.

f)   We use linear interpolation between the binary values as given in footnote a).

g)   We use linear interpolation between the effective masses in the (100) crystallographic direction.

h)   Following footnote b), we assume that 40% of the bandgap discontinuity is in the valence band.

We obtained Fig. 2b and Fig. 2d (which best agree with Fig. 2a and Fig. 2c) using γ_a = 0.18±0.03. This value is in close agreement with the results of Hayakawa et al (1988), Molenkamp et al (1988) and Shanabrook et al (1989). This value is quite far, however, from the theoretical calculations of Lawaetz (1971) which have been extensively used for modeling QWs in this material system.

To summarize, we report on optical spectroscopy of (110) oriented quantum wells. We account for both the energies and oscillator strengths of all the observed interband transitions in terms of an eight band k·p effective-mass theory. We used our results to independently determine the valence band anisotropy in GaAs.

References

Baraff G A and Gershoni D 1991 Phys. Rev. **B43** 4011

Dingle R, Wiegmann W and Henry C H 1974 Phys. Rev. Lett. **33** 827

Eppenga R, Schuurmans M F and Colak S 1987 Phys. Rev. **B36** 1554

Hayakawa T, Suyama T, Takahashi K, Kondo M, Suyama T, Yamamoto S, and Hijikata T, 1988 Phys. Rev. Lett. **60** 349

Hermann C and Weisbuch C 1977 Phys. Rev. **B15** 823, and **ibid** 826.

Lawaetz P 1971 Phys. Rev. **B4**, 3460

Luttinger J M 1956, Phys. Rev. **102**, 1030

Madelung O, Schulz M and Weiss H 1982 editors of Semiconductors, Landolt-Börnstein, New Series Group 3, Vol. 17a (Springer, Berlin, 1982).

Miller R C, Kleinman D A, and Gossard A C 1984, Phys. Rev. **B29** 7085

Molenkamp L W, Eppenga R, 't Hooft G W, Dawson P, Foxon C T and Moore K J, 1988, Phys. Rev. **B38** 4314

Nelson D F, Miller R C and Kleinman D A 1987, Phys. Rev. **B35** 7770

Pfeiffer L N, West K, Stormer H L, Eisenstein J, Baldwin K W, Gershoni D and Spector J 1990, Appl. Phys. Lett. **56** 1697

Schulman J N and Chang Y C 1985 Phys. Rev. **B31**, 2056

Shanabrook B V, Glembocki O J, and Broido D A 1989, Phys. Rev. **B39** 3411

Inst. Phys. Conf. Ser. No 120: Chapter 9
Paper presented at Int. Symp. GaAs and Related Compounds, Seattle, 1991

425

Ionized-impurity scattering of electrons in Si-doped GaAs/AlGaAs quantum wells in low and high electric fields

W. Ted Masselink

IBM Research Division,
T.J. Watson Research Center, Yorktown Heights, New York 10598 USA

Abstract: — The first measurements of low- and high-field transport properties of electrons in doped quantum wells are reported. We find that ionized-impurity scattering of electrons in quasi-two-dimensional systems is generally enhanced compared to that in similarly doped bulk material. This enhancement is evident not only at low electric fields, but continues to high fields, resulting in lower peak velocities. By delta-doping (rather than uniformly doping) the quantum wells, the low-field mobility is further reduced; at electric fields between 2 and 4 kV/cm, however, the differential mobility of the δ-doped quantum wells increases dramatically, also leading to higher peak velocities.

I. Introduction

The modulation-doped heterojunction transistor uses a two dimensional electron gas (2DEG) which is spatially separated from the parent donors resulting in very high mobilities. Although high mobilities can improve the ultimate speed performance of some FETs by decreasing access resistances and allowing higher electron velocities under the source-end of the gate, high-field mobility and peak velocity are much more important. For FETs with short gate lengths and for FETs using carriers with already high mobilities (Γ valley electrons), the entire velocity versus electric field behavior of the carriers is important. This fact along with the problems of the D-X center in AlGaAs and the difficulty in obtaining reproducible doping levels during epitaxial growth leads one to consider alternatives to modulation-doped FETs such as MESFETs [1] or MISFET-like structures with either undoped or doped channels. In these structures, high low-field mobility is not made such a high priority.

In previously published articles, we have reported the velocity versus electric field characteristics for electrons in doped GaAs and in modulation-doped GaAs [2-4]. These studies showed that, up to about 10 kV/cm, the entire velocity versus electric field characteristic of the 2DEG is higher than that of *heavily* doped GaAs; at very high electric fields, it appears that the velocities do not differ. Further data showed, however, that at both 300 and 77 K, the peak velocity of electrons in undoped bulk GaAs is higher than that of electrons in the 2DEG. This result is explained by a combination of enhanced transfer of electrons from the GaAs Γ valley to the L valleys along with the addition of real space transfer into both the indirect and direct valleys of the AlGaAs. An enhanced polar optical phonon scattering mechanism appears to add to this effect at 77 K.

In the present studies, the peak velocity of electrons in doped quantum wells is measured. In these quantum wells, dopant atoms are placed into the wells only; the barriers remain undoped. The present results indicate that although the peak velocity of these quasi-two-dimensional electrons is lower than those in bulk GaAs — qualitatively similar to the result for quasi-two-dimensional electrons in undoped channels, the mechanisms appear to be quite different. The lower peak velocity occurs instead in these structures because of ionized impurity scattering which is enhanced in these quasi-two-dimensional structures.

II. Samples and High-Field Results

The samples were grown by molecular beam epitaxy under identical conditions on (100) oriented undoped GaAs substrates. All sources were solid and a low substrate temperature of about 560° C was used. The bulk GaAs used for comparisons was 0.25 μm thick and doped with Si to a level of 6×10^{17} cm^{-3}. The quantum wells, which were in multiple-quantum-well samples with 25 to 50 periods, were 50 to 400 Å wide with 34 Å (17 Å in the case of the 50 Å wells) wide barriers of

$Al_{0.4}Ga_{0.6}As$. These barriers are narrow enough to ensure that the two dimensional concentration of traps in the AlGaAs is small compared to the two dimensional electron and dopant concentrations, yet are thick enough to contain the wavefunction in the GaAs quite well. All quantum wells are also doped to an average three dimensional concentration of 6×10^{17} cm^{-3}. The 100 Å wells and 50 Å wells were prepared with progressively narrower doping profiles, maintaining the same average doping concentration. The widest doping profile comprised 84% of the well width; the entire well was not doped in order to prevent D-X center formation near the interfaces. The narrow-profile limits of these sample series includes samples delta-doped [5, 6] with sheet densities of 6×10^{11} cm^{-2} in the 100 Å wells and 2.8×10^{11} cm^{-2} in the 50 Å wells. (In δ-doped samples, the sheet density of donor atoms is grown on one atomic plane; the final profile is determined by diffusion, but is still quite narrow.) Wells 100 Å wide were also delta doped off-center similar sheet density of 6×10^{11} cm^{-2}. The samples with 200 Å and 400 Å wells were doped 6×10^{17} cm^{-3} in the entire wells except the 6 Å adjacent to each interface.

Transport properties of the electron gas were measured at high fields using a sinusoidally varying electric field with frequency of 35 GHz [4] to avoid the formation of charge and field domains in the samples. This technique has been shown reliable in the measurement of electron velocity in bulk GaAs [7] and in a two-dimensional electron-gas [4]. Figure 1 shows the measured electron velocity as a function of the electric field for three samples, all with a three-dimensional doping density of 6×10^{17} cm^{-2}: the bulk sample of total thickness 0.25 μm, a sample with 100 Å quantum wells, and a sample with 50 Å quantum wells.

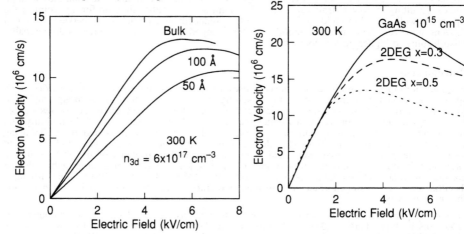

Figure 1: Measured electron velocities for bulk GaAs and 100 Å and 50 Å wide GaAs quantum wells uniformly doped with Si. The volume density in each is 6×10^{17} cm^{-3}.

Figure 2: Measured electron velocities comparing bulk GaAs with $n=1\times10^{15}$ cm^{-3} and two different two-dimensional electron gas samples with different AlGaAs composition from Ref. [2].

From Figure 1 we see that even with the same average 3-D electron density, the electron gas confined to narrower quantum wells has a lower peak velocity than in bulk GaAs. This result is qualitatively similar to lower peak velocity found in in the two-dimensional electron gas in undoped GaAs when compared to that of 3-D electrons in bulk undoped GaAs [2, 4]. (See Figure 2.) Comparing Figure 1 and Figure 2, however, we see that while in Figure 2, the mobility at low electric fields of the 2-D case is essentially the same as in the 3-D case, this is not true in Figure 1. In doped bulk GaAs, as the donor concentration is increased, both the peak electron velocity *and* the low field mobility decrease. The decrease of both of these values is due to an increase in ionized-impurity scattering resulting from a higher ionized impurity concentration. A fairer comparison to make regarding the data of Figure 1, then, is to compare the peak velocities as functions of low-field mobility. Comparing the peak velocity of the 100 Å wide sample along with its low-field mobility of 2536 cm^2/Vs and of the 50 Å wide sample along with its low-field mobility of 1767 cm^2/Vs to a summary of peak velocity versus low-field mobility [3, 4], we see that these velocities are consistent with the velocities found in bulk GaAs. From this we conclude that what limits the high-field velocity in these structures is *not* enhanced transfer of electrons out of the Γ valley into the GaAs L valleys and (real space transfer) into the AlGaAs as occurs in modulation-doped heterostructures, but, rather, simply an enhancement of the ionized-impurity scattering.

III. Ionized-Impurity Scattering in Two Dimensions

In this section we show how the dimensionality itself affects the ionized-impurity scattering in a semiconductor. In the case of a bulk semiconductor, the total wavefunction of the electrons overlaps with the ionized impurities which are the same in number as the electron density. If such a uniformly doped bulk semiconductor is confined in one dimension, the overlap of the electronic wavefunction with the ionized impurities (an effective impurity concentration) will be altered because the electronic wave function will be peaked in the center, but vanishing at the edges of the resulting quantum well.

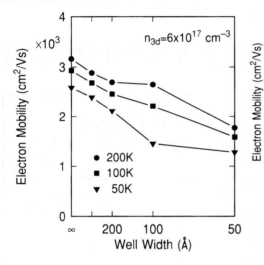

Figure 3: Measured Hall mobility as functions of quantum well width.

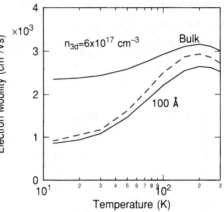

Figure 4: Measured Hall mobilities as functions of temperature for bulk GaAs and 100 Å wide quantum wells. The solid curves are measured data; the dashed curve is adjusted as described in the text in order to make an appropriate comparison with bulk.

Van der Pauw-Hall measurements were made on the samples between 10 and 300 K using low electric and magnetic fields. In heavily doped semiconductors, such as the samples studied here, the Hall factor is very close to unity, regardless of scattering mechanism. The curves of Figure 3 show low-field mobility of electrons at several temperatures at functions of quantum well width. It is clear that for a given electron density (and identical ionized impurity density), narrower quantum wells lead to lower mobilities. One explanation for such behavior can be the increased importance of interface roughness scattering. Previous data [8, 9] indicate that interface roughness scattering of a 2DEG in wells of 100 Å or wider is quite small. These same data, on the other hand, also indicate that in 50 Å wells it can be significant. Structures with 50 Å wells and AlAs barriers have mobilities limited to about 1000 cm²/Vs [8]; when the barrier alloy is $Al_xGa_{1-x}As$ with x between 0.30 and 0.35, mobilities of about 10^4 cm²/Vs were measured [9]. Electrons in 100 Å wells achieve mobilities of 10^5 cm²/Vs even with the rougher AlAs barriers [8]. From this we conclude that the interface roughness scattering almost certainly needs to be accounted for in the 50 Å well samples, but is not important compared to the ionized impurity scattering in our 100 Å (and wider) samples with $Al_{0.4}Ga_{0.6}As$ barriers. The solid curves of Figure 4 show the temperature dependence of the mobilities of two of the uniformly doped samples: bulk GaAs and 100 Å wells. The mobility of the bulk sample is what is typically measured at this doping level and is well understood theoretically [10]. Mobilities for the doped quantum wells are significantly lower than that observed in bulk GaAs for all temperatures. How the confinement of the electronic wavefunction affects the ionized impurity scattering can be understood theoretically following Ref. [11]. It is then straightforward to numerically obtain the theoretical mobilities as functions of dopant distribution, temperature, and well width.

Neither Figure 3 nor the solid curves of Figure 4 really provide a definitive comparison of ionized-impurity scattering in quasi-two-dimensional versus bulk semiconductors. This is because the quantum wells are not doped over 100% of the well, but only in the center. For example the 100 Å well is doped in the center 84% of the well. Additionally, the 50 Å sample's mobility has also been lowered by interface roughness scattering. Using the analysis above, however, we can accurately correct for the effect of the the less than 100% doping of the wells. Such corrected data for the 100 Å wells is displayed in the dashed curve on Figure 4. These data are slightly different from the raw data of the solid curve and demonstrate unambiguously that quasi-two-dimensional confinement results in a decrease of mobility and therefore an increase in ionized impurity scattering over a broad temperature range. Thus, we see that the lower mobility and the lower peak velocity of electrons in doped quantum wells when compared to electrons in doped bulk GaAs result in a fundamental way from the dimensionality of the channel. This decreased mobility is explained quantitatively [11] by a decrease in screening and an increase in the overlap between electrons and ionized impurities in quantum wells when compared to bulk.

IV. Delta-Doped Quantum Wells

From the analysis and data presented above, it is clear that uniformly doped quantum wells have a lower mobility than identically doped bulk GaAs. Figure 5 shows the effect of concentrating the dopants into the center of the well.

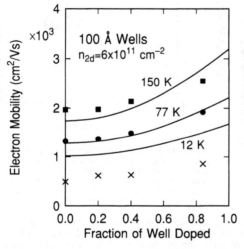

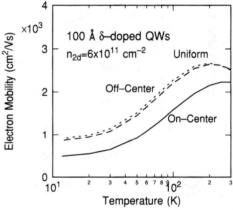

Figure 5: Electron mobility for 100 Å quantum wells as a function of temperature and fraction of the well which is doped. The curves are calculated as described in the text and the symbols are experimental data taken at 150 K (■), 77 K (•), and 12 K (×).

Figure 6: Measured Hall mobilities as functions of temperature for 100 Å wide GaAs quantum wells uniformly doped (dashed curve), δ-doped in the center (solid curve), and δ-doped midway between the center and an edge (dotted curve). The sheet density in each well is 6×10^{11} cm^{-2}.

In each case, the areal concentration is about 6×10^{11} cm^{-2}, but the distributions vary. When the fraction of the well which is doped reaches 0, the well is δ-doped in the center; when the fraction is 1.0, it is uniformly doped. The solid curves of Fig. 2 are calculated as described in Ref. [11] with no corrections or adjustable parameters. The symbols in Figure 5 are experimental data. We see that the doping profile has a significant impact on the ionized impurity mobility and that the model described above describes the physics adequately, except at low temperatures. At higher temperatures (and especially for larger fractions of the well which is doped), phonon scattering (not included in the calculation) becomes important. The mobility of the on-center δ-doped quantum wells is lower than when the same sheet concentration is uniformly doped in the wells because in the δ-doped wells all of the ionized impurities are located where the electronic wavefunction is maximum. By moving the δ-doping away from the center, we can increase the mobility. Figure 6 shows the mobility versus temperature for three samples. The solid curves are experimental. The one labeled "Uniform" is uniformly doped in the center 84% of the well. The lowest curve is for the center-delta-doped wells. The third curve is from a sample which is also δ-doped, but with the doping midway between the center and the edge of the well. From Fig. 6, we see that the mobility is higher than when the dopants are all at the center of the well. That the mobilities in the off-center δ-doped sample and in the uniformly doped sample are not significantly different is not surprising because in each case, the ionized impurities sample the average of the electronic wavefunction.

Electron velocity at high electric fields was also measured for the δ-doped samples [12]. Figure 7 depicts the measured electron velocity as a function of the electric field for the three samples. Although the on-center δ-doped quantum-well sample had the lowest mobility at low electric fields, the peak velocity of electrons in this sample is higher than in either the uniformly doped wells or in the off-center δ-doped wells. This higher velocity is possible because the differential mobility increases at electric fields between 0 and ε_p, the electric field at which the peak velocity occurs.

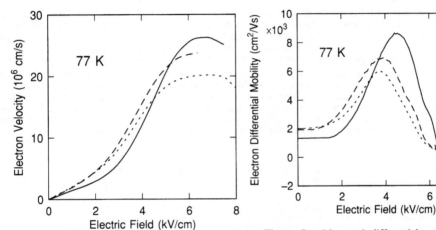

Figure 7: Measured velocities versus electric field for 100 Å wide GaAs quantum wells uniformly doped (dashed curve), δ-doped in the center (solid curve), and δ-doped midway between the center and an edge (dotted curve). The sheet density in each well is 6×10^{11} cm^{-2}.

Figure 8: Measured differential mobilities versus electric field for 100 Å wide GaAs quantum wells uniformly doped (dashed curve), δ-doped in the center (solid curve), and δ-doped midway between the center and an edge (dotted curve). The sheet density in each well is 6×10^{11} cm^{-2}.

Particularly noticeable about Figure 7 is that the velocity of the lowest mobility sample — the on-center δ-doped sample — is the highest. This is because a super-linear increase of velocity with increasing electric field over some range of electric field. This effect is more clearly seen in Figure 8 which depicts the derivatives of the velocity-field curves of Figure 7. From Figure 8 we see that the differential mobility of the electrons in the on-center δ-doped quantum wells increases much more than that of the electrons in uniformly doped quantum wells or in off-center δ-doped quantum wells. We expect that the ionized impurity scattering rate of the on-center δ-doped wells would decrease relatively more than that of the uniformly doped wells simply because the ionized impurity scattering rate is so much greater at low electric fields and, therefore, has much farther to fall. From Figure 8, however, we see that the differential mobility of the electrons in the on-center δ-doped wells actually increases enough to be significantly greater than that for the electrons in the uniformly doped wells. The large increase in differential mobility in the on-center δ-doped sample may also be contrasted with the smaller increase in the off-center δ-doped sample's mobility.

These data may be explained as follows: The dramatic increase in the differential mobility of the on-center δ-doped samples is due to the heating of the electrons from the ground state (even parity) quantum-well subband into the first excited (odd parity) subband. This odd-parity state has a node at the center of the quantum well, exactly where the ionized impurities are located. It follows, then, that heating electrons so that they begin to occupy the first excited subband should lead to higher mobility because the excited electrons experience much less ionized impurity scattering. The off-center δ-doped sample will not show this effect to the same extent because the δ-doping spike is located at one of the maxima of the first excited state instead of at the minimum.

V. Conclusions

The present studies represent the first measurements of the high-electric-field transport properties of electrons confined in *doped* GaAs/AlGaAs quantum wells. These measurements are made both at low electric fields, as well as up to about 9000 V/cm. Results are summarized as follows:

1. The electron mobility in doped quantum wells is lower than for electrons in similarly doped bulk GaAs, indicating an enhancement of the ionized impurity scattering rate of quasi-two-dimensional carriers. We have also calculated electron mobility in doped quantum wells and from these calculations learn that this enhanced scattering rate is characteristic of the confinement in one dimension and is due to a decrease in screening, an increase in the effective overlap of electronic wavefunction with the dopant atoms, and an increase in large-angle scattering in confined systems.

2. At high electric fields, the peak velocity of quasi-two-dimensional electrons in doped quantum wells is lower than what we measure in similarly doped bulk GaAs. Detailed transport measurements indicate that this lowering of the peak velocity is not primarily due to real space transfer as is the similar effect in modulation-doped heterostructures, but, rather, is also due to the enhanced ionized impurity scattering experienced by electrons confined in one dimension.

3. Delta-doping the quantum wells can lead to a further decrease in low-field mobility because the overlap of the electron wavefunction with the dopant atoms in increased. At high electric fields, however, the differential mobility dramatically increases, leading to a somewhat higher velocity than observed in uniformly doped quantum wells, in spite of the lower low-field mobility. This large increase in differential mobility may be the result of heating of the electrons out of the symmetric ground state into the anti-symmetric first excited state.

References

1) M. Feng, C.L. Lau, V. Eu, and C. Ito, Appl. Phys. Lett. **57**, 1233-1235 (1990).
2) W.T. Masselink, N. Braslau, W.I. Wang, and S.L. Wright, Appl. Phys. Lett. **51**, 1533-1535 (1987).
3) W.T. Masselink and T.F. Kuech, J. Electronic Materials **18**, 579-584 (1989).
4) W.T. Masselink, Semicond. Sci. Technol. **4**, 503-512 (1989).
5) C.E.C. Wood, G. Metze, J. Berry, and L.F. Eastman, J. Appl. Phys. **51**, 383-387 (1980).
6) E.F. Schubert and K. Ploog, Jap. J. Appl. Phys. Lett. **24**, L608-L610 (1985).
7) M.V. Fischetti, IEEE Trans. Electron Devices **38**, 634-649 (1991).
8) H. Sakaki, T. Noda, H. Hirakawa, M. Tanaka, and T. Matsusue, Appl. Phys. Lett. **51**, 1934-1936 (1987).
9) R. Gottinger, A. Gold, G. Abstreiter, G. Weimann, and W. Schlapp, Europhys. Lett. **6**, 183-188 (1988).
10) J.R. Meyer and F.J. Bartoli, Phys. Rev. B **36**, 5989-6000 (1987).
11) W.T. Masselink, Phys. Rev. Lett. **66**, 1513-1516 (1991).
12) W.T. Masselink, Appl. Phys. Lett. **59**, 694-696 (1991).

Inst. Phys. Conf. Ser. No 120: Chapter 9
Paper presented at Int. Symp. GaAs and Related Compounds, Seattle, 1991

431

Free carrier induced changes in the absorption and refractive index for intersubband optical transitions in $Al_xGa_{1-x}As/GaAs/Al_xGa_{1-x}As$ quantum wells

Gita U. Iyengar, Kelin J. Kuhn, and Sinclair Yee
Department of Electrical Engineering
University of Washington
Seattle, WA 98195

ABSTRACT

The changes in the real index of refraction and the optical absorption for conduction intersubband transitions in $Al_xGa_{1-x}As/GaAs/Al_xGa_{1-x}As$ quantum wells are examined as a function of the carrier density. Various values for the input optical field and quantum well width are considered in the calculations. The linear contribution due to $\chi^{(1)}$ as well as the non-linear contribution from $\chi^{(3)}$ is included. The relationship of the results to device applications such as waveguides and optical modulators is discussed.

I. INTRODUCTION

Quantum confinement of electrons in the conduction band of a semiconductor quantum well leads to the formation of discrete energy levels within the well. For a one-dimensional semiconductor quantum well grown in the z-direction, the energies in the z-direction are quantized as E_1, E_2 .... If light polarized in the z-direction is incident on the quantum well, photons whose energy corresponds to one of the transition energies $(E_{n+1} - E_n)$ can induce an intersubband optical transition. Such intersubband optical transitions have attracted much attention due to the large dipole matrix element and the resulting large changes in the complex dielectric constant.

The changes in the intersubband optical absorption as a function of the incident optical and DC electrical fields have been investigated theoretically by a number of researchers including the extensive efforts by Ahn and Chuang (1987a, 1987b) and Roan and Chuang (1991). Experimentally, the n=1 to n=2 intersubband transition in a modulation doped $Al_xGa_{1-x}As/GaAs/Al_xGa_{1-x}As$ quantum well was first observed by West and Eglash (1985) confirmed numerous other experiments. Levine first measured the large absorption possible in a waveguide structure (1987a) as well as the potential use of intersubband transitions as $10\mu m$ infrared detectors (1987b).

However, several unusual optical properties of intersubband transitions in semiconductor quantum wells have not been extensively investigated. One of the most interesting of these is the change in the complex dielectric constant as a function of the carrier density. This paper will present some initial calculations on the change in absorption and change in index of refraction for intersubband transitions as a function of the carrier density. More complete results will be given in a later paper Kuhn et al (1991).

II. THEORETICAL APPROACH

The conduction band of an $Al_xGa_{1-x}As/GaAs/Al_xGa_{1-x}As$ quantum well possesses several energy states $(n=1, n=2 ...)$. The energies of these states $(E_1, E_2 ...)$ are determined by solving the single-band effective mass Schrödinger equation for parabolic bands,

$$H \phi(z) = E_n \phi(z),$$ (1)

where H is the Hamiltonian and $\phi(z)$ is the envelope wavefunction. In this work, the BenDaniel and Duke Hamiltonian (1966),

$$H = \frac{-\hbar^2}{2} \frac{\partial}{\partial z} \left(\frac{1}{m^*(z)} \frac{\partial}{\partial z} \right) + V_o(z) + |q|F_{DC} \, z,$$ (2)

is used. In this Hamiltonian, m^* is the conduction band effective mass (m^*_{Ga} for the effective mass in the quantum well and m^*_{AlGa} for the effective mass in the $Al_xGa_{1-x}As$ barrier region), $V_o(z)$ is the quantum well potential for the conduction band, F_{DC} is the DC applied field, and $|q|$ is the absolute value of the electron charge.

The energies (E_1, E_2 ...) of the states in the quantum well are altered by application of the DC electrical field F_{DC}. The application of the electric field also changes the character of the envelope wavefunctions (ϕ_1, ϕ_2, ...) and thus alters the dipole matrix element between states approximated as given by Ahn and Chuang (1987a),

$$M_{ab} = |q| \langle a|z|b \rangle = |q| \int_{-\infty}^{+\infty} \phi_a z \phi_b dz. \tag{3}$$

In this paper, the effective mass equations (1) and (2) and the dipole matrix element equation (3) are solved numerically using a finite-difference approach. The numerical range is 707.5 Å wide and is discretized at 1.415 Å (2.83 Å/2) intervals. The quantum well is located in the center of the numerical range and the boundary conditions on the wavefunction $\phi(z)$ are assumed to be $\phi(-250\text{Å})=0$ and $\phi(+250\text{Å})=0$.

Injection of carriers in the quantum well strongly affects the carrier population in the various sub-bands and thus the location of the Fermi level. In most cases, the change in the location of the Fermi level is the principal factor responsible for the changes in the complex dielectric function with injected carriers. (The energies of the states and the structure of the electron wavefunction can also be altered by carrier injection (Li 1991). In addition at carrier densities on the order of $1 \cdot 10^{18} cm^{-3}$, electron screening effects must be taken into account (Roan and Chuang 1991). However these effects are small for the relatively low carrier densities typical of modulator devices,and will be neglected for the structures discussed in this paper.)

The energy of the Fermi level E_F for a conduction band quantum well with two states E_1 and E_2 (measured from the bottom of the conduction band) is related to the free electron concentration in the well by,

$$n_e = \frac{m^* k_B T}{\pi \hbar^2 L_{eff}} \left[\ln\left(1 + e^{\frac{E_F - E_1}{k_B T}} \right) + \ln\left(1 + e^{\frac{E_F - E_2}{k_B T}} \right) \right], \tag{4}$$

where k_B is Boltzman's constant, T is the temperature in degrees Kelvin,and L_{eff} is the effective width of the quantum well. In this work, the effective width of the quantum well L_{eff} will be assumed equal to the actual width of the quantum well L. In all cases discussed in this paper, it will be assumed that the entire carrier population is contained in the first two subbands.

Consider a quantum well under a DC applied field in the z-direction $F_{DC,z}$, and in the presence of an optical field $E(\omega t)$ where the polarization is along the z direction of the well. The electronic polarization will be approximated as,

$$P(t) \approx \epsilon_o \chi^{(1)} E(j\omega t) + \epsilon_o \chi^{(3)} E(j\omega t). \tag{5}$$

The analytical forms of the linear ($\chi^{(1)}$) and the nonlinear ($\chi^{(3)}$) susceptibilities have been previously calculated by Ahn and Chuang (1987a,1987b) by modeling the quantum well as a two-level system.

The susceptibility is related to the change in refractive index as,

$$\frac{\Delta n(\omega)}{n_r} = \mathcal{R}e \left(\frac{\chi(\omega)}{2n_r^2} \right), \tag{6}$$

where n_r is the refractive index and $\chi(\omega)$ is the Fourier component of $\chi(t)$ with $\exp(-i\omega t)$ dependence.

Using these relations, the analytic forms of the linear and nonlinear changes in the refractive index are given as follows. For the linear term,

$$\frac{\Delta n^{(1)}(\omega)}{n_r} = \frac{1}{2n_r^2\epsilon_o}|M_{fi}|^2\left(\frac{m^*k_BT}{L\pi\hbar^2}\right) \times$$

$$\ln\left(\frac{1 + \exp\left(\dfrac{E_F - E_i}{k_BT}\right)}{1 + \exp\left(\dfrac{E_F - E_f}{k_BT}\right)}\right)\left(\frac{E_f - E_i - \hbar\omega}{(E_f - E_i - \hbar\omega)^2 + \dfrac{\hbar^2}{\tau_{ij}^2}}\right). \tag{7}$$

For the third order term,

$$\frac{\Delta n^{(3)}(\omega)}{n_r} = \frac{-1}{2n_r^2\epsilon_o}|M_{fi}|^2\left(\frac{m^*k_BT}{L\pi\hbar^2}\right)\ln\left(\frac{1 + \exp\left(\dfrac{E_F - E_i}{k_BT}\right)}{1 + \exp\left(\dfrac{E_F - E_f}{k_BT}\right)}\right) \times$$

$$\frac{|E|^2}{\left((E_f - E_i - \hbar\omega)^2 + \dfrac{\hbar^2}{\tau_{ij}^2}\right)^2}\left[\frac{2}{\tau_{ij}^2}(E_f - E_i - \hbar\omega)(\tau_{ii} + \tau_{ff})|M_{fi}|^2\right.$$

$$-\frac{(M_{ff} - M_{ii})^2}{\left((E_f - E_i)^2 + \dfrac{\hbar^2}{\tau_{ij}^2}\right)}\left\{(E_f - E_i - \hbar\omega)\left((E_f - E_i)(E_f - E_i - \hbar\omega) - \frac{\hbar^2}{\tau_{ij}^2}\right)\right.$$

$$\left.\left. - \frac{\hbar^2}{\tau_{ij}^2}(2E_f - 2E_i - \hbar\omega)\right\}\right]. \tag{8}$$

The susceptibility χ is related to the absorption coefficient $\alpha(\omega, I)$ by,

$$\alpha(\omega) = \omega\left(\frac{\mu}{\epsilon_R}\right)^{1/2}\mathcal{I}m\left(\epsilon_o\chi(\omega)\right), \tag{9}$$

where μ is the permeability of the material and ϵ_R is the real part of the permittivity. The first and third order absorption coefficients $\alpha^{(1)}(\omega)$ and $\alpha^{(3)}(\omega, I)$ have also been calculated by Ahn and Chuang (1987a, 1987b).

III. RESULTS AND DISCUSSION

Example A: The change in the absorption as a function of the carrier density for a quantum well with an intersubband transition near $10.6\mu m$.

In optical device applications, a large induced absorption change with injected carrier concentration could be used to modulate the intensity in optical waveguides. Such modulation effects would be of special interest in the $10.6\mu m$ region, where the absorption change corresponds to the CO_2 laser lines.

Consider a 27 monolayer $Al_{0.3}Ga_{0.7}As/GaAs/Al_{0.3}Ga_{0.7}As$ quantum well (a single GaAs monolayer grown in the [100] direction is 2.83 Å thick or one-half the lattice constant of a GaAs conventional unit cell, thus a 27 monolayer well is 76.4 Å thick). Figure 1 illustrates the two components of the conduction band intersubband optical absorption (α_1 and α_3) as well as the total optical absorption ($\alpha_1 + \alpha_3$) for a 30 kV applied DC electric field at an incident optical intensity of 0.5 MW/cm^2 with a carrier concentration of $5 \cdot 10^{15}$ cm^{-3} at 300K. The large positive absorption is due to the α_1 contribution and this large absorption is significantly reduced by the α_3 contribution.

Figure 2 illustrates the total optical absorption ($\alpha_1 + \alpha_3$) a function of carrier density for the same conditions as Figure 1. Notice the rather large change in optical absorption possible with relatively small changes in the carrier density.

Example B: The change in the index of refraction as a function of the carrier density for a quantum well with an intersubband transition near $10.6\mu m$.

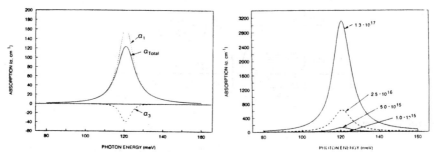

Figure 1. Example A: Intersubband absorption coefficients for a 27 monolayer QW with $n_e = 5.0 \times 10^{15} cm^{-3}$, $300K$, $30kV/cm$ and $0.5MW/cm^2$.

Figure 2. α_{total} Vs n_e for the QW in Figure 1.

Consider a 25 monolayer (70.7 Å) $Al_{0.3}Ga_{0.7}As/GaAs/Al_{0.3}Ga_{0.7}As$ quantum well with a 30 kV/cm applied DC electric field, at an incident optical intensity of 0.5 MW/cm^2 with a carrier concentration of $5 \cdot 10^{15}$ cm^{-3} at 300K.

Figure 3 illustrates the change in the two components of the index of refraction ($\Delta n_1/n$, and $\Delta n_3/n$) as well as the total change ($\Delta n_1/n + \Delta n_3/n$) for this quantum well. Notice that the largest change in the total refractive index is contributed by the linear $\chi^{(1)}$ term and is not at line center (as is true for the absorption) but is rather at a photon energy approximately 15% off line center. Unfortunately, the large linear change generated by the $\chi^{(1)}$ term is of the opposite sign of that generated by the $\chi^{(3)}$ term. Thus, calculation of the index of refraction change using only the linear $\chi^{(1)}$ term may be excessively optimistic for systems operating with a high optical intensity.

Figure 4 illustrates the total change in the index of refraction ($\Delta n_1/n + \Delta n_3/n$) as a function of carrier density for the same situation as in Figure 3. Notice the rather large refractive index change possible with a relatively small change in the carrier density. Notice also that the 25 monolayer structure locates the 10.6μm wavelength slightly lower in energy than the peak of the refractive index change. This has some potential advantages for reducing the sensitivity of the index of refraction change to well width.

The large change in refractive index as a function of carrier concentration is more readily apparent in Figure 5, where the refractive index change is plotted versus carrier density. As can be seen by comparison with Figure 4, the curves for photon energies of 0.088 → 0.120eV are for energies less than the peak of the refractive index change. The curve for the photon energy 0.128 eV is greater in energy than the the the peak of the refractive index change.

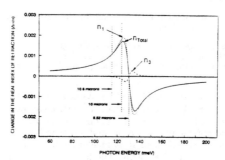

Figure 3. Example B: Changes in the index of refraction for a 25 monolayer quantum well in Figure 1.

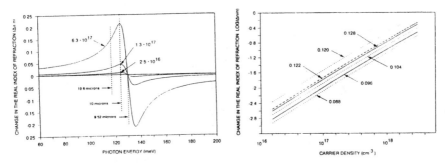

Figures 4 and 5. Δn_{total} Vs n_e for
the QW in Figure 3.

Example C: Bleaching and refractive index changes with incident optical fields.

At sufficiently high intensities the absorption at line center in an intersubband transition can be completely bleached. An equivalently large change in the refractive index would be detrimental if the induced refractive index change were being employed to modulate optical wave confinement in optical waveguides. Consider a 25 monolayer $Al_{0.3}Ga_{0.7}As/GaAs/Al_{0.3}Ga_{0.7}As$ quantum well (70.7 Å) possessing a carrier concentration of $1.0 \cdot 10^{16}$ cm^{-3} with a 30 kV/cm applied DC electric field at 300 K. Figure 6 illustrates the total optical absorption ($\alpha_1 + \alpha_3$) as a function of the input optical intensity. Notice the strong bleaching effect at line center for the 2.25 MW/cm^2 incident optical radiation. In contrast, Figure 7 illustrates the total change in the index of refraction ($\Delta n_1/n + \Delta n_3/n$) as a function of the optical intensity under the same circumstances as in Figure 6. Although the total change in the index of refraction is reduced by approximately 30% at line center, the effect is significantly less dramatic than that of the absorption.

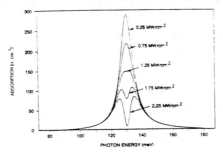

Figure 6. Example C: α_{total} Vs.
the optical intensity in a 25 monolayer
QW with $n_e = 1.0 \times 10^{16} cm^{-3}$, $300 K$
and $30 kV/cm$.

However, at some point the intensity of the input optical radiation will be sufficiently large for the peak $\Delta n_3/n$ contribution to exceed the peak $\Delta n_1/n$ contribution. This situation is illustrated in Figure 8 for similar conditions as in Figure 7. Figure 8A illustrates the change in the two components of the index of refraction ($\Delta n_1/n$, and $\Delta n_3/n$) as well as the total change ($\Delta n_1/n + \Delta n_3/n$) for an input optical power of 1.0 MW/cm^2. Figure 8B illustrates the change in the two components of the index of refraction as well as the total change of the index of refraction for an input optical power of 6.0 MW/cm^2. Notice that the factor of six increase in the input optical power has dramatically altered both the magnitude and sign of the refractive index change. This contrasts significantly with the situation in Figure 7, where a factor of six increase in the input optical power has merely reduced the index of refraction by approximately 30% at line center.

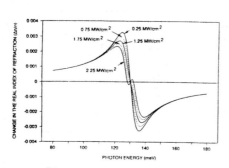

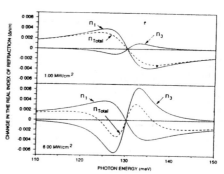

Figure 7. Δn_{total} Vs the optical intensity for the QW in Figure 6.

Figures 8A and 8B. Δn for the QW in Figure 6 at optical intensities of $1MW/cm^2$ and $6MW/cm^2$ respectively.

IV. SUMMARY

Intersubband transitions possess a rather large change in the complex dielectric constant as a function of the carrier density. This generates large increases in the intersubband absorption and index of refraction with increasing carrier density. The effect has both static applications (for example, increasing the absorption of a particular intersubband transition by doping) and dynamic applications (for example, modulating the real index of refraction, and thus the optical field confinement, in a waveguide structure by means of injected carriers).

V. ACKNOWLEDGEMENTS

The authors wish to express their thanks to Prof. L. Tsang and to Mr. T.L. Li for numerous stimulating discussions. This work was supported by the National Science Foundation and the Office of Naval Research under grant numbers ECS-8909082 and N00014-91-J-1662 respectively.

REFERENCES:

D Ahn and S L Chuang (1987a) J. Appl. Phys. **62**(7) 3052

D Ahn and S L Chuang (1987b) IEEE J. Quantum Electronics QE-23(12) 2196

D J BenDaniel and C B Duke (1966) Phys. Rev. **152** 683

K Kuhn G Iyengar and S Yee (1991) J. Appl. Phys. to be published

B F Levine R J Malik J Walker K K Choi C G Bethea D A Kleinman and J M Vandenberg (1987a) Appl. Phys. Lett. **50**(5) 273

B F Levine K K Choi C G Bethea J Walker and R J Malik (1987b) Appl. Phys. Lett. **50**(16) 1093

T L Li (1991) Masters Thesis Department of Electrical Engineering University of Washington

E J Roan and S L Chuang (1991) J. Appl. Phys. **69** (5) 3249

L C West and S J Eglash (1985) Appl. Phys. Lett. **46**(12) 1156

Inst. Phys. Conf. Ser. No 120: Chapter 9
Paper presented at Int. Symp. GaAs and Related Compounds, Seattle, 1991

437

Electro-absorption in InGaAs/AlGaAs quantum wells

B. Pezeshki, S. M. Lord, J. S. Harris Jr.

Solid State Laboratories, Stanford University, Stanford CA 94305

ABSTRACT: Theoretical and experimental work show that electro-absorption improves and the quantum-confined Stark effect becomes stronger in $In_{0.2}Ga_{0.8}As$ / $Al_xGa_{1-x}As$ quantum wells as the aluminum concentration (x) is increased in the barriers. This higher electro-absorption was used to fabricate a reflection modulator which exhibited a record reflectivity change of 77%. The enhanced electro-absorption is in contrast to material quality, as measured by photoluminescence intensity, which severely degrades with increasing aluminum composition.

1. INTRODUCTION

Electro-absorption effects in quantum wells are of interest for a variety of modulator structures. In the quantum-confined Stark effect, described by Miller et al (1985), an electric field causes a red-shift in the absorption edge of a quantum well and produces large changes in the absorption coefficient. These changes can be exploited directly in the normal incidence geometry to form a transmission modulator, as reported by Van Eck et al (1986), Whitehead et al (1988), and Woodward et al (1990) or used in a Fabry-Perot cavity to create a more efficient reflection device as reported earlier by us (Pezeshki et al 1990a, 1990b) and by Yan et al (1990). Although most of the early work focused on the GaAs/AlGaAs system, strained InGaAs/GaAs has also attracted attention primarily because of its longer wavelength operating regime.

However, electro-absorption in the InGaAs/GaAs material system tends to be poor. Apart from the problems associated with dislocations due to the lattice mismatch, the excitonic resonance is less intense than in the GaAs/AlGaAs system and tends to rapidly deteriorate as the electric field increases (Van Eck et al 1986). To reduce these detrimental effects, we investigated increasing the barrier height by adding aluminum to the barrier regions. Although there is some evidence in the literature that adding aluminum to the barriers may improve electro-absorption (Fujiwara et al 1990), such InGaAs/AlGaAs quantum wells are rarely used since conventionally this interface is considered poor. Consequently, in most instances, either the barriers contain no aluminum or buffer GaAs regions are inserted between the well and the AlGaAs barriers. This latter scheme does not substantially enhance the confinement of the exciton and therefore has negligible effect on the absorption or luminescence properties.

2. CALCULATIONS

We performed theoretical calculations to investigate the effect of higher aluminum barriers on $In_{0.2}Ga_{0.8}As$ / $Al_xGa_{1-x}As$ electro-absorption. The calculations followed the method of Bastard (1982) and Miller (1985). In brief, we used a variational technique using a hydrogenic $1s$ wavefunction to minimize the exciton binding energy. Since the strain shifts the light hole to much higher energies, we neglected bandmixing in the valence band. As the matrix element Ep between the electron and hole states is not accurately known in this system, and simply scales our results we adjusted this parameter to provide a best fit to the experimental data. The solid line in Figure 1 shows the results of our calculations, where the y-axis is the calculated size of the exciton resonance, given by the maximum absorption times the FWHM. As the aluminum concentration increases from 0 to 0.4, this value increases by almost 30 %. This enhancement arises from two factors. About 1/3 of this effect is due to the increase in the 1 D overlap integral between the electron and hole wavefunctions in the z or growth direction. The hole, due to its large effective mass, stays strongly confined in the well and is insensitive to the barrier composition. The electron, on the other hand, due to its lighter mass becomes better confined with increasing barrier height, and therefore the overlap integral increases for larger barriers. As the exciton wavefunction becomes better confined in the growth direction, the accompanying shrinkage of the wavefunction in the x-y plane accounts for the remaining 2/3 of the resonance enhancement.

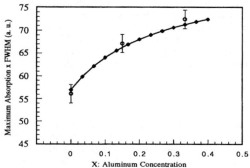

Figure 1: Plot of calculated (solid line) exciton absorption, expressed as the maximum absorption value times the full width at half maximum (FWHM), as a function of aluminum concentration in 75 Å $In_{0.2}Ga_{0.8}As$ / $Al_xGa_{1-x}As$ quantum wells at zero electric field. The open circles represent experimental data taken taken at low bias. Since the vertical scaling is fit to the data, only the relative increase is significant.

3. RESULTS

To check this prediction experimentally, we grew three multi-quantum well samples with different aluminum concentration in the barriers. The samples were grown by MBE on semi-insulating GaAs substrates at about 500 °C. After growing an n-doped contact region, twenty undoped 75Å $In_{0.2}Ga_{0.8}As$ quantum wells were grown with 200Å $Al_xGa_{1-x}As$ barriers with x=0, 0.15, and 0.33. A 0.5μm thick p-doped region was added to complete the p-i-n structure. After wet etching to form mesas, ohmic contacts were made to the p and n regions. The absorption coefficient was then calculated from the optical transmission (Pankove 1971). The experimental results are shown in Figure 2. The size of the exciton resonance in the low field case is then estimated and plotted in Figure 1 as the open circles. At zero reverse bias, there is still some field across the devices due to the built-in field of the p-i-n diode. However, as the exciton shape and position change slowly with field at small bias, we neglect these effects in the comparison with the calculations done at zero electric field. The error bars in the experimental data of Figure 1 arise due to the difficulty in differentiating the exciton absorption from the absorption due to transitions into the continuum.

We see close agreement between the theory and the experiment at zero bias, with the magnitude of the exciton absorption increasing with the aluminum concentration in the barriers. As an electric field is applied, the higher barriers also prevent the dissociation of

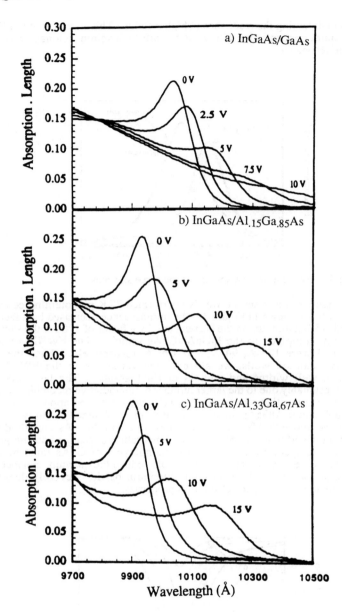

Figure 2: Experimental plots of absorption vs wavelength for various reverse bias voltages. The samples consisted of 20 x 75 Å $In_{0.2}Ga_{0.8}As$ quantum wells with 200Å $Al_xGa_{1-x}As$ barriers in a p-i-n configuration. Plots a, b, and c are for x=0, 0.15, and 0.33 respectively.

the exciton thus maintaining the resonance at higher fields. This is confirmed in Figure 2. The higher aluminum concentration allows us to shift the exciton to longer wavelengths without losing the resonance.

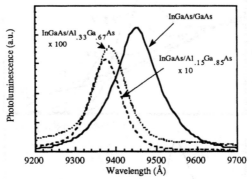

Figure 3: Photoluminescence spectra of the three samples.

To evaluate the quality of the material, we also performed 77 K photoluminescence on the samples. The results are shown in Figure 3. The luminescence decreased by a factor of 14 in the sample with 15% aluminum in the barriers and by a factor of 120 for the sample with the 33% aluminum compared to the InGaAs/GaAs sample. The FWHM of the x=0, x=0.15, and x=0.33 were 125 Å, 94 Å, and 109 Å, respectively. Since the samples contained twenty quantum wells, we attribute the variations in the FWHM to the nonuniformities in well widths and monolayer fluctuations. The large decrease in luminescence is presumably due to the larger number of nonradiative recombination centers in the AlGaAs, indicating that such quantum wells would not be appropriate for laser structures. Though better AlGaAs material quality can be obtained by growing at much higher temperatures (above 700 °C), this in incompatible with InGaAs growth, due to the high rate of indium desorbtion. If the substrate temperature is changed during growth to accommodate both temperature ranges, a few monolayers of GaAs must be inserted between the well and the barrier. This in turn would reduce the quantum enhancement of the higher barriers. So although using high aluminum barriers severely degrades the luminescence, it greatly improves the absorption properties.

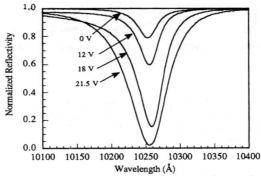

Figure 4: Normalized reflectivity versus wavelength for a reflective modulator incorporating 30 x 75Å $In_{0.2}Ga_{0.8}As/Al_{0.33}Ga_{0.67}As$ quantum wells.

To exploit this superior electro-absorption in InGaAs / AlGaAs, we constructed a reflection modulator similar to previous designs (Pezeshki et al 1990 a,b). The modulator was grown on an n-type substrate with an n-type 18.5 period back mirror of GaAs/AlAs and a top 5 period p-type mirror of the same composition. The cavity consisted of 30 quantum wells, each with 75 Å $In_{0.2}Ga_{0.8}As$ wells and 200 Å $Al_{0.33}Ga_{0.67}As$ barriers. The device was processed and characterized as described previously. Figure 4 shows the reflectivity versus wavelength of the modulator at different reverse bias voltages. We measured a maximum reflectivity change of 77% at the Fabry-Perot wavelength of 10255 Å and a contrast ratio of 37. This is by far the largest reflection modulation ever reported in such a device. Since the reflection modulation can be simply related to the absorption changes in the quantum wells (Pezeshki et al 1990a), we attribute the superior performance of our device to the large absorption changes in this material system.

4. DISCUSSION

Although we have clearly shown that the excellent electro-absorptive properties of InGaAs/AlGaAs allow large modulation to be obtained, this material system may have some disadvantages compared to the GaAs/AlGaAs system. The larger barrier heights in this system will probably lower the saturation intensity of the modulator as the carriers have more difficulty tunneling out of the wells and being collected at the contacts (Fox et al 1990). The longer time delay associated with this tunneling slows the photocurrent generation and may slow the operation of Self Electro-optic Effect Devices (SEEDs) that rely on photocurrent feedback (Boyd et al 1990). The operating voltages may also be higher in these structures due to the thicker barriers than in lattice matched systems. The 200 Å barriers used in this modulator are thicker than necessary and were chosen to permit comparison with an earlier test wafer. However, in general, barriers must be thick enough in this strained system to prevent relaxation due to the large average indium concentration in the strained quantum wells. Such uncontrolled relaxation broadens and reduces the exciton resonance, thereby severely degrading performance (Van Eck et al 1989).

In conclusion, we have shown both theoretically and experimentally that adding aluminum to the barriers in the InGaAs/AlGaAs system greatly improves the electro-absorption characteristics. We have used such optimized wells to fabricate a reflection modulator with the largest reflectivity change to date. This work was supported by ONR / DARPA contract N00014-90-J-4056 and ONR/SDIO contract N00014-89-K-0067.

5. REFERENCES

Bastard G, Mendez E E, Chang L L and Esaki L 1982 Phys. Rev. B **26** 1974
Boyd G D, Fox A M, Miller D A B, Chirovsky L M F, D'Asaro L A, Kuo J M, Kopf R F and Lentine A L 1990 Appl. Phys. Lett. **57** 1843
Fox A M, Miller D A B, Livescu G, Cunningham J E, Henry J E and Jan W Y 1990 Appl. Phys. Lett. **57** 2315
Fujiwara K, Kawashima K, Kobayashi K and Sano N 1990 Appl. Phys. Lett. **57** 2234
Miller D A B, Chemla D S, Damen T C, Gossard A C, Wiegmann W, Wood T H and Burrus C A 1985 Phys. Rev. B **32** 1043
Pankove J I 1971 Optical Processes in Semiconductors (New York: Dover) pp 103
Pezeshki B, Thomas D and Harris J S Jr 1990a Appl. Phys. Lett. **57** 1491
Pezeshki B, Thomas D and Harris J S Jr 1990b IEEE Photon. Tech. Lett. **2** 807
Van Eck T E, Chu P, Chang W S C and Wieder H H 1986 Appl. Phys. Lett. **49** 135
Van Eck T E, Niki S, Chu P, Chang W S C, Wieder H H, Mardingly A J, Aron K and Hansen G A 1989 Quantum Wells for Optics and Optoelectronics: Topical Meeting, Salt Lake City, Utah, (Washington DC: Optical Society of America) WA 3-1
Whitehead M, Stevens P, Rivers A, Parry G, Roberts J S, Mistry P, Pate M and Hill G 1988 Appl. Phys. Lett. **53** 956
Woodward T K, Sizer T, Sivco D L and Cho A Y 1990 Appl. Phys. Lett. **57** 548
Yan R H, Simes R J, Coldren L A and Gossard A C 1990 Appl. Phys. Lett. **56** 1626

Inst. Phys. Conf. Ser. No 120: Chapter 9
Paper presented at Int. Symp. GaAs and Related Compounds, Seattle, 1991

443

Studies of piezoelectric effects in [111] oriented strained $Ga_{1-x}In_xSb/GaSb$ quantum wells

M. Lakrimi, R.W. Martin, C. López, S.L. Wong, E.T.R. Chidley, R.M. Graham, R.J. Nicholas, N.J. Mason, and P.J. Walker.

Clarendon Laboratory, Department of Physics, The University of Oxford, Parks Road, OXFORD OX1 3PU, United Kingdom.

Abstract:

Magnetotransport and optical measurements have been performed on two-dimensional hole gases confined in [001] and piezoelectrically active [111] oriented strained $Ga_{1-x}In_xSb/GaSb$ quantum wells. Comparison between simultaneously grown [001], [111]A, and [111]B structures reveals carrier density enhancement for the <111> growth direction and 50-70 meV Stark energy shifts which are well accounted for by the estimated piezoelectric field, of order 1×10^5 V/cm. Double wells and superlattices were also investigated by inter-band magneto-optical experiments. The energy gap between the symmetric and anti-symmetric states is found to decrease with increasing indium content in double well structures.

1. Introduction:

The existence of a strain induced piezoelectric field in [111] oriented zincblende structures, which do not have a centre of inversion symmetry, has generated new research possibilities. In a series of papers (see Ref. [1] for a review), Smith and Mailhiot have shown that such fields could lead to bandgap shifts and large optical non-linearities. Subsequent optical experiments [2-7] confirmed the existence of such in-built fields of order 10^5 V/cm. Nicholas and co-workers [8-9] presented the first magnetotransport studies on piezoelectrically active structures. They confirmed that [111] oriented structures contained larger carrier densities than their [001] equivalents, in agreement with the calculations of Snow et al. [10] and Ekenberg et al. [11]. Significant absorption enhancement was also reported in [111] oriented InAs/GaSb superlattices [9] making piezoelectrically active structures potential alternatives for the fabrication of long wavelength optical detectors.

We report on magnetotransport and optical studies performed on two-dimensional holes confined in InGaSb/GaSb quantum wells.

2. Magnetotransport and photoluminescence studies:

We have studied single, double and multiple $In_xGa_{1-x}Sb/GaSb$ quantum wells grown at atmospheric pressure in a metal-organic vapour phase epitaxy reactor using TMGa, TMSb, and EDMIn or TMIn source materials. Structures were simultaneously grown on [001], [111]A, and [111]B oriented GaAs substrates. The growth conditions were optimised for the [001] orientation and details can be found in Ref. [12].

The single well structure consists of an $In_xGa_{1-x}Sb$ well sandwiched between a GaSb capping layer ranging from 100 to 2000 Å and 2-4 μm GaSb buffer layer. Growth temperature is 565 °C.

The piezoelectric field is estimated to be of order 1×10^5 V/cm for a well with an indium content of 12% [8]. In this system the well is under compression, therefore the piezoelectric field points from the A (cation) to the B (anion) face. Using [111]A rather than [111]B growth axis inverts the symmetry and thus reverses the direction of the in-built field. A systematic comparison between [001], [111]A, and [111]B orientations is given in Table I for a series of single well structure in which both the well widths and upper capping layer thicknesses are varied. The first series consisting of samples 589 and 593 were grown using TMIn and the remainder with EDMIn. The second series has a different V/III flow rates compared to the last one. Typical 4.2 K magnetotransport traces for the three orientations are shown in Figure 1.

Sam-ple	In %	Well Å	Cap Å	[001] p (10^{11} cm^{-2})	μ cm^2/Vs	[111]A p (10^{11} cm^{-2})	μ cm^2/Vs	[111]B p (10^{11} cm^{-2})	μ cm^2/Vs
589	14	80	400	1.0	2200	2.0	2400	6.8	2400
593	16	120	1200	1.6	7800	4.5	6700	5.8	2600
875	12	40	260	0.8	4900	high resistance		3.4	2600
866	12	80	100	4.8	1800	high resistance		6.7	5800
885	12	80	260	1.2	8300	high resistance		5.4	7100
881	14	80	260	2.9	11000	high resistance		7.8	9800
880	12	120	260	1.5	9100	high resistance		7.5	7500
911	12	80	260	2.1	6600	1.7	20000	7.9	6400
912	12	80	1040	1.1	2800	2.5	8300	2.1	5900

Table I: Growth parameters and electrical properties of three series of quantum wells, showing differences between [001], [111]A, and [111]B orientations.

For sufficiently thick cap layers such that the effects of any surface depletion field is negligible, [111]A and [111]B oriented samples exhibit comparable carrier densities which are enhanced over the [001] values. As the cap layer is reduced both [001] and [111]B structures show increased carrier concentrations, while for [111]A there are no, or very few, carriers. The effect of the cap layer thickness on the carrier density is shown in Figure 2 for a series of [001] oriented single wells. This yields convincing evidence that the depletion field points from the cap layer to the substrate, with the Fermi level at the surface pinned very close to the valence band edge as sketched in the inset to Figure 2. It thus follows that whilst it opposes the piezoelectric field in [111]B orientation it reinforces it in [111]A samples, leading to strong changes in the band profile, as illustrated in Figure 1.

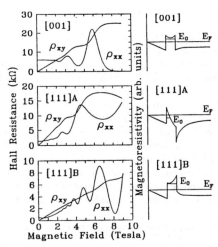

Figure 1: Magnetoresistivity (ρ_{xx}) and Hall effect (ρ_{xy}) for sample 911 along with the schematic profiles of the valence band edge for the three orientations. The carrier density is much higher in [111]B orientation and least in [111]A, as given in Table I.

In [111]A structures, this results in a sharper triangular well in which the confinement energies are higher, thus yielding fewer or no carriers. Conversely, in the [111]B samples, the partial cancellation of the piezoelectric field by the depletion field leads to a much flatter profile (than [111]A) in which the confinement energy is lower than that of the [001] counterpart, leading to the observed enhanced carrier density.

It is also apparent from Table I and the majority of structures studied that a number of [111] samples show signs of mobility enhancement compared to [001]. The mobility for the [111]A orientation is generally much higher than that of the other two orientations, such as the 20,000 cm^2/Vs measured in sample 911. It can be seen from Figure 1

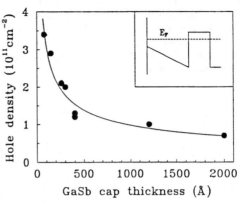

Figure 2: Variation of the hole density at 4.2 K as a function of cap layer thickness for [001] oriented single wells of identical well thickness and indium content.

that the piezoelectric field has displaced the hole wavefunction closer to the inverted interface (GaSb on GaInSb) for [111]A and closer to the normal interface (GaInSb on GaSb) for [111]B, and this may suggest that the inverted interface is more favourable. Notable differences between normal and inverted interfaces have already been reported in GaAs/AlGaAs [13] and InP/InGaAs [14] structures.

Further mobility enhancement has been found in strained double quantum wells (SDQW) and 10 period superlattices, more pronounced for thinner barriers. Comparison of Hall effect and Shubnikov-de Haas traces show that all the carriers fill a single (probably the last) well and give extremely good transport characteristics, as shown in Figure 3 and Table II; the effect of the thickness of the barrier on the mobility is obvious and the characterisation of these samples is to be reported elsewhere.

Sample	Type	In (%)	Well (Å)	Barrier (Å)	[001] p	[001] μ	[111]A p	[111]A μ	[111]B p	[111]B μ
938	SLS	15	40	50	1.3	380	5.0	33000	5.9	9300
932	SLS	15	40	100	high resist.		2.7	22500	6.0	1200
930	SLS	15	80	100	2.1	730	4.8	24400	12.0	450
931	SLS	15	80	500	8.7	900	4.3	3000	10.0	660
945	SDQW	15	40	30	2.1	9200	2.4	18200	6.1	6600
946	SDQW	15	40	50	1.4	3200	2.3	1900	4.8	5500
947	SDQW	15	80	20	1.9	8600	3.3	12200	6.9	6100
944	SDQW	15	80	30	1.6	7000	3.6	6100	6.5	6700
942	SDQW	15	80	200	1.4	5600	3.3	2200	6.3	7100

Table II: Hole density, p, and mobility, μ, for a series of strained double quantum wells (SDQW) and 10 period superlattices (SLS) with different well and barrier thicknesses. The cap thickness is 1740 Å. The units are as in Table I.

Further evidence for the presence of large piezoelectric fields is provided by photoluminescence studies at 4 K. Figure 4 shows spectra for sample 881 demonstrating a Stark shift, of order 50 meV, between the [111]A and [001] orientations at low laser power intensity. This corresponds to an in-built electric field of order 1×10^5 V/cm, in good agreement with our prediction. Raising the laser power increased the energy of the [111]A photoluminescence peak showing that the piezoelectric field is being screened by optically generated carriers. The [111]B sample shows a main peak at a similar energy to the [001] orientation. This is due to the oppositely directed piezoelectric and depletion fields producing a flatter well along with the extra band bending resulting from the very large carrier density.

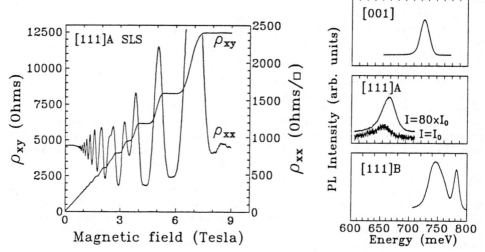

Figure 3: Magnetoresistance and Hall traces for a 10 period superlattice with 15% indium at 700 mK. The well and barrier thicknesses are 40 and 50 Å, respectively.

Figure 4: 4 K photoluminescence spectra for sample 881, showing the strong Stark shift for [111]A and the accompanying blue shift for high excitation power.

3. Interband magnetophotoconductivity experiments:

The study of coupled wells has generated much interest (for a review see Ref. [15]) because they allow controlled introduction of a degree of freedom in the z-direction through tuning of the barrier thickness, unlike the single well structure (or the decoupled multi-quantum wells) where the carriers are confined within the well. The double quantum well is the simplest case. It is found that when the barrier is sufficiently thin to allow interwell tunneling, the subband states of each well mix to form symmetric and antisymmetric states which are separated by an energy gap ΔSAS. In the presence of a magnetic field applied perpendicular to the layers, each state gives rise to a series of Landau levels.

In the following we concentrate on the interband magneto-optical studies performed at 1.8 K on a series of single, double and multiple quantum well structures with varying indium content. The magnetic field was applied both in the direction perpendicular $(B_\perp)$ and parallel $(B_\parallel)$ to the layers. Again these photoconductivity studies confirmed Stark energy shifts between [001] and [111]A of order 50 meV, in good agreement with theory and the photoluminescence measurements. Figure 5 shows the fan energy diagrams for a coupled double well and a 10 period superlattice in the $B_\perp$ configuration. We observe both the type-I E1-HH1 transition and a weaker E1-LH1 transition which is type-II.

An 80/50/80 Å GaSb/InGaSb SDQW shows a symmetric-antisymmetric state energy gap in excess of 13 meV, much larger than that of an equivalent GaAs/AlGaAs system

[16] which is smaller than 2 meV. This is expected since the GaSb barrier height for both the electrons and holes is of order 50 meV, much less than those of the AlGaAs barriers. The masses are also lighter in the InGaSb/GaSb system.

For indium contents of 10, 12, and 15%, the splitting between the symmetric and antisymmetric levels (ΔSAS) in the double wells is measured to be 17.7, 17.3, and 13.1 meV, respectively. This decrease in ΔSAS with increasing In content is expected since this results in an increase in well depth, leading to less coupling between the wells. The opposite behaviour is found in an equivalent series of 10 period superlattices, probably

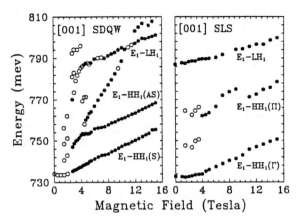

Figure 5: Fan energy diagrams for left) a double well and right) a 10 period superlattice with the same In content, 15%, and well/barrier ratio of 80 Å/50 Å.

due to some strain relaxation given the larger total thickness. For a layer with mean In content of 9.2%, the critical thickness is expected to be 460 Å. In this case the zone centre (Γ) and zone edge (Π) transitions are seen, suggesting that the miniband widths increase from 19 to 28 meV when the indium content is increased from 10 to 15%.

4. Conclusion:

We have shown that the existence of piezoelectric fields in GaSb/InGaSb strained structures leads to increased carrier density in [111] orientations compared to [001]. The observed photoluminescence shifts are in very good agreement with interband magnetoptical shifts and are well interpreted by the in-built electric fields. Finally, the symmetric-antisymmetric state energy gap (ΔSAS) is found to decrease with increasing indium content in double well structures but the opposite behaviour is seen in miniband width of superlattices, probably due to strain relaxation.

5. Acknowledgements:

The authors wish to thank Messieurs K.R. Belcher and S. Moulder for technical assistance. This work is supported by the Science and Engineering Research Council (U.K.).

References

[1] Smith D L and Mailhiot C 1990 *Rev. Mod. Phys.* **62** 173

[2] Laurich B K, Elcess K, Fonstad C G, Beery J G, Mailhiot C and Smith D L 1989 *Phys. Rev. Lett.* **62** 649

[3] Halsall M P, Nicholls J E, Davies J J, Wright P J and Cockayne B 1990 *Surf. Sci.* **228** 41

[4] Caridi E A, Chang T Y, Goossen K W and Eastman L F 1990 *Appl. Phys. Lett.* **56** 659

[5] Schwartz G P, Gualtieri G J and Sunder W A 1990 *J. Cryst. Growth* **102** 147

[6] Sela I, Watkins D E, Laurich B K, Smith D L, Subbanna S and Kroemer H 1991 *Appl. Phys. Lett.* **58** 684

[7] Xoo B S, Liu X C, Petrou A, Cheng J-P, Reeder A A, McCombe B D, Elcess K and Fonstad C 1989 *Superlattices and Microstruct.* **5** 363

[8] Martin R W, Lakrimi M, López C, Nicholas R J, Chidley E T R, Mason N J and Walker P J 1991 *Appl. Phys. Lett.* **59** 659

[9] Lakrimi M, López C, Martin R W, Summers G M, Sundaram G M, Dalton K S H, Nicholas R J, Mason N J and Walker P J 1991 *Proc. 9th. Int. Conf. on Electronic Properties of Two-Dimensional Systems (Nara, Japan 1991)*, to be published in *Surf. Sci.*

[10] Snow E S, Shanabrook B V and Gammon D C 1990 *Appl. Phys. Lett.* **56** 758

[11] Ekenberg U and Richards D 1991 *Proc. 9th. Int. Conf. on Electronic Properties of Two-Dimensional Systems (Nara, Japan 1991)*, to be published in *Surf. Sci.*

[12] Haywood S K, Chidley E T R, Mallard R E, Mason N J, Nicholas R J, Walker P J and Warburton R J 1989 *Appl. Phys. Lett.* **54** 922

[13] Kim D, Madhukar A, Hu K-Z and Chen W 1990 *Appl. Phys. Lett.* **56** 1874

[14] Kane M J, Anderson D A, Taylor L L and Bass S J 1986 *J. Appl. Phys.* **60** 657

[15] Fox A M, Miller D A B, Livescu G, Cunningham J E and Jan W Y 1991 submitted to *Phys. Rev. B.*

[16] Boebinger G S, Jiang H W, Pfeiffer L N and West K W 1990 *Phys. Rev. Lett.* **64** 1793

A long-wavelength PIN–FET receiver OEIC on GaAs substrate

K.Goto, E.Ishimura, T.Shimura, M.Miyashita, Y.Mihashi, T.Shiba,
Y.Okura, E.Omura and H.Kumabe

Optoelectronic and Microwave Devices Laboratory
Mitsubishi Electric Corp., 4-1,Mizuhara,Itami,Hyogo,664 Japan

ABSTRACT : A monolithic receiver OEIC, composed of an InGaAs PIN photodiode(PD) and a GaAs FET, has been successfully fabricated on 3-inch diameter GaAs substrate using InP-on-GaAs system. The quality of PD layers were improved by buffer layers with low temperature grown layer, thermal cyclic annealing, and InGaAs/InP strained layer superlattice. The receiver OEIC has sensitivity of -28 dBm at transmission rate of 622 Mb/s with a bit error rate of 10^{-9}, and is applicable for practical communication systems.

1. Introduction

Long wavelength optoelectronic integrated circuit (OEIC), which can realize low cost and high speed response, is one of the key devices of optical fiber communication systems(Koren 1982). However, InP-based long wavelength OEICs have a problem of immature InP-based electronic circuit technology.

Lattice mismatched systems, such as GaAs-on-InP or InP-on-GaAs are attractive in the OEICs for optical communication systems, because the long-wavelength InP based optical devices can be combined with well-established GaAs electronic components. Long wavelength receiver OEIC on a GaAs-on-InP heterostructure in which an InGaAs PD and a GaAs FET with GaAs buffer layer are grown on InP substrate has been reported(Suzuki 1987) as an approach for this purpose. For cost effective OEICs, InP-on-GaAs system seems more promising compared to the GaAs on InP system with an InP substrate, since low-cost large-diameter GaAs substrate with high quality are available, and the GaAs ICs can be formed directly onto GaAs substrates. In the InP-on-GaAs lattice mismatched system, reduction of dislocation density is essential to avoid the deterioration of the performance of optical devices(Dentai 1987). We have reported the low dark current and high efficiency InGaAs PD grown on GaAs substrate using lattice mismatched materials with buffer layers to reduce dislocation density(Ishimura 1990).

In this paper, we describe the first InGaAs /GaAs monolithic PIN-FET receiver OEIC fabricated on a 3-inch diameter GaAs substrate, which has a high sensitivity comparable with PIN-FET on InP substrate.

2. Reduction of Mismatch Dislocations by Buffer Layers

We employed the 5μm-thick buffer layers to reduce the dislocation density of PD layers. Epitaxial growth was performed by low pressure(150 torr) MOCVD with RF-heated horizontal reactor. First, 0.3 μm thick n-GaAs layer for n-electrodes of PD was grown on semi-insulating GaAs substrates at 650°C. Next, the substrates were cooled down to 500°C, and 30 nm thick low temperature buffer layer(LTB) was grown. Then the substrates were heated up to 600°C, and 2.7 μm thick InP layer was grown. Then the growth was interrupted, and a thermal cyclic annealing(TCA) of 4 cycles was performed in a PH₃ flow. The substrate were heated up to 700°C for 10 min, and cooled down to 200°C for each cycle. Then five periods of $In_{0.63}Ga_{0.37}As/InP$ (100Å/100Å) were grown in order to suppress the threading of the dislocations into upper epitaxial layers(Kimura 1991). Finally, 2 μm thick InP layer was grown. The dislocation density is estimated to be ~1×10^7cm^{-2} by EPD. This value is about 1/10 of the dislocation density of InP on GaAs with no buffer layer.

3. Structure and Fabrication

A planar type InGaAs PIN PD, a GaAs MESFET and a load resistance are monolithically integrated on a GaAs substrate. A schematic drawing and a photograph of the receiver OEIC chip are depicted in Fig.1 with the circuit diagram. The chip size of the OEIC is 500×800 μm². The diameter of the photosensitive area of the integrated PD is 50 μm. The FET gate length and gate width are 1 μm and 400 μm, respectively. The load resistance is 1 kΩ.

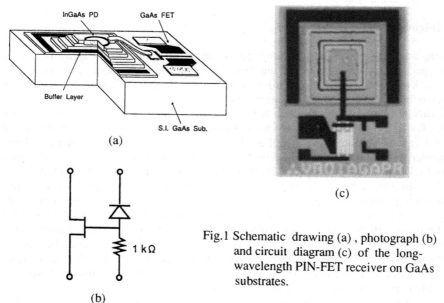

(a)

(c)

(b)

Fig.1 Schematic drawing (a) , photograph (b) and circuit diagram (c) of the long-wavelength PIN-FET receiver on GaAs substrates.

InGaAs/InP PIN PD layers were grown on a semi-insulating 3-inch diameter GaAs substrate with buffer layers described above. The photodiode consists of a 1 μm thick InP layer, a 3 μm thick (n=5×10^{15}cm^{-3}) InGaAs absorbing layer and a 1 μm thick (n=5×10^{15}cm^{-3}) InP window layer.

The total thickness of the epitaxial layers was about 10 μm. After the epitaxial growth, the layers were selectively etched to make a gently-sloped PD structure with 5 small steps. The graded structure facilitates the following fabrication processes such as the photolithographic processes or connecting the p-electrode of the PD and the gate electrodes of the FET. The p-region of the PD was formed by Be ion implantation, since conventional vapor phase Zn diffusion in a closed tube is not applicable for large diameter wafer. The active region of the FET were formed by Si ion implantation. The activation of the implanted ions were carried out at the same time at relatively low temperature 750°C in PH$_3$ atmosphere to prevent dissociation of the InGaAs/InP PD material. The photodiode p-electrode was formed by Ti/Mo/Au, and the photodiode n-electrode , the FET source and drain were formed by AuGe/Ni/Au. The Ti/Mo/Ti/Au gate was formed by a recess gate process. The photodiode p-electrode and the gate electrodes of the FET were connected by air-bridge using lift-off technique.

4. Device Characteristics

The integrated PDs have the dark current of as low as 10 nA at bias voltage of -5 V. The sensitivity for 1.3 μm-incident light is 0.9 A/W, which is as high as the sensitivity of the PDs on InP substrates. The quantum efficiency is 90%.

The transconductance of the FET was 125 mS/mm(Vgs=0V). Dynamic characteristics of the OEIC chip including feedback resistance of 1 kΩ were measured using a transimpedance amplifier comprising two GaAs HEMTs. The frequency response of the receiver OEIC is shown in Fig.2. The 3 dB bandwidth of the receiver OEIC is 1.4 GHz. The equivalent input current density as low as 7 pA/ $\sqrt{Hz}$ has been obtained. The input capacitance of the OEIC chip is estimated to be 1 pF from the frequency dependence of the noise characteristics. The transimpedance is 60 dB, and the dynanic range is over 27 dB.

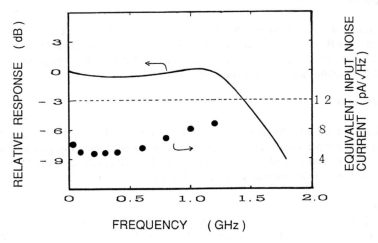

Fig.2 Frequency response of the receiver OEIC

The receiver sensitivity was measured at 1.3 μm wavelength. Bit error rate as a function of the average incident optical input is shown in Fig.3 for 622 Mbit/s-NRZ signals. The sensitivity of -28.1dBm was achieved with the bit error rate of 10^{-9}. Calculated sensitivity of the OEIC chip from the equivalent input noise current density is -31.2 dBm. The difference of the measured value and the calculated one is attributed to the noise in the amplifier.

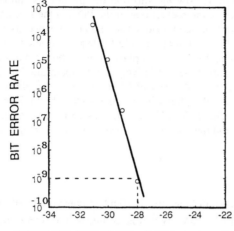

Fig.3 Bit error rate vs. average
incident optical power

There is apprehension that lattice mismatch might degrade the reliability of the OEIC. Therefore, the reliability of the PDs integrated in the OEICs were investigated by a bias temperature test. The dark current has been very stable under the condition of Vb=-10 V,T=175°C for 700 hours. The reliability of the integrated PD is estimated to be comparative with the PDs on InP substrate.

5. Conclusion

We have realized a long-wavelength PIN-FET receiver OEIC on 3-inch diameter GaAs substrate. The OEIC has a sufficient receiver sensitivity and is applicable for practical communication systems. The potential of the InP-on-GaAs heterostructure for long wavelength OEIC applications has been demonstrated.

Acknowledgements

The authors would like to thank Drs. T. Murotani, K. Ikeda, M. Otsubo and H. Namizaki for their supports and advices to this work. We also express our gratitude to Mr. K. Maemura, Mr. H. Nakano, Mr. H. Matsuoka, Mrs. H. Nishiguchi and Mr. M. Nogami for their help in design, fabrication and test of the OEIC.

References
Dentai A G, Campbell J C, Joyner C H and Qua J 1987 Electron. Lett. **23** 3
Ishimura E, Kimura T, Shiba T, Mihashi Y and Namizaki H 1990 Appl. Phy. Lett. **56** 644
Kimura T, Kimura T, Ishimura E, Uesugi F, Tsugami M, Mizuguchi K and Murotani T
    1991 J. Crystal Growth **107** 827
Koren U, Marglit S, Chen T R, Yu K L, Yaliv A, Barchaim N, Lau K Y and Ury I 1982
    IEEE J. Quantum Electron. **QE-10** 1653
Suzuki A, Itoh T, Terakado T, Kasahara K, Asano K, Inomoto Y Ishihara H, Torikai T and
    Fujita S 1987 Electron Lett. **23** 955

Inst. Phys. Conf. Ser. No 120: Chapter 10
Paper presented at Int. Symp. GaAs and Related Compounds, Seattle, 1991

453

Monolithically integrated optoelectronic transmitter by MOVPE

A.A. Narayanan, D. Yap, S.E. Rosenbaum, C.S. Chou, W.W. Hooper, R.H. Walden.
Hughes Research Laboratories, Malibu, CA 90265.

Abstract

We have grown a vertically integrated FET/Laser structure for optoelectronic transmitters. It is designed to operate at signal frequencies of several gigahertz. It combines a GaAs/GaAlAs ridge-waveguide laser with a GaAs MESFET driver circuit. The laser has one of its cavity mirrors formed by dry etching so that the die size of the transmitter is not limited to the laser cavity length. The step coverage problems that might result from this vertical integration are avoided by the use of air-bridge connections. The performance of the MESFET (g_m = 360 mS/mm), and the laser (I_{th} = 300 A/cm^2) grown by the one-step MOVPE technique will be discussed.

1. Introduction

The low pressure MOVPE technique[1] is used to grow the high performance MESFET[2] and the laser devices[3]. The transmitter consists of both of the devices stacked one above the other. The monolithically integrated optoelectronic transmitters and receivers operating at microwave frequencies are needed for optical interconnect systems. To complement our recently developed receivers, which operate at frequencies up to 6.5 GHz[4], we have been developing an integrated transmitter to operate at a center frequency of 3 GHz with a bandwidth of at least 2 GHz.

2. Integration Approach

The integrated transmitter combines a GaAs/GaAlAs ridge-waveguide laser[5], a GaAs MESFET amplifier circuit and various biasing components on a common semi-insulating GaAs substrate. GaAs based devices were chosen for our initial demonstration because of their comparatively mature fabrication technologies. Much of the technology developed for the transmitter can be extended to the InP based materials that would be well suited for long fiber optic links. The transmitter has a vertically integrated material structure in which all of the epilayers, for both the MESFET and the laser, are formed on a planar substrate. The epilayer design is shown in Fig. 1. All the layers are grown by the low-pressure MOVPE technique. Initially we have optimized the growth parameters by separately growing individual devices. The integration design we follow imposes several challenges to the fabrication process because of the resultant stepped surface. To fabricate the driver and bias-tee components, we must first selectively remove the laser epilayers. This produces steps that are as high as 5 μm between the top of the laser and the FET layers. The potential step-coverage problem is solved by using polyimide planarizing layers and air-bridge

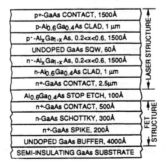

Figure 1. The epitaxial layer design of a vertically integrated transmitter structure.

metal connections, as illustrated in Fig. 2. Direct write electron-beam lithography is used to define the submicrometer FET gates.

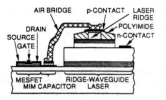

Figure 2. The schematic diagram of a transmitter showing the integration of a ridge waveguide laser, a MESFET driver and an MIM coupling capacitor.

3. MOVPE Growth

The reactor used for the growth is a commercial system[6] for low pressure operation. The pressure maintained during the growth is 76 Torr. The system has a four-way valve manifold to minimize the dead volume in the gas stream and to enhance the abrupt switching of the process gases. The exhaust from the reactor is burned and then passed through an activated charcoal bed. The source gases used for the transmitter growth are TMGa, TMAl, DEZn, SiH_4 and AsH_3. The bubblers are maintained at atmospheric pressure by means of a needle valve. The growth temperature for all the devices is 750°C. The various performances of the individual devices are given in the following sections.

4. Laser Performance

The laser of the integrated transmitter was designed with the following constraints. First, it must be a high performance device whose drive current, output power and modulation response adequately meet the requirements of the transmitter. Second, the laser should have a relatively simple structure and fabrication process so that it does not adversely affect the performance of the other circuit components. Finally, the material growth and device fabrication processes must give a reasonable yield of working circuits. Our GaAs/GaAlAs lasers have a GaAs single quantum well (SQW) active layer and linearly graded separate-confinement layers. Broad-area 100-μm-wide devices fabricated from material grown on conducting n-type substrates typically have threshold current densities of 250 to 300 A/cm^2 for 500-μm cavity lengths.

Single-transverse-mode ridge waveguide lasers on semi-insulating substrate material were developed for the transmitter. The structure is shown in Fig. 3. Pulsed threshold currents of 13 mA, and external differential quantum efficiencies above 80%, have been obtained. Output powers are greater than 20 mW, CW, for devices with uncoated facets. As shown in Fig. 4., the small signal modulation responses of these lasers show 3 dB bandwidths greater than 4 GHz when the lasers are biased at twice their threshold current.

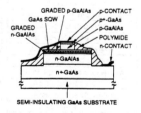

Figure 3. Structure of a polyimide planarized ridge waveguide laser fabricated on semi-insulating substrate.

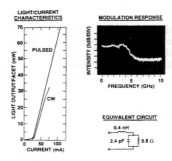

Figure 4. The ridge waveguide laser performance showing the modulation response of 3dB bandwidth greater than 4GHz.

5. Transistor Performance

The design of the MESFET was chosen to provide a high current-gain cutoff frequency. The FET structure shown in Fig. 5. has a conventional "square" gate of 0.2-μm length. The GaAs MESFET layers consist of an n^+ contact layer, a lightly doped Schottky layer, and a thin, heavily doped channel layer for high transconductance. A highly resistive buffer layer is grown directly below the channel layer. Resistivities above 1.5 E+5 Ω-cm, which give device isolation voltages greater than 500 V, have been achieved by adjusting the As/Ga ratio and lowering the growth temperature. Discrete 50-μm-wide FETs have transconductance measured at $I_{dss}(V_{gs}=0)$ of 360 mS/mm. Measurement of the short circuit current gain shows that the unity current-gain frequency, f_T, of these devices is greater than 50 GHz. In order to handle the 50- to 75-mA currents needed to drive the lasers, the FETs of the driver circuits have gate widths of 300 to 400 μm. These devices have multiple gate segments of 50-μm width, with metal air-bridge crossovers interconnecting the device contacts. The growth of the integrated transmitter involves a long (about 4 hours) growth time. Despite the prolonged exposure to high growth temperature (750°C), the performance of the FET is not compromised. The performance of the individual devices, grown separately, are repeated in the integrated structure.

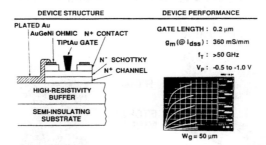

Figure 5. The GaAs MESFET and its performance showing g_m = 360 mS/mm.

6. Fabrication And Testing

The schematic diagram and the simulated frequency response of the driver circuit are shown in Fig. 6. The layout of the integrated transmitter with the driver circuit is shown in Fig. 7. We used a dry etching process to form the laser mirror, and a selective etching process to remove the unwanted laser material. The laser cavity mirror is etched in a low-pressure, low-voltage magnetron reactive-ion etching (RIE) system with a load locked chamber. Figure 8 shows SEM micrographs of the mirror region of the laser.

The FET structures are exposed by a selective etch process and then they are contacted. A mesa that is about 5 μm high remains after the selective etching. Air bridges are used in our circuit fabrication processes to form the cross-overs in the segmented-gate FETs, spiral inductors and interconnect lines. The SEM micrograph of a completed circuit, along with a receiver circuit on the same wafer is shown in Fig. 9. The completed circuits have been tested for frequency response, which is plotted in Fig. 10.

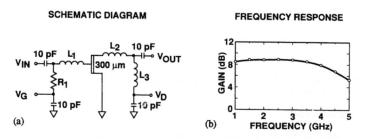

Figure 6. Schematic diagram (a) and simulated frequency response (b) of driver circuit, LDRVRO.

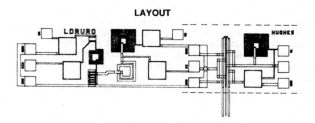

Figure 7. Layout of integrated transmitter with driver circuit LDRVRO.

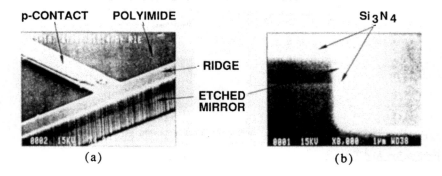

Figure 8. SEM micrograph of dry-etched mirror region of a ridge waveguide laser, showing (a) mirror surface smoothness and (b) cross-section profile. The mirror region has been coated with a thin layer of silicon nitride.

COMPLETED CIRCUITS   AIR-BRIDGE CONTACTED LASER

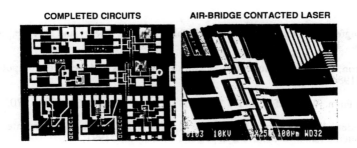

Figure 9. SEM photomicrographs of completed circuits showing the transmitter and receivers. An air-bridge contacted laser is also shown.

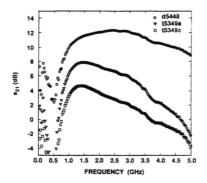

Figure 10. The frequency response measured on fabricated transmitters showing the response well over 4 GHz.

7. Summary

The low pressure MOVPE technique is used to grow the high performance MESFET and the ridge-waveguide laser for the monolithically integrated optoelectronic transmitter. We have achieved 360 mS/mm and $f_T > 50$ GHz for 0.2-μm gate MESFET devices. Single transverse mode ridge waveguide lasers for the transmitter have shown pulsed threshold currents of 13 mA, and external differential quantum efficiencies above 80% and output power greater than 20 mW, CW, with uncoated facets. The transmitter is fabricated using the dry etch process for the mirror, a selective etch process to expose the underlying device, and an air-bridge for contacting both devices. Our initial results showed the transmitter operating as expected with 3 GHz center frequency and a band width of 2 GHz.

Acknowledgments

The authors would like to acknowledge the encouragement of R. Blumgold, P.T. Greiling, G.S. Kamath, R.C. Lind, and H.W. Yen. The technical assistance of S.C. Ontiveros, D.M. Bohmeyer, S.L. Bourgholtzer, and R.E. Doty are appreciated. This work was supported in part by WRDC/EL, Wright-Patterson AFB, under contract number F33615-87-C-1546.

References

1. G.B. Stringfellow, "Organometallic Vapor-Phase Epitaxy: Theory and Practice," Academic Press, Inc., 1989.

2. M. Laviron, D. Delagebeaudeuf et al., Int. Symp. GaAs and Related Compounds, Biarritz (1984), 539.

3. R.D. Burnham, W. Streifer, and T.L. Paoli, J. Crystal Growth 68 (1984) 370.

4. R.H. Walden, W.W. Hooper, C.S. Chou, C. Ngo and R. Blumgold, presented at IEEE MTT-S International Microwave Symposium, Boston June, 1991, 491.

5. H. Lang, H.D. Wolf, L. Korte, H. Hedrich, C. Hoyler, C. Thanner, IEE Proceedings-J, Vol 138, No. 2. (1991) 117.

6. Spire MOCVD 450 System was used for the Growth.

A vertically integrated driver for light-emitting devices utilizing controllable electro-optical positive feedback

M. Selim Ünlü, S. Strite, A. Salvador, and H. Morkoç

I. Introduction

Recently, interest in new device technologies for digital optoelectronic circuits has been mounting. Optical switches will be necessary elements for optical signal processing and digital communication systems. An optoelectronic switch consisting of a bipolar inversion channel field-effect transistor was demonstrated [1]. An optical inverter consisting of a vertically integrated heterojunction phototransistor (HPT) and a light emitting diode (LED) has been realized [2]. A pulsed GaAs laser has been demonstrated also using a p-n-p-n structure [3]. Theoretical analysis of p-n-p-n devices have been carried out by several authors [4-6]. The operational principles of all of these two terminal optoelectronic devices are all similar to the Shockley diode. In this paper, we describe a heterojunction bipolar transistor (HBT) vertically integrated with a quantum well (QW) LED. The overall structure is an N-p-n-p diode with a heterojunction emitter. Switching is achieved by controlling the injection efficiency of the heterojunction emitter by external base current or the HBT can function as an HPT enabling the use of optical excitation. By modulating both the external base current and the optical excitation, this device can function as a high sensitivity optical switch or as an erasable programmable optical memory element.

II. Device Growth and Fabrication

Investigated structures were grown using a Perkin Elmer 430 molecular beam epitaxy machine on p^+-GaAs(100) substrates. Figure 1 shows the schematic device structure. From the substrate up is p-GaAs (0.5 μm, 5×10^{17} cm^{-3}), 2 50Å In$_{0.2}$Ga$_{0.8}$As quantum wells surrounded by 100Å GaAs barriers, an n-GaAs layer (0.1μm, 5×10^{17} cm^{-3}), an n-GaAs collector (0.5μm, 5×10^{16} cm^{-3}), a p-GaAs base (0.1μm, 5×10^{18} cm^{-3}), an N-Al$_{0.25}$Ga$_{0.75}$As emitter (0.2μm, 5×10^{17} cm^{-3}), and an n^+-GaAs cap (0.05μm, 5×10^{18} cm^{-3}). Standard photolithography and wet etching techniques were used to fabricate devices with various emitter sizes for electrical characterization. Circular devices with emitter windows were also fabricated for optical measurements. Contacts to n- and p-type layers were made by evaporating AuGe/Ni/Au and AuBe, respectively.

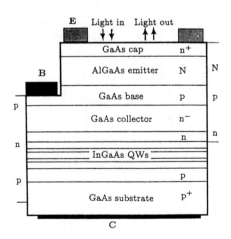

Fig.1. Schematic layer structure of the investigated devices.

III. Results and Discussion

Figure 2 gives the band diagram of the device (HBT-LED) showing various off-state current components. The structure can be viewed as interwoven N-p-n and p-n-p bipolar transistors. The base of the each is the collector of the other. The point denoted by **A** is the collector-base junction for both transistors (Fig.2).

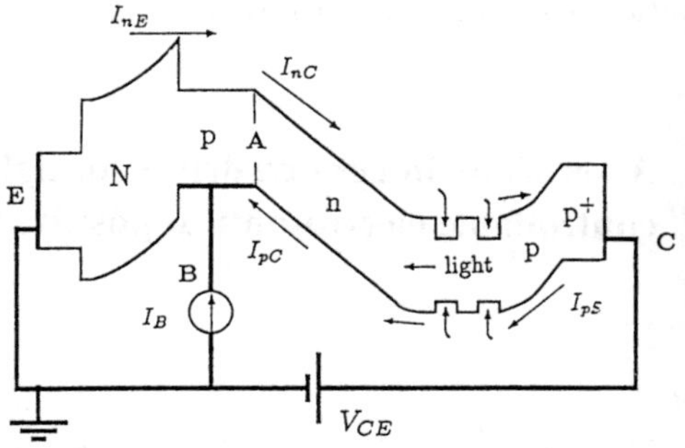

Common-emitter current-voltage (I-V) characteristics of the device are given in Figs. 3 (a) and (b) for constant base current (I_B) steps and for constant base-emitter voltage (V_{BE}) steps, respectively. As can be seen from Fig. 3(a), the device functions as a normal HBT in the low current regime (off-state). The measured collector current is supplied from the holes injected from the p+- substrate (I_{pS}). The electron current injected from the emitter into the collector of the N-p-n transistor (I_{nC}) recombines with the holes (I_{pS}) from the p-n-p device in the InGaAs quantum wells generating light (Fig. 2). The holes from the substrate which do not recombine in the quantum wells and the photo-regenerated carriers in the depleted collector and neutral base of the N-p-n device create additional base current (I_{pC}) constituting a positive feedback mechanism. The emission en-

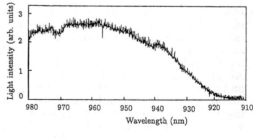

Fig.2. Band diagram of the investigated structure biased in OFF state. Electron and hole current components are denoted by subscripts *n* and *p*, respectively.

ergy of the LEDs was intentionally chosen to be smaller than the GaAs band gap to avoid an undesirably large feedback resulting from regeneration in the N-p-n HBT. In the investigated device, the entire spectrum of the emitted light is at energies below the GaAs band edge (Fig. 4). Therefore, the optical feedback is negligible. A small optical feedback allows the switching characteristics of the device to be controlled by the emitter-base junction bias of the N-p-n HBT. If desired, larger electro-optical feedback could be achieved by a different LED quantum well design.

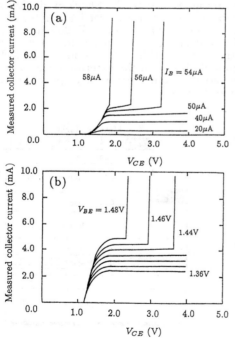

Fig.3. Current-voltage characteristics of the HBT-LED (emitter size is 10 × 40μm) for (a) constant base current steps and (b) constant base voltage steps.

Fig.4. Spectrum of light output from a 300 μm circular device in ON state (current=200 mA).

For the three terminal device the current relations can be derived as follows using the notation from Fig. 2 [6]:

$$I_{nE} = I_{pS} + I_B \tag{1}$$

and

$$I_{pS} = I_{nC}(A) + I_{pC}(A) + I_{C0} \tag{2}$$

where I_{C0} is the saturation current and

$$I_{nC}(A) = \alpha_N \cdot I_{nE} \quad \text{and} \quad I_{pC}(A) = \alpha_P \cdot I_{pS} \tag{3}$$

where α_N and α_P are the common-base current gains for N-p-n and p-n-p transistors, respectively. From Eqs. (1-3)

$$I_{pS} = \frac{\alpha_N I_B + I_{C0}}{1 - \alpha_N - \alpha_P} \tag{4}$$

Note that I_{pS} is the measured collector current which we use to emphasize that the hole current is supplied by the p$^+$-substrate. This notation was chosen to avoid confusion with the other current components in the collector of N-p-n HBT. As can be seen from Eq. (4), the switch on condition is reached when $\alpha_N + \alpha_P = 1$.

For the N-p-n HBT, the common base current gain α_N is given by [7]

$$\alpha_N = \gamma_N \alpha_T M \tag{5}$$

where γ_N is the emitter efficiency, α_T is the base transport factor, and M is the collector multiplication factor. Under small collector-emitter bias $M \approx 1$. For bipolar transistors with base widths less than one-tenth of the diffusion length, $\alpha_T > 0.995$; and the current gain is given almost entirely by the emitter efficiency, i.e., $\alpha_N \approx \gamma_N$. For the investigated structure, the base width is sufficiently small (0.1 μm) to justify this approximation.

The current gain of GaAs HBTs (α_N for our device) increases with increasing emitter current density, as the contribution of the surface and bulk recombination currents becomes less important compared with the useful diffusion current of minority carriers across the base region [7]. Therefore, the feedback can be controlled by external base bias or optical excitation. Controlling the switching by external base bias removes the need for a voltage bias exceeding the breakover voltage, which is more than 20 V in the absence of base current. We expect α_P to be dominated by the base transport factor (as given in Eq. (5)) and to be quite small due to recombination in the InGaAs QWs and the relatively thick base region of the p-n-p transistor. Moreover, the homojunction emitter, and the emitter-down configuration of the p-n-p transistor further degrades the current gain. Therefore, switching can be controlled by modulating α_N through the base bias, and the variation of the relatively smaller α_P is neglected in our model.

To determine the values of α_N and α_P at the switch-on point, we compare the I-V curves for constant base current and constant base voltage (Figs. 3 (a) and (b)). The value of α_N is a function of the total base current $(I_B + I_{pC})$ of the N-p-n HBT and we can expect that the oscillation condition occurs at a certain threshold value of $I_B + I_{pC}$. We will consider the transition at $V_{CE}=3$V. From Fig. 3(a) we obtain $I_B \approx 54.5$ μA and $I_{pC} = \alpha_P \cdot I_{pS} = \alpha_P \cdot 2.2$ mA from Eq. (2). Examining Fig. 3(b), we obtain $I_{pC} = \alpha_P \cdot 4.4$ mA, neglecting I_B which decreases as the switch-on condition is approached and is negligibly small at the switching threshold. Equating the total base currents for the constant base voltage and base current conditions

$$(0.0545 + 2.2\alpha_P) \text{ mA} = (4.4\alpha_P) \text{ mA} \tag{6}$$

Solving Eq. (6) gives $\alpha_P=0.025$. From $\alpha_N + \alpha_P = 1$ we deduce that $\alpha_N=0.975$ which is equivalent to a common emitter current gain of $h_{FE}=39$. That is, at $V_{CE}=3$V, switching occurs when the current gain of the N-p-n HBT exceeds 39.

This condition can be experimentally verified by plotting current gain versus measured collector current (I_{pS}) as the base current is swept at $V_{CE}=3V$ (Fig. 5). The curve for h_{FE}-I_C is similar to the typical HBT characteristics at low current levels. The slope of $\log(h_{FE})$–$\log(I_C)$ reflects a base-emitter ideality factor of 1.5 which is in good agreement with the measured value of 1.45. At higher current levels, unlike a typical HBT, the current gain increases rapidly and switching takes place. Just before switching occurs, the component of the base current supplied by the feedback mechanism (I_{pC}) becomes comparable to the external base current component (I_B). Therefore the measured current gain does not reflect the internal gain of the HBT. We estimate the actual current gain at the switching condition by extrapolating the gain curve from the low current regime as shown in Fig. 5. The current gain at switching obtained by this method is roughly 40, in very good agreement with the calculated $h_{FE}=39$.

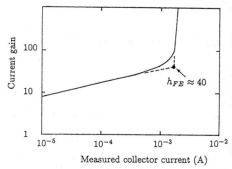

Fig.5. Current gain versus collector current for a $10 \times 40\mu$m emitter device. Dashed line is the extrapolation of the gain curve from low current region.

Figure 6 shows the Gummel plot of the investigated device. In this figure, the different switching conditions for constant base voltage and constant base current I_B can easily be identified. Note that base and collector currents are monitored as the V_{BE} is swept. The sudden increase in the collector current (arrow 2) corresponds to the switching condition at constant base voltage. Where I_B reaches its maximum (arrow 1) is the switching threshold at constant base current. As expected, these values for the switching condition are in good agreement with the current levels observed in Figs. 3(a) and (b). We interpret the downturn in I_B to result from the increasing effect of the parasitic resistances at the high current levels which reduce the fraction of the externally supplied V_{BE} which is actually dropped across the emitter-base junction. At the switching point denoted by arrow 2, the base current is an order of magnitude smaller than its maximum, and after switching it settles to a value determined by the external circuitry.

To further verify and understand the properties of this structure, we compared the current density and current gain at the switch-on condition in devices with various emitter perimeter to area ratios. As can be seen in Fig. 7, the current density at which switching occurs varies from 200 A/cm² to nearly 800 A/cm² whereas the critical current gain independent of the device geometry to within the accuracy of our measurement.

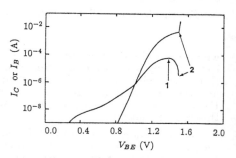

Fig.6. Gummel plot for the investigated device. At **1** $I_B = 55$ μA and $I_C = 2.2$ mA and at **2** $I_B < 5$ μA and $I_C = 4.4$ mA.

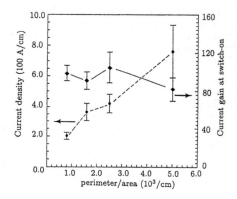

Fig.7. Dependence of critical current density and current gain for switch-on versus perimeter to area ratio.

The switching characteristics under optical excitation were also studied using a pulsed GaAs laser. The threshold light intensity for switching was observed to be controlled by the external base bias. We are unable to quantitatively analyze the optical input/output characteristics of the device due to our inability to measure absolute incident light intensities. However, intensities of the order of microwatts successfully induce switching in 300μm circular devices. If the devices are electrically biased just below the switching threshold, very weak optical excitations are capable of inducing the oscillatory ON state. The external circuit can be adjusted to achieve two different device functions. The load line can be adjusted so that the ON state can only be obtained by optical excitation. In this case the device functions as a high gain light amplifying switch. The load line can be adjusted so that two stable conditions are possible, one in the ON the other in the OFF state. In this configuration, if a device in the OFF state is illuminated, it will switch ON and remain ON until electrical bias condition is changed functioning as optical-write electrically erasable programmable read-only memory (optical E- EPROM). A complete analysis of optical device functions will be presented elsewhere [8].

IV. Summary

We have demonstrated an N-p-n-p device which is suitable for driving LEDs for digital optoelectronic applications. The vertical structure of the device should allow the application of this mechanism to surface emitting lasers. Controlling the switching conditions by modulating the base current removes the need for large voltages or excitations.The device can be switched by optical illumination promising useful applications for optical communications and computing.

V. Acknowledgements

This work is supported by Office of Naval Research under contract N00014-88-K-0724. We would like to thank B. Mazhari for valuable discussions and B. Bowdish for technical assistance. One of us (S. S.) wishes to acknowledge the support of an AFOSR fellowship.

References

1. G.W. Taylor, R.S. Mand, J.G. Simmons, and A.Y. Cho, Appl. Phys. Lett. 49, (1986), 1406-1408

2. F.R. Beyette Jun., S.A. Feld, X. An, K.M. Geib, M.J. Hafich, G.Y. Robinson, and C.W. Wilmsen, Electronics Lett. 27, (1991), 497-499

3. H.F. Lockwood, K.-F. Etzold, T.E. Stockton, and D.P. Marinelli, IEEE J.Quantum Electron., 10, (1974), 567-569

4. James F. Gibbons, Proc. IEEE, 55, (1967), 1366-1374

5. James F. Gibbons, IEEE Trans. Electron Dev., 11, (1964), 406-413

6. Finis E. Gentry, IEEE Trans. Electron Dev., 11, (1964), 74

7. S.M. Sze, Physics of Semiconductor Devices, (John Wiley& Sons, New York, 1981), 140-143

8. M.S. Ünlü, S. Strite, A. Salvador, A.L. Demirel, and H. Morkoç, IEEE Photon. Technol. Lett. (To be published)

Heavily p-doped GaAs/AlGaAs single quantum well lasers: growth, performance and integration with heterojunction bipolar transistors

J. Nagle[+], R. J. Malik, R. W. Ryan,
J. P. van der Ziel and L. C. Hopkins

AT&T Bell Laboratories
Murray Hill, NJ 07974

1. Introduction

The interest in p-doping of the active layer of quantum well lasers (QWLs) is motivated by high-speed and integration prospects. The high-speed modulation capability of QWLs is enhanced by heavy p-doping due to the steeper dependence of the gain as a function of the number of injected carriers (Uomi et al 1987, Zah et al 1990). In addition, heavily p-doped quantum wells sandwiched by two suitably graded n-doped confinement layers can simultaneously serve as the base of a double heterojunction bipolar transistor (DHBT) and the active layer of a quantum well laser, where the base contact of the transistor is used for the lateral injection of excess holes in the laser (Mori et al 1985, Hasumi et al 1987). A similar multifunctional approach combining FET and laser has been recently demonstrated (Taylor et al 1991, Honda et al 1991).

In this paper we demonstrate the successful operation of this QWL-HBT device and we present growth conditions by molecular beam epitaxy to obtain simultaneously good-quality lasers and HBTs. A device analysis of the QWL-HBT is presented showing the various tradeoffs involved in the structure design.

2. MBE Growth and Device Fabrication

The basic structure is that of a graded index separate confinement heterostructure (GRINSCH) quantum well laser. The samples are grown by molecular beam epitaxy using a Ta cracker As_2 source on n^+ GaAs substrates and include the following sequence of layers: 1.5 μm n- $Al_{0.4}Ga_{0.6}As$, 0.125 μm n- AlGaAs graded from 40% to 15% Al, 150Å p-doped GaAs quantum well, 0.125 μm n- AlGaAs graded from 15% to 40% Al, 1 μm n- $Al_{0.4}Ga_{0.6}As$ and 0.2 μm n^+-GaAs. The center of the quantum well is uniformly doped with Be at a nominal doping of $2 \times 10^{19} cm^{-3}$ to yield a total areal density of 2.7×10^{13} Be atoms/cm^2. Two control samples are grown to test the quality of the material. The first one is a conventional GRINSCH QWL with the same geometry

---

+ Present Address: Thomson-CSF/LCR, 91404 Orsay, FRANCE

and compositions but without doping in the well and with p-doping in the upper confinement layers. The second one has heavily p-doped quantum well and p-doping in the upper confinement layers. These control samples are processed as conventional top-injection stripe lasers of different lengths.

The traditional growth conditions to obtain high-quality GaAs/AlGaAs lasers and HBTs are very different: high growth temperature ($\approx 700°C$) for lasers to ensure good optical quality of AlGaAs and low growth temperatures ($\approx 550°C$) for HBTs to prevent unwanted diffusion of the base dopant (Be). We grow at intermediate temperatures of 580°C in the quantum well and 630°C in the AlGaAs barrier. We also use GaAs layers (7 monolayers) included every 500Å in the thick $Al_{0.4}Ga_{0.6}As$ confinement layers in order to improve the flatness of the interfaces. We confirm earlier findings (Foxon et al 1991) that low-threshold lasers can be obtained at lower temperatures when using the As_2 source. SIMS measurements on these samples show that there is no significant diffusion of the Be incorporated in the quantum well.

The laser-HBT samples are processed as shown on Figure 1. Ridge waveguides are formed by etching down ~ 1 μm AlGaAs through an SiO_2 protective layer. The overhang of the SiO_2 mask is then used to self-align the AuBe metallization. After sintering the sample to diffuse the p-contact to the quantum well, a 3 μm-wide window is opened in a new SiO_2 film to deposit the n-type AuGeNi top-contact. A TiAu overlay pad is then deposited for probing and bonding. In this first generation of devices, the collector contact is taken from the back of the substrate and there is no isolation between two adjacent ridges.

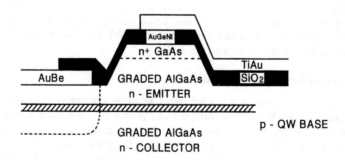

Figure 1. Schematic layout of the lateral injection ridge waveguide quantum well laser-HBT (cross-section).

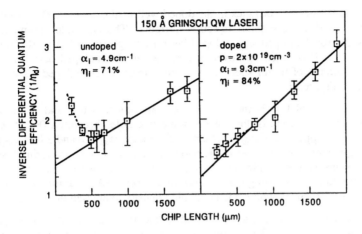

Figure 2. Measured inverse differential quantum efficiency as a function of the laser chip length for two GRINSCH GaAs/Al$_{0.15}$GaAs/Al$_{0.40}$GaAs single quantum well lasers grown at 580-630°C with or without p-doping in the well.

3. Device Performance

The results obtained on the top-injection stripe lasers made from the control samples are summarized in Figure 2. The internal quantum efficiency η_i and total internal losses α_i are determined by fitting the data to the usual expression

$$\frac{1}{\eta_d} = \frac{1}{\eta_i} \left[1 + \frac{2\,\alpha_i L}{\ln(1/R_1 R_2)} \right]$$

where L is the length of the laser chip and R_1 and R_2 the mirror facet reflectivities.

The lowest threshold current densities obtained for long lasers are 350 A/cm^2 for the p-doped well and 450 A/cm^2 for the undoped well. This result together with the higher internal quantum efficiency of the doped laser (84% vs 71%) seems surprising at first since the contribution of non-radiative Auger recombination is larger for doped wells and should diminish η_i. In fact, the observed experimental results reflect the contribution of additional non-radiative recombination paths probably related to the quality of the AlGaAs material. In the presence of this high density of holes, the radiative lifetime of minority electrons is reduced and hence the quantum efficiency is increased. Furthermore, the number of electrons at threshold is much lower for p-doped lasers and the spillover of excess electrons in the lesser-quality graded AlGaAs regions is larger for undoped wells and causes a reduction of η_i. This is due to the fact that few electrons are needed to reach high gain since a high population of holes is already present. Calculation shows that the quasi Fermi level of holes in the doped well is already 65 meV below the top of the valence band at very small injection currents. The calculated numbers of electrons at threshold for the doped and undoped laser are respectively $1.3 \times 10^{12}\,\text{cm}^{-2}$ and $5.5 \times 10^{12}\,\text{cm}^{-2}$. This larger spillover of electrons in the undoped laser explains why the observed decrease of η_i for short lasers is more marked for undoped lasers than for doped lasers (see Figure 2).

The measured internal loss is the sum of the scattering loss α_{sc} and of the free carrier absorption loss α_{fc} which is dominated by the free hole absorption (Henry et al 1983). We estimate an upper value $\alpha_{fc} = 9.3\,\text{cm}^{-1}$ supposing $\alpha_{sc} = 0$ and a lower value $\alpha_{fc} = 4.4\,\text{cm}^{-1}$ supposing that all losses in the undoped laser are due to scattering losses. This is converted by correcting with the optical mode confinement factor to an equivalent bulk free hole absorption at the lasing wavelength for a 150Å GaAs quantum well $\alpha_{fc} = (p/1 \times 10^{19})$ x $(110 \pm 29)\,\text{cm}^{-1}$ where p is the doping density expressed in cm^{-3}.

The common emitter I-V characteristics of a lateral injection ridge waveguide QWL-HBT is shown in Figure 3. The corresponding maximum common-emitter current gain is around 150. When the emitter-base junction is forward-biased with the collector floating, the same device is lasing at 890 nm with a pulsed threshold current of 220 mA at room temperature. The performance of these first devices can certainly be greatly improved by modifications of the process. In fact, conventional circular mesa HBTs processed from the same wafer have a higher maximum current gain of 500. The laser characteristics can also be improved by reducing the ridge width, increasing the facet reflectivity and replacing the AuBe p-contact. We have observed by SIMS that simultaneous diffusion of Be and Au into the sample occurs during sintering of the contact. The presence of Au atoms which are efficient non-radiative recombination centers is probably strongly increasing the threshold current of the laser.

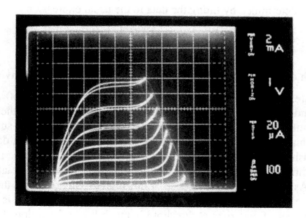

Figure 3. Common emitter I-V characteristics of a 520 μm long QWL-HBT. The width at the bottom of the ridge is 8 μm as measured by SEM.

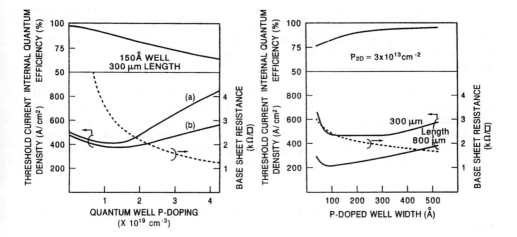

Figures 4 (left) and 5. Calculated dependence of the characteristics of the QWL-HBT as a function of the well doping for constant well width and as a function of the well width for constant well sheet doping. The curves labeled (a) and (b) correspond respectively to the maximum and minimum values of the free carrier absorption as determined in this paper.

4. Device Analysis of the QWL-HBT

There is obviously a tradeoff in the design of the QWL-HBT structure between reduction of the base intrinsic resistance for enhanced modulation bandwidth and improvement of the properties of the laser. Increasing the doping for a constant well width reduces the internal quantum efficiency because of the enhanced non-radiative Auger recombination and eventually also increases the threshold current because of Auger recombination current and of free hole absorption. The results of a detailed calculation of the influence of p-doping is presented in Figure 4. The calculation procedure is described by Nagle et al (1986) and includes a phenomenological Lorentzian broadening in order to account for the carrier collisions. No additional non-radiative recombination term is included except for Auger recombination $(C_A = 2 \times 10^{-30} \text{cm}^6 \text{s}^{-1})$. We use for the free carrier absorption, the values determined experimentally in the preceding section and we neglect scattering losses. It is important to note that contrary to the case of undoped lasers, the calculated threshold currents for heavily doped lasers are not sensitive to the exact value used for the broadening parameter. The energetically broad distribution of holes implies broad gain profiles and little effect of a relatively small additional broadening.

The influence of the width of the quantum well when keeping a constant hole sheet concentration is also calculated (Figure 5) for different laser lengths. The base sheet resistance decreases slightly for larger wells because the hole mobility is larger. Larger wells are to be preferred if the optical losses are high (short lasers) because of gain saturation for the narrow wells.

From these calculations it appears that low threshold and high quantum efficiency lasers can be obtained for base sheet resistance greater than 1-2 $k\Omega/\square$. In all cases considered, high values of the maximum current gain are expected since we deal with thin bases and dopings in the $10^{19}\,cm^{-3}$ range. This multifunctional epi-structure appears very promising since integration of high performance lasers and HBTs is possible through the use of standard process technologies.

Conclusions

Single quantum well lasers with heavy p-doping in the quantum well are grown by MBE. We show that, using a Ta cracker As$_2$ source, it is possible to grow high-quality lasers and HBTs at intermediate temperatures (580 - 630°C). Threshold current densities as low as 350A/cm^2 and internal quantum efficiencies of 84% are obtained for conventional top-injection 150Å single quantum well stripe lasers doped with Be at $2 \times 10^{19}\,cm^{-3}$. A value for the free hole absorption in a p-doped quantum well is estimated from the measurements.

We also demonstrate the first ridge waveguide QWL with lateral injection through the base, that also operates as an HBT. A theoretical model of the QWL-HBT predicts the best parameters of the structure for optimized device operation.

References

Foxon C T, Blood P, Fletcher E D, Hilton D, Hulyer P J and Vening M 1991 *J. Crystal Growth* **111** 1047

Hasumi Y, Kozen A, Temmyo J and Asahi H 1987 *IEEE Electron. Device Lett.* **EDL-6** 10

Henry C H, Logan R A, Merritt F R and Luongo J P 1983 *IEEE J. Quantum Electron.* **QE-19** 947

Honda Y, Suemune I and Yamanishi M 1991 *Appl. Phys. Lett.* **59** 621

Mori Y, Shibata J, Sasai Y, Serizawa H and Kajiwara T 1985 *Appl. Phys. Lett.* **47** 649

Taylor G W, Claisse P R and Cooke P 1991 *Appl. Phys. Lett.* **58** 666

Nagle J, Hersee S, Krakowski M, Weil T and Weisbuch C 1986 *Appl. Phys. Lett.* **49** 1325

Uomi K, Mishima T and Chinone N 1987 *Appl. Phys. Lett.* **51** 78

Zah C E, Bhat R, Menocal S G, Favire F, Andreadakis N C, Koza M A, Caneau C, Schwarz S A, Lo Y, and Lee T P 1990 *IEEE Photonics Technology Lett.* **2** 231

Inst. Phys. Conf. Ser. No 120: Chapter 10
Paper presented at Int. Symp. GaAs and Related Compounds, Seattle, 1991

471

MOCVD growth of vertical cavity surface-emitting lasers with graded-composition mirrors

S. Z. Sun, E. A. Armour, D. P. Kopchik, K. Zheng, P. Zhou, Julian Cheng, and C. F. Schaus

Center for High Technology Materials, University of New Mexico, Albuquerque, NM 87131 USA

ABSTRACT: A GaAs/AlGaAs vertical cavity surface-emitting laser (VCSEL) with a graded-index mirror composition has been grown using low pressure metalorganic chemical vapor deposition (LP-MOCVD). The graded compositions were obtained by ramping the reactant gas flows while maintaining a constant growth temperature. The devices have excellent electrical characteristics, including a low series resistance (22 Ω for 30 mm diameter), high overall power efficiency (6%), and low operating voltages. These characteristics are the best results ever reported for MOCVD-grown vertical cavity lasers and are comparable to, or better than, the best MBE-grown devices.

1. INTRODUCTION

Vertical cavity surface-emitting lasers are well-suited for two-dimensional optical switching applications [1,2] because of their low threshold currents, high output powers [3], and small beam divergence [4]. Low operating currents and voltages permit the integration of dense VCSEL arrays by minimizing thermal dissipation, thereby reducing its negative impact on output power and operating temperature range. However, while room-temperature CW operation of VCSELs with high output powers has been achieved, the series resistance and operating voltages of these devices have remained high. The large series resistances arise from a combination of contact resistance, spreading resistance, and the impedance to carrier transport from the energy barriers at the heterointerfaces of the distributed Bragg reflector (DBR) mirrors. These factors also contribute to increased operating voltages. Contact resistances are reduced by increasing the contact areas and surface doping. To reduce the contribution arising from the energy-band discontinuities at the heterointerfaces, various grading techniques have been proposed [5], including the use of a chirped superlattice, or the use of a piecewise-linear gradient with one or more $Al_xGa_{1-x}As$ layers of intermediate aluminum composition. Continuous grading can, in principle, eliminate the barriers altogether, however this is difficult to achieve with MBE. While these methods have significantly reduced the series resistance from the range of kΩ to several hundred ohms, these values remain too high for power-efficient devices. On the other hand, a continuous grading of an arbitrary composition profile can be readily achieved with MOCVD by regulating the gas flows while maintaining a constant temperature. In this paper, we report the details of MOCVD growth of high quality VCSELs with continuously-graded DBR mirrors.

2. VCSEL STRUCTURE DESIGN

The VCSEL epitaxial structure shown in Fig. 1 consists of an undoped active region bounded by p-doped and n-doped DBR mirror stacks, all of which were grown by a single growth sequence on a buffered (100) n-GaAs substrate. The upper p-doped DBR mirror contains 24 periods of graded $Al_xGa_{1-x}As$ layers. The Al content of the $Al_xGa_{1-x}As$ is 15 %, except at the heterointerfaces, where it is linearly graded from 15% to 100% over a distance of 12 nm. The lower n-doped DBR mirror contains 43.5 periods of graded $Al_xGa_{1-x}As$ layers with a graded structure. The active layer contains a symmetrical, graded-index (GRIN) heterostructure with four 8-nm quantum wells separated by 2-nm spacers. The total thickness of the GRIN MQW structure corresponds to a single wavelength of the cavity mode, which is designed to match the MQW gain peak at 850 nm.

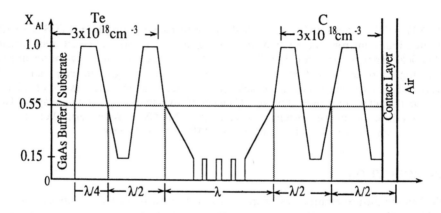

Fig. 1. Reduced structure of the VCSEL with graded-composition mirror. Graded layers are 12 nm thick between $X_{Al}=1.0$ to 0.15 and 6 nm thick for X_{Al} between 1.0 and 0.55. The active region consists of 4, 8-nm GaAs quantum wells spaced by 2-nm barrier layers of $X_{Al}=0.15$.

3. Material Growth

The VCSELs were grown in a horizontal MOCVD reactor [6] at 100 Torr and a temperature of 725° C. The carrier flow was 10 slm of palladium-diffused hydrogen. The reactant sources used were trimethylgallium (-10.0 ± 0.1° C), trimethylaluminum (17.0 ± 0.1° C), and 100 % arsine. The dopants used were diethyltellurium (74 ppm in H_2), diethylzinc (20.0 ± 0.1° C), and carbon tetrachloride (489 ppm in H_2). The growth rate for the structure was 500 Å/min. except for the quantum wells which were grown at 425 Å/min. The V/III ratio was maintained at 50 (58.8 for the quantum wells). The graded alloy compositions were obtained by ramping of the reactant gas flows with mass flow controllers as shown in Fig. 2. To maintain a constant carbon doping level of 3 x 10^{18} cm^{-3} throughout the graded aluminum mirrors, it is necessary to modulate the carbon tetrachloride flow rates due to increased carbon incorporation efficiency at high aluminum $Al_xGa_{1-x}As$ compositions [7,8]. Prior to the growths, a 760° C heat treatment was conducted under an arsine overpressure. The heavily

Zn-doped contact layer is 6-nm thick and grown at 580° C in order to achieve 1×10^{20} cm^{-3} hole concentration. Between each growth, an *in-situ* low pressure etching bakeout was conducted to clean the flow chamber using HCl vapor at 900° C.

The MOCVD reactor has several unique design features which help obtain tight thickness and composition tolerances necessary for these devices. The reactor features a fast-switching radial injection manifold consisting of twelve 3-way 5-port valves. The reactants enter the manifold body, where the position of the 3-way valves determine if the reactants are sent through the reaction chamber or through the bypass vent line. The symmetry of the manifold assures that the carrier gas flows are split evenly between the vent and reaction chamber and eliminates the need for elaborate pressure-balancing schemes. Dead-space is also reduced in the manifold, allowing for abrupt growth interfaces. The reactor is monitored and controlled in real-time using a multi-processing VAX computer connected to a CAMAC system. The speed and flexibility of the system allow precise timing and reproducible control with a cycle time of 10 ms.

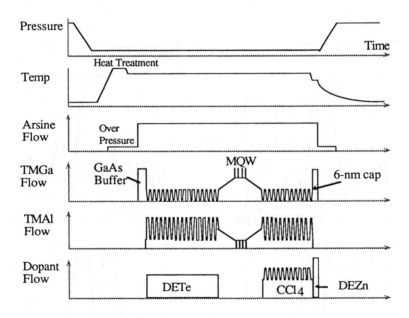

Fig. 2. A typical growth sequence for a graded-mirror VCSEL using LP-MOCVD. The number of mirror periods is less than that of an actual device.

To achieve the desired laser structure, several preliminary calibration structures were grown. The first calibration is the GRIN-MQW active layer section composed of four undoped 8-nm GaAs quantum wells separated by 2-nm Al$_{0.15}$Ga$_{0.85}$As barriers and sandwiched between the GRIN layers. Room-temperature photoluminescence is used to characterize quantum well emission wavelength. IR reflectometry is used to correct the overall structure thickness by changing the thickness of the GRIN layers. The results from the reflectometry are compared to calculated values, as shown in Figure 3a.

Once a satisfactory correlation between the reflectometry data and theory is achieved for the active region, ten periods of graded $Al_{0.15}Ga_{0.85}As/AlAs$ mirror layers are grown in order to achieve the correct the layer thicknesses and composition. The reflectance spectrum near 850 nm is used to assure a correct reflectivity profile from the mirrors. A typical reflectance profile for the multilayer reflector (MLR) calibration is shown in Fig 3b. Usually several iterations in layer thickness are needed to obtain a satisfactory reflectance peak at the lasing wavelength.

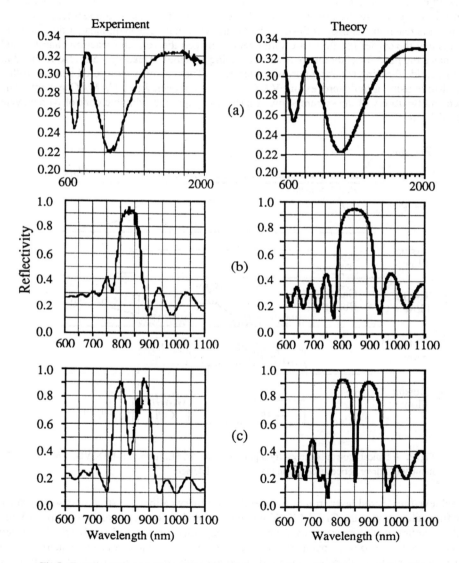

Fig.3 Experimental and theoretical reflectance spectra for calibration runs. (a) Active region , (b) 10 period multilayer reflector, and (c) 5 period mirror and 4 QW VCSEL.

Next, a reduced vertical cavity structure is grown consisting of a 5.5 period bottom (n-type) mirror, a GRIN-MQW active layer, and a 5 period top mirror. This structure is suitable for calibration purposes, as it contains a substantial mode in the reflectance peak which should be matched to the lasing wavelength. As the number of mirror periods for the full structure are increased, the measured mode width and size decrease. The reflectance spectrum is again used to assure correct layer thicknesses and proper interfaces between the mirror and active layers as shown in Fig. 3c.

4. RESULTS

The growth sequence for the entire structure is shown in Fig. 2. Room-temperature photoluminescence and reflectometry are again used to verify the structure against theoretical models prior to optical pumping and subsequent processing.The VCSEL device structure is a large contact area design described further in [9]. Current confinement is achieved by implantation of 310 keV protons at a dosage of 3×10^{14} cm^{-2} around a masked active area with diameters ranging from 5 to 35 μm. A post-implant anneal was conducted at 430° C for 30 seconds. Large area Ti-Pt-Au p-type and Au-Ge-Ni-Au n-type contacts were used. The room temperature CW electrical and optical characteristics were measured on VCSELs with diameters of 8, 11, 20, and 35 μm. The lasing for these devices occurs at thresholds of 2.2 mA, 2.7 mA, 6.0 mA, and 9.6 mA for the respective diameters, corresponding to current densities of 4.4, 2.8, 1.9, and 1.2 kA/cm^2, respectively. A current density of 780 A/cm^2 was measured for a 40 μm diameter device, which is one of the lowest values ever reported. The threshold voltage for all devices was at or below 3 V, which is significantly lower than other reported devices. The series resistances for the respective device diameters are 75, 50, 30 and 22 Ω (measured at 20 mA bias current), which is three times lower than comparably-sized devices as grown by MBE [3]. These low series resistances and threshold voltages can be attributed to the continuous grading of the heterointerfaces. Although a low series resistance was also recently reported for strained quantum well substrate-emitting devices [10], lasing was observed only under pulsed conditions and the quantum efficiencies were low (8% pulsed). The peak output optical powers of our devices range from 0.5 mW for the 8 μm ϕ devices to 2.0 mW for the 32 μm ϕ devices, which were not solder-bonded or optimally heat-sunk. The peak differential quantum efficiency is 80% for the 11 μm device, with a somewhat lower efficiency (~ 50%) for the larger devices. The overall power efficiency (electrical to optical power) is about 6% for the 11 μm device.

The lower operating voltages and the very low series resistances of the VCSELs indicate that the electrical characteristics have been substantially improved by the grading, and that thermal dissipation has been significantly reduced. The peak optical output is lowered as a result of some misalignment between the cavity mode and the gain peak and unoptimized heat-sinking. With improved epilayer alignment and better thermal management, higher optical powers and power efficiencies are possible. Nearly all the VCSELs showed good lasing characteristics with excellent local uniformity. Lasing was observed across the entire 2" wafer with a high yield.

Far above threshold, the spectrum of the smaller devices (11 and 14 μm) shows a single peak with a full width at half maximum (FWHM) of 0.7 Å. The best devices show a resolution-limited spectrum of less than 0.4 Å. Even below threshold, the spectrum is relatively narrow, indicating a very high Q for the cavity. For the larger devices, the spectrum shows a dominant peak at threshold, although vestiges of a second mode is already apparent. As the drive current increases, a discontinuity in the slope of the L-I characteristics can be observed, concurrent with the formation of a second dominant peak in the spectrum shifted towards higher energies. This corresponds to the appearance of a higher order transverse mode. At even higher currents, three or even four transverse modes appear.

5. CONCLUSIONS

The electrical and optical properties of these devices represent the best VCSEL characteristics for material grown by LP-MOCVD and are comparable to state-of-the-art MBE results. The excellent electrical characteristics that these devices display can be attributed to the grading of the heterojunction interfaces, which eliminated the band-edge discontinuities. These graded composition mirrors are easy to achieve with MOCVD by ramping the reactant gas flows while maintaining a constant growth temperature. Due to the lack of precise *in-situ* monitoring techniques presently available with MOCVD, several calibration steps were used to obtain a suitable VCSEL structure. These calibration runs were measured using photoluminescence and IR reflectance spectroscopy and compared against theoretical models to correct the layer thicknesses and alloy compositions. Good uniformity was achieved across the wafer, with nearly all devices showing good lasing characteristics.

6. ACKNOWLEDGEMENTS

The authors would like to acknowledge D.R. Myers and J. Puechner of Sandia National Laboratories for conducting the proton-implant isolation. This work was supported by AFOSR/AFWL, the National Science Foundation and DARPA.

7. REFERENCES

[1]   G. R. Olbright, R. P. Bryan, K. Lear, T. M. Brennan, G. Poirier, Y. H. Lee, and J. L. Jewell, Electron. Lett. 27 (1991) 216.
[2]   Julian Cheng, G. R. Olbright, R. P. Bryan, 1991 Technical Digest of Optical Computing 6 (1991) 10.
[3]   B. Tell, Y. H. Lee, K. F. Brown-Goebeler, J. L Jewell, R. E. Leibenguth, M. T. Asom, G. Livescu, L. Luther, and V. D. Materra, Appl. Phys. Lett. 57 (1990) 1855.
[4]   K. Tai, G. Hasnain, J. D. Wynn, R. J. Fischer, Y. H. Wang, B. Weir, J. Gawelin, and A. Y. Cho, Electron. Lett. 26 (1990) 1628.
[5]   K. Tai, L. Yang, Y. H. Wang, J. D. Wynn, and A. Y. Cho, Appl. Phys. Lett. 56 (1990) 2496.
[6]   H. E. Schaus, C. F. Schaus, S. Z. Sun, M. Y. A. Raja, A. Jacome-Torres, and J. G. McInerney, Proc. 16th Int. Symp. GaAs Related Compounds, Inst. Phys. Conf. Ser. (1990) 749.
[7]   B.T. Cunningham, J. E. Baker, and G. E. Stillman, J. Electron. Mat., 19 (1990) 331.
[8]   S. Z. Sun, E. A. Armour, K. Zheng, and C. F. Schaus, unpublished results.
[9]   P. Zhou, Julian Cheng, C. F. Schaus, S. Z. Sun, K. Zheng, E. A. Armour, C. Hains, Wei Hsin, D. R. Myers, and G. A. Vawter, to be published in IEEE Photon. Tech. Lett. (1991).
[10]   M. Orenstein, A. C. Von Lehmen, C. Chang-Hasnain, N. G. Stoffel, J. P. Harbison, and L. T. Florez, Electron Lett. 27 (1991) 438.

Inst. Phys. Conf. Ser. No 120: Chapter 10
Paper presented at Int. Symp. GaAs and Related Compounds, Seattle, 1991

Influence of gain saturation on the T_0 values of short AlGaAs–GaAs single and multiple quantum well lasers

H. Jung, E. Schlosser, and R. Deufel

Daimler-Benz Research Center, Wilhelm Runge Str. 11, D 7900 Ulm (Germany)

Abstract

In GaAs-AlGaAs Single and Multiple Quantum Well (SQW and MQW) ridge lasers, grown by molecular beam epitaxy, the temperature dependence of the threshold current, expressed by the characteristic temperature T_0, is investigated as a function of the cavity length (L) at different temperatures. SQW lasers, in contrast to MQW's, show a strong decrease in T_0 from 250 K to about 100 K when the length is reduced from 400 μm to 200 μm. We found that by further reducing L to about 130 μm, a strong increase in T_0 up to 250 K takes place and T_0 decreases again to 80 K for 60 μm SQW lasers. This T_0 behavior is directly correlated to the gain saturation of the $n=1$ transition and the change of the laser emission to the $n=2$ transition.

1. Introduction

During the last decade GaAs-AlGaAs Quantum Well (QW) lasers have attained increasing attention because of their superior properties which resulted in important applications [1]. In QW lasers the threshold current is drastically reduced compared to conventional Double Heterostructure (DH) lasers due to the decreased active volume of the quantum well. The optical absorption loss is lower and the differential gain coefficient is higher for QW lasers [2]. From these arguments, SQW lasers should have the lowest threshold current. However, in SQW lasers the threshold current increases drastically when the cavity length, L, is reduced below a critical value [3]. Simultaneously, the temperature dependence of the threshold current increases drastically [4]. Both effects can be attributed to gain saturation [5], which strongly influence the design of optimized QW lasers.

This paper reports the temperature dependence of the threshold current as a function of the cavity length of AlGaAs-GaAs SQW and MQW ridge lasers with separate confinement (SC) waveguide structure at various temperatures. From these measurements, the characteristic temperature, T_0, is determined as a function of L and temperature. These investigations reveal that, in contrast to results published to date [4], in very short SQW lasers the T_0 value increases again by further shortening L. This experimental observation will be explained by considering a shift of the laser emission to the $n=2$ transition.

2. Experimental

The studies were performed on GaAs-AlGaAs SQW and MQW ridge lasers. The epitaxial layers were grown by Molecular Beam Epitaxy (MBE) on n-doped ($n=1\cdot10^{18}$ cm^{-3}) (100)-oriented GaAs substrates at 700°C. For p- and n-doping we used Be and Si, respectively. The active GaAs QW layers have thicknesses of 10 nm and are sandwiched in between two 100 nm n- and p-doped (both $5\cdot10^{17}$ cm^{-3}) Al$_{0.2}$Ga$_{0.8}$As SC layers. Both, the n- ($1\cdot10^{18}$ cm^{-3}, 1.2 μm thick) and p- ($5\cdot10^{17}$ cm^{-3}, 1.2 μm thick) cladding layers consist of $Al_{0.5}Ga_{0.5}As$. The MQW structure contains four wells where the $Al_{0.2}Ga_{0.8}As$ barriers have a thickness of 5 nm. The uppermost contact layer consists of a highly p-doped ($5 \cdot 10^{18}$ cm^{-3}) 0.4 μm thick GaAs layer. The ridge structuring was performed by a wet chemical etching with a solution of $H_2SO_4:H_2O_2:H_2O$. Figure 1 shows (a) the schematical structure and (b) a scanning electron micrograph of the ridge laser. The ridge width was about 3.5 μm with a residual thickness of the p-cladding layer outside the ridge region of about 0.2 μm. The side region of the ridge was covered with SiO_2 for electrical insulation. For contacting we used annealed NiAuGe (n) and TiAu (p) metallization. The wafers were then

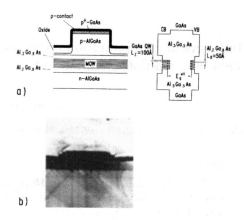

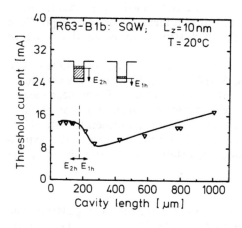

Fig. 1: a) Schematical structure and b) scanning electron micrograph of an AlGaAs-GaAs ridge laser

cleaved into laser chips with lengths from 60 μm to 1000 μm and mounted on heat-sinks. The light-current characteristics were measured in cw and pulse conditions at temperatures ranging from 10°C to 93°C.

3. Results and Discussion

3.1 Threshold current as a function of cavity length

Figure 2 represents the length dependence of threshold current, I_{th}, for SQW lasers at 20°C. In lasers longer than 300 μm, I_{th} increases linearly with L, according to the increase of the active volume. In this range, laser emission takes place via the first

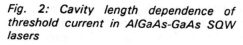

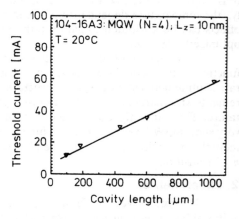

Fig. 2: Cavity length dependence of threshold current in AlGaAs-GaAs SQW lasers

Fig. 3: Cavity length dependence of threshold current in AlGaAs-GaAs MQW lasers

subband transition E_{1h}. The minimal threshold current is about 7 mA at a cavity length around 300 μm. When reducing L below 300 μm, I_{th} increases significantly, by nearly a factor of two, and remains nearly constant at lengths between 150 μm and 60 μm. This increase of I_{th} is directly correlated with a change of the emission energy from E_{1h} to the second subband transition E_{2h}. This change in the lasing transition is schematically indicated in the inset of Fig. 2, representing the population of the QW conduction band. For comparison, in Fig. 3, I_{th} is drawn as a function of L for the MQW laser also at 20°C. In contrast to the SQW behavior, I_{th} increases linearly within the total length range from 100 μm to 1000 μm.

3.2 Temperature dependence of threshold current

The temperature dependence of threshold current in semiconductore lasers can be approximated empirically by the following relation:

$$I_{th} \propto \exp(T/T_0)$$

T_o, which is the characteristic temperature can be determined according to the above relation by measuring I_{th} at different temperatures. In Fig. 4 the T_0 values are shown for SQW lasers as a function of L at different temperatures. For long cavities (above 500 μm), T_0 is nearly independent of L and reaches values of about 250 K determined between 20°C and 40°C, 180 K determined between 40°C and 60°C and 150 K determined between 80°C and 93°C. It can be seen that T_0 decreases with increasing temperature for long cavities. By reducing L to about 200 μm, T_0 decreases drastically to about 100 K.

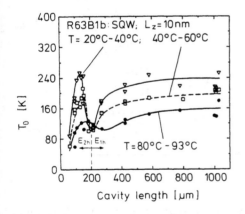

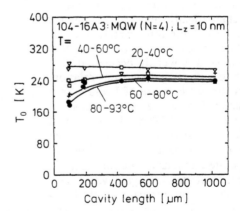

Fig. 4: T_0 behavior of AlGaAs-GaAs SQW lasers at different temperatures

Fig. 5: T_0 behavior of AlGaAs-GaAs MQW lasers at different temperatures

This minimum T_0 value is nearly independent of T between 20°C and 93°C. However, the corresponding cavity length increases slightly to longer L for the T_0 determined between 80°C and 93°C as can be seen from the curve. Reducing L to about 130 μm, T_0 increases again very strongly from 100 K to 250 K at 20°C and 40°C, to 200 K at 40°C and 60°C and to 130 K at 80°C and 93°C. In this length range, the laser emission changes from the E_{1h} to E_{2h} transition. A further reduction of L (below 130 μm) again yields a significant decrease in T_0 at all temperatures. This new T_0 behavior in short SQW lasers could be observed on a great number of lasers fabricated from different wafers. For comparison, Fig. 5 shows the T_0 dependence on L of MQW lasers at different temperatures. In these MQW lasers T_0 is nearly constant at about 250 K for lengths between 100 μm and 1000 μm. Only at higher temperatures does T_0 decrease slightly if L is reduced below 400 μm.

3.3 Discussion

Referring to Fig. 6, we can explain the length dependence of the T_0 behavior especially of SQW lasers. Both curves in the lower part show the behavior of the maximum modal gain as a function of the injection current at two temperatures. The modal gain, responsible for laser emission, is obtained by multiplying the volume gain with the confinement factor Γ. At low injection current I, an increase in I leads to an approximately linear increase in the peak volume gain and hence also in the peak modal gain [5]. This is shown in the lower linear region of the gain-current relation in Fig. 6. When I is further increased, the peak gain saturates because the spectral carrier

density, responsible for the maxi-maximum gain, cannot be further enhanced due to the constant 2-dimensional density of states (2D-DOS) in the QW structure. This can be understood by means of the upper part of Fig. 6, where the carrier population in the QW is schematically shown for 2 subbands and for 3 different excitation levels. At low excitation, only the first subband is weakly occupied yielding a linear gain-current relation. A further increase of I results in a saturation of the gain because of constant 2D-DOS. Finally, with further increase of current, the second subband system will be populated and, as a consequence, the gain will increase again with I. This is due to the relevant spectral carrier density which is now doubled. At threshold, the modal gain is equal to the total optical loss.

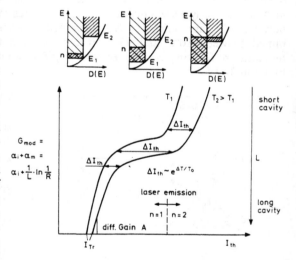

Fig. 6: Schematical gain behavior as a function of current density in SQW lasers at two temperatures

Thererefore, the curves in Fig. 6 describe threshold gain versus threshold current density for lasers with various optical losses at two temperatures. Lasers with long cavities L, having low mirror losses, are characterized by the lowet part of the curves where laser emission takes place via E_{1h}. Lasers with very short L having high mirror losses are represented in the upper part of the gain-current relation and laser emission occurs via E_{2h}. In both regions of the cavity length, the slopes of the curves are steep, but in the transition region the curves are flat due to gain saturation via E_{1h}. Comparing both curves, we see that the relative increase of I_{th} with temperature, given by the horizontal distance of the curves, is nearly constant at long L (small required gain) and increases more and more with decreasing L up to a maximum value in the saturation range. At further reduction of L the increase of I_{th} with temperature is again lower because laser transition now takes place via E_{2h} without gain saturation.

Since in MQW lasers the modal gain is N times the value of SQW lasers ($\Gamma \sim N$), the required threshold gain is distinctively below the saturation value (lower linear region of the gain-current relation in Fig. 6) for all lengths between 100 μm and 1000 μm. This leads to a nearly constant T_0 value — in agreement with the experimental results in Fig. 5. Only at high temperatures, where the gain saturation effects are more pronounced, a weak decrease of T_0 can be observed in Fig. 5.

In Fig. 7 the measured modal gain is shown as a function of the effective threshold current density for SQW ridge lasers with lenghts between 60 μm and 1000 μm at 20°C and 40°C. The curves are determined from the threshold current density measured as a function of cavity length at different temperatures. At threshold the modal gain is equal to the total optical losses, which is the sum of the intrinsic loss α_i and the mirror loss α_m. The intrinsic loss α_i is obtained from the length dependence of

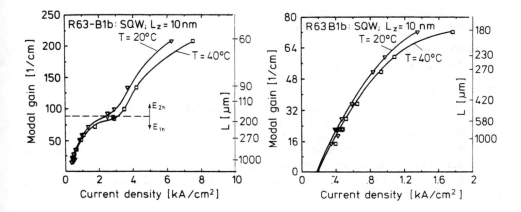

Fig. 7: Current density dependence of modal gain of AlGaAs-GaAs SQW lasers at two temperatures

Fig. 8: Current density dependence of modal gain of short AlGaAs-GaAs SQW lasers at two temperatures

the differential quantum efficiency. For α_i in SQW lasers, a value of 5 cm^{-1} is found and in MQW lasers a value of 7 cm^{-1}. The mirror loss is expressed by $(1/L) \cdot \ln(1/R)$. In Fig. 7 we obtained a saturation value for the modal gain of about 80 cm^{-1}. This represents a volume gain of about 2600 cm^{-1} calculated with a Γ factor of 3% (SQW). Figure 8 shows the saturation region in more detail (shorter cavities). Figure 9 exhibits the gain current behavior of MQW lasers for L between 100 μm and 1000 μm from 10°C to 93°C. These curves have all a constant slope indicating no gain saturation. The maximum modal gain of about 130 cm^{-1} (100 μm) represents a volume gain of only 1000 cm^{-1} with an assumed Γ of $4 \cdot 3\%$ ($N=4$). This value is well below the saturation value of about 2600 cm^{-1}, obtained for the SQW-laser.

4. Summary

The study of the threshold current in GaAs-AlGaAs SQW and MQW lasers as a function of the cavity length at various temperatures has led to the conclusion that by shortening the SQW lasers down to about 200 μm the temperature dependence of threshold current increases drastically resulting in a small T_0 due to gain saturation. Further reduction of L, however, leads to a significant increase of T_0 because of enhanced gain via the second subband transition. In contrast, MQW lasers show no marked gain saturation, even at temperatures up to 93°C within the same length range as in SQW lasers. As a consequence, in MQW lasers the temperature dependence of threshold current shows no length dependence and yields therefore nearly constant T_0 values of 250 K for all cavity length between 100 μm and 1000 μm.

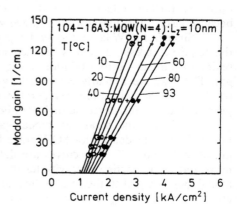

Fig. 9: Current density dependence of modal gain of AlGaAs-GaAs MQW lasers at various temperatures

5. Acknowledgement

This work was supported by the German Ministry of Research and Technology.

References

[1]   W. T. Tsang, IEEE J. Quantum Electron., vol. **QE-20**, (1984) 1119
[2]   P. L. Derry and A. Yariv, Appl. Phys. Lett. **50**, (1987) 1773
[3]   P. S. Zory, and A.R. Reisinger, Electron. Lett. **22**, (1986) 415
[4]   P. S. Zory, S.R. Chinn, and A. R. Reisinger, 11th IEEE International semiconductor laser conference in Boston, USA; Proc. (1988) 32
[5]   P. W. A. Mc Ilroy, A. Kurobe, and Y. Uematsu, IEEE J. Quantum Electron., vol.   **QE-21**, (1985) 1958

Inst. Phys. Conf. Ser. No 120: Chapter 10
Paper presented at Int. Symp. GaAs and Related Compounds, Seattle, 1991

483

Low threshold current density GaInAsSb/GaAlAsSb DH lasers emitting at 2.2 μm

J. L. Herrera-Pérez[+], M. B. Z. Morosini, A. C. F. da Silveira
and N. B. Patel[*]

LPD/DFA, Instituto de Física GLEB WATAGHIN-UNICAMP
C. P. 6165 Campinas S.P. Brazil

ABSTRACT

GaInAsSb/GaAlAsSb double-heterojunction injection lasers emitting at 2.2 μm with 27% Al in the confining layers have been prepared by liquid-phase-epitaxy, with very smooth hetero-interfaces. The best result of the threshold current density J_{th} = 2 KA/cm² at room temperature was a factor three lower than reported earlier for similar lasers.

1 Introduction

Low threshold injection lasers operating at room temperature in the mid infrared range are of great interest because of many potential applications. The work of Caneau *et al* (1986) with $Ga_{0.86}In_{0.14}As_{0.15}Sb_{0.85}/Ga_{0.73}Al_{0.27}As_{0.04}Sb_{0.96}$ DH lasers emitting at 2.2 μm reported best results of J_{th} (290 K) = 7-8 kA/cm² ($d_{act} \approx$ 0.8-1 μm) and it was suggested that the optical confinement with 27% Al in the confining layers was weak. By increasing the Al content to 34% a best result of J_{th}(290 K) = 2.6 kA/cm² was obtained by Zyskind *et al* (1989).

In this work we report the variation of J_{th} (295 K) as a function of active layer thickness of broad-area four-layer 2.2 μm GaInAsSb/GaAlAsSb DH lasers, with 27% Al in the confining layers.

2 Experimental

The lasers were grown by LPE under H_2 atmosphere. The liquid composition of the active layer melt were obtained by the dissolution method. (X_{Ga}^1 = 0.168, X_{Sb}^1 = 0.245, X_{In}^1 = 0.585 and X_{As}^1 = 0.00116). The composition of the confining layers melts was (X_{Al}^1 = 0.01257, X_{Sb}^1 = 0.003385, X_{As}^1 = 4.8 x 10⁻⁴). The GaInAsSb/GaAlAsSb heterostruture were grown at a temperature near 530 °C on (100) Te-doped GaSb (N_D-N_A 1 = 1-3 X 10¹⁷ cm⁻³) wafers, using a cooling rate of 0.3 °C/min. Small variations on As and Sb compositions were done for obtaining a morphology of high quality in the epitaxial heterostruture. The 2-3 μm thick AlGaAsSb layers were grown in 3 min, while a growth time of 10 s gave an active layer thickness of 0.9 μm. A 1 μm undoped GaSb layer is used as contact layer. A typical heterostruture is shown in Figure 1.

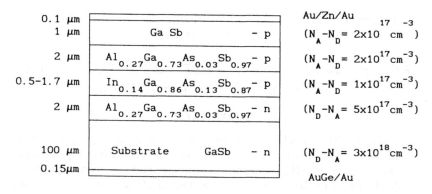

Fig. 1. Diagram of the DH heterostruture.

The solid composition of the layers were determined to be $Ga_{0.86}In_{0.14}As_{0.13}Sb_{0.87}$ and $Ga_{0.73}Al_{0.27}As_{0.03}Sb_{0.97}$ by microprobe analysis. Double-crystal X-ray diffraction rocking curve from layers grown showed narrow peaks ($\sim$ 30 arc-sec full widh at half power FWHP) with $\Delta a/a \leq 3 \times 10^{-4}$. Photoreflectance (PR) measurements at room temperature from the active layer doped with Te showed band-gap wavelength at 2.22 μm and for the confining layers at 1.18 μm. No signal of photoreflectance was observed in p type layers. The non intentionally doped active layers grown are p-type with residual doping of $\approx$ 2 x 10^{17} cm^{-3}, and the n-type GaAlAsSb layers were doped with Te at 5×10^{17} cm^{-3}.

The contacts were evaporated whith alloys of AuZn on the p-side and alloys of GeAu on the n-side. Then the wafers were alloyed at 375 °C for 3 min. to make low resistence ohmic contacts. The wafers were processed into individual broad area lasers obtained by cleaving and cutting them in cavities of 250 μm by 300 μm.

3   Results

The pulsed room-temperature L-I characteristic were measured , using a cooled PbS detector to determine their threshold current density. Figure 2 shows J_{th}(295 K) as a function of active layer thickness (d_{act}). J_{th} decreases with decreasing d_{act} and reaches a minimum value J_{th} (295 K) = 2.1 KA/cm^2 for d_{act} = 0.5-0.6 μm, which was the thinnest tested. We suppose that the low J_{th} for our lasers may be due fundamentally to the quality of the DH interfaces.

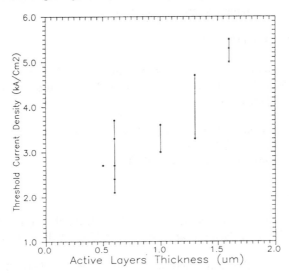

Fig. 2. Threshold current density J_{th} against active layer
thickness d_{act} for DH GaInAsSb/GaAlAsSb lasers

Absolute values of the refractive indices of GaInAsSb are not known. Using the laser far field data of Drakin *et al* (1987) and the recently published experimental data on the refractive index of GaAlAsSb by Alibert *et al* (1991), we estimated the refractive index step Δn for our lasers to be 0.18. This value of Δn should provide reasonably strong light confinement in the double heterostruture. The ratio between d_{act} and confinement factor Γ (d_{act}/Γ) has a minimum at d = 0.4 μm for this value of Δn. If we use the calculated values by Bhan *et al* (1987) of n for our active region together with results of Alibert *et al* (1991), we have a Δn of 0.1, which leads to a minimum of d/Γ at d=0.6 μm.

The minimun of J_{th} vs d_{act} should theoretically occur where d/Γ is minimum (Casey *et al* 1978) if extra losses like Auger recombination etc. are ignored. Our experimental J_{th} vs d_{act} is at d = 0.5-0.6 μm or less. These results suggest that the optical confinement with 27% Al in the confining layers is adequate and that the non-radiative Auger recombination rate in GaInAsSb material may not be as high for these lasers as a theoretic evaluation of Brosson *et al* (1987) has supposed.

4   Summary

Low threshold current densities as low as 2 KA/cm² can be obtained in high-quality GaInAsSb/GaAlAsSb DHs lasers emitting at 2.2 μm with 27% Al in the confining layer, by optimizing the growth conditions.

Acknowledgments

This research was suported by TELEBRAS and FAPESP (Brazil). The authors wish to thank to E. I. Piza, A. A. G. Von Zuben, M. A. Santos and F. Z. A. Von Zuben for help with devices fabrication.

[+]Now at Instituto de Física, Universidad Autónoma de Puebla
Apdo. Postal J-48, Puebla 72570 Mexico
[*]Also at CPqD, TELEBRAS; Campinas S.P. Brazil

References

Alibert C, Skouri C, Joullie A and Benouna M 1991 J. Appl. Phys, **69** 3208.

Bhan J, Joullie A, Mani H, Benot J and Brosson P 1987 SPIE *Mat. and Tech. for Comm. Opt.***866** 126.

Brosson P, Benoit J, Joullie A and Sermage B 1987 Electron. Lett. **23** 417.

Caneau C, Srivastava A K, Dentai A G, Zynkind J L and Pollak M A 1986 Electron. Lett. **22** 992.

Casey H C and Panish M B 1987 *Heterostruture Lasers* (Academic Press, New York).

Drakin A E, Peter G, Elieseev P G, Sverdlov B N, Bochkaren A E, Dolgienov L M and Druzhina L V 1987 IEEE J.Quantum Electr. QE **23** 1089.

Zyskind J L, Dewinter J C, Burrus C A, Centanni J C, Dentai A G and Pollack M A 1989 Electron. Lett. **25** 568.

Inst. Phys. Conf. Ser. No 120: Chapter 10
Paper presented at Int. Symp. GaAs and Related Compounds, Seattle, 1991

487

MBE growth, material properties, and performance of GaSb-based 2.2 μm diode lasers

S J Eglash and H K Choi

Lincoln Laboratory, Massachusetts Institute of Technology
Lexington, Massachusetts 02173-9108

ABSTRACT: Molecular beam epitaxy has been used to grow $Ga_{0.84}In_{0.16}As_{0.14}Sb_{0.86}/$ $Al_{0.75}Ga_{0.25}As_{0.06}Sb_{0.94}$ double-heterostructure diode lasers emitting at ~ 2.2 μm. The high-Al-content cladding layers provide improved carrier and optical confinement. For broad-stripe devices, the pulsed threshold current density was 0.94 kA/cm^2 for a cavity length of 1000 μm. Narrow-stripe devices were operated cw at temperatures up to 30°C, with maximum cw output power at 20°C of 4.6 mW/facet. These values are the best reported for semiconductor lasers emitting beyond 2 μm.

1. INTRODUCTION

High-performance 2- to 5-μm diode lasers would provide compact and reliable sources for applications such as laser radar, fluoride fiber communications, and spectroscopy. Double-heterostructure diode lasers incorporating a GaInAsSb active layer and AlGaAsSb cladding layers lattice matched to a GaSb substrate are being developed to meet this need. Room-temperature cw operation of index-guided devices grown by liquid phase epitaxy (LPE) has been achieved by two groups (Bochkarev *et al* 1988, Baranov *et al* 1988). We have recently reported (Choi and Eglash 1991A) the use of molecular beam epitaxy (MBE) to grow broad-stripe double-heterostructure $Ga_{0.84}In_{0.16}As_{0.14}Sb_{0.86}/Al_{0.50}Ga_{0.50}As_{0.04}Sb_{0.96}$ lasers emitting at 2.3 μm. These devices operated in the pulsed mode at room temperature with differential quantum efficiency η_d as high as 50 percent and pulsed output power as high as 900 mW/facet, the highest room-temperature values for semiconductor lasers emitting beyond 2 μm.

For GaInAsSb/AlGaAsSb diode lasers emitting near 2.3 μm, $Al_xGa_{1-x}As_ySb_{1-y}$ cladding layers with $x = 0.5$ provide conduction- and valence-band offsets of several kT for carrier confinement and a refractive index step of ~ 0.3 for optical confinement. Increasing the Al content increases both carrier and optical confinement, and can therefore be expected to improve laser performance. However, it is very difficult to grow lattice-matched $Al_xGa_{1-x}As_ySb_{1-y}$ layers with $x \geq 0.4$ by LPE (Motosugi and Kagawa 1980). Furthermore, it has not been possible to grow *n*-type lattice-matched AlGaAsSb layers with $x > 0.15$ by organometallic vapor phase epitaxy, since the layers contain high concentrations of acceptors, presumably C atoms incorporated from the organometallic sources (Palfrey 1991, Roth 1991). On the other hand, the use of MBE permits the growth of both *n*- and *p*-type AlGaAsSb over the entire range of lattice-matched compositions up to $x = 1$.

In this paper, we report the MBE growth of GaInAsSb/AlGaAsSb double-heterostructure diode lasers, emitting at ~ 2.2 μm, with approximately the same active layer composition used in our earlier devices but with $Al_xGa_{1-x}As_ySb_{1-y}$ cladding layers in which x has been increased to 0.75. This is the first report of GaInAsSb/AlGaAsSb diode lasers having cladding layers with $x > 0.55$. Narrow-stripe devices have been operated cw at temperatures up to 30°C, with maximum cw output power at 20°C of 4.6 mW/facet. For broad-stripe devices, the pulsed threshold current density J_{th} has been reduced to 0.94 kA/cm^2 for a cavity length $L =$ 1000 μm, compared with our earlier value of 1.5 kA/cm^2 for $L = 700$ μm.

2.   GROWTH AND MATERIAL PROPERTIES

2.1   Growth Technique

Conventional solid-source MBE was used for layer growth. The sources were the group III and group V elements, which yielded beams of Al, Ga, and In atoms and of As$_4$ and Sb$_4$ molecules. As expected, the incorporation efficiency was unity for the group III atoms and less than unity for the group V molecules. The group III fluxes were chosen to yield the desired Ga/In and Al/Ga ratios, with alloy growth rates of approximately 1 μm/h. An As beam-equivalent pressure of approximately 4 × 10^{-6} Torr was maintained to provide an excess As flux, and the Sb flux was adjusted to yield the desired As/Sb ratios. Because the sticking coefficient is much greater for Sb than for As, high concentrations of Sb can be incorporated even though the As flux is much greater than both the Sb flux and the total group III flux (Eglash *et al* 1991). The *n*-type dopant was Te provided by the sublimation of GaTe, and the *p*-type dopant was Be.

2.2   GaInAsSb Active Layers

The lattice constants of Ga$_x$In$_{1-x}$As$_y$Sb$_{1-y}$ alloys are related to the lattice constants of the constituent binary compounds by Vegard's Law. The condition for lattice matching to GaSb is

$$y = 0.91 \; (1 - x) \; / \; (1 + 0.05 \; x). \tag{1}$$

The bandgap of Ga$_x$In$_{1-x}$As$_y$Sb$_{1-y}$ alloys at 300 K, as estimated from the bandgaps of the ternary alloys (DeWinter *et al* 1985), is given by

$$\begin{aligned} E_g(x,y) = {} & 0.18 + 0.131 \, x - 0.4 \, y - 0.122 \, x \, y \\ & + 0.415 \, x^2 + 0.58 \, y^2 + 0.021 \, x^2 \, y + 0.62 \, x \, y^2. \end{aligned} \tag{2}$$

Simultaneous solution of Eqs. (1) and (2) yields the curve of E_g vs x shown in Fig. 1 for alloys lattice matched to GaSb. The E_g values range from 0.73 to 0.28 eV, corresponding to wavelengths from 1.7 to 4.4 μm. For this study, active layers with a nominal composition of Ga$_{0.84}$In$_{0.16}$As$_{0.14}$Sb$_{0.86}$ were chosen to provide laser emission at approximately 2.3 μm.

The peak wavelength of the active-layer photoluminescence (PL) measured at 4.5 K is 0.26 μm less than the laser emission wavelength at 300 K. The PL spectrum for a typical laser, as shown in Fig. 2, has a peak with full-width at half maximum of 7 meV. Such a narrow peak is indicative of uniform and well lattice-matched layers of GaInAsSb.

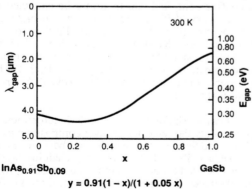

Fig. 1. Bandgap at 300 K vs composition parameter x for Ga$_x$In$_{1-x}$As$_y$Sb$_{1-y}$ lattice matched to GaSb.

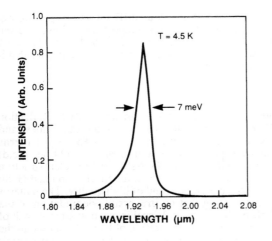

Fig. 2. Photoluminescence spectrum at 4.5 K from laser active layer.

2.3 AlGaAsSb Cladding Layers

The condition for lattice matching of $Al_xGa_{1-x}As_ySb_{1-y}$ is $y = 0.08\ x$. Cladding layers of $Al_{0.75}Ga_{0.25}As_{0.06}Sb_{0.94}$ were used for the present study. By careful calibration of the As and Sb fluxes it is possible to obtain closely lattice-matched layers. Figure 3 shows the double-crystal x-ray rocking curve for a sample consisting of the following layers (with x-ray peak positions): n-GaSb substrate (0 arc sec), 0.8-μm-thick GaSb buffer layer (+ 40 arc sec), 2.0-μm-thick $Al_{0.75}Ga_{0.25}As_{0.06}Sb_{0.94}$ (- 30 arc sec), and 0.05-μm-thick GaSb cap layer (negligible diffracted intensity). The AlGaAsSb layer exhibits a lattice mismatch at room temperature of only 1.2×10^{-4}. The GaSb layers are also slightly mismatched due to incorporation of a small amount of background As.

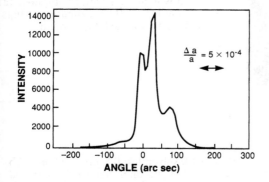

Fig. 3. Double-crystal x-ray rocking curve from an $Al_{0.75}Ga_{0.25}As_{0.06}Sb_{0.94}$ test sample.

We selected GaTe as the n-type dopant source for AlGaAsSb alloys because this compound had been employed successfully for doping MBE-grown GaSb (Furukawa and Mizuta 1988). By using GaTe we have grown n-GaSb layers with carrier concentrations up to 1×10^{18} cm^{-3} at 300 K with a mobility of 1900 cm^{2}V^{-1}s^{-1} (Eglash *et al* 1991). For a given GaTe source temperature, however, the electron concentration for n-$Al_{0.75}Ga_{0.25}As_{0.06}Sb_{0.94}$ is always lower than the value observed (Eglash *et al* 1991) for GaSb. Furthermore, the highest 300 K carrier concentration we have obtained in n-$Al_{0.75}Ga_{0.25}As_{0.06}Sb_{0.94}$ is only 6×10^{16} cm^{-3}, and the mobility at this level is only 10 cm^{2}V^{-1}s^{-1}. If the GaTe source temperature is increased above the value that gives this concentration, the measured carrier concentration decreases and the mobility rapidly increases. This anomalous high mobility may be due to the presence of metallic inclusions in the layers grown at the higher GaTe fluxes (Wolfe *et al* 1972).

The p-type doping of $Al_{0.75}Ga_{0.25}As_{0.06}Sb_{0.94}$ is not nearly as difficult as n-type doping. Although the hole concentration obtained for a given Be source temperature is lower for the alloy than for GaSb, we have obtained hole concentrations up to 4×10^{17} cm^{-3} at 300 K with a mobility of 140 cm^{2}V^{-1}s^{-1}. The observed dependence on Be source temperature indicates that it should be possible to obtain even higher concentrations.

2.4   Double Heterostructures

Laser-quality GaInAsSb/AlGaAsSb double heterostructures can be grown by MBE. Reflection high-energy electron diffraction (RHEED) has been used to characterize the smoothness and atomic order at the surface of the epitaxial layers. Figures 4(a,b) show RHEED patterns obtained for a 0.8-μm GaSb buffer layer that was grown on a GaSb substrate, and Figs. 4(c,d) show the patterns observed after the growth of a GaInAsSb/AlGaAsSb double heterostructure on this buffer layer. A 3×1 reconstruction is observed, as expected for an Sb-stabilized GaSb surface. The RHEED patterns show that the surface is smooth and of high quality. The patterns in Figs. 4(a,b) and 4(c,d) are very similar, indicating that no irreversible degradation of the surface occurred during growth of the double heterostructure. Visual inspection after growth of double heterostructures shows that the surfaces are specular and free of haze. As shown in the optical micrographs in Fig. 5, a crosshatch pattern is evident on structures that exhibit significant lattice mismatch (Fig. 5a), but adequately matched structures are very smooth (Fig. 5b). Some isolated defects that appear to be similar to GaAs oval defects are visible.

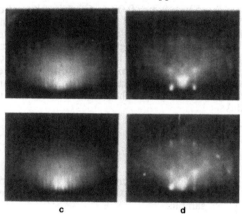

c   d

Fig. 4. RHEED patterns from (a,b) GaSb epitaxial layer on a GaSb substrate and (c,d) GaInAsSb/AlGaAsSb double heterostructure grown on the surface of the layer shown in Figs. 4(a,b). Figures (a) and (c) are for the $[1\bar{1}0]$ azimuth and Figs. (b) and (d) are for the [110] azimuth.

3.   DIODE LASERS

The following epitaxial layers were grown on (100) n-GaSb substrates: 0.8-μm-thick n-GaSb buffer, 2-μm-thick n-Al$_{0.75}$Ga$_{0.25}$As$_{0.06}$Sb$_{0.94}$ cladding, 0.4-μm-thick nominally undoped Ga$_{0.84}$In$_{0.16}$As$_{0.14}$Sb$_{0.86}$ active, 2-μm-thick p-Al$_{0.75}$Ga$_{0.25}$As$_{0.06}$Sb$_{0.94}$ cladding, and 0.05-μm-thick p^+-GaSb cap. Broad-stripe lasers 300 μm wide were fabricated by a lift-off process similar to the one reported previously (Eglash and Choi 1990). In order to obtain low threshold currents, narrow-stripe devices 30 μm wide were fabricated by etching mesas using a procedure described elsewhere (Choi and Eglash 1991B). Because of the difficulty of doping Al$_{0.75}$Ga$_{0.25}$As$_{0.06}$Sb$_{0.94}$, the series resistance in the forward direction is 12 Ω for the 30-μm-wide devices.

The lasers were first probe tested in the pulsed mode at room temperature. For the broad-stripe lasers the emission spectrum exhibits multiple longitudinal modes centered at 2.19 μm. The difference between this wavelength and the target wavelength of 2.3 μm is due to a deviation of the active-layer composition from the target value. Figure 6 shows the dependence of J_{th} on L. As L increases from 300 to 1000 μm, J_{th} decreases monotonically from 1.3 to 0.94 kA/cm^2.

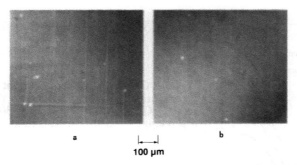

a   b
|———|
100 μm

Fig. 5. Nomarski interference micrographs showing surface morphology of (a) $Ga_{0.84}In_{0.16}As_{0.14}Sb_{0.86}/Al_{0.50}Ga_{0.50}As_{0.04}Sb_{0.96}$ laser structure with a crosshatch pattern resulting from lattice mismatch and (b) $Ga_{0.84}In_{0.16}As_{0.14}Sb_{0.86}/Al_{0.75}Ga_{0.25}As_{0.06}Sb_{0.94}$ laser structure with a very smooth surface.

The data previously reported (Choi and Eglash 1991A) for lasers with $Al_{0.50}Ga_{0.50}As_{0.04}Sb_{0.96}$ cladding layers are also shown. For the same L, J_{th} is about 30 percent lower for the devices with the higher Al content. The reduction in J_{th} was achieved not only by improved optical confinement, but also by an improvement in lattice matching.

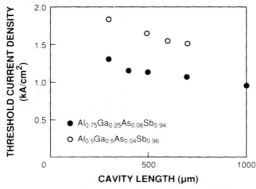

Fig. 6. Threshold current density vs cavity length for broad-stripe lasers with $Ga_{0.84}In_{0.16}As_{0.14}Sb_{0.86}$ active layers and either $Al_{0.50}Ga_{0.50}As_{0.04}Sb_{0.96}$ or $Al_{0.75}Ga_{0.25}As_{0.06}Sb_{0.94}$ cladding layers.

Figure 7 shows the cw output power vs current for a narrow-stripe device 300 μm long that was operated at several heatsink temperatures. Output powers as high as 10.5 and 4.6 mW/facet were obtained at heatsink temperatures of 5 and 20°C, respectively. Continuous operation was observed up to 30°C, where the maximum power was 0.9 mW/facet. The only previously reported cw power level for GaInAsSb/AlGaAsSb lasers is ~ 1 mW, which was obtained at room temperature for devices grown by LPE (Bochkarev *et al* 1988). The quasi-cw output of our narrow-stripe devices was measured by applying 100-μs pulses at 40 Hz. The peak output power at 24°C was 14 mW/facet.

4. CONCLUSION

Double-heterostructure $Ga_{0.84}In_{0.16}As_{0.14}Sb_{0.86}/Al_{0.75}Ga_{0.25}As_{0.06}Sb_{0.94}$ diode lasers emitting near 2.2 μm have been grown by MBE. Increasing the Al content of the cladding layers produced an improvement in the carrier and optical confinement. Narrow-stripe lasers have operated cw at heatsink temperatures up to 30°C. The maximum cw output power at 20°C was 4.6 mW/facet, and the pulsed threshold current density for broad-stripe devices was as low as 0.94 kA/cm². It should be possible to achieve still lower threshold current density with improvements in material quality, utilization of quantum-well and strained active layers, and optimization of device design.

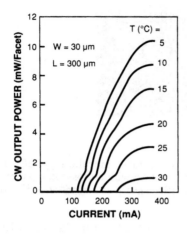

Fig. 7. CW output power vs current at several heatsink temperatures for $Ga_{0.84}In_{0.16}As_{0.14}Sb_{0.86}/Al_{0.75}Ga_{0.25}As_{0.06}Sb_{0.94}$ laser 30 μm wide by 300 μm long.

ACKNOWLEDGEMENT

We are grateful to D.R. Calawa, J.W. Chludzinski, D.F. Kolesar, W.L. McGilvary, J.V. Pantano, and G.D. Silva for expert technical assistance, to A.J. Strauss and G.W. Turner for many helpful discussions, and to I. Melngailis for his enthusiastic support. This work was sponsored by the Department of the Air Force.

NOTE

We have recently fabricated $InAs_{0.91}Sb_{0.09}/AlAs_{0.08}Sb_{0.92}$ double-heterostructure diode lasers emitting at 3.98 μm. These lasers exhibit cw operation up to 80 K and pulsed operation up to 155 K.

REFERENCES

Baranov A N, Danilova T N, Dzhurtanov B E, Imenkov A N, Konnikov S G, Litvak A M, Usmanskii V E, and Yakovlev Yu P 1988 *Sov. Tech. Phys. Lett.* **14** 727
Bochkarev A E, Dolginov L M, Drakin A E, Eliseev P G, and Sverdlov B N 1988 *Sov. J. Quantum Electron.* **18** 1362; paper PD-8 presented at *11th Int. Semiconductor Laser Conf.* Boston MA 1988
Choi H K and Eglash S J 1991A *IEEE J. Quantum Electron.* **27** 1555
Choi H K and Eglash S J 1991B *Appl. Phys. Lett.* **59**, 1165
DeWinter J C, Pollack M A, Srivastava A K, and Zyskind J L 1985 *J. Electron. Mater.* **14** 729
Eglash S J and Choi H K 1990 *Appl. Phys. Lett.* **57** 1292
Eglash S J, Choi H K, and Turner G W 1991 *J. Cryst. Growth* **111** 669
Furukawa A and Mizuta M 1988 *Electron. Lett.* **24** 1378
Motosugi G and Kagawa T 1980 *J. Cryst. Growth* **49** 102
Palfrey S L 1991 personal communication
Roth T J 1991 personal communication
Wolfe C M, Stillman G E, and Rossi J A 1972 *J. Electrochem. Soc.* **119** 250

Short wavelength operation of low threshold current AlGaInP strained quantum well laser diodes

I. Yoshida, T. Katsuyama, J. Shinkai, J. Hashimoto, and H. Hayashi

Optoelectronics Laboratories, Sumitomo Electric Industries, Ltd.

Taya-cho 1, Sakae-ku, Yokohama 244, Japan

ABSTRACT: Strained single quantum well, $(Al_{0.2}Ga_{0.8})_{0.43}In_{0.57}P$ ($\Delta a/a = 0.65\%$), was applied to visible laser diode to obtain short wavelength, low threshold devices. The laser was grown by low pressure OMVPE at 760°C on (100) GaAs substrate. High growth temperature (Tg) was used to grow high quality Al contained active layers. The high Tg also suppressed the formation of natural superlattice and kept the oscillation wavelength short without using off-angle substrates. The continuous wave (CW) oscillation wavelength was 637nm at 25°C for the index guided structure device (5x230um). The threshold current at the temperature was 52mA, which is the lowest value, to our knowledge, for the 630nm range lasers.

1. INTRODUCTION

Shortening the emission wavelength has been one of the main interests in the research of AlGaInP visible lasers [Valster et. al (1990), Kobayashi et al (1990), Hamada et al (1991), Ishikawa et al (1990), Kaneko et al (1990)]. However, several problems exist in obtaining short wavelength lasers of 630nm range or below.

(a) More carriers overflow from an active layer when the active layer bandgap is wider, due to the smaller bandgap difference between active and cladding materials [Itaya et al (1990)].

(b) The quality of an active layer material tends to degrade when the Al concentration is increased to obtain short wavelength [Kobayashi et al (1985), Yoshida et al (1990)].

(c) The use of off-angle substrate from (100), to suppress the formation of natural superlattice and to have the effective bandgap larger [Valster et al (1988)]], may cause the degradation of beam quality due to its asymmetricalness.

To solve these problems, we employed (a) compressively strained single quantum well (SSQW) structures, (b) high growth temperature, and (c) (100) oriented substrate. Compressively strained quantum well structures reduce the threshold current [Katsuyama et al (1990), Hashimoto et al (1991), Chang-Hasnain et al (1991)] and therefore is expected to reduce the carrier overflow. Although the use of tensilely strained GaInP well is also an attractive approach for short wavelength, low threshold lasers [Katsuyama et al (1991), Welch et al (1991)], the devices with tensile SSQW have operated only at pulse mode. In this study, we used a compressively strained AlGaInP well to obtain CW operation of lasers. High growth temperature improves the optical quality of AlGaInP active layer [Yoshida et al (1991), Domen et al (1991)], suppresses the formation of natural superlattice even on (100) GaAs substrate [Minagawa et al (1989), Yoshida et al (1991)], and suppresses also a development of hillocks [Ohba et al (1986), Bour and Shealy (1988)].

In the following, we report the growth temperature dependence of optical quality of undoped AlGaInP double hetero (DH) structures, an optical quality of undoped SSQW structures grown at a high temperature, and the device performance of SSQW lasers.

2. Growth and characterization of undoped structures

2.1 Double hetero-structures:

Epitaxial (epi) layers were grown on Si-doped ($n=2\times10^{18}cm^{-3}$) (100) GaAs substrate by low pressure (60Torr) organo-metallic vapor phase epitaxy. Background donor concentration of GaInP or AlGaInP grown in our reactor was high $10^{15}cm^{-3}$. Triethylgallium, trimethylaluminum, trimethylindium, and 20% phosphine diluted with H_2 were used to grow cladding and active layers. To see the effect of growth temperature, unstrained DH-structures, each consisting of an 900Å $(Al_{0.2}Ga_{0.8})_{0.52}In_{0.48}P$ active layer and two 4400Å $(Al_{0.7}Ga_{0.3})_{0.52}In_{0.48}P$ cladding layers, were grown at Tg=700-760°C. Similar structures but with $(Al_yGa_{1-y})_{0.52}In_{0.48}P$ active layers (y=0.0 - 0.3) were also grown at 700°C for comparison. Five to three ratio was 170 for active layers and 140 for cladding layers. PL was measured using an Ar+ ion laser (λ=5145Å) for the excitation at the power of 100mW. The spot diameter of the beam was approximately 100μm. The signal was detected by a photomultiplier placed after the monochromator. Wavelength dependence of the photomultiplier sensitivity was corrected after the measurements.

Photoluminescence spectra of the DH samples are shown in Figure 1. When the growth temperature is higher, the wavelength is shorter due to disordering of an active layer. The sample with y=0.2 and Tg=760 had a short wavelength, a narrow linewidth, and a large peak intensity. The sample with y=0.3 and Tg=700°C had about the same wavelength but the linewidth was wider and the peak intensity was smaller. The measurement of an excitation intensity dependence of PL intensity suggested the sample grown at 760°C had the least non-radiative recombination centers. [Yoshida et al (1991)]

2.2 Strained single quantum well structures:

Separate confinement hetero-structure (SCH) single quantum wells (SSQW) were grown at 760°C. The well composition was $(Al_xGa_{1-x})_{0.43}In_{0.57}P$ (x=0.2, 0.26, 0.3). The strain in these wells ($\Delta a/a = 0.65\%$) is the same as that of lasers with very low threshold current [Katsuyama et. al. 1990] or low threshold current density [Hashimoto et al (1991)]. Each well was sandwiched by two separate confinement layers of 800Å thick $(Al_{0.5}Ga_{0.5})_{0.52}In_{0.48}P$. A sample with $(Al_{0.1}Ga_{0.9})_{0.52}In_{0.48}P$ unstrained single quantum well (SQW) was also grown for comparison.

The PL spectra of SCH-(S)SQW structures are shown in Figure 2. Although, the PL intensity is decreased as the peak wavelength is shorter, it is independent of if the structure is strained or not. This indicate that the wells of SSQW structures are strained coherently without the generation of misfit dislocation which would significantly degrade the optical quality.

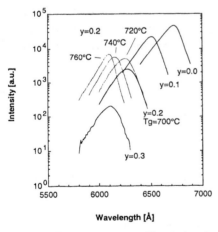

Figure 1 Room temperature PL spectra of undoped AlGaInP double hetero-structures

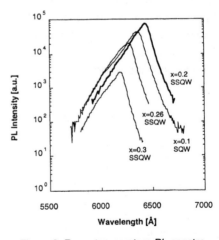

Figure 2 Room temperature PL spectra of undoped SSQW's

3. Fabrication and characterization of lasers

The SCH-SSQW laser was grown on a Si-GaAs (100) substrate at 760°C. Figure 3 shows the cross-section of the laser. It consists of a 0.23 μm Si doped

GaAs buffer layer, a 1.1 µm Se doped $(Al_{0.7}Ga_{0.3})_{0.52}In_{0.48}P$ cladding layer, a 100Å undoped $(Al_{0.2}Ga_{0.8})_{0.43}In_{0.57}P$ SSQW active layer ($\Delta a/a$=0.65%) sandwiched between 800Å $(Al_{0.5}Ga_{0.5})_{0.52}In_{0.48}P$ optical confinement layers, a 1.1 µm Zn doped $(Al_{0.7}Ga_{0.3})_{0.52}In_{0.48}P$ cladding layer, and 0.14 µm Zn doped $(Al_{0.1}Ga_{0.9})_{0.52}In_{0.48}P$ buffer layer. The P cladding layer was grown at lower temperature of 740°C to help the incorporation of Zn into the cladding layer and suppress the diffusion of Zn to the well. The SIMS analysis showed that little Zn was diffused to the well.

Index guided lasers with 5 µm stripe were fabricated by mesa etching and regrowth of a Si doped GaAs blocking layer ($n=1.5\times10^{18}cm^{-3}$), and a Zn-doped GaAs contact layer ($p=1.5\times10^{19}cm^{-3}$). The p-side and n-side electrodes were Ti/Pt/Au and AuGe/Ni/Au, respectively. Laser facets were made by cleaving and the devices were mounted on a copper heat sink with Au/Sn solder in the p-side down configuration.

Figure 4 shows the PL spectrum of the laser epi and the lasing spectrum of the laser device. PL of the epi was measured after removing the GaInP contact layer and some of the p-cladding layer. The sharp PL linewidth of 30meV suggests that the $(Al_{0.2}Ga_{0.8})_{0.43}In_{0.57}P$ SSQW layer is coherently strained without degradation. Continuous wave (CW) operation was obtained at 637nm. This is among the shortest wavelength reported for AlGaInP lasers continuously operating at room temperature. The emission wavelength is 13nm shorter than the peak wavelength of the PL spectrum. It is probably due to the band filling effects.

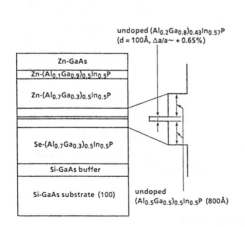

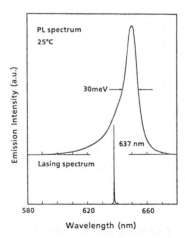

Figure 3 Cross-sectional view of the SSQW laser

Figure 4 Room temperature PL and the CW lasing spectrum of the SSQW laser at 25°C

Figure 5 shows the temperature dependence of current-output characteristics of SSQW laser (5x230µm). The threshold current at 25°C was 52mA, which is the lowest threshold current ever reported for CW operation of AlGaInP lasers with room temperature emission wavelength shorter than 640nm. Lasing was observed up to 42°C. The thermal resistance estimated by the pulse measurements was about 39°C/W. The slope efficiency was 0.24W/A and the characteristic temperature was 46K (15-30°C).

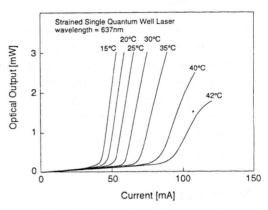

Figure 5 Temperature dependence of current-light output characteristics of the SSQW laser (5 x 230µm)

Similar laser but with $(Al_{0.26}Ga_{0.74})_{0.43}In_{0.57}P$ SSQW layer was also fabricated. Although, it did not oscillate in CW mode, it showed the pulse oscillation wavelength 621nm (Figure 6) with the threshold current below 200mA (180mA), which is also low for a device of this short wavelength.

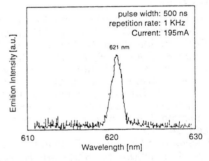

Figure 6 Pulse mode lasing spectrum of higher Al concentration (x=0.26) SSQW laser (5x270µm)

The advantage of SSQW structure lasers is the low threshold current. However, in SSQW structure lasers, band filling occurs easily due to its small volume of active region. When the band filling occurs, many carriers overflow from the well and the temperature dependence of the device becomes worse. In this study, we were successful in obtaining low threshold current devices, but not very successful in suppressing the carrier overflow because of band filling. One approach to solve this problems would be an introduction of strained multiple quantum well structures.

4. Conclusion

AlGaInP separate confinement hetero-structure lasers with AlGaInP compressive strained single quantum well have been fabricated. Successful room temperature CW operation at 637nm with a threshold current of 52mA was obtained. The combination of high temperature growth and the strained quantum structure is a promising approach in shortening the oscillation wavelength of AlGaInP lasers. Further improvement of the device performance at shorter wavelengths will be made by introducing strained multiple quantum well structures.

Acknowledgements

The authors would like to thank N. Ikoma for SIMS measurement. They also thank M. Koyama, K. Koe, and K. Yoshida for their continuous encouragement.

References

Bour D P and Shealy J R 1988 IEEE J. Quantum Elec 24 1856
Chang-Hasnain C J, Bhat R and Koza M A 1991 Confer. Lasers and Electro-Optics, Baltimore Maryland, Technical Digest Series 10 94
Domen K, Sugiura K, Anayama C, Kondo M, Sugawara M, Tanahashi T and Nakajima K 1991 Intern. Confer. Vapour Growth and Epitaxy, Nagoya, July 15pA09, also to be published in J Cryst. Growth
Hamada H, Shono M, Honda S, Hiroyama R, Matsukawa K, Yodoshi K and Yamaguchi T 1991 Elec Lett. 27 661
Hashimoto J, Katsuyama T, Shinkai J, Yoshida I and Hayashi H 1991 Appl. Phys. Lett. 58 879
Ishikawa M, Shiozawa H, Tsuburai Y and Uematsu Y 1990 Elec. Lett. 26 213
Itaya K, Ishikawa M, Shiozawa H, Nishikawa Y, Suzuki M, Sugawara H and Hatakoshi G (1990) Extended abs. 22nd Intern. Confer. Solid State Devices & Materials, Sendai, 565
Kaneko Y, Kikuchi A, Nomura I and Kishino K (1990) Elec. Lett. 26 658
Katsuyama T, Yoshida I, Shinkai J, Hashimoto J and Hayashi H 1990 17th Intern. Confer GaAs & Related Compounds, Jersey, Inst. Phys. Conf. Ser. No112 545
Katsuyama T, Yoshida I, Shinkai J, Hashimoto J and Hayashi H 1991 Confer. Lasers and Electro-Optics, Baltimore Maryland, Technical Digest Series 10 96
Kobayashi K, Ueno Y, Hotta H, Gomyo A, Tada K and Hara K (1990) Japan. J. Appl. Phys. 29 L1669
Kobayashi K, Hino I and Suzuki T, 1985 Appl. Phys. Lett. 46 7
Minagawa S and Kondow M 1989 Elec. Lett. 25 758
Ohba Y, Ishikawa M, Sugawara H, Yamamoto M and Nakanishi T 1986 J. Crystal Growth 77 374
Valster A, André J P, Dupont-Nivet E and Martin G M (1988) Elec Lett 24 326
Valster A, Liedenbaum C T H F, v.d.Heijden J M M, Finke M N, Severens A L G and Boermans M J B (1990) 12th IEEE Intern. Semiconductor Laser Confer. C-1
Yoshida I, Katsuyama T and Hayashi H 1990 17th Intern. Confer GaAs & Related Compounds Jersey Inst. Phys. Conf. Ser. No112 201
Yoshida I, Katsuyama T and Hayashi H 1991 Intern. Confer. Vapour Growth and Epitaxy NagoyaJuly 16aB05, also to be published in J Cryst. Growth
Welch D F, Wang T and Scifres D R 1991 Elec Lett. 27 694

Broad band side-emitting GaAs/AlGaAs/InGaAs single quantum well

LEDS

G. Vermeire, P. Demeester, K. Haelvoet, B. Van der Cruyssen, G. Coudenys and P. Van Daele
University of Gent - IMEC, Laboratory of Electromagnetism and Acoustics
Sint-Pietersnieuwstraat 41, B-9000 Gent, Belgium

ABSTRACT: Broad spectrum LED's have become very important for some specific applications like optical sensors and gyroscopes. In this paper we present the fabrication of such LED's using a special growth technique, called Shadow Masked Growth (SMG). This technique allows a spatial control and change in the thickness of quantum wells over the sample. Using this lateral variation of quantum well thickness, side-emitting single InGaAs quantum well GRINSCH-LED's were fabricated with a spectral width of already 60 nm. Higher spectral widths were obtained for multi section LED's.

1. INTRODUCTION

Recently side-emitting LED's with large spectral widths have become very usefull for some specific applications [Chen '90, Bradley '91]. Fiber optic sensor networks, based on Wavelength Division Multiplexing, are using broad spectrum LED's in combination with spectrographs to define the different windows in the spectrum corresponding to individual sensors. Other interesting applications are the fiber gyroscopes which, together with the ring-laser gyroscopes, tend to replace the mechanical gyroscopes. In this paper we present the fabrication of broad spectrum side-emitting LED's using a special growth technique called Shadow Masked Growth (SMG).

2. SHADOW MASKED GROWTH

The SMG technique is a non-planar growth technique which finds very promising and interesting applications in the field of monolithic integration of optoelectronic devices [Demeester '90, Demeester '91]. The principle of SMG is illustrated in figure 1. In a first growth run a thick (3 - 9 μm) AlGaAs spacer layer and a 1 μm GaAs mask layer are grown using an atmospheric pressure MOVPE system with a small horizontal reactor. The precursors for group III, V and doping elements are TMG, TMA, TMI, AsH$_3$, SiH$_4$ and DEZ. The growth temperature is 720 $^{\circ}$C for GaAs, 660 $^{\circ}$C for InGaAs and 760 $^{\circ}$C for AlGaAs with growth velocities in the range of 100 to 200 nm/min. In a second step the mask itself is defined using photolithography and consequent wet or dry etching of the mask layer, followed by wet selective etching of the spacer layer using KI:I$_2$. Using this process to define the crystalline mask has some major advantages such as a very accurate control on the thickness of the mask and spacer layers and also a well controlled growth behaviour on the mask itself. After growth of the device layers, the mask (together with the deposition on top of it) is lifted off by selective etching of the AlGaAs spacer layer using KI:I$_2$. The growth on the masked substrate is limited by the diffusion of group III species from the gas phase to the surface and this process can be split up into 3 different kinds of diffusion, as shown in figure 2. A typical profile of a layerstructure grown on a shadow masked substrate is given in figure 3.

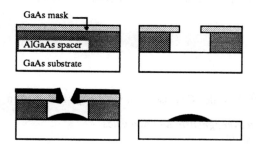

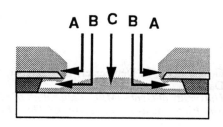

fig. 1 : Principle of Shadow Masked Growth

fig. 2 : Diffusion of group III species (A) to the mask edges, (B) under the mask, (C) to the substrate

Due to this diffusion limitation in the growth process a rather simple theoretical model could be developed using the Laplace equation and conformal transformation. In figure 4 the relative growth velocity in channels of a masked substrate is shown as a function of the channel width and for different spacer layer thicknesses. A very good agreement between experimental and theoretical results can be seen, confirming the model. Figure 4 also reveals the basic idea to use the SMG technique for the fabrication of broad spectrum LED's.

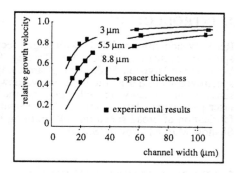

fig. 3 : Scanning Electron Micrograph of a cross-section of a GaAs/AlGaAs multilayer grown on a shadow masked substrate

fig. 4 : Comparison between experimental results and theoretical calculations on the influence of spacer thickness and channel width

Growing a Quantum Well (QW) on a masked substrate, the graph shows that a thickness variation of a factor 2.5 can be obtained when changing the channel widths, and this thickness variation will consequently also change the emission wavelength of the QW's. A change of 30 nm in emission wavelength between the different channels can be measured when growing GaAs/AlGaAs GRIN SQW LED's on a masked substrate with channel widths ranging from 10 to 100 μm [Demeester '91].

3. 2-SECTION SIDE-EMITTING LED's

Based on this reduction in growth velocity in masked channels, we have fabricated a 2-section side-emitting LED using the principle illustrated in figure 5. A schematic drawing of the topside of the masked sample before growth and of a cross-section after growth is given. The mask is defined in a 3 μm thick spacerlayer and the channels have different widths (5, 10, 15, 20, 50 & 100 μm). The QW on the planar part will be thicker and will emit light with a longer wavelength (lower energy) than the light coming from the QW's in the channels (the SMG part of the sample). Since we intend to fabricate side-emitting LED's it is important to notice that the QW's at the SMG part are transparant for the light generated at the planar part, but not vice versa. Using the strategy of figure 5, a double wavelength side-emitting LED can be fabricated where on one side light is emitted with a spectrum which is the superposition of the light emitted from the planar section and the SMG part.

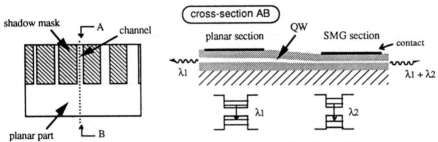

fig. 5 : principle of a 2-section broad spectrum side-emitting LED.

In our experiments a GRINSCH SQW LED structure was grown on a sample (topology as in figure 5) with a 10 nm (nominal thickness on the planar part) $In_{15}Ga_{85}As$ strained QW. A standard processing was carried out to define the side-emitting LED's with a cross section as shown in figure 5. Each LED has two separated contacts, one on the thick "planar" part and one on the thinner "SMG" part enabling us to inject different currents at the two sections of the LED. AuGe/Ni was used as back n-type contact and Zn/Au/Al as p-type top contact.

Injecting current at the SMG section only we could not detect any emission from the facet at the planar section, as expected. Measuring the spectrum at the other facet we see huge differences as we inject current in one or both sections of the LED's. Figure 6 shows spectra emitted by a LED with a 20 μm (figure 6a) and 10 μm (figure 6b) wide masked channel. The very flat and broad spectrum of figure 6a (FWHM = 76 nm) was obtained by applying a pulsed current of 250 mA to the planar section and 50 mA to the SMG section of the LED. This difference in current is caused by the coupling losses between the two sections and also by some propagation losses for light passing through the SMG section. The smaller channel width of figure 6b results in a higher reduction of the growth velocity and hence a larger spectral shift between the two sections. Consequently the spectrum of the LED is broader (FWHM = 103 nm) but it is not so flat anymore. In order to obtain a flat and broad spectrum at the same time, we changed the design to a multi-section LED.

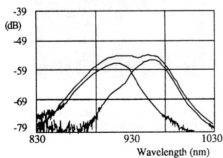

fig 6a: Spectra of a 2-section LED with a channel width of 20 mm. Individual spectra of the 2 sections are also given

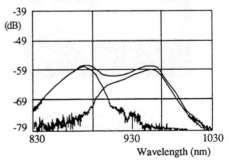

fig 6b: Spectra of a 2-section LED with a tchannel width of 10 mm. Individual spectra of the 2 sections are also given

4. MULTI-SECTION SIDE-EMITTING LED's

Taking into account the relation between the channel width and the emission wavelength of the QW one can understand that a gradual decrease in the width of a channel will result gradual decrease in the thickness of the QW and also a gradual decrease in the emission wavelength in the range of 860 to 800 nm in one single growth step. The principle is shown in figure 7 by a schematic topview and cross-section. It is clear that by defining several contacts on this structure, we can "compose" a broad spectrum by adjusting the different injection currents. Again the light generated at the different sections is only emitted at one facet of the LED. Also important to notice is the fact that for certain sections of the LED, the QW thickness will already change significantly under one single contact. As a result, the individual spectrum of these sections will already be very broad as will be shown later.

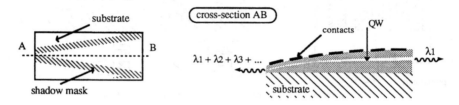

fig. 7 : Principle of a multi-section broad spectrum side-emitting LED.

First a 6 μm spacer layer and a 1 μm masking layer are grown. Then the mask is defined using photolithography and two etching steps resulting in rhombic shaped unmasked regions. A SEM picture of the masked substrate before growth is shown in figure 8a. The topangle and the height of the triangle are respectively 6° and 1 mm and the maximum width is 100 μm. The next step is the growth of a GRINSCH SQW LED with a GaAs QW with a thickness of 8.9 nm where the channel is 100 μm wide. The QW should be 4.7 nm thick at the top of the triangle, were the channel width is 20 μm. This reduction of the QW thickness should correspond with a wavelength shift of approximately 25 nm. Figure 8b shows a SEM picture of the overgrown masked substrate.

fig 8a: Scanning Electron Micrograph of the masked substrate

fig 8b: Scanning Electron Micrograph of the masked substrate after overgrowth

After removing the shadow mask we etch a 13 µm wide mesa ridge through the top GaAs layer as shown in figure 9. For each triangular LED 7 (100 µm x 7 µm) electrically isolated Zn/Au contacts are defined and after deposition and dry etching of a SiN$_x$ insulation layer, extra thicker and larger (100 µm x 100 µm) TiW/Au contacts are deposited to ensure easy contacting. Finally the LED's are polished and AuGe/Ni back contacts deposited. Before measuring, the samples are cleaved at different places to enable detection of light emitted from the smallest part of the LED.

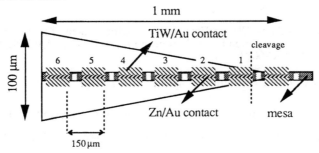

fig. 9 : final configuration of the multi-section side-emitting LED's

When we inject curent at contact 1, we already see a very broad spectrum of 88 nm as is shown in figure 10a. This is caused by the strong QW thickness reduction under the first contact and was expected because (figure 4) the relative growth velocity reduction becomes larger as the channel width decreases. In figure 10b we see the spectra obtained by contacting 1, 3 & 6 separately and in parallel. We get a narrower spectrum because only a very small current had to be applied to contact 1 to be still able to adjust the current at contact 6 to get the same optical power at the different emission wavelengths. This means that there are rather large propagation losses through the multi-section LED. It is also interesting to notice that the high energy side of the spectra falls very quikly off due to the absorption of the thinner QW's.

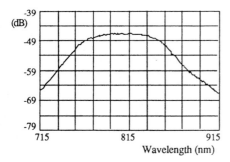

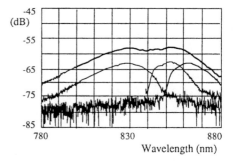

fig 10a: Spectrum obtained from the multi-section LED when using contact 1

fig 10b: Spectra obtained from the multi-section LED when using contact 1, 3, and 6 individually and in parallel

5. CONCLUSION.

The SMG technique is shown to be very promising for the fabrication of broad spectrum side-emitting LED's. Both the 2- and multi-section LED's exhibit broad and flat spectra while only one QW was used in the LED structure. The technique still leaves a lot of freedom to explore other methods for obtaining even broader spectra. It should be very interesting to combine the SMG technique with an AlGaAs multiple quantum well structure, where each QW has the same thickness but a different aluminum content [2]. The fact that the SMG technique is very easy to combine with other techniques, makes it a very powerfull technique for also a lot of other applications.

ACKNOWLEDGEMENTS

The authors wish to thank Claude Eeckhout and Dirk Lootens for the processing of the LED's. This work was supported by the European Project RACE 1069 EPLOT and the USA army project DAJA-90-C-0003.

REFERENCES

Bradley R. R., P. Bromley, J. P. Hall, A.J. Mosely; presented at "the IV European Workshop on MOVPE", may 1991

Chen T.R, L. Eng, Y.H. Zhuang, A. Yariv, N.S. Kwong, P.C. Chen; Appl. Phys. Lett. 56 (14), 2 april 1990,
   pp. 1345-1346

Demeester P., L. Buydens, P. Van Daele; Appl. Phys. Lett. 57 (2), 9 july 1990, pp. 168-170

Demeester P., L. Buydens, I. Moerman, D. Lootens, P. Van Daele; Journal of crystal growth 107 (1991); pp. 161-165

Inst. Phys. Conf. Ser. No 120: Chapter 10
Paper presented at Int. Symp. GaAs and Related Compounds, Seattle, 1991

505

Fabrication of circular gratings by FIB damage on GaAs

M. Fallahi , I.M. Templeton , R. Normandin

National Research Council & Solid-State Opto-Electronics Consortium
Ottawa, Ontario, K1A 0R6, Canada

Introduction:

In the past few years, there has been a growing interest in the study of circular grating resonators [1-3]. Circular gratings have potential applications in surface-emitting distributed feedback (S-E DFB) lasers and in multi-port couplers. In addition to the advantages of surface-emitting lasers such as 1) no cleavage process, 2) possibility of two-dimentional laser arrays, and 3) single mode operation, circular S-E DFB lasers can also provide low-divergence circular laser beams for efficient coupling into fibers. Despite these promising applications, little work has been done on their fabrication. Recently, we demonstrated the fabrication of circular gratings on InP by E-beam lithography and reactive ion etching [4]. In this paper, we demonstrate a simple technique for the fabrication of circular gratings using direct focused ion beam (FIB) damaging of GaAs and selective etching. This simple technique allows not only the fabrication of high quality circular gratings, but also good control over the tooth height and pitch which is essential for good optical performance of the grating.

FIB System:

The ion beam system is a JEOL JIBL-104 UHV FIB instrument. The accelerating voltage can be varied from 20 to 100 kV. An ion beam current of 2 to 600 pA can be obtained with this machine which corresponds to a beam diameter of 50 to 500 nm. The system is fully computer-controlled, allowing automatic stage motion and auto-acquisition of alignment marks. Two types of ion , Ga^+ from a Ga source and Au^+ from a Au-Si-Be source are used for the damage formation. The patterns are described in a simple (JEOL proprietary) format, JEOL01. Rings are assembled from arcs (maximum angle 45°) whose center, inner and outer radii, and start and finish angles must be defined. Within the conversion procedure, the rings are broken into rectangular and trapezoidal elements. Once converted, the pattern file may be used as often as needed, and the write time is only a few minutes for a 250 µm diameter grating.

Experiments:

In the fabrication of circular gratings, a GaAs substrate was exposed with a focused Au^+ ion beam. In this case an accelerating voltage of 100 kV and a beam current of 16 pA were used. 0.4 µm period circular gratings with 350 rings, each with 0.1 µm width / 0.3 µm spaces, were first designed as described above. In order to determine the influence of the exposure dose on the grating profile, the doses were varied from 2.5×10^{13} to 5×10^{13} ion/cm^2. After exposure, the exposed areas become amorphous and can be selectively etched in hot HCl. Figure 1 shows a scanning electron microscope (SEM) photograph of 0.4 µm period circular gratings at different doses.

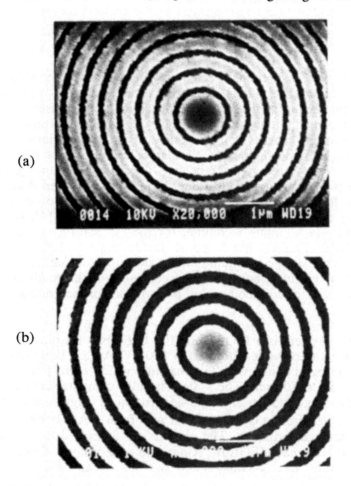

(a)

(b)

Fig. 1: SEM photographs of 0.4 µm period circular gratings in GaAs for two different exposure doses using Au^+ ions , (a) 2.5×10^{13} ion/cm^2, (b) 5×10^{13} ion/cm^2.

As shown in fig.1, width/space ratios varying from 1/4 for 2.5×10^{13} ion/cm^2 to 1/1 for 5×10^{13} ion/cm^2 were obtained. Very good definition of the gratings was successfully achieved in these cases. A combination 0.2 μm-0.4 μm period circular grating was also designed and fabricated. In this design, 100 rings of 0.4 μm are surrounded by 500 rings of 0.2 μm period. A width/space ratio of 1/1 was used for both periods. Figure 2 shows a SEM photograph of this combined structure at 2.5×10^{13} ion/cm^2. The grating depth in this case is in the order of 50 nm.

Fig. 2 : SEM photograph of a combined 0.2-0.4 μm period
circular grating in GaAs.

Since the optical response of gratings is strongly related to the depth of the gratings, it is important to fabricate deeper gratings. In order to increase the depth of the gratings to around 100 nm, we have also used a Ga source. In this experiment, an optical waveguide was grown by conventional MBE techniques on a GaAs substrate. The buffer layer consisted of 1.6 μm $Al_{0.3}Ga_{0.7}As$ and the waveguiding layer of 0.6 μm GaAs. The guide supported single mode TE and TM propagation at a wavelength of 1.06 μm

Propagation losses due to evanescent mode coupling to the substrate and scattering losses were negligible in our geometry. A second order circular grating with a period of 0.311 μm was then defined in the GaAs layer with Ga ions. The exposure doses were varied from 5×10^{13} to 1×10^{14} ion/cm^2. An exposure dose higher than 5×10^{13} ion/cm^2 was required to define the gratings. Figure 3 shows a SEM photograph of a grating exposed at 9×10^{13} ion/cm^2 and selectively etched. The total diameter of the grating is 260 μm.

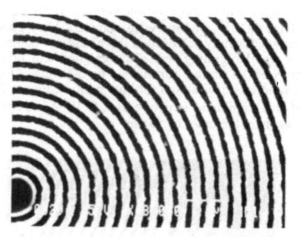

Fig. 3 : SEM photograph of a 0.311 µm period grating in
GaAs waveguide using Ga$^+$ ions at 9x10^{13} ion/cm^2.

The optical characteristics of the grating were then evaluated with a
1.06 µm YAG laser. The reflected light from the grating was observed with
an I.R. camera. Typical near field measurements are shown in Figure 4 for
the second order grating.

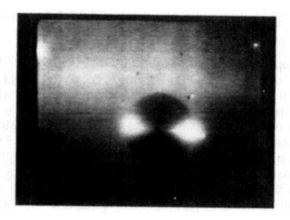

Fig. 4 : Near field scattered light at λ = 1.06 µm from a second order
circular grating in GaAs/AlGaAs waveguide.

The results are in good agreement with ref. [5] showing 90° scattering
interference. Indeed by translating the grating pattern in and out of the
collimated guided beam the interference condition could be varied in
addition to the built-in 3.75 waveshift at the center. This varied the relative

phase between the incoming and Bragg reflected components from the output side. The difference in scattering between the two components from each side gave constructive and destructive (in the center) interference conditions, indicative of the good optical performance of these gratings. A far field beam profile proved directional and exhibited a diffraction angular pattern consistent with a plane wave of the grating dimensions. Little light could be observed unless the camera was precisely positioned on axis. Thus little parasitic incoherent scattering, either from the grating structure or the waveguide mode propagation, could be observed.

Conclusion:

We have demonstrated the fabrication of circular gratings by FIB damaging and selective etching of GaAs . Two types of ion , Au and Ga, were used giving 50 and 100 nm deep gratings respectively. The variation of grating profile as a function of the exposure dose was also investigated. The optical characteristics of second order circular gratings on a GaAs/AlGaAs waveguide show good performance suitable for surface-emitting lasers.

Acknowledgements:

The authors would like to thank G. Champion for his technical assistance.

[1]. R.M. Schimpe, U.S. patent 4 743 083 , 1989.
[2]. X. Zheng , Electron. Letts. , vol. 25 , No 19, p. 1311, 1989.
[3]. T. Erdogan and D.G. Hall , J. Appl. Phys. , vol. 68 , No. 4 , p. 1435, 1990.
[4]. M. Fallahi, C. Wu, C. Maritan, K. Fox and C. Blaauw , Presented at LEOS 1991 Summer Topical Meetings: Microfabrication for Photonics and Optoelectronics, Newport Beach, California, July 31-Aug 2, 1991.
[5]. C.H. Henry, R.F. Kazarinov, R.A. Logan and R. Yen, IEEE J. Quant. Electron, QE-21, No 2, p. 151, 1985.

A new application for III–V quantum well systems—efficiency enhancement in solar cells

K.W.J.Barnham, J.Nelson, M.Paxman.
Physics Department, Imperial College of Science Technology and Medicine, London SW7 2BZ, U.K.
R.Murray.
Interdisciplinary Research Centre in Semiconductor Growth and Characterisation, ICSTM, London SW7 2BZ, U.K.
C.T.Foxon.
Philips Research Laboratories,Redhill, RH1 5HA, U.K.

1. Introduction

Given current environmental concerns it is important to consider afresh the fundamental limitations on solar cell efficiency. A conventional solar cell with a single, fixed band-gap E_g has a maximum power conversion efficiency which depends on E_g. GaAs solar cells are close to the optimum and have an upper efficiency limit of around 30% (Henry 1980) when operating in unconcentrated sunlight. Higher efficiencies are possible in principle if cells with different band-gaps are stacked in series but there are technological difficulties in connecting the cells electrically and optically. A new approach has recently been suggested (Barnham and Duggan 1990) in which Multi-Quantum-Well (MQW) or Super-Lattice (SL) systems are used as the active absorbers in the intrinsic region of a p-i-n solar cell. This is the first time that the wells themselves have been considered as active absorbers for solar cells.

We have recently reported a test of this new approach by studying $Al_{0.3}Ga_{0.7}As$ p-i-n photodiodes with GaAs MQWs in the intrinsic region (Barnham et al. 1991). These were compared with devices which were identical in all respects apart from having no wells in the AlGaAs i-region. In samples with low charged impurity levels in the undoped region it was observed that the power conversion efficiency of the MQW sample in a white light source was more than twice that of the control.

These first test samples were not solar cells but were based on successful MQW photodiodes (Whitehead et al 1988). In this paper we discuss the principles of the quantum well solar cell and ways in which the design of these test devices can be improved to increase the overall power conversion efficiency while preserving the enhancement due to the quantum wells. We report for the first time recent results obtained with new samples in which changes of geometry have doubled the efficiency in comparison with our published test samples.

2. Principle of the Quantum Well Solar Cell

The fundamental efficiency limit of a conventional solar cell can be appreciated from Fig. 1a. If light with energy $E\gamma$ greater than the band-gap E_g is absorbed within a diffusion length of the p-n junction the carriers are separated by the built-in field, producing a current. The smaller the band-gap the larger the absorption and the bigger the output current. However, at room temperature, carriers rapidly thermalise to the band edge as a result of phonon interactions. The higher the band-gap the less this thermalisation loss and the higher the output voltage V. In a conventional solar cell the power conversion efficiency is

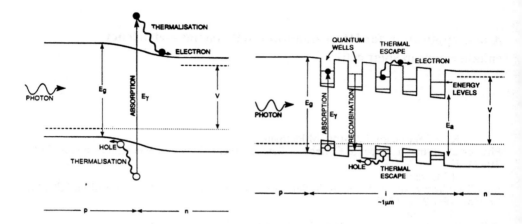

Fig. 1a) Conventional solar cell   b) Quantum well solar cell

therefore a compromise between the greater absorption available with a narrow band-gap and the greater output voltage achieved with a wide band-gap.

The quantum well solar cell in Fig. 1b behaves like a conventional p-i-n solar cell for photons with energy greater than E_g. However photons with energy below E_g and greater than E_a may be absorbed in the wells. At room temperature many of the resulting carriers escape from the wells and are separated by the built-in field. Hence the resulting output current is determined by the effective band-gap for absorption E_a and the output voltage depends primarily on the wider band-gap of the barrier region E_g leading to higher maximum efficiencies (Barnham and Duggan 1990).

The carriers photogenerated in the wells either recombine (radiatively or non-radiatively) or escape from the well by thermal emission, tunnelling or a combination of the two. In good quality, undoped, quantum well material the radiative recombination lifetime is at least 10 ns at room temperature (Dawson et al 1985) and the non-radiative lifetime is even longer. The escape time is longest in the flat-band limit when carriers must be thermally activated to the barrier band edge to escape. They will be separated, forming a current, if there is a residual built-in field present. Our calculations show that in typical Al$_3$Ga$_7$Al/GaAs wells this upper limit to the carrier escape time (for the electron) is approximately 0.2 ns in agreement with other authors (Schneider and von Klitzing 1988).

Hence, if the material quality is good, the carrier escape lifetime is much shorter than the recombination lifetime and the majority of the carriers will escape from the well before they recombine. In contrast to the carriers in the conventional solar cell in Fig. 1a the carriers from the wells in Fig. 1b are thermalising up to the band-gap. If the charged impurity level is low the residual field can separate these carriers forming a current which is additional to that from absorption above the barrier.

3. Optimisation of Structure Geometries

When first testing the principle of the quantum well solar cell we studied test devices based on successful MQW photodiode designs. At their reverse bias operating points such devices are known (Whitehead et al 1988) to have high internal quantum efficiencies for carriers photoexcited in the wells. Our study (Barnham et al 1991) demonstrated that, provided the intrinsic region charged impurity level (N_i) is sufficiently low that the built-in field is maintained across the i-region, high quantum efficiency can also be achieved in forward bias.

In order to maximise the efficiency enhancement available in a quantum well solar cell the additional current must be increased with least cost to the output voltage. The open circuit voltages of our 30 well test devices (Barnham et al 1991) were only slightly below those of the control samples without wells.

At photon energies below the barrier band-gap E_g higher absorption is clearly achieved by increasing the number of wells and hence the width of the i-region. To increase the absorption of above band-gap photons in the field bearing region the top surface (p-region) thickness should be reduced.

The new samples to be reported here had 65Å thick $Al_{.33}Ga_{.67}As$ barriers and 85Å GaAs quantum wells. Two samples have 50 wells, one has 30 wells and one (the control for the 30 well sample) has the same i-region thickness as the 30 well sample but no wells.

All samples have a $Al_{.33}Ga_{.67}As$ p-region (beryllium doped at 1.3×10^{18} cm^{-3}) of thickness 0.15 mm which is half that of the devices studied previously (Barnham et al. 1991). There are undoped $Al_{.33}Ga_{.67}As$ spacers 300Å wide on each side of the MQW structures. The $Al_{.33}Ga_{.67}As$ n-region (silicon doped at 1.3×10^{18} cm^{-3}) was grown on top of a n-type GaAs buffer and n^+ GaAs substrate.

The samples were grown by Molecular Beam Epitaxy at the Philips Research Laboratories, Redhill. Prior to the growth of the structures the system clean-up had been monitored by studying the minority carrier lifetime at room temperature of standard beryllium doped (2×10^{16} cm^{-3}) MQWs (60 periods of 55Å wells). It is known that for high quality low-threshold current lasers a minimum lifetime of 10 ns is required in such structures. Using As_4 this can only be achieved at rather high substrate temperatures (680°C) where the morphology of bulk AlGaAs is good. It has recently been established, however, that, using As_2, high quality material can be grown over a wide temperature range (Foxon et al 1991). This enables a comparatively low substrate temperature to be used (630°C) which prevents diffusion and segregation of impurities from the n and p regions into the i-region of the MQW structure. For the growth temperature chosen, the rate of change of doping level with distance is approximately 1 decade per 100Å for both silicon and beryllium dopants (Harris et al 1991). It should also be noted that extremely low background doping levels have been achieved using this particular As_2 source (Kean et al 1991).

Details of the processing into mesa diode structures and a description of the measurement procedures are given in the papers of Barnham et al. (1991) and Whitehead et al (1988). A low temperature study of the widths of excitonic features and their field variation suggests that $N_i \sim 10^{14}$ cm^{-3}.

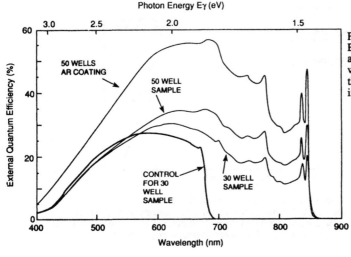

Fig. 2
External quantum efficiency as a function of photon wavelength at zero bias for the four samples described in the text.

The external quantum efficiency (QE) of these devices at zero applied bias is shown in Fig. 2. Comparison of the 30 well device with its control shows the large, extra contribution to short circuit current resulting from absorption below barrier band-gap in the quantum wells. Also it can be seen that the 30 quantum wells do not seriously degrade the QE for photons well above the barrier band-gap as the two curves are similar for wavelengths below 600 nm (energies greater than 2 eV).

Comparison of the 50 well and 30 well sample in Fig 2. clearly demonstrates the advantage of the wider i-region and increased number of quantum wells. This 50 well sample, the 30 well sample and the control all have a 200Å GaAs cap on the top surface. Results are also shown in Fig.2 for a similar 50 well sample with the cap removed and replaced by an anti-reflection coating. This demonstrates that much of the QE loss at short wavelengths (energies well above barrier band-gap) in the first three samples is due to absorption in the GaAs cap. A quantitative analysis of the effect of the reduced p-region thickness on above-barrier QE is in progress.

A high efficiency solar cell needs the QE enhancement to be maintained into forward bias. In Fig. 3a we compare the QE of the 50 well sample for absorption above the barrier band-gap (600 nm) and on the n=1 continuum in the wells (800nm). It should be noted that for the same sample the QE has a "cut-off" at a similar voltage for carriers excited in the wells and above the barrier.

Fig. 3b shows the QE for the 30 well sample and its control as a function of applied bias for an absorption wavelength of 600 nm near the peak in spectral response. The 30 quantum well sample has a forward bias "cut-off" at a similar voltage to that of the control. This confirms that for photons with energy greater than the barrier band-gap, where the MQW devices are acting as conventional solar cells, the quantum wells do not affect the cell

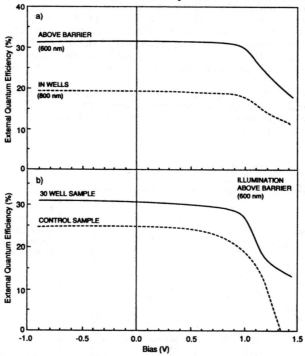

Fig.3 External quantum efficiency as a function of bias for:
a) the 50 well sample (no A-R coating) for photons with energy above barrier band-gap (600 nm) and for absorption on the n=1 continuum in the quantum wells,
b) the 30 well sample and its control for photons with energy above bulk band-gap (600 nm).

performance significantly. Comparison of the 600 nm results for the 50 well sample in Fig. 3a and the 30 well sample in Fig. 3b show that the widening of the i-region does not significantly degrade the QE in forward bias.

The open-circuit voltage and power conversion efficiency of these devices depend on a number of factors, in addition to the QE variation with bias, such as the dark-current and illumination level. Full details will be presented elsewhere but preliminary tests suggest that the efficiency of these new devices is double that of our first samples. The efficiency of the structure with anti-reflection coating would be around 10% in unconcentrated sunlight.

4.Conclusions

The results presented here on our new samples support the view expressed in our first study (Barnham et al 1991) that it should be possible to carry over the efficiency enhancement observed in our test devices to practical solar cells . With good material the change from 30 to 50 wells enhances the absorption but does not appear to degrade the voltage performance or the QE for photons with energy above barrier band-gap. We also believe that more wells can be added and further p-region thickness optimisations are possible. Furthermore, the addition of a high-band-gap top-surface window should improve the short wavelength QE.

Other III-V systems, such as GaAs barriers with strained InGaAs wells and InP barriers with lattice matched InGaAs wells, offer the prospect of higher efficiencies than the AlGaAs/GaAs system. It should also be noted that commercial solar cells are already produced by MOVPE and recent production with MBE has also been reported (Melloch et al 1991). It should therefore be possible to incorporate quantum wells with little or no cost increase.

We estimate that the first optimisation of our device structures reported here has more than doubled the overall efficiency in comparison with the first test results. There is clearly potential for further improvements of a similar magnitude. We believe that ultimately efficiencies well above 30% should be possible by this method.

5. Acknowledgements

We are most grateful to Geoff Hill and Malcolm Pate for processing these samples at the SERC III-V Facility, Department of Electrical Engineering at the University of Sheffield. We wish to acknowledge the financial support of the Greenpeace Trust and the SERC LDSD programme.

References

K.W.J.Barnham and G.Duggan, J.Appl.Phys. 67, 3490 (1990).
K.W.J.Barnham, B.Braun, J.Nelson, M.Paxman, C.Button, J. S.Roberts, C.T.Foxon, Appl. Phys. Lett. 59, 135 (1991).
P.Dawson, G.Duggan, H.I.Ralph and K.Woodbridge, Proc. 17th International Conference on the Physics of Semiconductors (Springer-Verlag, 1985) p.551.
C.T.Foxon, P.Blood, E.D.Fletcher, D.Hilton, P.J.Hulyer and M.Vening, J.Crystal Growth, 111, 1047 (1991).
J.J.Harris, J.B.Clegg, R.B.Beall, J.Castagne, K.Woodbridge and C.Roberts, J.Crystal Growth, 111, 239 (1991).
C.H.Henry, J.Appl.Phys. 51, 4494 (1980).
A.H.Kean, C.R.Stanley, M.C.Holland, J.L.Martin and J.N..Chapman, J.Crystal Growth, 111, 189 (1991).
M.R.Melloch, E.S.Harmon and K.A.Emery, IEEE Electron Device Lett. 12, 137 (1991).
H.Schneider and K. v. Klitzing, Phys. Rev. B38, 6160 (1988).
M.Whitehead, P.Stevens, A.Rivers, G.Parry, J.S.Roberts, P.Mistry, M.Pate, G.Hill, Appl. Phys. Lett. 53, 956 (1988).

Inst. Phys. Conf. Ser. No 120: Chapter 10
Paper presented at Int. Symp. GaAs and Related Compounds, Seattle, 1991

517

Theoretical studies of impact ionisation in pseudomorphic structures of InGaAlAs on GaAs and InP substrates

Yeesoo Keom, Vasu Sankaran and Jasprit Singh
Center for High Frequency Microelectronics
Department of Electrical Engineering
and Computer Science, 2234 EECS Bldg.,
University of Michigan, Ann Arbor MI 48109–2122.

Pseudomorphic semiconductor structures have been exploited for high speed electronic devices and low threshold lasers. In this paper the potential of utilizing strain for suppressing impact ionization is evaluated. It is found that if compressive strain is introduced <u>without</u> altering the bandgap (e.g. by using properly tailored InGaAlAs alloys) the threshold energy for electron impact ionization is increased for both GaAs based and InP based materials. This is expected to relax design constraints for high output power devices.

High speed electronic devices are currently being designed using pseudomorphic structures. For example, n-type In_xGa_{1-x}/AlGaAs (grown on GaAs) and $In_{0.53+x}Ga_{0.47-x}As$/$In_{0.52}Al_{0.48}As$ (grown on InP) MODFET structures are now routinely being used for high frequency applications. While the speed related improvements are well documented, strain has not been utilized to alter and control breakdown performance so far. Breakdown related to impact ionization is important in a variety of device applications and is usually tailorable only by altering bandgap. However, the large changes in valence bandstructure due to strain[1–4] suggests that electron impact ionization could be tailored by strain. This is because the severe constraints due to simultaneous satisfaction of energy and momentum conservation for the scattering particles are very much dependent on the bandstructure.

Two issues are examined in this paper: i) how are the impact ionization rates altered if excess In is added to $In_{0.53}Ga_{0.47}As$ which is initially lattice matched to $In_{0.52}Al_{0.48}As$; ii) how are the impact ionization rates affected if in addition to excess In we also add excess Al to maintain a <u>constant bandgap.</u> In the second case, the bandgap is maintained constant while the compressive strain is introduced by excess In. The motivation for examining the second possibility will become clear after we present our results. To study the effect of strain on impact ionization, the well known effects of strain on bandstructure are briefly reviewed[5]. In pseudomorphic growth, the in-plane lattice constant of the mismatched ovelayer is forced to match that of the substrate, while the lattice constant in the growth direction is altered by the Poisson effect. The conduction band is not significantly affected by strain since the symmetry of the conduction band states is primarily s-type for direct bandgap materials. The valence band is dramatically affected because the symmetry of the valence band is mainly p-type, and biaxial strain breaks the cubic symmetry. The compressive in-plane strain produced by addition of excess In in the channel reduces the valence band density of states mass considerably. These results have been reported in considerable detail both experimentally and theoretically for cases near the bandedge[6,7]. It is important to note that the reduction in the hole mass is dependent primarily upon the strain and <u>not</u> on the chemical compositon of the material.

The electron impact ionization rates are addressed since they are calculated to be affected much more due to bandstructure modifications. This process arises due to the electron-electron Coulomb interaction that acts as the perturbation driving transitions between the eigenstates of the one-electron Hamiltonian of the system. The transition rates may be obtained using first-order perturbation theory. A typical term in the matrix element for the process is:

$$M = \int d\vec{r}_1 \int d\vec{r}_2 \; \psi^\star_{n'_1}(\vec{k}'_1, \vec{r}_1)\, \psi^\star_{n'_2}(\vec{k}'_2, \vec{r}_2)\, V(\vec{r}_1, \vec{r}_2)\, \psi_{n_1}(\vec{k}_1, \vec{r}_1)\, \psi_{n_2}(\vec{k}_2, \vec{r}_2) \qquad (1)$$

where $\psi_n(\vec{k}, \vec{r})$ is the unperturbed eigenfunction belonging to band n and character-ized by two-dimensional wavevector $\vec{k}$. Choosing a representation for the heterostruc-ture stationary states, it can be shown that the threshold for impact ionization is determined by the conservation of energy, and in-plane momentum. The dispersion of the subbands in the quantum well is thus required to determine the threshold.

The valence bandstructure is calculated using a 6-band k.p Hamiltonian[8]:

$$H_{kp}(\mathbf{k}_\parallel, k_z) = \begin{bmatrix} H' & 0 \\ 0 & H' \end{bmatrix} \qquad (2)$$

$$H' = \begin{bmatrix} Lk_x^2 + M(k_y^2 + k_z^2) & Nk_xk_y & Nk_zk_x \\ Nk_xk_y & Lk_y^2 + M(k_z^2 + k_x^2) & Nk_yk_z \\ Nk_zk_x & Nk_yk_z & Lk_z^2 + M(k_x^2 + k_y^2) \end{bmatrix} \qquad (3)$$

where L, M and N are the valence band parameters. Spin-orbit interaction is included as a perturbation:

$$H_{so} = \Delta_0/3 \begin{bmatrix} 0 & -i & 0 & 0 & 0 & 1 \\ i & 0 & 0 & 0 & 0 & -i \\ 0 & 0 & 0 & -1 & i & 0 \\ 0 & 0 & -1 & 0 & i & 0 \\ 0 & 0 & -i & -i & 0 & 0 \\ 1 & i & 0 & 0 & 0 & 0 \end{bmatrix} \qquad (4)$$

The quantum well eigenstates are given in the envelope function approximation by:

$$\psi(\mathbf{k}_\parallel, \mathbf{r}) = \frac{e^{i\mathbf{k}_\parallel \cdot \mathbf{r}_\parallel}}{2\pi} \sum_\nu f_\nu(\mathbf{k}_\parallel, z)\; u_{\nu 0}(\mathbf{r}) \qquad (5)$$

The envelope functions $f_\nu(\mathbf{k}_{\parallel}, z)$ are obtained by the transformation $k_z \to -i\frac{d}{dz}$ in the k.p Hamiltonian and solving the resulting set of coupled differential equations numerically by finite difference. The effect of strain is incorporated using deformation potential theory, yielding a perturbation to the k.p Hamiltonian:

$$H_\epsilon = \begin{bmatrix} H'' & 0 \\ 0 & H'' \end{bmatrix} \tag{6}$$

$$H'' = \begin{bmatrix} l\epsilon_{xx} + m(\epsilon_{yy} + \epsilon_{zz}) & n\epsilon_{xy} & n\epsilon_{zz} \\ n\epsilon_{xy} & l\epsilon_{yy} + m(\epsilon_{zz} + \epsilon_{xx}) & n\epsilon_{yz} \\ n\epsilon_{zx} & n\epsilon_{yz} & l\epsilon_{zz} + m(\epsilon_{xx} + \epsilon_{yy}) \end{bmatrix} \tag{7}$$

where l,m,n are the independent valence band deformation potential.

While there has been considerable work reported in literature showing the effect of strain on the near bandedge effective mass, there have been no reports of how strain affects the bandstucture away from the bandedge. Since the impact ionisation process requires the hole to have a fairly large momentum to balance the high-energy incident electron's momentum, one has to examine the subband dispersion 100 to 200 meV away from the bandedge. The effect of strain is quite significant even $\sim$200 meV away from the bandedge. The dispersion for the first heavy hole subband is shown in Figures 1 and 2

Electron states in the conduction band are obtained using the tight-binding description. Impact ionization requires a high energy initial electron, for which tight-binding is a more accurate description than k.p formalism. Stationary states (subbands) of the heterostructure are obtained by solving the Schroedinger equation:

$$H\psi(\mathbf{r}) = E\psi(\mathbf{r}) \tag{8}$$

with the planar orbital form of the tight-binding representation:

$$\psi(\mathbf{r}) = \sum_m \sum_\nu c_\nu(m) \sum_n e^{i\mathbf{k}_{\parallel}\cdot\mathbf{R}_n} \xi_\nu(\mathbf{r} - [\mathbf{R}_n + \mathbf{r}_m]) \tag{9}$$

where m is an index over lattice planes, $\mathbf{r}_m$ an origin in each plane chosen to locate any one ion in the plane, and n an index over *in-plane* lattice translations $\mathbf{R}_n$. The Schroedinger equation is converted to a matrix equation for coefficients $c_\nu(m)$ by taking scalar product of either side successively with distinct orbitals on each plane. For the right hand side, we have

$$\int d\mathbf{r}\, \xi_\mu^*(\mathbf{r} - [\mathbf{R}_k + \mathbf{r}_p])\, \psi(\mathbf{r}) = c_\mu(p)\, e^{i\mathbf{k}_{\parallel}\cdot\mathbf{R}_k} \tag{10}$$

where $\mathbf{R}_k$ is an in-plane lattice translation in plane p. On the left hand side,

$$\int d\mathbf{r}\, \xi_\mu^*(\mathbf{r} - [\mathbf{R}_k + \mathbf{r}_p])\, H\, \psi(\mathbf{r}, 0) = \sum_m \sum_\nu c_\nu(m, 0) \sum_n e^{i\mathbf{k}_{\parallel}\cdot\mathbf{R}_n} \tag{11}$$

$$\times \int d\mathbf{r}\, \xi_\mu^*(\mathbf{r} - [\mathbf{R}_k + \mathbf{r}_p])\, H\, \xi_\nu(\mathbf{r} - [\mathbf{R}_n + \mathbf{r}_m])$$

The summation on the right is over those neighbouring orbitals with which the orbital centered at $[\mathbf{R}_k + \mathbf{r}_p]$ has nonzero matrix elements. The latter are the known tight-binding parameters of the bulk material. We have used the sp^3 orbital basis, retaining upto second nearest neighbour interactions.

The threshold energies for impact ionization in a 70 Å quantum well are listed in Table 1. The low energy final electron states in the conduction band have been described by a simple nonparabolic and isotropic dispersion. For a high energy electron incident along a chosen direction with calculated energy, we search for the initial valence band state and final conduction band states that satisfy energy and momentum conservation. The minimum energy along the incident direction that satisfies the conservation conditions yields the threshold. Results listed in the table are restricted to (100) incidence, as in source to drain transport in a FET. Two important points are to be noted. As excess In is added to make the structure pseudomorphic, the impact ionization threshold decreases. The decrease in bandgap in these structures decreases the threshold for breakdown, but the decreasing hole mass in partially compensates for this. However, in addition to introducing strain with excess In, if the bandgap is maintained by adding excess Al, we predict improvements in the impact ionization threshold as compared to the lattice-matched case due to the presence of strain.

We are now in the process of carrying out Monte Carlo simulations to find the electric field dependence of the impact ionisation rate and to predict breakdown voltages. We have seen that the inclusion of strain in the manner suggested here can improve the breakdown thresholds by ~10%. We expect this to result in power output improvements of ~20-30%. Our computations reflect an angular average of the breakdown, and conceal the angular dependence of impact ionisation. The threshold values obtained for the lattice-matched cases are quite comparable to reported values. The increased breakdown thresholds suggested by the theory for cases where compressive strain is introduced using excess In but where bandgap is maintained by adding Al have not yet been tested experimentally. Since a number of devices can benefit by this large tailoring of impact ionization rates it would be worthwhile to test these possibilities experimentally. It is important to note that the predicted benefits can arise only because of pseudomorphic epitaxy, since it is essential to maintain a constant bandgap and produce a uniaxial strain to lift the light hole– heavy hole degeneracy.

Acknowledgement: This work was supported by the Army URI program (Grant No. DAAL03-87-K-0007).

Table 1: Threshold energies for impact ionization in a 70 Å quantum well with $In_{0.52}Al_{0.48}As$ substrate.

Well Material	Bandgap (eV)	Threshold (eV)
$In_{0.53}Ga_{0.47}As$	0.72	1.104
$In_{0.73}Ga_{0.27}As$	0.54	0.947
$In_{0.73}Ga_{0.12}Al_{0.15}As$	0.72	1.210

REFERENCES:

1. E. Jones, I. Fritz, J. Schinber, M. Smith and T. Drummond, Proc. of the International Conference of GaAs and Related Compounds, 1986, pp. 227-233.

2. T. Drummond, T. Zipperian, I. Fritz, J. Schirber and T. Plut, Appl. Phys. Lett. vol. 49, pp 461-463, 1986.

3. J. Hinckley and J. Singh, Appl. Phys. Lett. vol. 53, 785-788, 1988.

4. C. Lee, H. Wang, G. Sullivan, N. Sheng, and D. Miller, IEEE Elect. Dev. Lett. vol. EDL-8, pp 85-87, 1987.

5. M. Jaffe and J. Singh, J. Appl. Phys. vol. 65, pp 329-338, 1989.

6. M. Jaffe, J. E. Oh, J. Pamulapati, J. Singh and P. Bhattacharya, Appl. Phys. Lett. vol. 54, pp 2345-2346, 1989.

7. T. E. Zipperian, L. R. Dawson, T. Drummond, J. E. Schinber, and I. J. Fritz, Appl. Phys. Lett. vol. 52, pp 975-977, 1988.

8. J. M. Hinckley, Ph.D. Thesis, University of Michigan, Ann Arbor (1990).

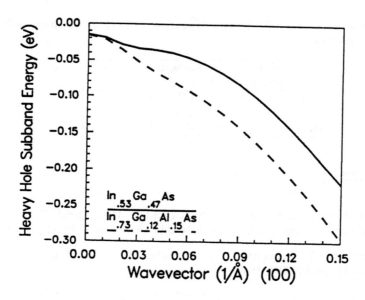

Figure 1: Heavy-hole in-plane dispersion in a 70 Å quantum well of 2 different alloys lattice-matched to $In_{0.52}Al_{0.48}As$

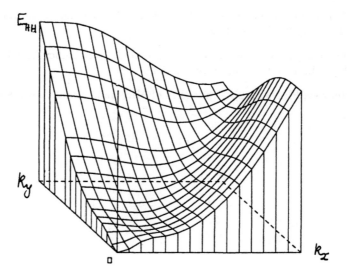

Figure 2: Schematic 2D dispersion in a 70 Å quantum well of $In_{0.53}Ga_{0.47}As$ lattice-matched to $In_{0.52}Al_{0.48}As$

Inst. Phys. Conf. Ser. No 120: Chapter 10
Paper presented at Int. Symp. GaAs and Related Compounds, Seattle, 1991

523

Evaluation of new multiple quantum well avalanche photodiode structures: the MQW, the doped barrier and doped quantum well

P. Aristin, A. Torabi, A.K. Garrison, H.M. Harris, and C.J. Summers

Physical Sciences Laboratory, Georgia Tech Research Institute, Atlanta, Georgia 30332

Abstract

A quantitative experimental investigation of AlGaAs/GaAs multiple-quantum-well avalanche photodiode (APD) structures, the superlattice APD; the doped barrier APD and the doped quantum well APD is reported. Diodes exhibiting self consistent C-V, I-V and breakdown voltage characteristics showed strong agreement between electron- and hole-ionization rates, as determined from gain and noise measurements, respectively. This study provides new data on the performance of doped barrier and quantum well APDs and establishes a comparison with the electron- and hole-ionization rates from $Al_xGa_{1-x}As/GaAs$ MQW-APDs. These devices exhibit gains of ~ 20 with excess-noise factors <5 at bias voltages <10V. The dependence of these properties on structure, doping concentration, bias and temperature is presented.

1. Introduction

Multiple quantum well structures are used in the design of high performance APDs because of their potential to minimize the excess noise of the avalanche process. Low noise is achieved if the ionization rates of the electrons and the holes, α and β respectively, are greatly different, equivalently α/β (or β/α) is high. Since the presentation of this original idea by Chin et al(1980) and the first experimental investigations by Capasso et al (1982a), several novel designs have been proposed: the doped multilayer APD (1982), the staircase APD (1982b), the channeling APD (1983), the pn-doped homojunction and heterojunction APD (1986) (or doped barrier APD) and the doped quantum well APD (1990). It is predicted that the last two designs could lead to the first solid state photomultiplier APD where the avalanche excess noise has been totally suppressed.

This paper presents detailed investigations on undoped MQW APDs having different geometries and aluminum compositions, doped barrier and doped well APDs. A comprehensive self-consistent methodology was used where current-voltage, capacitance-voltage, doping profile, and noise characteristics were extensively analyzed.

2. Experimental

The structures were grown by molecular beam epitaxy. Growth was initiated on a n^+ Si doped substrate followed by a short period superlattice to prevent propagation of dislocations and impurities. All the device structures were PINs where the I region was composed of the MQW structure with P and N contact layers of 1 μm and 1.5 μm doped to 1×10^{18} cm^{-3} with Be and Si, respectively. The MQW structures had 25 $Al_xGa_{1-x}As$ / GaAs multilayers with aluminum compositions of 0.30, 0.35 and 0.45. The aluminum composition, x, of the AlGaAs layers was calibrated using photoluminescence measurements. The samples showed high exciton recombination photoluminescence intensities with a half width of 5 meV. Growth interruption techniques were used to obtain well defined pn doped regions in the MQW

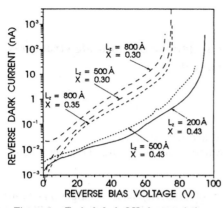

Figure 1. Typical dark I-V characteristics for undoped $Al_xGa_{1-x}As$ / GaAs MQW APDs.

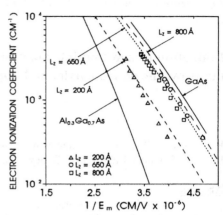

Figure 2. α_{MQW} for $Al_{0.3}Ga_{0.7}As$ / GaAs MQW APDs.

structures. The dopant concentration was calibrated by Hall effect and was kept below 6×10^{18} cm^{-3} to limit dopant diffusion.

The devices were fabricated into 2×10^{-4} cm^2 area mesa structures using standard photolithography techniques. The device configuration allows for electron or hole injection because both p^+ and n^+ layers can be illuminated. Precise control of the growth and fabrication procedures yields identical photodiode characteristics on the same wafer and from one run to another. A SiO_2 dielectric coating suppressed surface leakage currents and provided devices with very low dark currents. The photodiode characterization consisted of computer automated I-V, C-V and noise measurements. The dc I-V characteristics were measured in the dark and under illumination by a HeNe laser light with a 5 µm diameter spot. The photocurrent gain was calculated from the increase of the unmultiplied photocurrent and was verified to be independent of the light intensity. The C-V measurements were made on a LCZ meter between 80 Khz and 1 Mhz and at 300 K and 77 K. The apparent free carrier concentration profile was obtained from differentiation of the C-V data. Noise measurements consisted of measuring the variance of the photodiode output current for different gains. Absolute noise measurements were performed using a spectrum analyzer tuned to a frequency of 200 Khz. Several noise sources were used to calibrate the system and the noise measurements are accurate to within a few percent.

3. Results

A. Undoped MQW APDs

Investigations were performed on 2.5 µm thick MQW $Al_xGa_{1-x}As$ / GaAs structures with different well and barrier widths, L_B and L_Z respectively, and with a constant MQW period width of 1000 Å. The values of L_Z studied were: 200 Å, 350 Å, 500 Å, 650 Å and 800 Å. A 400 Å thick spacer layer was added prior to the first well on the P^+ contact side and also prior to the first barrier on the N^+ contact side to avoid trapped carriers at the first heterojunction for electric fields above 100 kV/cm. The MQW structures have an Al composition, x, of 0.3, 0.35 and 0.43. Typical dark I-V characteristics are shown in Figure 1 for electron injection. The dark current was below 10 Na at 80 % of the breakdown voltage for x = 0.30 and below 1 Na for x = 0.43. The dark current decreased with decreasing well width, and with increasing x.

The photocurrent increases slowly with the applied voltage, becomes constant between 25-35 V and finally increases exponentially above 50 V. As also shown in Figure 1, the breakdown voltage, V_B, increased from 70 to 85 V as the barrier width increased from 200 Å

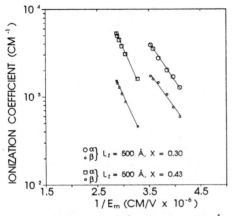

Figure 3. α_{MQW} and β_{MQW} for $L_z = 500$ Å, x = 0.30 and 0.43

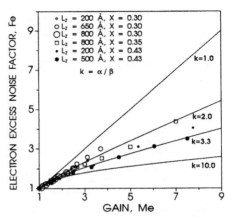

Figure 4. Electron excess noise versus gain for undoped MQW APDs.

to 800 Å. For a given geometry and Al composition, V_B is constant within 2 % for all photodiodes tested, and increases as the Al composition increases. Similarly, I-V measurements were taken for hole injection. These results indicate that the dark current is due to the generation-recombination of carriers in the narrow bandgap GaAs layer combined with a thermionic emission of the carriers over the barriers. The carriers are generated in the well since the dark current increases when the well width increases. However, the thick barriers do not permit carriers to tunnel and carriers need to gain enough energy from the applied field to be injected over the barriers and avoid being trapped at the AlGaAs / GaAs interface. Therefore MQW structures with high x values have lower dark currents since the barrier height increases with the Al composition. Similarly, at low applied field the photo-injected carriers are trapped at the first AlGaAs / GaAs interface unless they gain enough energy from the field to be injected over the barrier. Consequently, the minimum applied voltage, V_m, which allows a constant current to flow depends on the width, L_s, of the spacer layer plus the first well located prior to the first AlGaAs / GaAs hetero-interface for electron injection. Consequently, V_m is reduced as L_S is increased.

Carrier multiplication starts for bias voltages above 50 V and has a maximum value of 9. The electron gain, M_e is always greater than the hole gain, M_h, indicating that the electron ionization coefficient, α_{MQW}, is always greater than the hole ionization coefficient, β_{MQW}, for well widths between 200 Å and 800 Å.

C-V measurements confirm the low carrier concentration in the MQW region. The capacitance is constant (≈ 0.9 pF) from very low bias voltages (≈ 5 V) up to the breakdown voltage, for a wide range of frequencies (80 Khz - 1 Mhz), at 77 K and 300 K. This indicates that the 2.5 μm thick MQW ($N_D < 10^{14}$ cm^{-3}) is fully depleted even at low bias voltages.

The values of $\alpha_{MQW}(E_m)$ and $\beta_{MQW}(E_m)$ were calculated from the electron and hole gain, M_e and M_h, using the well known formula for a PIN structure, and were compared to $\alpha_{AV}(E_m)$ (Capasso 1985), defined as

$$\alpha_{AV}(E_m) = \frac{\alpha_{AlGaAs}(E_m) \times L_B + \alpha_{GaAs}(E_m) \times L_Z}{L_B + L_Z} \tag{1}$$

where α_{GaAs} and α_{AlGaAs} (x = 0.30) are the electron ionization coefficients for the bulk materials, from Bulman et al (1983) and Robbins et al (1988), respectively. $\alpha_{AV}(E_m)$ was calculated for all values of L_B and L_Z studied. Similar calculations were made for $\beta_{AV}(E_m)$ by replacing α by β in the above equation. If α_{MQW} is enhanced over the bulk material values, then it should be

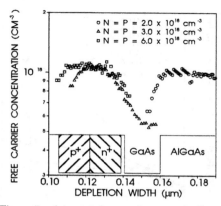

Figure 5. Apparent free carrier concentration of doped barrier APDs.

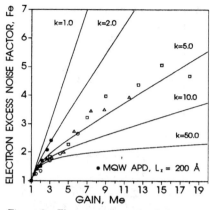

Figure 6. Electron excess noise factor vs gain, the solid circles are for the MQW APD, L_z = 200 Å, and the open symbols for the doped barrier APD.

larger than α_{AV}.

α_{MQW} is plotted versus $1 / E_m$ for MQW structures having 200 Å, 650 Å, and 800Å well widths and x = 0.30 in Figure 2 . The values of α_{AV} are also shown as dashed lines, along with the bulk values of α_{GaAs} and α_{AlGaAs} as solid lines. α_{MQW} is between 300 and 4000 cm^{-1} for an electric field between 222 Kv/cm and 333 Kv/cm. Finally, Figure 3 shows the results for α_{MQW} and β_{MQW} for a 500 Å well and two values of x.

The results presented in Figure 2 indicate that the values of $\alpha_{MQW}(E_m)$ agree with the calculated values of α_{AV} for all geometries studied. Therefore, for x = 0.30, the electron ionization α_{MQW} is obtained from (1) using the values of α from the bulk materials. Since α_{GaAs} >> α_{AlGaAs}, electron ionization occurs in the GaAs well. The conduction band edge discontinuity does not contribute to the enhancement of the electron ionization since α_{MQW} is not enhanced over its value in the bulk GaAs.

The results indicate also that for a higher Al composition, x, α_{MQW} stays in the same range [300-4000 cm^{-1}], but the corresponding range of electric fields is shifted to higher values. As x increases, the value of E_m at which multiplication starts is increased from 222 Kv/cm for x = 0.30 to 315 Kv/cm for x = 0.43. Consequently, as x increases, the breakdown voltage increases. This is indicated in Figure 3 by a translation of the data to higher electric field values. These results show that the AlGaAs layer of the MQW is "inactive" for x values of 0.30. However, for x = 0.43, the AlGaAs layer appears to reduce the average kinetic energy of the electrons which enter the GaAs layer since the device operates under higher applied electric fields. This energy loss is due to scattering in the X-band of the AlGaAs layer which increases exponentially with the Al composition.

The ratio $k_{MQW}(E_m) = \alpha_{MQW}(E_m) / \beta_{MQW}(E_m)$ increased from 1.72 to 2.5 when the electric field increased from 220 Kv/cm to 280 Kv/cm for the 5 geometries studied and x = 0.30, as shown in Figure 3. However $k_{MQW}(E_m)$ is reduced to a constant k = 2.5 for x = 0.43. The results agree with the predicted value using the relation $k_{MQW}(E_m) = \alpha_{AV}(E) / \beta_{AV}(E)$ for x = 0.30, but fail for x = 0.43 since the ratio obtained from the measurements is lower. These results indicate that both the electron and hole average energy are reduced due to the AlGaAs layer. The hole average energy is reduced by a larger amount since MQW structures with x = 0.43 give a higher k value.

Noise measurements indicate a k value (McIntyre 1966) between 1.7 and 2.5 for the x = 0.30 MQW and between 2.5 and 3.3 for the x = 0.43 MQW. Results for F_e are plotted versus M_e in Figure 4 for the 200 Å well MQW APD having x = 0.30 and 0.43. The solid lines correspond to the theoretical curves of McIntyre. These results agree with the k_{MQW} obtained from the coefficients $\alpha_{MQW}(E_m)$ and $\beta_{MQW}(E_m)$ which were calculated separately using the electron and hole gain measurements and demonstrate that the characterization techniques are self-consistent.

B. Doped MQW APDs: the pn junction doped barrier and doped well APDs.

The doped barrier APD was designed with the same unit cell as the undoped MQW APD and consists of a 800 Å barrier, 200 Å well $Al_{0.35}Ga_{0.65}As$ / GaAs MQW structure where a p^+n^+ equally doped junction was built in the barrier prior to the GaAs well. The fully depleted 150 Å p^+/ 150 Å n^+ junction, doped at 3 x 10^{18} cm^{-3}, locally enhances the electric field of the MQW structure by superimposing 0.5 eV on the MQW band potential.

The I-V characteristics indicate a low dark current and a low breakdown voltage of about -10 V. C-V characteristics indicate that the photodiode capacitance is higher (15 Pf) than measured for the MQW APD (0.9 Pf) and decreases for increasing diode bias voltage as for a one sided abrupt pn$^+$ junction. The free carrier concentration profile obtained from the analysis of the C-V data is presented in Figure 5. This modeling shows that only one period of the MQW structure is fully depleted at zero bias due to unbalanced doping concentrations in the 300 Å thick junctions. As the field is increased, the depletion width punches through the highest doped side of the 300 Å thick junction to deplete the second period and pn junction of the MQW structure. The results confirm that the depletion region is located close to the P$^+$ contact thus indicating that $p^+ \ll n^+$ in the 300 Å thick pn junctions. This is due to the difficulty of achieving equal p-and n- type dopant concentrations in the AlGaAs layer by using solid dopant sources. Although a qualitative variation of the electric field E is predicted, precise calculations of E are not available at present. Thus α and β can not be obtained separately from the gain measurements. Noise measurement results are presented in Figure 6 where F_e is plotted versus M_e and show that F_e is low for gains up to 5 with a corresponding k between 12.5 and 50. As M_e increases, F_e also increases with a corresponding k between 5 and 10.

These results indicate that k is reduced at low applied electric fields. Since the peak of the electric field is located prior to the well, the injected electrons are more likely to ionize in the well. However, holes generated from ionizations in the well travel in the opposite direction and enter the AlGaAs layer where their ionization probability is smaller compared to the ionization probability of the electrons in GaAs. Consequently, α is greatly different from β and a noise reduction is observed. However, k increases at higher gain indicating that the hole ionization probability is no longer small compared to the electrons and that the applied electric field is high enough to supply kinetic energy to the holes to impact ionize in the following well.

Similar results were obtained for a doped 500 Å well / 500 Å barrier $Al_{0.43}Ga_{0.57}As$ / GaAs MQW structure where a 300 Å thick pn junction was grown into the well immediately following the barrier. I-V and C-V measurements show that 2 to 3 periods were depleted when the bias voltage was increased from 0 to -12 V at the avalanche breakdown. For these structures high gains up to 400 were obtained. Noise measurements give a similarly low k value as reported for the doped barrier APDs. These results indicate that α / β is enhanced, consequently the noise is reduced, but no information was obtained on the separate magnitude of α and β. The comparable noise performance of both designs suggests that the location of the pn junctions, either in AlGaAs or GaAs material, has little consequence as long as the other parameters of the MQW remain constant (L_Z = 200 Å, L_B = 500 Å). However, the two designs have some differences in their characteristics since the gain is 10 times greater for the doped well than for the doped barrier APD. Further studies are needed to explain this difference.

4. Conclusion

    Investigations of undoped $Al_xGa_{1-x}As$ / GaAs MQW APDs show that α_{MQW} is not enhanced over its value in the bulk materials. Since $\alpha_{GaAs} >> \alpha_{AlGaAs}$, α_{MQW} in the MQW is obtained from an average calculation of the bulk values for low x values, $x \approx 0.3$. Thus, k_{MQW} = $\alpha_{MQW}(E)$ / $\beta_{MQW}(E)$ is between 1.72 and 2.5 which is higher than for GaAs in the same electric field range (k = 1.6). For higher Al composition, x = 0.43, higher electric fields are required for impact ionization. This indicates that both the electron and hole average kinetic energy is reduced due to the AlGaAs layer. However, low k values indicate a noise reduction. Even though the ratio α_{MQW} / β_{MQW} is enhanced, α_{MQW} is not enhanced over its value in the bulk GaAs material. These results have been demonstrated using self-consistent measurement and analysis techniques since similar values of α/β were obtained from gain and noise measurements.

    Results on doped barrier and doped well MQW APDs show that the noise of the doped structures is always lower than for the undoped structure having the same geometry. The noise reduction is due to a local enhancement of the built-in potential which confines the electron ionization in the GaAs well. High k values between 12.5 and 50 were obtained for gains up to 5, and values between 5 and 10 for gains above 5. The doped well APD gives the highest gain with values up to 400 with a corresponding k equal to 5. These new designs provide low noise, low breakdown voltage and high gain as required for optoelectronic applications.

Acknowledgements

We wish to thank the Polaroid Corporation for sponsoring this work, and we are grateful to K. F. Brennan, W. T. Wetterling and K. A. Maloney for technical discussions.

References

Blauvelt H, Margalit S and Yariv A 1982 Electron. Lett. **18** 375
Brennan K 1986 IEEE Trans. Electron Dev. ED-33 1683
Brennan K and Vetterling W T 1990 IEEE Trans. Electron Dev. 37 536
Bulman G E, Robbins V M, Brennan K, Hess K and Stillman G E 1983 IEEE Trans. Electron Dev. EDL-4 181
Capasso F, Tsang W T, Hutchinson A L and Williams G 1982a Appl. Phys. Lett. **40** 38
Capasso F, Tsang W T, Hutchinson A L and Foy P 1982b Conf. Ser.- Inst. Phys. 63 473
Capasso F, Tsang W T and Williams G P 1983 IEEE Trans. Electron Dev. ED-30 381
Capasso F 1985 Semiconductors and Semimetals - ed R K Willardson and A C Beer (New-York:Wiley) p121
Chin R, Holoniak N, Stillman G E, Tsang J and Hess K 1982 Appl. Phys. Lett. **16** 467
McIntyre R J 1966 IEEE Trans. Electron Dev. ED-13 164
Robbins V M, Smith S C and Stillman G E 1988 Appl. Phys. Lett. **52** 296

Inst. Phys. Conf. Ser. No 120: Chapter 11
Paper presented at Int. Symp. GaAs and Related Compounds, Seattle, 1991

529

The growth of $Ga_{0.52}In_{0.48}P$ and $Al_{0.18}Ga_{0.34}In_{0.48}P$ on lens-shaped GaAs substrates by metalorganic vapor phase epitaxy

N. Buchan, W. Heuberger, A. Jakubowicz and P. Roentgen

IBM Research Division, Zurich Research Laboratory,
Säumerstrasse 4, 8803 Rüschlikon, Switzerland

Abstract. The substrate orientation dependence of ordering of the group III-sublattice in epitaxial $Ga_{0.52}In_{0.48}P$ and $Al_{0.18}Ga_{0.34}In_{0.48}P$ has been studied on metalorganic vapor phase epitaxy (MOVPE) overgrown lens-shaped GaAs substrates centered around the (100). The photoluminescence peak wavelength from the $Ga_{0.52}In_{0.48}P$ and $Al_{0.18}Ga_{0.34}In_{0.48}P$ demonstrates a novel "saddle"-shaped mirror plane symmetry with respect to the $[011]$ and $[01\bar{1}]$ crystallographic axes. Ordering of the group-III sub-lattice is optimized on growth planes oriented (100) $5°$ toward $[01\bar{1}]$ and $[0\bar{1}1]$, as indicated by the maximum wavelength observed there. Disordering is indicated by a gradual decrease in the luminescence wavelength of approximately 40nm as the growth plane approaches orientations of $20°$ off the (100). Hence, the ordering *does not* demonstrate a simple monotonic decrease with orientation off the (100). These results are duplicated by growth on planar substrates of select orientations off the (100). The good compositional uniformity of In on the lenses is determined by x-ray diffractometry.

1. Introduction

$Ga_xIn_{1-x}P$ and $(Al_yGa_{1-y})_xIn_{1-x}P$, lattice matched to GaAs substrates, are important III-V materials that have the highest direct bandgaps among the commonly used III-V compounds. Epitaxial layers of $Ga_xIn_{1-x}P$, grown by metalorganic vapor phase epitaxy (MOVPE) at a constant value of x, have demonstrated a variation in bandgap that has been correlated to the degree of ordering of the group-III sub-lattice on (111) planes (Suzuki *et al* 1988). The degree of ordering has been shown to depend on numerous growth parameters: the group-V/group-III ratio introduced to the reactor (V/III) and the growth temperature (Gomyo *et al* 1986), the growth rate (Kurtz *et al* 1990) and the impurity concentration (Gomyo *et al* 1987). Most recently, the degree of ordering of both $Ga_{0.52}In_{0.48}P$ and $(Al_yGa_{1-y})_{0.52}In_{0.48}P$ (Minagawa *et al* 1989) and the variant of ordering of $Ga_{0.52}In_{0.48}P$ (Chen *et al* 1991) have also been shown to have a dependence on the growth plane orientation

The growth plane orientation dependence of III-V material parameters in both MOVPE and molecular beam epitaxy (MBE) has heretofore been dominantly performed on substrates oriented on or slightly off the (100) plane. However, improved device characteristics of $Al_xGa_{1-x}As/GaAs$ lasers grown by MBE have been demonstrated on substrates oriented (100) $4°$ toward $[011]$ (Chen *et al* 1987). Recently, the use of lens-shaped GaAs substrates has allowed the investigation of the substrate orientation dependence of photoluminescence (PL) and surface morphology of $Al_xGa_{1-x}As$ epilayers grown by MOVPE (Johnson *et al* 1987, Johnson 1991). We have extended the use of GaAs lens-shaped substrates to simultaneously and comprehensively investigate the dependence of ordering of the group-III sub-lattice in $Ga_{0.52}In_{0.48}P$ and $Al_{0.18}Ga_{0.34}In_{0.48}P$ on growth planes oriented up to 30° off the (100). The homogeneity of the In composition over the lens is measured by x-ray diffractometry (XRD) and the ordering is observed by its effect on the peak PL wavelength. The results on the lenses are duplicated, and hence confirmed, by growths performed on planar substrates of select orientations off the (100).

2. **Experimental**

The lens-shaped GaAs substrates used in these experiments were ground from 2.5 cm diameter semi-insulating (100) oriented GaAs discs into plano-convex lenses having a radius of curvature of 2.54 cm. Thus, growth orientations of up to 30° off the (100) are accessed. A narrow flat was ground on the side of each lens to indicate the [011] direction. The lenses were initially etched to remove polishing damage as previously described by Johnson (1991). Two lenses were investigated: the growth sequence involved the deposition of a 60nm GaAs buffer layer, followed by a nominally lattice matched layer of either $Ga_{0.52}In_{0.48}P$ (200nm) or $Al_{0.18}Ga_{0.34}In_{0.48}P$ (300nm). The precursors for the phosphorus containing layers were trimethyl-aluminum, -gallium and -indium, and PH_3. The layers were nominally undoped and were grown under conditions selected to produce ordering of the group III sub-lattice on growth planes near the (100) (670°C growth temperature, V/III = 220, reactor pressure 350 mbar and 25nm/min growth rate). The growths were performed with the [011] and [01$\bar{1}$] crystallographic axes of the lens rotated approximately 45° with respect to the reactor flow in order to distinguish the effect on the results of the crystallographic orientation from that of the fluid dynamics of the reactor. Growths were also performed on planar substrates of select orientations off the (100), at a composition of 1-X = 0.484±0.004, for comparison to the results obtained on the lenses.

The PL spectra were measured using 40mW argon laser excitation focussed to a 0.5mm diameter. Measurements were made every 1-2° off the (100) and every 15° in the axis of rotational symmetry of the lens using a refrigerated copper "cold-finger" mount that rotated the lenticular substrate on two orthogonal axes. Polar coordinate maps of the PL peak wavelength and intensity for both lenses were determined at RT and 100K for the $Ga_{0.52}In_{0.48}P$ and $Al_{0.18}Ga_{0.34}In_{0.48}P$ overgrown lenses, respectively.

The spatial variation of the In composition on the two lenses was investigated by XRD using a Philips analytical high resolution double crystal x-ray diffractometer. Narrow slits on the x-ray source and use of the (006) reflection (which employs a small angle of incidence and reflection with respect to the (100)) allowed a spatial resolution of less than 1mm. The substrate-epilayer diffraction peak splitting was then converted to the deviation of the group-III mole percent of In from the lattice matched condition (1-X≃0.482) after accounting for the dependence of the conversion on the relative orientations of the measurement plane, the growth plane and the direction in which the lattice elastically distorts (Hornstra *et al* (1978),Bartels *et al* 1978). Small compositional variations of Al/Ga are assumed to have an insignificant effect on the lattice constant relative to those of In/Ga due to the closely matched lattice constants of AlP to GaP compared to the wide lattice mismatch between InP and GaP (Krijn 1991). The spatial dependence of the In composition is converted to the PL wavelength expected for a disordered group-III sub-lattice:

$$E_{Ga_xIn_{1-x}P} (300K) = 1.295 + 1.151 \cdot X \quad \text{for} \quad 0.50 < X < 0.53 \tag{1}$$

$$E_{(Al_{0.35}Ga_{0.65})_xIn_{1-x}P} (100K) = 1.410 + 1.207\,X + 0.514\,X^2 \quad . \tag{2}$$

Eq.(1) is provided by Roberts *et al* (1981) and Eq.(2) is interpolated using Vegard's law from the temperature dependence of the direct bandgap for the constituent binary alloys (Madelung 1982).

3. **Results**

$Ga_{0.52}In_{0.48}P$:

The spatial variation of the peak PL wavelength and intensity on the $Ga_{0.52}In_{0.48}P$ overgrown lens is shown in Figs.1a and 1b. All figures show the flat indicating the [011] direction, while the [100] points out of the page. Aids to the eye are drawn in 5° increments off the (100). The peak PL wavelengths measured on the lens may be compared to those obtained for growth on select planar substrates oriented off the (100), as shown in Fig.1c.

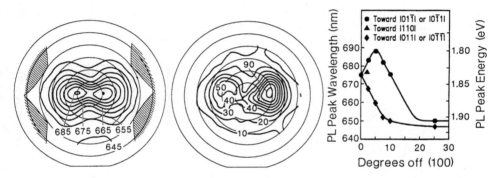

Fig. 1a: Peak PL wavelength (nm) at RT of the $Ga_{0.52}In_{0.48}P$ overgrown lens.

Fig. 1b: Peak PL intensity (arb units) of the $Ga_{0.52}In_{0.48}P$ overgrown lens at RT.

Fig. 1c: Peak PL wavelength (nm) at RT of $Ga_{0.52}In_{0.48}P$ on *planar* substrates.

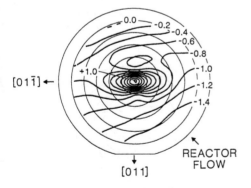

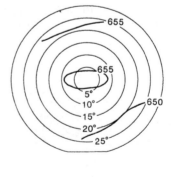

Fig. 1d: Compositional variation of the mole percent of In off the lattice-matched condition on the $Ga_{0.52}In_{0.48}P$ overgrown lens.

Fig. 1e: Peak PL wavelength (nm) on the $Ga_{0.52}In_{0.48}P$ overgrown lens, calculated for only the variation of In composition.

The peak PL wavelength and intensity demonstrate "saddle"-shaped profiles and mirror planes of symmetry with respect to [011] and [01$\bar{1}$]. The peak wavelength of 688nm and the maximum intensity coincide at an orientation of (100) 5° toward [01$\bar{1}$] and [0$\bar{1}$1], but fall off to a wavelength of 645nm and 10% of maximum intensity at growth planes of (100) 20°. The hashed regions in Fig.1a indicate that two distinct PL peaks are observed (approximately 645nm and 735nm). The same peak wavelengths and intensities observed on the lens are essentially duplicated for growth on the planar substrates.

The uniformity of the In composition, determined by XRD and shown in Fig.1d, demonstrates a steady increase of In in the direction of reactor flow. Additionally, a higher In concentration is centered at [$\bar{1}$00] and falls monotonically at orientations off [$\bar{1}$00]. The minimal effect of the In composition on the emission wavelength is calculated via eq.1 and shown in Fig.1e.

$Al_{0.18}Ga_{0.34}In_{0.48}P$:

The effect of exchanging approximately 35% of the Ga for Al in $Ga_{0.52}In_{0.48}P$ layer has little qualitative effect on the results. The spatial variation of the 100K peak PL wavelength and intensity and the peak PL wavelength measured on selected planar substrates are shown in Figs.2a, 2b, and 2c, respectively.

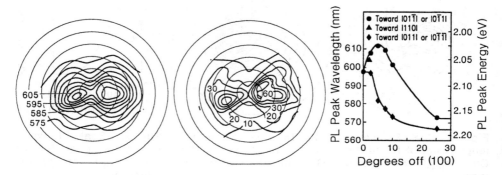

Fig. 2a: Peak PL wavelength (nm) at 100K of the $Al_{0.18}Ga_{0.34}In_{0.48}P$ overgrown lens.

Fig. 2b: Peak PL intensity (arb units) of the $Al_{0.18}Ga_{0.34}In_{0.48}P$ overgrown lens at 100K.

Fig. 2c: Peak PL wavelength (nm) at 100K of $Al_{0.18}Ga_{0.34}In_{0.48}P$ on *planar* substrates.

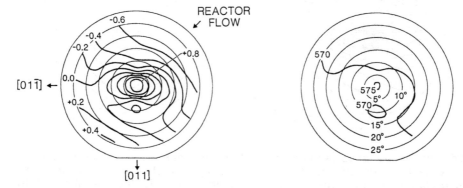

Fig. 2d: Compositional variation of the mole percent of In off the lattice-matched condition for the $Al_{0.18}Ga_{0.34}In_{0.48}P$ overgrown lens.

Fig. 2e: Peak PL wavelength (nm) on the $Al_{0.18}Ga_{0.34}In_{0.48}P$ overgrown lens at 100K calculated for only the variation of In composition.

The peak PL wavelength and intensity have the same "saddle"-profile and mirror planes of symmetry observed on the $Ga_{0.52}In_{0.48}P$ overgrown lens. The highest wavelength of 615nm and the maximum intensity coincide at an orientation of about (100) 5° toward [01$\bar{1}$] and [0$\bar{1}$1], but the values of the luminescence fall off to 575nm and 10% maximum intensity at 20° off the (100). The physical limitations of the low temperature PL apparatus prevented analysis of the $Al_{0.18}Ga_{0.34}In_{0.48}P$ lens at orientations of greater than (100) 20°. The same peak wavelengths and intensities observed on the lens are essentially duplicated for growth on the planar substrates. The exchange of Ga for Al also has little effect on the compositional variation of In which duplicates the dependence on reactor flow and crystallographic orientation previously observed for the $Ga_{0.52}In_{0.48}P$ lens, as shown in Fig.2c, and has only a weak effect on the calculated luminescence wavelength, as shown in Fig.2d.

4. Discussion and Summary

The dependence of the photoluminescence (PL) of MOVPE grown $Ga_{0.52}In_{0.48}P$ and $Al_{0.18}Ga_{0.34}In_{0.48}P$ epilayers on the growth plane orientation has been simultaneously investigated at orientations of up to 30° off the (100) by growing on lens-shaped GaAs substrates. Growth conditions were selected to produce an ordered group-III sub-lattice on growth planes near the (100). The peak PL wavelength and intensity from both the $Ga_{0.52}In_{0.48}P$ and $Al_{0.18}Ga_{0.34}In_{0.48}P$ lenses demonstrate a "saddle"-shaped spatial symmetry with respect to mirror planes in the [011] and [01$\bar{1}$] crystallographic axes, and reach a maximum at orientations of approximately (100) 5° toward [01$\bar{1}$] and [0$\bar{1}$1]. The uniquely high wavelength observed at these orientations correlates with the observation of highly developed ordering observed at (100) 6° toward [01$\bar{1}$] in $Ga_{0.52}In_{0.48}P$ (Gomyo et al 1989). Disordering of the group-III sub-lattice occurs on growth planes further off the (100) as evidenced by a decrease in the luminescence wavelength by approximately 40nm at 20° off the (100), and is accompanied by a decrease in luminescence intensity. The "saddle"-shaped symmetry of the luminescence wavelength shows that the ordering of $Ga_{0.52}In_{0.48}P$ and $Al_{0.18}Ga_{0.35}In_{0.48}P$ *does not* demonstrate a simple monotonic decrease with orientation off the (100), as may be inferred from the results of Minagawa et al (1989). The peak PL wavelengths observed on the lenses were reproduced on planar substrates at various orientations off the (100).

The utility of the lens-shaped GaAs substrates for studying ordering is indicated by the good agreement of the wavelength and intensities observed on the lens overgrown with $Ga_{0.52}In_{0.48}P$ and the planar substrates grown under identical growth conditions. The utility of the lenses is further shown by the good uniformity of the In composition in the layer and the minimal effect on the luminescence wavelength of the In composition relative to that of disordering. The composition of the In on both lenses demonstrates a minor increase along the flow direction and is symmetric with respect to the [011] and [01$\bar{1}$] crystallographic axes. The compositional variation of Al across the $Al_{0.18}Ga_{0.34}In_{0.48}P$ overgrown lens cannot be excluded as having an influence on the luminescence wavelength, however, due to the qualitatively analogous results obtained for both lenses, the effect of disordering is concluded to be dominant.

The observation of two distinct wavelengths in some regions of the $Ga_{0.52}In_{0.48}P$ lens located at orientations of greater than 20° off the (100) seemingly disagrees with previous observations of a single wavelength for growth on substrates oriented on (100) 55° toward [01$\bar{1}$] (Chen et al 1987) and these regions are too far removed from the edge to be attributed to "edge effects" during growth. The explanation may lie in terraced growth morphology (such as has been previously reported for growth of $Al_xGa_{1-x}As$ on GaAs lenses

(Johnson *et al* 1987, Johnson 1991) on which the surface diffusion of In (Arent *etal* 1989) can allow the formation of binary alloys of stoichiometry other than $Ga_{0.52}In_{0.48}P$.

References

Suzuki T, Gomyo A and Iijima S 1988 *J. Cryst. Growth* **93** 396

Gomyo A, Kobayashi K, Kawata S, Hino I and Suzuki T 1986 *J. Cryst. Growth* **77** 367

Kurtz S R, Olsen J M and Kibbler A 1990 *Appl. Phys. Lett.* **57** 1922

Gomyo A, Suzuki T, Kobayashi K, Kawata S, and Hino I 1987 *J. Cryst. Growth* **50** 673

Minagawa S and Kondow K 1989 *Electron. Lett.* **25** 758

Chen G S, Stringfellow G B 1991 *Appl. Phys. Lett.* **59** 324

Chen H Z, Ghaffari A, Markoc H and Yariv A 1987 *Appl. Phys. Lett.* **51** 2094

Johnson E S, Legg G E and Curless J A 1987 *J. Cryst. Growth* **85** 182

Johnson E S 1991 *J. Cryst. Growth* **107** 237

Hornstra J and Bartels W J 1978 *J. Cryst. Growth* **44** 513

Bartels W J and Nijman W 1978 *J. Cryst. Growth* **44** 518

Krijn M C P M 1991 *Semicond. Sci. Technol.* **6** 27

Roberts J S, Scott G B and Gowers J P 1981 *J. Appl. Phys.* **52** 4018

Numerical Data and Function Relationship in Science and Technology Landolt-Börnstein., editor Madelung O (Springer, Berlin, 1982) Vol. 17, Part A, pp 162, 205, 288

Gomyo A, Kawata S, Suzuki T, Iijima S and Hino I 1989 *Jap. J. Appl. Phys.* **28** L1728

Arent D J, Nilsson S, Galeuchet Y D, Meier H P and Walter W 1989 *Appl. Phys. Lett.* **55** 2611

Inst. Phys. Conf. Ser. No 120: Chapter 11
Paper presented at Int. Symp. GaAs and Related Compounds, Seattle, 1991

Incorporation of acceptor impurities in MOCVD and GSMBE InP

S. S. Bose, S. L. Jackson, A. P. Curtis, and G. E. Stillman
Center for Compound Semiconductor Microelectronics
University of Illinois at Urbana-Champaign, Urbana, Illinois-61801

Abstract: The incorporation of acceptor and donor impurities in MOCVD and GSMBE InP is investigated using Hall-effect, photoluminescence, magneto-photoluminescence, and photothermal ionization spectroscopy. These samples have 77 K electron concentrations in the range of 5×10^{13} - 2.8×10^{15} cm^{-3} and 77 K mobilities in the range of 38,000 - 300,000 cm^2/V-s. The dominant residual acceptor impurity species in the MOCVD and GSMBE samples are Zn and an unidentified species labeled A$_1$, respectively. C is not incorporated as a residual acceptor in InP grown by either of these techniques. Si and S are the dominant donor species detected in MOCVD InP. There is no indication of C donors in the layers that could be measured by either magnetophotoluminescence or photothermal ionization spectroscopy.

1. Introduction

The incorporation of residual impurities in InP has not been as extensively studied as in GaAs. The common acceptor impurities in epitaxial GaAs are C and Zn, although Zn has never been observed as a residual impurity in MBE GaAs. Zn has also been identified as a residual acceptor in epitaxial InP. The incorporation of C impurities in epitaxial InP is not well understood. There have been numerous reports on the presence of C impurities in InP. As cited by Bose, et al. (1990), the identification of C acceptor impurities was either incorrect or inconclusive. C acceptors have not been detected in LEC, LPE or VPE InP, based on the identification of C(D^0-A^0) and C(e-A^0) transitions in photoluminescence of C implanted high purity InP samples by Skromme et al. (1984a). Benchimol et al. (1990) have recently reported on the correlation of net electron concentration (measured by Hall-effect) with C concentration (measured by SIMS) in chemical beam epitaxy (CBE) InP grown at low substrate temperatures. This result suggests the possible incorporation of C as donors in CBE grown InP layers. Both Stillman et al. (1977) and Dean et al. (1984) have reported that photothermal ionization spectroscopy and magnetophotoluminescence measurements have been used to identify the residual donor species in high purity layers. The residual acceptor labeled A$_1$ has been detected in epitaxial InP, but the identification of the chemical species responsible for this acceptor has not yet been made. In this work, the incorporation of these acceptor impurities has been studied in MOCVD and GSMBE InP. The residual donor impurities incorporated in MOCVD InP are also identified.

2. Experimental

The samples studied in this work were grown by MOCVD and GSMBE. The MOCVD samples were grown on Fe-doped InP substrates in two different reactors under different conditions. Trimethylindium and phosphine were used as the In and P sources respectively. The GSMBE samples were grown using elemental Indium and phosphine, also on Fe-doped InP substrates. The V/III ratio was varied for the growth of the GSMBE samples.

The samples were characterized by Hall-effect and low temperature photoluminescence (PL). Magneto-photoluminescence (MPL) and photothermal ionization spectroscopic (PTIS) measurements were also performed to identify residual donor species. Hall-effect measurements on all samples utilized the Van der Pauw configuration at a magnetic field of 0.66 T. The ohmic contacts were alloyed Sn spheres. Photoluminescence measurements were used for the identification of acceptor impurity species. The samples were mounted in a strain-free manner in pumped liquid helium or gaseous helium and were optically excited by the defocused 5145 Å line from an argon ion laser. The sample luminescence was analyzed by a 1.0 m double spectrometer and detected by a liquid nitrogen cooled S-1 photomultiplier tube or a thermoelectrically cooled GaAs photomultiplier tube, depending on the spectral region of interest, using the photon counting technique. PL measurements were made over the temperature range 1.7 - 21 K with optical excitation power ranging from 2 to 20 mW/cm^2. A tunable dye laser was used to achieve resonant excitation for MPL measurements made at a magnetic field of 9.0 T. PTIS measurements were made near liquid helium temperature and at a magnetic field of 6.32 T. The 1s-2p donor transitions were recorded at this magnetic field by the detection of photoconductivity which results from the photothermal ionization of donor species.

3. Results and Discussion

The results of Hall-effect measurements on the samples studied here are given in the table. Both surface and interface depletion corrections were included in these results. The samples have 77 K electron concentrations in the range from 5×10^{13} to 2.8×10^{15} cm^{-3} and 77 K mobilities in the range from 38,000 to 300,000 cm^2/V-s. For GSMBE samples, the sample with the lowest carrier concentration was grown at the lowest V/III ratio.

Figure 1 shows PL spectra of donor/band-acceptor transitions for MOCVD InP sample 3, GSMBE InP sample 7, and C implanted LEC InP (as a reference), recorded at 1.7 K. These spectra were recorded with low optical excitation power and at low temperature for reliable identification of the acceptor impurities. The residual acceptor species in the MOCVD sample are Zn and A_1, while A_1 is the only acceptor species detectable in the GSMBE sample. PL measurements with variable excitation and variable temperature were performed to identify these transitions and impurity species. At higher temperatures, $C(e-A^0)$ transitions become more intense relative to $C(D^0-A^0)$ transitions due to the thermal ionization of donors. No residual C acceptors are detected in any of these samples. It should be noted, however, at higher excitation powers in MOCVD samples 1 and 2 photoluminescence peaks are observed at the same energies where $C(D^0-A^0)$ and $C(e-A^0)$ transitions for C acceptors occur, as determined from measurements on C implanted InP by Skromme et al. (1984b). As noted by Lee et al., these peaks have been shown to be due to LO and TO phonon replicas of free exciton recombination in the undoped MOCVD samples 1 and 2, rather than due to carbon acceptors.

Electrical Parameters and Impurities of InP Layers

Sample No.	Growth Techn.	n_{77} ($\times 10^{14}$ cm^{-3})	μ_{77} (cm^2/V-s)	Impurities	
	Donor			Acceptor	
1	MOCVD	0.5	297000	Zn	Si, S
2	MOCVD	0.5	305000	Zn	Si, S
3	MOCVD	12	75000	A_1,Zn	Si
4	MOCVD	2.4	130000	Zn	Si, S
5	MOCVD	1.4	122000	Zn	Si
6	MOCVD	1.1	79300	Zn	Si, S
7	GSMBE	13	49080	A_1	*
8	GSMBE	23	37880	A_1	*
9	GSMBE	28	40200	A_1	*

*Donor species could not be measured on these samples

The acceptor species detected in the samples studied are also listed in the table. The dominant acceptor species detected in MOCVD InP is Zn while the dominant acceptor species detected in GSMBE InP is the unknown species labeled A_1. Since the peak position of the (D^0-A^0) transition for the A_1 acceptor closely matches that of Mg and Be acceptors, as reported by Skromme et al. (1984b), the dominant acceptor detected in GSMBE could be either Be and/or Mg. If this acceptor is Be, this would indicate that there is a

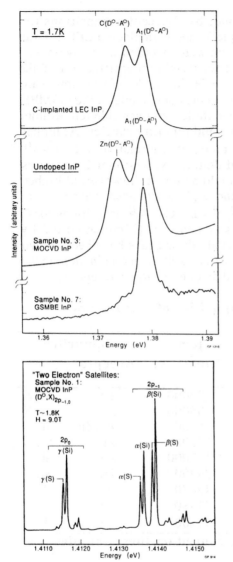

Fig. 1 PL spectra of $(D^0\text{-}A^0)$ transitions for MOCVD and GSMBE InP samples. Zn and A_1 are the acceptors identified in the MOCVD sample, while A_1 is identified as the only acceptor species in the GSMBE sample. C-implanted LEC InP is included at top as a reference.

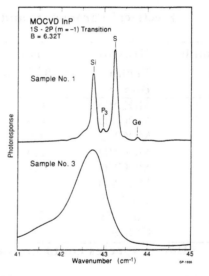

Fig. 2 MPL spectrum of the 'two electron' transitions for a high purity MOCVD InP sample. The donor species identified are Si and S.

Fig 3. PTIS spectra of two MOCVD InP samples. Si and S are idenified as the dominant donor species in sample 1 with trace amounts of Ge and an unknown donor P_3 (speculated to be Se) present. Si is the dominant donor species in the less pure sample 3.

memory effect for Be in InP growth as there is in GaAs growth. If the acceptor is Mg, the source of these impurities may be due to either the In source material, the hot Tungsten of the gas cracker, or possibly the AsH3 as has been previously observed for GaAs grown by GSMBE, noticed by Skromme et al. (1984c). The more likely candidate for the A1 acceptor in MOCVD InP is Mg since the Mg is detected at very·low levels in trimethylindium. C acceptors are not detected in GSMBE samples despite the fact that after growth, large C related peaks were detected with a quadrupole mass analyzer in the MBE chamber during the high temperature degassing of the indium containing PBN crucible. These results confirm that C is not incorporated as a residual acceptor in MOCVD and GSMBE InP and based on the previous work on InP by Benchimol et al. (1990), it can be concluded that C is not incorporated as a residual acceptor in InP grown by any of the LEC, LPE, VPE, MOCVD, and GSMBE growth techniques.

The MPL spectrum of 'two-electron' transitions ('two-electron' replicas of donor bound exciton transitions) for the MOCVD InP sample 1, recorded at 1.7 K, is shown in Figure 2. The dominant donor species is Si. S donors are also present in this sample at a lower concentration. No other donor species are detectable. The same donor species are identified in MOCVD samples 1, 3, 4, and 6. For the other samples, the 'two-electron' replicas are either very weak or not well resolved. Therefore, donor identification by MPL was not possible for these samples.

Photothermal ionization spectroscopy is a more sensitive technique that can be used to identify donor species present at very low levels in high purity samples. PTIS measurements were made on the MOCVD InP samples 1, 2, 3, 4, and 5. Figure 3 shows the spectra for two of these samples 1 and 3. Si is the only donor species detected in the less pure sample 3. For the higher purity sample 1, S is the dominant donor species, Si donors are present at a lower concentration, and a trace amount of Ge donors is also present. In addition, another yet unidentified donor, labeled P_3, is present. Lee et al. shows a comparison of the relative ordering of the chemical shifts of the donors in GaAs and InP indicates that Se is a likely candidate for the P_3 donor. There is an additional weak peak observed in the PTIS spectrum for the sample 2. It has been speculated by Lee et al. and others that the donor species associated with this peak is Te. The table lists the dominant donor species detected in MOCVD InP. Both MPL and PTIS measurements agree with each other regarding the dominant donor species. It should be noted that PTIS, like MPL, is not applicable to samples with carrier concentrations, $n \geq 10^{16}$ cm^{-3}. There is no spectroscopic indication of C donors in the samples that could be measured.

Whether C impurities are incorporated as residual donors in epitaxial InP is not clear. Results of MPL and PTIS measurements on the samples used in this work indicate that C donors are not present in these samples. However, recently Benchimol et al. (1990) have reported on the correlation between C concentration, measured by SIMS, and the net electron concentration in InP grown by CBE under certain conditions. Although the prescence of C donors is not conclusive, it is possible that different impurity incorporation mechanisms in various growth techniques lead to different incorporation behavior of impurity species.

4. Summary

Residual acceptor impurities in MOCVD and GSMBE InP samples are identified. C acceptors are not detected in any of these InP samples. Since it has been shown previously that C is not incorporated as a residual acceptor in LEC, LPE, and VPE InP, these results indicate that C is not incorporated as a residual acceptor in InP. Zn is identified as the dominant residual acceptor species in MOCVD InP. On the other hand, the dominant acceptor in GSMBE InP is A_1. At present, the A_1 acceptor is speculated to be Be and/or Mg. Si and S are identified as the dominant residual donor species in MOCVD InP. There is no indication of C donors in any of the samples that could be measured spectroscopically.

Acknowledgement

We would like to thank E. J. Thrush and M. Razeghi for some of the samples used in this work. We also thank B. Lee for photothermal ionization spectroscopic measurements. This work was supported by NSF-ECD 89-43166 under contract no. 1-5-50321.

References

Benchimol J L, Alaoui F, Gao Y, Le Roux G, Rao E V K and Alexandre F 1990 *J. Crystal Growth* **105** 135
Bose S S, Szafranek I, Kim M H and Stillman G E 1990 *Appl. Phys. Lett.* **56** 752
Dean P J, Skolnick M S and Taylor L L 1984 *J. Appl. Phys.* **55** 957
Lee B, Bose S S, Kim M H, McCollum M J and Stillman G E *unpublished*
Skromme B J, Stillman G E, Oberstar J D and Chan S S 1984a *Appl. Phys. Lett.* **44** 319
Skromme B J, Stillman G E, Oberstar J D and Chan S S 1984b *J. Electron. Mater.* **13** 463
Skromme B J, Stillman G E, Calawa A R and Metze G M 1984c *Appl. Phys. Lett.* **44** 240
Stillman G E, Wolfe C M and Dimmock J O 1977 *Semiconductors and Semimetals* (Academic: New York) vol. 12 p. 169

Inst. Phys. Conf. Ser. No 120: Chapter 11
Paper presented at Int. Symp. GaAs and Related Compounds, Seattle, 1991

541

MOVPE growth of wire-shaped InGaAs on corrugated InP

Takuya Fujii, Osamu Aoki, and Susumu Yamazaki
Fujitsu Laboratories Ltd.
10-1 Morinosato-Wakamiya, Atsugi 243-01, Japan

Abstract

   The InGaAs growth in valleys of corrugated InP substrates is vital to fabricating quantum wire structures for optical device applications. The shape of such corrugations is easily deformed at fairly low temperatures, however, we found that coating the corrugation with an ultra-thin GaAs pregrowth layer at a low temperature effectively preserves the shape. Adding this new process enabled us to grow high-quality wire-shaped InGaAs embedded in InP and observe intense photoluminescence from the InGaAs wires.

1. Introduction

   Two- or three-dimensional quantum-confined structures, such as quantum wires and quantum dots, have been attracting a great deal of attention because they present high-performance devices which operate in new ways [1-4]. For optical device applications, InGaAs wires embedded in InP crystals are considered the basic low-dimensional structure. Such InGaAs quantum wires have been fabricated combining InGaAs/InP quantum well growth on InP substrates and selective mesa-etching of the quantum wells [5,6]. However, it is difficult to fabricate the quantum wires without generating any defects during mesa-etching.

   Selective InGaAs growth in valleys of corrugated InP substrates is an attractive way to get quantum wire structures [7], because this method does not need etching after wire structures are grown. However, no success has been reported in growing high-quality InGaAs wires on the corrugation, because the InP corrugation is easily deformed at fairly low temperatures [8] and it is difficult to preserve its shape without generating new defects.

   We have developed a growth technique which preserves the shape of the corrugation without generating defects, and fabricated high-quality InGaAs

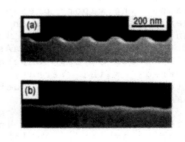

Figure 1. SEM cross sections of (a) an as-etched corrugated InP and (b) the corrugated InP after annealing at 600°C.

wires on the corrugated InP substrates.

2. Experiment

2.1 Prevention of thermal deformation

Figure 1 shows the scanning electron microscope (SEM) images for cross sections of (a) an as-etched corrugated InP substrate and (b) a corrugated InP substrate after annealing at 600°C in a phosphine ambient. The shape of the corrugation is thermally deformed as shown in the figure. According to SEM observations, thermal deformation started at about 450°C.

We consider that the basic deformation process consists of three steps (Figure 2): (1) evaporation of phosphorous atoms from the convex region of the corrugated InP substrate, (2) surface migration of indium atoms from the convex to the concave region, and (3) recombination of the migrated indium atoms at the surface and phosphorous atoms in the vapor at kink sites in valleys. According to this model, we can preserve the corrugation shape by preventing phosphorous from evaporating. Below the deformation temperature, we coated the InP corrugation with an additional pregrowth layer that keeps the phosphorous from evaporating.

We chose GaAs as the pregrowth coating material, because the vapor pressure of arsenic in GaAs is much lower than that of phosphorous in InP and single-crystal GaAs is easily grown at low temperatures compared with ternary and quaternary materials such as InGaAs and InGaAsP.

Figure 3 shows the growth procedure. First, in a phosphine ambient, an InP corrugated substrate is heated to the GaAs growth temperature, which is below the deformation temperature. After the temperature stabilizes, an ultra-thin GaAs layer is grown on the corrugation. Next, the GaAs-coated substrate is heated to the usual metalorganic vapor phase epitaxy (MOVPE) growth temperature in an arsine ambient. Last, an InGaAs crystal is grown on the GaAs-coated

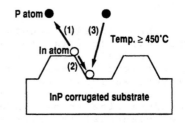

Figure 2. Model for thermal deformation of InP corrugations.

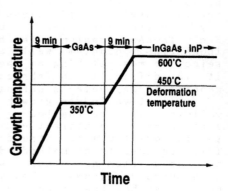

Figure 3. Low-temperature GaAs pregrowth.

corrugated substrate at this temperature.

In this pregrowth method, it is very important to optimize the GaAs thickness. If the GaAs is so thin that the InP corrugated surface is not fully covered, arsine attacks the exposed InP surface during heating to the InGaAs growth temperature and the reaction of arsine gas and InP generates defects at the InP surface. When GaAs is thicker than the critical layer thickness for maintaining the coherent growth of the strained GaAs layer on the InP substrate, misfit dislocations are generated at the GaAs/InP heterointerface and a high-quality InGaAs layer can not be grown on the GaAs-coated InP substrate.

Optimizing the thickness, the GaAs pregrowth layer prevents the phosphorous evaporation and ensures the quality of InGaAs layers grown on the GaAs-coated InP corrugation.

2.2 Growth conditions

InP corrugations as deep as 20-40 nm were prepared by interference lithography and chemical etching. The periodicity of the corrugation was 200 nm. InP/InGaAs/GaAs double hetero (DH) structures were grown directly on the InP corrugations by low-pressure MOVPE after treating InP surfaces with acid. Trimethylindium and triethylgallium were used as group III sources, and phosphine and arsine were used as group V sources. Hydrogen gas was used as the carrier gas. The operating pressure was 50 torr. Growth rates were 1 μm/h for InGaAs and InP layers and 0.1 μm/h for GaAs layers. We grew GaAs very slowly to obtain high-quality layers at low temperatures.

2.3 Procedure

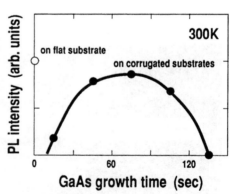

Figure 4. Surface morphologies of InP/InGaAs /GaAs grown on corrugated InP. GaAs growth time is (a) 15 sec, (b) 75 sec, and (c) 135 sec.

Figure 5. Dependence of InGaAs PL intensity on GaAs growth time.

First, we grew InP/InGaAs/GaAs DH structures on the InP corrugated substrates, varying the GaAs growth time to obtain high-quality InGaAs layers. Quality was analyzed by photoluminescence (PL) measurements and the optimized GaAs layer thickness was observed by transmission electron microscope (TEM). Using this pregrowth technique, we tried growing quantum-sized-thin InGaAs layers on the corrugated InP substrates to obtain wire-shaped InGaAs embedded in InP.

3. Results and Discussion

After coating the corrugated InP substrates by GaAs pregrowth layers at 350°C, we grew InP/InGaAs DH structures at 600°C, as shown in Figure 3. Ramping time from the room temperature to the GaAs growth temperature of 350°C and from the GaAs growth temperature to the the InGaAs growth temperature of 600°C was 9 min. This growth procedure enabled us to preserve well the shape of corrugation having a height as low as 20 nm and no defect was observed by SEM.

Figure 4 shows Nomarski interference-contrast images of the DH structures, where InGaAs and InP are 0.3 μm thick. GaAs growth time is (a) 15 sec, (b) 75 sec, and (c) 135 sec. Epilayer surface morphology depends on GaAs growth time. Optimizing the GaAs growth time at 75 sec produced the smooth surface in Figure 4-(b). At 30 sec (a) and 120 sec (c), however, surface morphologies were rough and many surface defects were observed.

Figure 5 shows the strong dependence of PL intensity for the DH structures on the GaAs growth time. The highest PL signal was obtained at 75 sec which corresponds to the time determined by the surface morphology in Figure 4. The PL

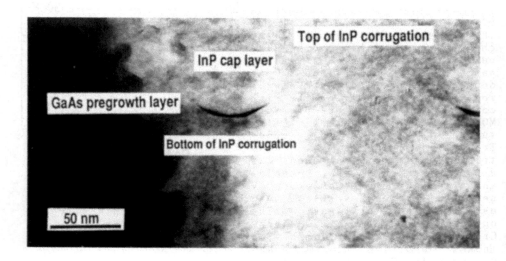

Figure 6.   Cross-sectional TEM image of the InP/GaAs structure grown on a corrugated InP.

intensity rapidly decreases as the GaAs growth time decreases from 60 sec or increases from 90 sec. This indicates that the InP corrugation has to be fully covered with the GaAs, i.e., not be exposed to the arsine ambient, and the GaAs must be thinner than the critical layer thickness. The PL spectra of the optimized DH-structure sample and a reference sample grown on a flat InP substrate without a GaAs pregrowth layer had almost the same intensities, as shown in the figure. Thus, the InGaAs layer grown on the optimized GaAs-coated InP corrugation is high-quality.

We grew an InP/GaAs structure on a corrugated InP substrate to measure the optimized GaAs thickness by TEM, because we can observe a high-contrast image of the cross section of the InP/GaAs/InP structure. Figure 6 shows the TEM bright-field image from the InP/GaAs/InP corrugated substrate. GaAs in the convex region of the corrugation is much thicker than that in the concave region. The GaAs in valleys is about 2 nm thick, and the GaAs at the top of the corrugation is 1 or 2

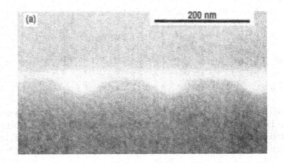

Figure 7. Cross-sectional SEM images of InP/InGaAs structures grown on GaAs-coated corrugated InP.

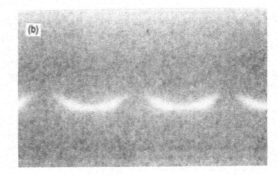

Figure 8. Low-temperature PL spectra of wire-shaped InGaAs crystals compared with the reference sample.

monolayers thick. Such an ultra-thin GaAs pregrowth layer keeps phosphorous evaporation away from the InP substrate and preserves the shape of the InP corrugation.

Applying this process to InGaAs quantum-sized thin-layer growth, we could grow wire-shaped InGaAs in valleys in the InP corrugation. Figure 7 is the SEM cross section of the DH structure of InGaAs of (a) 50 nm and (b) 13 nm thick. In Figure 7-(a) where InGaAs is thicker than the corrugation is high, the InGaAs surface is flat in the (100) direction and spreads over the substrate. However, in Figure 7-(b), where InGaAs is thinner than the corrugation in high, InGaAs grows selectively in valleys in the corrugation and InGaAs wires embedded in the InP crystals are spatially separated. We observed an intense PL signal from the wire-shaped InGaAs at 4.2K. Figures 8-(a) and (b) are PL spectra compared with the reference DH-structure sample grown on a flat substrate (c). These spectra show that few defects were generated at the InGaAs/GaAs and GaAs/InP hetero-interfaces.

The width of the InGaAs wires, determined by the corrugation periodicity, was about 100 nm, not narrow enough to observe quantum wire effects. However, combining this new technique and improving micro-lithography will enable us to fabricate quantum wire structures.

4. Conclusion

We have shown that a low-temperature GaAs pregrowth technique is an effective method in growing high-quality InGaAs on corrugated InP substrates without deforming the corrugation shape. Adding this new process, we could grow wire-shaped InGaAs crystals in the valleys in the InP corrugation and observed an intense PL signal from the InGaAs wires.

References

[1] Y.Arakawa and H.Sasaki, Appl. Phys. Letters 40 (1982) 939.
[2] M.Asada, Y.Miyamoto, and Y.Suematsu, IEEE J. Quantum Electron. QE-22 (1986) 1915.
[3] S.Schmitt-Rink, D.A.D.Miller, and D.S.Chemla, Phys. Rev. 35 (1987) 8113.
[4] H.Zarem, K.Vahala, and A.Yariv, IEEE J. Quantum Electron. QE-25 (1989) 705.
[5] P.Daste, Y.Miyake, M.Cao, Y.Miyamoto, S.Arai, and Y.Suematsu, J. Crystal growth 93 (1988) 365.
[6] M.Naganuma, M.Notomi, H.Iwamura, M.Okamoto, T.Nishida, and T.Tamamura, J. Crystal Growth 105 (1990) 254.
[7] F.S.Turco, S.Simhony, K.Kash, D.M.Hwang, T.S.Ravi, E.Kapon, and M.C.Tamargo, J. Crystal Growth 104 (1990) 766.
[8] H.Nagai, Y.Noguchi, and T.Matsuoka, J. Crystal Growth 71 (1985) 225.

Inst. Phys. Conf. Ser. No 120: Chapter 11
Paper presented at Int. Symp. GaAs and Related Compounds, Seattle, 1991

547

Atomic layer epitaxy growth of InAs/GaAs heterostructures and quantum wells

Shu Goto, Keiichi Higuchi and Hideki Hasegawa

Department of Electrical Engineering and Research Center for Interface
Quantum Electronics, Hokkaido University, Sapporo, 060 Japan

ABSTRACT: The atomic layer epitaxy (ALE) growth and characterization of InAs/GaAs heterostructures and quantum wells (QW) are presented. Trimethylgallium, Trimethylindium and AsH_3 were used in a vertical atmospheric pressure metalorganic vapor phase epitaxy (MOVPE) system. Conditions for the successful ALE growth for heterostructures and QWs by switching the gas and the growth temperature are clarified. The structures were characterized by the Hall effect, quantum Hall effect and photoluminescence (PL) measurements.

1. INTRODUCTION

Recent advances in crystal growth technique have made it possible to produce very fine artificial structures. Lattice-mismatched quantum wells and superlattices can be grown without defects if their layers are sufficiently thin. The atomic layer epitaxy (ALE) is a powerful technique to realize such structures because of its built-in self-limiting mechanism for the monolayer (ML) thickness control.

This paper investigates the ALE growth and properties of InAs/GaAs heterostructures and quantum wells (QWs).

TMG, TMI and AsH_3 were used in a vertical atmospheric pressure MOVPE system. Conditions for the successful ALE growth for heterostructures and QWs by switching the gas and the growth temperature are presented.

The electrical properties of heterostructures having thick InAs ALE layers on GaAs were investigated by the Hall effect and quantum Hall effect measurements. The results indicate presence of the interface Fermi level pinning at the relaxed interface.

Single InAs quantum wells with various well widths were successfully grown by ALE, and they exhibited clear photoluminescence (PL) at 9-77K. From the PL data, the critical thickness is concluded to be 3 MLs.

2. EXPERIMENTAL

The growth system used in the present work was a conventional atmospheric pressure MOVPE system with a vertical reactor. The vertical reactor consisted of a 50mm inner-diameter quartz tube with a water-cooling jacket and a carbon susceptor. The susceptor was heated by an RF induction coil. Substrates with the size of 10mm x 10mm were placed on the carbon susceptor with their surfaces inclined at about 30° to the gas flow so as not to disturb the gas flow pattern. The source materials were TMG, TMI and 10% AsH_3 in H_2. Hydrogen was used as the carrier gas and the total flow rate was kept to 6.0 SLM. GaAs and InAs epitaxial layers were grown on semi-insulating (SI) GaAs substrates with (001) orientation.

Prior to growth, surface oxides on the substrates were thermally removed by heating in an AsH_3 ambient at $650^{\circ}C$. ALE growth of GaAs and InAs were done by alternately supplying TMIII(III=Ga or In) and AsH_3 according to the timing chart shown in Fig.1. In the standard growth, t_{III} for TMG was 5sec, t_{III} for TMI was 10sec and t_A for AsH_3 was 5sec, respectively. A purge time t_p of 5sec was always inserted between the change of the source material.

For thickness measurements, selective etching of epitaxial layers were made for step height measurements. GaAs layers were etched by $H_2SO_4:H_2O_2:H_2O=3:1:1$ at $20^{\circ}C$ and InAs layers were removed by HCl at $20^{\circ}C$.

The structures of the samples used in the present study are shown in Fig.2(a) and (b). For the heterostructure having a thick InAs ALE layer on GaAs shown in Fig.2(a), a thick InAs epitaxial layer was directly grown on the semi-insulating GaAs substrate. Their electrical properties were investigated by both Hall effect and quantum Hall effect measurements.

Samples having a QW structure shown in Fig.2(b) were grown by using both of the conventional MOVPE mode and the ALE mode on semi-insulating GaAs (001) substrates. The undoped GaAs buffer layer with the thickness of 5500A was grown by the conventional MOVPE mode where both of the AsH_3 and TMG are supplied simultaneously to the substrate at $700^{\circ}C$. Then, the QW structures were made by growing GaAs layers and InAs layers by using the ALE mode. The growth temperature was $450^{\circ}C$ for the GaAs layer and $305^{\circ}C$ for the InAs layer. During the gas and temperature switching period between TMG and TMI which took about approximately 5 minutes, the sample was kept under an AsH_3 flow. The QWs were characterized by the PL measurements at 77K (1q.N_2) and 9-12K (1q.He) using an Ar^+ laser light (λ =514.5nm) as the excitation source.

3.RESULTS AND DISCUSSION

3.1 ALE conditions

The observed growth rate of GaAs per cycle is plotted in Fig.3(a) as a function of the injected amount of TMG per cycle for a growth

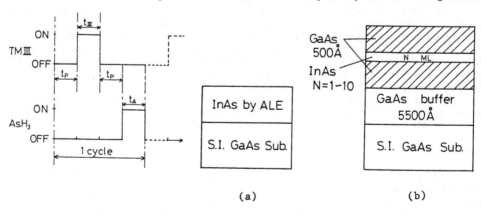

(a)   (b)

Fig.1 Timing chart for ALE mode
   t_{III} : TMIII injection period
   t_A : AsH_3 injection period
   t_p : purge period

Fig.2 Sample structures (a)thick
   InAs/GaAs heterostructures
   (b)ultrathin QW having InAs
   well with of N monolayers
   (N=1-10)

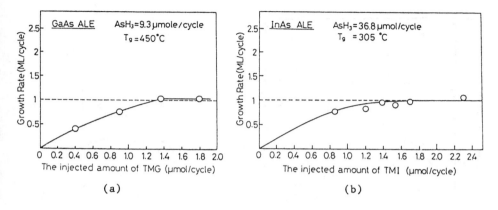

Fig.3 Growth rates vs. injected amount of TMIII
(a) GaAs   (b) InAs

temperature of 450°C. The growth rate showed a clear saturation at 1 ML/cycle, indicating occurrence of the ALE growth mode. The measurement of the temperature dependence of the growth rate indicated that the ALE mode is possible within a narrow temperature range of 450°C ± 5°C.

Similarly, Fig.3(b) shows the observed growth rate of InAs per cycle as a function of the injected amount of TMI per cycle at a growth temperature 305°C. Growth experiment was done on GaAs substrates. Again, a clear saturation to 1ML/cycle is seen. The temperature range for the ALE mode was found to be even narrower for InAs, being in the range of 305°C ± 4°C. Thus, to grow both GaAs and InAs in the ALE mode, a switching of temperature was necessary in our system. When the ALE conditions were met, both GaAs and InAs had mirror-like excellent surface morphology, whereas severe pitting occurred due to direct decomposition of TMG and TMI into Ga and In droplets when the growth temperature was too high.

3.2 Properties of thick InAs/GaAs heterostructures

The undoped GaAs ALE layers grown on SI GaAs substrates were of highly p-type with a typical carrier concentration of 1×10^{18} cm^{-3} possibly due to a high rate of carbon incorporation.

On the other hand, the undoped InAs ALE layers grown on SI GaAs substrates were of n-type, and their electrical properties depended strongly on the layer thickness. The measured carrier concentration and Hall mobility are shown in Fig.4(a) as a function of the InAs layer thickness. As seen in Fig.4(a), carrier density is reduced and the Hall mobility is increased, as the InAs layer thickness increased. The behavior is similar to the case of the MBE growth of InAs (Holmes 1989), and can be explained by the formation of accumulation layers at surface and interfaces as shown in Fig.4(b). Namely, due to presence of high density of surface states and interface states at the relaxed heterointerface, there is a strong tendency that the surface and interface Fermi level is pinned at the hybrid orbital charge neutrality energy E_{HO} of InAs which lies within the conduction band (Hasegawa 1986). Then, the data in Fig.4(a) can be explained by applying the well-known parallel conduction model.

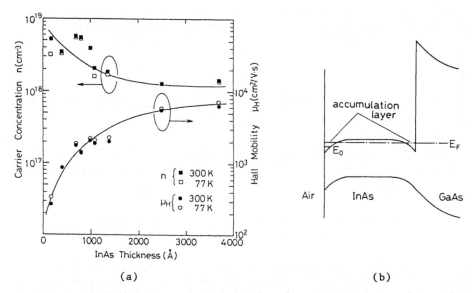

Fig.4 (a) Carrier concentration and Hall mobility of thick ALE InAs layers on GaAs vs. thickness of InAs , (b) a model to explain the data

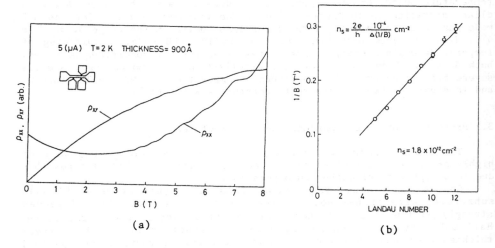

Fig.5 (a) quantum Hall effect data and
(b) determination of 2DEG concentration

The formation of the accumulation layer having the two-dimensional electron gas (2DEG) was directly confirmed by the quantum Hall effect measurements. As shown in Fig.5(a) for a sample having an InAs layer of 900Å, well defined quantum Hall plateau and Shubnikov-de Hass (SdH) oscillations appeared under strong magnetic fields. From a standard plot shown in Fig.5(b), the 2DEG sheet carrier concentration is obtained to be 1.8×10^{12} cm^{-2}. The sample having 200Å thick InAs layer did not show

quantum Hall plateau, whereas the sample
having 400Å thick InAs showed the quantum
Hall plateau at higher fields only. The
result seems to indicate that the observed
2DEG behavior is from the InAs surface,
whereas 2DEG at the InAs/GaAs interface
does not show up due to its low mobility.

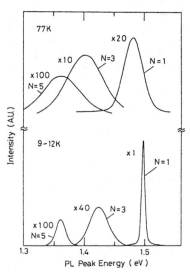

3.3 Properties of single quantum wells

The properties of the single quantum
well by ALE having the structure shown in
Fig.2(b) were investigated by
photoluminescence at 77K and 9-12K. The
number of InAs monolayers in the well N
was changed from 1 to 10. Figure 6 shows
the observed PL spectra for various QWs.
As seen in Fig.6, the ultra narrow QWs in
the present study show well defined PL
peaks with reasonably narrow FWHM values
of a few ten meV even at 77K. The
narrowest FWHM of 6 meV was observed from
a QW of N=1. Generally, the PL intensity
decreased with the increase of N. At N=5,
PL intensity became very much reduced, and

Fig.6 Observed PL spectra
from ultra narrow
InAs QWs by ALE

for N=6-10, no luminescence was observed possibly due to introduction of
high density of dislocations, which produce efficient SRH recombination
centers.

The observed PL peak energies are plotted in Fig.7 vs. the numbers of
monolayers in the well, N. Data by other workers MBE (Brandt 1991, Marzin
1989), MEE(Marzin 1989), ALE(Tischler 1986, Ahopelto 1990), FME(Sato 1989)
and conventional MOVPE (Taira 1988) methods were included in Fig.7. The
shifts of the PL peak energy between 77K and 9-12K were only about 10meV

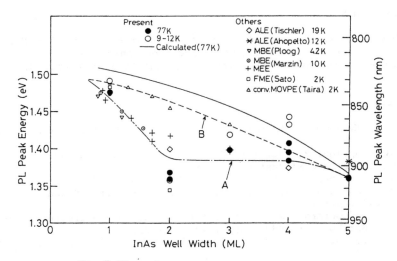

Fig.7 PL peak energy vs. InAs well width

at most which is consistent with the result by Sato (1989). As indicated by the dashed curves A and B in Fig.7, the present result is similar to MBE, MEE, FME and other ALE results (dashed curve A), but is distinctly different from the conventional MOVPE result (dashed curve B). The theoretical curve on the PL peak energy position assuming a complete relaxation at interface is also shown by the solid curve in Fig.7. As is clear in Fig.7, the behavior of the dashed curve A is very different from that of the solid curve up to N=2 and the approaches the solid curve for N=3-5. This can be best interpreted as two-dimensional pseudomorphic growth takes place up to N=2, and start to relax at N=3. Thus, the critical thickness of the InAs in the present ALE is estimated to be N=3.

The difference between the dashed curve A for ALE, FME, MBE and MEE results and the dashed curve B for MOVPE result seems to reflect the difference in the microscopic details of the growth mode. The former involves two-dimensional layer-by-layer growth while the latter involves formation of small islands whose size is below the exciton wavelength.

4.CONCLUSION

Growth of InAs/GaAs heterostructures and QWs by ALE in a vertical atmospheric pressure MOVPE system using TMG, TMI and AsH_3 was investigated. Main conclusions are the following.

(1)ALE growth for heterostructures and QWs is possible just by switching the growth temperature.

(2)The electrical properties of thick InAs layers on GaAs showed strong dependence on the thickness of InAs and this is due to surface and interface Fermi level pinning which produces accumulation layers.

(3)Ultrathin QWs by present ALE show clear photoluminescence at 9-77K. The observed PL behavior indicates that ALE growth mode involves two-dimensional layer-by-layer growth and the critical thickness is estimated to be 3 MLs.

ACKNOWLEDGMENTS

This work was supported in part by a Grant-in-Aid for Scientific Research on Priority Areas on Mesoscopic Electronics (No.03237103) and by a Grant-in-Aid for Scientific Research (#03456147), all from Ministry of Education.

REFERENCES

Ahopelto J, Kattelus H P, Saarilahti J and Suni I 1990 J.Cryst.Growth **99** 550
Brandt O, Cingolani R, Tapfer L, Scamarcio G and Ploog K 1991 Superlattices and Microstructures **9** 147
Hasegawa H and Ohno H 1986 J.Vac.Sci.Technol. **4** 1130
Holmes S, Strandling R A, Wang P D, Droopad R, Parker S D and Williams R L 1989 Semicond.Sci.Technol. 4 303
Marzin J Y and Gerard J M 1989 Superlattices and Microstructures **Vol.5** 51
Sato M and Horikoshi Y 1989 J.Appl.Phys. **66** 851
Taira K, Kawai H, Hase I, Kaneko K and Watanabe N 1988 Appl.Phys.Lett. **53** 495
Tischler M A, Anderson N G and Bedair S M 1986 Appl.Phys.Lett. **49** 1199

Inst. Phys. Conf. Ser. No 120: Chapter 11
Paper presented at Int. Symp. GaAs and Related Compounds, Seattle, 1991

553

Radical-assisted organometallic vapor-phase epitaxy

S. H. Li*, C. H. Chen, and G. B. Stringfellow
Dept. of Materials Science and Engineering
University of Utah, Salt Lake City, UT 84112

*Current Address: Solid-State Electronics Laboratory
Dept. of Electrical Engineering and Computer Science
University of Michigan, Ann Arbor, MI 48109

ABSTRACT: Supplemental radicals are demonstrated to be a method of enhancing the low temperature growth rate in organometallic vapor-phase epitaxy. This may prove an attractive alternative to the use of lasers and plasmas, reported earlier. t-C$_4$H$_9$ radicals, produced from the pyrolysis of azo-t-butane ((t-C$_4$H$_9$)$_2$N$_2$), have been used to assist the growth of GaAs from trimethylgallium and arsine. The epilayer growth rates at temperatures below 500°C are significantly enhanced by the presence of the radicals; at 450°C, the growth rate is increased by a factor of 6. The surface morphologies are also improved. X-ray diffraction results confirm that the layers are single-crystalline GaAs.

1. INTRODUCTION

Low temperature epitaxial growth of semiconductor materials is desired for the production of many electronic device structures, particularly those containing extremely fine composition and/or doping modulations. Low temperature processing also reduces the concentrations of residual impurities and defects, and the probability of wafer warpage in thermally mismatched systems. Growth rates at low temperatures are usually low in chemical vapor transport systems due to the slow pyrolysis rates of the precursor molecules, leading to long growth times. In the past, lasers and plasmas have been utilized to assist the growth at low temperatures (Putz et al 1984, Heinecke et al 1986). In this work, supplemental radicals were employed to serve the same purpose. This is the first report of the radical-assisted growth of III/V semiconductors, although H radicals were previously used by Yasui et al (1990) to assist the growth of SiN films in a low pressure system.

Radicals are thought to play an important role in organometallic vapor-phase epitaxy (OMVPE) (Butler et al 1986, Speckmen and Wendt 1990). In ersatz reactors, at relatively high input concentrations, radical reaction pathways are common in the decomposition of a number of OMVPE precursors (Price 1972). It is believed that in a multi-precursor system (trimethylgallium (TMGa) plus arsine for example), radicals produced from the pyrolysis of one precursor will assist the decomposition of other precursors . This, coupled with the generation of supplemental radicals at low temperatures demonstrated by Li et al 1989, leads to the idea that OMVPE growth can be assisted at low temperatures by providing the proper radicals.

Radicals such as CH_3, C_2H_5, and C_4H_9 at typical OMVPE growth conditions are reactive in the gas phase. They have a tendency to abstract an H from the precursors TMGa and arsine, as well as from the ambient H_2 (Li 1991). After H abstraction, the remaining fragments of the precursor molecules are immediately destablized. Abstracting H from H_2 produces H radicals, which themselves abstract ligands from TMGa and arsine, leading to rapid pyrolysis. In both cases, the decomposition rates of both TMGa and arsine are increased; thus, the GaAs growth rate is increased.

The chemistry of radical-assisted OMVPE has been described in more detail by Buchan et al (1991) and Li et al (to be published). This paper emphasizes the growth results.

2. EXPERIMENTAL

Azo-t-butane ((t-C_4H_9)$_2$$N_2$ or ATB) was selected as the radical source. ATB pyrolyzes at approximately 200-300°C yielding t-C_4H_9 radicals and inert N_2 (Blackham and Eatough 1962). The ATB purity was 97.7%; the impurities were mainly ATB isomers. The Ga and As precursors, TMGa and arsine, were of electronic grade.

The growth apparatus is a conventional, atmospheric-pressure, IR-heated, horizontal OMVPE reactor. The carrier gas was Pd-purified H_2. The partial pressures of TMGa, arsine, and ATB were approximately 6.6×10^{-5}, 1×10^{-2}, and 2×10^{-3} atm, respectively. The total flow rate was 2000 sccm. The substrates were (100)-oriented semi-insulating GaAs and InP. Before growth, the substrates were degreased using trichloroethane, acetone, and methanol sequentially. Then, they were etched in H_2SO_4 for 3 minutes and in $1H_2O:1H_2O_2:4H_2SO_4$ for 4 minutes. After etching, the substrates were rinsed in deionized water. Before transferring into the reactor, they were dried using compressed N_2.

The growth rate was determined by measuring the epilayer thickness (normally between 1-3 μm) on a cleaved cross-section. A DIANO 8000 diffractometer was used for the x-ray diffraction measurements using Cu $K\alpha_1$ and $K\alpha_2$ radiation.

3. RESULTS AND DISCUSSION

GaAs growth using TMGa and arsine without the addition of ATB was first studied to provide reference data for comparison with the ATB-assisted growth results. For substrate temperatures above 500°C, reasonably good surface morphologies were obtained at a V/III ratio of 150. At temperatures of less than 500°C, the surface morphology began to show whisker-like features, as shown in Fig.1a, indicative of vapor-liquid-solid (VLS) growth. This is attributed to an insufficient As concentration at the surface giving Ga-rich conditions. This is due to incomplete pyrolysis of arsine under these conditions. Addition of ATB leads to improved surface morphologies, as seen in Fig.1b, because an increase in the rate of arsine decomposition was caused by the added t-C$_4$H$_9$ radicals (Li et al, to be published).

(a)   (b)

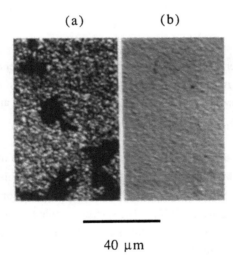

40 μm

Fig.1 Surface morphologies for GaAs epilayers grown on InP substrates for 1 hour at 500°C. (a) without ATB, and (b) with ATB. The ATB partial pressure was 2×10^{-3} atm.

The temperature dependence of the GaAs growth efficiency (growth rate/TMGa molar flow rate) with and without ATB is shown in Fig.2. The solid data are for TMGa/arsine/ATB partial pressures of $6.6\times10^{-4}/2\times10^{-3}/2\times10^{-3}$ atm. The precursor partial pressures for the open data are $6.6\times10^{-5}/1\times10^{-2}/2\times10^{-3}$ atm. In both cases, the growth efficiency was found to be increased by the addition of ATB at temperatures of less than 500°C. The activation energy for the radical-assisted growth is lower than for the regular OMVPE growth. Obviously, the reaction mechanism has been changed by the addition of ATB.

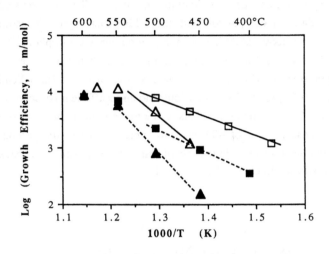

Fig.2 Temperature dependence of GaAs growth efficiency with and without ATB. Ratios of concentrations of TMGa:arsine:ATB = 1:3:0 (▲), 1:3:3 (■), 1:150:0 (Δ), and 1:150:30 (□).

The growth efficiency for the low TMGa input concentrations (open data points) was higher than for the high TMGa concentrations (solid data points). This is because the low temperature growth rate saturates for high TMGa input concentrations as demonstrated by Reep and Ghandhi (1983). The surface morphologies for the samples grown with ATB at temperatures between 380 and 460°C have an orange-peel texture similar to the morphology shown in Fig.1b.

Figs. 3a and 3b show the x-ray diffraction scans for the samples grown at 460 and 420°C using ATB. The samples were grown on InP substrates so that the peaks from the substrate and the epilayer could be distinguished. The two peaks with 2θ values between 63.5 and 64.0 are the $K\alpha_1$ and $K\alpha_2$ peaks from the InP substrate. The peaks with 2θ values near 66.5 are from the GaAs epilayer. Even though there is a large lattice mismatch between InP and GaAs, the $K\alpha_1$ and $K\alpha_2$ peaks are still well-resolved. The results indicate that the addition of ATB does not harm the crystallinity of the GaAs.

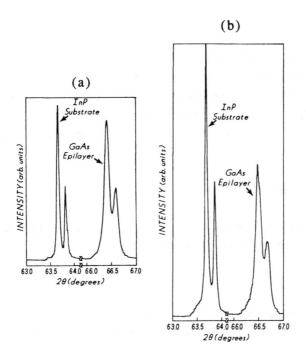

Fig.3 X-ray diffraction scans for GaAs epilayers grown on
(100) InP at (a) 460°C, and (b) 420°C. The doublet is due to
the Kα₁ and Kα₂ radiation.

Attempts have been made to characterize the samples with the Van der Pauw technique.
However, the epilayers are highly resistive (resistivity in excess of 100 Ω-cm),
preventing meaningful measurements. The carrier concentration in these samples is
probably below 10^{14} cm^{-3}. Similar results have been reported by Dapkus et al. (1981)
for GaAs grown using TMGa and arsine at temperatures below 550°C.

The carbon doping level was not determined for these epitaxial layers. However,
Buchan et al. (1991) showed that the use of ATB in the OMVPE growth of GaAs using
ordinary growth conditions results in reduced carbon levels. Incorporation of other
impurities in the samples may be possible, since low growth temperatures are known to
increase the incorporation of volatile impurities (Stringfellow 1986).

Low temperature photoluminescence (PL) measurements were carried out at 10 K. The
spectra show a single, very broad peak (the full widths at half maximum were more
than 100 meV) near the band edge. The origin of the broadening is unclear.

5. SUMMARY

t-C4H9 radical-assisted OMVPE growth of GaAs using TMGa and arsine as the precursors has been demonstrated. The radicals were produced from the pyrolysis of the co-reactant ATB. The OMVPE growth rate was significantly increased in the range from 380 to 500°C: The growth rate was enhanced by a factor of 6 at 450°C. The resulting epilayers are single-crystalline. The surface morphology was improved by the addition of ATB.

6. ACKNOWLEDGEMENTS

Thanks to D. H. Jaw for his help in characterization. This work was supported by the U. S. Air Force, contract number of AFOSR-87-0233.

7. REFERENCES

Blackham A U and Eatough N L 1962 J. Am. Chem. Soc. **84** 2922

Buchan N I, Kuech T F, Beach D, Scilla G, and Cardone F 1991 J. Appl. Phys. **69** 2156

Butler J E, Bottka N, Sillmon R S, and Gaskill D K 1986 J. Crystal Growth **77** 163

Dapkus P D, Manasevit H M, Hess K L, Low T S, and Stillman G E 1981 J. Crystal Growth **55** 10

Heinecke H, Brauers A, Luth H, and Balk P 1986 J. Crystal Growth **77** 241

Li S H, Buchan N I, Larsen C A, and Stringfellow G B 1989 J. Crystal Growth **98** 309

Li S H 1991 Ph.D. Dissertation, University of Utah

Li S H, Chen C H, Jaw D H, and Stringfellow G B (to be published) Appl. Phys. Lett.

Price S J W 1972 in *Decomposition of Inorganic and Organometallic Compounds*, eds. C H Bamford and C F Tipper(Amsterdam: Elsevier Pub.) p.254

Putz N, Heinecke H, Veuhoff E, Arens G, Heyen M, Luth H, and Balk P 1984 J. Crystal Growth **68** 194

Reep D H and Ghandhi S K 1983 J. Electrochem. Soc. **130** 675

Speckman D M and Wendt J P 1990 J. Electronic Materials **19** 495

Stringfellow G B 1986 J. Crystal Growth **75** 91

Yasui K, Nasu M, and Kaneda S 1990 Jpn. J. Appl. Phys. **29** 2822

Inst. Phys. Conf. Ser. No 120: Chapter 11
Paper presented at Int. Symp. GaAs and Related Compounds, Seattle, 1991

559

MOVPE growth of InGaAsP laser diodes using tertiarybutylarsine and tertiarybutylphosphine

A. Kuramata, H. Kobayashi, S. Ogita, and S. Yamazaki
Fujitsu Laboratories Ltd., 10-1 Morinosato-Wakamiya, Atsugi
243-01, Japan

1. Introduction

MOVPE growth is a very useful technique both for making the abrupt heterointerfaces used in advanced III-V semiconductor devices, such as HEMTs and MQW laser diodes, and for growing on large-scale and multiple wafers. MOVPE enables devices to be mass-produced. Given that large number of MOVPE apparatuses are used in manufacturing plants, a problem has arisen in the potential danger in the handling of group V sources, AsH_3 and PH_3, because these are toxic and stored under high pressure. One way to solve this problem would be to use safe group V sources. Among the organic group V sources that have been tried are tertiarybutylarsine (TBA) and tertiarybutylphosphine (TBP), which show promise because of the high-quality GaAs[1], AlGaAs[2], InP[3], InGaAsP[4], GaInP[5] and AlGaInP[6] crystals grown using them. Devices such as FETs[7], HBTs[8] and HEMTs[2] have been produced using TBA, and its usefulness has already been proven at the device level. Few reports have yet to be made on device fabrication using TBP, however. Duncan et. al. made an InGaAsP laser diode using TBA and TBP[9], but the lasing performance was poor.

The fabrication of InGaAsP optical devices using TBA and TBP requires that we study doping conditions into crystals and growth conditions for making InP/InGaAsP/InP double heterostructures. This paper covers these issues and the 1.3-μm-wavelength InGaAsP laser diodes we fabricated. The laser diodes grown using TBA and TBP showed the same performance as those grown using AsH_3 and PH_3.

2. Experiments

A low-pressure vertical reactor with four gas injectors[10] was used to grow InP and InGaAsP crystals at an operating pressure of 76 or 20 torr at 600°C. Hydrogen was used as the carrier gas, with the total flow adjusted to 6 slm. Trimethylindium (TMI) and triethylgallium (TEG) were used as the group III sources and TBA and TBP as the group V sources. The typical V/III ratio was between 20 and 50 for both InP and InGaAsP. To determine the difference between organic and hydrides sources, we also grew crystals using AsH_3 and PH_3. The growth rate was about 1 μm/h with the total group III source gas flow being 0.25 sccm. We tried using silane (SiH_4) and dimethylzinc (DMZ) as the n-type and p-type dopants. Carrier concentrations of the crystals were measured by the C-V

method using a Polaron etch profiler.   Photoluminescence (PL)
intensity was measured at room temperature using an Ar^+ laser.
The surface morphology was observed by Nomarsky microscopy and
scanning electron microscopy (SEM).

3. Results and Discussions
3.1 Doping into InP
Doping into the crystal is an important factor in device
fabrication.   We studied the surface morphology and doping
efficiency in intentional doping when we grew InP crystals
using TBP as the group V source, and compared them with the
case using PH_3.

Figures 1 and 2 show the growth-pressure dependence of
surface morphology for InP in SiH_4 doping and DMZ doping.   When
we used SiH_4 as a n-type dopant and doped it to a carrier
concentration of about $5x10^{17}$ cm^{-3}, a large number of small
pits were observed at 76 torr (Fig. 1(a)).   InP cyrstal was not
grown in these pits, some of which contain small particles as
shown in the SEM image in Fig. 1(d).   At 20 torr, no pits were
observed even when doping up to carrier concentration of $1x10^{18}$
cm^{-3} (Fig. 1(b)).   Pits did not appear, ether, on InP surfaces
with a carrier concentration of $1x10^{18}$ cm^{-3} when we grew InP
using PH_3 as the group V source and SiH_4 as the n-type dopant
at 76 torr (Fig. 1(c)).   These results suggest an undesirable
side reaction between TBP and SiH_4 at 76 torr, resulting in the
surface pits.   The small particles observed in the pits are
probably products of the side reaction.   This reaction is
suppressed when the pressure is 20 torr because of the high gas
velocity and low partial pressure of TBP and SiH_4, which leads
to a low molecule collision probability.   When DMZ was used as
the p-type dopant, no surface degradation was observed at
either 76 or 20 torr (Fig. 2).   Thus, we assume that no side
reaction occurs between TBP and DMZ in this pressure range.

Figure 3 shows the relationship between the carrier
concentration and dopant flow at 20 torr.   Both SiH_4 and DMZ
showed an almost linear relationship, indicating normal doping
up to a carrier concentration of about $1x10^{18}$ cm^{-3}.   Silicon is
the n-type dopant substituting for group III atoms, so the
doping efficiency should be independent of the group V source
in the simplest case where the dopant and group V source do not
interact.   The doping efficiency of SiH_4 with TBP is about 10
times that with PH_3, suggesting the interaction between the
group V sources and SiH_4.   No difference in doping efficiency
was observed between TBP and PH_3 for DMZ, however.   Zinc is
also a dopant substituting for group III atoms, and this case
corresponds to the simplest case with no interaction between
the dopant and group V sources.

We confirmed the possibility of n-type and p-type doping up
to a carrier concentration of about $1x10^{18}$ cm^{-3} without surface
degradation when we grew InP using TBP as the group V source.
This was done using SiH_4 and DMZ as the dopant and by reducing
the pressure to 20 torr.

3.2 Growth of InP/InGaAsP/InP double heterostructures
The InP/InGaAsP/InP double heterostructure is the basic
structure for optical devices such as laser diodes and photodiodes.   When we grew such structures using TBA and TBP, we

found some peculiar surface defects. We checked the effects of gas-switching conditions on the density of surface defects in InP/InGaAsP/InP heterostructures and the relationship between defect density and PL intensity. Growth was at 20 torr.

Figure 4 shows typical surface defects produced when we grew double heterostructures using TBA and TBP. We alternatively performed surface observation and selective etching, and found no defects on the surface of either the InP buffer layer or the InGaAsP layer. This indicates that the defects observed in double heterostructures were generated from the InGaAsP/InP heterointerfaces.

The quality of the heterointerface changes if we change gas-switching conditions. Changing these gas-switching conditions at the InGaAsP/InP heterointerface, we studied the defect density and PL intensity. Figure 5 shows the gas-switching sequence. Four quantities were changed as parameters, i.e., the total flow of TBA and TBP in InGaAsP growth, V_1; the TBP flow in InP growth, V_2; the TBA and TBP purge time in growth interruption, t_1; and the TBP purge time in growth interruption, t_2. As a standard condition, we set V_1, V_2, t_1 and t_2 at 20 sccm, 20 sccm, 30 seconds, and 20 seconds. When one parameter was changed, the other parameters were made equal to the standard condition.

Figure 6 shows the dependence of defect density on parameters. Two parameters, V_1 and t_2, affected the defect density (Figs. 6(b) and (c)) but not parameters, V_2 and t_1 (Figs. 6(a) and (d)). The defect density first decreased from 5×10^4 cm^{-2} when t_2 was changed from 1 second to 10 seconds, then remained at less than 2×10^3 cm^{-2} with t_2 between 10 seconds and 30 seconds, and finally increased above 1×10^5 cm^{-2} when t_2 exceeded 40 seconds (Fig. 6(b)). The influence of V_1 was rather small. The defect density increased from 6×10^2 cm^{-2} to 1.5×10^4 cm^{-2} when V_1 was changed from 20 sccm to 50 sccm (Fig. 6(c)). These results indicate that arsenic molecules remaining after InGaAsP growth are related to defect generation during InP growth, because the concentration of residual arsenic molecules on the surface at the beginning of InP growth is decided by both the total flow of TBA and TBP in InGaAsP growth, V_1, and the TBP purge time from the end of TBA supply up to the beginning of InP growth, t_2. When the defect density is minimum on t_2, we considered the arsenic concentration on the surface at the beginning of InP growth on the InGaAsP crystal to be suitable.

The PL intensity reflects the amount of nonradiative recombination in the crystal. Figure 7 shows the dependence of PL intensity on defect density. PL intensity remained unchanged with the defect density and equalled that of the double heterostructure grown using AsH$_3$ and PH$_3$, even when the defect density was 5×10^5 cm^{-2}. Two possible explanations can be posited: first, such defects do not operate as a nonradiative recombination center in the crystal, and, second, the amount of defects is too low to reduce the PL intensity even if the defects operate as such. Whichever the case, the fact that the defect density in this range does not affect PL intensity favors the application to laser diodes.

We clarified that the generation of surface defects was minimized to a density of 6×10^2 cm^{-2} under proper gas-switching conditions and that defects of this type do not affect PL intensity when the defect density is less than 5×10^5 cm^{-2}.

3.3 Laser diode performance

Based on our findings, we fabricated 1.3-μm-wavelength Fabry-Perot InGaAsP FBH laser diodes. The first growth of the active and cladding layers was by MOVPE using TBA and TBP, and the second growth of embedded layers and the third growth of contact layers were by LPE. The cavity was 300 μm long and the the active layer was 1.6 μm wide. Figure 8 shows the light output dependence of the drive current. The threshold current was 15 mA, the maximum light output exceeded 25 mW, and slope efficiency was 0.27 mW/mA. These values were the same as for laser diodes grown using AsH$_3$ and PH$_3$. Lasing wavelength uniformity and diode yield were also very good. The usefulness of TBA and TBP for the fabrication of optical devices is thus confirmed by these results.

4. Conclusion

We studied doping into the InP crystal and the growth of InP/InGaAsP/InP double heterostructures by MOVPE using TBA and TBP. We found both n-type and p-type doping into InP possible when we used SiH$_4$ and DMZ at 20 torr. Surface defects typically appearing in growth using TBA and TBP were minimized when we optimized the TBP purge time in growth interruption and the total TBA and TBP flow in InGaAsP growth, even though such defects did not affects the PL intensity. The 1.3-μm-wavelength laser diodes we fabricated showed the same performance as those grown using AsH$_3$ and PH$_3$. This confirms the usefulness of TBA and TBP in the fabrication of such optical devices as laser diodes.

References
[1] G. Haacke, S. P. Watkins, and H. Burkhard, Appl. Phys. Lett. 56, 478· (1990)
[2] T. Kikkawa, H. Tanaka, and J. Komeno, J. Appl. Phys. 67, 7576 (1990)
[3] A. Kuramata, S. Yamazaki, and K. Nakajima, Ninth Symposium Record of Alloy Semicondutor Physics and Electronics, Izunagaoka, 1990, pp. 49
[4] A. Kuramata, S. Yamazaki, and K. Nakajima, Mater. Res. Soc. Proc. 216, Boston, 1990, pp. 365
[5] S. R. Kurtz, J. M. Olson, and A. Kibbler, J. Electron. Mater. 18, 15 (1989)
[6] D. S. Cao, and G. B. Stringfellow, J. Electron. Mater. 20, 97 (1991)
[7] R. M. Lum, J. K. Klingert, F. Ren, and N. J. Shah, Appl. Phys. Lett. 56, 379 (1990)
[8] T. S. Kim, B. Bayraktaroglu, T. S. Henderson, and D. L. Plumton, Appl. Phys. Lett. 58, 1997 (1991)
[9] W. J. Duncan, D. M. Baker, M. Harlow, A. English, A. L. Burness, and J. Haigh, Electron. Lett. 25, 1603 (1989)
[10] A. Kuramata, S. Yamazaki, and K. Nakajima, Inst. Phys. Conf. Ser. 96 (Inst. Phys., London, 1989) 113

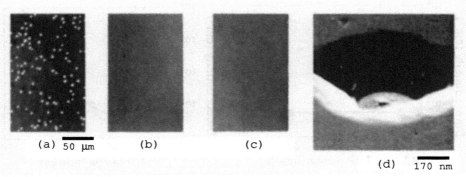

Fig. 1 Surface morphology of SiH4-doped InP: (a) TBP at 76 torr; (b) TBP at 20 torr; (c) PH3 at 76 torr; and (d) SEM image of the pit shown in (a).

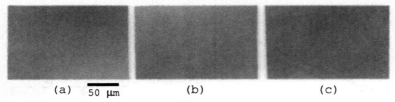

Fig. 2 Surface morphology of DMZ-doped InP: (a) TBP at 76 torr; (b) TBP at 20 torr; (c) PH3 at 76 torr.

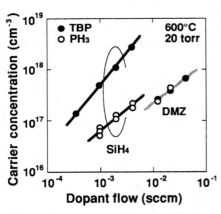

Fig. 3 Dependence of the InP carrier concentration on the dopant flow.

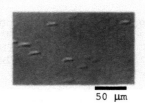

Fig. 4 Defects tipically appearing in InP/InGaAsP/InP double heterostructures.

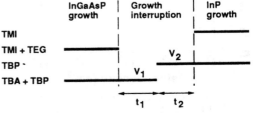

Fig. 5 Gas-switching sequence at the InGaAsP/InP heterointerface.

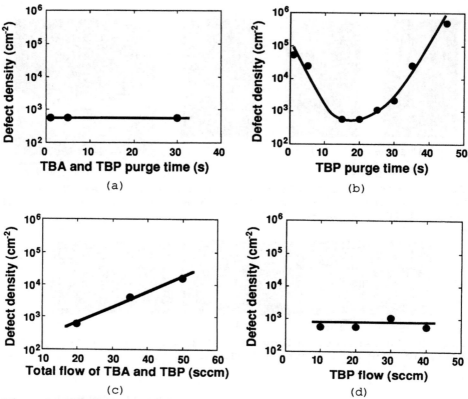

Fig. 6 Dependence of the defect density on the parameter in gas-switching: (a) TBA and TBP purge time in growth interruption, t1; (b) TBP purge time in growth interruction, t2; (c) total TBA and TBP flow in InGaAsP growth, V1; (d) TBP flow in InP growth, V2.

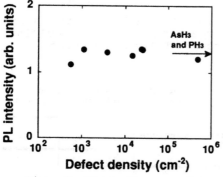

Fig. 7 Dependence of PL intensity on the defect density.

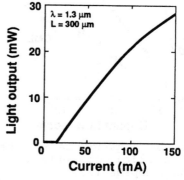

Fig. 8 I-L characteristics of 1.3-μm-wavelength InGaAsP laser diodes fabricated using TBA and TBP.

Extremely uniform growth on 3-inch diameter InP substrates by OMVPE for N-AlInAs/GaInAs HEMT application

Michio Murata, Hiroshi Yano, Goro Sasaki, Hidenori Kamei,
and Hideki Hayashi

Optoelectronics R&D Lab., Sumitomo Electric Industries,Ltd.
1,Taya-cho,Sakae-ku,Yokohama 244,Japan

ABSTRACT: Highly uniform GaInAs, AlInAs and GaInAs/InP MQW structures with excellent electrical and optical qualities were grown on 3-inch diameter InP substrates by low pressure OMVPE. The undoped GaInAs layer showed an electron mobility of 12,000 cm^2/Vs with a carrier density of 7×10^{14} cm^{-3} at room temperature. The variation of lattice mismatch across 3-inch diameter substrates was less than $\pm 3 \times 10^{-5}$ both for GaInAs and AlInAs layers. An excellent thickness uniformity with a fluctuation within $\pm 1\%$ was obtained for the GaInAs/InP MQW structure. The doping uniformity of Si-doped AlInAs was $\pm 5\%$. Si-doped AlInAs/GaInAs HEMTs grown on a 3-inch diameter InP substrate showed the uniform threshold voltage with a standard deviation of 5%.

1. Introduction

N-AlInAs/GaInAs HEMTs on an InP substrate have the potential of ultra-high-speed operation (Mishra et al. 1988). Furthermore, they have the merit of applicability to monolithic integration with optical devices such as GaInAs *pin*-photodiodes and lasers operating in the wavelength from 1.3 to 1.5 μm (Yano et al 1991). Many works related to N-AlInAs/GaInAs HEMTs have been done using molecular beam epitaxy (MBE). However, organometallic vapor phase epitaxy (OMVPE) is preferable to MBE in terms of integration with optical devices, because of its higher growth rate and aptitude for growths of phosphorus compounds.

Scaling up of a wafer size is a very effective way to expand applications. Actually, 3-inch diameter epitaxial wafers are commonly used for GaAs-based devices (Komeno et al 1989). However, few works have been done for epitaxial growth on 3-inch diameter InP at present (Kondo et al 1990). The only work on N-AlInAs/GaInAs HEMTs grown on a 3-inch diameter InP substrate by MBE was recently reported (Imanishi et al 1991). It becomes more difficult to obtain uniform epitaxial layers with increasing wafer size. A precise composition control is very important for InP-based materials to avoid lattice mismatch. A good uniformity of thickness and carrier density is also required, because the threshold voltage of HEMTs strongly depends on the sheet carrier density. Furthermore, residual impurities in the GaInAs channel layer and/or the AlInAs donor layer degraded the device performance.

In this paper, we describe the properties of epitaxial layers grown on 3-inch diameter InP substrates by OMVPE in detail. First, optical and electrical properties of undoped GaInAs layers are described. Next, the growth rate uniformity of GaInAs/InP multiple quantum well (MQW) structure is discussed. Finally, uniformity of composition and doping density of Si-doped AlInAs is mentioned. For the application of these materials, Si-doped AlInAs/GaInAs HEMTs were fabricated. Good characteristics with highly uniform threshold voltage was obtained.

2. Experiment

A conventional OMVPE growth system with a vertical quartz reactor was used. Palladium-diffused hydrogen with a total flow rate of 5 standard liters per minute was supplied from the top of the reactor. A 3-inch diameter InP substrate was placed perpendicular to the gas flow on a SiC coated carbon susceptor. The susceptor was heated with a carbon resistive heater and rotated at 12 rpm. Growth pressure was kept at 60 Torr. A vent-run system was used to obtain abrupt hetero-interfaces.

76 mm-diameter Fe-doped InP substrates oriented (100) 2° off toward <110> were used. Triethylgallium (TEG), trimethylindium (TMI), trimethylaluminum (TMA), arsine (AsH_3) and phosphine (PH_3) were used as source materials. Disilane (Si_2H_6) was used as an n-type dopant. Growth temperature was 650 °C to 675 °C. The growth rate was about 2 μm/h.

Lattice mismatch of layers and the period of the GaInAs/InP MQW structure were determined by a double crystal X-ray diffraction using (400) reflection. Optical quality of GaInAs was examined by photoluminescence at 4 K. The 514.5 nm line of an Ar ion laser were used for excitation. Sheet carrier density and electron mobility of GaInAs and Si-doped AlInAs layers were measured using Van der Pauw method at 77 K and room temperature.

3. Characterization of epitaxial layers

3-1. undoped GaInAs

A 5μm-thick GaInAs layer was grown upon a 0.2 μm-thick InP buffer layer to evaluate properties of undoped GaInAs. This layer was closely lattice matched to the InP substrate within the lattice mismatch of 3×10^{-4}. This layer showed electron mobilities of 12,000 cm^2/Vs and 74,000 cm^2/Vs at room temperature and 77 K, respectively. The background carrier density was 7×10^{14} cm^{-3} at both temperatures. This value is comparable to the best value obtained for layers grown on 2-inch diameter substrates.

Figure 1 shows the photoluminescence spectrum measured at 4 K. A sharp band edge emission with a full width at half maximum of 1.6 meV was observed, and donor-acceptor (DA) pair related luminescence was very weak. The high electron mobility and the sharp PL spectrum indicated that the purity of GaInAs was good enough for the channel layer of HEMTs.

Figure 2 shows the uniformity of the photoluminescence peak wavelength. The peak wavelength of the photoluminescence was 1540.7±0.7 nm across the 3-inch diameter substrate. The compositional uniformity was evaluated for another sample. Figure 3 shows the lattice mismatch variation of a 1 μm-thick intentionally lattice-mismatched GaInAs layer across the 3-inch diameter substrate. The variation of lattice mismatch is less than ±3×10⁻⁵ in the 60mm diameter area. The fluctuation of InAs concentration in this area was estimated to be less than ±0.02%, assuming that the layer was strained without misfit dislocations. The lattice constant decrement of 2×10⁻⁴ at the edge is equivalent to the InAs concentration decrement of 0.15%.

3-2. GaInAs/InP MQW

In order to examine the growth rate uniformity, the period thickness of a GaInAs/InP MQW structure on a 3-inch diameter InP substrate was investigated. The MQW structure consists of 40 periods of a 12 nm-thick GaInAs well and a 12 nm-thick InP barrier layer. The X-ray diffraction rocking curve of this sample showed the sharp satellite peaks at all areas measured across the

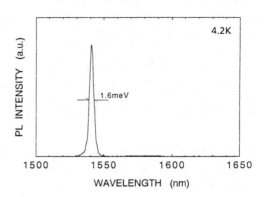

Fig. 1. PL Spectrum of undoped GaInAs.

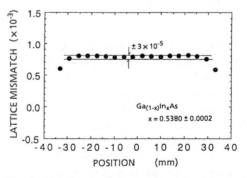

Fig. 2. PL peak wavelength variation of GaInAs across wafer.

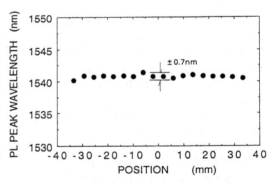

Fig. 3. Lattice mismatch variation of intentionally lattice mismatched GaInAs across wafer.

3-inch diameter substrate. Figure 4 shows the variation of the period thickness which was deduced from a separation of satellite peaks across the 3-inch diameter. An almost flat profile was obtained. The period was 24.4 nm±1%. The thickness fluctuation was less than ±1 monolayer.

3-3. Si-doped AlInAs

A 1 μm-thick Si-doped AlInAs layer was grown. The lattice mismatch of this sample across the 3-inch diameter substrate is shown in figure 5. The variation of lattice mismatch was less than ±3×10⁻⁵ across 3-inch diameter substrate. To our best knowledge, this is the best value ever reported for this material grown either on 2-inch diameter or 3-inch diameter substrates.

The uniform carrier density of the AlInAs layer is essential to the uniformity of threshold voltage of HEMTs. The activation energy of Si doping into AlInAs using Si_2H_6 is about 2.5 eV. This value is much higher than that of Si doping into GaAs (Kamei et al 1988). Therefore, a good uniformity of substrate temperature is very important to achieve a highly uniform carrier density of Si-doped AlInAs layer. Figure 6 shows the variation of carrier density of Si-doped AlInAs layer across the 3-inch diameter substrate. The carrier density was 1.21×10^{18} cm^{-3} with a variation of ±5%. This doping variation corresponds to a variation of substrate temperature of less than 2 °C, assuming the activation energy of 2.5eV.

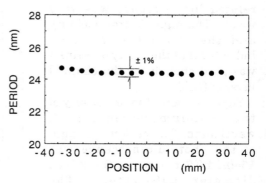

Fig. 4. Variation of period thickness of GaInAs/InP MQW across wafer.

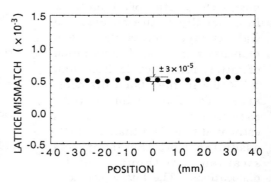

Fig. 5. Lattice mismatch variation of Si-doped AlInAs across wafer.

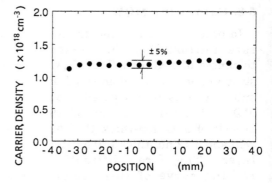

Fig. 6. Carrier density variation of Si-doped AlInAs across wafer.

3-4. Performance of HEMTs

A HEMT structure was grown on a 3-inch diameter InP substrate to confirm the applicability of highly uniform OMVPE grown layers to device applications.

First, a 200 nm-thick undoped InP buffer was grown, followed by a 45 nm-thick undoped GaInAs channel layer, a 5 nm-thick undoped AlInAs spacer layer, an 11 nm-thick Si-doped AlInAs (2×10^{18} cm^{-3}) layer, and a 42 nm-thick Si-doped AlInAs (5×10^{17} cm^{-3}) layer, capped by a 13.5 nm-thick Si-doped GaInAs layer. The higher doped AlInAs layer supplies a high density two-dimensional electron gas (2DEG). The lower doped AlInAs layer enhances the Schottky barrier height. Figure 7 shows the uniformity of electrical properties. The electron mobility and sheet carrier concentration were 10,200 cm^2/Vs $\pm 1\%$ and 2.59×10^{12} cm^{-2} $\pm 4\%$, respectively, in the 3-inch diameter HEMT wafer at room temperature. They were 58,200 cm^2/Vs and 2.46×10^{12} cm^{-2}, respectively, at 77K. The 2DEG was confirmed by Shubnikov-de Haas oscillation at 4.2 K after removing the cap layer. The sheet carrier densities at the ground state and the first exited state were determined to be 1.82×10^{12} cm^{-2} and 4.6×10^{11} cm^{-2}, respectively. HEMTs with a 1.0 μm-gate-length were fabricated using a conventional wet recess etching. Figure 8 and figure 9 show typical DC drain I-V and transfer characteristics of a 40

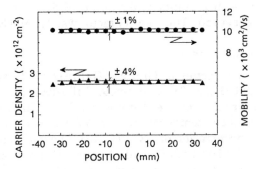

Fig. 7. Uniformity of electrical properties of HEMT structure over 3-inch diameter substrate.

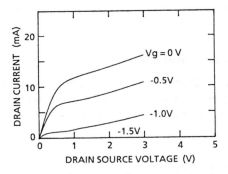

Fig. 8. DC drain I-V characteristics for a 40μm-gate-width device.

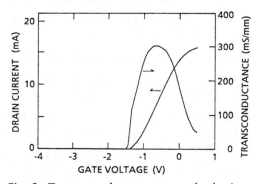

Fig. 9. Transconductance and drain current characteristics for a 40 μm-gate-width HEMT at V_{ds} = 2V.

μm-gate-width device. The maximum transconductance of 300 mS/mm was obtained at $V_{gs}=$ -0.7 V and $V_{ds}=$ 2 V. A good pinch-off characteristic was obtained. The threshold voltage of 120 devices over 3-inch diameter was -1.44 V with a standard deviation of 5%. The Schottky characteristics were also measured. The ideality factor was 1.1 and Schottky barrier height was 0.55eV.

4. Summary

Properties of GaInAs, GaInAs/InP MQW and Si-doped AlInAs grown by OMVPE on 3-inch diameter InP substrate have been investigated. The GaInAs layer showed the electron mobility as high as 12,000 cm^2/Vs and 74,000 cm^2/Vs at room temperature and 77 K, respectively, with a back ground carrier concentration of 7×10^{14} cm^{-3} The variation of lattice mismatch was less than $\pm 3 \times 10^{-5}$ both for GaInAs and AlInAs layers. The period of GaInAs/InP MQW showed a uniformity of $\pm 1\%$. The doping uniformity of the Si-doped AlInAs layer was $\pm 1\%$. HEMTs showed a uniform threshold voltage with a standard deviation of 5%, and a transconductance of 300 mS/mm. These results show that the properties and uniformity of OMVPE-grown layers on a 3-inch diameter substrate is good enough for large area fabrication of InP-based devices.

Acknowledgement

The authors would like to acknowledge T. Katsuyama for fruitful discussions. They also wish to thank M. Koyama and K. Koe for their continuous encouragements throughout this work.

References

Imanishi K, Ishikawa T, Higuchi M, Kondo K, Katakami T, Kuroda S 1991
    Digest 3rd Int. Conf. on InP and related materials, Cardiff, Wales, UK 652

Kamei H, Hashizume K, Murata M, Kuwata N, Ono K, Yoshida K 1988
    J. Crystal Growth **93** 329

Komeno J, Tanaka H, Tomesakai N, Ohori T 1989 Inst Phys. Conf. Ser. **106** 889

Kondo M, Fujii T, Kuramata A, Tanahashi T, Yamazaki S, Nakajima K 1990
    Inst. Phys. Conf. Ser. **112** 187

Mishra U K, Brown A S, Rosenbaum S E, Hooper C E, Pierce M W,
    Delaney M J, Vaughn S, White K 1988 IEEE Electron Device Lett. **9** 647

Yano H, Aga K, Murata M, Kamei H, Sasaki G, Hayashi H 1991
    OFC'91 Tech. Dig., San Diego, CA, 2

Inst. Phys. Conf. Ser. No 120: Chapter 11
Paper presented at Int. Symp. GaAs and Related Compounds, Seattle, 1991

571

MOCVD growth of high-quality delta-doped $Al_xGa_{1-x}As$–GaAs structures

S.M. Vernon, V.E. Haven, A.L. Mastrovito, M.M. Sanfacon, P.A. Sekula-Moisé,
Spire Corporation, Bedford, MA 01730

J. Ramdani, *North Carolina State Univ., ECE Dept. Raleigh, NC 27695*

M.M. Al-Jassim, *SERI, Golden, CO, 80401.*

ABSTRACT

Metalorganic chemical vapor deposition has been used to form Si delta-doped structures of GaAs-$Al_xGa_{1-x}As$, which have been characterized by mobility and sheet-carrier density measurements. At 20 K, mobility values of up to 300,000 cm^2/V-sec, with carrier densities of $\approx 1\times10^{12}$ cm^{-2}, are reported. The mobility at 77 K is 131,000 cm^2/V-sec. The Si delta-doped region has been imaged by cross-sectional transmission electron microscopy, and is seen to have a width of approximately 30 Å. Si delta doping in $Al_xGa_{1-x}As$ (x=0.23) has also resulted in high carrier levels, with an unexplained amphoteric behavior having been observed.

INTRODUCTION

The use of planar doping, also known as "delta" doping (Ploog, 1987), was first demonstrated by Wood *et al* (1980); it is commonly used in MBE-grown devices. Delta doping by MOCVD has been reported (Hobson *et al*, 1989; Hong *et al*, 1989; Kim *et al*, 1990; Mori *et al*, 1986; Nakajima *et al.* 1990; Pan *et al*, 1990; Pearton *et al*, 1989; Schmidt *et al*, 1991) and by others, but its use is less well developed than in the MBE method. Delta doping improves the performance of modulation-doped structures by placing the donor atoms closer to the active region, where they can more completely transfer their electrons to the two-dimensional electron gas (2DEG), while effectively reducing the scattering from the donor impurity atoms.

EXPERIMENTAL PROCEDURES

The structures are grown by low-pressure metalorganic chemical vapor deposition (MOCVD), using the SPI-MO CVD 450™ reactor described previously (Daly and Roberts, 1989). Trimethylgallium (TMG), trimethylaluminum (TMA), and AsH_3 (100%) are the source reagents, SiH_4 in H_2 (100 ppm) is the dopant, and palladium-purified H_2 is the carrier gas. The reactor pressure is varied between 53 and 760 torr, the growth temperatures explored are 650 to 720°C, the growth rate is ≈ 2Å/sec, and the V-III ratio is maintained at 60 to 115. For the growth of the test wafers studied by Hall-effect measurements, the growth conditions are as follows: temperature equals 650°C, pressure equals 53 torr, and the V-III ratio is 75 for the GaAs layers, and 115 for the $Al_xGa_{1-x}As$.

The structure of the delta-doped test samples is shown in Figure 1. For the $Al_xGa_{1-x}As$ test wafers, the structure is similar, but with the epi layers being 0.2 μm thick each, and the composition being $x \approx 0.25$. For the mobility test structures, the layers are as shown in Figure 2. Comparison structures were also grown with the $Al_xGa_{1-x}As$ having conventional bulk doping throughout the 400 Å thick layer.

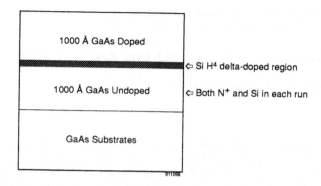

Figure 1   *Test structure used for the calibration of delta doping in GaAs.*

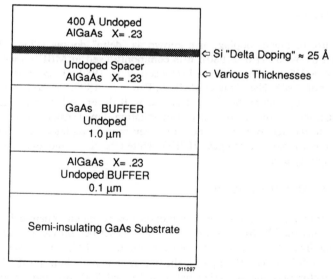

Figure 2   *Structure of delta-doped GaAs - $Al_xGa_{1-x}As$ test wafers.*

Growth has been characterized by Polaron profiling, variable-temperature Hall measurements, and cross-sectional transmission electron microscopy (XTEM). Secondary ion mass spectrometry (SIMS) analysis was also performed, using a 10 keV Cesium beam, a primary ion current of 150 nA, and a sputtering rate of $\approx$ 3Å/sec.

CHARACTERIZATION RESULTS

Delta-Doped GaAs

The test structure employed in these experiments is shown in Figure 1. Depletion-mode Polaron profiles reveal that the peak doping increases significantly as the doping flow times increase. Peak values observed are up to 5.5×10^{18} cm^{-3}. The other variables studied (purge time, growth temperature, and reactor pressure), appear to have less of an influence on the peak height. The full width half maximum (FWHM) of the peak is strongly dependent on both the purge time and pressure, with longer purge times and lower pressures serving to give the narrowest FWHM values. The minimum FWHM measured, by depletion-mode Polaron profiling, in these samples is 83 Å. The SIMS measurement conditions used were not able to accurately resolve such narrow layers; the roughness due to sputtering damage caused the peak widths to appear broader, in the range of 150 to 250 Å.

The sheet carrier densities have been determined by Hall measurements at room temperature. Values up to 1.53×10^{13} cm^{-2} have been obtained. The peak values, measured by Polaron, and the sheet carrier values, measured by Hall, are well correlated, as shown in Figure 3. The Polaron peak values are essentially proportional to the SIMS peak values, although there is considerable scatter in the data, with the Polaron values being typically a factor of two less than those measured by SIMS.

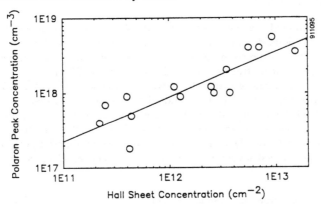

Figure 3   *Plot of peak dopant concentration, measured by Polaron, versus sheet-carrier density, measured by Hall effect, for the delta-doped GaAs test structures.*

Delta-Doped Al$_x$Ga$_{1-x}$As

Delta-doped Al$_x$Ga$_{1-x}$As, x=0.23, was also analyzed by Polaron and Hall measurements. The Al$_x$Ga$_{1-x}$As layers grown at low SiH$_4$ flows are all *P type*. This has been confirmed both by Polaron and Hall measurement results. Peak Polaron values range up to 2×10^{19} cm^{-3}, and Hall results show carrier densities up to 1.93×10^{13} cm^{-2}, with 300 K mobilities of 115 to 181 cm^2/V-sec. Increasing the SiH$_4$ flow causes the films to be N type, with mobilities (at room temperature) up to $\approx$ 1600 cm^2/V-sec, and sheet densities up to 6×10^{12} cm^{-2}.

The peak widths (on the Polaron) generally are narrower in the $Al_xGa_{1-x}As$ samples than in the GaAs samples. The depletion-mode Polaron of the narrowest-FWHM sample is shown in Figure 4. The peak height is 5×10^{19} cm^{-3}, and the FWHM is only ≈ 25 Å.

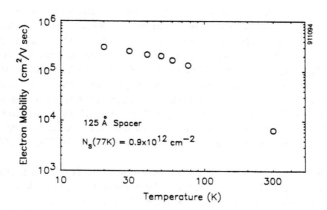

Figure 4 *Depletion-mode Polaron profile of delta-doped $Al_{0.23}Ga_{0.77}As$ test structure.*

Delta-Doped Test Structures

Test structures have been grown, with the $Al_xGa_{1-x}As$ top layer having bulk doping, at 2×10^{18} cm^{-3}, or delta doping, using the same conditions described directly above. The delta-doped structure is shown in Figure 2. The undoped GaAs layer, grown under the same conditions as in the test structure (and with the same AsH$_3$ tank and TMG bubbler) in previous test runs, is characterized to be N type, with $N = 2 \times 10^{13}$ cm^{-3}, and mobility (77 K) of 140,000 cm^2/V-sec. The structures studied have undoped $Al_xGa_{1-x}As$ layer thicknesses of 50 to 200 Å. Table I shows the Hall measurement results, at 77 K, for these structures. For each run, the data shown are averages of four Hall samples cut from a two-inch wafer. We see from these data that the delta-doped structures show superior performance compared to those with bulk-doped layers, so long as the undoped $Al_xGa_{1-x}As$ layer is over 100 Å thick.

For the test structures shown in Table I, the standard deviation (SD) in 77 K mobility is 3-9% for the delta-doped samples, and is 6-18% for the bulk-doped ones. For the sheet carrier values at 77 K, the SD range is 1.7-6.3% for the delta-doped structures, and is 3.9-10.7% for the bulk-doped samples. Thus, it appears that the delta-doping process yields wafers with improved uniformity, as compared with the bulk-doped samples.

One delta-doped test structure, with an undoped $Al_xGa_{1-x}As$ thickness of 125 Å, has been measured by variable-temperature Hall measurements, down to a temperature of 20 K. The results are shown in Figure 5. The maximum mobility (at 20 K) is 300,000 cm^2/V-sec, and the sheet carrier density, for all measurement temperatures, is 9×10^{11} cm^{-2}.

The delta-doping region in one test sample has been imaged by cross-sectional transmission electron microscopy. The micrograph is shown in Figure 6. The delta-doped layer appears to be approximately 30-35 Å wide in this photo. This value agrees well with our Polaron measurement.

Table I *77 K Hall results for test structures (averaged over 4 spots per wafer).*

Run No.	Structure	$Al_xGa_{1-x}As$ Thickness ($\mathring{A}$)	Carrier Density (cm^{-2})	Mobility (cm^2/V-sec)
1865	Conventional	50	1.03×10^{12}	66,500
1866	Conventional	100	6.51×10^{11}	104,200
1877	Conventional	50	1.22×10^{12}	71,040
1875	Delta-doped	50	2.08×10^{12}	6110
1887	Delta-doped	125	1.14×10^{12}	116,500
1940	Delta-doped	200	7.05×10^{11}	123,300

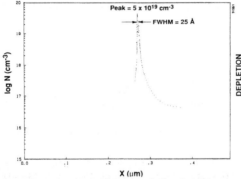

Figure 5 *Mobility versus temperature for the delta-doped test structure with a spacer thickness of 125 $\mathring{A}$ (run #1887).*

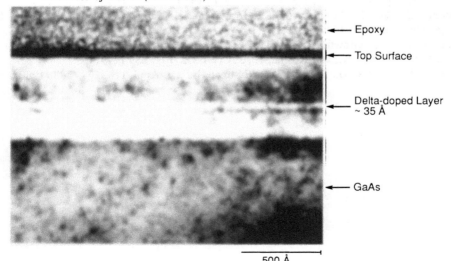

Figure 6 *Cross-sectional transmission electron micrography of a delta-doped $Al_{0.23}Ga_{0.77}As$ test structure.*

SUMMARY

Low-pressure MOCVD has been used to produce Si-delta-doped GaAs, $Al_xGa_{1-x}As$, and GaAs-$Al_xGa_{1-x}As$ test structures. In GaAs, delta doping has produced sheet carrier densities up to 1.53×10^{13} cm^{-2} and peaks with a FWHM value as narrow as 83 Å. In $Al_xGa_{1-x}As$ (x=0.23), delta doping has resulted in n-type layers with sheet carrier densities up to 6×10^{12} cm^{-2}, Polaron peak heights up to 5×10^{19} cm^{-3} and FWHM values down to 25 Å. The width of the delta-doped layer has also been confirmed by XTEM. In $Al_xGa_{1-x}As$, lower amounts of SiH_4 in the delta-doping process result in P-type behavior. The use of delta doping leads to high mobilities and high carrier densities, up to 131,000 cm^2/V-sec and 9×10^{11} cm^{-2} at 77 K. These structures, at 20 K, show mobility values up to 300,000 cm^2/V-sec.

ACKNOWLEDGMENTS

This work was supported by a program funded by SDIO/IST and managed by the Defense Nuclear Agency.

REFERENCES

Daly, J.T., and Roberts, C.B., Mat. Res. Soc. Symp. Proc. 145, 239 (1989).

Hobson, W.S., Pearton, S.J., Schubert, E.F., and Cabaniss, G., Appl. Phys. Lett. 55, 1546, (1989).

Hong, W-P., DeRosa, F., Bhat, R., Allen, S.J., and Hayes, J.R., Appl. Phys. Lett. 54, 457 (1989).

Kim, Y., Kim, M.S., Min, S-K., Lee, C., Yoo, K-H., J. Appl. Phys. 68, 2747 (1990).

Mori, Y., Nakamura, F., and Watanabe, N., J. Appl. Phys. 60, 334 (1986).

Pan, N., Carter, J., Jackson, G.S., Hendriks, H., Huang, J.C., and Zheng, X.L., at GaAs and Related Compounds Symp., Jersey, U.K., Sept., 1990.

Pearton, S.J., Ren, F., Abernathy, C.R., Hobson, W.S., Chu, S.N.G., and Kovalchick, J., Appl. Phys. Lett. 55, 1342 (1989).

Ploog, K., J. Cryst. Growth 81, 304 (1987).

Schmidt, P., Deppert, K., and Kostial, H., J. Cryst. Gr. 107, 259 (1991).

Wood, C.E.C., Metze, G., Berry, J., and Eastman, L.F., J. Appl. Phys. 51, 383 (1980).

1.3 μm luminescence in $(InAs)_n/(GaAs)_n$ strained quantum-well structures grown on GaAs

E. J. Roan, K. Y. Cheng, P. J. Pearah, X. Liu, K. C. Hsieh, and S. G. Bishop

Department of Electrical and Computer Engineering, and Center for Compound Semiconductor Microelectronics, University of Illinois at Urbana-Champaign, Urbana, Illinois 61801

ABSTRACT: Optical emission from pseudomorphic InGaAs-on-GaAs quantum-well heterostructures has been extended to useful wavelengths (1.3 μm) by replacing an $In_xGa_{1-x}As$ random alloy quantum well with an $(InAs)_n/(GaAs)_n$ short-period superlattice (SPS). With the same quantum-well width, the photoluminescence peak wavelength of the SPS structure is always longer than that of the $In_{0.5}Ga_{0.5}As$ random-alloy structure. Strong photoluminescence was observed in $(InAs)_n/(GaAs)_n$ SPS quantum wells with thicknesses up to 72 Å. The longest optical-emission wavelength observed at room temperature in $(InAs)_n/(GaAs)_n$ SPS quantum-well structures was 1.34 μm.

1. Introduction

In order to realize GaAs-based long-wavelength optoelectronic integrated circuits for optical fiber communication applications, a compatible direct-bandgap meterial with optical emission wavelengths greater than 1.3 μm must be developed. To this end, strained $In_xGa_{1-x}As$/GaAs heterostructure quantum-well laser technologies have been investigated previously using metalorganic chemical vapor deposition (Feketa et al. 1986) and molecular beam epitaxy (MBE) (Laidig et al. 1984) techniques. Due to the lattice mismatch inherent in this material system, the maximum quantum-well width in heterostructures useful for photonic device applications is limited by the critical thickness of the strained $In_xGa_{1-x}As$ active region. For $x \geq 0.5$, theoretical (Matthews and Blakeslee 1974, People and Bean 1985) and experimental (Orders and Usher 1987, Anderson et al. 1987, Elman et al. 1989, Weng 1989, Elman et al. 1990) studies have clearly demonstrated that the critical thickness is less than 40 Å. The concomitant blue shift (due to the quantum size effect) has thus far prevented researchers from obtaining wavelengths suitable for optical fiber communication application; the longest wavelength repoorted to date is 1.1 μm (York et al. 1989).

In order to extend the optical emission from the $In_xGa_{1-x}As$/GaAs material system to longer wavelengths, the bandgap-narrowing effect due to ordering on the atomic scale has been exploited. This effect has been observed in short-period superlattices (SPS's) in several material systems, such as $(InAs)_m/(GaAs)_n$ on InP (Fukui and Saito 1984, Wei and Zunger 1990), $(AlAs)_m/(GaAs)_n$ on GaAs (Isu et al. 1987, Marzin et al. 1986), and $(GaP)_2/(InP)_2$ on GaAs (Hsieh et al. 1990). In each instance, the bandgap of the ordered system was determined to be narrowed with respect to the random ternary alloy of the same average composition. This red shift was the greatest (150 meV) for $(InAs)_1/(GaAs)_5$ on GaAs (Marzin et al. 1986). Similarily, using $(InAs)_m/(GaAs)_n$ SPS, instead of $In_xGa_{1-x}As$ random alloy as the quantum-well material, increases the emission wavelength of $In_xGa_{1-x}As/GaAs$ strained quantum-well structures.

For $GaAs/[(InAs)_m/(GaAs)_n$ SPS$]/GaAs$ pseudomorphic quantum-well structures, there are two kinds of strain, the strain between $(InAs)_m$ and $(GaAs)_n$ layers and the strain between $(InAs)_m/(GaAs)_n$ SPS and GaAs cladding layers, that should be taken into account. The critical thickness of InAs layers on GaAs substrates has been widely discussed (Brandt et al. 1991, Kudo et al. 1991). One monolayer of InAs can be grown on GaAs to form a dislocation-free InAs/GaAs heterojunction (Brandt et al. 1991). In this study, the strain effect between $(InAs)_m/(GaAs)_n$ SPS and GaAs layers is investigated for SPS (m,n < 2) quantum-well structures through photoluminescence (PL) measurements. The $(InAs)_n/(GaAs)_n$ (n~1) SPS was utilized in order to suppress the generation of misfit dislocations. Low temperature (77K) PL measurements were made on both $GaAs/[(InAs)_n/(GaAs)_n$ SPS$]/GaAs$ and $GaAs/In_{0.5}Ga_{0.5}As/GaAs$ quantum-well structures. The peak energies, linewidths, and intensities of the PL spectra are presented and discussed for both random alloy and SPS quantum wells. Room-temperature 1.3-μm luminescence has been achieved in $GaAs/[(InAs)_n/(GaAs)_n$ SPS$]/GaAs$ quantum-well structures.

2.   Experimental

Undoped $GaAs/[(InAs)_1/(GaAs)_1$ SPS$]/GaAs$ and $GaAs/In_{0.5}Ga_{0.5}As/GaAs$ pseudomorphic quantum-wells were grown on (100) semi-insulating GaAs substrates by MBE. The structures consist of an ordered $(InAs)_1/(GaAs)_1$ SPS or $In_{0.5}Ga_{0.5}As$ random alloy quantum well with a 0.4-μm GaAs buffer layer and a 0.15-μm GaAs cap layer. The growth rates of InAs and GaAs layers were calibrated using reflection high-energy electron diffraction intensity oscillation. The widths of both SPS and random alloy quantum wells and the InAs molar fraction of the $In_xGa_{1-x}As$ alloy were estimated from the growth rates. The $(InAs)_1/(GaAs)_1$ SPS's were formed by growing alternating InAs and GaAs layers approximately one monolayer at a time. The substrate temperatures used for the growth of In-containing ordered and random-alloy materials were less than 530°C. The room-temperature and 77K PL spectra were measured in a cold finger cryostat with synchronous detection electronics. Luminescence was excited with the 5145-Å line of an Ar^+ ion laser, dispersed with a 1.26-m Spex spectrometer and detected with a liquid-nitrogen cooled Ge detector.

3.   Results and Discussion

3.1.   PL Peak Energies:

The 77K luminescence peak energies of $GaAs/In_{0.5}Ga_{0.5}As/GaAs$ and $GaAs/[(InAs)_1/(GaAs)_1$ SPS$]/GaAs$ quantum-well structures are shown in Fig. 1. Due to quantum-size effect and strain, these transition energies are all much higher than that of the nearly lattice matched $In_{0.5}Ga_{0.5}As/InP$ system (0.767 eV) (Fuhui and Saito 1984). For $In_{0.5}Ga_{0.5}As$ quantum wells between 12 and 48 Å thick, the luminescence peak energies are in the range of 1.109 eV (1.12 μm) to 1.395 eV (0.89 μm). No luminescence can be detected in those structures with $In_{0.5}Ga_{0.5}As$ quantum-wells wider than 48 Å. For $(InAs)_1/(GaAs)_1$ SPS quantum wells between 12 and 96 Å, the luminescence peak energies are in the range of 1.003 eV (1.24 μm) to 1.309 eV (0.95 μm). With the same quantum-well width, the ordered $(InAs)_1/(GaAs)_1$ SPS structure shows a lower transition energy as compared to that of the $In_{0.5}Ga_{0.5}As$ random alloy. The differences in transition energies between ordered $(InAs)_1/(GaAs)_1$ SPS quantum wells and their corresponding random-alloy $In_{0.5}Ga_{0.5}As$ quantum wells range from 65 meV to 120 meV. For nearly lattice-matched bulk $In_{0.5}Ga_{0.5}As$ random alloy and thick $(InAs)_1/(GaAs)_1$ SPS grown on InP substrates (Fukui and Saito 1984), the energy shift between the random alloy and the SPS at 77K has been determined to be 35 meV. This value is much smaller than those of the strained InGaAs quantum-well structures grown on GaAs substrates observed in this study. One possible explanation is that the strain in random alloy and/or ordered quantum wells was partially relaxed, with less strain effect at $[(InAs)_1/(GaAs)_1$ SPS]/GaAs interfaces than at $In_{0.5}Ga_{0.5}As$/GaAs interfaces. Therefore, the GaAs/$[(InAs)_1/(GaAs)_1$ SPS]/GaAs quantum-well structures can still emit readily detectable luminescence up to a quantum-well width of 96 Å, which is twice the maximum thickness of the random alloy $In_{0.5}Ga_{0.5}As$ quantum-well.

Fig. 1 The 77K PL peak energies of GaAs/$In_{0.5}Ga_{0.5}As$/GaAs quantum-well structures (solid circles) and GaAs/$[(InAs)_1/(GaAs)_1$ SPS]/GaAs quantum-well structures (open circles).

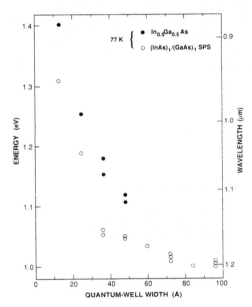

3.2. PL Linewidths:

Figure 2 shows the full widths at half maximum (FWHM) of 77K PL spectra of GaAs/$[(InAs)_1/(GaAs)_1$ SPS] /GaAs and GaAs/ $In_{0.5}Ga_{0.5}As$/GaAs quantum-well structures. For GaAs / $In_{0.5}Ga_{0.5}As$ / GaAs quantum-well structures, the FWHM is 12.5 meV for a 12-Å well and increases to 30-45 meV for wider wells. In random alloy quantum-well structures, the broadening of the PL spectrum is an indication of lattice relaxation. The critical thickness of the GaAs/$In_{0.5}Ga_{0.5}As$/GaAs quantum-well structure is between 12 Å and 24 Å, which is consistent with previous reported data (Elman et al. 1990). The FWHM's of the GaAs/$[(InAs)_1/(GaAs)_1$ SPS]/GaAs quantum-well structures range from 40 to 50 meV. For closely lattice-matched $(InAs)_1/(GaAs)_1$ SPS grown on InP, the FWHM has been reported to be about 30 meV (Fukui and Saito 1984). The PL peak broadening is probablly due to strain between InAs and GaAs layers. For GaAs/$[(InAs)_1/(GaAs)_1$ SPS]/GaAs quantum-well structures, the strain is much larger than that of the lattice-matched $[(InAs)_1/(GaAs)_1$ SPS]/InP structure. Therefore, its FWHM should be larger than 30 meV. Unlike GaAs/$In_{0.5}Ga_{0.5}As$/GaAs quantum-well structures, the FWHM does not vary much with quantum-well width in GaAs/$[(InAs)_1/(GaAs)_1$ SPS]/GaAs quantum-well structures. The reason for this characteristic is not fully understood at this stage; compositional nonuniformity and lattice relaxation are among the probable causes.

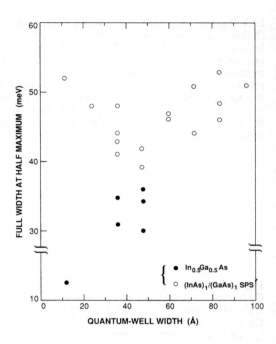

Fig. 2 The full widths at half maximum of 77K PL spectra of GaAs/In$_{0.5}$Ga$_{0.5}$As/GaAs quantum-well structures (solid circles) and GaAs/[(InAs)$_1$/(GaAs)$_1$ SPS]/GaAs quantum-well structures (open circles).

3.3.   PL Intensities:

In Fig. 3, the 77K PL intensities of GaAs/In$_{0.5}$Ga$_{0.5}$As/GaAs and GaAs/[(InAs)$_1$/(GaAs)$_1$ SPS]/GaAs quantum-well structures are shown. Luminescence only could be observed from random-alloy quantum-well structures, with quantum-well width less than

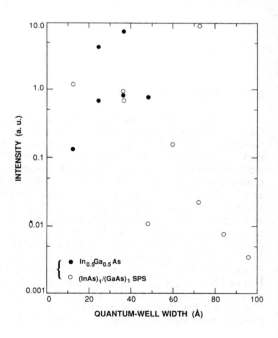

Fig. 3 The 77K PL intensities of GaAs/In$_{0.5}$Ga$_{0.5}$As/GaAs quantum-well struc-tures (solid circles) and GaAs/[(InAs)$_1$/(GaAs)$_1$ SPS]/GaAs quantum-well struc-tures (open circles).

or equal to 48 Å. The PL intensities of those samples are of the same order of magnitude. For GaAs/[(InAs)$_1$/(GaAs)$_1$ SPS]/GaAs quantum-well structures, 77K luminescence could be observed with the quantum-well widths up to 96 Å. The PL intensity varies irregularly, independent of quantum-well width. The PL intensity also strongly depends on the growth conditions such as substrate temperature and V/III ratio. Samples with the same quantum-well width grown under different conditions gave similar luminescence peak energy and FWHM but very different PL intensities. Further study is needed to examine this phenomenon.

The room-temperature PL spectrum of a 72-Å GaAs/[(InAs)$_1$/(GaAs)$_1$ SPS]/GaAs quantum-well structure is shown in Figure 4. The luminescence peak energy and the FWHM are 0.925 eV (1.34 µm) and 54 meV, respectively. The emission peak wavelength of 1.34 µm is the longest ever achieved in the strained InGaAs/GaAs quantum-well system grown on GaAs substrate. The well width of this (InAs)$_1$/(GaAs)$_1$ SPS quantum well is much thicker than the theoretical critical thickness of the In$_{0.5}$Ga$_{0.5}$As alloy calculated with Matthews and Blakeslee's model (1974) for a single quantum well structure.

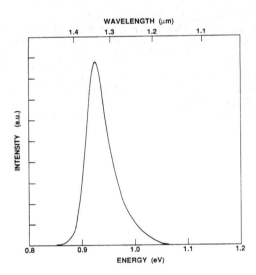

Fig. 4 The room-temperature photoluminescence spectrum of a 72-Å GaAs/[(InAs)$_n$/(GaAs)$_n$]/GaAs quantum-well structure.

4.   Conclusion

At the same quantum-well width, a red shift of 65 to 120 meV has been observed in the emission peak wavelength of the GaAs/[(InAs)$_1$/(GaAs)$_1$ SPS]/GaAs quantum-well structure compared to that of the corresponding GaAs/In$_{0.5}$Ga$_{0.5}$As/GaAs quantum-well structure. For GaAs/ In$_{0.5}$Ga$_{0.5}$As/GaAs quantum-well structures, the FWHM increases from 12.5 meV to 30-45 meV as the quantum-well width increases from 12 Å. The FWHM's of GaAs/ [(InAs)$_1$/(GaAs)$_1$ SPS]/GaAs quantum-well structures are in the range of 40 to 50 meV, independent of the quantum-well width. Compositional nonuniformity and lattice relaxation are possible reasons for this behavior. Even though lattice relaxation may occur in thick GaAs/[(InAs)$_1$/(GaAs)$_1$ SPS]/GaAs quantum-well structures, strong photoluminescence still could be observed. Room-temperature 1.3-µm luminescence has been observed, for the first time, in GaAs/[(InAs)$_1$/(GaAs)$_1$ SPS]/GaAs pseudomorphic quantum-well structures grown on GaAs substrates.

Acknowledgements

The authors would like to thank B. W. Bowdish III for technical assistance. This work was supported by the National Science Foundation grants (ECD-89-43166 and DMR-89-20538) and the Joint Service Electronics Program (N0014-90-J-1270).

References

Andersson T G, Chen Z G, Kulakovskii V D, Uddin A and Vallin T J 1987 *Appl Phys Lett* **51** 752
Brandt O, Tapfer L, Ploog K, Hohenstein M and Phillipp F 1991 *J Cryst Growth* **111** 383
Elman B, Koteles E S, Melman P, Jagannath C, Lee J and Dugger D 1989 *Appl Phys Lett* **55** 1659, and references therein.
Elman B, Koteles E S, Melman P and Rothman M 1990 *SPIE* **1361** 362
Feketa D, Chan K T, Ballantyne J M and Eastman L F 1986 *Appl Phys Lett* **49** 1659
Fukui T and Saito H 1984 *Jpn J Appl Phys* **23** L521
Hsieh K C, Baillargeon J N and Cheng K Y 1990 *Appl Phys Lett* **57** 2244
Isu T, Jiang D S and Ploog K 1987 *Appl Phys A* **43** 75
Kudo K, Lee J S, Tanaka K, Makita Y and Yamada A 1991 *J Cryst Growth* **111** 402
Laidig W D, Caldwell P J, Lin Y F and Peng C K 1984 *Appl Phys Lett* **44** 653
Marzin J Y, Goldstein L, Glas F and Quillec M 1986 *Surface Science* **174** 586
Matthews J W and Blakeslee A E 1974 *J Cryst Growth* **27** 118
Orders P J and Usher B F 1987 *Appl Phys Lett* **50** 980
People R and Bean J C 1985 *Appl Phys Lett* **47** 332
Wei S H and Zunger A 1990 *Appl Phys Lett* **56** 662, and references therein.
Weng S-L 1989 *J Appl Phys* **66** 2217
York P K, Beernink K J, Fernandez G E and Coleman J J 1989 *Appl Phys Lett* **54** 499

Inst. Phys. Conf. Ser. No 120: Chapter 12
Paper presented at Int. Symp. GaAs and Related Compounds, Seattle, 1991

583

Valence-band energy dispersion in modulation-doped quantum wells: effect of strain and confinement on heavy- and light-hole mixing

S. K. Lyo and E. D. Jones

Sandia National Laboratories, Albuquerque, New Mexico 87185

ABSTRACT. We present theory and data for the valence-band structure for lattice-matched, strained, symmetrically- and asymmetrically-doped quantum wells (QW's). Our self-consistent k•p calculation using a plane-wave basis (PWB) yields accurate results and can be applied to arbitrary potential profiles. The PWB is convenient for shallow as well as type II QW's (with the light-hole QW inverted due to strain). Excellent agreement is obtained between the theoretical dispersion curves and the data.

1. INTRODUCTION

A large hole mobility is desirable in many device applications. As is well known, one way to enhance the mobility is to reduce the mass as well as collisions with the impurities at low temperatures. Recently light in-plane mass was achieved in layered structures by introducing biaxial strain in the quantum well (QW) through crystal growth (Osbourn et al. 1987). For example, in a strained $In_xGa_{1-x}As/GaAs$ QW, the light-hole (LH) band ($|3/2,\pm1/2\rangle$) is shifted down away from the valence-band edge while the heavy-hole (HH) band ($|3/2,\pm3/2\rangle$) is shifted up due to the compression in the QW. As a result, only the HH's with light in-plane mass are populated in typical modulation-doped QW's. Another way of achieving this kind of HH-LH separation is through a confinement effect. Namely, when the holes are confined in a QW, LH quantum levels move down relative to the HH levels in the valence band. A lattice-matched $GaAs/Al_xGa_{1-x}As$ QW is expected to show only the confinement effect with negligible strain effect. Therefore it is interesting to see the effect of the interplay between the strain and the confinement.

Recently valence-band structures of modulation-doped strained semiconductor single-quantum-wells (SQW's) such as $In_xGa_{1-x}As/GaAs$ were determined from magneto-luminescence (Jones et al. 1989). The data indicate the importance of strain in determining the degree of HH-LH mixing and the hole masses for the motion in the QW plane (Jones et al. 1989). The magneto-luminescence data from lattice-matched $GaAs/Al_xGa_{1-x}As$ QW's to be presented below clearly indicate the HH-LH separation through a confinement effect. The purpose of this paper is to present theory for the valence-band energy dispersion for lattice-matched and strained QW's and compare with the data. We consider symmetrically-doped and asymmetrically-doped SQW's and study how the strain and confinement affect the valence-band mass.

The low-energy valence-band structure is calculated by using a self-consistent k·p calculation. This method which can be applied to arbitrarily-shaped potential profiles is based on a plane-wave basis and yields numerically accurate results. The plane-wave basis is especially convenient for studying narrow or shallow QW's with only a few confined levels as well as partially type II QW's (without any confined level) where the LH QW is inverted relative to the HH QW due to strain-induced shift of the LH-band bottom. For example, the LH band changes from type I to type II around $x \approx 0.2$ for $In_xGa_{1-x}As/GaAs$. The effect of the above mentioned LH QW inversion on the HH-LH mixing is studied for this system. Also, the conduction as well as valence bands are severely distorted by the fields from the dopant ions and the majority carriers. The plane-wave basis is convenient for studying the effects of band bending on the valence-band structure. Excellent agreement is obtained between the theoretical valence-band energy dispersion curves and the data for both lattice-matched and strained QW's.

2. SELF-CONSISTENT K·P CALCULATION OF VALENCE-BAND ENERGY

The valence-band energy dispersion curve is calculated by diagonalizing the 4×4 block-diagonal Hamiltonian obtained from the Luttinger Hamiltonian through a unitary transformation (Broido and Sham 1985) in the absence of an external magnetic field:

$$
\begin{bmatrix} H^U & 0 \\ 0 & H^L \end{bmatrix} = \begin{bmatrix} F_+ & |I|+i|H| & 0 & 0 \\ |I|-i|H| & F_- & 0 & 0 \\ 0 & 0 & F_- & |I|+i|H| \\ 0 & 0 & |I|-i|H| & F_+ \end{bmatrix} \begin{matrix} |u_1> \\ |u_2> \\ |u_3> \\ |u_4> \end{matrix} \qquad (1)
$$

where the orthogonal HH (LH) basis functions u_1 and u_4 (u_2 and u_3) are linear combinations $|3/2,\pm3/2>$ ($|3/2,\pm1/2>$) and

$$
F_\pm = E_{v\pm}(z) + \frac{1}{2}(-\gamma_1 \pm 2\gamma_2)k_z^2 - \frac{1}{2}(\gamma_1 \pm \gamma_2)k^2, \qquad (2)
$$

$$
H = \sqrt{3}\gamma_3 k_- k_z , \quad I = \frac{\sqrt{3}}{2}(-\bar{\gamma}k_-^2 + \mu k_+^2). \qquad (3)
$$

In (2) $E_{v\pm}(z)$ are the valence-band edges for HH's (+) and LH's (-), $k^2 = k_x^2 + k_y^2$, $k_z = \frac{\partial}{i\partial z}$ and $\mu = m_o - 1$. Here m_o is the free electron mass. In (3) $k_\pm = k_x \pm ik_y$, $\bar{\gamma} = (\gamma_2 + \gamma_3)/2$, $\mu = (\gamma_2 - \gamma_3)/2$. In a lattice-matched SQW, the valence-band edges for the HH and LH are degenerate (i.e., $E_{v+} = E_{v-}$) as shown in Fig. 1a. In a strained SQW such as $In_xGa_{1-x}As/GaAs$ with the growth direction (i.e., the z-axis) in the [100] direction, for example, the effect of the compression inside the QW is to raise $E_{v+}(z)$ and lower $E_{v-}(z)$ inside the QW by Δ_+ (> 0) and Δ_- (> 0), respectively, as shown in Fig. 1b. Therefore the well is deeper for the HH than for the LH. The LH QW may even become inverted under a large strain as indicated by a dashed line in the figure.

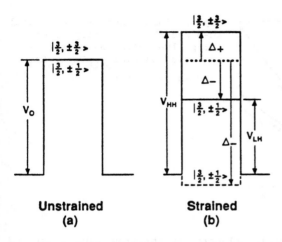

Fig. 1 HH and LH valence-band edges for (a) lattice-matched and (b) strained QW's.

Unstrained   **Strained**
   (a)   **(b)**

When the band-edge potentials have an inversion symmetry (i.e., $E_{v\pm}(-z) = E_{v\pm}(z)$), the 2×2 Hamiltonians H^U and H^L are equivalent, so that the HH and LH band edges are degenerate (Broido and Sham 1985, Eckenberg and Altarelli 1985, Ando 1985). The degeneracies are lifted in the absence of the inversion symmetry. In order to diagonalize H^U (or H^L) we introduce a sinusoidal-wave basis set:

$$\psi_{en} = (2/d)^{1/2} \cos k_n z, \qquad k_n = (2n-1)\pi/d$$

$$\psi_{on} = (2/d)^{1/2} \sin k_n z, \qquad k_n = 2n\pi/d$$

(4)

for the HH as well as LH states where $n = 1, 2, \cdots, n_{max}$ and $z = 0$ is the center of the QW. The wave functions in (4) are assumed to vanish for $|z| > d/2$. The quantity d is arbitrary but is chosen to be much larger than the QW width b in order to avoid the cut-off effect. Also, the final result is independent of n_{max} for large n_{max}. When the γ-parameters depend on z, the kinetic-energy terms in (2) are rewritten in a Hermitian form (Morrow and Brownstein 1984): $k_z(-\gamma_1 \pm \gamma_2)k_z/2$. This expression insures the continuity of the current. Also, the quantity H in (3) is replaced by $H = (\sqrt{3}k_-/2)(\gamma_3 k_z + k_z \gamma_3)$.

If we ignore the interactions of the hole with the dopant ions and the majority carriers, the valence-band energy-dispersion curve is obtained by diagonalizing H^U (or H^L). For typical QW structures considered here with b < 120 Å, a very rapid convergence is obtained for d = 1000 Å and $n_{max} = 150$. However, the QW is severely distorted by Coulomb fields arising from the electrons (for n-type samples considered in this paper) and dopant ions, complicating the problem. For an n-type system we find the potential profile of the conduction band first. For this purpose we evaluate the matrix elements of the Hamiltonian $H_c = k_z \gamma_c k_z/2 + E_c(z)$ for the conduction band in terms of the basis set given in (4). Here γ_c

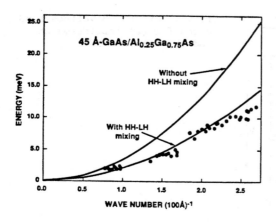

Fig. 2 Comparison of theoretical in-plane HH energy dispersion along [100] with the data from 45 Å-wide GaAs/Al$_{0.25}$Ga$_{0.75}$As. Band warping is small.

is the inverse mass of the conduction electron in the growth direction and $E_c(z)$ is the square-well potential of depth V_c. The electron distribution is found from the eigen-functions of H_c at 0 K and the Coulomb potential from the electrons and ions is calculated. We add these Coulomb potential to the original H_c and diagonalize the new H_c and recalculate the new Coulomb potential. Iterating this procedure, we find that the charge distribution and the potential profile converge very fast to self-consistent values usually after three iterations. This process yields at the same time the Coulomb-potential profile for the holes, which is added to $E_{v\pm}(z)$ in (2). The valence-band energy dispersion is then obtained by diagonalizing H^U (H^L).

3. NUMERICAL RESULTS AND COMPARISON WITH DATA

The γ-parameters used for applications are $\gamma_1 = 6.85$, $\gamma_2 = 2.1$, $\gamma_3 = 2.9$ for GaAs and $\gamma_1 = 20.4$, $\gamma_2 = 8.3$, $\gamma_3 = 9.1$ for InAs (Madelung 1987). A small difference between the gamma parameters of Al$_x$Ga$_{1-x}$As and those of GaAs is ignored because the wave functions penetrate little into the barriers. The compositional-average γ-values are used for In$_x$Ga$_{1-x}$As/GaAs QW. The samples to be studied below are of n-type. However, the magneto-luminescence technique can measure the energy-dispersion curves for both the conduction and valence bands (Jones et al. 1989). We first study a 45 Å-wide GaAs/Al$_{0.25}$Ga$_{0.75}$As SQW with an electron density $N = 6.6 \times 10^{11}$ cm^{-2}. Only one side of the QW is doped with Si following a 70 Å-wide undoped Al$_{0.25}$Ga$_{0.75}$As spacer layer. The ions are calculated to be uniformly spread out within a 51 Å-wide layer at 0 K. A more detailed description of the sample is given elsewhere (Jones et al. 1991). The conduction and valence band offsets are estimated from the recent work of Fu et al. (1990) and equal approximately 284 meV and $V_o = 135$ meV (see Fig. 1), respectively. The theoretical energy-dispersion curve for the HH agrees well with the data as shown in Fig. 2. The LH energies are separated away from the HH energies through confinement effect and are not shown there. It is seen that the HH and LH mixing alters the energy dispersion

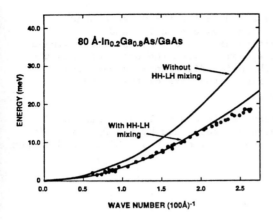

Fig. 3 Comparison of theoretical in-plane HH energy dispersion along [100] with the data from 80 Å-wide $In_{0.2}Ga_{0.8}$/GaAs. Band warping is small.

significantly. The degree of HH-LH mixing depends on the QW width, which determines the HH and LH separations and their wave functions.

We now consider a symmetric 80 Å-wide Si-doped GaAs/$In_{0.2}Ga_{0.8}$As/GaAs QW with N = 5×10^{11} cm^{-2} (Jones et al. 1989). The width of the undoped GaAs spacer layers on both sides of the QW equals 100 Å. The ions are distributed over 17 Å-wide layers at 0 K. The conduction and HH band offsets equal approximately 128 meV and V_{HH} = 67 meV (cf. Fig. 1), respectively. The LH band is assumed to lie approximately $\Delta = \Delta_+ + \Delta_-$ = 57 meV below the HH band and is a shallow QW (of type I) of depth V_{LH} = 10 meV (cf. Fig. 1). The theoretical energy dispersion curve is compared with the data (Jones et al. 1989) in Fig. 3. Again the agreement is excellent. The amount of HH-LH mixing and therefore the mass are much smaller here than in Fig. 2, because the HH-LH separation in $In_{0.2}Ga_{0.8}$As/GaAs is mainly strain-induced and, as a result, is much larger than that in GaAs/$Al_{0.25}Ga_{0.75}$As, which arises from the confinement effect alone. Because there are uncertainties in Δ, we increased Δ to Δ = 67 meV and Δ = 77 meV corresponding to a flat LH band (i.e., V_{LH} = 0) and an inverted (i.e., type II) LH band (indicated by a dashed line in Fig. 1), respectively, in order to see if there is any significant change in the fitting. No significant change was found except that the HH-LH mixing diminished somewhat and the curves were shifted up slightly relative to that of Δ = 57 meV in Fig. 3 for large wave numbers.

The theoretical energy dispersion curves in Figs. 2 and 3 are obtained by fully taking into account the band bending. We calculated the dispersion curves ignoring Coulomb fields, namely for simple square-well potentials. Apart from a uniform shift of a few meV of the HH ground level, nearly the same dispersion curves were obtained for both cases, indicating that band-bending effects are not important. This is consistent with the fact that only a slight shift to the electron charge distribution was found even when an asymmetric band-bending (as for the above lattice-matched case) is introduced. Namely, the electrons are rigidly locked in the QW unless the QW is very wide (> 500 Å). We find negligible spin-splitting for the asymmetrically doped QW.

In summary we presented theory and data for the valence-band structure for lattice-matched, strained, symmetrically- and asymmetrically-doped QW's. Our self-consistent k.p calculation using a plane-wave basis yielded excellent agreement with the data for both the strained and unstrained systems. The major difference between the strained and unstrained systems is simply the size of the HH-LH separations in the strained system.

The authors acknowledge valuable conversations with I. J. Fritz, G. C. Osbourn, and C. P. Tigges. This work was supported by the Division of Materials Science, Office of Basic Energy Science, U.S. DOE under Contract No.DE-AC04-76DP00789

References

Ando T 1985 J. Phys. Soc. Jpn 54 1528
Broido D A and Sham L J 1985 Phys. Rev. B 31 888
Eckenberg U and Altarelli M 1985 Phys. Rev. B 32 3712
Fu W S, Olbright G R, Owyoung A, Klem J F, Biefeld R M, and Hadley G R 1990 Appl. Phys. Lett. 57 1404
Jones E D, Lyo S K, Fritz I J, Klem J F, Schirber J E, Tigges C P, and Drummond T J 1989 Appl. Phys. Lett. 54 2227
Jones E D, Lyo S K, Klem J F, Schirber J E, and Lin S Y 1991 Proceedings: Gallium Arsenide and Related Compound-1991.
Madelung O 1987 Landolt-Bónstein Series Vol. 22 pp 82-120 Springer (New York)
Morrow R A and Brownstein K R 1984 Phys. Rev. B 30 678
Osbourn G C, Gourley P L, Fritz I J, Biefeld R M, Dawson L R, and Zipperian T E 1987 in Semiconductors and Semimetals, Vol. 24 459, Eds. R. K. Willardson and A. C. Beer, Academic Press (New York)

Inst. Phys. Conf. Ser. No 120: Chapter 12
Paper presented at Int. Symp. GaAs and Related Compounds, Seattle, 1991

589

Formation of lateral quantum-wells in vertical short-period superlattices

K. Y. Cheng, K. C. Hsieh, J. N. Baillargeon, and A. Mascarenhas*

Department of Electrical and Computer Engineering, and Center for Compound
Semiconductor Microelectronics, University of Illinois at Urbana-Champaign, Urbana,
Illinois 61801 U.S.A.
*Solar Energy Research Institute, Golden, Colorado 80401 U.S.A.

ABSTRACT: Lateral quantum well (QW) structures with periodicities as small as ~200Å
were formed in vertical $(GaAs)_n/(InAs)_n$ and $(GaP)_n/(InP)_n$ short-period-superlattices
(SPS) grown on nominally (100) InP and GaAs substrates using solid and gas source
molecular beam epitaxy techniques, respectively. Both InGaAs and InGaP ordered vertical
superlattice layers were found to have a growth-induced lateral periodic modulation of the
composition along the [$\bar{1}$10] direction and form parabolic-shaped lateral QWs. When the
InGaP QW structure was excited with laser light along [110] and [$\bar{1}$10] directions, strong
anisotropy was observed in the intensity ratio of the electron-to-light-hole and electron-to-
heavy-hole transitions. This measurement confirms the existence of lateral two-
dimensional quantum confinement in vertical SPS structures.

1. Introduction:

One-dimensional confinement of carriers in quantum-wire structures holds great
promise for novel electronic and photonic device applications (Sakaki 1980, Asada et al.
1986). Limited success has been achieved in structures prepared by a number of techniques
utilizing fine-line lithographical (Kash et al. 1986) and tilted superlattice (TSL) on
misoriented substrates methods (Fukui and Saito 1988). The lithographical techniques have
been limited by several intrinsic difficulties such as lithography resolution and process related
damage. The TSL method has the potential to achieve quantum-wire structures. However, it
requires a misoriented substrate to provide a uniform tilt angle over the entire wafer (Petroff
et al. 1991). It is also necessary to maintain sharp interfaces between the QW's and the
cladding layers during TSL growth (Petroff et al. 1991). In this study, as a first step toward
the realization of a true quantum-well-wire structure, we have obtained *lateral* InGaAs and
InGaP multiple quantum well (MQW) structures with periodicities as small as ~200Å on
nominally (100) oriented substrates. These lateral MQWs were formed in vertical
$(GaAs)_n/(InAs)_n$ and $(GaP)_n/(InP)_n$ short-period-superlattices (SPS) grown on InP and
GaAs substrates. Transmission electron microscopy (TEM) and low temperature polarization
dependent photoluminescence (PL) techniques were used to examine the microstructure of
the SPS and to determine its lateral two-dimensional characteristics, respectively.

2. Experiment:

The epitaxial growth of $(GaP)_n/(InP)_n$ short-period superlattice (SPS) structures was
performed in a gas source molecular beam epitaxy (GSMBE) system. Elemental indium and

gallium were used for the column III sources and P_2 generated by cracking PH_3 injected through a tantalum-baffled PBN cracker maintained at 950°C served as the column V source. Detailed information about the design of the GSMBE and the characteristics of the gas cracker has been published elsewhere (Baillargeon et al. 1990). After desorption of the protective surface oxide, an undoped ternary $Ga_xIn_{1-x}P$ buffer layer with a lattice constant close to that of GaAs was first grown on the on-axis (001) GaAs substrate. During the growth, the substrate temperature was held constant at 510°C, and a V/III ratio equal to 16 was used with a PH_3 flow rate of 3 sccm. The growth rate was measured to be 1.45 μm/hr. Growth of the buffer layer was immediately followed by the growth of a 110-pair $(GaP)_n/(InP)_n$ SPS layer. Using *in-situ* reflection high-energy electron diffraction (RHEED) intensity oscillation to calibrate the growth rate, approximately one or two monolayer thick InP and GaP superlattice layers were grown alternatingly.

The growth of $(GaAs)_n/(InAs)_n$ SPS on InP was carried out in a solid source MBE system. A 3000Å $Ga_xIn_{1-x}As$ buffer layer and a 100 pair $(GaAs)_n/(InAs)_n$ SPS layer were successively grown on the on-axis (100) InP substrate. The composition of the buffer layer was adjusted to close to that of the InP substrate. The periodicity of the SPS layers, n = 1 or 2, was calibrated through *in-situ* RHEED intensity oscillations. The substrate temperature was held at ~500°C throughout the growth of the entire structure.

After the growth, the composition of both the random alloy buffer layer and the SPS layer were determined using the double-crystal x-ray diffraction (DCXD) technique. Both cross-sectional TEM and plan-view TEM have been employed to examine the microstructure of the buffer layer and the superlattice layer. The lateral (parallel to the heterojunction) composition variation of the SPS was examined with energy dispersive x-ray spectroscopy (EDS). The low temperature PL polarization measurements were performed with the sample held in an optical cryostat in the backscattering geometry. The 4545Å line of a linearly polarized Ar^+-ion laser was used to excite the sample. The PL was measured using a double grating spectrometer placed after a linear polarization analyzer and scrambler and detected using the photon counting technique.

3.   Results and Discussion:

3.1.   Microstructures:

The microstructure of both $(GaP)_n/(InP)_n$ and $(GaAs)_n/(InAs)_n$ SPS structures was first examined by cross-sectional TEM. From the average composition obtained from DCXD and the measured periodicity from the TEM diffraction pattern, the periodicity of the superlattices can be determined. Samples with nearly integral monolayer periodicity showed similar uniform image contrast in both [110] and [$\bar{1}$10] directions. When the monolayer periodicity is slightly nonstoichiometric, i.e. the binary layer thickness, expressed in monolayers, does not assume an exact integer value, there is a marked difference in image contrast in the [110] and [$\bar{1}$10] directions. Figure 1 shows such cross-section images of a $(GaP)_2/(InP)_2$ SPS structure. It is estimated that this two monolayer superlattice is slightly nonstoichiometric, i.e. the InP layer is exactly two monolayers thick, but the GaP layer averages about 2.2 monolayers (Hsieh et al. 1990). Using (002) dark field imaging, a uniform image appears in [$\bar{1}$10] cross section, and a modulated image composed of weak dark and light fringes, each approximately 100Å wide, in parallel to the growth direction appears in the [110] cross section. Similar cross-sectional TEM images were also observed in $(GaAs)_n/(InAs)_n$ SPS structures grown on InP substrates. Figure 2 shows the (002) dark-field image of a $(GaAs)_1/(InAs)_1$ SPS structure in which the InAs layers are several percent thicker than one monolayer. Most often, growth induced compositional modulation generated by either the instability of the source flux or substrate temperature fluctuation and

Fig. 1.   Dark field cross-sectional
transmission electron micrographs of the
(GaP)₂/(InP)₂ short-period superlattice
grown on the Ga₀.₅₂₅In₀.₄₇₅P buffer
layer along the [110] and [Ī10] cross
sections. Note that within the SPS layer
a uniform image appears in the [Ī10]
cross section while weak dark and light
fringes are observable in the [110] cross
section.

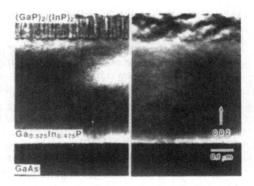

Fig. 2.   Dark field cross-sectional
transmission electron micrographs of the
(GaAs)₁/(InAs)₁ short-period superlattice
grown on the Ga₀.₅In₀.₅As buffer layer
along the [110] and [Ī10] cross sections.
Note that within the SPS layer a uniform
image appears in the [Ī10] cross section
while weak dark and light fringes are
observable in the [110] cross section.

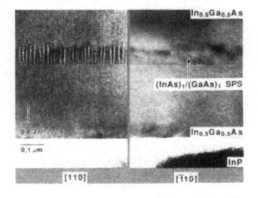

consequent strain relaxation is found perpendicular to the growth direction (Fraser et al.
1989). The lateral contrast modulations in Figs. 1 and 2, however, are different from those
commomly observed vertical modulations. In general, image contrast can result from
compositional difference and/or strain relaxation, such as structure factor contrast and strain
contrast, respectively. The vertical fringes formed in these SPS layers could be a result of
compositional modulation accompanied by local strain (Hsieh et al. 1990).

To separate the effect of structure factor contrast and strain contrast, two
perpendicular but crystallographically equivalent g vectors, in this case g = [002] and [200],
were used to image the structure (Fraser et al. 1989). Since both g vectors are
crystallographically equal, a constant structure factor contrast will be observed under these
imaging conditions. On the other hand, by aligning g vectors lying in the undistorted planes,
the strain contrast can be confined to a minimum value in all images. We can then suppress
the strain contrast by choosing proper g vectors while maintaining the structure factor
contrast. The image contrast shown in Fig. 1 was imaged with g = [002] and the incident
electron beam close to the [110] direction. A similar image was also obtained when g =
[200] was used. Since the modulating fringes shown in Figs. 1 and 2 are parallel to the
growth direction, it suggests that strain relaxation should have taken place along the same
direction if the image contrast was entirely due to the strain contrast. However, imaging with
g = [002] should exhibit little or no strain contrast, i.e. **g·b** = 0. Then, most of the
observed fringe contrast shown in Fig. 1 should be largely due to structure factor contrast,
and indicating that there exists compositional modulation parallel to the growth direction.

This result is very different from the compositional modulations in GeSi heteroepitaxial layers reported by Fraser et al. (1989).

To further confirm the orientation of the modulation fringes and to estimate how far each dark/light fringe extends into the [110] direction, we also examined the $(GaP)_2/(InP)_2$ SPS layer in plan-view TEM. Figure 3 shows nearly parallel fine fringes lie in the [110] direction. Therefore, the alternating dark/light fringes strongly indicate a growth-induced [$\bar{1}$10] superlattice, i.e. each dark/light fringe stacked along the [$\bar{1}$10] direction represents a thin stripe of $Ga_xIn_{1-x}P$ with different In and Ga content. The average width of each fringe is about 100Å, which is exactly the same periodicity that was measured in the modulation fringes observed in the [110] cross section. Unlike an intentionally grown superlattice having uniform layer thickness extending to the whole substrate surface, each dark/light fringe in this growth-induced superlattice has a finite width in the [$\bar{1}$10] direction and an extended longitudinal dimension in [110] direction. The average longitudinal size in [110] direction is around 3000Å. In general, this dimension is comparable to or larger than the thickness of a TEM sample. Therefore, it is likely that a dark/light fringe observed in a thinned TEM [110] cross section can be entirely composed of just one composition.

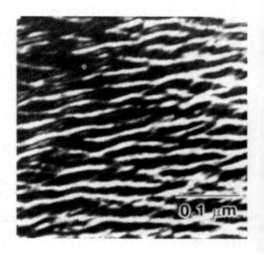

Fig. 3.   Dark field plan-view transmission electron micrograph of the $(GaP)_2/(InP)_2$ short-period superlattice grown on nominally (100) GaAs substrate. The nearly parallel fringes lie in the [110] direction.

The composition of the modulating fringes was also examined with EDS. An electron probe of 10 Å in diameter generated in a VG HB-5 TEM system operating at 100 KV has been used to obtain x-ray spectra from dark/light fringes in a thin [110] sectional TEM sample. We found a higher In content than the average value in dark fringes, and a higher Ga content than the average value in light fringes. Extrenal compositions of $Ga_{0.44}In_{0.56}P$ and $Ga_{0.58}In_{0.42}P$ have been measured among the In-rich fringes and the Ga-rich fringes, respectively, in the $(GaP)_2/(InP)_2$ SPS layer. Since the image contrast gradually changes across the dark/light fringe, we expect the lateral composition in each fringe should also vary gradually. The measured In or Ga content represents only the possible extremal average value in the fringes. Due to possible absorption of x-ray signals from lighter elements by heavier atoms and the finite dimension of each fringe, the measured composition may still underestimate the maximum compositional modulation. Nevertheless, the measured complementary Ga and In compositional modulation is consistent with the observed bright-field (g = 002) image contrast, namely In-rich and Ga-rich composition corresponding to dark and light fringes, respectively.

3.2.   Photoluminescence Measurements:

As mentioned already, the (001)-ordered $(GaP)_n/(InP)_n$ and $(GaAs)_n/(InAs)_n$ SPS's can actually be treated as a growth induced [$\bar{1}$10] $Ga_xIn_{1-x}P/Ga_yIn_{1-y}P$ and $Ga_xIn_{1-x}As/Ga_yIn_{1-y}As$ superlattice layers, respectively, exhibiting the dark/light modulation fringes parallel to the growth direction. Although EDS data show that among these layers the In content can vary between 0.42 and 0.56, the highest In content in each fringe may have been underestimated due to the influence of sample geometry and electron scattering effects. Since a higher In content implies a lower band gap, this growth-induced lateral composition modulation forms a lateral MQW superlattice structure. The continuous but oscillating image contrast indicates that this lateral MQW superlattice has a sinusoidal-like well-barrier structure instead of the more popular square-well superlattice. To verify this finding, low temperature PL polarization measurements were performed. Figure 4 (a) and (b) show the 140K PL spectra of the $(GaP)_2/(InP)_2$ SPS sample measured with the axis of TE mode of the laser beam polarized in the [110] and [$\bar{1}$10] directions, respectively. The parallel ($\parallel$) and perpendicular ($\perp$) symboles on each curve denote the orientation of the polarization analyzer for the PL measurement. Three major peaks were generated from the $Ga_{0.525}In_{0.475}P$ buffer layer (~1.963 eV), the $(GaP)_2/(InP)_2$ SPS layer (~1.818 eV), and the GaAs substrate (~1.510 eV) (Hsieh et al. 1990). In these spectra, the GaAs substrate serves as a good control sample for depicting the PL behaviour of a zinc-blende type cubic symmetry system.

(a)

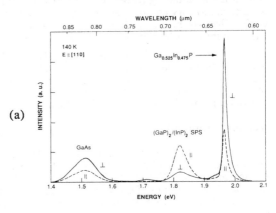

Fig. 4.   Low temperature (140K) polarization photoluminescence spectra of the $(GaP)_2/(InP)_2$ short-period superlattice grown on the $Ga_{0.525}$ $In_{0.475}P$ buffer layer and (100) GaAs substrate. The sample was measured with the axis of TE mode of the laser beam been polarized in the (a) [110] and (b) [$\bar{1}$10] directions. The parallel ($\parallel$) and perpendicular ($\perp$) symboles on each curve denote the orientation of the polarization analyser for the PL measurement.

(b)

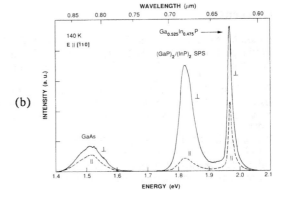

One must keep in mind that the PL intensity polarized in the perpendicular configuration for a system with T_d point group symmetry is always larger than that polarized in the parallel configuration because of the anisotropic momentum distribution of photocreated carriers. This is verified in the PL spectrum from the substrate for both orientations. On the other hand, for one geometry of excitation [Fig. 4 (a)] the PL intensity of the SPS layer in the parallel configuration is always greater than that in the perpendicular configuration, whereas in the other excitation geometry [Fig. 4 (b)] the opposite holds true. In this MQW system,

PL emission polarized in the [$\bar{1}$10] direction is allowed only for transitions from the conduction band to the light hole band whereas PL emission polarized in the [110] direction is allowed for transitions from the conduction band to both the heavy and light hole transitions (Yamanishi and Suemune 1984). In addition, at low temperature the Fermi distribution factor for heavy holes is larger than that for light holes. If, together with this information, one keeps in mind that the experimental polarization analysis is not perfectly exact because of small misorientation errors, then the data can be interpreted as confirming the electronic behaviour of lateral MQW's in the $(GaP)_2/(InP)_2$ SPS.

4. Conclusion:

Both $(GaAs)_n/(InAs)_n$ and $(GaP)_n/(InP)_n$ ordered vertical SPS layers grown on InP and GaAs substrates, respectively, were found to have a growth-induced lateral periodic modulation of the composition along the [$\bar{1}$10] direction. The periodicity was ~200Å. Within the modulating fringes of the $(InP)_n/(GaP)_n$ SPS layer the In and Ga compositions were found to vary complementarily between 42 and 58% and form parabolic-shaped lateral QWs. The existence of such growth induced *lateral MQW structures* was confirmed by low temperature photoluminesence polarization measurements. When the MQW structure was

excited with polarized laser light along [110] and [$\bar{1}$10] directions, strong anisotropy was observed in the intensity ratio of the electron-to-light-hole and electron-to-heavy-hole transitions. These results represent the first evidence of lateral two-dimensional quantum confinement in MQWs formed in vertical $(GaAs)_n/(InAs)_n$ and $(GaP)_n/(InP)_n$ SPS structures grown on nominally (100) InP and GaAs substrates, respectively.

Acknowledgment:

The authors are grateful to Mrs. M. Mochel for assistance in composition analysis, and to Prof. G.E. Stillman and Prof. N. Holonyak, Jr. for support. We would also like to acknowledge the Microanalysis Center at the University of Illinios for the use of the microanalysis facilities. This work has been supported by the National Science Foundation (ECD-89-43166 and DMR-89-20538) and the Joint Service Electronics Program (N00014-90-J-1270).

References

Asada M, Miyamoto Y and Suemetsu Y 1986 *IEEE Quantum Electron* **QE-22** 1915
Baillargeon J N, Cheng K Y, Hsieh K C and Stillman G E 1990 *J Appl Phys* **68** 2133
Fraser H L, Maher D M, Knoell R V, Eaglesham D J, Humphreys C J and Bean J C 1989 *J Vac Sci Technol* **B7** 210
Fukui T and Saito H 1988 *J Vac Sci Technol* **B6** 1373
Hsieh K C, Baillargeon J N and Cheng K Y 1990 *Appl Phys Lett* **57** 2244
Kash K, Scherer A, Worlock J M, Craighead H G and Tamargo M C 1986 *Appl Phys Lett* **49** 1043
Petroff P M, Miller M S, Lu Y T, Chalmers S A, Metin H, Kroemer H and Gossard A C 1991 *J Crystal Growth* **111** 360
Sakaki H 1980 *Jpn J Appl Phys* **19** 95
Yamanishi M and Suemune I 1984 *Jpn J Appl Phys* **23** L35

Inst. Phys. Conf. Ser. No 120: Chapter 12
Paper presented at Int. Symp. GaAs and Related Compounds, Seattle, 1991

595

An overview of the electronic properties of InAs$_{1-x}$Sb$_x$ strained-layer superlattices (0 $\leqslant x \leqslant$ 1)

S. R. Kurtz, R. M. Biefeld, L. R. Dawson, and B. L. Doyle

Sandia National Laboratories, Albuquerque, NM 87185

Abstract: Long wavelength infrared photoluminescence, magneto-transmission, and detector spectral response measurements have been used to characterize the electronic properties of InAs$_{1-x}$Sb$_x$/InAs$_{1-y}$Sb$_y$ strained-layer superlattices (SLSs) ranging throughout much of the In(AsSb) ternary system. We find that there is a large type II band offset in the Sb-rich (x, y $\geq$ 0.8) SLSs. Ordering and further bandgap reduction is observed in As-rich (x, y = 0.5 - 0.4) SLSs. In either Sb-rich or As-rich structures, optical absorption is readily extended to wavelengths greater than 10 μm.

1. Introduction

There have been few attempts to "bandgap engineer" In(AsSb) semiconductors although In(AsSb) alloys have the smallest bandgaps and effective masses among the III-V semiconductors. In(AsSb) may prove useful for high speed devices, and this system is an obvious starting point to extend the optical response of III-V semiconductors to longer wavelength for infrared detectors and lasers. In this paper we summarize studies of the electronic and optical properties of In(AsSb) strained-layer superlattices (SLSs) for a wide range of compositions. Previously, long wavelength infrared response was observed at low As content in these SLSs, and we concluded that a type II (or staggered) band offset occurs in the In(AsSb) system [1,2]. Prototype infrared detectors have been fabricated from these structures [3].

2. Growth of In(AsSb) Alloys and SLSs

The In(AsSb) structures were grown by metal-organic chemical vapor deposition (MOCVD) or molecular beam epitaxy (MBE). (Growth temperatures of 475 °C (MOCVD) and 425 °C (MBE) were used. A review of growth procedures and a list of references is found in the chapter by Biefeld et al.[4].) Each technique offers potential advantages in material quality and device manufacturability. Sb-rich MBE material has displayed background doping levels ($\geq$ 1x10^{14} / cm^3) compatible with bipolar device applications. However, MOCVD growth allows for easier control of group V fluxes which determine In(AsSb) composition and lattice constant, and to date, lifetimes of carriers in MOCVD SLSs appear superior to those in MBE structures. For both MBE and MOCVD grown SLSs, double crystal x-ray analysis is required to determine the resulting SLS composition due to difficult group V flux control and complex growth kinetics. Also, considerable effort has gone into the development of strain-relief buffers to accommodate lattice mismatch between the substrate and these In(AsSb) structures. Material quality is critical in the performance of narrow bandgap devices, and dislocations and cracks resulting from lattice mismatched layers exceeding the critical layer thickness should be eliminated from the electrically and optically active regions of devices.

3. Sb-Rich In(AsSb) SLS Overview

A. Long Wavelength Infrared Absorption and Photoluminescence

Representative of the optical properties of Sb-rich In(AsSb) SLSs, we examine two InAs$_{0.13}$Sb$_{0.87}$ / InSb SLS structures grown using MOCVD. These two samples are designated the "100 Å" (96 Å / 104 Å layer thicknesses) and "250 Å" (251 Å / 258 Å layer thicknesses) SLSs. Low temperature (4.1 K) infrared absorption spectra for these SLSs are shown in Fig. 1. The absorption spectra are quantum size shifted relative to each other, and several absorption edge-like features, corresponding to optical transitions involving various quantum confinement states, are observed in each spectrum. Comparing the spectra in Fig. 1, the magnitude of the lowest energy absorption edge increases with decreasing layer thickness. This feature and the "soft" shape of the absorption edges, approximating $(\omega - \omega_0)^{3/2}$, are characteristic of type II, interband transitions [5].

A double-modulation technique [6] was used to obtain infrared photoluminescence spectra shown in Figure 2. At the higher laser powers, multiple photoluminescence peaks can be resolved, and the energies of the photoluminescence peaks correspond to edges in the absorption spectra (Fig. 1). Indicative of the low dislocation density in these SLSs, the linewidth of the main photoluminescence peak at low pump power was 11 meV and 18 meV FWHM for the 250 Å and 100 Å SLSs, respectively. Photoluminescence and the onset of

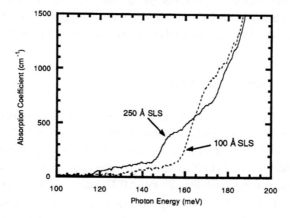

Figure 1 - Infrared absorption spectra for the 100 Å (dotted line) and 250 Å (solid line) InAs$_{0.13}$Sb$_{0.87}$/InSb SLSs at 4.1 K.

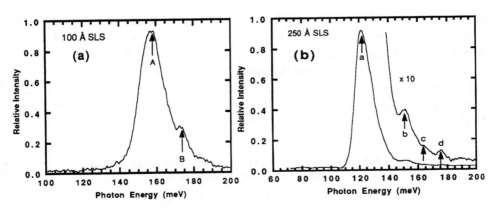

Figure 2 - Photoluminescence spectrum for the (a) 100 Å SLS and (b) 250 Å SLS with 15 K sample temperature.

optical absorption occurs at noticeably lower energy than the bandgap of the unstrained InAs$_{0.13}$Sb$_{0.87}$ constituent (180 meV). Observed transitions are significantly lower in energy than the strain-shifted, low energy transitions that would occur in type I SLSs. Therefore, these In(AsSb) SLSs must have a type II band offset which results in "spatially indirect", low energy optical transitions [1,2].

B. SLS Electronic Properties

Infrared magneto-transmission experiments were performed to identify the multiple transitions observed in photoluminescence and absorption and to study in-plane electronic properties of the InAs$_{0.13}$Sb$_{0.87}$ / InSb SLSs [6]. All of the transitions observed in photoluminescence are seen in magneto-optical spectra covering a wide range in magnetic field, and one finds that the zero-field magneto-transmission energies are in excellent agreement with the photoluminescence peak positions. Linear field dependences of the magneto-optical transitions reveal in-plane reduced masses ($1/\mu = 1/m^{*}_{e} + 1/m^{*}_{h}$) corresponding to interband electron - heavy hole (e-H) and electron - light hole (e-L) transitions of $\mu_{e-H} = 0.011 \pm 0.001$ and $\mu_{e-L} = 0.016 \pm 0.002$. For a range of in-plane light hole mass spanning k•p and cyclotron resonance values, $m^{*}_{L} = 0.05 - \infty$ [6,7], the reduced masses predict in-plane electron and heavy hole masses of $m^{*}_{e} = 0.024 - 0.016$ and $m^{*}_{H} = 0.021 - 0.035$, respectively.

The energies of the interband transitions and identification of those transitions based on magneto-optical data are summarized in Table 1. Note that (1e-1H) is the lowest energy transition observed in both SLSs. This proves that the hole quantum well is located in the biaxially compressed InSb layer, and these SLSs have a type II band offset. SLS quantum confinement energies were calculated for different band offset values using an envelope function model which corrects for band nonparabolicity [8]. The strain-induced energy shifts were determined for each layer of the SLS using deformation potentials found in the literature [9], SLS strain and composition data , and Vegard's law. An unstrained, valence band offset of 47 meV produced the fit of the optical data for the InAs$_{0.13}$Sb$_{0.87}$ / InSb SLSs shown in Table 1. The predicted quantum well structure for the InAs$_{0.13}$Sb$_{0.87}$ / InSb SLSs is shown in Figure 3. By examining a variety of SLS compositions, we find that an InAs$_{1-x}$Sb$_x$ unstrained, valence band energy of $(0.36\pm0.04 \text{ eV})\bullet x$ produces an offset that satisfactorily describes optical data for SLSs with $x \geq 0.8$. (Zero energy corresponds to the valence band energy of InAs.) The projected InAs /InSb valence band offset (0.36 eV) is in rough agreement with the value predicted by pseudopotential calculations (0.58 eV) [10].

Table 1 - Experimental energies and predicted quantum confinement transition energies for the InAs$_{0.13}$Sb$_{0.87}$ / InSb SLSs

SLS	Peak	Energy (meV)		SLS Heterojunction Model	
		Photolum. Peak	Magneto-Op. Transition, B=0	Quantum Confinement Transition	Energy (meV)
	a	122	121	(1e-1H)	117
	b	151	148	(1e-1L)	153
250 Å	c	163	165	(1e-2L)	164
	d	175	177	(2e-1H)	180
	e		194	(1e-ΓH)	191
100 Å	A	158	164	(1e-1H)	163
	B	174	177	(1e-2H)	181

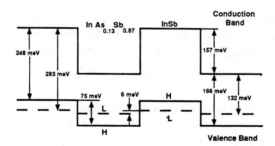

Figure 3 - Quantum well energies (0 K) of a InAs$_{0.13}$Sb$_{0.87}$/InSb SLS. Out-of-plane heavy-hole (H, solid line) and light-hole (L, dashed line) wells are indicated. Quantum confinement states have been omitted from the drawing.

4. Optical Properties of As-Rich In(AsSb) Structures

Preliminary results have been obtained for a high quality, $InAs_{0.62}Sb_{0.38}$ / $InAs_{0.54}Sb_{0.46}$ (83 Å / 83 Å layer thicknesses) SLS. This In(AsSb) SLS was grown using MBE on GaSb substrates with (AlIn)Sb / AlSb SLS strain relief buffers. Double-crystal x-ray rocking curve measurements with position (angle) sensitive 2θ detection [11] was used to determine the composition of the infrared active In(AsSb) SLS. A photoluminescence spectrum and a photoconductive response spectrum for this SLS are shown in Figure 4. Notice that the photoluminescence peak and the lowest energy absorption edge coincide at an energy of 116 meV (10.7 μm). These results demonstrate that long wavelength infrared photoresponse can easily be obtained with As-rich In(AsSb) SLSs. Surprisingly, the photoresponse was approximately 80 meV lower in energy than that predicted from accepted In(AsSb) ternary bandgaps and the band offsets observed in Sb-rich SLSs. Our preliminary electron diffraction results and previous reports [12] indicate that ordering is occurring in the In(AsSb) ternary layers of the SLS, and we are working to determine the effect that ordering has on the optical properties of As-rich In(AsSb) alloys and SLSs.

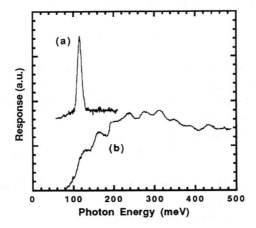

Figure 4 - (a) Photoluminescence spectrum (15 K) of a $InAs_{0.62}Sb_{0.38}$ / $InAs_{0.54}Sb_{0.46}$ (83 Å / 83 Å layer thicknesses) SLS. (b) Spectral response (77 K) of a photoconductive detector fabricated from this SLS.

5. Summary

The energies and reduced effective masses of several interband optical transitions in Sb-rich, $InAs_{0.13}Sb_{0.87}$/InSb SLSs were determined from photoluminescence and magneto-transmission experiments. SLSs with different layer thicknesses produced self-consistent results. With these data, the type II band offset was accurately determined.

Consistent with a type II offset, an extremely small, in-plane effective mass hole ground state was observed, thus proving that the hole quantum well is located in the biaxially compressed InSb layer. InAsSb SLS holes display an in-plane effective mass which is lower than that reported for ground state holes in any other III-V structure.

As-rich SLSs also display long wavlength photoresponse. We speculate that ordering in the In(AsSb) ternary may be an important factor in extending the photoresponse of these structures beyond 10 μm.

In(AsSb) alloys and SLSs display a variety of novel electronic properties which can be controlled through "band structure engineering" and altering growth conditions. Most significantly, we have identified a number of In(AsSb) structures that extend the photoresponse of III-V materials to long wavelength. We anticipate that these materials will ultimately find applications as III-V based infrared sources and detectors and as high speed devices.

References

1. S. R. Kurtz, G. C. Osbourn, R. M. Biefeld, L. R. Dawson, and H. J. Stein, Appl. Phys. Lett. 52, 831 (1988).
2. S. R. Kurtz, G. C. Osbourn, R. M. Biefeld, and S. R.Lee, Appl. Phys. Lett. 53, 216 (1988).
3. S. R. Kurtz, L. R. Dawson, T. E. Zipperian, and R. D. Whaley, Jr., IEEE Elect. Dev. Lett. 11, 54 (1990).
4. R. M. Biefeld, S. R. Kurtz, L. R. Dawson, and G. C. Osbourn, Crystal Properties and Preparation (Trans Tech Pub., Zurich, 1989), Vol. 21, pp. 141-164.
5. Gerald Bastard, Wave Mechanics Applied to Semiconductor Heterostructures, Les 'Editions de Physique, Paris, 1988.
6. S. R. Kurtz and R. M. Biefeld, Phys. Rev. B 44, 1143 (1991).
7. S. Y. Lin, D. C. Tsui, L. R. Dawson, C. P. Tigges, and J. E. Schirber, Appl. Phys. Lett. 57, 1015 (1990).
8. G. Bastard and J. A. Brum, IEEE J. Quantum Elect. QE-22, 1625 (1986), and G. Bastard, Phys. Rev. B 24, 5693 (1981).
9. Landolt-Börnstein, Numerical Data and Functional Relationships in Science and Technology, edited by O. Madelung, M. Schulz, and H Weiss (Springer, New York), Vols. 17a and 17b, 1982.
10. Chris G. Van de Walle, Phys. Rev. B 39, 1871 (1989).
11. S. T. Picraux, B. L. Doyle, and J. Y. Tsao, Semiconductors and Semimetals, edited by T. P. Pearsall (Academic, New York, 1991), Vol. 33, p. 139.
12. H. R. Jen, K. Y. Ma, and G. B. Stringfellow, Appl. Phys. Lett. 54, 1154 (1989).

Inst. Phys. Conf. Ser. No 120: Chapter 12
Paper presented at Int. Symp. GaAs and Related Compounds, Seattle, 1991

601

Resonant tunneling in a novel coupled-quantum-well base transistor

Takao Waho, Koichi Maezawa and Takashi Mizutani

NTT LSI Laboratories
3-1, Morinosato Wakamiya, Atsugi
Kanagawa 243-01, Japan

ABSTRACT: Resonant tunneling in a novel coupled-quantum-well (CQW) base transistor is investigated with marked modulation in collector current being observed in common-emitter output characteristics. Transconductance measurements reveal that collector current is attributable to resonant tunneling transfer from emitter to collector through the modified energy states in the CQW-base. Current gain as large as 5.5 is observed for the first time with respect to the resonant tunneling current. This indicates 85% of the electrons injected from the emitter are transferred into the collector without recombination in the CQW-base.

1. INTRODUCTION

Resonant tunneling is expected to provide functional operations as well as ultra-high speed switching in future devices (Capasso 1990, Bate *et al* 1987). High potential has been demonstrated in resonant tunneling diodes (Brown *et al* 1991) and transistors such as RHET (Yokoyama *et al* 1985) and RTBT (Capasso *et al* 1986). Although these transistors successfully utilize resonant tunneling, the current amplifying process is the same as the one used in conventional HETs and HBTs. Recently, novel types of transistors have been proposed in which resonant tunneling current is modulated by directly controlling the quantum well potential (Reed *et al* 1989, Haddad *et al* 1989, Schulman and Waldner 1988, Bonnefoi *et al* 1985).

In these devices, a single quantum well (SQW) is generally used as the tunneling barrier structure. Employing coupled quantum wells (CQW), in which the energy state can be designed by adjusting quantum well dimensions and coupling strength, would make it possible to enhance or modify resonant properties, thereby opening up a whole range of new operations. Although some diodes using CQW have already been proposed (Nakagawa *et al* 1986, Tanoue *et al* 1988, Collins *et al* 1989), no transistor structures with CQW have been realized yet.

We have fabricated a novel transistor structure employing the CQW as a base and observed an enhancement in resonance properties. Observation of a clear peak in transconductance corresponding to the energy states in CQW leads to the conclusion that collector current in this transistor results from resonant tunneling transfer from emitter to collector through the modified energy states in the CQW-base. A current gain β as large as 5.5 is achieved, which is the first observed gain with respect to resonant tunneling current.

2. EXPERIMENTAL

All epitaxial layers were grown by MBE. First, a 200-nm n^+-doped GaAs buffer layer, a 300-nm n^--GaAs collector layer and an undoped 2-nm AlAs barrier layer were grown on n^+-GaAs(001) substrates. Coupled quantum wells (CQW) consisting of three 5-nm GaAs wells separated by a 1-nm AlAs barrier were then grown. This was followed by the growth of an undoped 2-nm AlAs barrier layer, a 270-nm N-$Al_xGa_{1-x}As$ ($x = 0.1$, $n = 5 \times 10^{17}$ cm^{-3}) emitter layer and a 50-nm n^+-GaAs cap layer. Si was used as an n-type dopant.

In order to use the CQW as a p-type base, Be was planar-doped at the middle of each well. Sheet concentration was 1×10^{13} cm^{-2} for each of the three wells, making total concentration 3×10^{13} cm^{-2}. This value was selected because lower concentrations resulted in a depleted base region, which prevented successful contact formation. At higher concentrations, Be atoms abnormally diffused out of the well into the emitter and collector layers, which seriously deteriorated device characteristics. The confinement of Be atoms within the base region was confirmed by secondary ion mass spectroscopy (SIMS) analysis.

A three terminal structure, as shown in Figure 1, was fabricated using conventional photolithography, wet-etching and metalization techniques. The emitter layer was removed partly by wet etching while monitoring the process electrically. After the etching, Ni/Zn/Au/Ti/Au and AuGe/Ni/Ti/Au were deposited on the p-type CQW-base layers and the emitter and collector layers, respectively. Annealing was then carried out to form ohmic contacts. Contact and sheet resistances for the CQW-base were estimated through the TLM method. Contact resistance of 1.2 Ω mm and sheet resistance of 6 kΩ/$\square$ were obtained. Although sheet resistance was rather high, this value was anticipated by taking account of the mobility of holes and the CQW thickness. The size of the emitter mesas ranged from 12×40 to 100×100 μm^2. For a reference, a transistor using a 14-nm single-quantum-well (SQW) base planar-doped at the same sheet carrier concentration (3×10^{13} cm^{-2}) was also fabricated.

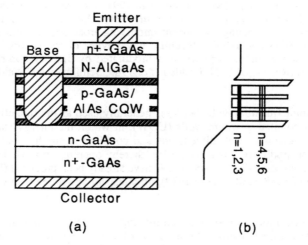

(a)   (b)

Figure 1   Structure of the coupled-quantum-well (CQW) base transistor (a) and schematic band diagram with energy states ($n = 1 \sim 6$) under an operating bias condition (b).

3. RESULTS AND DISCUSSION

3.1. Three terminal characteristics

Common-emitter output characteristics at 77 K were measured for both CQW- and SQW-base transistors with emitter mesa size of 12×40 μm^2. Marked modulations were found in collector current I_C, as shown in Figure 2, with an increase in base current I_B from 0 to 10 mA by 1-mA step. In the reference SQW-base transistor, the rate of I_C increase became small at $I_B = 3$ mA and then became large again at $I_B = 6$ mA (Figure 2a). This behavior can be interpreted as follows: with increasing I_B, an energy state in the SQW-base is lowered. When it aligns with the conduction band minimum (CBM) in the emitter, resonant current flows from emitter to collector. A further increase in I_B quenches this flow and turns on the next resonance. This interpretation was first proposed by Reed *et al* (1989) for their transistor called BiQuaRTT.

In the CQW-base transistor, the I_C increase rate also became small at $I_B = 3$ mA. However, it did not recover for larger I_B (Figure 2b). This can be interpreted in terms of modified energy states in the CQW. By coupling three quantum wells with thin tunneling barriers, the first three states ($n = 1, 2, 3$) in the CQW-base move closer together and effectively behave as one resonance state (Figure 1b). In addition, a large gap arises between the $n = 3$ and 4 states that reduces the chance for electrons in the emitter to tunnel through the CQW-base via $n > 4$ states. Being estimated to be 270 meV in this case, the gap was too large to turn on the next resonance. The unique properties in this I_C modulation due to the modified energy states will be discussed later with respect to transconductance.

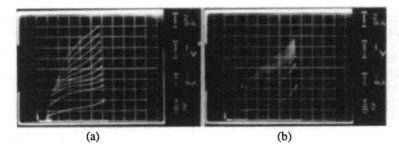

(a)   (b)

Figure 2 Common-emitter output characteristics at 77 K of the reference SQW-base transistor (a) and the CQW-base transistor (b). The emitter size was 40×12 μm^2.

3.2. I_B and I_C behaviors as functions of V_{BE}

In order to further examine resonant tunneling behavior in the CQW-base transistor, base current I_B and collector current I_C as functions of base-emitter voltage V_{BE} were investigated in detail in the common-emitter configuration (Figure 3). For small V_{BE}, I_B increased as an exponential function of V_{BE}. The ideal factor was 1.9 at room temperature. For larger V_{BE}, the I_B variation showed an ohmic behavior. Estimated resistance (42 Ω) was nearly equal to the calculated value (40 Ω) using the sheet and contact resistance mentioned above. Thus I_B increased with V_{BE} in a manner similar to that in usual HBTs except for the lack of a diffusion region, which can be explained by the fact that the CQW-base is too thin to permit carrier diffusion.

Contrary to I_B, I_C behavior was found to be completely different from that in usual HBTs. With an increase in V_{BE}, I_C increased abruptly at $V_{BE} = 1.5$ V. This was followed by a slow increase in I_C or almost saturation (Figure 3). As V_{BE} increases, there is a possibility that thermionic current from emitter to collector increases because the tunnel barriers are lowered by applying V_{BE}. However, since this increase should be monotonic, it is obvious that the I_C behavior observed here is not attributed to the thermionic emission from emitter to collector. The origin of I_C at $V_{BE} > 1.5$ V will be discussed in the following section. The small I_C for $V_{BE} < 1.5$ V was due to reverse leakage current in the base-collector junction.

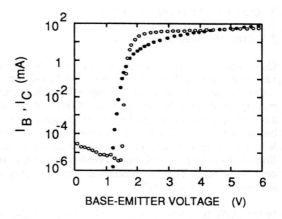

Figure 3 I_C and I_B as functions of V_{BE} for a CQW-base transistor at 77 K. Open circles and closed circles are I_C and I_B, respectively. The emitter size was 100×100 μm^2 and $V_{CE} = 6$ V.

3.3. Origin of collector current

In order to identify I_C origin clearly, the first derivative of I_C vs V_{BE}, i. e., transconductance of the CQW-base and SQW-base transistors was compared. As shown in Figure 4, one clear peak in the transconductance ($g_m = dI_C/dV_{BE}$) was observed for the CQW-base transistor, while there were two peaks for the SQW-base transistor. Peak voltages agree reasonably well with predicted values shown by arrows in Figure 4. These voltages were estimated for V_{BE} that aligned the emitter CBM to the CQW- or SQW-levels. The base sheet resistance was taken into account in the estimation. The good agreement provides clear evidence that the collector current is attributed to the resonant tunneling transfer from emitter to collector. Furthermore, it is concluded that the outstanding peak is attributed to the modified energy states in the CQW-base.

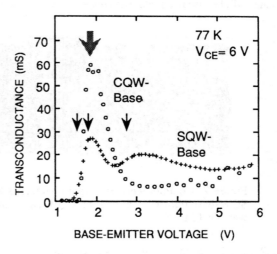

Figure 4 Transconductance as a function of V_{BE} for the CQW-base transistor with emitter size of 100×100 μm^2. The values for the SQW-base transistor are also plotted for comparison. One large arrow and three small arrows show calculated resonance voltages for the first three states (n = 1, 2, 3) in the CQW- and SQW-base, respectively.

The maximum current gain β in this case was 5.5 at $V_{BE} = 2.0$ V (Figure 5). As far as we know, this is the first clear evidence of an amplifying operation with respect to the resonant tunneling current. The current transfer ratio from emitter to collector α ($= \beta / (\beta+1)$) is derived to be 0.85. It is worth noting that 85% of the electrons injected from the emitter reach the collector without recombination in the CQW-base. The ratio α is represented as a product of emitter injection efficiency α_E, base transport efficiency α_T and collector collection efficiency α_C. The transport efficiency α_T could be described as $1 - \tau_{tun} / \tau_{rec}$, where τ_{rec} is the recombination time and τ_{tun} is the tunneling transit time through the CQW-base, i. e., phase time. Since τ_{tun} is calculated to be on the order of a few ps, and τ_{rec} is estimated to be at least a few hundred ps, α_T can be nearly equal to unity. If scattering effects due to high-concentration Be in the CQW-base could be suppressed and if emitter-base structure could be optimized to obtain $\alpha_E = 1$, then α of almost unity and therefore larger current gain β would be expected.

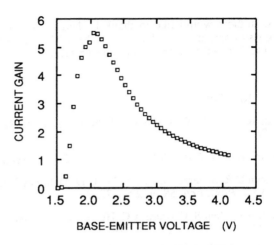

Figure 5 Current gain β as a function of V_{BE}. The maximum value was 5.5 at $V_{BE} = 2.0$ V. This corresponds to a current transfer ratio α ($= I_C / I_E$) of 0.85.

Seabough *et al* (1989) observed dI_C/dV_{BE} - V_{BE} for the BiQuaRTT, and obtained inflections that were claimed to represent the resonant transfer. The reference SQW-base transistor here has a structure similar to their device except for AlAs-barrier thickness (2 nm in the present device versus 5 nm in theirs). The thinner barrier drastically increases the resonant current and probably causes two peaks instead of the inflections for the SQW-base.

4. CONCLUSION

A novel transistor structure employing coupled quantum wells (CQW) as a base was proposed, and an enhancement in resonance properties was observed. Transconductance analysis reveals that the collector current is attributed to the resonant tunneling current through the modified energy states in the CQW-base. Current gain β as large as 5.5 was observed for the first time with respect to the resonant tunneling current. This indicates that 85% of electrons injected from the emitter were transferred into the collector without recombination in the CQW-base.

ACKNOWLEDGMENTS

The authors would like to thank K. Nagata for his helpful suggestions in fabricating transistor structures, K. Honma for the SIMS analysis, Y. Hasuike for the MBE growth, and K. Hirata for his continuous encouragement.

REFERENCES

Bate R T, Frazier G A, Frensley W R, Lee J W and Reed M A 1987 *Proc. SPIE* **792** pp 26-35

Bonnefoi A R, Chow D H and McGill T C 1985 *Appl. Phys. Lett.* **47** pp 888-890

Brown E R, Söderström J R, Parker C D, Mahoney L J, Molvar K M and McGill T C 1991 *Appl. Phys. Lett.* **58** pp 2291-2293

Capasso F, Sen S, Gossard A C, Hutchinson A L and English J H 1986 *IEEE Electron Dev. Lett.* **EDL-7** pp 573-576

Capasso F 1990 (Ed.) *Physics of Quantum Electron Devices* (Berlin: Springer-Verlag)

Collins D A, Chow D H, Ting D Z -Y, Yu E T, Söderström J R and McGill T C 1989 *Solid-State Electron.* **32** pp 1095-1099

Haddad G I, Mains R K, Reddy U K and East J R 1989 *Superlattices and Microstr.* **5** pp 437-441

Nakagawa T, Imamoto H, Kojima T and Ohta K 1986 *Appl. Phys. Lett.* **49** pp 73-75

Reed M A, Frensley W R, Matyi R J, Randall J N and Seabaugh A C 1989 *Appl. Phys. Lett.* **54** pp 1034-1036

Seabaugh A C, Frensley W R, Randall J N, Reed M A, Farrington D L and Matyi R J 1989 *IEEE Trans. Electron Devices* **36** pp 2328-2334

Schulman J N and Waldner M 1988 *J. Appl. Phys.* **63** pp 2859-2861

Tanoue T, Mizuta H and Takahasi S 1988 *IEEE Electron Device Lett.* **9** pp 365-367

Yokoyama N, Imamura K, Muto S, Hiyamizu S and Nishi H 1985 *Jpn. J. Appl. Phys.* **24** L853-L854

Inst. Phys. Conf. Ser. No 120: Chapter 12
Paper presented at Int. Symp. GaAs and Related Compounds, Seattle, 1991

607

Tunneling spectroscopic study of miniband break-up and coherence in finite superlattices

Mark A. Reed
Department of Electrical Engineering, Yale University, New Haven, CT, USA

Ranji Aggarwal
Department of Electrical Engineering and Computer Science, Massachusetts Institute of Technology, Cambridge, MA, USA

Yung-Chung Kao
Central Research Laboratories, Texas Instruments, Dallas, TX, USA

ABSTRACT: We have utilized tunneling spectroscopy to investigate the miniband structure of MBE-grown finite superlattice samples, and have observed for the first time the transition from a superlattice miniband to a tightly-coupled sequential well structure far below the Stark localization threshold. The transition from an indistinguishable miniband to a coupled well structure in these samples was experimentally found to be 2.5 meV < W(miniband width)/n(# periods) < 10.5 meV. By varying the number of periods until the individual eigenstates are no longer visible, the superlattice coherence length can be directly measured, which for these structures is ~1000A.

1. INTRODUCTION

Semiconductor superlattices have received renewed interest for the design and fabrication of novel electronic structures utilizing perpendicular transport. A central issue for the design, utilization, and analysis of superlattice structures is the nature of the electronic states. We present here a tunneling density of states study of the transition of a finite superlattice from a superlattice miniband to a coupled well structure. We can use this technique to directly measure the coherence length of superlattice states in a superlattice, far under the Stark localization threshold.

2. EXPERIMENTAL APPROACH

A generic superlattice tunnel diode structure[1] was utilized to study the density of states in a series of superlattices. Table 1 illustrates the series of superlattice structures investigated. The samples consisted of a 10-period superlattice on either side of a tunnel barrier, except sample 3808 which only has a superlattice on one side, and 3969 which has 20-period superlattices. Structural details are given elsewhere.[2] Structural parameters were verified by cross-section transmission electron microscopy, and photoluminescence of nominally identical superlattices (grown without doping and contact structures) was used to verify superlattice bandgap and aluminum content.

Sample	d_{GaAs}/d_{AlGaAs}	$E_{SL,0}\text{-}Ec, GaAs$	Width	$E_{F,SL}\text{-}Ec, GaAs$
3488	140Å / 30Å	17, 66, 139 meV	5, 19, 50 meV	22 meV
3379	60Å / 30Å	53 meV	30 meV	62 meV
3450	40Å / 50Å	90 meV	25 meV	96 meV
3486	49Å / 14Å	45 meV	105 meV	57 meV
3445	40Å / 10Å	43 meV	190 meV	55 meV
3808 (asymm.)	40Å / 10Å	43 meV	190 meV	55 meV
3969 (20-p)	40Å / 10Å	43 meV	190 meV	55 meV

Table 1. Summary of the superlattice samples investigated.

The superlattice structure 3379 is presented to compare to previous work.[1] Structures 3488, 3450, 3486, 3445, and 3808 are also 10 period superlattices, designed such that the superlattice miniband widths span the available range in the conduction band. 3808's superlattice is identical to 3445, except the asymmetry allows one to investigate injection into a superlattice from a 3D system, and visa versa. The superlattice of 3969 is identical to 3445, except the number of periods has been increased to 20.

3. RESULTS

Figure 1 shows a self-consistent band diagram at resonant bias (a), along with the experimental current (I) (and conductance (G)) versus voltage (V) characteristics (b), of the type of structures investigated in this study. This specific example is a structure identical to the initial work of Davies *et al.*[1] The band diagram is determined from a self-consistent finite temperature Thomas-Fermi zero-current calculation, with the superlattice structure determined from an envelope function calculation superimposed. When the top of the first collector miniband crosses the bottom of the available emitter electron supply, a decrease in current occurs due to the requirement to conserve both energy and momentum. It should be emphasized that realistic band diagrams are necessary for an accurate understanding of resonant effects, and that an accurate determination of the Fermi level in the superlattices must be done.[3]

The superlattice Fermi levels were calculated by assuming free electrons in the transverse directions and Bloch states for the vertical direction. The Fermi level was then inferred as the chemical potential which leads to a miniband-occupied carrier density corresponding to the average carrier concentration. It should be noted that determination of the superlattice Fermi level in general produces a higher Fermi level than that for a bulk system of the same density.[3]

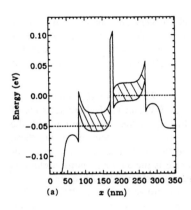
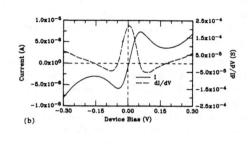

(a) z (nm)

(b)   Device Bias (V)

Figure 1. (a) Self-consistent Γ-point energy band versus epitaxial dimension of sample 3379 at resonant bias. The hatched regions denote the 25 meV wide lowest superlattice minibands, and the dotted lines the Fermi level. T=4.2K. (b) Experimental current (solid) and conductance (dashed) versus voltage characteristics of S1. T=4.2K.

Sample 3448 exhibits tunneling into a spectrum of superlattice states, as shown from Figure 2. The superlattice has 3 available minibands. NDR is seen when each is brought into resonance.

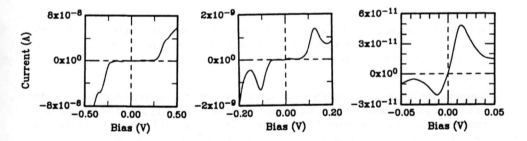

Figure 2. (a) Current versus voltage characteristics of the 3-miniband sample, in increasingly higher magnification from left to right. The leftmost graph exhibits tunneling into the n=3; the next shows the n=2; and the furthest right graph shows tunneling into the n=1. T=1.4.2K.

The I-V and G-V characteristics of samples 3379 and 3450 are very similar, exhibiting well-defined negative differential resistance (NDR) at low temperature. Aside from the major resonance (Figure 1), there is no apparent additional structure in the conductance greater than the 1mV (i.e., 12K) experimental resolution for either 3379 or 3450, at a sample temperature of 4.2K (immersed).

We now experimentally increase the superlattice miniband from 25 meV to 105 meV and examine the vertical transport. Figure 3 shows the low voltage I-V and G-V characteristics of 3486 at 4.2K. The 120mV major peak corresponds to the line-up of the first minigap with the emitter. A series of peaks on the low bias side of the major peak is apparent. Note that these biases correspond to electric fields well below that expected for Stark localization. The condition for Stark localization of a superlattice is $eEd > W$, where E is the applied electric field, d is the superlattice period, and W the width of the miniband under consideration. At the biases considered here, the Stark splitting is < 10meV, compared to a miniband width of 105meV. The "sub-resonant series" starts to degrade above 20K, and is unobservable (except for the highest sub-resonance peak) above 50K.

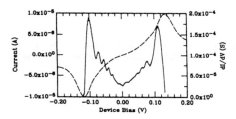

Figure 3. Low voltage I-V (solid) and G-V (dashed) characteristics of sample 3486 (105 meV wide superlattice miniband) at 4.2K. The 120mV major resonances corresponds to the alignment line-up of the first minigap with the emitter.

To explore the origin of the "sub-resonant series", we vary the superlattice miniband width while keeping the number of superlattice periods constant. Figure 4(a) shows the I-V and G-V characteristics of a superlattice miniband experimentally increased to 190 meV (3445), showing a very pronounced sub-resonance series. The higher bias peaks of this series are evident even at room temperature. To determine if the structure is due to the finite extent of the superlattice, we calculate the single electron transmission coefficient of the 10-period superlattice/coupled quantum well system, and map these 10 resonant peaks onto the self-consistent band structure. Figure 4(b) shows the calculated resonant crossings of the collector finite superlattice transmission peaks with the emitter Fermi level, compared with the experimental resonant peaks. The calibration of the top resonance is determined by the number of periods in the finite superlattice, and the low peak cut-off is determined from the superlattice Fermi level. The agreement between calculated and experimental peak position is qualitatively and quantitatively good. High voltage deviation may indicate a zero-current model is no longer valid.

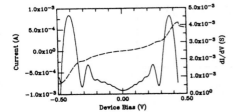
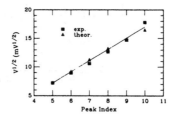

Figure 4. (a) Low voltage I-V (dashed) and G-V (solid) characteristics of sample 3445 (190 meV wide superlattice miniband) at 10K. (b) Experimental (square) and theoretical (circle) resonant crossings of the collector finite superlattice transmission peaks with the emitter superlattice Fermi level. The calculated resonant crossings were determined from mapping the finite superlattice transmission peaks onto the self-consistent band structure and determining the bias at which they cross the emitter Fermi level. To check that the resonances are indeed arising from the collector density of states, a sample (3808) identical to 3445 but with bulk GaAs on one side of the superlattice was investigated. As expected, oscillations in the conductance were observed for injection from the bulk GaAs into the finite superlattice, but none were observed in the opposite polarity.

The absence of structure in 3379 and 3450 implies that we have experimentally observed the transition (in this system) from a indistinguishable miniband to a coupled well structure. In energy, this implies the transition occurs between state splittings of 4 meV (the maximum in 3379) and 8 meV (the minimum observable in 3486), when kT < the state splitting $E(i+1)$ - $E(i)$. Note that this is a function of the position of eigenstate i within the miniband. In rationalized units, this corresponds to 2.5 meV < W(miniband width)/n(# periods) < 10.5 meV. The origin of the eigenstate broadening mechanism (such as epitaxial or alloy fluctuations) is not known.

Finally, we can use this technique to directly measure the coherence length of superlattice states in a superlattice. Sample 3969 has the same superlattice structure as 3445, except now the number of periods has been increased to 20. In this situation, if all the peaks are observed, it implies that these superlattice states are coherent over the entire superlattice. Figure 5 shows the conductance-voltage characteristics at 4K of this sample (with a background subtracted). The oscillations are weak, as the spacing of the states is approaching the line broadening limit. However, there appears to be all the peaks within this voltage range corresponding to the superlattice states. Over the entire voltage range, 14 peaks are observed, and 14-15 are expected (the uncertainty is due to the precision of knowing the Fermi level).

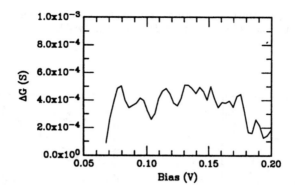

Figure 5. G-V characteristics of sample 3969 at 4K (with a background subtracted). Over the entire voltage range, 14 peaks are observed, and 14-15 are expected (the uncertainty is due to the precision of knowing the Fermi level).

4. SUMMARY

In summary, we have observed by vertical tunneling transport the eigenstates of a finite superlattice system far below the Stark localization threshold. We find for the samples investigated, when kT is lower than the splitting, the transition from an indistinguishable miniband to a coupled well structure experimentally is 2.5 meV < W(miniband width)/n(# periods) < 10.5 meV. We have also observed, using this technique, a coherence length of ~1000Å of superlattice states.

5. ACKNOWLEDGEMENTS

We are thankful to W. R. Frensley and J. L. Luscombe for modeling support, and to R. T. Bate, D. C. Collins, and C. Fonstad for constant support and encouragement, to W. M. Duncan for the photoluminescence measurements, to J. N. Randall for discussions, to H.-L. Tsai for cross section TEMs, and to R. K. Aldert, P. Q. Montague, E. D. Pijan, P. F. Stickney, F. H. Stovall, and J. R. Thomason for technical assistance. This work was done under the MIT-TI VI-A Internship Program, and was supported in part by the Office of Naval Research.

REFERENCES

1. R. A. Davies, M. J. Kelly, and T. M. Kerr, *Phys. Rev. Lett.* **55**, 1114 (1985).

2. R. J. Aggarwal, M. A. Reed, W. R. Frensley, Y.-C. Kao, and J. L. Luscombe, *Appl. Phys. Lett.* **57**, 707 (1990).

3. J. H. Luscombe, R. J. Aggarwal, M. Luban, and M. A. Reed, to be published, *Phys. Rev. B*, 15 September.

Inst. Phys. Conf. Ser. No 120: Chapter 12
Paper presented at Int. Symp. GaAs and Related Compounds, Seattle, 1991

613

Ultrathin GaAs p–n junction wires

K. Hiruma, K. Haraguchi, T. Katsuyama, M. Yazawa, and H. Kakibayashi
Central Research Laboratory, Hitachi Ltd., Kokubunji, Tokyo 185, Japan

Abstract :

Ultrathin GaAs needle-shaped microcrystals (whiskers) as thin as 20 nm and 2 to 3 μm long are grown by organometallic vapor phase epitaxy. The whiskers are selectively grown within a SiO_2 patterned substrate surface, and the growth axis is parallel to the [111] direction. From transmission electron microscopic analysis, the crystal structure coincides with that of zincblende at a growth temperature of 460℃, but changes to that of wurtzite above 500℃. A p-n junction along the wire is grown by changing dopants from silicon to carbon during growth. Light emission from the p-n junction is observed at a wavelength of 920 nm in continuous operation at room temperature.

1. INTRODUCTION

Quantum wires and boxes have recently attracted much attention, because such structures raise the possibility of developing a low-power consuming laser with an extremely narrow emission spectrum and high gain (Asada et al 1986). To obtain such a device, the device active region should be as small as 100 nm.

For the past several years, GaAs quantum wires have been fabricated by ion etching (Scherer et al 1987), molecular beam epitaxy (MBE) (Tsuchiya et al 1989, Petroff et al 1984), or organometallic vapor phase epitaxy (OMVPE) (Fukui et al 1988, Asai et al 1987). With these methods, the microstructures with thicknesses of 20 to 100 nm are formed on a substrate plane and cannot be grown outside of the substrate.

In this paper, we present a new approach for making quantum wires using needle-shaped microcrystals (whiskers) by OMVPE. The advantage of a whisker is that a p-n junction and heterostructure can be grown along the wire. We also discuss p-n junction growth as well as the crystal structure of the whisker.

2. EXPERIMENTAL

Whiskers were grown on GaAs substrates whose surface orientations were (100), (110), (111)A, (111)B, (211)B, and (411)B. The substrate surface was covered with CVD SiO_2 (thickness: 0.3 μm), and window patterns were made by standard photolithographic techniques. The window patterns were square, rectangular, and other shapes ranging from 3 to 300 μm. Prior to whisker growth, the substrate surface was degreased, rinsed in a buffered HF solution (HF:H_2O = 1:100) for 30 sec, and then rinsed in deionized water for 1 min.

Trimethylgallium (TMG) and arsine (AsH_3;10% in H_2) were used as souece materials, and disilane (Si_2H_6;150ppm in H_2) was used for an n-type dopant. These source materials were carried by palladium diffused H_2. The substrate was placed on a graphite base in a vertical SiO_2 reactor, and the temperature was controlled by rf induction heating. In this study, the temperature range for whisker growth was between 380 and 520℃.

The shape and size of the whiskers grown on the substrate were observed by scanning electron microscopy (SEM) and transmission electron microscopy (TEM). The crystal structure was analyzed by transmission electron diffraction pattern and the chemical composition was analyzed by energy dispersive X-ray spectroscopy (EDX).

3. RESULTS

3.1 WHISKER GROWTH

An SEM photograph of whiskers grown at 460℃ on a SiO_2 patterned (100)GaAs substrate is shown in Fig.1(a). The surface configuration of the substrate atoms is also shown in Fig.1(b). The whiskers were grown within the SiO_2 window area, and their growth axis is parallel to the [111] direction which is 35.3 degrees tilted from the substrate surface, as seen in Fig.1. When whiskers were grown on

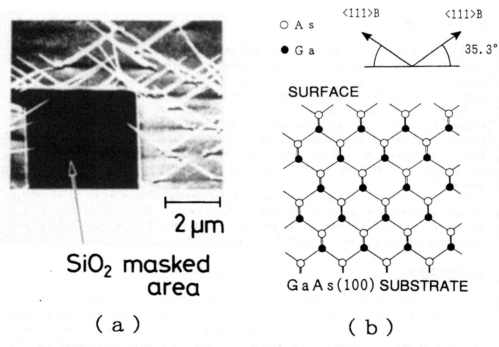

Fig.1 SEM photograph of whiskers. (a) Whiskers grown on (100) GaAs substrate. (b) Cross-sectional view of substrate viewed along [01$\bar{1}$] and surface configuration of substrate atoms.

the (111)B substrate, the growth axis was perpendicular to the substrate surface, as seen in Figs.2(a) and 2(b). The growth axis was also parallel to the [111] direction when whiskers were grown on the (110), (211)B, and (411)B substrate. However, the growth axis was random when grown on the (111)A substrate. From these growth experiments, it is concluded that the growth axis is parallel to the [111] arsenic dangling bond directon on the substrate surface.

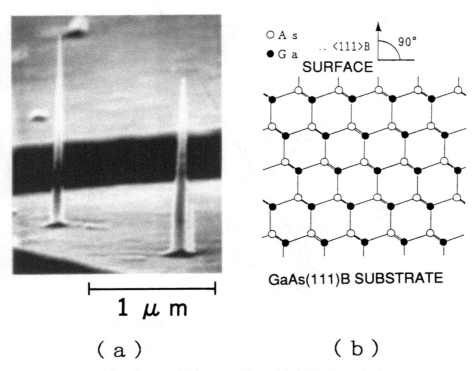

1 μ m

(a) **(b)**

Fig.2 SEM photograph of whiskers. (a) Whiskers grown on (111)B GaAs substrate. (b) Cross-sectional view of substrate viewed along [01$\bar{1}$] and surface configuration of substrate atoms.

The width and length of the whiskers were dependent on the substrate temperature and growth time. The whisker width grown on the (111)B GaAs substrate is shown in Fig.3 as a function of growth time. In this figure, the width is tentatively defined by the width at a half length of the whisker, as indicated by the drawing in the graph. From Fig.3, it is known that the width increases as the growth time as well as the substrate temperature increase. However, the width did not depend on the AsH₃-to-TMG molar flow ratio under present growth conditions.

The growth mechanism of the whisker has not yet been clarified, but it is similar to the vapor-liquid-solid (VLS) growth reported by Wagner and Ellis (1965) on the whisker growth of Si.

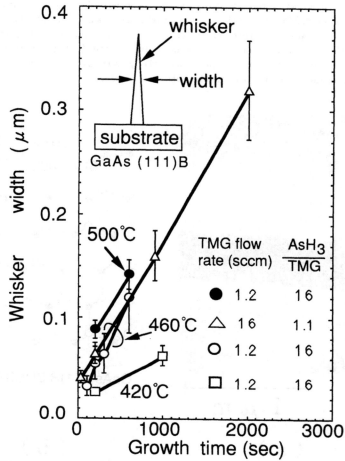

Fig. 3 Whisker width as a function of growth time.

3.2 CRYSTAL STRUCTURE ANALYSIS

A TEM image and transmission electron diffraction pattern of a whisker grown at 500℃ on the (111)B GaAs substrate are shown in Figs.4(a) and 4(b). In this measurement, the incident electron beam was focused on the cleaved edge of a substrate and was parallel to the [1$\bar{1}$0] direction. From the TEM image shown in Fig. 4(a), a stripe pattern with a periodicity of 10 to 50 nm can be seen along the wire. The diffraction pattern shown in Fig.4(b) coincides with the wurtzite structure, in contrast with the zincblende structure of a GaAs substrate. However, the structure changed to that of the zincblende when the whisker was grown at 460℃.

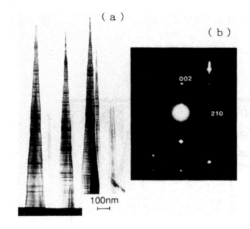

(a)

(b)

002

210

100nm

Fig.4 TEM photograph of whiskers.
(a) TEM image of whiskers , and
(b) transmission electron diffraction
pattern.

The chemical composition of the whisker determined by EDX analysis showed a Ga-to
-As ratio of nearly 50:50 at the tip and base of the whisker within a resolution
limit of ±1.5%.

4. P-N JUNCTION CHARACTERISTICS

Whiskers with a p-n junction along the wire were successfully grown on a Si
-doped (111)B GaAs substrate by changing the dopants from Si (using Si_2H_6) to
carbon (C;using TMG). The substrate temperature was 460℃. The whisker size was 3
to 5μm long and about 90 nm wide. The density of the whiskers was around 3 x 10^6 /
mm^2. The doping levels of Si and C of the whisker have not been evaluated, because
they are difficult to measure. The doping levels were thus approximated by those of
epitaxial layers grown under the same conditions, and they were 1 x 10^{18} cm^{-3}
for Si and 6 x $10^{18} cm^{-3}$ for C by Hall measurement.

The shape of the p-n junction interface was observed by SEM using a stain-etched
facet of a cleaved substrate. The Si-doped (n-type) whisker was put into the sheath
of the C-doped (p-type) whisker. The thickness of the p-type sheath was estimated
at 10 to 30 nm. This value is smaller than the depletion width at a doping level of
6 x 10^{18} cm^{-3}. Thus, the p-n junction can be approximated by the contact point
around the tip of the n-type whisker. The whiskers were partly buried in spin-on
glass (SOG).This was followed by making an ohmic contact by evaporating Au-Zn at
the tip and Au-Ge at the back side of the substrate. For electrical and optical
measurement, the substrate was cut down to 1 x 1 mm^2 and soldered onto a metal
mount.

The p-n junction showed diode I-V characteristics with a breakdown voltage of
around -3 volts. The light emitted from the p-n junction under forward biasing
conditions was detected by a GaAs photocathode (S-1). An emission spectrum is
shown in Fig.5 together with a cross-sectional view of the device.

The emission intensity was maximum around 920 nm in continuous operation at room temperature.   The full width at half maximum (FWHM) was 77 meV.   At 77K, the emission peak wavelength shifted to 835 nm with a FWHM of 51 meV, and the peak intensity increased by two orders of magnitude greater than that at room temperature.

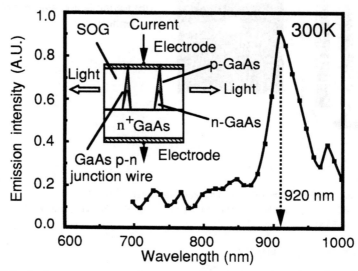

Fig.5 Emission spectrum and cross-sectional view of p-n junction diodes.

5. CONCLUSION

A novel method for making ultrathin wires has been developed using whiskers by OMVPE.   The crystal structure, analyzed by transmission electron microscopy, coincides with the zincblende structure at a growth temperature of 460 ℃, but changes to that of wurtzite above 500℃. Light emission from the p-n junction has been observed, showing promise for making quantum wire lasers.

REFERENCES

Asada M, Miyamoto Y and Suematsu Y 1986 IEEE J. Quantum electron. QE-22 1915.
Asai H, Yamada S and Fukui T 1987 Appl. Phys. Lett. 51 1518.
Fukui T and Saito H 1988 J. Vac. Sci. Technol. B6 1373.
Petroff P M, Gossard A C and Wiegmann W 1984 Appl. Phys. Lett. 45 620.
Scherer A, Roukes M L, Craighead H G, Ruthen R M, Beebe E D and Harbison J P
    1987 Appl. Phys. Lett. 51 2133.
Tsuchiya M, Graines J M, Yan R H, Simes R J, Holtz P O, Coldren L A and
    Petroff P M 1989 Phys. Rev. Lett. 62 466.
Wagner R S and Ellis W C 1965 Appl. Phys. Lett. 4 89.

Electron wave interference in fractional layer superlattice (FLS) quantum wires

Kotaro Tsubaki   Takashi Honda   Hisao Saito   Takashi Fukui

NTT Basic Research Laboratories,
Musashino-shi, Tokyo 180, JAPAN.

I. Introduction

A new concept of device action is desired to achieve high-performance electron devices. The electron wave interference effect[1-7] and mobility modulation effect[8] are enhanced in quantum confinement materials, presenting possibilities for such a new device concept. By using the electron wave interference effect, the current is controlled by the interference between incident electron waves and electron waves reflected by the potential in the channel. Since the phase of the electron wave modulates the current (in contrast to the conventional electron device where the wave amplitude modulates the current) an electron wave interference effect is promising for a high-performance electron device.

The electron wave interference effect, where the potential has sub-micron periodicity, can change the small drain current.[3-5] However, electron wave interference devices with AlAs/GaAs fractional layer superlattices (FLSs) have demonstrated an increases of drain current.[6] This is because an AlAs/GaAs FLS has periodicities of 5-20nm perpendicular to the growth direction. The FLS periodicity is adjusted to the order of the electron wavelength by changing the tilt angle. Current oscillations due to the electron wave interference effect has been observed in FLS electron wave interference devices having periodicities of 8, 11, and 16nm.[7]

FLS electron wave interference does not lead to negative transconductance which is important for applications. Negative transconductance, however, is expected in an FLS quantum wire, which is produced by reducing the channel width of the FLS electron wave interference device to the submicron level.

This letter reports the fabrication of FLS quantum wires 0.25μm wide and observation of drain current oscillation with negative transconductance. The temperature dependence of the drain current oscillation shows that the oscillation valley corresponds to the side-lobes which are spaced equally between consecutive Bragg reflection peaks shown in a reflection spectrum of the grating waveguide.

II. Fabrication Procedure

The cross-sectional view of a FLS quantum wire is shown in Fig.1. The modulation-doped AlGaAs/GaAs heterojunction structure with FLSs was grown by low-pressure metalorganic chemical vapor deposition on a Cr-doped semi-insulating GaAs substrate. (001) GaAs substrate tilted $2.0°$ towards the [$\bar{1}$10] direction were used for growing the 8-nm-period FLS. The epitaxial layers are 180-nm-thick undoped GaAs, 6-nm-thick undoped

$(Al-As)_{1/2}(GaAS)_{1/2}$ FLS, 100nm-thick Si-doped $Al_{0.3}Ga_{0.7}As$, and 10nm-thick Si-doped GaAs cap-layer. The growth procedure has already been reported in detail.[9]

The current flow was towards the $[\bar{1}10]$ direction, as shown in Fig.1. The sample was produced as follows. The channel was defined by electron beam lithography to a width of 0.25 μm and fabricated by wet-etching the GaAs cap-layer and AlGaAs layer 40nm. Since the wet-etched region has a high resistivity for the whole range of gate voltage at 4.2K, electrons go through the narrow channel. The effective channel width is 0.1 $\sim$ 0.13μm obtained from the relation between the conductance and structure width. The source and drain electrodes were formed by vacuum-evaporating eutectic GeAu and Ni, and alloying. The sample were completed by depositing Ti and Au for the gate electrode by vacuum evaporation.

The FLSs act as a spacer layer to separate the carrier-supplying layer from the conducting layer, as shown in Fig.1. It has been calculated that electrons confined at the GaAs/FLS heterojunction interface penetrate into the GaAs "teeth" of the FLSs.[10] This penetration modulates the electron distribution along the $[\bar{1}10]$ direction. The subband spacing between the grand and the first subband is about 20 meV.[10] It has been reported that the density of states (DOS) below the top of the barrier in the periodic potential is quasi-one-dimensional and that above the top of the barrier is two-dimensional.[6]

III. Experimental Results

Figure 2 shows the drain current and transconductance vs. gate voltage at 4.2K. The drain current clearly oscillates. Since the transconductance is very small between 0.2 and 0.3V, the transconductances divided by the drain current is shown instead of the transconductance. Negative transconductance is observed at 0.40 and 0.58V. The threshold voltage which is defined as the voltage where transconductance divided by drain current suddenly become large, is measured to be 0.2V at 4.2K.

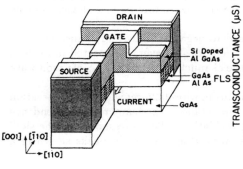

Fig.1
Cross-sectional view of the FLS quantum wire. The structure width of the quantum wire is 0.25 μm.

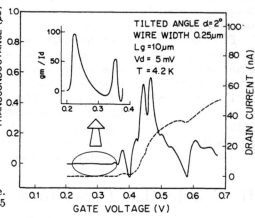

Fig.2
Transconductance (solid line) and drain current (dash-dotted line) at 4.2K as a function of gate voltage.

Since the transverse wavevector is quantized, conductivity oscillations associated with coincidence of the longitudinal wavevector and the FLS period are naturally enhanced as compared to the unconfined FLS case. Hence the conductivity oscillation as a function of gate voltage shows negative transconductance. Further, large transverse sub-level separation (on the order of sub-meV) which results from the wire width of $0.1 \sim 0.15 \mu m$ inhibits inter-sublevel scattering and contribute to the sharpness of the drain current oscillation features.

Due to the complicated density of states caused by low-dimensionality, the precise longitudinal wavelength associated with a given conductivity oscillation cannot be easily deduced from the electron concentration. Therefore, the Fourier analysis of drain current oscillation, which gave us the detailed mechanism of the electron wave interference,[6] cannot be performed.

The temperature dependence of the oscillations clarify the origin of electron wave interference effect. Figure 3 shows the temperature dependence of transconductance as a function of gate voltage measured from the threshold voltage. The oscillation can be observed up to 80K due to the large subband separation. The oscillation has an amplitude which diminishes and a full width at half maximum (FWHM) which grows as temperature is increased. This broadening, which follows from the degradation of the monochoromaticity of the electron wave (the usual thermal broadening the Fermi surface), is accompanied by a shift of peaks and valleys to higher gate voltages (*i.e.* shorter longitudinal wavelengths). This latter effect indicates that a valley in the transconductance oscillation is associated with the sidelobes in a reflected spectrum of a grating waveguide, which will be discussed in next section.

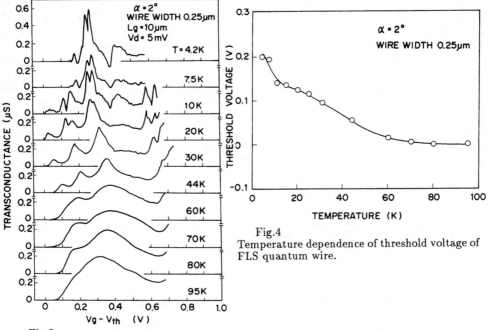

Fig.3
Temperature dependence of transconductance oscillation.

Fig.4
Temperature dependence of threshold voltage of FLS quantum wire.

The threshold voltage (V_{th}) as a function of the temperature is shown in Fig.4. the rapid decrease shown at 10 and 40K are only observed in FLS quantum wire. The decrease of V_{th} with increasing temperature thought to be due to the increasing number of mobile electrons. The DOS calculated from the temperature dependence of V_{th} will be discussed in Section V.

IV. Electron Wave Interference

Figure 5 shows schematic representations of an interference of a light wave in a grating waveguide and of an electron wave in a periodic potential. The drain current oscillation due to electron wave interference can be associated with the reflection spectrum of an optical grating waveguide.

Using the analogy to light waves, the superposition of electron waves, such that a multiple of half the wavelength is close to the periodicity of the periodic potential, causes the valleys in the drain current oscillation. Therefore, the valleys whose positions only determined by the periodicity of the periodic potential are thought to correspond to the Bragg reflection peaks.

On the other hand, the sidelobes in the reflection spectrum of the optical waveguide originate as follows. The sidelobes are the reflection peaks resulting from the interference between the incident light wave and the light wave reflected at the gratings shown in Fig.5. When the path difference between the two light waves is close to an odd multiple of half wavelength, the waves are superposed in opposite phase resulting in the reflection peak. The sidelobes in the reflection spectrum are spaced equally between consecutive Bragg reflection peaks, due to the path differences being multiples of the grating period, and their positions depend on the size of the grating waveguide. The maximum path difference between two light waves is the twice the size of the grating waveguide.

The above mechanism also causes the superposition of the electron waves. A path difference between the incident and reflected electron waves close to an odd multiple of half the wavelength produces a reduction of the drain current resulting in valleys in the drain current oscillation. The maximum path difference between electron waves is the phase coherence length, within which the phase of a traveling wave is preserved. The coherence length is thought to correspond to the the size of grating waveguide, because the coherence length and twice the size of the grating waveguide are the maximum path differences of the electron and light waves. Therefore, valleys whose positions depend on the coherence length correspond to the sidelobes.

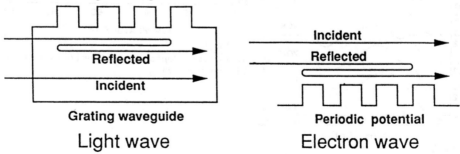

Fig.5
Interference of light waves in a grating waveguide and a electron waves in a periodic potential.

Valleys whose positions change with temperature are sidelobes because the coherence length deceases as increasing the temperature. The experimental results in Fig.3 indicate that the valleys correspond to the sidelobes and agree with the result of Fourier analysis in Ref.6.

V. Density of States

The DOS is calculated from the temperature dependence of V_{th} as follows. The electrons within $k_B T$ of Fermi level are mobile and can move from source to drain electrode when their energy is higher than the top of the barrier in the periodic potential. Therefore, the energy difference between the top of the periodic potential and the Fermi level is $k_B T/2$ at threshold. When the temperature increases by ΔT, FWHM of the electron distribution (the derivative of the Fermi distribution function) increases by $k_B \Delta T$ resulting in an increase in the number of mobile electrons. The increase of mobile electrons is expressed by $DOS \cdot k_B \Delta T/2$. The mobile electron increase is also expressed by $\Delta V_{th} \cdot C_{gs}/e$, where ΔV_{th} is the V_{th} shift corresponding to ΔT, and C_{gs}, the gate capacitance, which is $1.83 \times 10^{-7} F/cm^2$ as measured by a 1 MHz capacitance meter, is roughly equal to the AlGaAs layer capacitance. Therefore, the DOS between the bottom and top of the periodic potential is calculated using,

$$DOS = \Delta V_{th} \cdot C_{gs}/(e k_B \Delta T/2).$$

The DOS so-obtained is shown in Fig.6, is more complicated than that of the unconfined FLS modulation doped structure.[11]

The FWHM of DOS peaks are about 0.7meV. The electron scattering broadens the peaks in the DOS, whose FWHM is roughly estimated by $\hbar/\tau$ (τ : scattering time). This FWHM, which is estimated to be 0.8meV from the electron mobility at 4.2K, is close to the measured data. On the other hand, the DOS peak separations are 0.5 and 1.0 meV. Therefore, the sub-level separations are 0.5 and 1.0 meV, which are explained by the effective width of $0.1\mu m$ and the periodic potential amplitude of 4meV.

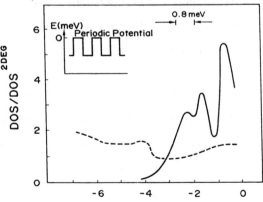

Fig.6
DOS of the FLS quantum wire (solid curve) and DOS of the FLS modulation doped structure (dashed curve). The position of 0 corresponds to the top of the barrier in the periodic potential as indicated in the inset. DOS^{2DEG} is the DOS of a 2-dimensional electron gas system.

Since the measured FWHM of the DOS peaks as well as the sub-level separation energy are on the order of the calculated values, a 1-dimensional state is thought to be obtained.

The reduction in the degree of freedom due to the sub-micron confinement can create the other lower-dimension state. In contrast to the FLS modulation doped structure consisting of 1- and 2-dimensional states, the FLS quantum wire is expected to have lower dimensionality (0- and 1- dimensionality).

V. Conclusion

Electron wave interference in AlAs/GaAs FLSs quantum wires was observed. The drain current oscillation with negative transconductance due to electron wave interference was first observed below 7.5K. The oscillation with negative transconductance originated from the reduction of wire width. The dimensionality of this system deceased due to submicron wire width. This oscillation became small as the temperature increases and diminishes over 95K.

Since the oscillation valleys moved to higher gate voltages as temperature increased, oscillation valleys corresponding to the sidelobes were displaced by the coherence length reduction concomitant to the temperature increase.

The DOS is estimated from the temperature dependence of V_{th}. The DOS has sharper peaks compared with that of the FLS modulation-doped structure. The DOS is found to consist of 1-dimensional and 0-dimensional states. The reduction in the degree of freedom of FLS modulation-doped structures results from sub-micron confinement.

References

[1] H. Sakaki, K. Wagatsuma, J. Hamasaki, and S. Saito, Thin Solid Films, **36**, 497 (1976).

[2] M. J. Kelly, J. Phys. C: Solid State Phys., 1984, **17**, L781 (1984).

[3] Y. Tokura and K. Tsubaki, Appl. Phys. Lett. **51**, 1807 (1987).

[4] K. Tsubaki and Y. Tokura, Appl. Phys. Lett. **53**, 859 (1988).

[5] K. Ismail, W. Chu, D. A. Antoniadis, and H. I. Smith, Appl. Phys. Lett., **52**, 1071 (1988).

[6] K. Tsubaki, T. Fukui, Y. Tokura, H. Saito, and N. Susa, Electron. Lett. **25**, 728 (1989).

[7] K. Tsubaki, T. Honda, H. Saito, and T. Fukui, Appl. Phys. Lett. **58**, 376 (1991).

[8] K. Tsubaki, Y. Tokura, and N. Susa, Appl. Phys. Lett. **57**, 804 (1990).

[9] T. Fukui and H. Saito, Jpn. J. Appl. Phys. **29**, L731 (1990).

[10] Y. Tokura, K. Tsubaki, and N. Susa, Appl. Phys. Lett. **55**, 1403 (1989).

[11] K. Tsubaki, Y. Tokura, and N. Susa, Appl. Phys. Lett. **57**, 2101 (1990).

Inst. Phys. Conf. Ser. No 120: Chapter 12
Paper presented at Int. Symp. GaAs and Related Compounds, Seattle, 1991

625

Second harmonic generation in AlInAs/GaInAs asymmetric coupled quantum wells

Carlo Sirtori, Federico Capasso,
Deborah L. Sivco, and Alfred Y. Cho

AT&T Bell Laboratories
Murray Hill, NJ 07974

1. Introduction

Recently optical nonlinearities associated with intersubband transitions at $\lambda \approx 10\ \mu m$ in doped AlGaAs/GaAs quantum wells have been experimentally investigated (Fejer *et al* 1989; Yoo *et al* 1991; Rosencher *et al* 1989a; Rosencher *et al* 1989b; Rosencher *et al* 1990) In particular, large nonlinear susceptibilities have been deduced from second harmonic (Fejer *et al* 1989; Yoo *et al* 1991; Rosencher *et al* 1989a) and optical rectification (Rosencher *et al* 1989b; Rosencher *et al* 1990) experiments. Physically the enhancement of these nonlinearities with respect to the bulk values results from the large intersubband matrix elements ($<z> \sim 10\ \text{Å}$).

In this paper we summarize our recent results on the investigation of second harmonic generation associated with intersubband transitions at $\lambda \sim 10\ \mu m$ in the AlInAs/GaInAs material system (Sirtori *et al* 1991a; Sirtori *et al* 1991b). The lower effective mass of GaInAs compared to GaAs has the advantage of larger matrix elements for the same intersubband transition wavelengths.

2. Sample Structure and Absorption (FTIR) Measurements

Our $Al_{0.48}In_{0.52}As/Ga_{0.47}In_{0.53}As$ structures, grown lattice matched to a semi-insulating (100) InP substrate, consists of 40 coupled well periods separated from each other by 150 Å undoped AlInAs barriers. Each period consists of a 64 Å GaInAs well, doped with $Si(n \approx 2\times10^{17}\,cm^{-3})$ separated from a 28 Å undoped GaInAs well by a 16 Å undoped AlInAs barrier. Undoped, 100 Å thick GaInAs spacer layers separate the multiquantum well structure from 4000 Å thick n^+ GaInAs layers. All thicknesses were checked by transmission electron microscopy. The band diagram and corresponding $|\psi|^2$s of each period are schematically illustrated in the inset of Fig. 1. To achieve enhancement of second harmonic generation (at $\lambda \approx 5.3\ \mu m$) the asymmetric structure is designed in such a way that the energy separation ΔE_{12} is within the tuning range of a CO_2 laser. In addition, the compositional profile is tailored in a way to maximize the product of the dipole matrix elements of the three transitions, since the second order susceptibility $\chi^{(2)}_{2\omega}$ is proportional to $<z_{12}> <z_{23}> <z_{13}>$ (Rosencher *et al* 1989a). In our structure our calculations (for the AlInAs/GaInAs parameters we used $\Delta E_c = 0.53$ eV; $m_e^* = 0.043\ m_0$; γ (nonparabolicity coefficient) $= 1.03\times10^{-18}m^2$) give: $<z_{12}> = 15.4\ \text{Å}$, $<z_{23}> = 22.3\ \text{Å}$ and $<z_{13}> = 12.1\ \text{Å}$. Previous structures used for second harmonic

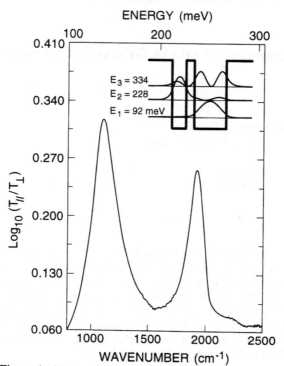

Figure 1. Measured absorbance at room temperature of the sample. The absorbance for polarization in the plane of the layers ($-\log_{10}T_\parallel$) is subtracted so as to remove some of the contribution of free electron absorption from the heavily doped regions. The inset shows a schematic band diagram of the coupled-well structure with its energy levels and the corresponding $|\psi|^2$.

generation consisted of asymmetric step graded wells (Yoo *et al* 1991; Rosencher *et al* 1989a) or of rectangular wells with an asymmetry induced by an electric field (Fejer *et al* 1989).

The sample transmission was measured at room temperature using an infrared Fourier transform interferometer and a multipass waveguide geometry (Levine *et al* 1987). The absorbance (= $-\log_{10}$ transmission) spectrum of the structure is shown in Fig. 1. The peaks at 137.3 meV and 238.2 meV are due to the $1 \rightarrow 2$ and $1 \rightarrow 3$ transitions, respectively; their position is in excellent agreement with our calculations. By integrating the area under the absorption peaks one obtains the integrated absorption strength I_A (West and Eglash 1985).

For the peak related to the $1 \rightarrow 2$ transition one obtains from the data of Fig. 1, $I_A = 9.5$ meV. From this value, the parameters of the structure and the sample geometry, one finds $<z_{12}> = 15.5$ Å in good agreement with the calculated value. From the $1 \rightarrow 3$ transition peak one finds from Fig. 1, $I_A = 5$ meV. From the ratio of the integrated absorption strength (I_A) for the two transitions one obtains $<z_{12}>/<z_{13}> = 1.38$. Thus

$<z_{13}> \simeq 11.2 \, \text{Å}$, in good agreement with the theoretical value. Note that the $1 \rightarrow 3$ transition is forbidden in a symmetric quantum well.

3. Zero Field Nonlinear Optical Properties

For these experiments, the samples were processed into circular mesas (350μm diameter) and ohmic contacts were provided to the two n^+ GaInAs cladding layers. They were then cleaved in narrow strips and the cleaved edges were polished at 45° to provide a two pass waveguide for the pump beam. In this geometry the beam from a stabilized CO_2 laser, after entering the sample at normal incidence (on one of the polished edges) and traversing the superlattice, is reflected off the top surface of the mesa and passes a second time through the multilayers. The second harmonic radiation is collected by a lens, followed by a sapphire window to cut the 10 μm radiation, an analyzer and filters to select the second harmonic. The signal is synchronously detected using a high detectivity calibrated InSb detector followed by a lock-in.

The second harmonic power is expected to increase quadratically with the power of the primary beam. In addition, only the component normal to the layer of the electric field of the incident wave at ω contributes to second harmonic generation (Fejer *et al* 1989; Yoo *et al* 1991; Rosencher *et al* 1989a). To maximize the latter we polarized the primary beam so as to maximize the component of the electric field normal to the layers and we adjusted the laser photon energy $\hbar\omega$ at 118 meV. At this energy $2\hbar\omega$ is very close to the separation between the third and first energy level ($E_3 - E_1 = 242$ meV).

The second harmonic power accurately follows the expected square law dependence on the pump power $P(\omega)$ (Fig. 2).

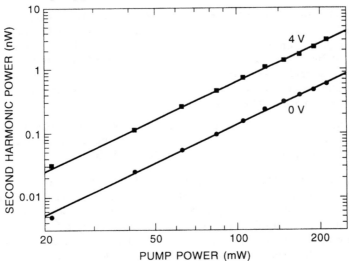

Figure 2. Second harmonic power at $\lambda = 5.3$ μm as a function of pump power under different bias conditions.

From the solid curve of Fig. 2 at zero bias, which represents the best fit to the data and Eq. (1) of Sirtori *et al* (1991a) for the second harmonic power, which includes the effects of absorption of the pump and the second harmonic (Singh 1984), one then obtains $|\chi^{(2)}(2\omega)| = 2.8\times10^{-8}$ m/V. Note that typical bulk values are $\approx 4\times10^{-10}$ m/V for GaAs, InAs and InP (Flytzanis and Ducuing 1969). Note that P(2ω) arises in general from the coherent superposition of the nonlinear polarization of the well and that of the bulk (dominated by the InP substrate) (Fejer *et al* 1989; Yoo *et al* 1991). The bulk and interference terms are negligible since $\chi^{(2)}_{\text{well}} \gg \chi^{(2)}_{\text{bulk}}$ and since in our geometry the component normal to the layers of the electric field of the pump is maximized. The dependence of P(2ω) on the polarization of the pump beam has been investigated elsewhere (Sirtori *et al* 1991a).

4. Electric Field Dependence of Second Harmonic Generation: Resonant Stark Tuning

For the type of 3-level structure shown in Fig. 1 and other asymmetric quantum wells (Rosencher *et al* 1989a) $\chi^{(2)}_{2\omega}$ is well approximated by the expression

$$\chi^{(2)}_{2\omega} = \frac{e^3}{\varepsilon_0} n \frac{<z_{12}> <z_{23}> <z_{31}>}{(\hbar\omega-\Delta E_{12}-i\Gamma)(2\hbar\omega-\Delta E_{13}-i\Gamma)} \qquad (1)$$

where n is the electron density, e is the electronic charge, $<z_{ij}>$ is the matrix element of the ij transition, $\Delta E_{ij} = E_j-E_i$ are the intersubband transition energies and Γ the level broadening. From Eq. (2) it is clear that, for a given structure, $\chi^{(2)}_{2\omega}$ exhibits a strong resonant enhancement if $\Delta E_{12} = \Delta E_{32}$ and $\hbar\omega \cong \Delta E_{12}$. We shall see that in our structures (Fig. 1) the $1 \rightarrow 2$ and $2 \rightarrow 3$ transitions can be strongly Stark shifted in opposite directions by an amount equal to the potential drop between the centers of the two wells, so that a pronounced peak is observed in $|\chi^{(2)}_{2\omega}|$ as a function of the electric field.

The second harmonic signal as a function of the CO_2 laser incident power at $\hbar\omega = 122.2$ meV is shown in Fig. 2 for 4 V positive bias, (i.e. close to the position of the peak power, Fig. 4). From the analysis of the data of Fig. 2 we then find $|\chi^{(2)}_{2\omega}| = 2.8\times10^{-8}$ m/V at zero bias and $|\chi^{(2)}_{2\omega}| = 7.5\times10^{-8}$ m/V at 4.0 V, corresponding to a field $F \approx 4.0\times10^4$ V/cm. The measured second harmonic power displays a pronounced peak as a function of positive bias, i.e. when the thin well is lowered in energy with respect to the thick well.

Figs. 3 and 4 summarize the results of our experiments at various pump photon energies. The second harmonic power and the corresponding $\chi^{(2)}_{2\omega}$ show a pronounced peak for positive polarity at fields in the range $3.5\times10^4 - 4.5\times10^4$ V/cm.

The enhancement of $\chi^{(2)}_{2\omega}$ by an electric field of the appropriate polarity can be understood in terms of the Stark shifts of the intersubband transitions calculated in Fig. 5. The intersubband energy separation ΔE_{13} is found to be weakly dependent on the electric field while the other two transitions (ΔE_{12} and ΔE_{23}) are strongly affected. Their shift with field is given, with excellent approximation, by the potential drop between the centers of the two wells. This behavior is physically understood by noting that the first and third states are confined by the thick well while the second state is confined by the thin well (see Fig. 1). It is well known that within a first order approximation the bound state energy of a quantum well with respect to the well center is independent of the electric field as long as the potential drop across the well is small compared to the bound state energy (Yuh and Wang 1989). Thus as the electric field is increased, the first and third energy levels track the center of the thick well while the second energy level tracks

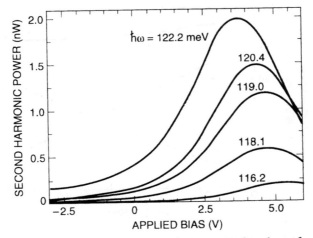

Figure 3. Second harmonic power as a function of applied bias for various pump photon energies at room temperature.

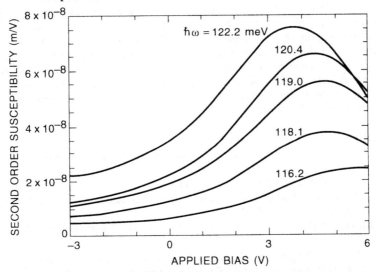

Figure 4. Measured $|\chi^{(2)}_{2\omega}|$ at room temperature as a function of applied bias at different pump photon energies $\hbar\omega$.

the center of the thin well. The net effect is that the energy of the $1 \rightarrow 3$ transition is weakly dependent on the electric field while the energy of the $1 \rightarrow 2$ and $2 \rightarrow 3$ transitions are shifted by an amount equal to the potential drop between the centers of the wells. The latter behavior is similar to that of the so called "local-to-global state" transitions in step quantum wells (Yuh and Wang 1989; Mii *et al* 1990). Our calculations also show that the dipole matrix elements appearing in Eq. (1) vary weakly with the electric field in the range of the experiments.

We are now in position to understand the data of Fig. 4. For the photon energies shown $\hbar\omega < \Delta E_{12}$ so that a positive bias must be applied to achieve the resonance

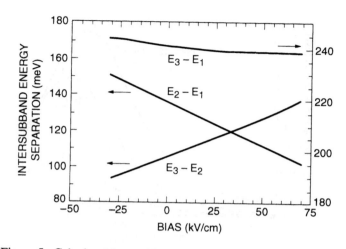

Figure 5. Calculated intersubband transition energies vs. electric field.

condition $\hbar\omega = \Delta E_{12}$ and a peak in $|\chi_{2\omega}^{(2)}|$ (see Eq. (2)). The higher $\hbar\omega$ the lower the field required to achieve resonance so that the peak of $|\chi_{2\omega}^{(2)}|$ will shift to lower bias; this trend is clearly manifested in the data of Fig. 4. The maximum value of $|\chi_{2\omega}^{(2)}|$ will occur when $\Delta E_{12} = \Delta E_{23} = \hbar\omega$. Our calculations predict that this should occur at $\hbar\omega \approx 120$ meV and for an electric field $\approx 3.2\times10^4$ V/cm (Fig. 5), in good agreement with the experimental values (122.2 meV and 3.7×10^4 V/cm). Using for Γ the experimental value of 15 meV obtained from the FTIR measurements and the calculated values $<z_{12}> = 16.1$ Å, $<z_{23}> = 20.4$ Å, $<z_{31}> = 12.4$ Å at $F = 3.5\times10^4$ V/cm, one then finds from Eq. (2) the maximum value $|\chi_{2\omega}^{(2)}| = 1\times10^{-7}$ m/V, in good agreement with the experimental data (0.75×10^{-7} m/V).

References

Fejer M M, Yoo S J B, Byer R L, Harwit A and Harris J S Jr. 1989 *Phys. Rev. Lett.* **62** 1041.

Flytzanis Chr and Ducuing J 1969 *Phys Rev.* **178** 1218.

Levine B F, Malik R J, Walker J, Choi K K, Bethea C G, Kleinman D A and Vandenberg J M 1987 *Appl. Phys. Lett.* **50** 273.

Mii Y J, Karunasiri R P G, Wang K L, Chen M and Yuh P F 1990 *Appl. Phys. Lett.* **56** 1986.

Rosencher E, Bois P, Nagle J and Delaitre S 1989a *Electron. Lett.* **25** 1063.

Rosencher E, Bois P, Nagle J, Costard E and Delaitre S 1989b *Appl. Phys. Lett.* **55** 1597.

Rosencher E, Bois P, Vinter B, Nagle J and Kaplan D 1990 *Appl. Phys. Lett.* **56** 1822.

Singh S 1984 *CRC Handbook Series of Laser Science and Technology, Vol. III: Optical Materials Part A* (Boca Raton: CRC Press).

Sirtori C, Capasso F, Sivco D L, Chu S N G and Cho A Y 1991a *Appl. Phys. Lett.*, in press.

Sirtori C, Capasso F, Sivco D L, Hutchinson A L and Cho A Y 1991b, submitted for publication.

West L C and Eglash S E 1985 *Appl. Phys. Lett.* **46** 1156.

Yoo S J B, Fejer M M, Byer R L and Harris J S Jr. 1991 *Appl. Phys. Lett.* **58** 1724.

Yuh P F and Wang K L 1989 *IEEE J. Quantum Electron.* **25** 1671.

Inst. Phys. Conf. Ser. No 120: Chapter 13
Paper presented at Int. Symp. GaAs and Related Compounds, Seattle, 1991

631

Incorporation of interstitial carbon during growth of heavily carbon-doped GaAs by MOVCD and MOMBE

G. E. Höfler[1], J. N. Baillargeon[1], J. L. Klatt[2], K. C. Hsieh[1], R. S. Averback[2] and K. Y. Cheng[1]

[1]Electrical and Computer Engineering Department; [2]Materials Research Laboratory University of Illinois at Urbana-Champaign.

Carbon is an attractive dopant for GaAs-based devices that require high doping concentrations and low atomic diffusivities. Hole mobilities reported for heavily carbon-doped samples, moreover, are at least 30% higher than those in beryllium or zinc-doped layers. However there appears to be a discrepancy between the total amount of carbon incorporated during growth,[1,2,3] as measured by Secondary Ion Mass Spectrometry (SIMS) concentration profiles, and the effective hole concentration as measured by Hall effect and Polaron electrochemical capacitance-voltage depth profiles. These measurements suggest, that at high carbon concentrations, some of the carbon atoms in the sample are either self-compensating or occupy neutral interstitial sites. Interstitial carbon could be detrimental to device performance if its diffusion out of the doped layer is rapid. It was the objective of this work to determine whether or not carbon occupies interstitial sites. We addressed this issue by directly measuring the amount of interstitial carbon in as-grown samples by nuclear reaction analysis (NRA) in conjunction with ion channeling, and by SIMS depth profiling of the redistribution of carbon in MOMBE grown GaAs n-i-p$^+$-i-n structures.

Rutherford Backscattering Spectrometry (RBS) has been widely used to determine the composition of matrix elements and impurities in near surface layers of many semiconductor materials, including GaAs.[4] When the ion beam is aligned perpendicular to a low index plane of a single crystal sample, however, the backscattering yield can decrease by $\approx$ two orders of magnitude. This effect, which can be used to probe structural defects in crystals, is denoted as ion channeling. However, RBS methods are not sensitive to impurities with low mass, such as carbon, because the backscattered yield from such impurities is small and overlaps with the larger signals from the Ga and As in the substrate. Consequently, NRA, using the ^{12}C(d,p)^{13}C, was employed since it yields a proton with an energy of 3.1 MeV that is detected without interference from the Ga and As signals. By comparing the signals obtained when the beam was either randomly or critically aligned with axial channeling directions in the sample, direct determinations of interstitial carbon present in the epilayers is possible.

Several GaAs n-i-p$^+$-i-n structures were grown by MOMBE using carbon and silicon as dopants. The SIMS depth profile of these structures clearly show the redistribution of carbon into the upper n-type layer. This effect has been previously reported for Be-doped GaAs grown by MBE[5] and for Zn-doped GaAs grown by MOCVD[6,7] and has been attributed to rapid diffusion of interstitial dopants. It is believed that this redistribution may affect the performance and reliability of HBT's when the carbon concentration in the base region exceeds $\approx 10^{20}$ cm^{-3}.

The first set of samples to be analyzed was grown by LP-MOCVD using TMGa, (100%) AsH$_3$ and CCl$_4$ as an external carbon source. The growth temperature was kept at 580°C and the carrier concentration was controlled by varying the flow of CCl$_4$ into the chamber and decreasing the V/III ratio to $\approx$ 10. Hole concentrations, determined by Hall effect measurements, were between 4 - 6 x 10^{19} cm^{-3}. A second set of samples was grown by MOMBE using elemental arsenic and TMGa. The amount of carbon incorporated into the epitaxial layers was controlled by adjusting the As$_2$/Ga ratio at the substrate from 5 to 2 while maintaining a sufficient arsenic overpressure to stabilize the surface. The growth temperature was measured using a calibrated infrared pyrometer, and it was maintained at 600°C. The range of carrier concentration, as determined by Hall effect and Polaron measurements, was 6 - 14 x 10^{19} cm^{-3}. SIMS depth profiles were obtained using a CAMECA IMS 3f in the negative ion detection mode with a Cs sputter beam. A standard with a peak concentration of 9.4 x 10^{19} cm^{-3} was used to calibrate the total concentration of carbon in the sample.

Standard ion beam analysis techniques were employed for the channeling study.[4] A beam of 1.3 MeV deuterium with a current of ≈ 150 nA was obtained from a Van der Graaff accelerator. The final spot size was set by a 3mm aperture placed at the entrance of the sample chamber. The current incident on the sample during the measurement was monitored by measuring the yield of ions backscattered from a rotating gold wire which intercepted the beam. Although radiation damage to the specimen was unlikely under these irradiation conditions, transmission electron microscopy (TEM) analysis was performed on the sample; no evidence of structural damage in the epilayer was observed. Figure 1 shows the backscattering spectra obtained from a MOCVD grown sample with a hole concentration of 4 x 10^{19} cm^{-3} when the ion beam was aligned with <100> surface normal. A minimum yield of 0.026 was determined by comparing the height of the aligned and random signals obtained from the GaAs substrate. A second spectra along the <110> direction was also obtained. Since backscattering techniques are based on statistical events, the experimental uncertainty is minimized by increasing the total number of events. For this reason, the backscattered intensity was integrated over the 1μm thick carbon-doped layer. The ratio of counts in the channeling to random directions, or "normalized yield," unambiguously showed that at least 27% (± 8%) of the C was located in interstitial sites.

Figure 2 shows the carbon yield obtained for a sample grown by MOMBE with a carrier concentration of 1.4 x 10^{20} cm^{-3}. A minimum yield of 0.05 was obtained for this and most other samples grown by MOMBE. The interstitial concentration in these samples was 28% (± 6%) of the total carbon concentration. However, as the carbon concentration increases much beyond 2.5 x 10^{20} cm^{-3} this techniques becomes less reliable as the minimum yield increases rapidly due to grown-in structural defects.

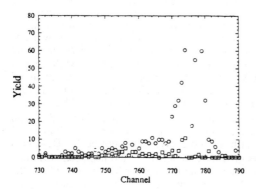

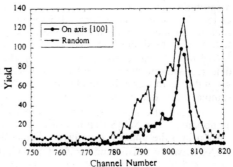

Figure 1   Random and <100> aligned backscattered yield obtained from a carbon-doped epilayer grown by LP-MOCVD with a hole concentration of 4 x 10^{19} cm^{-3}.

Figure 2   Random and <100> aligned backscattered yield obtained from a carbon-doped epilayer grown by MOMBE with a hole concentration of 1.4 x 10^{20} cm^{-3}.

An important application of a heavily carbon-doped layer is the base region of an HBT device, where the lower resistivity obtained is expected to improve the high frequency characteristics of the device. However, the presence of interstitial carbon can cause a non-abrupt emitter-base junction which will increase the space-charge combination and decrease the gain of the device. In order to determine the behavior of interstitial carbon during growth of a heavily carbon-doped layer, a set of n-i-p$^+$-i-n structures was grown by MOMBE using similar growth conditions to those employed for the samples analyzed by NRA. The growth structure, layer thickness and sources used are shown schematically in Fig.3. At a carbon concentration $\approx 1 \times 10^{20}$ cm^{-3} the redistribution of carbon into the upper n-region of the n-i-p$^+$-i-n structure is clearly observed in the SIMS depth profile shown in Figure 4. A similar redistribution of carbon into both the upper undoped and n-doped region was observed in all samples with carbon concentrations $\geq 6 \times 10^{19}$ cm^{-3}. The shape of the silicon signal in the upper undoped region [see Fig. 4] is not an artifact of the SIMS measurement caused by ion mixing but a real diffusion profile; this was verified in a series of samples with different silicon concentration. In fact, as the silicon level increased beyond 4×10^{18} cm^{-3} the amount of silicon diffusing into the undoped region increased. A systematic increase of the silicon concentration in the n-region, with carbon concentrations of $\approx 7 \times 10^{19}$ cm^{-3}, did not have an effect on the relative amount of carbon redistributing into the upper layers. However, an increase in the carbon content in the p-region, while the silicon concentrations was kept constant at $\approx 2.5 \times 10^{18}$ cm^{-3}, increased the relative amount of carbon diffusing into both upper regions.

Sources:

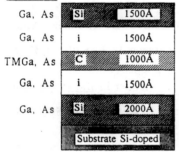

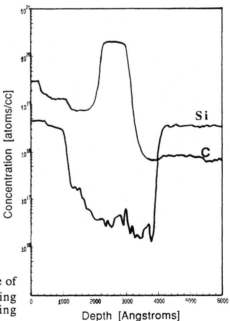

Figure 3 Schematic drawing of the n-i-p$^+$-i-n structures grown by MOMBE. Solid As and Ga were used throughout the structure except during the growth of the carbon-doped layer which required TMGa and solid As .

Figure 4 SIMS depth profile of a n$^+$-i-p$^+$-i-n$^+$ structure showing redistribution of carbon during growth of adjacent layers.

At present the mechanisms responsible for the anomalous redistribution of carbon into the upper layers are not fully understood. Similar results observed in p-type samples doped with beryllium or zinc have been attributed to the presence[6,7] or the thermodynamically favored creation[8,7] of interstitials during growth. Previous studies showed the absence of interstitial diffusion at carbon levels $\approx 10^{18}$ cm^{-3} in n-doped and p-doped materials.[9,10] This study shows that at carbon concentrations $\geq 6 \times 10^{19}$ cm^{-3} the operating diffusion mechanisms are different from those observed for lower carbon concentrations. Thus, at very high doping levels, the behavior obtained for carbon-doped layers is similar to that of zinc or beryllium-doped layers. This suggests that an upper limit exists on the total carbon concentration that can be usefully employed, as the incorporation of interstitials will change the favorable diffusion properties previously observed in carbon-doped epilayers.[9,10]

In conclusion, we have directly measured the presence of interstitial carbon in carbon-doped epilayers grown by MOCVD and MOMBE. At least 27% of the total carbon in an epilayer grown by MOCVD resides on interstitial sites. Similar trends were observed for samples grown by MOMBE of which, the one with the highest dopant concentration exhibit an interstitial concentration $\approx 28\%$ of the total carbon concentration. The detrimental effect that interstitial carbon may pose on the growth of HBT structures was demonstrated by the redistribution of carbon into the subsequent layers of a n-i-p$^+$-i-n structure. Carbon redistribution into subsequently grown layers was observed in all samples with total carbon concentration $\geq 6 \times 10^{19}$ cm^{-3}.

ACKNOWLEDGEMENTS: This work was supported by the National Science Foundations, grants Nos. DMR-89-20538 and ECD-89-43166. The authors wish to thank Dr. H. J. Höfler and S. A. Stockman for numerous technical discussions.

REFERENCES

1. T. J. de Lyon, J. M. Woodall, M. S. Goorsky, and P. D. Kirchner, Appl. Phys. Lett. **64**, 3975 (1990).
2. S. A. Stockman, G. E. Fernandez-Höfler, J. N. Baillargeon, G. E. Stillman, K. Y. Cheng, and K. C. Hsieh, to be published.
3. G. E. Höfler, H. J. Höfler, and K. C. Hsieh, to be published.
4. For a review see W. K. Chu, J. W. Meyer, M. A. Nicollet in <u>Backscattering Spectrometry</u>, (Academic Press Inc., New York, 1978).
5. P. M. Enquist, G. W. Wicks, and L. F. Eastman, J. Appl. Phys. **58**, 4130 (1985).
6. P. M. Enquist, J. Cryst. Growth **93**, 637 (1988).
7. W. S. Hobson, S. Pearton, and A. S. Jordan, Appl. Phys. Lett. **56**, 1251 (1990).
8. D. G. Deppe, Appl. Phys. Lett. **56**, 371 (1990).
9. B. T. Cunningham, L. J. Guido, J. E. Baker, J. S. Major, Jr., N. Holonyak, Jr. and G. E. Stillman, Appl. Phys. Lett. **55**, 687 (1989).
10. L. J. Guido, J. S. Major, Jr., J. E. Baker, N. Holonyak, Jr., B. T. Cunningham, and G. E. Stillman, Appl. Phys. Lett. **56**, 572 (1990).

Inst. Phys. Conf. Ser. No 120: Chapter 13
Paper presented at Int. Symp. GaAs and Related Compounds, Seattle, 1991

635

Optical determination of electric field and carrier concentration in nanometric epitaxial GaAs by photoreflectance

Ali Badakhshan*, C. Durbin, R. Glosser, *Center for Applied Optics and Physics Programs, UT-Dallas, Richardson, TX 75083;* **Kambiz Alavi,** *Dept. of Elec. Eng., UT-Arlington, TX;* **S. Lambert,** *VARO/IMO Inc., Garland, TX;* **R.S. Sillmon, P.E. Thompson,** *Nav. Res. Lab., Washington, DC;* and **K. Capuder,** *Emcore Corp., Somerset, NJ*

ABSTRACT

In this paper, we describe the application of photoreflectance (PR) to obtain the surface electric field and carrier concentration in moderately and heavily doped nanoscale epilayers of GaAs in a contactless manner. The technique exploits the PR line width of the E_1 response (~2.9eV). This is described by the empirical broadening parameter Γ_1, which we find increases linearly with the logarithm of the carrier concentration and the surface electric field beyond $\approx 10^{17}$ cm^{-3}. The slope of the linear relation for n-GaAs is nearly twice of that of p-GaAs. We explain the observed effect by the Schottky relation and Fermi level pinning and show that this effect is a function of the field intensity near the surface. We also show that the PR line shape at E_1 is applicable in determining the field intensity and the carrier concentration of a nanometric GaAs layer in the vicinity of the surface.

INTRODUCTION

For characterization of many advanced high density multilayer devices there is a great need for new methods of measurement which are nondestructive and applicable to very thin layers. Photoreflectance (PR) spectroscopy provides detailed information about the electronic properties of epitaxial compound semiconductors and different interfaces in multilayer structures. PR studies of compound semiconductors, which have received much attention in the last few years are strongly focused on the bandgap transition, E_0 (at 1.4 eV)[1-3]. The PR responses at E_0 in most thin multilayer device structures of GaAs and its dilute alloys are often very complicated[4] because of the large penetration depth of light at this energy. Recently a new approach in PR has been introduced[5,6] which helps to simplify the interpretation of such responses. In multilayer structures, we have demonstrated [7] that the PR measurement at $E_1 (2.9eV)$ has advantages over that of E_0 because of the smaller optical penetration depth at E_1 $(\alpha^{-1} \approx 17nm$ vs $\approx 1\mu m)$ and more limited lineshape distortion when the doping is increased. In more recent work, we showed[8,9] a linear relation between the broadening parameter at E_1, Γ_1 and the logarithm of the carrier concentration. The relation depends only on whether the material is n- or p-GaAs. We present a theoretical approach based on the Schottky relation and Fermi level pinning which explains the observed effect on Γ_1.

RESULTS AND DISCUSSION

A wide range of n- and p-GaAs samples were investigated by PR measurement at the E_1 transition. They were grown by metalorganic chemical vapor deposition (MOCVD) (Si, Se and Zn doped), low pressure chemical vapor deposition (LPCVD) (Si and Be doped) or molecular beam epitaxy (MBE)(Si, Be and Si+Be doped [This last set of dopants was chosen as a test for compensation effects]).

Fig. 1 (a) and (b) show Γ_1 vs. logarithm of carrier concentration for all n- and p-GaAs samples.For clarity, the results for those 16 Si:GaAs and 15 p-GaAs (Zn and Be doped) samples which had been recently reported[9], are merely presented by the best linear fit to these data. Γ_1 as a function of carrier concentration for Se:GaAs and Si+Be:GaAs as well as the reported values for S:GaAs[10] are also shown in Fig 1 (a). As seen in this figure the increase in Γ_1 is linear within the range of $\sim 1 \times 10^{17}$ to $\sim 4 \times 10^{18} cm^{-3}$. The best linear fit to Γ_1's within the above range is referred to as the "linear relation". Fig. 1 (a) clearly shows that a single linear relation with a slope of 56 ± 5 *meV per decade* describes all of our results for n-GaAs. This is further supported by Dittrich's results[10]. The reason the Γ_1's for S:GaAs seem to be somewhat lower than other n-GaAs samples has been explained elsewhere[9]. Our recalculation of the linear relation for S:GaAs[10] yields a value of $56 meV / decade$, which is in excellent agreement with our results for n-GaAs. This is shown by the dashed line in Fig. 1 (a). Based on this consideration we believe that the observed linear relation is the same for all n-GaAs.

Fig. 1 (b) shows Γ_1 versus carrier concentration for p-GaAs samples, where the linear region extends up to $1 \times 10^{19} cm^{-3}$. The slope of this range is 30 ± 3 *meV per decade*, which is considerably smaller than the slope for n-GaAs. Fig. 1 (b) clearly shows that a single linear relation describes all Zn: and Be:GaAs. This includes ion implanted Be:GaAs reported by Brown and co-workers[11] in their electroreflectance (ER) study. However, we do not observe any significant shift in E_1 for n- or p-GaAs, while Brown et al report a large shift (up to 80 *meV*).

Behavior of Γ_1 versus carrier concentration in Figs. 1 and 2 seems to be quite general, because Γ_1 is independent of both the growth apparatus and technique and it is also independent of doping element, compensation of carriers in the sample (as shown for Si:GaAs, or for the method of introducing the doping element [epitaxial vs. ion implantation, as demonstrated for Be:GaAs]). We believe that what is being observed in Γ_1 is associated with the electric field broadening as described by some authors[12-15]. In the present case, the broadening is a result of the increased electric field F within the depletion layer produced by the ionized impurities. Qualitatively, the broadening which correlates with increased doping, is comparable to that obtained in ER measurements of GaAs[14] and germanium[15] with increased applied F.

Recently Batchelor and co-workers[16,17] presented a theoretical treatment of the effect of the field gradient on electroreflectance (ER). Fig. 2, shows the reported[17] changes in ER line shape at E_1 with increased reverse bias for two different carrier concentrations in p-GaAs. They clearly show the increase in Γ_1 with increase in the electric field and carrier concentration, which is similar to what we observe in PR for Γ_1 as the carrier concentration is increased[7,8].

We now discuss the correlation between the increase in the doping concentration and the electric field within the depletion layer. The electric field in the semiconductor as a function of distance from the surface is[18]:

$$F(z) = F_s(1 - z/W)$$

where F_s is the surface electric field and W is the depletion width. W is related to the parameters of the sample as:

$$W = (2\epsilon_s V_b/eN)^{1/2}$$

where V_b is the built-in potential and N is the carrier concentration. As it is well known the Fermi level is pinned for doped GaAs. Though there are uncertainties involved in the value of V_b for air exposed GaAs, reasonable values for V_b are about $0.75 \pm .05$ and $0.5V$ (less certain) for n- and p-GaAs respectively[19,20]. As it relates to PR at E_1, the probe light samples only a portion of the depletion width (α^{-1} about $17nm$ compared with the depletion widths for doped GaAs, $W_D \approx 180$ and $W_A \approx 150nm$ for $norp = 5 \times 10^{17} cm^{-3}$). Therefore, $F(\alpha^{-1}) \approx F_s$ is a reasonable approximation at least up to about $5 \times 10^{17} cm^{-3}$. F_s is defined as[1,18]:

$$F_s \approx [(2eN/\epsilon_s)(V_b)]^{1/2} \tag{1}$$

and ϵ_s is the static dielectric constant of the semiconductor. As the doping is increased and degeneracy becomes more important, a more detailed examination is required.

Based on our experimental results and Eqn. 1, we propose that:

$$\Gamma_1 \approx C\log F_s = (1/2)C[\log(\xi N)] \qquad (2),$$

where $\xi = 2eV_b/\epsilon_s$. Eqn. 2, which holds for both donors and acceptors, is in agreement with the observed linear relation. Since ξ is larger for n-GaAs, Eqn. 2 also qualitatively explains the larger slope for n-GaAs with respect to p-GaAs.

In conclusion, for moderately and highly doped GaAs, there exist a linear relation between the width of PR line shape at the E_1 transition and the logarithm of the carrier concentration. The linear relation is considerably larger for n-GaAs compared with p-GaAs. On the other hand this line shape broadening is proportional to the electric field within the optical penetration depth, as we formulated. Consequently, upon proper calibration, the linear relation provides a contactless measurement of carrier concentration as well as the electric field within a $20nm$ distance from the surface/interface of the epilayer. In addition, this approach has important advantages over the existing techniques. Among these are application to nanometer scale layers within the depletion region and avoiding a composite response from a few hundred nm layer, which is the case for most common techniques. This characteristic makes the proposed technique particularly valuable in thin multilayer structures.

We are very grateful to Dr. Richard A. Batchelor for providing the copies of figures from his PhD thesis. This research is supported in part by the Texas Advanced Technical Program under project #0033656 − 141 and Texas Advanced Research Program under project #003656 − 107.

REFERENCES

*- Present address: Department of Physics, University of Minnesota-Duluth, 371 Marshall Hall, Duluth, MN 55812

1. N. Bottka, D. K. Gaskill, R. S. Sillmon, R. Henry, and R. Glosser, J. Electron. Mater. 17, 161 (1988).
2. L. Peters, L. Phaneuf, L. W. Kapitan, and W. M. Theis, J. Appl. Phys. 62, 4558 (1987).
3. U. Behn and H. Roppischer, J. Phys. C: Solid State Phys. 21, 5507 (1988).
4. See for example, Fred H. Pollak and H. Shen J. Crystal Growth 98, 53 (1989).
5. M. Sydor, Ali Badakhshan and J. R. Anghlom and D. A. Dale, Appl. Phys. Lett 58, 948 (1991).
6. M. Sydor and Ali Badakhshan, J. Appl. Phys. 70, 2322 (1991).
7. A. Badakhshan, C. Durbin, A. Giordana, R. Glosser, S. Lambert and J. Liu, in *Nanostructure Physics and Fabrication*, edited by M.A. Reed and W.P. Kirk (Academic, New York, 1989), p.485.
8. Ali Badakhshan, R. Glosser and S. Lambert, J. Appl. Phys. 69, 2525 (1991) and SPIE, Int. Soc. Opt. Eng. 1286, 382 (1990).
9. Ali Badakhshan, R. Glosser and S. Lambert, M. Anthony R. S. Sillmon, P. E. Thompson and K. Alavi, Appl. Phys. Lett, 59, 1218 (1991).
10. T. Dittrich, Phys. Stat. Sol. (a) 119, 479 (1990).
11. R.L. Brown, L. Schoonveld, L.L. Abels, S. Sundraram and P.M. Raccah, J. Appl. Phys. 52, 2950 (1981).
12. D.E. Aspnes and A.A. Studna, Phys. Rev. B 7, 4605 (1973).
13. A. Frova and D. E. Aspnes, Phys. Rev., 182, 795 (1969).
14. R. Del Sole and D. E. Aspnes, Phys. Rev. B 17, 3310 (1978).
15. D. E. Aspnes, Solid State Comm. 8, 267 (1970).
16. R. A. Batchelor, A. C. Brown, and A. Hamnett, Phys. Rev. B 41, 1401 (1990).
17. R.A. Batchelor, Ph.D. thesis Oxford University 1990.
18. See for example S. M. Sze, *Physics of Semiconductor Devices*, 2nd ed. (Wiley, New York, 1981).
19. D. K. Gaskill, N. Bottka, and R. S. Sillmon, J. Vac. Sci. Technol. B, 6, 1497 (1988) and references there in.
20. C. Van Hoof, K. Deneffe, J. Deboeck, D.J. Arent, and G. Borghs, Appl. Phys. Lett. 54, 603 (1989) and references there in.

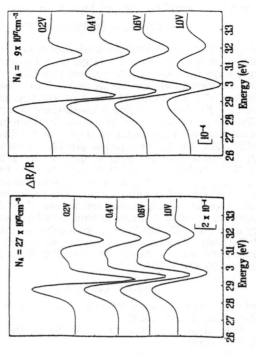

Fig. 1- Γ_1 as a function of carrier concentration for different n-GaAs (a) and p-GaAs samples (b). The solid lines are the best linear fit through Si:GaAs (16 samples), and Zn: or Be:GaAs (15 samples) which were previously reported (reference 9) and are not shown in these figures. Si+Be:GaAs, which has more donors than acceptors is called mixed doped sample. The best linear fit for S:GaAs samples, which are reported in Ref 10, is shown by the dashed line. In (b) Γ_1 for 5 Be:GaAs samples are shown. The Ion implanted samples has been reported in Ref 11. The error bars are typical for all of our data points in these figures.

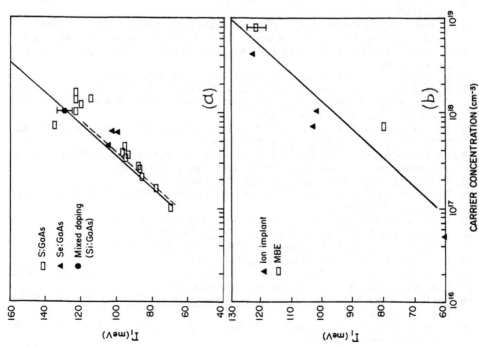

Fig. 2- The change in the electroreflectance line shape of p-GaAs with the change in reverse bias as reported by Batchelor et al., reference 17. The effect of increase in the electric field which is due to increase in bias voltage on the apparent lineshape broadening Γ_1 is noticeable. We observed similar results in Γ_1 due to increased carrier concentration.

Inst. Phys. Conf. Ser. No 120: Chapter 13
Paper presented at Int. Symp. GaAs and Related Compounds, Seattle, 1991

639

Magnetotransport properties of modulation doped pseudomorphic $Al_{0.30}Ga_{0.70}As/In_{0.17}Ga_{0.83}As$ heteroepitaxial layers grown by MBE on GaAs and $Al_{0.30}Ga_{0.70}As$ buffer layers

Juan M. Fernández and H. H. Wieder

Department of Electrical and Computer Engineering, 0407, University of California, San Diego, La Jolla, CA 92093-0407

Abstract: We have investigated at 1.6 K the Shubnikov de Haas (SdH) magnetoresistance oscillations of Si-modulation-doped, pseudomorphic, $Al_{0.30}Ga_{0.70}As/In_{0.17}Ga_{0.83}As$ heteroepitaxial layers grown on both GaAs and $Al_{0.30}Ga_{0.70}As$ buffers deposited on (100) GaAs substrates by Molecular Beam Epitaxy and also determined their electrical and low field galvanomagnetic properties between 300 K and 1.6 K. A comparison between the total scattering rate relaxation time τ_q obtained from the amplitude of the SdH magnetoresistance oscillations, and the conductivity relaxation time τ_c, suggests a strong large angle scattering contribution from the buffer interface of $In_{0.17}Ga_{0.83}As/Al_{0.30}Ga_{0.70}As$.

1. Introduction

Heteroepitaxial AlGaAs/InGaAs structures which employ pseudomorphic modulation doped epitaxial layers grown on GaAs substrates provide major advantages for high speed, high frequency field effect transistor applications (Moll 1987, Chao 1987). Such devices require large sheet electron concentrations in their two-dimensional electron gas channels with a high degree of spatial confinement. Pseudomorphic modulation doped AlGaAs/InGaAs/AlGaAs with doping from both sides of the quantum well channel layer may fit this requirement.

The electrical properties of double barrier heterostructures, however, were found to be inferior to those of single barrier structures in the AlGaAs/GaAs system (Fisher 1984, Pfeiffer 1991). A three period superlattice of graded average composition at the heterointerfaces was needed to obtain a 160 fold-increase in the photoluminescence efficiency (Fisher 1984). Similar problems might be expected with double barrier pseudomorphic AlGaAs/InGaAs/AlGaAs structures. It is, therefore, desirable to determine the relations between their structural and electrical properties.

We have determined the scattering times of such pseudomorphic modulation-doped doped $Al_{0.30}Ga_{0.70}As/In_{0.17}Ga_{0.83}As$ structures grown by MBE either on an undoped GaAs buffer layer or on an undoped $Al_{0.30}Ga_{0.70}As$ buffer layer. The conductivity scattering time, $\tau_c = \mu m^*/e$, where μ is the carrier mobility, m^* is the carrier effective mass and e the electronic charge, weighted by the factor $(1 - \cos\theta)$, differs from the total scattering time τ_q. This can be determined from the semi-classical expression for the low magnetic field-dependent transverse component of the resistivity matrix, ρ_{xx}, (Coleridge 1989)

$$\rho_{xx}(B) = \rho_0 \left[1 + 4D(X)\exp(-\pi/\omega_c\tau_q)\cos(\frac{2\pi E_f}{\hbar\omega_c} - \pi) \right] \tag{1}$$

where ρ_0 is the zero field resistivity, $\omega_c = eB/m^*$ is the cyclotron frequency, k is Boltzmann's constant, $\hbar$ is the reduced Planck's constant, T the temperature, E_f is the Fermi energy, and $D(X) = X/\sinh(X)$ where $X = 2\pi^2 kT/\hbar\omega_c$. This expression assumes single band occupancy and is applicable in low magnetic fields such that $\Delta\rho_{xx}/\rho_0 \ll 1$.

2. Experimental

The samples used for our experimental investigations were grown in a modified Varian Gen II MBE reactor on (100)-oriented GaAs substrates with an As_4/Ga flux ratio of ~ 30 and a GaAs growth rate of 0.9 μm/hr. Figure 1a is a schematic cross-section of sample MBE1632. It consists of a 1 μm thick GaAs buffer layer grown on a semi-insulating GaAs substrate with a substrate temperature of 580°C with a superposed 17.5 nm thick undoped $In_{0.17}Ga_{0.83}As$ layer, grown at a substrate temperature of 535°C. It contains the two-dimensional electron gas (2DEG) produced by modulation doping from a superposed 80 nm thick $Al_{0.30}Ga_{0.70}As$ barrier doped with Si to $2\times10^{18}/cm^3$ separated from the 2DEG channel by a 10 nm thick, undoped, $Al_{0.30}Ga_{0.70}As$ spacer layer. A 2.5 nm thick undoped GaAs cap layer covers the barrier layer reducing its degradation. The barrier, spacer and the cap layers were grown at a substrate temperature of 535°C. The thickness of the $In_{0.17}Ga_{0.83}As$ layer was deliberately chosen near the critical thickness limit, d = 17.7 nm (Fritz 1985, Weng 1989). Figure 1b represents sample MBE1633; it is in every respect the same as MBE1632 except that an undoped 0.1 μm thick $Al_{0.30}Ga_{0.70}As$ buffer layer is inserted between the GaAs buffer and the $In_{0.17}Ga_{0.83}As$ layer.

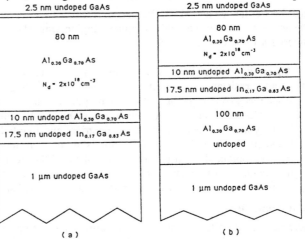

Fig. 1. Schematic cross section of heteroepitaxial samples MBE1632 (a) and MBE1633 (b), employed for Hall bar structures. Growth details are summarized in the text.

Photolithographic and etching procedures were used to define six arm Hall bar configurations with length to width ratio of 3. Ohmic contacts were made by means of alloying at 450°C vacuum deposited Ni/AuGe/Ni contact layers. Measurements were carried out in an Oxford superconducting magnet which provided a transverse, variable magnetic field, up to 8T, and stable temperatures variable from 300 K to 1.6 K.

3. Results and Discussion

Sheet electron concentrations and mobilities derived from Hall-effect measurements are included in Table 1. Shubnikov de Haas (SdH) oscillations measurements made at 1.6 K using an excitation current of 10 μA are shown in Figures 2a and 2b. The sheet electron densities, $n_s = 7.4\times10^{11}/cm^2$, determined from the Fast Fourier power spectra of the SdH oscillations, agree within 1% with the values determined from Hall-effect measurements.

Table I: Charge carrier transport parameters

Sample	μ (cm^2/V–s)		n_s (cm^{-2})		τ_c (ps)	τ_q (ps)	τ_c/τ_q
	300 K	1.6 K	300 K	1.6 K	1.6 K	1.6 K	1.6 K
MBE 1632	9500	95000	9.1×10^{11}	7.4×10^{11}	3.73	0.40	9.3
MBE 1633	5900	12000	1.0×10^{12}	7.4×10^{11}	0.47	0.13	3.6

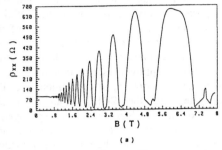
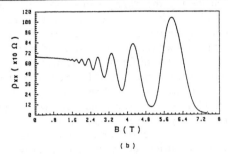

Fig. 2. Shubnikov de Haas oscillatory magnetoresistance as a function of magnetic field for samples MBE1632 (a), and MBE1633 (b) measured at T = 1.6 K with an excitation current of $10\,\mu A$.

The low field, oscillatory components of the magnetoresistance were extracted by digitally filtering the raw SdH data through a bandpass filter (Coleridge 1989, Kaiser 1978). The extrema of the oscillations were then normalized to $2\rho_0 D(X)$. The logarithm of the theoretically derived normalized extrema should be linear with respect to 1/B with an intercept for 1/B = 0 of 0.69315 and a slope of $(-\pi/\omega_c\tau_q)$. Given the experimentally measured n_s and the temperature, T, the Fermi level is defined; the effective electron mass, m^* and the total scattering time, τ_q, are then chosen to fit, in eq.(1), the experimentally measured SdH oscillation extrema.

Assuming a linear interpolation between the effective masses of unstrained GaAs and InAs, m^*(In$_{0.17}$Ga$_{0.83}$As) $= 0.0595m_0$. The effective mass corrected for strain is, $m^*_{st} = 0.0644m_0$. Consequently, the increase of ~ 8% in m^*/m_0 due to strain is added to an additional 8% attributed to the non-parabolicity at the Fermi level. The effective mass ratio in our case, $m^*/m_0 = 0.069$ is considered to include both corrections (Liu 1988). With this value we obtain,

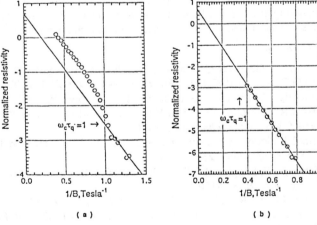

Fig. 3. Normalized extrema (open circles) of the SdH oscillations for samples MBE1632 (a) and MBE1633 (b) expressed as Ln $(\Delta\rho_{xx}/2\rho_0 D(X))$ as a function of reciprocal magnetic field. The solid lines are plots of equation (1) with $m^* = 0.069$ and the τ_q values listed in Table I.

by using a least square fit for τ_q, the data shown in Figures 3a and 3b and the values listed in Table 1, where $\tau_c = \mu m^*/e$.

The results presented in Figure 3 are in accord with the observations of Coleridge et. al. (Coleridge 1989). They indicate that eq. (1) may apply not only for $\Delta\rho_{xx}/\rho_0 \ll 1$ but for even higher magnetic fields for which $\omega_c\tau_q < 1$. The arrows in Figures 3a and 3b show the positions of $\omega_c\tau_q = 1$ in each case; it appears that the experimental data fit the model to this limit.

Considering the τ_c/τ_q ratios listed in Table 1, it appears that low angle scattering, associated with long range, screened ionized impurity scattering is present in the 2DEG of both sample types (Harrang 1985). The large difference between the τ_c/τ_q ratios of the sample grown on a GaAs buffer and that grown on an AlGaAs buffer indicates, however, the presence of different total scattering processes. In the latter, this ratio is closer to unity suggesting the presence of large angle scattering. It is tempting to ascribe this to a form of non-transitive interfacial roughness and to lattice defects formed, in consequence, at the AlGaAs/InGaAs heterointerface which propagate into the 2DEG channel if the $In_{0.17}Ga_{0.83}As$ layer is grown on the ternary alloy rather than on GaAs. Appropriate interface grading and or increased substrate temperature during growth of the backside AlGaAs barrier, however, may improve interface quality and therefore the transport properties of these double barrier, modulation doped pseudomorphic heterostructures.

4. Conclusions

A comparison is presented of the synthesis, electrical and galvanomagnetic properties of pseudomorphic $Al_{0.30}Ga_{0.70}As/In_{0.17}Ga_{0.83}As/GaAs$ and $Al_{0.30}Ga_{0.70}As/In_{0.17}Ga_{0.83}As$ /$Al_{0.30}Ga_{0.70}As$ modulation-doped heterostructures and of Shubnikov de Haas oscillatory magnetoresistance measurements made on the two-dimensional electron gas present in the $In_{0.17}Ga_{0.83}As$ layer. It is shown that the experimental measurements fit a theoretical expression for the amplitude extrema of the oscillations, developed for low magnetic fields provided that the applied magnetic fields are restricted to $\omega_c\tau_q \leq 1$. Although the sheet electron densities determined from the Fourier spectra of the SdH oscillations are the same for both heterojunction types the nearly factor of 3 difference in their classical to quantum scattering time is attributed to the non-transitivity of the $Al_{0.30}Ga_{0.70}As/In_{0.17}Ga_{0.83}As$ heterojunction and to lattice defects generated at its interface.

References

Chao P C, Tiberio R C, Duh K G, Smith P M, Balingall J M, Lester L F, Lee B R, Jabra A and Gifford G G 1987 *IEEE Electron Device Lett.* **EDL-8** 4899.
Coleridge P T, Stoner R and Fletcher R 1989 *Phys. Rev. B* **39** 1120.
Fisher R, Masselink W T, Sun Y L, Drummond T J, Chang Y C, Klein M V and Morkoç H 1984 *J. Vac. Sci. Technol.* **B2**(2) 170.
Fritz I J, Picraux S T, Dawson L R, Drummond T J, Laidig W D and Anderson N G 1985 *Appl. Phys. Lett.* **46** 967.
Harrang J P, Higgins R J, Goodall R K, Jay P R, Laviron M and Delescluse P 1985 *Phys. Rev. B* **32** 8126.
Kaiser J F and Reed W A 1978 *Rev. Sci. Instrum.* **49** 1103.
Liu C T, Lin S Y, Tsui D C, Lee H and Ackley D 1988 *Appl. Phys. Lett.* **53** 2510.
Moll N, Hueschen M R and Fisher-Colbrie A 1987 *IEEE Trans. Electron Devices* **ED-35** 879.
Pfeiffer L, Schubert E F and West K W 1991 *Appl. Phys. Lett.* **58** 2258.
Weng S L 1989 *J. Appl. Phys.* **66** 2217.

Inst. Phys. Conf. Ser. No 120: Chapter 13
Paper presented at Int. Symp. GaAs and Related Compounds, Seattle, 1991

643

High temperature (>150°C) and low threshold current operation of AlGaInP/Ga$_x$In$_{1-x}$P strained multiple quantum well visible laser diodes

T. Katsuyama, I. Yoshida, J. Shinkai, J. Hashimoto, and H. Hayashi

Optoelectronics R&D Laboratories, Sumitomo Electric Industries Ltd.,

1, Taya-cho, Sakae-ku, Yokohama 244, Japan

Abstract: High temperature and very low threshold current operation of separate confinement heterostructure AlGaInP/Ga$_x$In$_{1-x}$P (x = 0.43) strained multiple quantum well lasers has been achieved. Continuous wave (cw) operation was observed up to at least 150°C with an output power of more than 7mW, which is the highest cw operating temperature ever reported for devices operating in the visible wavelength region. The characteristic temperature was 130K (20 - 80°C) and the threshold current at 25°C was 13.9mA.

1. Introduction

AlGaInP visible lasers have attracted much attention as light sources in optical information processing systems such as magneto-optic disks and laser printers. Recently, significant developments in the reduction of the emission wavelength (Ishikawa et al 1990, Valster et al 1990, Kobayashi et al 1990, Hamada et al 1991), and the increase of the optical power (Bour et al 1987, Itaya et al 1990, Ueno et al 1990, Serreze et al 1991) have been made. However, high temperature operation of the lasers, which is also a very important characteristic, especially for the application to laser arrays and optoelectronic integrated circuits (OEICs), has been difficult. This is mainly due to the relatively large thermal resistance of AlGaInP and their high threshold current densities compared to those of AlGaAs/GaAs lasers. Reduction of the threshold current density, enhancement of the hetero-barrier height, and suppression of band-filling effects are very effective methods for improving the temperature characteristics of the devices, and much effort has been made including the optimization of the active layer thickness (Tanaka et al 1991) and the application of a multiple quantum barrier (Kishino et al 1991). Recently, we demonstrated the effective reduction of the threshold current (density) by introducing a strained quantum well structure to the active region of an AlGaInP laser (Katsuyama et al 1990, Hashimoto et al 1991). Increasing the In content of GaInP, which results in a compressively strained structure, provides a higher effective barrier height than that of a lattice matched structure. Moreover, multiple quantum well structures suppress the band-filling effects. Therefore, the improvement of temperature characteristics of devices by using compressively strained multiple quantum well structures can be expected.

In this paper, we report on the high performance of AlGaInP/Ga$_x$In$_{1-x}$P strained multiple quantum well (SMQW) lasers. An effective reduction of the threshold current density and suppression of carrier overflow by incorporating SMQW structures into the visible laser diode remarkably improved the temperature characteristics of the devices.

2. Fabrication of SCH-SMQW lasers

A separate confinement heterostructure (SCH) AlGaInP/Ga$_x$In$_{1-x}$P SMQW laser was grown on Si-doped (100) GaAs by low pressure (60Torr) organometallic vapor phase epitaxy (OMVPE) at 700°C. Triethylgallium, trimethylaluminum, trimethylindium, and PH$_3$ were used as sources. H$_2$Se and diethylzinc were used as an n-type and a p-type dopants, respectively. Trimethylgallium, AsH$_3$ and Si$_2$H$_6$ were also used for the growth of GaAs as sources and as an n-type dopant, respectively.

Figure 1 shows the cross-sectional view of the SMQW laser. The epitaxial structure consists of a 0.23 μm Si-doped GaAs buffer layer (n=1.5×10^{18} cm^{-3}), a 1.1 μm Se-doped (Al$_{0.7}$Ga$_{0.3}$)$_{0.5}$In$_{0.5}$P cladding layer (n=2.0×10^{17} cm^{-3}), the active region of the SMQW sandwiched between 800Å undoped (Al$_{0.4}$Ga$_{0.6}$)$_{0.5}$In$_{0.5}$P optical confinement layers, a 1.1 μm Zn-doped (Al$_{0.7}$Ga$_{0.3}$)$_{0.5}$In$_{0.5}$P cladding layer (p=3.8×10^{17} cm^{-3}), and a 0.14 μm Zn-doped Ga$_{0.5}$In$_{0.5}$P buffer layer (p=1×10^{18} cm^{-3}). The active region is formed by three 100Å undoped Ga$_{0.43}$In$_{0.57}$P strained quantum wells (Δa/a=0.65%) separated by 80Å (Al$_{0.4}$Ga$_{0.6}$)$_{0.5}$In$_{0.5}$P barrier layers.

The epitaxial wafer was processed into 80μm-wide and 5μm-wide stripe index guided structures by mesa etching and regrowth of a Si-doped GaAs blocking layer (n=1.5×10^{18} cm^{-3}) and a Zn-doped GaAs contact layer (p=1.5×10^{19} cm^{-3}). The p-side and n-side electrodes consist of Ti/Pt/Au and AuGe/Ni/Au, respectively. Laser facets were made by cleaving to various cavity lengths and the devices were mounted in the p-side down configuration directly onto copper heat sinks with a Au/Sn solder.

3. Characteristics of SCH-SMQW lasers

Figure 2 shows the temperature dependence of output power against cw current characteristics for the SMQW laser (5×500μm, uncoated). The device operated up to at least 150°C with optical output power exceeding 7mW. The

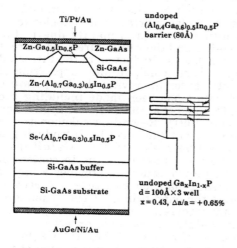

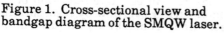

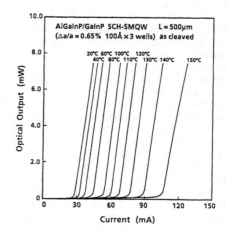

Figure 1. Cross-sectional view and bandgap diagram of the SMQW laser.

Figure 2. Temperature dependence of optical power-current characteristics for the SMQW laser (5×500μm).

operating temperature of 150°C, which is the limit of our measurement system, is the highest value ever observed in lasers operating at the visible wavelength region. Further improvement of the T_{max} can be made by facet coating and the use of a heat sink having a high thermal conductivity, such as diamond.

Figure 3 shows the dependence of the threshold current, I_{th}, on the operating temperature for 160, 300, 500, 700μm-long devices. All the devices, except for the 160μm-long device, showed a similar $I_{th}(T)$ dependence. The characteristic temperatures, T_0, were about 130K between 20°C and 80°C and about 100K between 80°C and 120°C. For the 160μm-long device, the T_0 was 110K between 20°C and 80°C and it decreased to 64K between 80°C and 120°C. Since the threshold gain is inversely proportional to the cavity length, L, a shorter cavity device requires higher gain to overcome the internal loss and mirror loss. Therefore, as L decreases, the threshold current density increases as does the thermal resistance. This enhances the band-filling effects which increase carrier loss by nonradiative recombination and leakage over the hetero-barriers. Although the 160μm-long device showed slightly smaller T_0 than those of the longer cavity devices, the device operated up to 130°C without showing the anomalous $I_{th}(T)$ dependence observed in the GaInAs/GaAs (Blood et al 1990, Van der Ziel et al 1991) and AlGaAs/GaAs (Zory et al 1986) single quantum well lasers, which is attributed to the shift in the optical transition levels due to a significant band-filling. It should also be noted that the threshold current decreased linearly with decreasing L and the threshold current of the 160μm-long device at 25°C was 13.9mA, which is the lowest value ever reported for visible lasers.

These results suggest that the multiple quantum well structure provides a high gain, thus substantially suppressing the band-filling effects even in the short cavity device. In addition, the increase of the In content of GaInP enhances the effective hetero-barrier height between the active layers and the cladding layers, which suppresses the electron overflow from the active region to the p-type cladding layer. Moreover, the threshold current density is reduced by the strain induced effects. The minimum threshold current density obtained in a $80 \times 770\mu$m device at 25°C was 430A/cm^2, which is among the lowest value ever reported for a device consisting of this material system (Serreze et al 1991).

The long-term reliability of these devices are currently under an

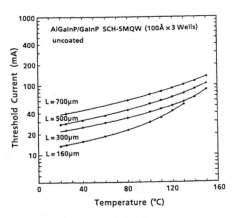

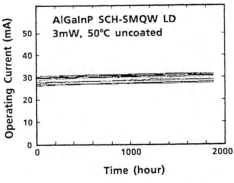

Figure 3. Threshold current dependence on the operating temperature for 160, 300, 500, 700μm-long devices

Figure 4. Lifetime test result of the SMQW lasers. (50°C, 3mW APC)

investigation. Eleven uncoated devices with initial operating currents ranging from 25 to 30mA were chosen for the lifetest. Figure 4 shows the result of the lifetest. These devices have been operating at 50°C with 3mW optical power under automatic power controlled (APC) mode for more than 1900 hours without a significant degradation.

4. Conclusion

We have grown an AlGaInP/Ga$_x$In$_{1-x}$P SMQW laser by low pressure OMVPE. Although the device structure was not optimized, high temperature (>150°C) and very low threshold current (I_{th} = 13.9mA) operation has been achieved. These remarkable improvements in the characteristics of the SMQW device are attributed to the high gain of the MQW structure, the enhancement of hetero-barrier height by the increase of the In content of GaInP quantum well, and the reduction of the threshold current density by the strain induced effects. These results show that the use of strained multiple quantum well structures is a very effective approach for improving the high temperature characteristics of AlGaInP/GaInP visible lasers.

Acknowledgements

The authors would like to acknowledge H. Kamei, M. Murata, and N. Kuwata for their useful discussions and I. Nakagaki for his technical assistance. They also thank M. Koyama, A. Ishida, K. Koe, and K. Yoshida for their continuous encouragement.

References

Blood P., Fletcher E. D., Woodbridge K., and Vening M., *Appl. Phys. Lett.*, 57, 1482 (1990)

Bour D. P., and Shealy R., *Appl. Phys. Lett.*, 51, 1658 (1987)

Hamada H., Shono M., Honda S., Hiroyama R., Matsukawa K., Yodoshi K., and Yamaguchi T., *Electron. Lett.*, 27, 662 (1991)

Hashimoto J., Katsuyama T., Shinkai J., Yoshida I., and Hayashi H., *Appl. Phys. Lett.*, 58, 879 (1991)

Ishikawa M., Shinozawa H., Tsuburai Y., and Uematsu Y., *Electron. Lett.*, 26, 211 (1990)

Itaya K., Hatakoshi G., Watanabe Y., Ishikawa M., and Uematsu Y., *Electron. Lett.*, 26, 214 (1990)

Katsuyama T., Yoshida I., Shinkai J., Hashimoto J., and Hayashi H., *Electron. Lett.*, 26, 1375 (1990)

Kishino K., Kikuchi A., Kaneko Y., and Nomura I., *Appl. Phys. Lett.*, 58, 1822 (1991)

Kobayashi K., Ueno Y., Hotta H., Gomyo A., Tada K., Hara K., and Yuasa T., *Jpn. J. Appl. Phys.*, 29, L1669 (1990)

Serreze H. B., Chen Y. C., and Waters R. G., *Appl. Phys. Lett.*, 58, 2464 (1991)

Tanaka T., Ooishi A., Kajimura T., and Minagawa S., *Extended Abstracts of the 22nd Conference on Solid State Devices and Materials* Sendai, 1177 (1990)

Ueno Y., Fujii H., Kobayashi K., Endo K., Gomyo A., Hara K., Kawata S., Yuasa T., and Suzuki T., *Jpn. J. Appl. Phys.*, 29, L1666 (1990)

Valster A., Liedenbaum C. T. H. F., Van der Heijden J. M. M., Finke M. N., Severens A. L. G., and Boermans M. J. B., *12th IEEE Int. Semiconductor Laser Conf., Davos, Conference Digest* 28 IEEE LEOS (1990).

Van der Ziel J. P., and Chand N., *Appl. Phys. Lett.*, 58 1437 (1991)

Zory P. S., Reisinger A. R., Water R. G., Mawst L. J., Zmudzinski C. A., Emanuel M. A., Givens M. A., and Coleman J. J., *Appl. Phys. Lett.*, 49, 16 (1986)

Inst. Phys. Conf. Ser. No 120: Chapter 13
Paper presented at Int. Symp. GaAs and Related Compounds, Seattle, 1991

MBE growth and characterization of high gain AlGaAs/GaAsSb/GaAs NpN HBTS

G.J. Sullivan, W.J. Ho, R.L. Pierson, M.K. Szwed, M.D. Lind, and R.L.Bernescut

Rockwell International Science Center, Thousand Oaks, CA 91358

Abstract
The growth and characterization of high gain NpN HBTs on GaAs with a pseudomorphic $GaAs_{.92}Sb_{.08}$ base layer are reported. The layers were grown in an elemental source MBE system, using tetramer As_4 and Sb_4 sources. The emitter to base heterojunction was abrupt. The HBTs fabricated were large area, mesa isolated transistors, with a beta of 150 at a current density of 2 KA/cm^2, and high gain down to low current densities. The specific contact resistance of the refractory base contact has been measured to be as low as 6×10^{-7} Ωcm^2. The turn-on voltage, V_{BE} at $J_E = 2$ A/cm^2, is 0.98V. Perhaps most significant, the inclusion of Sb in the base dramatically reduces the diffusion of Be from the base into the wider band gap AlGaAs emitter.

1. Introduction

Very high performance has been attained in AlGaAs/GaAs HBTs over the past several years. The AlGaAs/GaAsSb ternary system has not been examined for device applications because of the large lattice mismatch (~8%) of the AlGaAs/GaSb systems. However, a GaAsSb base layer in an AlGaAs/GaAsSb/GaAs double heterojunction bipolar transistor (DHBT) has some significant advantages, compared with a lattice matched GaAs base. The smaller band gap of the GaAsSb should reduced the turn-on voltage, which can reduce power dissipation in circuits. Specific contact resistance to P type GaAsSb is lower than contact resistance to GaAs. There is the potential for improved hole mobility and therefore higher base conductivity, because GaSb has a hole mobility three times larger than GaAs. The transistors displayed several other advantages which were less expected. The gain in the HBT remains high to low current densities, which can be used to reduce power dissipation in circuits. Perhaps most remarkably, inclusion of Sb in the base of an HBT dramatically reduces the diffusion of Be from the base into the wider band gap AlGaAs emitter. In this paper, we report on the growth and characterization of high gain DHBTs on GaAs with a pseudomorphic $GaAs_{.92}Sb_{.08}$ base layer.

2. MBE Growth of GaAsSb

The HBT layers were grown in a solid source MBE system, using tetramer beams of As_4 and Sb_4. GaAsSb is not lattice matched to GaAs, which limits how much Sb can be incorporated without forming dislocations. The Sb alloy composition of the base was chosen such that it would be close to this pseudomorphic limit at the thickness of the base layer. Small increases (4%) in the Sb alloy content caused crosshatching to appear in the surface morphology, indicating the formation of dislocations. This 500Å of 8% GaAsSb is in good agreement with published values for the critical layer thickness (Ikossi-Anastasiou et al., 1990). The GaAsSb alloy composition was graded on the collector side to avoid a conduction band edge spike which could block efficient collection of the electrons.

The (400) X-ray reflection of a strained GaAsSb layer doped with Be to 1×10^{20} cm^{-3} is shown in Figure 1. There is 100Å of graded GaAsSb on either side of the 400Å doped layer and 100Å undoped GaAsSb layers. This layer is identical to the GaAsSb layer used as the base of the HBTs. The peak from the GaAsSb layer has a full width at half max (FWHM) of 375 arcsec. This FWHM is very close to the 250 arcsec. intrinsic line width calculated for a perfectly homogeneous layer of this thickness. If the GaAsSb layer is assumed to be fully pseudomorphic, the alloy composition should be GaAs$_{.92}$Sb$_{.08}$.

The low temperature (15K) photoluminescense of this same 100Å undoped GaAsSb layer and 400Å GaAsSb layer doped with Be to 1×10^{20} cm^{-3} is shown in Figure 2. The peak of the luminescence is at 0.929 µm, which is estimated to be 8% GaAsSb, in very good agreement with the X-ray data. The very sharp drop in intensity on the high energy side of the peak is probably due to recombination in the undoped GaAsSb, while the wide sholder on the low energy side is probably due to recombination in the heavily doped GaAsSb. The FWHM of the total peak is only 22 meV.

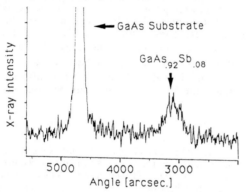

Fig. 1. Double crystal X-ray diffraction of pseudomorphic GaAsSb layer.

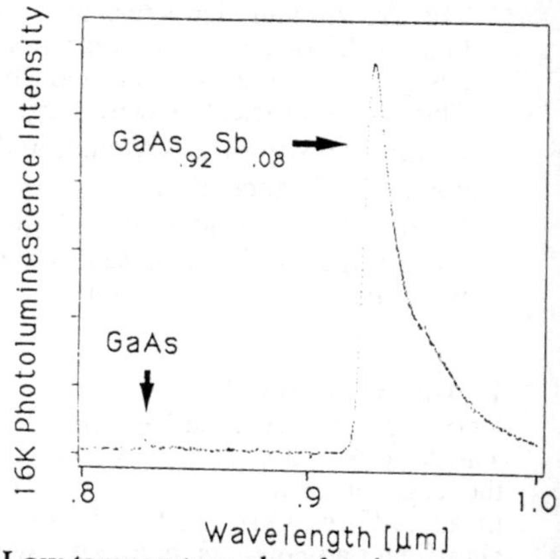

Fig. 2. Low temperature photoluminescense of heavily doped 500Å GaAsSb layer.

The room temperature Hall mobility of several Be doped pseudomorphic GaAsSb layers is plotted in Figure 3, along with some typical GaAs data. The hole concentration indicates that the Be is an efficient acceptor, but the hole mobility in GaAsSb is much worse than hole mobility in GaAs. The cause of the poor mobility is not clear. The X-ray and photoluminescence results suggest that the GaAsSb layers are not heavily dislocated. A low density of dislocations might not degrade these results. Similarly, the gain of HBTs should not be severely degraded by a low density of dislocations (Kroemer et al., 1989). Whether this poor Hall mobility can be improved by better source material or growth conditions is under investigation. This poor mobility does significantly degrade the base sheet conductivity, and would degrade the high frequency performance of HBTs.

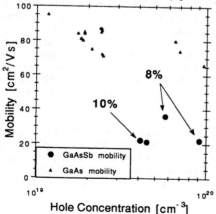

Fig. 3. Mobility vs. Hole Concentration

Table I. The epitaxial layer structure for AlGaAs/GaAsSb/GaAs NpN HBT on GaAs.

Layer	Thickness (Å)	Alloy Comp.	Doping (cm^{-3})
n+GaAs	500	0	N_D=5E18
n+GaAs	1000	0	N_D=2E18
down ramp	400	25 to 0%	N_D=2E18
AlGaAs	550	25%	N_D=2E18
GaAsSb	100	8 to 0%	undoped
p+GaAsSb	400	8%	N_A=5E19
GaAsSb	100	8%	N_D=4E16
Sb up ramp	100	0 to 8%	N_D=4E16
GaAs	3000	0	N_D=4E16
GaAs	4000	0	N_D=3E17
GaAs	6000	0	N_D=6E18

(100) oriented semi-insulating GaAs substrate

The typical HBT layer structure is listed in Table I. The base Be density was varied in sequential growths from 5×10^{19} to 1×10^{20} cm^{-3}. The growth temperature was 570°C, as measured by a pyrometer. The abrupt AlGaAs emitter was used for ease of growth.

3. Device Fabrication

A self-aligned, dual mesa process was employed to fabricate the HBTs. This procedure simplifies the device fabrication and produces fully isolated, large area transistors. The base layer was exposed using selective wet chemical etching. The base metal is a refractory Ti/Pt/Au contact, and the emitter and collector contacts are AuGe/Ni/Au. The devices were alloyed at 350°C for 25 sec in forming gas. The specific contact resistance of the refractory base contact has been measured to be as low as 6×10^{-7} Ωcm^2, suggesting that the Sb is improving contact to the p-type base. The base sheet resistance is 780 Ω/sq for a 475Å thick base doped at 5×10^{19} cm^{-3} and 550Ω/sq. at 1×10^{20}cm^{-3} doping level. This high sheet resistance is due to the low hole mobility in the p-GaAsSb, as shown in Figure 3.

4. Device Characterization

To measure the DC current gain characteristics, large area transistors were characterized, for which recombination at the emitter edge has negligible effect. Figure 4 shows the common emitter characteristics of a typical HBT with GaAsSb base doping level of 5×10^{19} cm^{-3}. No short term degradation of device characteristics was observed while characterizing the HBTs on a curve-tracer. The offset voltage is only 0.1 volts. The devices show significant output conductance. This is not due to modulation of the base width (Early effect), but is believed due to a blocking spike forming in the conduction band on the collector side of the base, due to inadequate grading distance for the Sb. The collector breakdown voltage, BV_{ceo}, is greater than 15V.

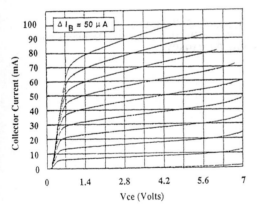

Fig. 4. Common Emitter Characteristics

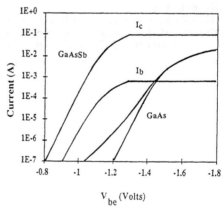

Fig. 5. Gummel plots for HBTs, with and without Sb in the base.

Figure 5 shows a representative Gummel plot of an HBT with base doping density of 5×10^{19} cm^{-3}. The turn-on voltage (V_{BE} at $J_C = 2$A/cm^2) is only 0.98 volts, and the DC gain is 150 ($J_C = 2$ kA/cm^2). The base current has an ideality factor very close to unity, so that the HBTs retain high gain at low current densities. This high gain down to low current densities can be used to minimize the power dissipation in circuits. HBTs with a base doping of 1×10^{20} cm^{-3} had very similar Gummel plots, with a gain of 80 ($J_C = 2$ kA/cm^2), almost identical turn-on voltage, and high gain to low current densities. The lack of significant change in the turn-on voltage at the heavier doping indicates that Be diffusion into the emitter is still negligible at the heavier doping.

Also plotted in Figure 5 is the Gummel plot for an HBT which is identical to the first device except that there is no Sb in the base. The HBT without Sb suffers from severe diffusion of the Be into the emitter. This suppression of Be diffusion is especially dramatic because GaAsSb base HBTs doped to twice this doping density and grown under similar conditions ($N_A=1\times10^{20}$ cm^{-3} mentioned above) do not show indications of diffusion. This reduction in Be diffusion due to Sb incorporation has been reproduced with almost identical results in sets of layers grown 9 months apart, where MBE sources had been refilled between sets. It is therefore believed to be real and significant. The cause of this reduction in diffusion is not clear. Hypotheses that the diffusion is suppressed by the small Be atom compensating the lattice strain caused by the large Sb atom fail to explain why a similar reduction in diffusion is not seen for InGaAs base HBTs, where the strain is very similar. We have seen no dramatic reductions in Be diffusion when InGaAs base layers are used in the base of HBTs. This would seem to indicate that the reduction in diffusion due to GaAsSb is due to a "chemical" effect, perhaps modification of the Ga vacancy population.

5. Summary
In summary, high gain, pseudomorphic GaAsSb base NpN HBTs have been demonstrated on GaAs substrates. The HBTs on GaAs show low turn-on voltage, low base contact resistance, and a dramatic reduction in Be diffusion into the AlGaAs emitter.

Acknowledgements
The authors thank C. Farley, M. Mondry, E. Peterson, S. Pittman, and S. Skylstad for technical assistance. The constant support and encouragement from D.T. Cheung and P. Asbeck are greatly appreciated.

References
Ikossi-Anastasiou K. et al.,Elect. Lett., V. 27 (2), pp.965-967, (1990).
Kroemer H, T.Y. Lui and P.M. Petroff, "GaAs on Si and Related Systems; Problems and Prospects," J Cryst. Growth, V.95, pp. 96-102, (1989).

Subject Index

Author Index